W9-DIW-255

ENZYME NOMENCLATURE
1978

NOMENCLATURE COMMITTEE
OF THE
INTERNATIONAL UNION OF BIOCHEMISTRY

Heinz Bielka (German Democratic Republic)
Bernard L. Horecker (U.S.A.)
William B. Jakoby (U.S.A.)
Peter Karlson (Federal Republic of Germany), Chairman
Borivoj Keil (France)
Claude Liébecq (Belgium)
Bengt Lindberg (Sweden)
Edwin C. Webb (Australia)

In consultation with the following members of the Joint Commission of Biochemical Nomenclature and its Predecessor, the Commission on Biochemical Nomenclature, jointly sponsored by the International Union of Pure and Applied Chemistry (IUPAC) and the International Union of Biochemistry (IUB).

Alexander E. Braunstein (U.S.S.R.)
Waldo E. Cohn (U.S.A.)
Henry B. F. Dixon (United Kingdom)
(France)
Otto Hoffmann-Ostenhof (Austria)
Kurt L. Loening (U.S.A.)
G. P. Moss (United Kingdom)

REF
QP601
I54
1979

ENZYME NOMENCLATURE
1978

RECOMMENDATIONS OF THE NOMENCLATURE COMMITTEE
OF THE INTERNATIONAL UNION OF BIOCHEMISTRY ON THE
NOMENCLATURE AND CLASSIFICATION OF ENZYMES

This edition is a revision of the Recommendations (1972) of the IUPAC–IUB
Commission on Biochemical Nomenclature, and has been approved for
publication by the Executive Committee of the International Union of
Biochemistry

NO LONGER THE PROPERTY
OF THE
UNIVERSITY OF R. I. LIBRARY

ACADEMIC PRESS   New York   San Francisco   London   1979
Published for the International Union of Biochemistry by Academic Press, Inc.

COPYRIGHT © 1979, BY ACADEMIC PRESS, INC.
ALL RIGHTS RESERVED.
NO PART OF THIS PUBLICATION MAY BE REPRODUCED OR
TRANSMITTED IN ANY FORM OR BY ANY MEANS, ELECTRONIC
OR MECHANICAL, INCLUDING PHOTOCOPY, RECORDING, OR ANY
INFORMATION STORAGE AND RETRIEVAL SYSTEM, WITHOUT
PERMISSION IN WRITING FROM THE PUBLISHER.

ACADEMIC PRESS, INC.
111 Fifth Avenue, New York, New York 10003

United Kingdom Edition published by
ACADEMIC PRESS, INC. (LONDON) LTD.
24/28 Oval Road, London NW1 7DX

Library of Congress Cataloging in Publication Data

International Union of Biochemistry. Nomenclature committee.
 Enzyme nomenclature, 1978.

 Revision of Enzyme nomenclature; recommendations (1972) of
the Commission on Biochemical Nomenclature published in 1973.
    1. Enzymes--Nomenclature. 2. Enzymes--Classification. I.
Commission on Biochemical Nomenclature. Enzyme nomenclature
II. Title.
QP601.I54 1979   574.1'925'014   79–1466
ISBN 0–12–227160–2

PRINTED IN THE UNITED STATES OF AMERICA
79 80 81 82   9 8 7 6 5 4 3 2 1

Errata

Enzyme Nomenclature

Page 603, line 10 *should read:*
5.3.2 Ferredoxins (abbreviation: Fd).

Page 603, line 2 of footnote 7 *should read:*
structure of the active center of these proteins . . . .

Page 605, line 5 *should read:*
Chromatium vinosum Hipip: $[4Fe-4S]^{2+(3+:2+)}$

Page 605, Table II, column 1, line 2 *should read:*
Azotobacter vinelandii . . . .

Page 605, Table II, column 2, line 1 *should read:*
Spinach chloroplast $[2Fe-2S]^{2+}$Fd . . . .

Page 605, Table II, column 2, line 2 *should read:*
specific content: $[2Fe-2S]^{2+(2+:1+)}$

ABBREVIATIONS USED IN THE ENZYME LIST

The spellings used throughout are those common in Great Britain, but those current in the U.S. may be used instead in recommended names as in the following examples: glycollate, glycolate; sulphur, sulfur; oestradiol, estradiol.

ADP	adenosine 5′-diphosphate
AMP	adenosine 5′-monophosphate
ATP	adenosine 5′-triphosphate
dATP	deoxyadenosine 5′-triphosphate
CDP	cytidine 5′-diphosphate
CMP	cytidine 5′-monophosphate
dCMP	deoxycytidine 5′-monophosphate
CoA	coenzyme A
CTP	cytidine 5′-triphosphate
dCTP	deoxycytidine 5′-triphosphate
DFP	diisopropyl fluorophosphate
DNA	deoxyribonucleic acid
FAD	flavin-adenine dinucleotide
FMN	flavin mononucleotide (riboflavin 5′-monophosphate)
GDP	guanosine 5′-diphosphate
dGDP	deoxyguanosine 5′-diphosphate
GMP	guanosine 5′-monophosphate
dGMP	deoxyguanosine 5′-monophosphate
GTP	guanosine 5′-triphosphate
dGTP	deoxyguanosine 5′-triphosphate
IDP	inosine 5′-diphosphate
IMP	inosine 5′-monophosphate
ITP	inosine 5′-triphosphate
NAD^+	oxidized nicotinamide-adenine dinucleotide

NADH	reduced nicotinamide-adenine dinucleotide
NADP+	oxidized nicotinamide-adenine dinucleotide phosphate
NAD(P)+	indicates either NAD+ or NADP+
NADPH	reduced nicotinamide-adenine dinucleotide phosphate
NAD(P)H	indicates either NADH or NADPH
NDP	nucleoside 5′-diphosphate
NMN	nicotinamide mononucleotide
NMP	nucleoside 5′-monophosphate
dNMP	deoxynucleoside 5′-monophosphate
NTP	nucleoside 5′-triphosphate
P	phosphate residues
poly(C)	synthetic polynucleotide composed of cytidylate residues
poly(G)	synthetic polynucleotide composed of guanylate residues
RNA	ribonucleic acid
tRNA	transfer ribonucleic acid
TDP	ribothymidine 5′-diphosphate
dTDP	thymidine 5′-diphosphate
TMP	ribothymidine 5′-monophosphate
dTMP	thymidine 5′-monophosphate
TTP	ribothymidine 5′-triphosphate
dTTP	thymidine 5′-triphosphate
UDP	uridine 5′-diphosphate
UMP	uridine 5′-monophosphate
dUMP	deoxyuridine 5′-monophosphate
UTP	uridine 5′-triphosphate
dUTP	deoxyuridine 5′-triphosphate

CHAPTER 1

HISTORICAL INTRODUCTION

The rapid growth in the science of enzymology, and the great increase in the number of enzymes known, have given rise to many difficulties of terminology in recent years. By about 1955 it had become evident that the nomenclature of the subject, in the absence of any guiding authority, was getting out of hand. The naming of enzymes by individual workers had proved far from satisfactory in practice. In many cases the same enzymes became known by several different names, while conversely there were cases in which the same name was given to different enzymes. Many of the names conveyed little or no idea of the nature of the reactions catalysed, and similar names were sometimes given to enzymes of quite different types. To meet this situation, various attempts to bring order into the general nomenclature of enzymes, or into that of particular groups of enzymes, were made by individuals or small groups of specialists. But none of the resulting nomenclatures met with general approval.

Furthermore, no general agreement had been reached on the nomenclature of the coenzymes, on which so many names of enzymes inevitably depend; in the equations of enzyme kinetics different systems of mathematical symbols were used by different workers; and the standardization of enzymes was in a chaotic state, owing to the multiplicity of arbitrarily defined units of enzyme activity.

In view of this state of affairs, the General Assembly of the International Union of Biochemistry (IUB) decided, during the third International Congress of Biochemistry in Brussels in August, 1955, to set up an International Commission on Enzymes. This step was taken in consultation with the International Union of Pure and Applied Chemistry (IUPAC).

The International Commission on Enzymes was established in 1956 by the President of the International Union of Biochemistry, Professor M. Florkin, with the advice of an *ad hoc* Committee. The following members were appointed by the Bureau of the International Union of Biochemistry:

> A.E. Braunstein, U.S.S.R.; S.P. Colowick, U.S.A.; P.A.E. Desnuelle, France; M. Dixon, U.K. *(Chairman)*; W.A. Engelhardt, U.S.S.R.; E.F. Gale, U.K.; O. Hoffmann-Ostenhof, Austria; A.L. Lehninger, U.S.A.; K. Linderstrøm-Lang, Denmark; F. Lynen, Germany.

> *Corresponding Members*: F. Egami, Japan; L.F. Leloir, Argentina.

In 1959, on the death of K. Linderstrøm-Lang, E.C. Webb (United Kingdom, later Australia) joined the Commission.

The terms of reference of the Enzyme Commission, as laid down by the *ad hoc* Committee, were as follows:

> *'To consider the classification and nomenclature of enzymes and coenzymes, their units of activity and standard methods of assay, together with the symbols used in the description of enzyme kinetics.'*

The Enzyme Commission faced many difficulties arising from the uncontrolled naming of the rapidly increasing number of known enzymes. Some of the names in use were definitely misleading; others conveyed little or nothing about the nature of the reaction catalysed, as for example, *diaphorase, Zwischenferment, catalase*. Enzymes catalysing essentially similar reaction had sometimes names implying that they belong to different groups, while some enzymes of different types had been placed in the same group, for example, the *pyrophosphorylases* had included both glycosyl-transferases and phosphotransferases. In some cases a name which had been well established for many years with a definite meaning, such as the term *synthetase*, had been later employed with different meanings, causing confusion.

One of the main tasks given to the Commission was therefore to see how the nomenclature of enzymes could best be brought into a satisfactory state, and whether a code of systematic rules could be devised that would serve as a guide for the consistent naming of new enzymes in the future. At the same time, the Commission realised the difficulties that would be caused by a large number of changes of well-known enzyme names, and the desirability of retaining the existing names wherever there was no good reason for making an alteration. Nevertheless, the overriding consideration was to reduce the confusion and prevent futher confusion from arising. Its task could not have been accomplished without causing some inconvenience, for this was the inevitable result of having allowed the problem to drift for a considerable time.

Throughout its work, the Enzyme Commission was in close touch with the Biological Chemistry Nomenclature Commission of IUPAC. In addition, it considered many comments and suggestions from various experts in the field; 52 formal documents were circulated and discussed in several meetings. Finally, the Commission prepared a report, which was presented to the General Assembly of the International Union of Biochemistry at their meeting in Moscow, 1961, and was adopted. The nomenclature set out in that report has been widely used in scientific journals, textbooks, *etc.* since 1961.

Subsequently, the Council of IUB dissolved the Commission on Enzymes and set up a Standing Committee on Enzymes consisting of S.P. Colowick, O. Hoffmann-Ostenhof, A.L. Lehninger and E.C. Webb *(Secretary)*. This Standing Committee discussed the comments and criticisms received on the published report of the Enzyme Commission and prepared in 1964 a second version, the *Recommendations (1964) of the International Union of Biochemistry on the Nomenclature and Classification of Enzymes.*

The function of the Standing Committee on Enzymes was then taken over by the IUPAC/IUB Commission on Biochemical Nomenclature (CBN). This Commission was originally set up to deal with the nomenclature of various compounds of biochemical interest. At a meeting in September, 1969, it was decided that the *Recommendations on Enzyme Nomenclature* should be revised, mainly to include the many enzymes discovered in recent years, and an Expert Committee on Enzymes was formed, consisting of A.E. Braunstein, J.S. Fruton, O. Hoffmann-Ostenhof, B.L. Horecker, W.B. Jakoby, P. Karlson, B. Keil, E.C. Slater, E.C. Webb *(convenor)* and W.J. Whelan. With the help of a number of expert subcommittees, and comments and suggestions solicited from authors and editors, a completely revised version of Enzyme Nomenclature was prepared and published as *Recommendations (1972) of the International Union of Pure and Applied Chemistry and the International Union of Biochemistry.*

After the publication of the third version of the complete report and enzyme list, the Commission on Biochemical Nomenclature decided that it would be appropriate to publish from time to time, a supplement to the Enzyme List, containing new entries, deletions and corrections. The first supplement was prepared during 1974-5 and published in Biochimica et Biophysica Acta (Vol. 429, pages 1-45 (1976).

During 1977 there was a reorganization of responsibility for consideration of biochemical nomenclature, and the responsibility for enzyme nomenclature passed to the new Nomenclature Committee of I.U.B.. At the same time the International Union of Biochemistry was able to arrange with the National Institutes of Health at Bethesda to enter the enzyme list on computer tape and prepare future versions of the list as a computer print-out. This phase of the operation was under the direction of Richard J. Feldmann of the Division of Computor Research and Technology, National Institutes of Health. The present version of Enzyme Nomenclature is the first to be produced in this way. It includes changes and additions approved by the Nomenclature Committee of I.U.B. at meetings in June 1977 and June 1978.

The number of enzymes in the various versions of the enzyme list are as follows:

Report of the Enzyme Commission (1961)	712
Enzyme Nomenclature (1964)	875
Enzyme Nomenclature (1972)	1770
Enzyme Nomenclature (1972)plus	
Supplement I (1975)	1974
Present version	2122

Many people have contributed to the steady growth of the Enzyme List. Particular mention should be made of Otto Hoffmann-Ostenhof, who was Secretary of the original Enzyme Commission, and as Chairman of the Commission on Biochemical Nomenclature from 1965 to 1976, has been largely responsible for continuity of the efforts in this field. Until 1976, Alexander E. Braunstein had a similarly long association with this work. Until the transfer of the enzyme list to the computer in 1977, it has been kept in the form of a card index in the office of Edwin Webb, first at the University of Queensland, and more recently at Macquarie University, Sydney. In both places, for a period of 15 years, Miriam Armstrong has been responsible for the management of the list and typed the many versions of Enzyme Nomenclature (1964) and its subsequent edition and supplement.

CHAPTER 2

THE CLASSIFICATION AND NOMENCLATURE OF ENZYMES

1. General principles

Because of their close interdependence, it is convenient to deal with the classification and nomenclature together in one chapter.

The first general principle of these 'Recommendations' is that names purporting to be names of enzymes, especially those ending in *-ase,* should be used only for single enzymes, *i.e.* single catalytic entities. They should not be applied to systems containing more than one enzyme. When it is desired to name such a system on the basis of the overall reaction catalysed by it, the word *system* should be included in the name. For example, the system catalysing the oxidation of succinate by *molecular oxygen,* consisting of succinate dehydrogenase, cytochrome oxidase, and several intermediate carriers, should not be named *succinate oxidase,* but it may be called the *succinate oxidase system.* Other examples of systems consisting of several structurally and functionally linked enzymes (and cofactors) are the pyruvate dehydrogenase system, the similar 2-oxoglutarate dehydrogenase system, and the fatty acid synthetase system.

In this context it is appropriate to express disapproval of a loose and misleading practice that is currently rather frequent in biological literature. It consists in designation of a natural substance (or even of an hypothetical active principle), responsible for a physiological or biophysical phenomenon that cannot be described in terms of a definite chemical reaction, by the name of the phenomenon in conjugation with the suffix *-ase,* which implies an individual enzyme. Some recent examples of such *phenomenase* nomenclature, which should be discouraged even if there are reasons to suppose that the particular agent may have enzymic properties, are: *permease, translocase, reparase, joinase, replicase, codase, etc.* .

The second general principle is that enzymes are classified and named according to the reaction they catalyse. The chemical reaction catalysed is the specific property that distinguishes one enzyme from another, and it is logical to use it as the basis for the classification and naming of enzymes.

Several alternative bases for classification and naming had been considered, *e.g.* chemical nature of the enzyme (whether it is a flavoprotein, a haemoprotein, a pyridoxal-phosphate-protein, a cuproprotein, and so on), or chemical nature of the substrate (nucleotides, carbohydrates, proteins, *etc.*). The first cannot serve as a general basis, for only a minority of enzymes have such identifiable prosthetic groups. The chemical nature of the enzyme has however, been used exceptionally in this revised report in certain cases where classification based on specificity is difficult, for example with the proteinases (sub-sub-groups 3.4.21-24). The second basis for classification is hardly practicable, owing to the great variety of substances acted upon and because it is not sufficiently informative unless the type of reaction is also given. It is the overall reaction, as expressed by the formal equation, that should be taken as the basis. Thus, the intimate mechanism of the reaction, and the formation of intermediate complexes of the reactants with the enzyme, is not taken into account, but only the observed chemical change produced by the complete enzyme reaction. For example, in those cases in which the enzyme contains a prosthetic group that serves to catalyse transfer from a donor to an acceptor (*e.g.* flavin, biotin, or pyridoxal-phosphate enzymes) the name of the prosthetic group is not included in the name of the enzyme. Nevertheless, where alternative names are possible, the mechanism may be taken into account in choosing between them.

A consequence of the adoption of the chemical reaction as the basis for naming enzymes is that a systematic name cannot be given to an enzyme until it is known what chemical reaction it catalyses. This applies, for example, to a few enzymes that have so far not been shown to catalyse any chemical reaction, but only isotopic exchanges; the isotopic exchange gives some idea of one step in the overall chemical reaction, but the reaction as a whole remains unknown.

A second consequence of this concept is that a certain name designates not a single enzyme protein but a group of proteins with the same catalytic property. Enzymes from different sources (various bacterial, plant or animal species) are classified as one entry. The same applies to isoenzymes (see below). However, there are exceptions to this general rule. Some are justified because the mechanism of the reaction or the substrate specificity is so different as to warrant different entries in the enzyme list. This applies, for example, to the two cholinesterases, EC 3.1.1.7 and 3.1.1.8, the two citrate hydro-lyases, EC 4.2.1.3 and 4.2.1.4, and the two amine oxidases, EC 1.4.3.4 and 1.4.3.6. Others are mainly historical, *e.g.* acid and alkaline phosphatase.

A third general principle adopted is that the enzymes are divided into groups on the basis of the *type of reaction* catalysed, and this, together with the name(s) of the substrate(s) provides a basis for naming individual enzymes. It is also the basis for classification and code numbers.

Special problems attend the classification and naming of enzymes catalysing complicated transformations that can be resolved into several sequential or coupled intermediary reactions of different types, all catalysed by a single enzyme (not an enzyme system). Some of the steps may be spontaneous non-catalytic reactions, while one or more intermediate steps depend on catalysis by the enzyme. Wherever the nature and sequence of intermediary reactions is known or can be presumed with confidence, classification and naming of the enzyme should be based on the *first enzyme-catalysed* step, that is essential to the subsequent transformations, which can be indicated by a supplementary term in parentheses, *e.g.* L-malate glyoxylate-lyase (CoA-acetylating) (EC 4.1.3.2, *cf.* section 3).

For the classification according to the type of reaction catalysed, it is occasionally necessary to choose between alternative ways of regarding a given reaction. Some considerations of this type are outlined in section 3 of this chapter. In general, that alternative should be selected which fits in best with the general system of classification and reduces the number of exceptions.

One important extension of this principle is the question of the direction in which the reaction is written for the purposes of classification. To simplify the classification, the direction chosen should be the same for all enzymes in a given class, even if this direction has not been demonstrated for all. Thus the systematic names, on which the classification and code numbers are based, may be derived from a written reaction, even though only the reverse of this has been actually demonstrated experimentally.

2. Systematic and trivial names

The first Enzyme Commission gave much thought to the question of a systematic and logical nomenclature for enzymes, and finally recommended that there should be two nomenclatures for enzymes, one systematic, and one working or trivial. The systematic name of an enzyme, formed in accordance with definite rules, showed the action of an enzyme as exactly as possible, thus identifying the enzyme precisely. The trivial name was sufficiently short for general use, but not necessarily very systematic; in a great many cases it was a name already in current use. The introduction of (often cumbersome) systematic names has been strongly criticised. It has been pointed out that in many cases the reaction catalysed, given in parentheses, is not much longer than the systematic name and can serve just as well for identification, especially in conjunction with the code number.

The Commission for Revision of Enzyme Nomenclature has discussed this problem at length. It was decided to give the *trivial* names more prominence in the enzyme list; they now follow immediately after the code number, and are described as *Recommended Name*. Also, in the index the recommended names appear in bold roman. Nevertheless, it was decided to retain the systematic names as the basis for classification as well as for identification (to be given only once in the paper) for the following reasons:

(i)   the code number alone is only useful for identification of an enzyme when a copy of the Enzyme List is at hand, whereas the systematic name is self-explanatory;

(ii)   the systematic name stresses the type of reaction, the reaction equation does not;

(iii)   systematic names can be formed for new enzymes by the discoverer, by application of the rules, but **code numbers should not be assigned by individuals;**

(iv)   recommended names for new enzymes are generally formed as a condensed version of the systematic name; therefore, the systematic names are helpful in finding recommended names that are in accordance with the general pattern.

It is recommended that for enzymes that are not the main subject of a paper or abstract, the recommended names should be used, but they should be identified at their first mention by their code numbers and source. Where an enzyme is the main subject of a paper or abstract, its code number, systematic name, or, alternatively, the reaction equation, and source should be given at its first mention; thereafter the recommended name may be used. In the light of the fact that enzyme names and code numbers refer to reactions catalysed rather than to discrete proteins, *it is of special importance to give also the source of the enzyme* for full identification; in cases where multiple forms are known to exist, knowledge of this should be included where available.

When a paper deals with an enzyme that is not yet in the enzyme list, the author may introduce a new name and, if desired, a new systematic name, both formed according to the recommended rules. A number should be assigned only by the Nomenclature Committee of IUB.

The enzyme list at the end of this volume, in contrast to the earlier versions, contains one or more references for each enzyme. It should be stressed that no attempt has been made to provide a complete bibliography, or to refer to the first description of an enzyme. The references are intended to provide sufficient evidence for the existence of an enzyme catalysing the reaction as set out. In those cases where there is a major paper describing the purification and specificity of an enzyme, this has been quoted to the exclusion of earlier and later papers. In some cases separate references are given for animal, plant and bacterial enzymes.

3. Scheme of classification and numbering of enzymes

The first Enzyme Commission, in its report in 1961, devised a system for classification of enzymes that also serves as a basis for assigning code numbers to them. These code numbers, prefixed by EC, which are now widely in use, contain four elements separated by points, with the following meaning:

(i)   **the first number shows to which of the six main divisions (classes) the enzyme belongs,**

(ii)   **the second figure indicates the sub-class,**

(iii)   **the third figure gives the sub-sub-class,**

(iv)   **the fourth figure is the serial number of the enzyme in its sub-sub-class.**

The sub-classes and sub-sub-classes are formed according to principles indicated below; the full key to the classification is given in section 5.

The main divisions and sub-classes are:

1. OXIDOREDUCTASES. To this class belong all enzymes catalysing oxidoreduction reactions. The substrate that is oxidized is regarded as hydrogen donor. The systematic name is based on *donor:acceptor oxidoreductase.* The recommended name will be *dehydrogenase,* wherever this is possible; as an alternative, *reductase* can be used. *Oxidase* is only used in cases where O_2 is the acceptor.

The second figure in the code number of the oxidoreductases indicates the group in the hydrogen donor which undergoes oxidation: 1 denotes a -CHOH- group, 2 an aldehyde- or keto-group, and so on, as listed in section 5.

The third figure, except in sub-groups 1.11 and 1.15, indicates the type of acceptor involved: 1 denotes NAD(P), 2 a cytochrome, 3 molecular oxygen, 4 a disulphide, 5 a quinone or related compound, *etc.*.

It should be noted that in reactions with a nicotinamide coenzyme, this is always regarded as acceptor, even if this direction of the reaction is not readily demonstrated. The only exception is the sub-class 1.6, in which NAD(P)H is the donor; some other redox catalyst is the acceptor.

Although not used as a criterion for classification, the two hydrogen atoms at carbon-4 of the dihydropyridine ring of nicotinamide nucleotides, are not equivalent in that the hydrogen is transferred stereospecifically. The stereospecificity of a large number of dehydrogenases has been summarized*

2. TRANSFERASES. Transferases are enzymes transferring a group, *e.g.* the methyl group or a glycosyl group, from one compound (generally regarded as donor) to another compound (generally regarded as acceptor). The systematic names are formed according to the scheme *donor:acceptor grouptransferase.* The recommended names are normally formed according to *acceptor grouptransferase* or *donor grouptransferase.* In many cases, the donor is a cofactor (coenzyme) charged with the group to be transferred. A special case is that of the aminotransferases (see below).

Some transferase reactions can be viewed in different ways. For example, the enzyme-catalysed reaction

$$X - Y + Z = X + Z - Y$$

may be regarded either as a transfer of the group Y from X to Z, or as a breaking of the X-Y bond by the introduction of Z. Where Z represents phosphate or arsenate, the process is often spoken of as 'phosphorolysis' or 'arsenolysis', respectively, and a number of enzyme names based on the pattern of *phosphorylase* have come into use. These names are not suitable for a systematic nomenclature, because there is no reason to single out these particular enzymes from the other transferases, and it is better to regard them simply as Y-*transferases.*

---

* You, K., Arnold, L.J., Jr., Allison, W.S. & Kaplan, N.O. *Trends Biochem. Sciences,* Vol. 3, December 1978; a fuller compilation is available as TIBS Databank No. 2.

Another problem is posed in the enzyme-catalysed transamination reactions. They involve the transfer of one electron pair and a proton, together with the NH$_2$ group, from a primary amine to an oxo compound, according to the general equation

$$R^1—CHNH_2—R^2 + R^3—CO—R^4 \xrightarrow{\text{enzyme}} R^1—CO—R^2 + R^3—CHNH_2—R^4.$$

The reaction can formally be considered as oxidative deamination of the donor (*e.g.* amino acid) linked with reductive amination of the acceptor (*e.g.* oxo acid), and the transaminating enzymes (pyridoxal-phosphate-proteins) might be classified as oxidoreductases. However, the unique distinctive feature of the reaction is the transfer of the amino group (by a well-established mechanism involving covalent substrate-coenzyme intermediates), which justifies allocation of these enzymes among the transferases as a special sub-group (2.6.1, aminotransferases).

The second figure in the code number of transferases indicates the group transferred: a one-carbon group in 2.1, an aldehydic or ketonic group in 2.2, a glycosyl group in 2.3 and so on.

The third figure gives further information on the group transferred: *e.g.* sub-class 2.1 is subdivided into methyltransferases (2.1.1), hydroxymethyl and formyltransferases (2.1.2) and so on; only in sub-class 2.7, does the third figure indicate the nature of the acceptor group.

3. HYDROLASES. These enzymes catalyse the hydrolytic cleavage of **C-O, C-N, C-C** and some other bonds, including phosphoric anhydride bonds. Although the systematic name always includes *hydrolase,* the recommended name is, in many cases, formed by the name of the substrate with the suffix *...ase.* It is understood that the name of the substrate with this suffix means a hydrolytic enzyme.

A number of hydrolases acting on ester, glycosyl, peptide, amide or other bonds are known to catalyse not only hydrolytic removal of a particular group from their substrates, but likewise the transfer of this group to suitable acceptor molecules. In principle, all hydrolytic enzymes might be classified as transferases, since hydrolysis itself can be regarded as transfer of a specific group to water as the acceptor. Yet, in most cases, the reaction with water as the *acceptor* was discovered earlier and is considered as the main physiological function of the enzyme. This is why such enzymes are classified as hydrolases rather than as transferases.

Some hydrolases (especially among the esterases and glycosidases) pose problems because they have a very wide specificity and it is not easy to decide if two preparations described by different authors (perhaps from different sources) have the same catalytic properties, or if they should be listed under separate entries. An example is vitamin A esterase (formerly EC 3.1.1.12, now believed to be identical with EC 3.1.1.1). To some extent the choice must be arbitrary; however, separate entries should be given only when the specificities are sufficiently different.

Another problem are the so-called 'esterolytic' proteases, which hydrolyse ester bonds in appropriate substrates even more rapidly than natural peptide bonds. In this case, classification among the peptide hydrolases was based on historical priority and presumed physiological function.

The second figure in the code number of the hydrolases indicates the nature of the bond hydrolysed: 3.1 are the esterases, 3.2 the glycosidases, and so on (*cf.* section 5).

The third figure normally specifies the nature of the substrate, *e.g* in the esterases the carboxylic ester hydrolases (3.1.1), thiol ester hydrolases (3.1.2), phosphoric monoesterases (3.1.3); in the glycosidases the *O*-glycosidases (3.2.1), *N*-glycosidases (3.2.2), *etc.*. Exceptionally, in the case of the peptidyl-peptide hydrolases, the third figure is based on the catalytic mechanism as shown by active centre studies or the effect of pH (*cf.* section 5 for the full key).

4. LYASES. Lyases are enzymes cleaving **C-C, C-O, C-N,** and other bonds by elimination, leaving double bonds, or conversely adding groups to double bonds. The systematic name is formed according to the pattern *substrate group-lyase.* The hyphen is an important part of the name, and to avoid confusion should not be omitted, *e.g. hydro-lyase* not 'hydrolyase'. In the recommended names, expressions like *decarboxylase, aldolase, dehydratase* (in case of elimination of water) are used. In cases where the reverse reaction is much more important, or the only one demonstrated, *synthase* (not synthetase) may be used in the name. Various sub-classes of the lyases include pyridoxal-phosphate enzymes that catalyse the elimination of β- or γ-substituent from an α-amino acid followed by a replacement of this substituent by some other group. In the overall replacement reaction, no unsaturated end-product is formed; therefore, these enzymes might formally be classified as alkyl-transferases (EC 2.5.1. ..). However, there is ample evidence that the replacement is a two-step reaction involving the transient formation of enzyme-bound α,β(or β, γ)-unsaturated amino acids. According to the rule that the first reaction is indicative for classification, these enzymes are correctly classified as lyases. Examples are tryptophan synthase (EC 4.2.1.20) and cystathionine β-synthase (EC 4.2.1.22).

The second figure in the code number indicates the bond broken: 4.1 are carbon-carbon-lyases, 4.2 carbon-oxygen-lyases and so on.

The third figure gives further information on the group eliminated (*e.g.* CO_2 in 4.1.1, H_2O in 4.2.1).

5. ISOMERASES. These enzymes catalyse geometric or structural changes within one molecule. According to the type of isomerism, they may be called *racemases, epimerases,* cis-trans-*isomerases, isomerases, tautomerases, mutases* or *cyclo-isomerases.*

In some cases, the interconversion in the substrate is brought about by an intramolecular oxidoreduction (5.3); since hydrogen donor and acceptor are the same molecule, and no oxidized product appears, they are not classified as oxidoreductases, even if they may contain firmly bound $NAD(P)^+$.

The sub-classes are formed according to the type of isomerism, the sub-sub-classes to the type of substrates.

6. LIGASES (SYNTHETASES). Ligases are enzymes catalysing the joining together of two molecules coupled with the hydrolysis of a pyrophosphate bond in ATP or a similar triphosphate. The bonds formed are often *high energy* bonds. The systematic names are formed on the system *X:Y ligase (ADP-forming).* In the recommended nomenclature the term *synthetase* may be used, if no other short term (*e.g.* carboxylase) is available. Names of the type 'X-activating enzyme' should not be used.

The second figure in the code number indicates the bond formed: 6.1 for C-O bonds (enzymes acylating tRNA), 6.2 for C-S bonds (acyl-CoA derivatives) *etc.*. Sub-sub-classes are only in use in the C-N ligases (*cf.* section 5).

In a few cases it is necessary to use the word *other* in the description of sub-classes and sub-sub-classes. They have been provisionally given the figure 99, in order to leave space for new subdivisions. Actually, in the revised Enzyme List presented here, a number of new sub-classes and sub-sub-classes have been introduced.

Some enzymes have been deleted from the List, some others have been renumbered. However, the old numbers have *not* been allotted to new enzymes; rather the place has been left vacant and cross-reference is made according to the following scheme:

> [n.m.o.p.   *Deleted entry: old name]*

or

> [n.m.o.p.   *Transferred entry: now EC q.r.s.t. - recommended name]*

Entries for reclassified enzymes transferred from one position in the List to another are followed, for reference, by a comment indicating the former number.

It is regarded as important that the same policy be followed in future revisions and extensions of the Enzyme List, which will become necessary from time to time, and will have to be carried out by future enzyme commissions or Expert Committees of NC-IUB.

4. Rules for classification and nomenclature

(a) GENERAL RULES AND GUIDELINES

Guidelines for recommended names

1. Generally accepted trivial names of substrates may be used in enzyme names. The prefix D- should be omitted for common sugars and L- for individual amino acids, unless ambiguity would be caused. If desired, α,β,γ may be used instead of numbers to indicate positions where such usage is widely established at present; in general, it is not necessary to indicate positions of substituents in recommended names, unless it is necessary to prevent two different enzymes having the same name. The prefix *keto* is used in the non-systematic names for derivatives of sugars in which -CHOH- has been replaced by -CO-.

Rules for systematic nomenclature, on which the classification is based

1. To produce usable systematic names, accepted trivial names of substrates forming part of the enzyme names should be used. Where no accepted and convenient trivial names exist, the official IUPAC rules of nomenclature should be applied to the substrate name. The 1,2,3 system of locating substituents should be used instead of the α,β,γ system; α,β should be used for indicating configuration, although group names such as β-aspartyl-, γ-glutamyl-, and also β-alanine, γ-lactone are permissible. For nucleotide radicals, *adenylyl* (not *adenyl*) *etc.* should be the form used. The name oxo acids (not keto acids) may be used as a class name, and for individual compounds in which -CH$_2$- has been replaced by -CO-, *oxo* should be used.

2. Where the substrate is normally in the form of an anion, its name should end in *-ate* rather than *-ic; e.g. lactate dehydrogenase,* not 'lactic dehydrogenase' or 'lactic acid dehydrogenase'.

Guidelines for recommended names

3. Commonly used abbreviations for substrates, *e.g.* ATP, may be used in names of enzymes, but the use of new abbreviations (*not listed in recommendations of the IUPAC-IUB Commission on Biochemical Nomenclature*) should be discouraged. Chemical formulae should not be used instead of names of substrates. Abbreviations for names of enzymes, *e.g.* GDH, should not be used.

4. Names of substrates composed of two nouns, such as glucose phosphate, which are normally written with a space, should be hyphenated when they form part of the enzyme names, and thus become adjustives, *e.g.* glucose-6-phosphate dehydrogenase (EC 1.1.1.49*).

5. The use as enzyme names of *descriptions* such as *condensing enzyme, Zwischenferment, acetate-activating enzyme, pH 5 enzyme* should be discontinued as soon as the catalysed reaction is known. The word *activating* should not be used in the sense of converting the substrate into a substance that reacts further; all enzymes act by activating their substrates, and the use of the word in this sense may lead to confusion.

6. If it can be avoided, a recommended name should not be based on a substance that is not a true substrate, *e.g.* enzyme EC 4.2.1.17 should not be called *crotonase*, since it does not act on crotonate.

7. Where a name in common use gives some indication of the reaction and is not incorrect or ambiguous, its continued use is recommended. In other cases a recommended name is based on the same general principles as the systematic name (see opposite), but with a minimum of detail, to produce a name short enough for convenient use. A few names of proteolytic enzymes ending in *-in* are retained; all other enzymes names should end in *-ase. [The termination -ese should not be used.]*

7. Systematic names consist of two parts. The first contains the name of the substrate or, in the case of a bimolecular reaction, of the two substrates separated by a colon, with small and equal spaces before and after the colon. The second part, ending in *-ase*, indicates the nature of the reaction. Where additional information is needed to make the reaction clear, a phrase indicating the reaction or a product should be added in parentheses after the second part of the name *e.g. (ADP-forming), (dimerizing), (CoA-acylating).*

8. A number of generic words indicating a type of reaction may be used in either recommended or systematic names: *oxidoreductase, oxygenase, transferase* (with a prefix indicating the nature of the group transferred), *hydrolase, lyase, racemase, epimerase, isomerase, mutase, ligase, synthetase.*

9. A number of additional generic words indicating reaction types are used in recommended names, but not in the systematic nomenclature, *e.g. dehydrogenase, reductase, oxidase, peroxidase, kinase, tautomerase, deaminase, dehydratase, etc..*

* This follows standard practice in phrases where two nouns qualify a third.

14

10. The direct attachment of *-ase* to the name of the substrate will indicate that the enzyme brings about hydrolysis.

10. The suffix *-ase* should never be attached directly to the name of the substrate.

11. The name *dehydrase* which has been used for both dehydrogenating and dehydrating enzymes, should not be used. *Dehydrogenase* will be used for the former and *dehydratase* for the latter.

12. Where possible, recommended names should normally be based on a reaction direction that has been demonstrated, *e.g. dehydrogenase* or *reductase, decarboxylase* or *carboxylase.*

12. In the case of reversible reactions, the direction chosen for naming should be the same for all the enzymes in a given class, even if this direction has not been demonstrated for all. Thus, systematic names may be based on a written reaction, even though only the reverse of this has been actually demonstrated experimentally.

13. When the overall reaction includes two different changes, *e.g.* an oxidative demethylation, the classification and systematic name should be based, whenever possible, on the one (or the first one) catalysed by the enzyme; the other function(s) should be indicated by adding a suitable participle in parentheses, *e.g.* sarcosine:oxygen oxidoreductase (*demethylating*) (EC 1.5.3.1); D-aspartate:oxygen oxidoreductase (*deaminating*) (EC 1.4.3.1); L-serine hydro-lyase (adding indolglycerol-phosphate) (EC 4.2.1.20).

Other examples of such additions are (*decarboxylating*), (*cyclizing*), (*acceptor-acylating*), (*isomerizing*).

14. When an enzyme catalyses more than one type of reaction, the name should normally refer to one reaction only. Each case must be considered on its merit, and the choice must be, to some extent, arbitrary. Other important activities of the enzyme may be indicated in the List under 'Reaction' or 'Comments'.

Similarly, when an enzyme acts on more than one substrate (or pair of substrates), the name should normally refer only to one substrate (or pair of substrates), although in certain cases it may be possible to use a term that covers a whole group of substrates, or an alternative substrate may be given in parentheses.

15. A group of enzymes with rather similar specificities should normally be described by a single entry. However, when the specificity of two enzymes catalysing the same reactions is sufficiently different (the degree of difference being a matter of arbitrary choice) two separate entries may be made, *e.g.* EC 1.2.1.4 and EC 1.2.1.7. Separate entries are also appropriate for enzymes having similar catalytic functions, but known to differ basically with regard to reaction mechanism or to the nature of the catalytic groups, *e.g.* amine oxidase (flavin-containing) (EC 1.4.3.4) and amine oxidase (pyridoxal-containing) (EC 1.4.3.6).

(b) RULES AND GUIDELINES FOR PARTICULAR CLASSES OF ENZYMES

Guidelines for recommended names

Rules for systematic nomenclature, on which the classification is based

Class 1

16. The terms *dehydrogenase* or *reductase* will be used much as hitherto. The latter term is appropriate when hydrogen transfer from the substance mentioned as donor in the systematic name is not readily demonstrated. *Transhydrogenase* may be retained for a few well-established cases. *Oxidase* is used only for cases where O_2 acts as an acceptor, and *oxygenase* only for those cases where the O_2 molecule is directly incorporated into the substrate. *Peroxidase* is used for enzymes using H_2O_2 as acceptor. *Catalase* must be regarded as exceptional. Where no ambiguity is caused, the second reactant is not usually named; but where required to prevent ambiguity, it may be given in parentheses, *e.g.* EC 1.1.1.1, *alcohol dehydrogenase* and EC 1.1.1.2, *alcohol dehydrogenase* ($NADP^+$).

16. All enzymes catalysing oxidoreductions should be named *oxidoreductases* in the systematic nomenclature, and the names formed on the pattern 'donor:acceptor oxidoreductase.'

17. For oxidoreductases using NAD^+ or $NADP^+$, the coenzyme should always be named as the acceptor except for the special case of Section 1.6 (*enzymes whose normal physiological function is regarded to be reoxidation of the reduced coenzyme*). Where the enzyme can use either coenzyme, this should be indicated by writing $NAD(P)^+$.

18. Where the true acceptor is unknown and the oxidoreductase has only been shown to react with artificial acceptors, the word *acceptor* should be written in parentheses, *e.g.* EC 1.3.99.1, succinate:(*acceptor*) oxidoreductase.

19. Oxidoreductases that bring about the incorporation of molecular oxygen into one donor or into either or both of a pair of donors are named *oxygenase*. If only one atom of oxygen is incorporated the term *monooxygenase* is used; if both atoms of O_2 are incorporated, the term *dioxygenase* is used.

19. Oxidoreductases bringing about the incorporation of oxygen into one of paired donors should be named on the pattern 'donor,donor:oxygen oxidoreductase (hydroxylating)'.

Guidelines for recommended names	Rules for systematic nomenclature, on which the classification is based

Class 2

20. Only one specific substrate or reaction product is generally indicated in the recommended names, together with the group donated or accepted.

The forms, *aminotransferase, etc.* may be replaced if desired by the corresponding forms *transaminase, etc.*.

A number of special words are used to indicate reaction types, *e.g. kinase* to indicate a phosphate transfer from ATP, to the named substrate (not *phosphokinase*), *pyrophosphokinase* for a similar transfer of pyrophosphate, *phosphomutase* for an apparent intramolecular phosphate transfer.

20. Enzymes catalysing group-transfer reactions should be named *transferase* and the names formed on the pattern 'donor:acceptor *group-transferred-transferase*', *e.g.* ATP:acetate phosphotransferase (EC 2.7.2.1). A figure may be prefixed to show the position to which the group is transferred, *e.g.* ATP:D-fructose 1-phosphotransferase (EC 2.7.1.3). The spelling *transphorase* should not be used. In the case of the phosphotransferases, ATP should always be named as the donor. In the case of the aminotransferases involving 2-oxoglutarate, the latter should always be named as the acceptor.

21. The prefix denoting the group transferred should, as far as possible, be non-committal with respect to the mechanism of the transfer, *e.g. phospho-*.

Class 3

22. The direct addition of *-ase* to the name of the substrate generally denotes an hydrolase. Where this is difficult, *e.g.* for EC 3.1.2.1, the word *hydrolase* may be used. Enzymes should not normally be given separate names merely on the basis of optimal conditions for activity. The acid and alkaline phosphatases (EC 3.1.3.12) should be regarded as special cases and not as examples to be followed. The recommended name *lysozyme* is also exceptional.

22. Hydrolysing enzymes should be systematically named on the pattern *substrate hydrolase*. Where the enzyme is specific for the removal of a particular group, the group may be named as a prefix, *e.g.* adenosine *amino*hydrolase (EC 3.5.4.4). In a number of cases this group can also be transferred by the enzyme to other molecules, and the hydrolysis itself might be regarded as a transfer of the group to water.

Guidelines for recommended names	Rules for systematic nomenclature, on which the classification is based

Class 4

23. The old names *decarboxylase, aldolase, etc.,* are retained; and *dehydratase* (not *dehydrase*) is used for the hydro-lyases. *Synthetase* should not be used for any enzymes in this class. The term *synthase* may be used instead for any enzyme in this class (or any class other than Class 6) when it is desired to emphasise the synthetic aspect of the reaction.

23. Enzymes removing groups from substrates non-hydrolytically, leaving double bonds (or adding groups to double bonds) should be called *lyases* in the systematic nomenclature. Prefixes such as *hydro-, ammonia-* should be used to denote the type of reaction, *e.g.* L- malate *hydro*-lyase (EC 4.2.1.2). Decarboxylases should be regarded as carboxy-lyases. A hyphen should always be written before *lyase* to avoid confusion with hydrolases, carboxylases, *etc.*.

24. Where the equilibrium warrants it, or where the enzyme has long been named after a particular substrate, the reverse reaction may be taken as the basis of the name, using *hydratase, carboxylase, etc., e.g. fumarate hydratase* for EC 4.2.1.2 (in preference to *fumarase*, which suggests an enzyme hydrolysing fumarate).

24. The complete molecule, not either of the parts into which it is separated, should be named as the substrate.

The part indicated as a prefix to *-lyase* is the more characteristic and usually, but not always, the smaller of the two reaction products. This may either be the removed (saturated) fragment of the substrate molecule, as in *ammonia-, hydro-, thiol-*lyases, *etc.* or the remaining unsaturated fragment, *e.g.* in the case of *carboxy-, aldehyde-* or *oxo acid-*lyases.

25. Various sub-classes of the lyases include a number of strictly specific or group-specific pyridoxal-5-phosphate enzymes that catalyse *elimination* reactions of β- or γ-substituted α-amino acids. Some closely related pyridoxal-5-phosphate-containing enzymes, *e.g.* tryptophan synthase (EC 4.2.1.20) and cystathionine β-synthase (EC 4.2.1.22) catalyse *replacement* reactions in which a β- or γ-substituent is replaced by a second reactant without creating a double bond. Formally, these enzymes appear to be transferases rather than lyases. However, there is evidence that in these cases the elimination of the β- or γ-substituent and the formation of an unsaturated intermediate is the first step in the reaction. Thus, applying rule 13, these enzymes are correctly classified as lyases.

Class 5

26. In this class, the recommended names are, in general, similar to the systematic names which indicate the basis of classification.

27. *Isomerase* will be used as a general name for enzymes in this class. The types of isomerization will be indicated by prefixes, *e.g.* maleate *cis-trans*-isomerase (EC 5.2.1.1), phenylpyruvate keto–enol-isomerase (EC 5.3.2.1), 3-oxosteroid Δ^5–Δ^4-isomerase (EC 5.3.3.1). Enzymes catalysing an aldose-ketose interconversion will be known as *ketolisomerases, e.g.* L-arabinose ketol-isomerase (EC 5.3.1.4). When the isomerization consists of an intramolecular transfer of a group, the enzyme is named a *mutase, e.g.* EC 5.4.1.1, and when it consists of an intramolecular lyase-type reaction, *e.g.* EC 5.5.1.1, it is named a *lyase (decyclizing)*.

28. Isomerases catalysing inversions at asymmetric centres should be termed *racemases* or *epimerases*, according to whether the substrate contains one, or more than one, centre of asymmetry:compare, for example, EC 5.1.1.5 with EC 5.1.1.7. A numerical prefix to the word *epimerase* should be used to show the position of the inversion.

18

Guidelines for recommended names

Rules for systematic nomenclature, on which the classification is based

Class 6

29. Since the enzymes of this class are still generally known as *synthetases,* this designation is retained, in order to avoid extensive changes in the existing nomenclature. *Synthetase* should not be used for enzymes that do not involve nucleoside triphosphates. [See rule 23]

29. The class of enzymes catalysing the linking together of two molecules, coupled with the breaking of a pyrophosphate link in ATP, *etc.* should be known as *ligases.* These enzymes have previously been known as *synthetases;* however, this terminology differs from all other systematic enzyme names in that it is based on the product and not on the substrate. For these reasons, a new systematic class name was necessary.

30. The recommended names should be formed on the pattern X-Y *synthetase,* where X-Y is the substance formed by linking X and Y. Exceptionally where Y is CO_2 the name X *carboxylase may be used. Names of the type X-activating enzyme* are used. [See rule 5]

30. The systematic names should be formed on the pattern *X:Y ligase (ADP-forming),* where X and Y are the two molecules to be joined together. The phrase shown in parentheses indicates both that ATP is the triphosphate involved, and also that the terminal pyrophosphate link is broken. Thus, the reaction is $X + Y + ATP = X - Y + ADP + Pi.$

31. In the special case where glutamine acts as an ammonia-donor, this is indicated by adding in parentheses (*glutamine-hydrolysing*).

31. In this case, the name *amido-ligase* should be used in the systematic nomenclature.

5. Key to numbering and classification of enzymes

1. OXIDOREDUCTASES

1.1 *Acting on the CH-OH group of donors*

1.1.1 With NAD$^+$ or NADP$^+$ as acceptor
1.1.2 With a cytochrome as acceptor
1.1.3 With oxygen as acceptor
1.1.99 With other acceptors

1.2 *Acting on the aldehyde or oxo group of donors*

1.2.1 With NAD$^+$ or NADP$^+$ as acceptor
1.2.2 With a cytochrome as acceptor
1.2.3 With oxygen as acceptor
1.2.4 With a disulphide compound as acceptor
1.2.7 With an iron-sulphur protein as acceptor
1.2.99 With other acceptors

1.3 *Acting on the CH-CH group of donors*

1.3.1 With NAD$^+$ or NADP$^+$ as acceptor
1.3.2 With a cytochrome as acceptor
1.3.3 With oxygen as acceptor
1.3.7 With an iron-sulphur protein as acceptor
1.3.99 With other acceptors

1.4 *Acting on the CH-NH$_2$ group of donors*

1.4.1 With NAD$^+$ or NADP$^+$ as acceptor
1.4.2 With a cytochrome as acceptor
1.4.3 With oxygen as acceptor
1.4.4 With a disulphide compound as acceptor
1.4.7 With an iron-sulphur protein as acceptor
1.4.99 With other acceptors

1.5 *Acting on the CH-NH group of donors*

1.5.1 With NAD$^+$ or NADP$^+$ as acceptor
1.5.3 With oxygen as acceptor
1.5.99 With other acceptors

1.6   *Acting on NADH or NADPH*

1.6.1   With NAD$^+$ or NADP$^+$ as acceptor
1.6.2   With a cytochrome as acceptor
1.6.4   With a disulphide compound as acceptor
1.6.5   With a quinone or related compound as acceptor
1.6.6   With a nitrogenous group as acceptor
1.6.7   With an iron-sulphur protein as acceptor
1.6.99   With other acceptors

1.7   *Acting on other nitrogenous compounds as donors*

1.7.2   With a cytochrome as acceptor
1.7.3   With oxygen as acceptor
1.7.7   With an iron-sulphur protein as acceptor
1.7.99   With other acceptors

1.8   *Acting on a sulphur group of donors*

1.8.1   With NAD$^+$ or NADP$^+$ as acceptor
1.8.2   With a cytochrome as acceptor
1.8.3   With oxygen as acceptor
1.8.4   With a disulphide compound as acceptor
1.8.5   With a quinone or related compound as acceptor
1.8.7   With an iron-sulphur protein as acceptor
1.8.99   With other acceptors

1.9   *Acting on a haem group of donors*

1.9.3   With oxygen as acceptor
1.9.6   With a nitrogenous group as acceptor
1.9.99   With other acceptors

1.10   *Acting on diphenols and related substances as donors*

1.10.1   With NAD$^+$ or NADP$^+$ as acceptor
1.10.2   With a cytochrome as acceptor
1.10.3   With oxygen as acceptor

1.11   *Acting on hydrogen peroxide as acceptor*

1.12   *Acting on hydrogen as donor*

1.12.1   With NAD$^+$ or NADP$^+$ as acceptor
1.12.2   With a cytochrome as acceptor
1.12.7   With an iron-sulphur protein as acceptor

1.13   *Acting on single donors with incorporation of molecular oxygen (oxygenases)*

1.13.11   **With incorporation of two atoms of oxygen**
1.13.12   **With incorporation of one atom of oxygen (internal monooxygenases or internal mixed function oxidases)**
1.13.99   **Miscellaneous (requires further characterization)**

1.14   *Acting on paired donors with incorporation of molecular oxygen*

1.14.11   **With 2-oxoglutarate as one donor, and incorporation of one atom each of oxygen into both donors**
1.14.12   **With NADH or NADPH as one donor, and incorporation of two atoms of oxygen into one donor**
1.14.13   **With NADH or NADPH as one donor, and incorporation of one atom of oxygen**
1.14.14   **With reduced flavin or flavoprotein as one donor, and incorporation of one atom of oxygen**
1.14.15   **With a reduced iron-sulphur protein as one donor, and incorporation of one atom of oxygen**
1.14.16   **With reduced pteridine as one donor, and incorporation of one atom of oxygen**
1.14.17   **With ascorbate as one donor, and incorporation of one atom of oxygen**
1.14.18   **With another compound as one donor, and incorporation of one atom of oxygen**
1.14.99   **Miscellaneous (requires further characterization)**

1.15   *Acting on superoxide radicals as acceptor*

1.16   *Oxidizing metal ions*

1.16.3   **With oxygen as acceptor**

1.17   *Acting on -CH_2 groups*

1.17.1   **With NAD^+ or $NADP^+$ as acceptor**
1.17.4   **With a disulphide compound as acceptor**

1.97   *Other oxidoreductases*

2.   TRANSFERASES

2.1   *Transferring one-carbon groups*

2.1.1   **Methyltransferases**

3. HYDROLASES

3.1 Acting on ester bonds

3.1.1 Carboxylic ester hydrolases
3.1.2 Thiolester hydrolases
3.1.3 Phosphoric monoester hydrolases
3.1.4 Phosphoric diester hydrolases
3.1.5 Triphosphoric monoester hydrolases
3.1.6 Sulphuric ester hydrolases
3.1.7 Diphosphoric monoester hydrolases
3.1.11 Exodeoxyribonucleases producing 5'-phosphomonoesters
3.1.13 Exoribonucleases producing 5'-phosphomonoesters
3.1.14 Exoribonucleases producing other than 5'-phosphomonoesters
3.1.15 Exonucleases active with either ribo- or deoxyribonucleic acids and producing 5'-phosphomonoesters
3.1.16 Exonucleases active with either ribo- or deoxyribonucleic acids and producing other than 5'-phosphomonoesters
3.1.21 Endodeoxyribonucleases producing 5'-phosphomonoesters
3.1.22 Endodeoxyribonucleases producing other than 5'-phosphomonoesters
3.1.23 Site-specific endodeoxyribonucleases: cleavage is sequence-specific
3.1.24 Site-specific endodeoxyribonucleases: cleavage is not sequence-specific
3.1.25 Site-specific endodeoxyribonucleases: specific for altered bases
3.1.26 Endoribonucleases producing 5'-phosphomonoesters
3.1.27 Endoribonucleases producing other than 5'-phosphomonoesters
3.1.30 Endonucleases active with either ribo- or deoxyribonucleic acids and producing 5'-phosphomonoesters
3.1.31 Endonucleases active with either ribo- or deoxyribonucleic acids and producing other than 5'-phosphomonoesters

3.2 Acting on glycosyl compounds

3.2.1 Hydrolysing *O*-glycosyl compounds
3.2.2 Hydrolysing *N*-glycosyl compounds
3.2.3 Hydrolysing *S*-glycosyl compounds

3.3 Acting on ether bonds

3.3.1 Thioether hydrolases
3.3.2 Ether hydrolases

24

3.4   *Acting on peptide bonds (peptide hydrolases)*

3.4.11   **α-Aminoacylpeptide hydrolases**
3.4.12   **Peptidylamino-acid or acylamino-acid hydrolases**
3.4.13   **Dipeptide hydrolases**
3.4.14   **Dipeptidylpeptide hydrolases**
3.4.15   **Peptidyldipeptide hydrolases**
3.4.16   **Serine carboxypeptidases**
3.4.17   **Metallo-carboxypeptidases**
3.4.21   **Serine proteinases**
3.4.22   **Thiol proteinases**
3.4.23   **Carboxyl (acid) proteinases**
3.4.24   **Metalloproteinases**
3.4.99   **Proteinases of unknown catalytic mechanism**

3.5   *Acting on carbon-nitrogen bonds, other than peptide bonds*

3.5.1   **In linear amides**
3.5.2   **In cyclic amides**
3.5.3   **In linear amidines**
3.5.4   **In cyclic amidines**
3.5.5   **In nitriles**
3.5.99   **In other compounds**

3.6   *Acting on acid anhydrides*

3.6.1   **In phosphoryl-containing anhydrides**
3.6.2   **In sulphonyl-containing anhydrides**

3.7   *Acting on carbon-carbon bonds*

3.7.1   **In ketonic substances**

3.8   *Acting on halide bonds*

3.8.1   **In C-halide compounds**
3.8.2   **In P-halide compounds**

3.9   *Acting on phosphorus-nitrogen bonds*

3.10   *Acting on sulphur-nitrogen bonds*

3.11   *Acting on carbon-phosphorus bonds*

4.   LYASES

4.1   Carbon-carbon lyases

      4.1.1   **Carboxy-lases**
      4.1.2   **Aldehyde-lyases**
      4.1.3   **Oxo-acid-lyases**
      4.1.99 **Other carbon-carbon lyases**

4.2   Carbon-oxygen lyases

      4.2.1   **Hydro-lyases**
      4.2.2   **Acting on polysaccharides**
      4.2.99 **Other carbon-oxygen lyases**

4.3   Carbon-nitrogen lyases

      4.3.1   **Ammonia-lyases**
      4.3.2   **Amidine-lyases**

4.4   Carbon-sulphur lyases

4.5   Carbon-halide lyases

4.6   Phosphorus-oxygen lyases

4.99   Other lyases

5.   ISOMERASES

5.1   Racemases and epimerases

      5.1.1   **Acting on amino acids and derivatives**
      5.1.2   **Acting on hydroxy acids and derivatives**
      5.1.3   **Acting on carbohydrates and derivatives**
      5.1.99 **Acting on other compounds**

5.2   Cis-trans isomerases

5.3   Intramolecular oxidoreductases

ENZYME LIST

NOTES ON THE ENZYME LIST

(1) *Other Names*. To enable the list to be used for finding recommended names, as many 'other names' that have been used as possible have been included. Some of these may only have been used to a small extent, and many of the names reported are unacceptable.

(2) *References*. It is not intended that the list should provide a complete bibliography, nor to give an indication of priority in the discovery of each enzyme. The references included for each entry are believed to be sufficient to justify the entry, showing the definite existence of an enzyme catalysing the reaction set out.

1. OXIDOREDUCTASES

To this class belong all enzymes catalysing oxido-reductions. The substrate oxidized is regarded as hydrogen or electron donor. The classification is based on 'donor:acceptor oxidoreductase'. The recommended name is 'dehydrogenase', wherever this is possible; as an alternative, 'acceptor reductase' can be used. 'Oxidase' is used only in cases where O_2 is an acceptor. Classification is difficult in some cases because of the lack of specificity towards the acceptor.

1.1 ACTING ON THE CH-OH GROUP OF DONORS

This sub-class contains all dehydrogenases acting on primary alcohols, secondary alcohols and hemiacetals. They are further classified according to the acceptor which can be NAD^+ or $NADP^+$ (sub-class 1.1.1), cytochrome (1.1.2), oxygen (1.1.3), or another acceptor (1.1.99).

1.1.1 WITH NAD⁺ OR NADP⁺ AS ACCEPTOR

When an enzyme can use either NAD^+ or $NADP^+$, the symbol $NAD(P)^+$ is used.

Number	Recommended Name	Reaction
1.1.1.1	Alcohol dehydrogenase	Alcohol + NAD^+ = aldehyde or ketone + NADH
1.1.1.2	Alcohol dehydrogenase ($NADP^+$)	Alcohol + $NADP^+$ = aldehyde + NADPH
1.1.1.3	Homoserine dehydrogenase	L-Homoserine + $NAD(P)^+$ = L-aspartate β-semialdehyde + NAD(P)H
1.1.1.4	D(—)-Butanediol dehydrogenase	D(—)-2,3-Butanediol + NAD^+ = acetoin + NADH
1.1.1.5	Acetoin dehydrogenase	Acetoin + NAD^+ = diacetyl + NADH
1.1.1.6	Glycerol dehydrogenase	Glycerol + NAD^+ = dihydroxyacetone + NADH
1.1.1.7	Propanediol-phosphate dehydrogenase	1,2-Propanediol 1-phosphate + NAD^+ = hydroxyacetone phosphate + NADH

Number	Other Names	Basis for classification (Systematic Name)	Comments	Reference
1.1.1.1	Aldehyde reductase	Alcohol:NAD$^+$ oxidoreductase	Acts on primary or secondary alcohols or hemiacetals; the animal, but not the yeast, enzyme acts also on cyclic secondary alcohols	340, 632, 2206, 2384, 3292, 3396, 3533, 871
1.1.1.2		Alcohol:NADP$^+$ oxidoreductase	Some members of this group oxidize only primary alcohols; others act more rapidly on secondary alcohols	691, 2755, 3362
1.1.1.3		L-Homoserine: NAD(P)$^+$ oxidoreductase	The yeast enzyme acts most rapidly with NAD$^+$; the *Neurospora* enzyme with NADP$^+$. The enzyme from *E. coli* is a multifunctional protein, which also catalyses the reaction of EC 2.7.2.4	309, 3823
1.1.1.4	Butyleneglycol dehydrogenase	D(—)-2,3-Butanediol: NAD$^+$ oxidoreductase	Also converts diacetyl to acetoin with NADH as reductant	3257, 3378
1.1.1.5	Diacetyl reductase	Acetoin:NAD$^+$ oxidoreductase	NADP$^+$ also acts	3257, 723
1.1.1.6		Glycerol:NAD$^+$ 2-oxidoreductase	Also acts on 1,2-propanediol	123, 437, 1969
1.1.1.7		1,2-Propanediol-1-phosphate:NAD$^+$ oxidoreductase		3003

Number	Recommended Name	Reaction
1.1.1.8	Glycerol-3-phosphate dehydrogenase (NAD⁺)	sn-Glycerol 3-phosphate* + NAD⁺ = dihydroxyacetone phosphate + NADH
1.1.1.9	D-Xylulose reductase	Xylitol + NAD⁺ = D-xylulose + NADH
1.1.1.10	L-Xylulose reductase	Xylitol + NADP⁺ = L-xylulose + NADPH
1.1.1.11	D-Arabinitol dehydrogenase	D-Arabinitol + NAD⁺ = D-xylulose + NADH
1.1.1.12	L-Arabinitol dehydrogenase	L-Arabinitol + NAD⁺ = L-xylulose + NADH
1.1.1.13	L-Arabinitol dehydrogenase (ribulose-forming)	L-Arabinitol + NAD⁺ = L-ribulose + NADH
1.1.1.14	L-Iditol dehydrogenase	L-Iditol + NAD⁺ = L-sorbose + NADH
1.1.1.15	D-Iditol dehydrogenase	D-Iditol + NAD⁺ = D-sorbose + NADH
1.1.1.16	Galactitol dehydrogenase	Galactitol + NAD⁺ = D-tagatose + NADH
1.1.1.17	Mannitol-1-phosphate dehydrogenase	D-Mannitol 1-phosphate + NAD⁺ = D-fructose 6-phosphate + NADH
1.1.1.18	myo-Inositol 2-dehydrogenase	myo-Inositol + NAD⁺ = 2,4,6/3,5-pentahydroxycyclohexanone + NADH
1.1.1.19	Glucuronate reductase	L-Gulonate + NADP⁺ = D-glucuronate + NADPH

---

*This is R-glycerol 3-phosphate

Number	Other Names	Basis for classification (Systematic Name)	Comments	Reference
1.1.1.8		*sn*-Glycerol-3-phosphate:NAD$^+$ 2-oxidoreductase	Also acts on 1,2-propanediol phosphate and dihydroxyacetone sulphate (but with a much lower affinity)	199, 200, 2226, 3388
1.1.1.9		Xylitol:NAD$^+$ 2-oxidoreductase (D-xylulose-forming)	Also acts as an L-erythrulose reductase	521, 1303, 1492
1.1.1.10		Xylitol:NADP$^+$ 4-oxidoreductase (L-xylulose-forming)		1303, 1347, 3452
1.1.1.11		D-Arabinitol:NAD$^+$ 4-oxidoreductase		1968, 3727
1.1.1.12		L-Arabinitol:NAD$^+$ 4-oxidoreductase (L-xylulose-forming)		521, 522
1.1.1.13		L-Arabinitol:NAD$^+$ 2-oxidoreductase (L-ribulose-forming)		522
1.1.1.14	Polyol dehydrogenase, Sorbitol dehydrogenase	L-Iditol:NAD$^+$ 5-oxidoreductase	Also acts on D-glucitol (giving D-fructose) and other closely related sugar alcohols	1375, 2148, 3688
1.1.1.15		D-Iditol:NAD$^+$ 5-oxidoreductase	Also converts xylitol to L-xylulose and L-glucitol to L-fructose	3028
1.1.1.16		Galactitol:NAD$^+$ 3-oxidoreductase	Also converts other alditols containing an L-*threo*-configuration adjacent to a primary alcohol group to the corresponding sugars	3028
1.1.1.17		D-Mannitol-1-phosphate:NAD$^+$ 2-oxidoreductase		2090, 3718, 3719
1.1.1.18		*myo*-Inositol:NAD$^+$ 2-oxidoreductase		277, 1879, 3551
1.1.1.19		L-Gulonate:NADP$^+$ 1-oxidoreductase	Also reduces D-galacturonate	3093, 3799

Number	Recommended Name	Reaction
1.1.1.20	Glucuronolactone reductase	L-Gulono-γ-lactone + NADP$^+$ = D-glucurono-γ-lactone + NADPH
1.1.1.21	Aldose reductase	Alditol + NADP$^+$ = aldose + NADPH
1.1.1.22	UDPglucose dehydrogenase	UDPglucose + 2 NAD$^+$ + H$_2$O = UDPglucuronate + 2 NADH
1.1.1.23	Histidinol dehydrogenase	L-Histidinol + 2 NAD$^+$ = L-histidine + 2 NADH
1.1.1.24	Quinate dehydrogenase	Quinate + NAD$^+$ = 3-dehydroquinate + NADH
1.1.1.25	Shikimate dehydrogenase	Shikimate + NADP$^+$ = 3-dehydroshikimate + NADPH
1.1.1.26	Glyoxylate reductase	Glycolate + NAD$^+$ = glyoxylate + NADH
1.1.1.27	Lactate dehydrogenase	L-Lactate + NAD$^+$ = pyruvate + NADH
1.1.1.28	D-Lactate dehydrogenase	D-Lactate + NAD$^+$ = pyruvate + NADH
1.1.1.29	Glycerate dehydrogenase	D-Glycerate + NAD$^+$ = hydroxypyruvate + NADH
1.1.1.30	3-Hydroxybutyrate dehydrogenase	D-3-Hydroxybutyrate + NAD$^+$ = acetoacetate + NADH
1.1.1.31	3-Hydroxyisobutyrate dehydrogenase	3-Hydroxyisobutyrate + NAD$^+$ = methylmalonate semialdehyde + NADH
1.1.1.32	Mevaldate reductase	Mevalonate + NAD$^+$ = mevaldate + NADH
1.1.1.33	Mevaldate reductase (NADPH)	Mevalonate + NADP$^+$ = mevaldate + NADPH
1.1.1.34	Hydroxymethylglutaryl-CoA reductase (NADPH)	Mevalonate + CoA + 2 NADP$^+$ = 3-hydroxy-3-methylglutaryl-CoA + 2 NADPH

Number	Other Names	Basis for classification (Systematic Name)	Comments	Reference
1.1.1.20		L-Gulono-γ-lactone:NADP+ 1-oxidoreductase		3307
1.1.1.21	Polyol dehydrogenase (NADP+)	Alditol:NADP+ 1-oxidoreductase	Wide specificity	1292, 131, 2950
1.1.1.22		UDPglucose:NAD+ 6-oxidoreductase	Also acts on UDP-2-deoxyglucose	2136, 3267, 3268, 759
1.1.1.23		L-Histidinol:NAD+ oxidoreductase	Also oxidizes L-histidinal	15, 16, 2002, 3815
1.1.1.24		Quinate:NAD+ 3-oxidoreductase		989, 2249
1.1.1.25		Shikimate:NADP+ 3-oxidoreductase		186, 2249, 3779
1.1.1.26		Glycolate:NAD+ oxidoreductase	Reduces glyoxylate to glycolate or hydroxypyruvate to D-glycerate	3838, 3839
1.1.1.27	Lactic acid dehydrogenase	L-Lactate:NAD+ oxidoreductase	Also oxidizes other L-2-hydroxymonocarboxylic acids. NADP+ also acts, more slowly	696, 1030, 1808, 3250
1.1.1.28	Lactic acid dehydrogenase	D-Lactate:NAD+ oxidoreductase		696
1.1.1.29		D-Glycerate:NAD+ oxidoreductase		1354, 3193
1.1.1.30		D-3-Hydroxybutyrate: NAD+ oxidoreductase	Also oxidizes other 3-hydroxymonocarboxylic acids	270, 688, 1912
1.1.1.31		3-Hydroxyisobutyrate: NAD+ oxidoreductase		2819
1.1.1.32		Mevalonate:NAD+ oxidoreductase		2956
1.1.1.33		Mevalonate:NADP+ oxidoreductase		576
1.1.1.34		Mevalonate:NADP+ oxidoreductase (CoA-acylating)		419, 772, 1623

34

Number	Recommended Name	Reaction
1.1.1.35	3-Hydroxyacyl-CoA dehydrogenase	L-3-Hydroxyacyl-CoA + NAD$^+$ = 3-oxoacyl-CoA + NADH
1.1.1.36	Acetoacetyl-CoA reductase	D-3-Hydroxyacyl-CoA + NADP$^+$ = 3-oxoacyl-CoA + NADPH
1.1.1.37	Malate dehydrogenase	L-Malate + NAD$^+$ = oxaloacetate + NADH
1.1.1.38	Malate dehydrogenase (oxaloacetate-decarboxylating)	L-Malate + NAD$^+$ = pyruvate + CO_2 + NADH
1.1.1.39	Malate dehydrogenase (decarboxylating)	L-Malate + NAD$^+$ = pyruvate + CO_2 + NADH
1.1.1.40	Malate dehydrogenase (oxaloacetate-decarboxylating) (NADP$^+$)	L-Malate + NADP$^+$ = pyruvate + CO_2 + NADPH
1.1.1.41	Isocitrate dehydrogenase (NAD$^+$)	*threo*-D$_S$-Isocitrate† + NAD$^+$ = 2-oxoglutarate + CO_2 + NADH
1.1.1.42	Isocitrate dehydrogenase (NADP$^+$)	*threo*-D$_S$-Isocitrate + NADP$^+$ = 2-oxoglutarate + CO_2 + NAD(P)H
1.1.1.43	Phosphogluconate dehydrogenase	6-Phospho-D-gluconate + NAD(P)$^+$ = 6-phospho-2-keto-D-gluconate + NAD(P)H
1.1.1.44	Phosphogluconate dehydrogenase (decarboxylating)	6-Phospho-D-gluconate + NADP$^+$ = D-ribulose 5-phosphate + CO_2 + NADPH
1.1.1.45	L-Gulonate dehydrogenase	L-Gulonate + NAD$^+$ = 3-keto-L-gulonate + NADH
1.1.1.46	L-Arabinose dehydrogenase	L-Arabinose + NAD$^+$ = L-arabinono-γ-lactone + NADH

† Following the stereochemical convention put forward by H. B. Vickery (1962) *J. Biol. Chem.* 237, 1739

Number	Other Names	Basis for classification (Systematic Name)	Comments	Reference
1.1.1.35	β-Hydroxyacyl dehydrogenase, β-Keto-reductase	L-3-Hydroxyacyl-CoA: NAD$^+$ oxidoreductase	Also oxidizes S-3-hydroxyacyl-N-acylthioethanolamine and S^6-3-hydroxyacyl-hydrolipoate. Some enzymes act, more slowly, with NADP$^+$	1911, 3208, 3586, 1312
1.1.1.36		D-3-Hydroxyacyl-CoA: NADP$^+$ oxidoreductase		3585
1.1.1.37	Malic dehydrogenase	L-Malate:NAD$^+$ oxidoreductase	Also oxidizes some other 2-hydroxydicarboxylic acids	652, 830, 1735, 3251, 3716
1.1.1.38	'Malic' enzyme, Pyruvic-malic carboxylase	L-Malate:NAD$^+$ oxidoreductase (oxaloacetate-decarboxylating)	Also decarboxylates added oxaloacetate	1619
1.1.1.39	'Malic' enzyme, Pyruvic-malic carboxylase	L-Malate:NAD$^+$ oxidoreductase (decarboxylating)	Does not decarboxylate added oxaloacetate	2932
1.1.1.40	'Malic' enzyme, Pyruvic-malic carboxylase	L-Malate:NADP$^+$ oxidoreductase (oxaloacetate-decarboxylating)	Also decarboxylates added oxaloacetate	1192, 2468, 2876, 3260, 3588
1.1.1.41	Isocitric dehydrogenase, β-Ketoglutaric-isocitric carboxylase	threo-D$_S$-Isocitrate: NAD$^+$ oxidoreductase (decarboxylating)	Does not decarboxylate added oxalosuccinate	1771, 2620, 2622, 2706, 1226
1.1.1.42		threo-D$_S$-Isocitrate: NADP$^+$ oxidoreductase (decarboxylating)	Also decarboxylates added oxalosuccinate	27, 2306, 2620, 3065
1.1.1.43	6-Phosphogluconic dehydrogenase	6-Phospho-D-gluconate: NAD(P)$^+$ 2-oxidoreductase		920
1.1.1.44	Phosphogluconic acid dehydrogenase, 6-Phosphogluconic dehydrogenase, 6-Phosphogluconic carboxylase	6-Phospho-D-gluconate:NADP$^+$ 2-oxidoreductase (decarboxylating)	Certain preparations also reduce NAD$^+$ as well as NADP$^+$	716, 2634, 2986, 2987
1.1.1.45	L-3-Aldonate dehydrogenase	L-Gulonate:NAD$^+$ 3-oxidoreductase	Also oxidizes other L-3-hydroxyacids	727, 3105
1.1.1.46		L-Arabinose:NAD$^+$ 1-oxidoreductase		3641

Number	Recommended Name	Reaction
1.1.1.47	Glucose dehydrogenase	β-D-Glucose + NAD(P)$^+$ = D-glucono-δ-lactone + NAD(P)H
1.1.1.48	Galactose dehydrogenase	D-Galactose + NAD$^+$ = D-galactono-γ-lactone + NADH
1.1.1.49	Glucose-6-phosphate dehydrogenase	D-Glucose 6-phosphate + NADP$^+$ = D-glucono-δ-lactone 6-phosphate + NADPH
1.1.1.50	3α-Hydroxysteroid dehydrogenase	Androsterone + NAD(P)$^+$ = 5α-androstane-3,17-dione + NAD(P)H
1.1.1.51	β-Hydroxysteroid dehydrogenase	Testosterone + NAD(P)$^+$ = 4-androstene-3,17-dione + NAD(P)H
1.1.1.52	3α-Hydroxycholanate dehydrogenase	3α-Hydroxy-5β-cholanate + NAD$^+$ = 3-oxo-5β-cholanate + NADH
1.1.1.53	20β-Hydroxysteroid dehydrogenase	17,20β,21-Trihydroxypregn-4-ene-3,11-dione + NAD$^+$ = cortisone + NADH
1.1.1.54	Allyl-alcohol dehydrogenase	Allyl alcohol + NADP$^+$ = acrolein + NADPH
1.1.1.55	Lactaldehyde reductase (NADPH)	1,2-Propanediol + NADP$^+$ = L-lactaldehyde + NADPH
1.1.1.56	Ribitol dehydrogenase	Ribitol + NAD$^+$ = D-ribulose + NADH
1.1.1.57	Fructuronate reductase	D-Mannonate + NAD$^+$ = D-fructuronate + NADH
1.1.1.58	Tagaturonate reductase	D-Altronate + NAD$^+$ = D-tagaturonate + NADH
1.1.1.59	3-Hydroxypropionate dehydrogenase	3-Hydroxypropionate + NAD$^+$ = malonate semialdehyde + NADH
1.1.1.60	Tartronate-semialdehyde reductase	D-Glycerate + NAD(P)$^+$ = tartronate semialdehyde + NAD(P)H
1.1.1.61	4-Hydroxybutyrate dehydrogenase	4-Hydroxybutyrate + NAD$^+$ = succinate semialdehyde + NADH
1.1.1.62	Oestradiol 17β-dehydrogenase	Oestradiol-17β + NAD$^+$ = oestrone + NADH

Number	Other Names	Basis for classification (Systematic Name)	Comments	Reference
1.1.1.47		β-D-Glucose:NAD(P)$^+$ 1-oxidoreductase	Also oxidizes D-xylose	387, 3258, 3415, 192, 2577
1.1.1.48		D-Galactose:NAD$^+$ 1-oxidoreductase		675, 1389
1.1.1.49	Zwischenferment, Robison ester dehydrogenase	D-Glucose-6-phosphate: NADP$^+$ 1-oxidoreductase	Also acts slowly on β-D-glucose and other sugars. Certain preparations also reduce NAD$^+$ as well as NADP$^+$	829, 1057, 1545, 2430
1.1.1.50		3α-Hydroxysteroid: NAD(P)$^+$ oxidoreductase	Also acts on other 3α-hydroxysteroids	1731, 2080, 3441
1.1.1.51		3 (or 17) β-Hydroxysteroid: NAD(P)$^+$ oxidoreductase	Also acts on other 3β- or 17β-hydroxysteroids	624, 2022, 2080, 3356
1.1.1.52		3α-Hydroxy-5β-cholanate: NAD$^+$ oxidoreductase	Also acts on other 3α-hydroxysteroids with an acidic side-chain	1240
1.1.1.53	Cortisone reductase	17,20β,21-Trihydroxysteroid: NAD$^+$ oxidoreductase	Also acts on other 17,20,21-trihydroxysteroids	1394, 1395, 2022, 2734
1.1.1.54		Allyl-alcohol:NADP$^+$ oxidoreductase	Also acts on saturated primary alcohols	2525
1.1.1.55		1,2-Propanediol: NADP$^+$ oxidoreductase		1148
1.1.1.56		Ribitol:NAD$^+$ 2-oxidoreductase		1347, 2437, 3727
1.1.1.57		D-Mannonate:NAD$^+$ 5-oxidoreductase	Also reduces D-tagaturonate	1304, 1683
1.1.1.58		D-Altronate:NAD$^+$ 3-oxidoreductase		1304
1.1.1.59		3-Hydroxypropionate: NAD$^+$ oxidoreductase		694
1.1.1.60		D-Glycerate: NAD(P)$^+$ oxidoreductase		1088
1.1.1.61		4-Hydroxybutyrate: NAD$^+$ oxidoreductase		2414
1.1.1.62		Oestradiol-17β:NAD$^+$ 17-oxidoreductase		1622, 1874

Number	Recommended Name	Reaction
1.1.1.63	Testosterone 17β-dehydrogenase	Testosterone + NAD$^+$ = 4-androstene-3,17-dione + NADH
1.1.1.64	Testosterone 17β-dehydrogenase (NADP$^+$)	Testosterone + NADP$^+$ = 4-androstene-3,17-dione + NADPH
1.1.1.65	Pyridoxine 4-dehydrogenase	Pyridoxine + NADP$^+$ = pyridoxal + NADPH
1.1.1.66	ω-Hydroxydecanoate dehydrogenase	10-Hydroxydecanoate + NAD$^+$ = 10-oxodecanoate + NADH
1.1.1.67	Mannitol dehydrogenase	D-Mannitol + NAD$^+$ = D-fructose + NADH

[1.1.1.68 Transferred entry: now EC 1.1.99.15, 5,10-Methylenetetrahydrofolate reductase (FADH$_2$)]

1.1.1.69	Gluconate 5-dehydrogenase	D-Gluconate + NAD(P)$^+$ = 5-keto-D-gluconate + NAD(P)H

[1.1.1.70 Deleted entry: now included with EC 1.2.1.3, Aldehyde dehydrogenase]

1.1.1.71	Alcohol dehydrogenase (NAD(P)$^+$)	Alcohol + NAD(P)$^+$ = aldehyde + NAD(P)H
1.1.1.72	Glycerol dehydrogenase (NADP$^+$)	Glycerol + NADP$^+$ = D-glyceraldehyde + NADPH
1.1.1.73	Octanol dehydrogenase	1-Octanol + NAD$^+$ = 1-octanal + NADH

[1.1.1.74 Deleted entry: D-Aminopropanol dehydrogenase (reaction due to EC 1.1.1.4)]

1.1.1.75	L-Aminopropanol dehydrogenase	L-1-Aminopropan-2-ol + NAD$^+$ = aminoacetone + NADH
1.1.1.76	L(+)-Butanediol dehydrogenase	L(+)-2,3-Butanediol + NAD$^+$ = acetoin + NADH
1.1.1.77	Lactaldehyde reductase	D(or L)-1,2-Propanediol + NAD$^+$ = D(or L)-lactaldehyde + NADH
1.1.1.78	D-Lactaldehyde dehydrogenase	D-Lactaldehyde + NAD$^+$ = methylglyoxal + NADH

Number	Other Names	Basis for classification (Systematic Name)	Comments	Reference
1.1.1.63		17β-Hydroxysteroid:NAD$^+$ 17-oxidoreductase		825, 3316, 3552
1.1.1.64		17β-Hydroxysteroid:NADP$^+$ 17-oxidoreductase	Also oxidizes 3-hydroxyhexobarbital to 3-oxohexobarbital	825, 3316, 3552
1.1.1.65	Pyridoxine dehydrogenase	Pyridoxine:NADP$^+$ 4′-oxidoreductase	Also oxidizes pyridoxine phosphate	1355
1.1.1.66		10-Hydroxydecanoate: NAD$^+$ oxidoreductase	Also acts, more slowly, on 9-hydroxynonanoate and 11-hydroxyundecanoate	1580, 2251
1.1.1.67		D-Mannitol:NAD$^+$ 2-oxidoreductase		2102
1.1.1.69	5-Keto-D-gluconate 5-reductase	D-Gluconate:NAD(P)$^+$ 5-oxidoreductase		2490, 690
1.1.1.71	Retinal reductase	Alcohol:NAD(P)$^+$ oxidoreductase	Reduces aliphatic aldehydes of carbon chain length from 2 to 14, with greatest activity on C-4, C-6 and C-8 aldehydes; also reduces retinal to retinol	876
1.1.1.72		Glycerol:NADP$^+$ oxidoreductase		1762, 3433
1.1.1.73		Octanol:NAD$^+$ oxidoreductase	Acts, less rapidly, on other long-chain alcohols	2823
1.1.1.75		L-1-Aminopropan-2-ol: NAD$^+$ oxidoreductase	Requires K$^+$	3492, 3493
1.1.1.76		L(+)-2,3-Butanediol: NAD$^+$ oxidoreductase		3378
1.1.1.77		D(or L)-1,2-Propanediol: NAD$^+$ oxidoreductase		3424
1.1.1.78		D-Lactaldehyde: NAD$^+$ oxidoreductase		3423

Number	Recommended Name	Reaction
1.1.1.79	Glyoxylate reductase (NADP$^+$)	Glycollate + NADP$^+$ = glyoxylate + NADPH
1.1.1.80	Isopropanol dehydrogenase (NADP$^+$)	2-Propanol + NADP$^+$ = acetone + NADPH
1.1.1.81	Hydroxypyruvate reductase	D-Glycerate + NAD(P)$^+$ = hydroxypyruvate + NAD(P)H
1.1.1.82	Malate dehydrogenase (NADP$^+$)	L-Malate + NADP$^+$ = oxaloacetate + NADPH
1.1.1.83	D-Malate dehydrogenase (decarboxylating)	D-Malate + NAD$^+$ = pyruvate + CO_2 + NADH
1.1.1.84	Dimethylmalate dehydrogenase	3,3-Dimethyl-D-malate + NAD$^+$ = 2-oxoisovalerate + CO_2 + NADH
1.1.1.85	3-Isopropylmalate dehydrogenase	2-Hydroxy-4-methyl-3-carboxyvalerate + NAD$^+$ = 2-oxo-4-methyl-3-carboxyvalerate + NADH
1.1.1.86	Ketol-acid reductoisomerase	2,3-Dihydroxyisovalerate + NADP$^+$ = 2-acetolactate + NADPH
1.1.1.87	2-Hydroxy-3-carboxyadipate dehydrogenase	2-Hydroxy-3-carboxyadipate + NAD$^+$ = 2-oxoadipate + CO_2 + NADH
1.1.1.88	Hydroxymethylglutaryl-CoA reductase	Mevalonate + CoA + 2 NAD$^+$ = 3-hydroxy-3-methylglutaryl-CoA + 2 NADH

[1.1.1.89 Deleted entry: Dihydroxyisovalerate dehydrogenase (isomerizing). Now included with EC 1.1.1.86]

1.1.1.90	Aryl-alcohol dehydrogenase	An aromatic alcohol + NAD$^+$ = an aromatic aldehyde + NADH
1.1.1.91	Aryl-alcohol dehydrogenase (NADP$^+$)	An aromatic alcohol + NADP$^+$ = an aromatic aldehyde + NADPH

Number	Other Names	Basis for classification (Systematic Name)	Comments	Reference
1.1.1.79		Glycollate:NADP$^+$ oxidoreductase	Also reduces hydroxypyruvate to glycerate; has some affinity for NAD$^+$	480
1.1.1.80		2-Propanol:NADP$^+$ oxidoreductase	Also acts on other short-chain secondary alcohols, and, slowly on primary alcohols	1376, 1377, 2756
1.1.1.81		D-Glycerate:NAD(P)$^+$ 2-oxidoreductase		1737
1.1.1.82		L-Malate:NADP$^+$ oxidoreductase	Activated by light	1220, 1519, 1520
1.1.1.83		D-Malate:NAD$^+$ oxidoreductase (decarboxylating)		3216
1.1.1.84		3,3,-Dimethyl-D-malate: NAD$^+$ oxidoreductase (decarboxylating)	Requires K$^+$ or NH$_4$$^+$ and Mn^{2+} or Co^{2+}; also acts on D-malate	2040
1.1.1.85		2-Hydroxy-4-methyl-3-carboxyvalerate:NAD$^+$ oxidoreductase	The product decarboxylates to 2-oxo-4-methylvalerate	435, 2560
1.1.1.86	Dihydroxyisovalerate dehydrogenase (isomerizing), Acetohydroxy acid isomeroreductase	2,3-Dihydroxy-isovalerate: NADP$^+$ oxidoreductase (isomerizing)	Also catalyses the reduction of 2-aceto-2-hydroxybutyrate to 2,3-dihydroxy-3-methylvalerate	103, 1696, 2921
1.1.1.87		2-Hydroxy-3-carboxyadipate: NAD$^+$ oxidoreductase (decarboxylating)		3247
1.1.1.88		Mevalonate:NAD$^+$ oxidoreductase (CoA-acylating)		879
1.1.1.90		Aryl-alcohol:NAD$^+$ oxidoreductase	Enzyme uses a range of primary alcohols with an aromatic or cyclohex-l-ene ring as substrates, but does not act on short-chain aliphatic alcohols	3283
1.1.1.91		Aryl-alcohol:NADP$^+$ oxidoreductase	Also acts on some aliphatic aldehydes but cinnamaldehyde was the best substrate found	1131

Number	Recommended Name	Reaction
1.1.1.92	Oxaloglycollate reductase (decarboxylating)	D-Glycerate + NAD(P)$^+$ + CO$_2$ = oxaloglycollate + NAD(P)H
1.1.1.93	Tartrate dehydrogenase	Tartrate + NAD$^+$ = oxaloglycollate + NADH
1.1.1.94	sn-Glycerol-3-phosphate dehydrogenase (NAD(P)$^+$)	sn-Glycerol 3-phosphate + NAD(P)$^+$ = dihydroxyacetone phosphate + NAD(P)H
1.1.1.95	Phosphoglycerate dehydrogenase	3-Phosphoglycerate + NAD$^+$ = 3-phosphohydroxypyruvate + NADH
1.1.1.96	Diiodophenylpyruvate reductase	β-(3,5-Diiodo-4-hydroxyphenyl) lactate + NAD$^+$ = β-(3,5-diiodo-4-hydroxyphenyl) pyruvate + NADH
1.1.1.97	3-Hydroxybenzyl-alcohol dehydrogenase	3-Hydroxybenzyl alcohol + NADP$^+$ = 3-hydroxybenzaldehyde + NADPH
1.1.1.98	D-2-Hydroxy-fatty-acid dehydrogenase	D-2-Hydroxystearate + NAD$^+$ = 2-oxostearate + NADH
1.1.1.99	L-2-Hydroxy-fatty-acid dehydrogenase	L-2-Hydroxystearate + NAD$^+$ = 2-oxostearate + NADH
1.1.1.100	3-Oxoacyl-[acyl-carrier-protein] reductase	D-3-Hydroxyacyl-[acyl-carrier protein] + NADP$^+$ = 3-oxoacyl-[acyl-carrier protein] + NADPH
1.1.1.101	Palmitoyldihydroxyacetone-phosphate reductase	1-Palmitoylglycerol 3-phosphate + NADP$^+$ = palmitoyldihydroxyacetone phosphate + NADPH
1.1.1.102	3-Dehydrosphinganine reductase	Sphinganine + NADP$^+$ = 3-dehydrosphingosine + NADPH
1.1.1.103	L-Threonine 3-dehydrogenase	L-Threonine + NAD$^+$ = L-2-amino-3-oxobutyrate + NADH

Number	Other Names	Basis for classification (Systematic Name)	Comments	Reference
1.1.1.92		D-Glycerate:NAD(P)$^+$ oxidoreductase (carboxylating)	Also reduces hydroxypyruvate to D-glycerate and glyoxylate to glycollate	1736
1.1.1.93		Tartrate:NAD$^+$ oxidoreductase	meso-Tartrate and L$^+$-tartrate act as substrates; requires Mn^{2+} and a monovalent cation	1739
1.1.1.94		sn-Glycerol-3-phosphate: NAD(P)$^+$ 2-oxidoreductase		1705
1.1.1.95		3-Phosphoglycerate:NAD$^+$ 2-oxidoreductase		3280, 3696
1.1.1.96		β-(3,5-Diiodo-4-hydroxyphenyl)-lactate:NAD$^+$ oxidoreductase	Substrates contain an aromatic ring with a pyruvate side chain. The most active substrates are halogenated derivatives. Compounds with hydroxyl or amino groups in the 3 or 5 position are inactive	3833
1.1.1.97		3-Hydroxybenzyl-alcohol: NADP$^+$ oxidoreductase		917
1.1.1.98		D-2-Hydroxystearate: NAD$^+$ oxidoreductase		1936
1.1.1.99		L-2-Hydroxystearate: NAD$^+$ oxidoreductase		1391
1.1.1.100		D-3-Hydroxyacyl-[acyl-carrier-protein]:NADP$^+$ oxidoreductase	Exhibits a marked preference for acyl-carrier-protein derivatives over CoA derivatives as substrates	3447, 2660
1.1.1.101		1-Palmitoylglycerol-3-phosphate:NADP$^+$ oxidoreductase	Also acts on alkylglycerol 3-phosphate	1171, 1853
1.1.1.102		D-erythro-Dihydrosphingosine: NADP$^+$ 3-oxidoreductase		3236, 3237
1.1.1.103		L-Threonine: NAD$^+$ oxidoreductase	The product spontaneously decarboxylates to aminoacetone	1106, 1206

Number	Recommended Name	Reaction
1.1.1.104	4-Oxoproline reductase	4-Hydroxy-L-proline + NAD$^+$ = 4-oxoproline + NADH
1.1.1.105	Retinol dehydrogenase	Retinol + NAD$^+$ = retinal + NADH
1.1.1.106	Pantoate dehydrogenase	D-Pantoate + NAD$^+$ = D-2-hydroxy-3,3-dimethyl-3-formylpropionate + NADH
1.1.1.107	Pyridoxal dehydrogenase	Pyridoxal + NAD$^+$ = 4-pyridoxolactone + NADH
1.1.1.108	Carnitine dehydrogenase	Carnitine + NAD$^+$ = 3-dehydrocarnitine + NADH

[1.1.1.109 Transferred entry: now EC 1.3.1.28, 2,3-Dihydro-2,3-dihydroxybenzoate dehydrogenase]

1.1.1.110	Indolelactate dehydrogenase	Indolelactate + NAD$^+$ = indolepyruvate + NADH
1.1.1.111	Imidazol-5-yl-lactate dehydrogenase	Imidazol-5-yl-lactate + NAD(P)$^+$ = imidazol-5-yl-pyruvate + NAD(P)H
1.1.1.112	Indanol dehydrogenase	1-Indanol + NAD(P)$^+$ = indanone + NAD(P)H
1.1.1.113	L-Xylose dehydrogenase	L-Xylose + NADP$^+$ = L-xylono-γ-lactone + NADPH
1.1.1.114	D-Apiose reductase	D-Apiitol + NAD$^+$ = D-apiose + NADH
1.1.1.115	D-Ribose dehydrogenase (NADP$^+$)	D-Ribose + NADP$^+$ + H$_2$O = D-ribonate + NADPH
1.1.1.116	D-Arabinose dehydrogenase	D-Arabinose + NAD$^+$ = D-arabinono-γ-lactone + NADH
1.1.1.117	D-Arabinose dehydrogenase (NAD(P)$^+$)	D-Arabinose + NAD(P)$^+$ = D-arabinono-γ-lactone + NAD(P)H
1.1.1.118	Glucose dehydrogenase (NAD$^+$)	D-Glucose + NAD$^+$ = D-glucono-δ-lactone + NADH
1.1.1.119	Glucose dehydrogenase (NADP$^+$)	D-Glucose + NADP$^+$ = D-glucono-δ-lactone + NADPH
1.1.1.120	Galactose dehydrogenase (NADP$^+$)	D-Galactose + NADP$^+$ = D-galactonolactone + NADPH

Number	Other Names	Basis for classification (Systematic Name)	Comments	Reference
1.1.1.104		4-Hydroxy-L-proline: NAD+ oxidoreductase		3128
1.1.1.105		Retinol:NAD+ oxidoreductase		1733
1.1.1.106		D-Pantoate:NAD+ 4-oxidoreductase		1079
1.1.1.107		Pyridoxal:NAD+ oxidoreductase		427
1.1.1.108		Carnitine:NAD+ oxidoreductase		139, 2970
1.1.1.110		Indolelactate:NAD+ oxidoreductase		1506
1.1.1.111		Imidazol-5-yl-lactate: NAD(P)+ oxidoreductase		584, 591
1.1.1.112		1-Indanol:NAD(P)+ oxidoreductase		296
1.1.1.113		L-Xylose:NADP+ 1-oxidoreductase	Also oxidizes D-arabinose and D-lyxose	3504
1.1.1.114		D-Apiitol:NAD+ 1-oxidoreductase		2381, 1182
1.1.1.115		D-Ribose:NADP+ 1-oxidoreductase	Also acts more slowly on D-xylose and other pentoses	2950, 2954
1.1.1.116		D-Arabinose:NAD+ 1-oxidoreductase		2546, 2954
1.1.1.117		D-Arabinose:NAD(P)+ 1-oxidoreductase	Also acts on L-galactose, 6-deoxy- and 3,6-dideoxy-L-galactose	544, 545, 546
1.1.1.118		D-Glucose:NAD+ 1-oxidoreductase		1389
1.1.1.119		D-Glucose:NADP+ 1-oxidoreductase	Also oxidizes D-mannose, 2-deoxy-D-glucose and 2-amino-2-deoxy-D-mannose	140
1.1.1.120		D-Galactose:NADP+ 1-oxidoreductase	Also acts on L-arabinose, 6-deoxy- and 2-deoxy-D-galactose	544, 545, 546, 2953

Number	Recommended Name	Reaction
1.1.1.121	Aldose dehydrogenase	D-Aldose + NAD$^+$ = D-aldonolactone + NADH
1.1.1.122	L-Fucose dehydrogenase	L-Fucose + NAD$^+$ = L-fucono-1,5-lactone + NADH
1.1.1.123	Sorbose dehydrogenase (NADP$^+$)	L-Sorbose + NADP$^+$ = 5-keto-D-fructose + NADPH
1.1.1.124	Fructose 5-dehydrogenase (NADP$^+$)	D-Fructose + NADP$^+$ = 5-keto-D-fructose + NADPH
1.1.1.125	2-Deoxy-D-gluconate dehydrogenase	2-Deoxy-D-gluconate + NAD$^+$ = 2-deoxy-3-ketogluconate + NADH
1.1.1.126	2-Keto-3-deoxy-D-gluconate dehydrogenase	2-Keto-3-deoxy-D-gluconate + NADP$^+$ = 2-keto-3-deoxy-D-gluconaldehyde + NADPH
1.1.1.127	2-Keto-3-deoxy-D-gluconate dehydrogenase (NAD(P)$^+$)	2-Keto-3-deoxy-D-gluconate + NAD(P)$^+$ = 3-deoxy-D-*glycero*-2,5-hexodiulosonate + NAD(P)H
1.1.1.128	L-Idonate dehydrogenase	L-Idonate + NADP$^+$ = 5-keto-D-gluconate + NADPH
1.1.1.129	L-Threonate dehydrogenase	L-Threonate + NAD$^+$ = 3-keto-L-threonate + NADH
1.1.1.130	3-Keto-L-gulonate dehydrogenase	3-Keto-L-gulonate + NAD(P)$^+$ = 2,3-diketo-L-gulonate + NAD(P)H
1.1.1.131	Mannuronate reductase	D-Mannonate + NAD(P)$^+$ = D-mannuronate + NAD(P)H
1.1.1.132	GDPmannose dehydrogenase	GDPmannose + NAD$^+$ + H$_2$O = GDPmannuronate + NADH
1.1.1.133	dTDP-4-ketorhamnose reductase	dTDP-6-deoxy-L-mannose + NADP$^+$ = dTDP-4-keto-6-deoxy-L-mannose + NADPH

Number	Other Names	Basis for classification (Systematic Name)	Comments	Reference
1.1.1.121		D-Aldose:NAD$^+$ 1-oxidoreductase	Acts on D-glucose, 2-deoxy- and 6-deoxy-D-glucose, D-galactose, 6-deoxy-D-galactose, 2-deoxy-L-arabinose and D-xylose	544, 545, 546
1.1.1.122		6-Deoxy-L-galactose:NAD$^+$ 1-oxidoreductase	Product is unstable and is hydrolysed to L-fuconate. Enzyme also acts on L-galactose, D-arabinose and 3-amino-3-deoxy-D-arabinose	2942
1.1.1.123	5-Keto-D-fructose reductase	L-Sorbose:NADP$^+$ 5-oxidoreductase		831
1.1.1.124	5-Keto-D-fructose reductase (NADP$^+$)	D-Fructose:NADP$^+$ 5-oxidoreductase		143
1.1.1.125		2-Deoxy-D-gluconate:NAD$^+$ 3-oxidoreductase		798
1.1.1.126		2-Keto-3-deoxy-D-gluconate: NADP$^+$ 6-oxidoreductase		2654
1.1.1.127		2-Keto-3-deoxy-D-gluconate: NAD(P)$^+$ 5-oxidoreductase		2655
1.1.1.128	5-Keto-D-gluconate 2-reductase	L-Idonate:NADP$^+$ 2-oxidoreductase		3341
1.1.1.129		L-Threonate:NAD$^+$ oxidoreductase		124
1.1.1.130		3-Keto-L-gulonate:NAD(P)$^+$ 2-oxidoreductase		3566
1.1.1.131		D-Mannonate:NAD(P)$^+$ 6-oxidoreductase		860
1.1.1.132		GDPmannose:NAD$^+$ 6-oxidoreductase	Also uses the corresponding deoxynucleoside diphosphate derivative as a substrate	2651
1.1.1.133		dTDP-6-deoxy-L-mannose: NADP$^+$ 4-oxidoreductase	In the reverse direct reduction on the 4-position of the hexose moiety takes place only while the substrate is	2195

Number	Recommended Name	Reaction
1.1.1.134	dTDP-6-deoxy-L-talose dehydrogenase	dTDP-6-deoxy-L-talose + $NADP^+$ = dTDP-4-keto-6-deoxy-L-mannose + NADPH
1.1.1.135	GDP-6-deoxy-D-talose dehydrogenase	GDP-6-deoxy-D-talose + $NAD(P)^+$ = GDP-4-keto-6-deoxy-D-talose + NAD(P)H
1.1.1.136	UDP-N-acetylglucosamine dehydrogenase	UDP-2-acetamido-2-deoxy-D-glucose + 2 NAD^+ + H_2O = UDP-2-acetamido-2-deoxy-D-glucuronate + 2 NADH
1.1.1.137	Ribitol-5-phosphate dehydrogenase	D-Ribitol 5-phosphate + $NAD(P)^+$ = D-ribulose 5-phosphate + NAD(P)H
1.1.1.138	Mannitol dehydrogenase ($NADP^+$)	D-Mannitol + $NADP^+$ = D-fructose + NADPH

[1.1.1.139 Deleted entry: now included with EC 1.1.1.21, Aldose reductase]

Number	Recommended Name	Reaction
1.1.1.140	D-Sorbitol-6-phosphate dehydrogenase	D-Sorbitol 6-phosphate + NAD^+ = D-fructose 6-phosphate + NADH
1.1.1.141	15-Hydroxyprostaglandin dehydrogenase	11α,15-Dihydroxy-9-oxoprost-13-enoate + NAD^+ = 11α-hydroxy-9,15-dioxoprost-13-enoate + NADH
1.1.1.142	D-Pinitol dehydrogenase	5D-5-O-Methyl-*chiro*-inositol + $NADP^+$ = 5D-5-O-methyl-2,3,5/4,6-pentahydroxycyclohexanone + NADPH
1.1.1.143	Sequoyitol dehydrogenase	5-O-Methyl-*myo*-inositol + NAD^+ = 5D-5-O-methyl-2,3,5/4,6-pentahydroxycyclohexanone + NADH
1.1.1.144	Perillyl-alcohol dehydrogenase	Perillyl alcohol + NAD^+ = perillyl aldehyde + NADH

Number	Other Names	Basis for classification (Systematic Name)	Comments	Reference
			bound to another enzyme which catalyses epimerization at C-3 and C-5	
1.1.1.134		dTDP-6-deoxy-L-talose:NADP$^+$ 4-oxidoreductase	Oxidation on the 4-position of the hexose moiety takes place only while the substrate is bound to another enzyme which catalyses epimerization at C-3 and C-5	999
1.1.1.135		GDP-6-deoxy-D-talose:NAD(P)$^+$ 4-oxidoreductase		2088, 3707
1.1.1.136		UDP-2-acetamido-2-deoxy-D-glucose:NAD$^+$ 6-oxidoreductase		857
1.1.1.137		D-Ribitol-5-phosphate:NAD(P)$^+$ 2-oxidoreductase		1052
1.1.1.138		D-Mannitol:NADP$^+$ 2-oxidoreductase		790, 3265
1.1.1.140		D-Sorbitol-6-phosphate:NAD$^+$ 2-oxidoreductase		760, 1986
1.1.1.141		11α,15-Dihydroxy-9-oxoprost-13-enoate:NAD$^+$ 15-oxidoreductase		86
1.1.1.142		5D-5-O-Methyl-*chiro*-inositol:NADP$^+$ oxidoreductase		2872
1.1.1.143		5-O-Methyl-*myo*-inositol:NAD$^+$ oxidoreductase		2872
1.1.1.144		Perillyl-alcohol:NAD$^+$ oxidoreductase	Oxidizes a number of primary alcohols with the alcohol group allylic to an endocyclic double bond, and a 6-membered ring, either aromatic or hydroaromatic	189

Number	Recommended Name	Reaction
1.1.1.145	3β-Hydroxy-Δ^5-steroid dehydrogenase	3β-Hydroxy-Δ^5-steroid + NAD$^+$ = 3-oxo-Δ^4-steroid + NADH
1.1.1.146	11β-Hydroxysteroid dehydrogenase	11β-Hydroxysteroid + NADP$^+$ = 11-oxosteroid + NADPH
1.1.1.147	16α-Hydroxysteroid dehydrogenase	16α-Hydroxysteroid + NAD(P)$^+$ = 16-oxosteroid + NAD(P)H
1.1.1.148	Oestradiol 17α-dehydrogenase	Oestradiol-17α + NAD(P)$^+$ = oestrone + NAD(P)H
1.1.1.149	20α-Hydroxysteroid dehydrogenase	17α,20α-Dihydroxypregn-4-en-3-one + NAD(P)$^+$ = 17α-hydroxyprogesterone + NAD(P)H
1.1.1.150	21-Hydroxysteroid dehydrogenase	Pregnan-21-ol + NAD$^+$ = pregnan-21-al + NADH
1.1.1.151	21-Hydroxysteroid dehydrogenase (NADP$^+$)	Pregnan-21-ol + NADP$^+$ = pregnan-21-al + NADPH
1.1.1.152	Etiocholanolone 3α-dehydrogenase	Etiocholanolone + NAD$^+$ = 5β-androstane-3,17-dione + NADH
1.1.1.153	Sepiapterin reductase	7,8-Dihydrobiopterin + NADP$^+$ = sepiapterin + NADPH
1.1.1.154	Ureidoglycollate dehydrogenase	(S)-Ureidoglycollate + NAD(P)$^+$ = oxalureate + NAD(P)H
1.1.1.155	Homoisocitrate dehydrogenase	1-Hydroxy-1,2,4-butanetricarboxylate + NAD$^+$ = 2-oxoadipate + CO$_2$ + NADH
1.1.1.156	Glycerol 2-dehydrogenase (NADP$^+$)	Glycerol + NADP$^+$ = dihydroxyacetone + NADPH
1.1.1.157	3-Hydroxybutyryl-CoA dehydrogenase	L-3-Hydroxybutyryl-CoA + NADP$^+$ = 3-acetoacetyl-CoA + NADPH
1.1.1.158	UDP-N-acetylenolpyruvoylglucosamine reductase	UDP-N-acetylmuramate + NADP$^+$ = UDP-2-acetamido-2-deoxy-3-enolpyruvoylglucose + NADPH

Number	Other Names	Basis for classification (Systematic Name)	Comments	Reference
1.1.1.145	Progesterone reductase	3β-Hydroxy-Δ^5-steroid:NAD$^+$ 3-oxidoreductase	Acts on 3β-hydroxyandrost-5-en-17-one to form androst-4-ene-3,17-dione and 3β-hydroxypregn-5-en-20-one to form progesterone	510, 1761, 2401
1.1.1.146		11β-Hydroxysteroid:NADP$^+$ 11-oxidoreductase		439
1.1.1.147		16α-Hydroxysteroid:NAD(P)$^+$ 16-oxidoreductase		2181
1.1.1.148		17α-Hydroxysteroid:NAD(P)$^+$ 17-oxidoreductase		2776
1.1.1.149		20α-Hydroxysteroid:NAD(P)$^+$ 20-oxidoreductase		1798, 3046
1.1.1.150		21-Hydroxysteroid:NAD$^+$ 21-oxidoreductase		2266
1.1.1.151		21-Hydroxysteroid:NADP$^+$ 21-oxidoreductase		2266
1.1.1.152		3α-Hydroxy-5β-steroid:NAD$^+$ 3-oxidoreductase		2828
1.1.1.153		7,8-Dihydrobiopterin:NADP$^+$ oxidoreductase		2119, 1609
1.1.1.154		(S)-Ureidoglycollate: NAD(P)$^+$ oxidoreductase		3530
1.1.1.155		1-Hydroxy-1,2, 4-butanetricarboxylate:NAD$^+$ oxidoreductase (decarboxylating)		2862
1.1.1.156	Dihydroxyacetone reductase	Glycerol:NADP$^+$ 2-oxidoreductase (dihydroxyacetone-forming)		253
1.1.1.157		L-3-Hydroxybutyryl-CoA: NADP$^+$ oxidoreductase		2037
1.1.1.158		UDP-N-acetylmuramate: NADP$^+$ oxidoreductase	A flavoprotein (FAD). Sodium thionite, sodium borohydride and, to a lesser extent, NADH can replace NADPH	3355, 3354

Number	Recommended Name	Reaction
1.1.1.159	7α-Hydroxysteroid dehydrogenase	3α,7α,12α-Trihydroxy-5β-cholanate + NAD⁺ = 3α,12α-dihydroxy-7-oxo-5β-cholanate + NADH
1.1.1.160	Dihydrobunolol dehydrogenase	DL-5-[(*tert*-Butylamino)-2′-hydroxypropoxy]-1,2,3,4-tetrahydro-1-naphthol + NADP⁺ = DL-5-[(*tert*-butylamino)-2′-hydroxypropoxy]-3,4-dihydro-1(2H)-naphthalenone + NADPH
1.1.1.161	Cholestanetetraol 26-dehydrogenase	5β-Cholestane-3α,7α,12α,26-tetrol + NAD⁺ = 5β-cholestane-3α,7α,12α-triol-26-al + NADH
1.1.1.162	D-Erythrulose reductase	Erythritol + NADP⁺ = D-erythrulose + NADPH
1.1.1.163	Cyclopentanol dehydrogenase	Cyclopentanol + NAD⁺ = cyclopentanone + NADH
1.1.1.164	Hexadecanol dehydrogenase	Hexadecanol + NAD⁺ = hexadecanal + NADH
1.1.1.165	2-Alkyn-1-ol dehydrogenase	2-Butyne-1,4-diol + NAD⁺ = 4-hydroxy-2-butynal + NADH
1.1.1.166	Hydroxycyclohexanecarboxylate dehydrogenase	(1S,3R,4S)-3,4-Dihydroxycyclohexane-1-carboxylate + NAD⁺ = (1S,4S)-4-hydroxy-3-oxocyclohexane-1-carboxylate + NADH
1.1.1.167	Hydroxymalonate dehydrogenase	Hydroxymalonate + NAD⁺ = oxomalonate + NADH
1.1.1.168	2-Oxopantoyl-lactone reductase	Pantoyl lactone + NADP⁺ = 2-oxopantoyl lactone + NADPH

Number	Other Names	Basis for classification (Systematic Name)	Comments	Reference
1.1.1.159		7α-Hydroxysteroid:NAD$^+$ 7-oxidoreductase	Catalyses the oxidation of the 7α-hydroxyl group of bile acids, alcohols, and sulphates, both in their free and conjugated forms. The *Bacteroides fragilis* enzyme can also utilize NADP$^+$	2028, 2029
1.1.1.160	Bunolol reductase	DL-5-[(*tert*-Butylamino)-2'-hydroxypropoxy]-1,2,3,4-tetrahydro-1-naphthol:NADP$^+$ oxidoreductase	Also acts more slowly with NAD$^+$	1916
1.1.1.161		5β-Cholestane-3α,7α,12α,26-tetraol:NAD$^+$ 26-oxidoreductase		2113
1.1.1.162		Erythritol:NADP$^+$ oxidoreductase	NAD$^+$ is also utilized, more slowly	3505
1.1.1.163		Cyclopentanol:NAD$^+$ oxidoreductase		1120
1.1.1.164		Hexadecanol:NAD$^+$ oxidoreductase	The liver enzyme acts on long-chain alcohols from C-8 to C-16. The *Euglena* enzyme also oxidizes the aldehydes to fatty acids	3238, 1751
1.1.1.165		2-Butyne-1,4-diol:NAD$^+$ 1-oxidoreductase	Acts on a variety of 2-alkyne-1-ols, and also 1,4-butanediol. NADP$^+$ also acts as acceptor, more slowly	2253
1.1.1.166		(1S,3R,4S)-3,4-Dihydroxycyclohexane-1-carboxylate:NAD$^+$ 3-oxidoreductase	Acts on (—)-hydroxycyclohexanecarboxylates having an equatorial carboxyl group at C-1, an axial hydroxyl group at C-3 and an equatorial hydroxyl or carbonyl group at C-4, including (—)-quinate and (—)-shikimate	3676
1.1.1.167		Hydroxymalonate:NAD$^+$ oxidoreductase		1544, 2602
1.1.1.168		Pantoyl lactone:NADP$^+$ oxidoreductase		1691

Number	Recommended Name	Reaction
1.1.1.169	2-Oxopantoate reductase	D-Pantoate + NADP$^+$ = 2-oxopantoate + NADPH
1.1.1.170	3β-Hydroxy-4β-methylcholestenoate dehydrogenase (decarboxylating)	3β-Hydroxy-4β-methyl-5α-cholest-7-en-4α-oate + NAD$^+$ = 4α-methyl-5α-cholest-7-en-3-one + CO$_2$ + NADH
1.1.1.171	5,10-Methylenetetrahydrofolate reductase (NADPH)	5-Methyltetrahydrofolate + NADP$^+$ = 5,10-methylenetetrahydrofolate + NADPH
1.1.1.172	2-Ketoadipate reductase	2-Hydroxyadipate + NAD$^+$ = 2-ketoadipate + NADH
1.1.1.173	L-Rhamnose dehydrogenase	L-Rhamnofuranose + NAD$^+$ = L-rhamno-γ-lactone + NADH
1.1.1.174	Cyclohexan-1,2-diol dehydrogenase	*trans*-Cyclohexan-1,2-diol + NAD$^+$ = 2-hydroxy-cyclohexan-1-one + NADH
1.1.1.175	D-Xylose dehydrogenase	D-Xylose + NAD$^+$ = D-xylonolactone + NADH
1.1.1.176	12α-Hydroxysteroid dehydrogenase	3α,7α,12α-Trihydroxy-5β-cholanate + NADP$^+$ = 3α,7α-dihydroxy-12-oxo-5β-cholanate + NADPH

1.1.2 WITH A CYTOCHROME AS ACCEPTOR

[1.1.2.1   Transferred entry: now EC 1.1.99.5, Glycerol-3-phosphate dehydrogenase]

1.1.2.2	Mannitol dehydrogenase (cytochrome)	D-Mannitol + ferricytochrome = D-fructose + ferrocytochrome
1.1.2.3	Lactate dehydrogenase (cytochrome)	L-Lactate + 2 ferricytochrome c = pyruvate + 2 ferrocytochrome c
1.1.2.4	D-Lactate dehydrogenase (cytochrome)	D-Lactate + 2 ferricytochrome c = pyruvate + 2 ferrocytochrome c

1.1.3 WITH OXYGEN AS ACCEPTOR

1.1.3.1	Glycollate oxidase	Glycollate + O$_2$ = glyoxylate + H$_2$O$_2$

[1.1.3.2   Transferred entry: now EC 1.13.12.4, Lactate 2-monooxygenase]

Number	Other Names	Basis for classification (Systematic Name)	Comments	Reference
1.1.1.169		D-Pantoate:NADP⁺ 2-oxidoreductase		1691
1.1.1.170		3β-Hydroxy-4β-methyl-5α-cholest-7-en-4α-oate:NAD⁺ oxidoreductase (decarboxylating)	Also acts on 3β-hydroxy-5α-cholest-7-en-4α-oate	2700
1.1.1.171		5-Methyltetrahydrofolate: NADP⁺ oxidoreductase	A flavoprotein (FAD)	1848
1.1.1.172		2-Hydroxyadipate:NAD⁺ oxidoreductase		3278
1.1.1.173		L-Rhamnofuranose:NAD⁺ 11-oxidoreductase		2790
1.1.1.174		*trans*-Cyclohexan-1,2-diol:NAD⁺ oxidoreductase	Acts, more slowly, on the *cis* isomer and on 2-hydroxycyclohexanone	644
1.1.1.175		D-Xylose:NAD⁺ 1-oxidoreductase		3763
1.1.1.176		12α-Hydroxysteroid:NAD⁺ 12α-oxidoreductase	Catalyses the oxidation of the 12α-hydroxyl group of bile acids, both in their free and conjugated form. Also acts on bile alcohols	2054, 2027
1.1.2.2		D-Mannitol:ferricytochrome 2-oxidoreductase	Also oxidizes erythritol, D-glucitol, D-arabinitol and ribitol	100
1.1.2.3	Lactic acid dehydrogenase	L-Lactate:ferricytochrome *c* oxidoreductase	Identical with cytochrome *b₂*; a flavohaemoprotein (FMN)	95, 96, 160, 2465
1.1.2.4	Lactic acid dehydrogenase	D-Lactate:ferricytochrome *c* oxidoreductase	A flavoprotein (FAD)	1115, 1116, 2464, 2465
1.1.3.1		Glycollate:oxygen oxidoreductase	A flavoprotein (FMN). Also oxidizes L-lactate and glyoxylate, more slowly	1824, 2783, 3840

Number	Recommended Name	Reaction
1.1.3.3	Malate oxidase	L-Malate + O_2 = oxaloacetate + (?)
1.1.3.4	Glucose oxidase	β-D-Glucose + O_2 = D-glucono-δ-lactone + H_2O_2
1.1.3.5	Hexose oxidase	β-D-Glucose + O_2 = D-glucono-δ-lactone + H_2O_2
1.1.3.6	Cholesterol oxidase	Cholesterol + O_2 = 4-cholesten-3-one + (?)
1.1.3.7	Aryl-alcohol oxidase	An aromatic primary alcohol + O_2 = an aromatic aldehyde + H_2O_2
1.1.3.8	L-Gulonolactone oxidase	L-Gulono-γ-lactone + O_2 = L-*xylo*-hexulonolactone + H_2O_2
1.1.3.9	Galactose oxidase	D-Galactose + O_2 = D-*galacto*-hexodialdose + H_2O_2
1.1.3.10	Pyranose oxidase	D-Glucose + O_2 = D-glucosone + H_2O_2
1.1.3.11	L-Sorbose oxidase	L-Sorbose + O_2 = 5-keto-D-fructose + H_2O_2
1.1.3.12	Pyridoxine 4-oxidase	Pyridoxine + O_2 = pyridoxal + H_2O_2
1.1.3.13	Alcohol oxidase	Primary alcohol + O_2 = aldehyde + H_2O_2

Number	Other Names	Basis for classification (Systematic Name)	Comments	Reference
1.1.3.3		L-Malate:oxygen oxidoreductase		555
1.1.3.4	Notatin, Glucose oxyhdrase	β-D-Glucose:oxygen 1-oxidoreductase	A flavoprotein (FAD)	256, 597, 1636, 1637
1.1.3.5		D-Hexose:oxygen 1-oxidoreductase	A cupro glycoprotein. Also oxidizes D-galactose, D-mannose, maltose, lactose and cellobiose	234, 235, 3287
1.1.3.6		Cholesterol:oxygen oxidoreductase		3189
1.1.3.7		Aryl-alcohol:oxygen oxidoreductase	Oxidizes many primary alcohols containing an aromatic ring; best substrates are β-naphthylcarbinol and 3-methoxybenzyl alcohol	861
1.1.3.8		L-Gulono-γ-lactone:oxygen 2-oxidoreductase	The product spontaneously isomerizes to L-ascorbate	508, 1454
1.1.3.9		D-Galactose:oxygen 6-oxidoreductase	A cuproprotein. Oxidizes at the 6-position of D-galactose	141
1.1.3.10		Pyranose:oxygen 2-oxidoreductase	Also oxidizes D-xylose, L-sorbose and D-glucono-1,5-lactone, which have the same ring conformation and configuration at C-2, C-3 and C-4	1503, 2870
1.1.3.11		L-Sorbose:oxygen 5-oxidoreductase	Also acts on D-glucose, D-galactose and D-xylose, but does not act on D-fructose. 2,6-Dichloroindophenol could act as acceptor	3752
1.1.3.12	Pyridoxin 4-oxidase	Pyridoxine:oxygen 4'-oxidoreductase	A flavoprotein. Can also use 2,6-dichloroindophenol as an acceptor	3294
1.1.3.13		Alcohol:oxygen oxidoreductase	A flavoprotein (FAD); acts on lower primary alcohols and unsaturated alcohols but branched-chain and secondary alcohols are not attacked	1502

Number	Recommended Name	Reaction
1.1.3.14	Catechol oxidase (dimerizing)	4 Catechol + 3 O_2 = 2 dibenzo[1,4]dioxin-2,3 dione + 6 H_2O
1.1.3.15	L-2-Hydroxyacid oxidase	L-2-Hydroxyacid + O_2 = 2-oxo acid + H_2O_2
1.1.3.16	Ecdysone oxidase	Ecdysone + O_2 = 3-dehydroecdysone + H_2O_2
1.1.3.17	Choline oxidase	Choline + O_2 = betaine aldehyde + H_2O_2

1.1.99   WITH OTHER ACCEPTORS

Number	Recommended Name	Reaction
1.1.99.1	Choline dehydrogenase	Choline + acceptor = betaine aldehyde + reduced acceptor
1.1.99.2	2-Hydroxyglutarate dehydrogenase	L-2-Hydroxyglutarate + acceptor = 2-oxoglutarate + reduced acceptor
1.1.99.3	Gluconate 2-dehydrogenase	D-Gluconate + acceptor = 2-keto-D-gluconate + reduced acceptor
1.1.99.4	Ketogluconate dehydrogenase	2-Keto-D-gluconate + acceptor = 2,5-diketo-D-gluconate + reduced acceptor
1.1.99.5	Glycerol-3-phosphate dehydrogenase	sn-Glycerol 3-phosphate + acceptor = dihydroxyacetone phosphate + reduced acceptor
1.1.99.6	D-2-Hydroxyacid dehydrogenase	D-Lactate + acceptor = pyruvate + reduced acceptor
1.1.99.7	Lactate—malate transhydrogenase	Lactate + oxaloacetate = malate + pyruvate
1.1.99.8	Alcohol dehydrogenase (acceptor)	Primary alcohol + acceptor = aldehyde + reduced acceptor

Number	Other Names	Basis for classification (Systematic Name)	Comments	Reference
1.1.3.14		Catechol:oxygen oxidoreductase (dimerizing)		2340
1.1.3.15		L-2-Hydroxyacid:oxygen oxidoreductase	Acts on various aliphatic 2-hydroxyacids. A flavoprotein (FMN)	2365
1.1.3.16		Ecdysone:oxygen 3-oxidoreductase	2,6-Dichloroindophenol can also act as an acceptor	1758
1.1.3.17		Choline:oxygen 1-oxidoreductase	A flavoprotein (FAD). Also oxidizes betaine aldehyde to betaine	1436
1.1.99.1		Choline: (acceptor) oxidoreductase		785, 2774
1.1.99.2		L-2-Hydroxyglutarate: (acceptor) oxidoreductase		3639
1.1.99.3		D-Gluconate: (acceptor) 2-oxidoreductase		2707
1.1.99.4		2-Keto-D-gluconate: (acceptor) 5-oxidoreductase		640
1.1.99.5		sn-Glycerol-3-phosphate: (acceptor) oxidoreductase	Formerly EC 1.1.2.1	2796
1.1.99.6		D-2-Hydroxyacid: (acceptor) oxidoreductase	A flavoprotein (FAD) containing zinc. Acts on a variety of D-2-hydroxyacids	1116, 2464, 3484
1.1.99.7		Lactate:oxaloacetate oxidoreductase	Catalyses hydrogen transfer from C-3 or C-4 L-2-hydroxyacids to 2-oxo acids. It contains tightly bound nicotinamide nucleotide in its active centre. This prosthetic group cannot be removed without denaturation of the protein	59, 60
1.1.99.8		Alcohol: (acceptor) oxidoreductase	Acts on a wide range of primary alcohols, including methanol	92

Number	Recommended Name	Reaction
1.1.99.9	Pyridoxine 5-dehydrogenase	Pyridoxine + acceptor = isopyridoxal + reduced acceptor
1.1.99.10	Glucose dehydrogenase (acceptor)	D-Glucose + acceptor = D-glucono-δ-lactone + reduced acceptor
1.1.99.11	D-Fructose 5-dehydrogenase	D-Fructose + acceptor = 5-keto-D-fructose + reduced acceptor
1.1.99.12	Sorbose dehydrogenase	L-Sorbose + acceptor = 5-keto-D-fructose + reduced acceptor
1.1.99.13	D-Glucoside 3-dehydrogenase	Sucrose + acceptor = 3-keto-α-D-glucosyl-β-D-fructofuranoside + reduced acceptor
1.1.99.14	Glycollate dehydrogenase	Glycollate + acceptor = glyoxylate + reduced acceptor
1.1.99.15	5,10-Methylenetetrahydrofolate reductase (FADH$_2$)	5-Methyltetrahydrofolate + acceptor = 5,10-methylenetetrahydrofolate + reduced acceptor
1.1.99.16	Malate dehydrogenase (acceptor)	L-malate + acceptor = oxaloacetate + reduced acceptor

1.2 ACTING ON THE ALDEHYDE OR OXO GROUP OF DONORS

This sub-class contains all enzymes oxidizing aldehydes to the corresponding acids; when this acid is concomitantly phosphorylated or acetylates CoA, this is indicated in parentheses. Oxo groups may be oxidized either with addition of water and cleavage of a carbon-carbon bond or, in the case of ring compounds, by addition of the elements of water and dehydrogenation. The sub-class is subdivided according to the acceptor, which may be NAD(P)$^+$, a cytochrome, oxygen, a disulphide compound or some other acceptor.

1.2.1 WITH NAD$^+$ OR NADP$^+$ AS ACCEPTOR

1.2.1.1	Formaldehyde dehydrogenase	Formaldehyde + glutathione + NAD$^+$ = S-formylglutathione + NADH

Number	Other Names	Basis for classification (Systematic Name)	Comments	Reference
1.1.99.9	Pyridoxol-5-dehydrogenase	Pyridoxine: (acceptor) 5'-oxidoreductase	A flavoprotein (FAD)	3294
1.1.99.10	Glucose dehydrogenase (*Aspergillus*)	D-Glucose: (acceptor) 1-oxidoreductase	2,6-Dichloroindophenol can act as acceptor; enzyme contains one mole of FAD per mole of enzyme and is a glycoprotein	181
1.1.99.11		D-Fructose: (acceptor) 5-oxidoreductase	The enzyme uses 2,6-dichloroindophenol as an acceptor	3751
1.1.99.12		L-Sorbose: (acceptor) 5-oxidoreductase	2,6-Dichloroindophenol was used as acceptor	2919
1.1.99.13		D-Aldohexoside: (acceptor) 5-oxidoreductase	A flavoprotein (FAD). The enzyme acts on D-glucose, D-galactose, D-glucosides and D-galactosides, but D-glucosides react more rapidly than D-galactosides	1244
1.1.99.14		Glycollate: (acceptor) oxidoreductase	Also acts on D-lactate. 2,6-Dichloroindophenol and phenazine methosulphate can act as acceptor	2004
1.1.99.15		5-Methyltetrahydrofolate: (acceptor) oxidoreductase	The acceptor is free FAD	1611
1.1.99.16		L-Malate: (acceptor) oxidoreductase	A flavoprotein (FAD)	1689, 1444, 1441
1.2.1.1	Formic dehydrogenase	Formaldehyde:NAD+ oxidoreductase (glutathione-formylating)	Some other 2-ketoaldehydes are also oxidized. In the reverse	1487, 3263, 2841, 3515

Number	Recommended Name	Reaction
1.2.1.2	Formate dehydrogenase	Formate + NAD$^+$ = CO_2 + NADH
1.2.1.3	Aldehyde dehydrogenase	Aldehyde + NAD$^+$ + H_2O = acid + NADH
1.2.1.4	Aldehyde dehydrogenase (NADP$^+$)	Aldehyde + NADP$^+$ + H_2O = acid + NADPH
1.2.1.5	Aldehyde dehydrogenase (NAD(P)$^+$)	Aldehyde + NAD(P)$^+$ + H_2O = acid + NAD(P)H
[1.2.1.6	Deleted entry: Benzaldehyde dehydrogenase]	
1.2.1.7	Benzaldehyde dehydrogenase (NADP$^+$)	Benzaldehyde + NADP$^+$ + H_2O = benzoate + NADPH
1.2.1.8	Betaine-aldehyde dehydrogenase	Betaine aldehyde + NAD$^+$ + H_2O = betaine + NADH
1.2.1.9	Glyceraldehyde-phosphate dehydrogenase (NADP$^+$)	D-Glyceraldehyde 3-phosphate + NADP$^+$ +H_2O = 3-phospho-D-glycerate + NADPH
1.2.1.10	Acetaldehyde dehydrogenase (acylating)	Acetaldehyde + CoA + NAD$^+$ = acyl-CoA + NADH
1.2.1.11	Aspartate-semialdehyde dehydrogenase	L-Aspartate β-semialdehyde + orthophosphate + NADP$^+$ = L-β-aspartyl phosphate + NADPH
1.2.1.12	Glyceraldehyde-phosphate dehydrogenase	D-Glyceraldehyde 3-phosphate + orthophosphate + NAD$^+$ = 3-phospho-D-glyceroyl phosphate + NADH
1.2.1.13	Glyceraldehyde-phosphate dehydrogenase (NADP$^+$) (phosphorylating)	D-Glyceraldehyde 3-phosphate + orthophosphate + NADP$^+$ = 3-phospho-D-glyceroyl phosphate + NADPH
1.2.1.14	IMP dehydrogenase	Inosine 5′-phosphate + NAD$^+$ + H_2O = xanthosine 5′-phosphate + NADH

Number	Other Names	Basis for classification (Systematic Name)	Comments	Reference
			direction, NADPH can replace NADH	
1.2.1.2	Formate hydrogenlyase	Formate:NAD⁺ oxidoreductase		659, 2670
1.2.1.3		Aldehyde:NAD⁺ oxidoreductase	Wide specificity, including oxidation of D-glucuronolactone to D-glucarate. Formerly EC 1.1.1.70	1487, 2686
1.2.1.4		Aldehyde:NADP⁺ oxidoreductase		1487, 2367, 2991
1.2.1.5		Aldehyde:NAD(P)⁺ oxidoreductase		305, 1487, 1693, 3205, 3362
1.2.1.7		Benzaldehyde:NADP⁺ oxidoreductase		1144, 3180
1.2.1.8		Betaine-aldehyde:NAD⁺ oxidoreductase		2857
1.2.1.9	Triosephosphate dehydrogenase	D-Glyceraldehyde-3-phosphate:NADP⁺ oxidoreductase		2847
1.2.1.10		Acetaldehyde:NAD⁺ oxidoreductase (CoA-acylating)	Also acts, more slowly, on glycolaldehyde, propionaldehyde and butyraldehyde	438
1.2.1.11		L-Aspartate-β-semialdehyde:NADP⁺ oxidoreductase (phosphorylating)		308, 1487
1.2.1.12	Triosephosphate dehydrogenase	D-Glyceraldehyde-3-phosphate:NAD⁺ oxidoreductase (phosphorylating)	Also acts very slowly on D-glyceraldehyde and some other aldehydes; thiols can replace phosphate	468, 587, 1158, 3545, 3613
1.2.1.13	Triosephosphate dehydrogenase (NADP⁺)	D-Glyceraldehyde-3-phosphate:NADP⁺ oxidoreductase (phosphorylating)		381, 1023, 2847
1.2.1.14		IMP:NAD⁺ oxidoreductase		2039, 3491

Number	Recommended Name	Reaction
1.2.1.15	Malonate-semialdehyde dehydrogenase	Malonate semialdehyde + NAD(P)$^+$ + H_2O = malonate + NAD(P)H
1.2.1.16	Succinate-semialdehyde dehydrogenase (NAD(P)$^+$)	Succinate semialdehyde + NAD(P)$^+$ + H_2O = succinate + NAD(P)H
1.2.1.17	Glyoxylate dehydrogenase (acylating)	Glyoxylate + CoA + NADP$^+$ = oxalyl-CoA + NADPH
1.2.1.18	Malonate-semialdehyde dehydrogenase (acetylating)	Malonate semialdehyde + CoA + NAD(P)$^+$ = acetyl-CoA + CO_2 + NAD(P)H
1.2.1.19	Aminobutyraldehyde dehydrogenase	4-Aminobutyraldehyde + NAD$^+$ + H_2O = 4-aminobutyrate + NADH
1.2.1.20	Glutarate-semialdehyde dehydrogenase	Glutarate semialdehyde + NAD$^+$ + H_2O = glutarate + NADH
1.2.1.21	Glycolaldehyde dehydrogenase	Glycolaldehyde + NAD$^+$ + H_2O = glycollate + NADH
1.2.1.22	Lactaldehyde dehydrogenase	L-Lactaldehyde + NAD$^+$ + H_2O = lactate + NADH
1.2.1.23	2-Oxoaldehyde dehydrogenase	2-Oxoaldehyde + NAD(P)$^+$ + H_2O = 2-oxo acid + NAD(P)H
1.2.1.24	Succinate-semialdehyde dehydrogenase	Succinate semialdehyde + NAD$^+$ + H_2O = succinate + NADH
1.2.1.25	2-Oxoisovalerate dehydrogenase (acylating)	2-Oxoisovalerate + CoA + NAD$^+$ = isobutyryl-CoA + CO_2 + NADH
1.2.1.26	2,5-Dioxovalerate dehydrogenase	2,5-Dioxovalerate + NADP$^+$ + H_2O = 2-oxoglutarate + NADPH
1.2.1.27	Methylmalonate-semialdehyde dehydrogenase (acylating)	Methylmalonate semialdehyde + CoA + NAD$^+$ = propionyl-CoA + CO_2 + NADH
1.2.1.28	Benzaldehyde dehydrogenase (NAD$^+$)	Benzaldehyde + NAD$^+$ + H_2O = benzoate + NADH
1.2.1.29	Aryl-aldehyde dehydrogenase	An aromatic aldehyde + NAD$^+$ + H_2O = an aromatic acid + NADH

Number	Other Names	Basis for classification (Systematic Name)	Comments	Reference
1.2.1.15		Malonate-semialdehyde: NAD(P)$^+$ oxidoreductase		2356
1.2.1.16		Succinate-semialdehyde: NAD(P)$^+$ oxidoreductase		1487, 1495, 2414
1.2.1.17		Glyoxylate:NADP$^+$ oxidoreductase (CoA-oxalylating)		2672
1.2.1.18		Malonate-semialdehyde: NAD(P)$^+$ oxidoreductase (decarboxylating, CoA- acetylating)		1237, 3745, 1487
1.2.1.19		4-Aminobutyraldehyde: NAD$^+$ oxidoreductase	1-Pyrroline was used as source of substrate	1493, 1487
1.2.1.20		Glutarate-semialdehyde: NAD$^+$ oxidoreductase		1422
1.2.1.21		Glycolaldehyde:NAD$^+$ oxidoreductase		651
1.2.1.22		L-Lactaldehyde:NAD$^+$ oxidoreductase		3176
1.2.1.23		2-Oxoaldehyde:NAD(P)$^+$ oxidoreductase		2265
1.2.1.24		Succinate-semialdehyde: NAD$^+$ oxidoreductase		38
1.2.1.25		2-Oxoisovalerate: NAD$^+$ oxidoreductase (CoA- isobutyrylating)	Also acts on L-2-oxo-3-methylvalerate and 2-oxoisocaproate	2370
1.2.1.26		2,5-Dioxovalerate: NADP$^+$ oxidoreductase		20
1.2.1.27		Methylmalonate-semialdehyde:NAD$^+$ oxidoreductase (CoA-propionylating)	Also converts propionaldehyde to propionyl-CoA	3145
1.2.1.28		Benzaldehyde:NAD$^+$ oxidoreductase		1144
1.2.1.29		Aryl-aldehyde:NAD$^+$ oxidoreductase	Oxidizes a number of aromatic aldehydes, but not aliphatic aldehydes	2702

Number	Recommended Name	Reaction
1.2.1.30	Aryl-aldehyde dehydrogenase (NADP+)	An aromatic aldehyde + NADP+ + ADP + orthophosphate + H_2O = an aromatic acid + NADPH + ATP
1.2.1.31	Aminoadipate-semialdehyde dehydrogenase	L-2-Aminoadipate 6-semialdehyde + NAD(P)+ + H_2O = L-2-aminoadipate + NAD(P)H
1.2.1.32	Aminomuconate-semialdehyde dehydrogenase	2-Aminomuconate 6-semialdehyde + NAD+ + H_2O = 2-aminomuconate + NADH
1.2.1.33	D-Aldopantoate dehydrogenase	D-2-Hydroxy-3,3-dimethyl-3-formylpropionate + NAD+ + H_2O = 3,3-dimethyl-D-malate + NADH
1.2.1.34	D-Mannonate dehydrogenase (NAD(P)+)	D-Mannonate + NAD(P)+ + H_2O = D-mannuronate + NAD(P)H
1.2.1.35	Uronate dehydrogenase	D-Galacturonate + NAD+ + H_2O = D-galactarate + NADH
1.2.1.36	Retinal dehydrogenase	Retinal + NAD+ + H_2O = retinoate + NADH
1.2.1.37	Xanthine dehydrogenase	Xanthine + NAD+ + H_2O = urate + NADH
1.2.1.38	N-Acetyl-γ-glutamyl-phosphate reductase	N-Acetyl-L-glutamate 5-semialdehyde + NADP+ + orthophosphate = N-acetyl-5-glutamyl phosphate + NADPH
1.2.1.39	Phenylacetaldehyde dehydrogenase	Phenylacetaldehyde + NAD+ + H_2O = phenylacetate + NADH
1.2.1.40	Cholestanetriol-26-al 26-dehydrogenase	5β-Cholestane-3α,7α,12α-triol-26-al + NAD+ + H_2O = 5β-cholestane-3α,7α,12α-triol-26-oate + NADH
1.2.1.41	Glutamate-semialdehyde dehydrogenase	L-Glutamate γ-semialdehyde + orthophosphate + NADP+ = L-γ-glutamyl phosphate + NADPH
1.2.1.42	Hexadecanal dehydrogenase (acylating)	Hexadecanal + CoA + NAD+ = hexadecanoyl-CoA + NADH
1.2.1.43	Formate dehydrogenase (NADP+)	Formate + NADP+ = CO_2 + NADPH

Number	Other Names	Basis for classification (Systematic Name)	Comments	Reference
1.2.1.30		Aryl-aldehyde:NADP$^+$ oxidoreductase (ATP-forming)		1130
1.2.1.31		L-2-Aminoadipate-6-semialdehyde:NAD(P)$^+$ oxidoreductase		447
1.2.1.32		2-Aminomuconate-6-semialdehyde:NAD$^+$ oxidoreductase	Also acts on 2-hydroxymuconate semialdehyde	1428
1.2.1.33		D-2-Hydroxy-3,3-dimethyl-3-formylpropionate:NAD$^+$ 4-oxidoreductase		2040
1.2.1.34		D-Mannonate:NAD(P)$^+$ 6-oxidoreductase (D-mannuronate-forming)		860
1.2.1.35		Uronate:NAD$^+$ 1-oxidoreductase	Also acts on D-glucuronate	1684
1.2.1.36		Retinal:NAD$^+$ oxidoreductase	A metalloflavoprotein (FAD); acts on both the 11-*trans*- and 13-*cis*-forms of retinal	2260
1.2.1.37		Xanthine:NAD$^+$ oxidoreductase	Acts on a variety of purines and aldehydes	2562, 2703
1.2.1.38		*N*-Acetyl-L-glutamate-5-semialdehyde:NADP$^+$ oxidoreductase (phosphorylating)		171, 1050
1.2.1.39		Phenylacetaldehyde:NAD$^+$ oxidoreductase		964, 965
1.2.1.40		5β-Cholestane-3α,7α,12α-triol-26-al:NAD$^+$ oxidoreductase		2113
1.2.1.41	γ-Glutamylphosphate reductase	L-Glutamate-γ-semialdehyde:NADP$^+$ oxidoreductase (phosphorylating)		170
1.2.1.42	Fatty acyl-CoA reductase	Hexadecanal:NAD$^+$ oxidoreductase (CoA-acylating)	Also acts more slowly on octadecanoyl-CoA	1525
1.2.1.43		Formate:NADP$^+$ oxidoreductase	Probably contains selenium and tungsten	82, 1995

Number	Recommended Name	Reaction
1.2.1.44	Cinnamoyl-CoA reductase	Cinnamaldehyde + CoA + $NADP^+$ = cinnamoyl-CoA + NADPH
1.2.1.45	2-Hydroxy-4-carboxymuconate-6-semialdehyde dehydrogenase	2-Hydroxy-4-carboxy-*cis-cis*-muconate 6-semialdehyde + $NADP^+$ = 2-hydroxy-4-carboxy-*cis-cis*-muconate + NADPH

1.2.2   WITH A CYTOCHROME AS ACCEPTOR

| 1.2.2.1 | Formate dehydrogenase (cytochrome) | Formate + ferricytochrome b_1 = CO_2 + ferrocytochrome b_1 |
| 1.2.2.2 | Pyruvate dehydrogenase (cytochrome) | Pyruvate + ferricytochrome b_1 + H_2O = acetate + CO_2 + ferrocytochrome b_1 |

1.2.3   WITH OXYGEN AS ACCEPTOR

1.2.3.1	Aldehyde oxidase	Aldehyde + H_2O + O_2 = acid + superoxide
1.2.3.2	Xanthine oxidase	Xanthine + H_2O + O_2 = urate + superoxide
1.2.3.3	Pyruvate oxidase	Pyruvate + orthophosphate + O_2 + H_2O = acetyl phosphate + CO_2 + H_2O_2
1.2.3.4	Oxalate oxidase	Oxalate + O_2 = 2 CO_2 + H_2O_2
1.2.3.5	Glyoxylate oxidase	Glyoxylate + H_2O + O_2 = oxalate + H_2O_2
1.2.3.6	Pyruvate oxidase (CoA-acetylating)	Pyruvate + CoA + O_2 = acetyl-CoA + CO_2 + H_2O_2

1.2.4   WITH A DISULPHIDE COMPOUND AS ACCEPTOR

| 1.2.4.1 | Pyruvate dehydrogenase (lipoamide) | Pyruvate + lipoamide = S^6-acetyldihydrolipoamide + CO_2 |

Number	Other Names	Basis for classification (Systematic Name)	Comments	Reference
1.2.1.44		Cinnamaldehyde:NADP$^+$ oxidoreductase (CoA-cinnamoylating)	Acts also on a number of substituted cinnamoyl esters of coenzyme A	3659
1.2.1.45		2-Hydroxy-4-carboxy-*cis-cis*-muconate-6-semialdehyde: NADP$^+$ oxidoreductase	The enzyme does not act on unsubstituted aliphatic or aromatic aldehydes or glucose; NAD$^+$ can replace NADP$^+$, but with lower affinity	2104
1.2.2.1		Formate:ferricytochrome b_1 oxidoreductase		985
1.2.2.2	Pyruvate dehydrogenase, Pyruvic dehydrogenase	Pyruvate:ferricytochrome b_1 oxidoreductase	A flavoprotein requiring thiamin diphosphate	3692
1.2.3.1		Aldehyde:oxygen oxidoreductase	A flavohaemoprotein containing Mo. Also oxidizes quinoline and pyridine derivatives	1084, 1725, 2051
1.2.3.2	Hypoxanthine oxidase, Schardinger enzyme	Xanthine:oxygen oxidoreductase	A flavoprotein (FAD) containing Fe and Mo. Also oxidizes hypoxanthine, some other purines and pterins, and aldehydes. The *Micrococcus* enzyme can use ferredoxin as acceptor	146, 378, 680
1.2.3.3	Pyruvic oxidase	Pyruvate:oxygen oxidoreductase (phosphorylating)	A flavoprotein requiring thiamin diphosphate	1162, 3691
1.2.3.4		Oxalate:oxygen oxidoreductase	A flavoprotein	642
1.2.3.5		Glyoxylate:oxygen oxidoreductase		1600
1.2.3.6		Pyruvate:oxygen oxidoreductase (CoA-acetylating)	A flavoprotein (FAD). May be identical with EC 1.2.7.1	3353, 2760
1.2.4.1	Pyruvate dehydrogenase, Pyruvic dehydrogenase	Pyruvate:lipoamide oxidoreductase (decarboxylating and	Requires thiamin diphosphate; component of the multienzyme	2467, 2989

Number	Recommended Name	Reaction

1.2.4.2   Oxoglutarate dehydrogenase

2-Oxoglutarate + lipoamide = S^6-succinyldihydrolipoamide + CO_2

[1.2.4.3   Deleted entry: now included with EC 1.2.4.4]

1.2.4.4   2-Oxoisovalerate dehydrogenase (lipoamide)

2-Oxoisovalerate + lipoamide = S^6-isobutyryldihydrolipoamide + CO_2

1.2.7   WITH AN IRON-SULPHUR PROTEIN AS ACCEPTOR

1.2.7.1   Pyruvate synthase

Pyruvate + CoA + oxidized ferredoxin = acetyl-CoA + CO_2 + reduced ferredoxin

1.2.7.2   2-Oxobutyrate synthase

2-Oxobutyrate + CoA + oxidized ferredoxin = propionyl-CoA + CO_2 + reduced ferredoxin

1.2.7.3   2-Oxoglutarate synthase

2-Oxoglutarate + CoA + oxidized ferredoxin = succinyl-CoA + CO_2 + reduced ferredoxin

1.2.99   WITH OTHER ACCEPTORS

1.2.99.1   Uracil dehydrogenase

Uracil + acceptor = barbiturate + reduced acceptor

1.3   ACTING ON THE CH-CH GROUP OF DONORS

In this sub-class are listed all enzymes introducing a double-bond into the substrate by direct dehydrogenation at a carbon-carbon single bond. Sub-sub-classes are based on the acceptor; 1.3.1 with NAD(P)$^+$, 1.3.2 with cytochrome, 1.3.3 with oxygen, 1.3.7 with an iron-sulphur protein and 1.3.99 with other acceptors.

1.3.1   WITH NAD$^+$ OR NADP$^+$ AS ACCEPTOR

1.3.1.1   Dihydrouracil dehydrogenase

5,6-Dihydrouracil + NAD$^+$ = uracil + NADH

1.3.1.2   Dihydrouracil dehydrogenase (NADP$^+$)

5,6-Dihydrouracil + NADP$^+$ = uracil + NADPH

1.3.1.3   Cortisone β-reductase

4,5β-Dihydrocortisone + NADP$^+$ = cortisone + NADPH

Number	Other Names	Basis for classification (Systematic Name)	Comments	Reference
		acceptor-acetylating)	pyruvate dehydrogenese complex	
1.2.4.2	α-Ketoglutaric dehydrogenase	2-Oxoglutarate:lipoamide oxidoreductase (decarboxylating and acceptor-succinylating)	Requires thiamin diphosphate; component of the multienzyme 2-oxoglutarate dehydrogenase system	2108, 2467, 2906
1.2.4.4	Branched-chain α-keto acid dehydrogenase	2-Oxoisovalerate:lipoamide oxidoreductase (decarboxylating and acceptor-isobutyrylating)	Requires thiamin diphosphate; apparently a multienzyme complex. Also acts on 2-oxo-isocaproate and 2- oxo-3-methylvalrate	570
1.2.7.1		Pyruvate-ferredoxin oxidoreductase (CoA-acetylating)		848, 3520, 1003
1.2.7.2		2-Oxobutyrate:ferredoxin oxidoreductase (CoA-propionylating)		417
1.2.7.3		2-Oxoglutarate:ferredoxin oxidoreductase (CoA-succinylating)		418, 1003
1.2.99.1		Uracil: (acceptor) oxidoreductase	Also oxidizes thymine	1235
1.3.1.1		5,6-Dihydrouracil:NAD^+ oxidoreductase		452
1.3.1.2		5,6-Dihydrouracil:$NADP^+$ oxidoreductase	Also acts on dihydrothymine	946, 1126
1.3.1.3		4,5β-Dihydrocortisone: $NADP^+$ Δ^4-oxidoreductase		403, 1948, 3442

Number	Recommended Name	Reaction
1.3.1.4	Cortisone α-reductase	4,5α-Dihydrocortisone + NADP$^+$ = cortisone + NADPH
1.3.1.5	Cucurbitacin Δ^{23}-reductase	23,24-Dihydrocucurbitacin + NAD(P)$^+$ = cucurbitacin + NAD(P)H
1.3.1.6	Fumarate reductase (NADH)	Succinate + NAD$^+$ = fumarate + NADH
1.3.1.7	*meso*-Tartrate dehydrogenase	*meso*-Tartrate + NAD$^+$ = dihydroxyfumarate + NADH
1.3.1.8	Acyl-CoA dehydrogenase (NADP$^+$)	Acyl-CoA + NADP$^+$ = 2,3-dehydroacyl-CoA + NADPH
1.3.1.9	Enoyl-[acyl-carrier-protein] reductase	Acyl-[acyl-carrier protein] + NAD$^+$ = 2,3-dehydroacyl-[acyl-carrier protein] + NADH
1.3.1.10	Enoyl-[acyl-carrier-protein] reductase (NADPH)	Acyl-[acyl-carrier protein) + NADP$^+$ = 2,3-dehydroacyl-[acyl-carrier protein] + NADPH
1.3.1.11	Melilotate dehydrogenase	2-Hydroxyphenylpropionate + NAD$^+$ = *o*-coumarate + NADH
1.3.1.12	Prephenate dehydrogenase	Prephenate + NAD$^+$ = 4-hydroxyphenylpyruvate + CO_2 + NADH
1.3.1.13	Prephenate dehydrogenase (NADP$^+$)	Prephenate + NADP$^+$ = 4-hydroxyphenylpyruvate + CO_2 + NADPH
1.3.1.14	Orotate reductase	L-5,6-Dihydroorotate + NAD$^+$ = orotate + NADH
1.3.1.15	Orotate reductase (NADPH)	L-5,6-Dihydroorotate + NADP$^+$ = orotate + NADPH
1.3.1.16	β-Nitroacrylate reductase	3-Nitropropionate + NADP$^+$ = 3-nitroacrylate + NADPH

Number	Other Names	Basis for classification (Systematic Name)	Comments	Reference
1.3.1.4		4,5α-Dihydrocortisone: NADP$^+$ Δ^4-oxidoreductase		2166
1.3.1.5		23,24-Dihydrocucurbitacin: NAD(P)$^+$ Δ^{23}-oxidoreductase	Requires Mn^{2+}; Fe^{2+} or Zn^{2+} can replace Mn^{2+} to some extent	2937, 2938
1.3.1.6		Succinate:NAD$^+$ oxidoreductase		1358
1.3.1.7		*meso*-Tartrate:NAD$^+$ oxidoreductase		1738
1.3.1.8	Enoyl-CoA reductase	Acyl-CoA:NADP$^+$ oxidoreductase	Acts on enoyl-CoA derivatives of carbon chain length 4 to 16, with optimum activity on 2-hexenoyl-CoA	3013
1.3.1.9		Acyl-[acyl-carrier-protein]: NAD$^+$ oxidoreductase	Catalyses the reduction of enoyl-acyl-[acyl-carrier-protein] derivatives of carbon chain lengths from 4 to 16	3636
1.3.1.10		Acyl-[acyl-carrier-protein]: NADP$^+$ oxidoreductase	Catalyses the reduction of enoyl-acyl-[acyl-carrier-protein] derivatives of carbon chain length from 4 to 16	3636
1.3.1.11		2-Hydroxyphenylpropionate: NAD$^+$ oxidoreductase		1945
1.3.1.12		Prephenate:NAD$^+$ oxidoreductase (decarboxylating)	This enzyme in the enteric bacteria also possesses chorismate mutase activity and converts chorismate into prephenate	445, 1730
1.3.1.13		Prephenate:NADP$^+$ oxidoreductase (decarboxylating)		990
1.3.1.14		L-5,6-Dihydroorotate:NAD$^+$ oxidoreductase	A flavoprotein (FAD, FMN)	940, 941, 1961
1.3.1.15		L-5,6-Dihydroorotate:NADP$^+$ oxidoreductase	A flavoprotein	3385, 3506
1.3.1.16		3-Nitropropionate:NADP$^+$ oxidoreductase		3032

Number	Recommended Name	Reaction
1.3.1.17	3-Methyleneoxindole reductase	3-Methyloxindole + NADP$^+$ = 3-methyleneoxindole + NADPH
1.3.1.18	Kynurenate-7,8-dihydrodiol dehydrogenase	7,8-Dihydro-7,8-dihydroxykynurenate + NAD$^+$ = 7,8-dihydroxykynurenate + NADH
1.3.1.19	*cis*-1,2-Dihydrobenzene-1,2-diol dehydrogenase	*cis*-1,2-Dihydrobenzene-1,2-diol + NAD$^+$ = catechol + NADH
1.3.1.20	*trans*-1,2-Dihydrobenzene-1,2-diol dehydrogenase	*trans*-1,2-Dihydrobenzene-1,2-diol + NADP$^+$ = catechol + NADPH
1.3.1.21	7-Dehydrocholesterol reductase	Cholesterol + NADP$^+$ = cholesta-5,7-dien-3β-ol + NADPH
1.3.1.22	Cholestenone 5α-reductase	5α-Cholestan-3-one + NADP$^+$ = cholest-4-en-3-one + NADPH
1.3.1.23	Cholestenone 5β-reductase	5β-Cholestan-3-one + NADP$^+$ = cholest-4-en-3-one + NADPH
1.3.1.24	Biliverdin reductase	Bilirubin + NAD(P)$^+$ = biliverdin + NAD(P)H
1.3.1.25	3,5-Cyclohexadiene-1,2-diol-1-carboxylate dehydrogenase	3,5-Cyclohexadiene-1,2-diol-1-carboxylate + NAD$^+$ = catechol + CO$_2$ + NADH
1.3.1.26	Dihydrodipicolinate reductase	2,3,4,5-Tetrahydrodipicolinate + NAD(P)$^+$ = 2,3-dihydrodipicolinate + NAD(P)H
1.3.1.27	2-Hexadecenal reductase	Hexadecanal + NADP$^+$ = 2-*trans*-hexadecenal + NADPH
1.3.1.28	2,3-Dihydro-2,3-dihydroxybenzoate dehydrogenase	2,3-Dihydro-2,3-dihydroxybenzoate + NAD$^+$ = 2,3-dihydroxybenzoate + NADH
1.3.1.29	*cis*-1,2-Dihydro-1,2-dihydroxynaphthalene dehydrogenase	*cis*-1,2-Dihydro-1,2-dihydroxynaphthalene + NAD$^+$ = 1,2-dihydroxynaphthalene + NADH
1.3.1.30	Progesterone 5α-reductase	5α-Pregnan-3,20-dione + NADP$^+$ = progesterone + NADPH

Number	Other Names	Basis for classification (Systematic Name)	Comments	Reference
1.3.1.17		3-Methyloxindole:NADP$^+$ oxidoreductase		2305
1.3.1.18		7,8-Dihydro-7,8-dihydroxykynurenate:NAD$^+$ oxidoreductase		3370
1.3.1.19		*cis*-1,2-Dihydrobenzene-1,2-diol:NAD$^+$ oxidoreductase		1031, 148
1.3.1.20		*trans*-1,2-Dihydrobenzene-1,2-diol:NADP$^+$ oxidoreductase		156
1.3.1.21		Cholesterol:NADP$^+$ Δ^7-oxidoreductase		692
1.3.1.22		3-Oxo-5α-steroid:NADP$^+$ Δ^4-oxidoreductase		3037
1.3.1.23		3-Oxo-5β-steroid:NADP$^+$ Δ^4-oxidoreductase		915
1.3.1.24		Bilirubin:NAD(P)$^+$ oxidoreductase		3089
1.3.1.25		3,5-Cyclohexadiene-1,2-diol-1-carboxylate:NAD$^+$ oxidoreductase (decarboxylating)		2764
1.3.1.26		2,3,4,5-Tetrahydrodipicolinate: NAD(P)$^+$ oxidoreductase		3358, 859
1.3.1.27	2-Alkenal reductase	Hexadecanal:NADP$^+$ oxidoreductase	Enzyme is specific for long chain 2-*trans* and 2-*cis*-alkenals, with chain length optimum around 14 to 16 carbon atoms	3232
1.3.1.28		2,3-Dihydro-2,3-dihydroxybenzoate:NAD$^+$ oxidoreductase	Formerly EC 1.1.1.109	3813
1.3.1.29		*cis*-1,2-Dihydro-1,2-dihydroxynaphthalene:NAD$^+$ 1,2-oxidoreductase	Also acts, at half the rate, on *cis*-anthracene dihydrodiol and *cis*-phenanthrene dihydrodiol	2564
1.3.1.30		5α-Pregnan-3,20-dione:NADP$^+$ oxidoreductase	Testosterone and 20α-hydroxy-4-pregnen-3-one can act in place of progesterone	516, 517

Number	Recommended Name	Reaction

1.3.2 WITH A CYTOCHROME AS ACCEPTOR

[1.3.2.1 Transferred entry: now EC 1.3.99.2, Butyryl-CoA dehydrogenase]

[1.3.2.2 Transferred entry: now EC 1.3.99.3, Acyl-CoA dehydrogenase]

1.3.2.3 Galactonolactone dehydrogenase

L-Galactono-γ-lactone + 2 ferricytochrome c = L-ascorbate + 2 ferrocytochrome c

1.3.3 WITH OXYGEN AS ACCEPTOR

1.3.3.1 Dihydroorotate oxidase

L-5,6-Dihydroorotate + O_2 = orotate + H_2O_2

1.3.3.2 Lathosterol oxidase

5α-Cholest-7-en-3β-ol + O_2 = cholesta-5,7-dien-3β-ol + H_2O_2

1.3.3.3 Coproporphyrinogen oxidase

Coproporphyrinogen-III + O_2 = protoporphyrinogen-IX + 2 CO_2

1.3.3.4 Protoporphyrinogen oxidase

Protoporphyrinogen-IX + O_2 = protoporphyrin-IX + H_2O

1.3.7 WITH AN IRON-SULPHUR PROTEIN AS ACCEPTOR

1.3.7.1 6-Hydroxynicotinate reductase

1,4,5,6-Tetrahydro-6-oxo-nicotinate + oxidized ferredoxin = 6-hydroxynicotinate + reduced ferredoxin

1.3.99 WITH OTHER ACCEPTORS

1.3.99.1 Succinate dehydrogenase

Succinate + acceptor = fumarate + reduced acceptor

1.3.99.2 Butyryl-CoA dehydrogenase

Butyryl-CoA + acceptor = crotonoyl-CoA + reduced acceptor

1.3.99.3 Acyl-CoA dehydrogenase

Acyl-CoA + acceptor = 2,3-dehydroacyl-CoA + reduced acceptor

Number	Other Names	Basis for classification (Systematic Name)	Comments	Reference
1.3.2.3		L-Galactono-γ-lactone: ferricytochrome *c* oxidoreductase		2077, 2078
1.3.3.1		L-5,6-Dihydroorotate:oxygen oxidoreductase	Ferricyanide can act as acceptor. A flavoprotein (FAD, FMN)	941, 3385
1.3.3.2		5α-Cholest-7-en-3β-ol:oxygen Δ^5-oxidoreductase		692
1.3.3.3	Coproporphyrinogenase	Coproporphyrinogen:oxygen oxidoreductase (decarboxylating)		227
1.3.3.4		Protoporphyrinogen-IX:oxygen oxidoreductase	Also slowly oxidizes mesoporphyrinogen-IX: oxygen	2642, 2641
1.3.7.1	6-Oxotetrahydro-nicotinate dehydrogenase	1,4,5,6-Tetrahydro-6-oxo-nicotinate:ferredoxin oxidoreductase		1342
1.3.99.1	Succinic dehydrogenase, Fumarate reductase, Fumaric hydrogenase	Succinate: (acceptor) oxidoreductase	A flavoprotein (FAD) containing iron	3082, 3083, 3619
1.3.99.2	Butyryl dehydrogenase, Ethylene reductase, Unsaturated acyl-CoA reductase	Butyryl-CoA: (acceptor) oxidoreductase	A flavoprotein; forms with another flavoprotein ('electron-transferring flavoprotein') a system reducing ubiquinone and other acceptors. Formerly EC 1.3.2.1	246, 1104, 1227, 2049
1.3.99.3	Acyl dehydrogenase	Acyl-CoA: (acceptor) oxidoreductase	A flavoprotein; forms with another flavoprotein ('electron-transferring flavoprotein') a system reducing ubiquinone and	246, 604, 1227

Number	Recommended Name	Reaction
1.3.99.4	3-Oxosteroid Δ^1-dehydrogenase	A 3-oxosteroid + acceptor = a Δ^1-3-oxosteroid + reduced acceptor
1.3.99.5	3-Oxo-5α-steroid Δ^4-dehydrogenase	A 3-oxo-5α-steroid + acceptor = a 3-oxo-Δ^4-steroid + reduced acceptor
1.3.99.6	3-Oxo-5β-steroid Δ^4-dehydrogenase	A 3-oxo-5β-steroid + acceptor = a 3-oxo-Δ^4-steroid + reduced acceptor
1.3.99.7	Glutaryl-CoA dehydrogenase	Glutaryl-CoA + acceptor = crotonoyl-CoA + CO_2 + reduced acceptor
1.3.99.8	2-Furoyl-CoA dehydrogenase	2-Furoyl-CoA + H_2O + acceptor = 5-hydroxy-2-furoyl-CoA + reduced acceptor
1.3.99.9	β-Cyclopiazonate dehydrogenase	β-Cyclopiazonate + acceptor = α-cyclopiazonate + reduced acceptor
1.3.99.10	Isovaleryl-CoA dehydrogenase	Isovaleryl-CoA + acceptor = 3-methylcrotonoyl-CoA + reduced acceptor

1.4 ACTING ON THE CH-NH₂ GROUP OF DONORS

These are the amino-acid dehydrogenases and the amine oxidases. In most cases the imine formed is hydrolysed to give an oxo-group and NH_3. This is indicated as (deaminating). Sub-sub-classes are formed according to acceptor: 1.4.1 with NAD(P)$^+$, 1.4.2 with a cytochrome, 1.4.3 with oxygen, 1.4.4 with a disulphide compound, 1.4.7 with an iron-sulphur protein as acceptor, and 1.4.99 with other acceptors.

1.4.1 WITH NAD⁺ OR NADP⁺ AS ACCEPTOR

1.4.1.1	Alanine dehydrogenase	L-Alanine + H_2O + NAD$^+$ = pyruvate + NH_3 + NADH
1.4.1.2	Glutamate dehydrogenase	L-Glutamate + H_2O + NAD$^+$ = 2-oxoglutarate + NH_3 + NADH

Number	Other Names	Basis for classification (Systematic Name)	Comments	Reference
			other acceptors. Formerly EC 1.3.2.2	
1.3.99.4		3-Oxosteroid: (acceptor) Δ^1-oxidoreductase		1949
1.3.99.5		3-Oxo-5α-steroid: (acceptor) Δ^4-oxidoreductase		1949
1.3.99.6		3-Oxo-5β-steroid: (acceptor) Δ^4-oxidoreductase		649
1.3.99.7		Glutaryl-CoA: (acceptor) oxidoreductase (decarboxylating)	A flavoprotein	289
1.3.99.8	Furoyl-CoA hydroxylase	2-Furoyl-CoA: (acceptor) oxidoreductase (hydroxylating)	A cuproprotein. The oxygen atom of the -OH produced is derived from water, not O_2; the actual oxidative step is probably C-CHOH- to C=C(OH)-. The product of the reaction tautomerizes non-enzymically to form 5-oxo-Δ^2-dihydro-2-furoyl-CoA. Methylene blue, nitro blue tetrazolium and a membrane fraction from *P. putida* can act as acceptors	1704
1.3.99.9	β-Cyclopiazonate oxidocyclase	β-Cyclopiazonate: (acceptor) oxidoreductase (cyclizing)	A flavoprotein (FAD). Cytochrome *c* and various dyes can act as acceptor	2939, 2936
1.3.99.10		Isovaleryl-CoA: (acceptor) oxidoreductase		159, 3359
1.4.1.1		L-Alanine:NAD$^+$ oxidoreductase (deaminating)		2471, 2604, 3800
1.4.1.2	Glutamic dehydrogenase	L-Glutamate:NAD$^+$ oxidoreductase (deaminating)		201, 424, 847, 932, 2423

Number	Recommended Name	Reaction
1.4.1.3	Glutamate dehydrogenase (NAD(P)$^+$)	L-Glutamate + H_2O + NAD(P)$^+$ = 2-oxoglutarate + NH_3 + NAD(P)H
1.4.1.4	Glutamate dehydrogenase (NADP$^+$)	L-Glutamate + H_2O + NADP$^+$ = 2-oxoglutarate + NH_3 + NADPH
1.4.1.5	L-Amino-acid dehydrogenase	An L-amino acid + H_2O + NAD$^+$ = a 2-oxo acid + NH_3 + NADH
1.4.1.6	D-Proline reductase	5-Aminovalerate + NAD$^+$ = D-proline + NADH
1.4.1.7	Serine dehydrogenase	L-Serine + H_2O + NAD$^+$ = 3-hydroxypyruvate + NH_3 + NADH
1.4.1.8	Valine dehydrogenase (NADP$^+$)	L-Valine + H_2O + NADP$^+$ = 3-methyl-2-oxobutyrate + NH_3 + NADPH
1.4.1.9	Leucine dehydrogenase	L-Leucine + H_2O + NAD$^+$ = 4-methyl-2-oxopentanoate + NH_3 + NADH
1.4.1.10	Glycine dehydrogenase	Glycine + H_2O + NAD$^+$ = glyoxylate + NH_3 + NADH
1.4.1.11	L-erythro-3,5-Diaminohexanoate dehydrogenase	L-erythro-3,5-Diaminohexanoate + H_2O + NAD$^+$ = 5-amino-3-oxohexanoate + NH_3 + NADH
1.4.1.12	2,4-Diaminopentanoate dehydrogenase	2,4-Diaminopentanoate + H_2O + NAD(P)$^+$ = 2-amino-4-oxopentanoate + NH_3 + NAD(P)H
1.4.1.13	Glutamate synthase (NADPH)	2 L-Glutamate + NADP$^+$ = L-glutamine + 2-oxoglutarate + NADPH
1.4.1.14	Glutamate synthase (NADH)	2 L-Glutamate + NAD$_+$ = L-glutamine + 2-oxoglutarate + NADH
1.4.1.15	Lysine dehydrogenase	Lysine + NAD$^+$ = 1-didehydro-piperidine-2-carboxylate + NH_3 + NADH

Number	Other Names	Basis for classification (Systematic Name)	Comments	Reference
1.4.1.3	Glutamic dehydrogenase	L-Glutamate:NAD(P)$^+$ oxidoreductase (deaminating)		2498, 3130, 3254
1.4.1.4	Glutamic dehydrogenase	L-Glutamate:NADP$^+$ oxidoreductase (deaminating)		24, 25, 3569, 1127
1.4.1.5		L-Amino-acid:NAD$^+$ oxidoreductase (deaminating)	Acts on aliphatic amino acids	2424
1.4.1.6		5-Aminovalerate:NAD$^+$ oxidoreductase (cyclizing)		3186, 3190
1.4.1.7		L-Serine:NAD$^+$ oxidoreductase (deaminating)		1803
1.4.1.8		L-Valine:NADP$^+$ oxidoreductase (deaminating)		1555
1.4.1.9		L-Leucine:NAD$^+$ oxidoreductase (deaminating)	Also acts on isoleucine, valine, norvaline and norleucine	2909, 3851
1.4.1.10		Glycine:NAD$^+$ oxidoreductase (deaminating)		1068
1.4.1.11		L-*erythro*-3,5-Diaminohexanoate: NAD$^+$ oxidoreductase (deaminating)		182
1.4.1.12		2,4-Diaminopentanoate: NAD(P)$^+$ oxidoreductase (deaminating)	Also acts, more slowly, on 2,5-diaminohexanoate forming 2-amino-5-oxohexanoate, which then cyclizes non-enzymically to 1-pyrroline-2-methyl-5-carboxylate	3188, 3149, 3474
1.4.1.13		L-Glutamate:NADP$^+$ oxidoreductase (transaminating)	An iron-sulphur flavoprotein. In the reverse reaction ammonia can act instead of glutamine but more slowly. Formerly EC 2.6.1.53	3389, 2227
1.4.1.14		L-Glutamate:NAD$^+$ oxidoreductase (transaminating)	A flavoprotein (FMN)	556
1.4.1.15		L-Lysine:NAD$^+$ oxidoreductase (deaminating, cyclizing)		430

Number	Recommended Name	Reaction

1.4.2 WITH A CYTOCHROME AS ACCEPTOR

1.4.2.1   Glycine dehydrogenase (cytochrome)

Glycine + H_2O + 2 ferricytochrome c = glyoxylate + NH_3 + 2 ferrocytochrome c

1.4.3 WITH OXYGEN AS ACCEPTOR

1.4.3.1   D-Aspartate oxidase

D-Aspartate + H_2O + O_2 = oxaloacetate + NH_3 + H_2O_2

1.4.3.2   L-Amino-acid oxidase

An L-amino acid + H_2O + O_2 = a 2-oxo acid + NH_3 + H_2O_2

1.4.3.3   D-Amino-acid oxidase

A D-amino acid + H_2O + O_2 = a 2-oxo acid + NH_3 + H_2O_2

1.4.3.4   Amine oxidase (flavin-containing)

$RCHNH_2$ + H_2O + O_2 = RCHO + NH_3 + H_2O_2

1.4.3.5   Pyridoxaminephosphate oxidase

Pyridoxamine phosphate + H_2O + O_2 = pyridoxal phosphate + NH_3 + H_2O_2

1.4.3.6   Amine oxidase (copper-containing)

RCH_2NH_2 + H_2O + O_2 = RCHO + NH_3 + H_2O_2

1.4.3.7   D-Glutamate oxidase

D-Glutamate + H_2O + O_2 = 2-oxoglutarate + NH_3 + H_2O_2

1.4.3.8   Ethanolamine oxidase

Ethanolamine + H_2O + O_2 = glycolaldehyde + NH_3 + H_2O_2

1.4.3.9   Tyramine oxidase

Tyramine + O_2 + H_2O = 4-hydroxyphenylacetaldehyde + NH_3 + H_2O_2

Number	Other Names	Basis for classification (Systematic Name)	Comments	Reference
1.4.2.1	Glycine—cytochrome c reductase	Glycine:ferricytochrome c oxidoreductase (deaminating)		2907
1.4.3.1	Aspartic oxidase	D-Aspartate:oxygen oxidoreductase (deaminating)	A flavoprotein (FAD)	3226, 3227, 729 3226, 3227, 729
1.4.3.2	Ophio-amino-acid oxidase (for the snake enzyme only)	L-Amino-acid:oxygen oxidoreductase (deaminating)	A flavoprotein (FAD). The enzyme from liver and kidney also oxidizes 2-hydroxyacids; that from snake venom does not	321, 2190, 2364, 3655
1.4.3.3		D-Amino-acid:oxygen oxidoreductase (deaminating)	A flavoprotein (FAD). Wide specificity for D-amino acids	2111, 2190, 730
1.4.3.4	Monoamine oxidase, Tyramine oxidase, Tyraminase, Amine oxidase, Adrenalin oxidase	Amine:oxygen oxidoreductase (deaminating) (flavin-containing)	A flavoprotein (FAD). Acts on primary, secondary and tertiary amines	322, 842, 3426, 3625, 3749, 3841
1.4.3.5		Pyridoxaminephosphate:oxygen oxidoreductase (deaminating)	A flavoprotein (FAD). Also oxidizes pyridoxine 5-phosphate and pyridoxin	3573
1.4.3.6	Diamine oxidase, Diamino oxhydrase, Histaminase, Amine oxidase (Pyridoxal containing)	Amine:oxygen oxidoreductase (deaminating) (copper-containing)	A copper-protein; may also contain pyridoxal phosphate. A group of enzymes oxidizing primary monoamines and diamines, including histamine	322, 323, 423, 2161, 2267, 3747, 3841
1.4.3.7	D-Glutamic oxidase	D-Glutamate:oxygen oxidoreductase (deaminating)		2821, 3517
1.4.3.8		Ethanolamine:oxygen oxidoreductase (deaminating)	A cobamide-protein	2373
1.4.3.9		Tyramine:oxygen oxidoreductase (deaminating)	A flavoprotein (FAD). Also acts on dopamine; slowly oxidizes secondary amines	1817

Number	Recommended Name	Reaction
1.4.3.10	Putrescine oxidase	Putrescine + O_2 + H_2O = 4-aminobutyraldehyde + NH_3 + H_2O_2
1.4.3.11	L-Glutamate oxidase	2 L-Glutamate + O_2 + H_2O = 2 2-oxoglutarate + 2 NH_3 + H_2O
1.4.3.12	Cyclohexylamine oxidase	Cyclohexylamine + O_2 + H_2O = cyclohexanone + NH_3 + H_2O_2

1.4.4   WITH A DISULPHIDE COMPOUND AS ACCEPTOR

1.4.4.1	D-Proline reductase (dithiol)	5-Aminovalerate + lipoate = D-proline + dihydrolipoate

1.4.7   WITH AN IRON-SULPHUR PROTEIN AS ACCEPTOR

1.4.7.1	Glutamate synthase (ferredoxin)	2 L-Glutamate + 2 oxidized ferredoxin = L-glutamine + 2-oxoglutarate + 2 reduced ferredoxin

1.4.99   WITH OTHER ACCEPTORS

1.4.99.1	D-Amino-acid dehydrogenase	A D-amino acid + H_2O + acceptor = a 2-oxo acid + NH_3 + reduced acceptor
1.4.99.2	Taurine dehydrogenase	Taurine + H_2O + acceptor = sulphoacetaldehyde + NH_3 + reduced acceptor
1.4.99.3	Amine dehydrogenase	RCH_2NH_2 + H_2O + acceptor = RCHO + NH_3 + reduced acceptor

1.5   ACTING ON CH-NH GROUP OF DONORS

Enzymes dehydrogenating secondary amines, introducing a $C=N$ double bond as the primary reaction. In some cases, this is later hydrolysed. Sub-sub-classes are 1.5.1 with $NAD(P)^+$, 1.5.3 with oxygen as acceptor, and 1.5.99 with other acceptors.

Number	Other Names	Basis for classification (Systematic Name)	Comments	Reference
1.4.3.10		Putrescine:oxygen oxidoreductase (deaminating)	A flavoprotein (FAD). 4- Aminobutyraldehyde condenses non-enzymically to 1-pyrroline	3746, 701
1.4.3.11		L-Glutamate:oxygen oxidoreductase (deaminating)	Can also use phenazine methosulphate and ferricyamide as electron acceptors	1550
1.4.3.12		Cyclohexylamine:oxygen oxidoreductase (deaminating)	A flavoprotein (FAD). Some other cyclic amines can act instead of cyclohexylamine, but not simple aliphatic and aromatic amides	3434
1.4.4.1		5-Aminovalerate:lipoate oxidoreductase (cyclizing)	Other dithiols can function as reducing agents; the enzyme contains bound pyruvate which acts as a cofactor	1331, 1332
1.4.7.1		L-Glutamate:ferredoxin oxidoreductase (transaminating)	An iron-sulphur flavoprotein	2219
1.4.99.1		D-Amino-acid: (acceptor) oxidoreductase (deaminating)	A flavoprotein (FAD) Acts to some extent on all D-amino acids except D-aspartate and D-glutamate	3477
1.4.99.2		Taurine: (acceptor) oxidoreductase (deaminating)		1756
1.4.99.3		Primary amine:acceptor oxidoreductase (deaminating)	Also on short-chain primary amines	779, 780

Number	Recommended Name	Reaction

1.5.1 WITH NAD⁺ OR NADP⁺ AS ACCEPTOR

1.5.1.1	Pyrroline-2-carboxylate reductase	L-Proline + $NAD(P)^+$ = 1-pyrroline-2-carboxylate + $NAD(P)H$
1.5.1.2	Pyrroline-5-carboxylate reductase	L-Proline + $NAD(P)^+$ = 1-pyrroline-5-carboxylate + $NAD(P)H$
1.5.1.3	Tetrahydrofolate dehydrogenase	5,6,7,8-Tetrahydrofolate + $NADP^+$ = 7,8-dihydrofolate + $NADPH$

[1.5.1.4   Deleted entry: Dihydrofolate dehydrogenase. Now included with EC 1.5.1.3]

1.5.1.5	Methylenetetrahydrofolate dehydrogenase (NADP⁺)	5,10-Methylenetetrahydrofolate + H^+ + $NADP^+$ = 5,10-methenyltetrahydrofolate + H_2O + $NADPH$
1.5.1.6	Formyltetrahydrofolate dehydrogenase	10-Formyltetrahydrofolate + $NADP^+$ + H_2O = tetrahydrofolate + CO_2 + $NADPH$
1.5.1.7	Saccharopine dehydrogenase (NAD⁺, lysine-forming)	N^6-(1,3-Dicarboxypropyl)-L-lysine + NAD^+ + H_2O = L-lysine + 2-oxoglutarate + $NADH$
1.5.1.8	Saccharopine dehydrogenase (NADP⁺, lysine-forming)	N^6-(1,3-Dicarboxypropyl)-L-lysine + $NADP^+$ + H_2O = L-lysine + 2-oxoglutarate + $NADPH$
1.5.1.9	Saccharopine dehydrogenase (NAD⁺, L-glutamate-forming)	N^6-(1,3-Dicarboxypropyl)-L-lysine + NAD^+ + H_2O = L-glutamate + 2-aminoadipate 6-semialdehyde + $NADH$
1.5.1.10	Saccharopine dehydrogenase (NADP⁺, L-glutamate-forming)	N^6-(1,3-Dicarboxypropyl)-L-lysine + $NADP^+$ + H_2O = L-glutamate + 2-aminoadipate 6-semialdehyde + $NADPH$
1.5.1.11	Octopine dehydrogenase	N^2-(1-Carboxyethyl)-L-arginine + NAD^+ + H_2O = L-arginine + pyruvate + $NADH$
1.5.1.12	1-Pyrroline-5-carboxylate dehydrogenase	1-Pyrroline-5-carboxylate + NAD^+ + H_2O = L-glutamate + $NADH$
1.5.1.13	Nicotinate dehydrogenase	Nicotinate + H_2O + $NADP^+$ = 6-hydroxynicotinate + $NADPH$

Number	Other Names	Basis for classification (Systematic Name)	Comments	Reference
1.5.1.1		L-Proline:NAD(P)$^+$ 2-oxidoreductase	Reduces 1-pyrroline-2-carboxylate to L-proline, or Δ^1-piperidine-2-carboxylate to L-pipecolate	2189
1.5.1.2		L-Proline:NAD(P)$^+$ 5-oxidoreductase	Also reduces 1-pyrroline-3-hydroxy-5-carboxylate to L-hydroxyproline	17, 2189, 3121, 3822
1.5.1.3	Dihydrofolate reductase	5,6,7,8-Tetrahydrofolate: NADP$^+$ oxidoreductase	The animal enzyme also slowly oxidizes 7,8-dihydrofolate to folate	3830, 320, 1613, 1668
1.5.1.5		5,10-Methylenetetrahydrofolate: NADP$^+$ oxidoreductase		1224, 2515, 3786, 2710
1.5.1.6		10-Formyltetrahydrofolate: NADP$^+$ oxidoreductase		1849
1.5.1.7	Lysine-2-oxo glutaryl reductase	N^6-(1,3-Dicarboxypropyl)-L-lysine:NAD$^+$ oxidoreductase (L-lysine-forming)		966, 2922
1.5.1.8		N^6-(1,3-Dicarboxypropyl)-L-lysine:NADP$^+$ oxidoreductase (L-lysine-forming)		1415
1.5.1.9		N^6-(1,3-Dicarboxypropyl)-L-lysine:NAD$^+$ oxidoreductase (L-glutamate-forming)		1416
1.5.1.10		N^6-(1,3-Dicarboxypropyl)-L-lysine:NADP$^+$ oxidoreductase (L-glutamate-forming)		1528
1.5.1.11		N^2-(1-carboxyethyl)-L-arginine:NAD$^+$ oxidoreductase (L-arginine-forming)		3407
1.5.1.12		1-Pyrroline-5-carboxylate:NAD$^+$ oxidoreductase	Oxidizes other 1-pyrrolines, *e.g.* 3-hydroxy-1-pyrroline-5-carboxylate	18, 3255
1.5.1.13		Nicotinate:NADP$^+$ 6-oxidoreductase (hydroxylating)	A flavoprotein containing non-haem iron. Also oxidizes NADPH	1341

Number	Recommended Name	Reaction
1.5.1.14	1,2-Didehydropipecolate reductase	L-Pipecolate + NADP$^+$ = 1,2-didehydropipecolate + NADPH
1.5.1.15	Methylenetetrahydrofolate dehydrogenase (NAD$^+$)	5,10-Methylenetetrahydrofolate + H$^+$ + NAD$^+$ = 5,10-methenyltetrahydrofolate + H$_2$O + NADH
1.5.1.16	D-Lysopine dehydrogenase	N^2-(1-Carboxyethyl)-L-lysine + NADP$^+$ + H$_2$O = L-lysine + pyruvate + NADPH

1.5.3   WITH OXYGEN AS ACCEPTOR

1.5.3.1	Sarcosine oxidase	Sarcosine + H$_2$O + O$_2$ = glycine + HCHO + H$_2$O$_2$
1.5.3.2	*N*-Methylamino-acid oxidase	An *N*-methyl-L-amino acid + H$_2$O + O$_2$ = an L-amino acid + HCHO + H$_2$O$_2$
[1.5.3.3	*Deleted entry: Spermine oxidase]*	
1.5.3.4	N^6-Methyl-lysine oxidase	N^6-Methyl-L-lysine + H$_2$O + O$_2$ = L-lysine + HCHO + H$_2$O$_2$
1.5.3.5	6-Hydroxy-L-nicotine oxidase	6-Hydroxy-L-nicotine + H$_2$O + O$_2$ = [(6-hydroxypyridyl(3))]-(3-*N*-methylaminopropyl)-ketone + H$_2$O$_2$
1.5.3.6	6-Hydroxy-D-nicotine oxidase	6-Hydroxy-D-nicotine + H$_2$O + O$_2$ = [(6-hydroxypyridyl(3))]-(3-*N*-methylaminopropyl)-ketone + H$_2$O$_2$

1.5.99   WITH OTHER ACCEPTORS

1.5.99.1	Sarcosine dehydrogenase	Sarcosine + acceptor + H$_2$O = glycine + HCHO + reduced acceptor
1.5.99.2	Dimethylglycine dehydrogenase	*N,N*-Dimethylglycine + acceptor + H$_2$O = sarcosine + HCHO + reduced acceptor
1.5.99.3	L-Pipecolate dehydrogenase	L-Pipecolate + acceptor + H$_2$O = 1,6-didehydropipecolate + reduced acceptor
1.5.99.4	Nicotine dehydrogenase	Nicotine + acceptor + H$_2$O = 6-hydroxynicotine + reduced acceptor

Number	Other Names	Basis for classification (Systematic Name)	Comments	Reference
1.5.1.14		L-Pipecolate:NADP+ 2-oxidoreductase		494
1.5.1.15		5,10-Methlenetetrahdrofolate: NAD+ oxidoreductase		2274
1.5.1.16		N^2-(1-Carboxyethyl)- L-lysine:NADP+ oxidoreductase (L-lysine forming)	In the reverse reaction, a number of L-amino acids can act instead of L-lysine, and 2-oxobutyrate and, to a lesser extent, glyoxylate, can act instead of pyruvate	2526
1.5.3.1		Sarcosine:oxygen oxidoreductase (demethylating)		1179, 1378, 943
1.5.3.2		N-Methyl-L-amino-acid:oxygen oxidoreductase (demethylating)	A flavoprotein	2290, 2291
1.5.3.4		N^6-Methyl-L-lysine:oxygen oxidoreductase (demethylating)		1686
1.5.3.5		6-Hydroxy-L-nicotine:oxygen oxidoreductase	A flavoprotein (FAD)	682, 631
1.5.3.6		6-Hydroxy-D-nicotine:oxygen oxidoreductase	A flavoprotein (FAD)	682, 407
1.5.99.1		Sarcosine: (acceptor) oxidoreductase (demethylating)	A flavoprotein (FMN)	944, 1379
1.5.99.2		N,N-Dimethylglycine: (acceptor) oxidoreductase (demethylating)	A flavoprotein	944, 1379
1.5.99.3		L-Pipecolate: (acceptor) oxidoreductase	A flavoprotein (FAD)	166
1.5.99.4		Nicotine: (acceptor) 6-oxidoreductase (hydroxylating)	Enzyme is a metalloflavoprotein (FMN). Acts on both D and L isomers	682, 1329

Number	Recommended Name	Reaction
1.5.99.5	Methylglutamate dehydrogenase	N-Methyl-L-glutamate + acceptor + H_2O = L-glutamate + formaldehyde + reduced acceptor
1.5.99.6	Spermidine dehydrogenase	Spermidine + acceptor + H_2O = 1,3-diaminopropane + 4-aminobutyraldehyde + reduced acceptor
1.5.99.7	Trimethylamine dehydrogenase	Trimethylamine + H_2O + acceptor = dimethylamine + formaldehyde + reduced acceptor

1.6 ACTING ON NADH OR NADPH

In general, enzymes using NADH or NADPH to reduce a substrate are classified according to the reverse reaction, in which NAD^+ or $NADP^+$ is formally regarded as acceptor. In class 1.6, only those enzymes are listed in which some other redox catalyst is the acceptor. This can be either $NAD(P)^+$, in sub-sub-class 1.6.1; a cytochrome in 1.6.2; a disulphide in 1.6.4; a quinone in 1.6.5; a nitrogenous group in 1.6.6; an iron-sulphur protein in 1.6.7; or some other acceptor in 1.6.99.

1.6.1 WITH NAD+ OR NADP+ AS ACCEPTOR

1.6.1.1	NAD(P)$^+$ transhydrogenase	NADPH + NAD^+ = $NADP^+$ + NADH

1.6.2 WITH A CYTOCHROME AS ACCEPTOR

[1.6.2.1   Transferred entry: now EC 1.6.99.3 - NADH dehydrogenase]

1.6.2.2	Cytochrome b_5 reductase	NADH + 2 ferricytochrome b_5 = NAD^+ + 2 ferrocytochrome b_5

[1.6.2.3   Deleted entry: Cytochrome reductase (NADPH)]

Number	Other Names	Basis for classification (Systematic Name)	Comments	Reference
1.5.99.5		*N*-Methyl-L-glutamate: (acceptor) oxidoreductase (demethylating)	A number of *N*-methyl-substituted amino acids can act as donor; 2,6-dichloroindophenol is the best acceptor	1296
1.5.99.6		Spermidine: (acceptor) oxidoreductase	A flavohaemoprotein (FAD). Ferricyanide, dichloroindophenol and 2,6-cytochrome *c* can act as acceptor. 4-Aminobutyraldehyde condenses non-enzymically to 1-pyrroline condenses non-enzymically to 1-pyrroline	3329, 3327
1.5.99.7		Trimethylamine: (acceptor) oxidoreductase (demethylating)	A number of alkyl-substituted derivatives of trimethylamine can also act as electron donor; phenazine methosulphate and 2,6-dichloroindophenol can act as electron acceptors	558
1.6.1.1	Pyridine nucleotide transhydrogenase, Transhydrogenase	NADPH:NAD$^+$ oxidoreductase	The enzyme from *Azotobacter vinelandii* is a flavoprotein (FAD). Also acts with deamino coenzymes	408, 1436, 1210
1.6.2.2		NADH:ferricytochrome b_5 oxidoreductase	A flavoprotein (FAD)	2052, 3262, 3264

Number	Recommended Name	Reaction
1.6.2.4	NADPH-cytochrome reductase	NADPH + 2 ferricytochrome = NADP$^+$ + 2 ferrocytochrome
1.6.2.5	NADPH—cytochrome c_2 reductase	NADPH + 2 ferricytochrome c_2 = NADP$^+$ + 2 ferrocytochrome c_2

1.6.4   WITH A DISULPHIDE COMPOUND AS ACCEPTOR

1.6.4.1	Cystine reductase (NADH)	NADH + L-cystine = NAD$^+$ + 2 L-cysteine
1.6.4.2	Glutathione reductase (NAD(P)H)	NAD(P)H + oxidized glutathione = NAD(P)$^+$ + 2 glutathione
1.6.4.3	Dihydrolipoamide reductase (NAD$^+$)	NADH + lipoamide = NAD$^+$ + dihydrolipoamide
1.6.4.4	Protein-disulphide reductase (NAD(P)H)	NAD(P)H + protein disulphide = NAD(P)$^+$ + protein dithiol
1.6.4.5	Thioredoxin reductase (NADPH)	NADPH + oxidized thioredoxin = NADP$^+$ + reduced thioredoxin
1.6.4.6	CoAS—Sglutathione reductase (NADPH)	NADPH + CoAS—Sglutathione = NADP$^+$ + CoA + glutathione
1.6.4.7	Asparagusate reductase (NADH)	NADH + asparagusate = NAD$^+$ + dihydroasparagusate

1.6.5   WITH A QUINONE OR RELATED COMPOUND AS ACCEPTOR

[1.6.5.1   Deleted entry: Quinone reductase]

[1.6.5.2   Transferred entry: now EC 1.6.99.2, NAD(P)H dehydrogenase]

[1.6.5.3   Deleted entry: Ubiquinone reductase]

1.6.5.4	Monodehydroascorbate reductase (NADH)	NADH + 2 monodehydroascorbate = NAD$^+$ + 2 ascorbate

1.6.6   WITH A NITROGENOUS GROUP AS ACCEPTOR

1.6.6.1	Nitrate reductase (NADH)	NADH + nitrate = NAD$^+$ + nitrite + H$_2$O

Number	Other Names	Basis for classification (Systematic Name)	Comments	Reference
1.6.2.4	NADPH-cytochrome c reductase, TPNH-cytochrome c reductase, Ferrihaemoprotein P_{450} reductase	NADPH:ferricytochrome oxidoreductase	Physiological acceptor is probably cytochrome P-450. Isolated enzyme reacts more readily with cytochrome c. A flavoprotein (FMN)	1151, 1366, 2010, 2112, 3690
1.6.2.5		NADPH:ferricytochrome c_2 oxidoreductase	A flavoprotein (FAD)	2887
1.6.4.1		NADH:L-cystine oxidoreductase		2832
1.6.4.2		NAD(P)H:oxidized-glutathione oxidoreductase	A flavoprotein (FAD)	2689, 2873, 3306, 3534, 1431
1.6.4.3	Diaphorase, Lipoyl dehydrogenase, Lipoamide dehydrogenase (NADH), Lipoamide reductase (NADH)	NADH:lipoamide oxidoreductase	A flavoprotein (FAD) Component of the multienzyme pyruvate dehydrogenase complex and 2-oxoglutarate dehydrogenase complex	2109, 2110, 2924, 3249
1.6.4.4		NAD(P)H:protein-disulphide oxidoreductase		1223
1.6.4.5		NADPH:oxidized-thioredoxin oxidoreductase	A flavoprotein (FAD)	2271, 3164
1.6.4.6		NADPH:CoAS—Sglutathione oxidoreductase	A flavoprotein	2501
1.6.4.7	Asparagusate dehydrogenase	NADH:asparagusate oxidoreductase	Also acts on lipoate	3771
1.6.5.4		NADH:monodehydroascorbate oxidoreductase		1662, 1666, 2377, 2980, 3203
1.6.6.1	Assimilatory nitrate reductase	NADH:nitrate oxidoreductase	A flavoprotein (FAD or FMN) containing a metal	875, 2375, 2404, 3162

Number	Recommended Name	Reaction
1.6.6.2	Nitrate reductase (NAD(P)H)	$NAD(P)H + nitrate = NAD(P)^+ + nitrite + H_2O$
1.6.6.3	Nitrate reductase (NADPH)	$NADPH + nitrate = NADP^+ + nitrite + H_2O$
1.6.6.4	Nitrite reductase (NAD(P)H)	$3\ NAD(P)H + nitrite = 3\ NAD(P)^+ + NH_4OH + H_2O$

[1.6.6.5 Transferred entry: now EC 1.7.99.3, Nitrite reductase]

1.6.6.6	Hyponitrite reductase	$2\ NADH + hyponitrite = 2\ NAD^+ + 2\ NH_2OH$
1.6.6.7	Azobenzene reductase	$NADPH + dimethylaminoazobenzene = NADP^+ + dimethyl\text{-}p\text{-}phenylenediamine + aniline$
1.6.6.8	GMP reductase	$NADPH + GMP = NADP^+ + IMP + ammonia$
1.6.6.9	Trimethylamine-*N*-oxide reductase	$NADH + trimethylamine\text{-}N\text{-}oxide = NAD^+ + trimethylamine + H_2O$
1.6.6.10	Nitroquinoline-*N*-oxide reductase	$2\ NAD(P)H + 4\text{-}nitroquinoline\ N\text{-}oxide = 2\ NAD(P)^+ + 4\text{-}hydroxyaminoquinoline\ N\text{-}oxide$
1.6.6.11	Hydroxylamine reductase (NADH)	$NADH + hydroxylamine = NAD^+ + ammonia + H_2O$

[1.6.7.1 Transferred entry: now EC 1.18.1.2, Ferredoxin-NADP$^+$ reductase]

[1.6.7.2 Transferred entry: now EC 1.18.1.3, Rubredoxin-NAD$^+$ reductase]

[1.6.7.3 Transferred entry: now EC 1.18.1.1, Ferredoxin-NAD reductase]

1.6.99 WITH OTHER ACCEPTORS

1.6.99.1	NADPH dehydrogenase	$NADPH + acceptor = NADP^+ + reduced\ acceptor$
1.6.99.2	NAD(P)H dehydrogenase (quinone)	$NAD(P)H + acceptor = NAD(P)^+ + reduced\ acceptor$
1.6.99.3	NADH dehydrogenase	$NADH + acceptor = NAD^+ + reduced\ acceptor$

Number	Other Names	Basis for classification (Systematic Name)	Comments	Reference
1.6.6.2	Assimilatory nitrate reductase	NAD(P)H:nitrate oxidoreductase	A flavoprotein (FAD or FMN)	2375, 2552
1.6.6.3		NADPH:nitrate oxidoreductase	A flavoprotein (FAD) containing Mo	2375, 2376, 2405, 3368
1.6.6.4		NAD(P)H:nitrite oxidoreductase	A flavoprotein containing a metal	1892, 2403, 3368
1.6.6.6		NADH:hyponitrite oxidoreductase	A metalloprotein	2175
1.6.6.7		NADPH:dimethylaminoazobenzene oxidoreductase		2311
1.6.6.8		NADPH:GMP oxidoreductase (deaminating)		2031, 2041, 2772
1.6.6.9		NADH:trimethylamine-*N*-oxide oxidoreductase		3511
1.6.6.10		NAD(P)H:4-nitroquinoline-*N*-oxide oxidoreductase		3450
1.6.6.11		NADH:hydroxylamine oxidoreductase	Also acts on some hydroxamates	282, 283
1.6.99.1	'Old Yellow enzyme', NADPH diaphorase	NADPH: (acceptor) oxidoreductase	A flavoprotein (FMN in yeast, FAD in plants)	33, 3398, 3399, 1485, 147
1.6.99.2	Menadione reductase, Phylloquinone reductase, DT-diaphorase, Quinone reductase	NAD(P)H: (quinone-acceptor) oxidoreductase	A flavoprotein. Inhibited by dicoumarol. Formerly EC 1.6.5.2	714, 1048, 2084, 2103, 2240, 3732
1.6.99.3	Cytochrome *c* reductase	NADH: (acceptor) oxidoreductase	A flavoprotein containing iron. After preparations have been subjected to certain treatments cytochrome *c* may act as	1590, 2601, 1328, 13

Number	Recommended Name	Reaction

[1.6.99.4 Transferred entry: now EC 1.6.7.1 - Ferredoxin-NADP+ reductase]

1.6.99.5	NADH dehydrogenase (quinone)	NADH + acceptor = NAD^+ + reduced acceptor
1.6.99.6	NADPH dehydrogenase (quinone)	NADPH + acceptor = $NADP^+$ + reduced acceptor
1.6.99.7	Dihydropteridine reductase	NADPH + 6,7-dihydropteridine = $NADP^+$ + 5,6,7,8-tetrahydropteridine
1.6.99.8	Aquacobalamin reductase	NADH + aquacob(III)alamin = NAD^+ + cob$_{(II)}$ alamin
1.6.99.9	Cob (II) alamin reductase	NADH + cob(II)alamin = NAD^+ + cob(I)alamin
1.6.99.10	Dihydropteridine reductase (NADH)	NADH + 6,7-dihydropteridine = NAD^+ + 5,6,7,8-tetrahydropteridine

1.7 ACTING ON OTHER NITROGENOUS COMPOUNDS AS DONORS

A small group of enzymes oxidizing diverse nitrogenous products with a cytochrome (1.7.2), oxygen (1.7.3), or with other acceptors (1.7.99).

1.7.2 WITH A CYTOCHROME AS ACCEPTOR

1.7.2.1	Nitrite reductase (cytochrome)	Nitric oxide + H_2O + 2 ferricytochrome *c* = nitrite + 2 ferrocytochrome *c*

1.7.3 WITH OXYGEN AS ACCEPTOR

1.7.3.1	Nitroethane oxidase	Nitroethane + H_2O + O_2 = acetaldehyde + nitrite + H_2O

Number	Other Names	Basis for classification (Systematic Name)	Comments	Reference
			acceptor. Formerly EC 1.6.2.1	
1.6.99.5		NADH: (quinone-acceptor) oxidoreductase	Menaquinone can act as acceptor. Inhibited by AMP and 2,4-dinitrophenol but not by dicoumarol or folic acid derivatives	1752
1.6.99.6		NADPH: (quinone-acceptor) oxidoreductase	A flavoprotein. Menaquinone can act as acceptor. Inhibited by dicoumarol and folic acid derivatives but not by 2,4-dinitrophenol	1752
1.6.99.7		NADPH:6,7-dihydropteridine oxidoreductase	Substrate is quinoid form of dihydropteridine; not identical with dihydrofolate reductase	1616, 1971
1.6.99.8		NADH:aquacob (III) alamin oxidoreductase	A flavoprotein	3590
1.6.99.9	Vitamin B_{12r} reductase	NADH:cob (II) alamin oxidoreductase	A flavoprotein	3590
1.6.99.10		NADH:6,7-dihydropteridine oxidoreductase	This enzyme has been separated from EC 1.6.99.7 from bovine liver	1210, 2361
1.7.2.1		Nitric-oxide:ferricytochrome c oxidoreductase	A cuproprotein. Cytochrome c-552 or cytochrome c-553 from *Pseudomonas denitrificans* acts as acceptor	2252
1.7.3.1		Nitroethane:oxygen oxidoreductase	Acts on some other aliphatic nitro-compounds	1990

Number	Recommended Name	Reaction
1.7.3.2	Acetylindoxyl oxidase	N-Acetylindoxyl + O_2 = N-acetylisatin + (?)
1.7.3.3	Urate oxidase	Urate + O_2 = unidentified products
1.7.3.4	Hydroxylamine oxidase	Hydroxylamine + O_2 = nitrite + H_2O

1.7.7 WITH AN IRON-SULPHUR PROTEIN AS ACCEPTOR

1.7.7.1	Ferredoxin—nitrate reductase	Ammonia + 3 oxidized ferredoxin = nitrite + 3 reduced ferredoxin

1.7.99 WITH OTHER ACCEPTORS

1.7.99.1	Hydroxylamine reductase	Ammonia + acceptor = hydroxylamine + reduced acceptor
1.7.99.2	Nitric-oxide reductase	Nitrogen + acceptor = 2 nitric oxide + reduced acceptor
1.7.99.3	Nitrite reductase	2 Nitric oxide + 2 H_2O + acceptor = 2 nitrite + reduced acceptor
1.7.99.4	Nitrate reductase	Nitrite + acceptor = nitrate + reduced acceptor

1.8 ACTING ON A SULPHUR GROUP OF DONORS

A small group of enzymes acting either on inorganic substrates or organic thiols. Sub-sub-groups again depend on the acceptor: 1.8.1, with NAD(P)$^+$; 1.8.2 with a cytochrome; 1.8.3 with oxygen; 1.8.4 with a disulphide compound; 1.8.5 with a quinone; 1.8.6 with a nitrogenous compound; 1.8.99 with other acceptors. Note that some such enzymes with NAD$^+$ as acceptor are listed under EC 1.6.4.1-4.

1.8.1 WITH NAD$^+$ OR NADP$^+$ AS ACCEPTOR

[1.8.1.1   Deleted entry:Cysteamine dehydrogenase]

Number	Other Names	Basis for classification (Systematic Name)	Comments	Reference
1.7.3.2		*N*-Acetylindoxyl:oxygen oxidoreductase		243
1.7.3.3	Uricase	Urate:oxygen oxidoreductase		1999, 2050, 2799
1.7.3.4		Hydroxylamine:oxygen oxidoreductase		2744
1.7.7.1		Ammonia:ferredoxin oxidoreductase	Contains iron	1540, 2712, 3858
1.7.99.1		Ammonia: (acceptor) oxidoreductase	A flavoprotein. Reduced pyocyanine, methylene blue or flavins act as donor for the reduction of hydroxylamine	3367, 3368, 3592
1.7.99.2		Nitrogen: (acceptor) oxidoreductase	A flavoprotein. Reduced pyocyanine acts as donor for the reduction of NO	537, 874
1.7.99.3		Nitric-oxide: (acceptor) oxidoreductase	A cuproprotein; the *Pseudomonas* enzyme also contains FAD. Reduced pyocyanine, flavins, *etc.* act as donor for the reduction of nitrite. Formerly EC 1.6.6.5	536, 3591, 1468
1.7.99.4	Respiratory nitrate reductase	Nitrite: (acceptor) oxidoreductase	The *Pseudomonas* enzyme is a cytochrome, but the enzyme from *Micrococcus halodentrificans* is a non-haem-iron protein containing Mo. Reduced benzyl viologen and other dyes bring about the reduction of nitrate	2375, 2694, 2852

Number	Recommended Name	Reaction
1.8.1.2	Sulphite reductase (NADPH)	Hydrogen sulphide + 3 $NADP^+$ + 3 H_2O = sulphite + 3 NADPH
1.8.1.3	Hypotaurine dehydrogenase	Hypotaurine + H_2O + NAD^+ = taurine + NADH

1.8.2   WITH A CYTOCHROME AS ACCEPTOR

1.8.2.1	Sulphite dehydrogenase	Sulphite + 2 ferricytochrome c + H_2O = sulphate + 2 ferrocytochrome c

1.8.3   WITH OXYGEN AS ACCEPTOR

1.8.3.1	Sulphite oxidase	Sulphite + O_2 + H_2O = sulphate + H_2O_2
1.8.3.2	Thiol oxidase	4 R'C(R)SH + O_2 = 2 R'C(R)S-S(R)CR' + $2H_2O$

1.8.4   WITH A DISULPHIDE COMPOUND AS ACCEPTOR

1.8.4.1	Glutathione—homocystine transhydrogenase	2 Glutathione + homocystine = oxidized glutathione + 2 homocysteine
1.8.4.2	Protein—disulphide reductase (glutathione)	2 Glutathione + protein-disulphide = oxidized glutathione + protein-dithiol
1.8.4.3	Glutathione—CoAS-SG transhydrogenase	CoA + oxidized glutathione = CoAS-SG + glutathione
1.8.4.4	Glutathione—cystine transhydrogenase	2 Glutathione + cystine = oxidized glutathione + 2 cysteine

1.8.5   WITH A QUINONE OR RELATED COMPOUND AS ACCEPTOR

1.8.5.1	Glutathione dehydrogenase (ascorbate)	2 Glutathione + dehydroascorbate = oxidized glutathione + ascorbate

1.8.6   WITH A NITROGENOUS GROUP AS ACCEPTOR

[1.8.6.1   Deleted entry: Nitrate-ester reductase. Now included with EC 2.5.1.18]

1.8.7   WITH AN IRON-SULPHUR PROTEIN AS ACCEPTOR

1.8.7.1	Sulphite reductase (ferredoxin)	Hydrogen sulphide + 3 oxidized ferredoxin + 3 H_2O = sulphite + 3 reduced ferredoxin

Number	Other Names	Basis for classification (Systematic Name)	Comments	Reference
1.8.1.2		Hydrogen-sulphide:NADP$^+$ oxidoreductase	A ferroflavoprotein (FAD and FMN)	1313, 3809, 3067
1.8.1.3		Hypotaurine:NAD$^+$ oxidoreductase		3289
1.8.2.1		Sulphite:ferricytochrome c oxidoreductase		500
1.8.3.1		Sulphite:oxygen oxidoreductase	A molybdohaemoprotein	2033, 3338, 1669
1.8.3.2		Thiol:oxygen oxidoreductase	R may be =S or =O, or a variety of other groups. The enzyme is not specific for R′	2394, 137
1.8.4.1		Glutathione:homocystine oxidoreductase	The reactions catalysed by this enzyme and by others in this sub-group may be similar to those catalysed by EC 2.5.1.18	2690
1.8.4.2	Glutathione-insulin transhydrogenase, Insulin reductase	Glutathione:protein-disulphide oxidoreductase	Reduces insulin and some other proteins	1612
1.8.4.3		Coenzyme A: oxidized-glutathione oxidoreductase		493
1.8.4.4		Glutathione:cystine oxidoreductase		2328
1.8.5.1		Glutathione:dehydroascorbate oxidoreductase		609
1.8.7.1		Hydrogen-sulphide:ferredoxin oxidoreductase		2961

Number	Recommended Name	Reaction

1.8.99 WITH OTHER ACCEPTORS

| 1.8.99.1 | Sulphite reductase | Hydrogen sulphide + acceptor + 3 H_2O = sulphite + reduced acceptor |

| 1.8.99.2 | Adenylylsulphate reductase | AMP + sulphite + acceptor = adenylylsulphate + reduced acceptor |

1.9 ACTING ON A HAEM GROUP OF DONORS

These are the cytochrome oxidases and nitrate reductases. Sub-sub-groups: 1.9.3 with oxygen; 1.9.6 with a nitrogenous compound as acceptor and 1.9.99 with other acceptors.

1.9.3 WITH OXYGEN AS ACCEPTOR

| 1.9.3.1 | Cytochrome c oxidase | 4 Ferrocytochrome c + O_2 = 4 ferricytochrome c + 2 H_2O |

| 1.9.3.2 | *Pseudomonas* cytochrome oxidase | 4 Ferrocytochrome c_2 + O_2 = 4 ferricytochrome c_2 + 2 H_2O |

1.9.6 WITH A NITROGENOUS GROUP AS ACCEPTOR

| 1.9.6.1 | Nitrate reductase (cytochrome) | Ferrocytochrome + nitrate = ferricytochrome + nitrite |

1.9.99 WITH OTHER ACCEPTORS

| 1.9.99.1 | Iron—cytochrome c reductase | Ferrocytochrome c + ferric ions = ferricytochrome c + ferrous ions |

1.10 ACTING ON DIPHENOLS AND RELATED SUBSTANCES AS DONORS

These enzymes oxidize diphenols or ascorbate. There are three sub-sub-groups: 1.10.1 with NAD^+ or $NADP^+$, 1.10.2 with cytochromes, and 1.10.3 with oxygen as acceptor. Some enzymes oxidizing phenols are oxygenases (sub-sub-group 1.14.18).

1.10.1 WITH NAD^+ OR $NADP^+$ AS ACCEPTOR

| 1.10.1.1 | *trans*-Acenaphthene-1,2-diol dehydrogenase | $(\pm)$-*trans*-Acenaphthene 1,2-diol + $NADP^+$ = acenaphthenequinone + NADPH |

Number	Other Names	Basis for classification (Systematic Name)	Comments	Reference
1.8.99.1		Hydrogen-sulphide: (acceptor) oxidoreductase	An iron-protein. A stoichiometry of six molecules of reduced methyl viologen per molecule of sulphide formed was found	115, 116, 3808
1.8.99.2		AMP,sulphite: (acceptor) oxidoreductase	Methyl viologen can act a ferroflavoprotein as acceptor	2216
1.9.3.1	Cytochrome oxidase, Cytochrome a_3, Indophenolase, Indophenol oxidase Atmungsferment	Ferrocytochrome c:oxygen oxidoreductase	A cytochrome of the a type containing Cu	1640, 1641, 3581, 3796, 3797
1.9.3.2	Cytochrome cd	Ferrocytochrome c_2:oxygen oxidoreductase	A cytochrome cd. Nitrite and hydroxylamine can act as acceptor	3120, 3427, 3764, 3085
1.9.6.1		Ferrocytochrome:nitrate oxidoreductase		2889
1.9.99.1		Ferrocytochrome c:iron oxidoreductase		3784
1.10.1.1		$(\pm)$-trans-Acenaphthene-1,2-diol:NADP$^+$ oxidoreductase	Some preparations also utilize NAD$^+$	1359

Number	Recommended Name	Reaction

1.10.2   WITH A CYTOCHROME AS ACCEPTOR

1.10.2.1   L-Ascorbate—cytochrome b_5 reductase

L-Ascorbate + ferricytochrome b_5 = monodehydroascorbate + ferrocytochrome b_5

1.10.2.2   Ubiquinol-cytochrome c reductase

QH_2 + 2 ferricytochrome c = Q + 2 ferrocytochrome c

1.10.3   WITH OXYGEN AS ACCEPTOR

1.10.3.1   Catechol oxidase

2 Catechol + O_2 = 2 1,2-benzoquinone + $2H_2O$

1.10.3.2   Laccase

4 Benzenediol + O_2 = 4 benzosemiquinone + $2H_2O$

1.10.3.3   Ascorbate oxidase

2 L-Ascorbate + O_2 = 2 dehydroascorbate + 2 H_2O

1.10.3.4 o-Aminophenol oxidase

o-Aminophenol + $\frac{3}{2}O_2$ = isophenoxazine + 3 H_2O

1.10.3.5   3-Hydroxyanthranilate oxidase

3-Hydroxyanthranilate + O_2 = 1,2-benzoquinoneimine-3-carboxylate + H_2O_2

1.11   ACTING ON HYDROGEN PEROXIDE AS ACCEPTOR

This single sub-sub-class (1.11.1) contains the peroxidases.

1.11.1.1   NAD$^+$ peroxidase

NADH + H_2O_2 = NAD$^+$ + 2 H_2O

Number	Other Names	Basis for classification (Systematic Name)	Comments	Reference
1.10.2.1		L-Ascorbate:ferricytochrome b_5 oxidoreductase		850
1.10.2.2		Ubiquinol:ferricytochrome c oxidoreductase	Contains cytochromes b-562, b-566 and c_1 and a 2-iron ferridoxin	2788
1.10.3.1	Diphenol oxidase, o-Diphenolase, Tyrosinase	1,2-Benzenediol:oxygen oxidoreductase	A group of copper proteins, that act also on a variety of substituted catachols, and many of which also catalyse the reaction listed under EC 1.14.18.1; this is especially true for the classical tyrosinase	399, 663, 900, 1117, 2105, 2631
1.10.3.2	Phenolase, Polyphenol oxidase, Urishiol oxidase	Benzenediol:oxygen oxidoreductase	A group of copper proteins of low specificity acting on both o-and p-quinones, and often acting also on aminophenols and phenylenediamine. The semiquinone may react further either enzymically or non-enzymically	663, 2103, 1644, 2068, 2357
1.10.3.3	Ascorbase	L-Ascorbate:oxygen oxidoreductase	A copper-protein	3198
1.10.3.4	Isophenoxazine synthase	o-Aminophenol:oxygen oxidoreductase	Requires Mn^{2+}; a flavoprotein. Isophenoxazine may be formed by a secondary condensation from the initial oxidation product	2338, 2341, 3275
1.10.3.5		3-Hydroxyanthranilate:oxygen oxidoreductase		2282
1.11.1.1		NADH:hydrogen-peroxide oxidoreductase	A flavoprotein (FAD). Ferricyanide, quinones, $etc.$ can replace H_2O_2	741, 2259, 3589

Number	Recommended Name	Reaction
1.11.1.2	NADP+ peroxidase	$NADPH + H_2O_2 = NADP^+ + 2 H_2O$
1.11.1.3	Fatty-acid peroxidase	$Palmitate + 2 H_2O_2 = pentadecanal + CO_2 + 3 H_2O$

[1.11.1.4 Transferred entry: now EC 1.13.11.11, Tryptophan 2,3-dioxygenase]

1.11.1.5	Cytochrome peroxidase	2 Ferrocytochrome $c + H_2O_2 = 2$ ferricytochrome $c + 2 H_2O$
1.11.1.6	Catalase	$H_2O_2 + H_2O_2 = O_2 + 2 H_2O$
1.11.1.7	Peroxidase	$Donor + H_2O_2 = oxidized\ donor + 2 H_2O$
1.11.1.8	Iodide peroxidase	$Iodide + H_2O_2 = iodine + 2 H_2O$
1.11.1.9	Glutathione peroxidase	2 Glutathione $+ H_2O_2 = oxidized$ glutathione $+ 2 H_2O$
1.11.1.10	Chloride peroxidase	$2 RH + 2 Cl^- + H_2O_2 = 2 RCl + 2 H_2O$

1.12 ACTING ON HYDROGEN AS DONOR

Hydrogenases using iron-sulphur compounds as donor for the reduction of H^+ to H_2 are listed in sub-group 1.18. Other hydrogenases have been provisionally left in this sub-group.

1.12.1   WITH NAD+ OR NADP+ AS ACCEPTOR

[1.12.1.1 Transferred entry: now EC 1.18.3.1 - Hydrogenase]

1.12.1.2	Hydrogen dehydrogenase	$H_2 + NAD^+ = H^+ + NADH$

1.12.2   WITH A CYTOCHROME AS ACCEPTOR

1.12.2.1	Cytochrome c_3 hydrogenase	$H_2 + 2$ ferricytochrome $c_3 = 2 H^+ +$ ferrocytochrome c_3

[1.12.7.1 Transferred entry: now EC 1.18.3.1, Hydrogenase]

Number	Other Names	Basis for classification (Systematic Name)	Comments	Reference
1.11.1.2		NADPH:hydrogen-peroxide oxidoreductase		569
1.11.1.3		Palmitate:hydrogen-peroxide oxidoreductase	Acts on long-chain fatty acids from lauric to stearic acid	2100
1.11.1.5		Ferrocytochrome c:hydrogen-peroxide oxidoreductase	A haemoprotein	63, 3765, 3795
1.11.1.6		Hydrogen-peroxide:hydrogen-peroxide oxidoreductase	A haemoprotein. Several organic substances, especially ethanol, can act as hydrogen donor	1288, 1639, 2406, 3291, 1495
1.11.1.7		Donor:hydrogen-peroxide oxidoreductase	A haemoprotein	1660, 2297, 2573, 3337, 3397
1.11.1.8	Iodotyrosine deiodase, Iodinase	Iodide:hydrogen-peroxide oxidoreductase	A haemoprotein	598, 613, 1381, 3008
1.11.1.9		Glutathione:hydrogen-peroxide oxidoreductase	A selenium-protein. Steroid hydroperoxides can also act as acceptor	2233
1.11.1.10		Chloride:hydrogen-peroxide oxidoreductase	Brings about the chlorination of a range of organic molecules, forming stable C—Cl bonds	2294
1.12.1.2	Hydrogenase	Hydrogen:NAD$^+$ oxidoreductase	This may be a system involving EC 1.18.3.1	339, 2966
1.12.2.1	Hydrogenase	Hydrogen:ferricytochrome c_3 oxidoreductase	Requires iron. Methylene blue and other acceptors can be reduced with H$_2$.	2791, 2890, 3740, 700

1.13 ACTING ON SINGLE DONORS WITH INCORPORATION OF MOLECULAR OXYGEN (OXYGENASES)

The enzymes in this sub-group differ from all those in earlier sub-groups in that oxygen is actually incorporated from O_2 into the substance oxidized; they differ from those in 1.14 in that a second hydrogen donor is not required. Sub-sub-groups are 1.13.11, when two atoms of oxygen are incorporated; 1.13.12 when only one oxygen atom is used, and 1.13.99 for cases where information is not complete. Recommended names in this sub-group are of the form 'monooxygenase' and 'dioxygenase'.

[1.13.1.1   Transferred entry: now EC 1.13.11.1, Catechol 1,2-dioxygenase]

[1.13.1.2   Transferred entry: now EC 1.13.11.2, Catechol 2,3-dioxygenase]

[1.13.1.3   Transferred entry: now EC 1.13.11.3, Protocatechuate 3,4-dioxygenase]

[1.13.1.4   Transferred entry: now EC 1.13.11.4, Gentisate 1,2-dioxygenase]

[1.13.1.5   Transferred entry: now EC 1.13.11.5, Homogentisate 1,2-dioxygenase]

[1.13.1.6   Transferred entry: now EC 1.13.11.6, 3-Hydroxyanthranilate 3,4-dioxygenase]

[1.13.1.7   Transferred entry: now EC 1.13.11.7, 3,4-Dihydroxyphenylacetate 3,4-dioxygenase]

[1.13.1.8   Transferred entry: now EC 1.13.11.8, Protocatechuate 4,5-dioxygenase]

[1.13.1.9   Transferred entry: now EC 1.13.11.9, 2,5-Dihydroxypyridine 5,6-dioxygenase]

[1.13.1.10 Transferred entry: now EC 1.13.11.10, 7,8-Dihydroxykynurenate 8,8a-dioxygenase]

[1.13.1.11 Transferred entry: now EC 1.13.99.1, myo-Inositol oxygenase]

[1.13.1.12 Transferred entry: now EC 1.13.11.11, Tryptophan 2,3-dioxygenase]

[1.13.1.13 Transferred entry: now EC 1.13.11.12, Lipoxygenase]

1.13.11 WITH INCORPORATION OF TWO ATOMS OF OXYGEN

Number	Recommended Name	Reaction
1.13.11.1	Catechol 1,2-dioxygenase	Catechol + O_2 = *cis, cis*-muconate
1.13.11.2	Catechol 2,3-dioxygenase	Catechol + O_2 = 2-hydroxymuconate semialdehyde
1.13.11.3	Protocatechuate 3,4-dioxygenase	Protocatechuate + O_2 = 3-carboxy-*cis,cis*-muconate
1.13.11.4	Gentisate 1,2-dioxygenase	Gentisate + O_2 = maleylpyruvate

Number	Other Names	Basis for classification (Systematic Name)	Comments	Reference
1.13.11.1	Pyrocatechase	Catechol:oxygen 1,2-oxidoreductase (decyclizing)	Contains ferric ion. Formerly EC 1.99.2.2 and EC 1.13.1.1	1234, 3092, 1233
1.13.11.2	Metapyrocatechase	Catechol:oxygen 2,3-oxidoreductase (decyclizing)	Contains ferrous ion. Formerly EC 1.13.1.2	1748, 2457, 1233
1.13.11.3	Protocatechuate oxygenase	Protocatechuate:oxygen 3,4-oxidoreductase (decyclizing)	Contains ferric ion. Formerly EC 1.99.2.3 and EC 1.13.1.3	968, 1133, 3197
1.13.11.4	Gentisate oxygenase	Gentisate:oxygen 1,2-oxidoreductase (decyclizing)	Requires ferrous ion. Formerly EC 1.99.2.4 and EC 1.13.1.4	3282, 1233

Number	Recommended Name	Reaction
1.13.11.5	Homogentisate 1,2-dioxygenase	Homogentisate + O_2 = 4-maleylacetoacetate
1.13.11.6	3-Hydroxyanthranilate 3,4-dioxygenase	3-Hydroxyanthranilate + O_2 = 2-amino-3-carboxymuconate semialdehyde
1.13.11.7	3,4-Dihydroxyphenylacetate 3,4-dioxygenase	3,4-Dihydroxyphenylacetate + O_2 = 3-carboxymethylmuconate
1.13.11.8	Protocatechuate 4,5-dioxygenase	Protocatechuate + O_2 = 2-hydroxy-4-carboxymuconate semialdehyde
1.13.11.9	2,5-Dihydroxypyridine 5,6-dioxygenase	2,5-Dihydroxypyridine + O_2 = maleamate + formate
1.13.11.10	7,8-Dihydroxykynurenate 8,8a-dioxygenase	7,8-Dihydroxykynurenate + O_2 = 5-(γ-carboxy-γ-oxo) propenyl-4,6-dihydroxypicolinate
1.13.11.11	Tryptophan 2,3-dioxygenase	L-Tryptophan + O_2 = L-formylkynurenine
1.13.11.12	Lipoxygenase	Linoleate + O_2 = 13-hydroperoxyoctadeca-9,11-dienoate
1.13.11.13	Ascorbate 2,3-dioxygenase	Ascorbate + O_2 = oxalate + threonate
1.13.11.14	2,3-Dihydroxybenzoate 3,4-dioxygenase	2,3-Dihydroxybenzoate + O_2 = 3-carboxy-2-hydroxymuconate semialdehyde
1.13.11.15	3,4-Dihydroxyphenylacetate 2,3-dioxygenase	3,4-Dihydroxyphenylacetate + O_2 = 2-hydroxy-5-carboxymethylmuconate semialdehyde
1.13.11.16	3-Carboxyethylcatechol 2,3-dioxygenase	β-(2,3-Dihydroxyphenyl) propionate + O_2 = 2-hydroxy-6-oxonona-2,4-diene-1,9-dioate
1.13.11.17	Indole 2,3-dioxygenase	Indole + O_2 = 2-formylaminobenzaldehyde
1.13.11.18	Sulphur dioxygenase	Sulphur + O_2 + H_2O = sulphite

Number	Other Names	Basis for classification (Systematic Name)	Comments	Reference
1.13.11.5	Homogentisicase, Homogentisate oxygenase	Homogentisate:oxygen 1,2-oxidoreductase (decyclizing)	Requires ferrous ion. Formerly EC 1.99.2.5 and EC 1.13.1.5	12, 603, 1727, 2728, 1233
1.13.11.6	3-Hydroxyanthranilate oxygenase	3-Hydroxyanthranilate:oxygen 3,4-oxidoreductase (decyclizing)	Requires ferrous ion. Formerly EC 1.13.1.6	683, 1233
1.13.11.7	Homoprotocatechuate oxygenase	3,4-Dihydroxyphenyl-acetate:oxygen 3,4-oxidoreductase (decyclizing)	Requires ferrous ion. Formerly EC 1.13.1.7	1702
1.13.11.8	Protocatechuate 4,5-oxygenase	Protocatechuate:oxygen 4,5-oxidoreductase (decyclizing)	Requires ferrous ion. Formerly EC 1.13.1.18	3463
1.13.11.9	2,5-Dihydroxypyridine oxygenase	2,5-Dihydroxypyridine:oxygen 5,6-oxidoreductase (decyclizing)	Requires ferrous ion. Formerly EC 1.13.1.9	245, 1000
1.13.11.10	7,8-Dihydroxy-kynurenate oxygenase	7,8-Dihydroxykynurenate:oxygen 8,8a-oxidoreductase (decyclizing)	Requires ferrous ion. Formerly EC 1.13.1.10	1837
1.13.11.11	Tryptophan pyrrolase, Tryptophanase, Tryptophan oxygenase	L-Tryptophan:oxygen 2,3-oxidoreductase (decyclizing)	A cuprohaemoprotein. Formerly EC 1.11.1.4 and EC 1.13.1.12	1239, 3361, 366
1.13.11.12	Lipoxidase, Carotene oxidase (probably identical with)	Linoleate:oxygen oxidoreductase	Contains iron. Also oxidizes other methylene-interrupted polyunsaturated fatty acids. Formerly EC 1.99.2.1 and EC 1.13.1.13	534, 3400, 3846
1.13.11.13		Ascorbate:oxygen 2,3-oxidoreductase (bond-cleaving)	Requires ferrous ion	3671
1.13.11.14	o-Pyrocatechuate oxygenase	2,3-Dihydroxybenzoate:oxygen 3,4-oxidoreductase (decyclizing)		2779
1.13.11.15		3,4-Dihydroxyphenylacetate:oxygen 2,3-oxidoreductase (decyclizing)	Contains ferrous ion	14
1.13.11.16		β-(2,3-Dihydroxyphenyl)propionate:oxygen 1,2-oxidoreductase (decyclizing)	Requires ferrous ion	620
1.13.11.17		Indole:oxygen 2,3-oxidoreductase (decyclizing)	A cupro-flavoprotein	2337
1.13.11.18		Sulphur:oxygen oxidoreductase	A non-haem-iron protein	3305

Number	Recommended Name	Reaction
1.13.11.19	Cysteamine dioxygenase	Cysteamine + O_2 = hypotaurine
1.13.11.20	Cysteine dioxygenase	L-Cysteine + O_2 = cysteine sulphinate
1.13.11.21	β-Carotene 15,15′-dioxygenase	β-Carotene + O_2 = 2 retinal
1.13.11.22	Caffeate 3,4-dioxygenase	3,4-Dihydroxy-*trans*-cinnamate + O_2 = 3-carboxy-ethen-*cis,cis*-muconate
1.13.11.23	2,3-Dihydroxyindole 2,3-dioxygenase	2,3-Dihydroxyindole + O_2 = anthranilate + CO_2
1.13.11.24	Quercetin 2,3-dioxygenase	Quercetin + O_2 = 2-protocatechuoylphloroglucinol carboxylate + CO
1.13.11.25	3,4-Dihydroxy-9,10-secoandrosta-1,3,5(10)-triene-9,17-dione 4,5-dioxygenase	3,4-Dihydroxy-9,10-secoandrosta-1,3,5(10)-triene-9,17-dione + O_2 = 3-hydroxy-5,9,17-trioxo-4,5:9,10-disecoandrosta-1(10),2-dien-4-oate
1.13.11.26	Peptidyltryptophan 2,3-dioxygenase	Peptidyltryptophan + O_2 = peptidylformylkynurenine
1.13.11.27	4-Hydroxyphenylpyruvate dioxygenase	4-Hydroxyphenylpyruvate + O_2 = homogentisate + CO_2
1.13.11.28	2,3-Dihydroxybenzoate 2,3-dioxygenase	2,3-Dihydroxybenzoate + O_2 = 2-carboxy-*cis,cis*-muconate
1.13.11.29	Stizolobate synthase	3,4-Dihydroxyphenylalanine + O_2 = 2-hydroxy-4-alanine muconic-6-semialdehyde
1.13.11.30	Stizolobinate synthase	3,4-Dihydroxyphenylalanine + O_2 = 2-hydroxy-5-alanine muconic-6-semialdehyde

Number	Other Names	Basis for classification (Systematic Name)	Comments	Reference
1.13.11.19		Cysteamine:oxygen oxidoreductase	A non-haem-iron protein	487, 3723
1.13.11.20		L-Cysteine:oxygen oxidoreductase	Requires ferrous ion and NAD(P)H	1998
1.13.11.21		β-Carotene:oxygen 15,15′-oxidoreductase (bond-cleaving)	Requires bile salts and ferrous ion	1080, 1081
1.13.11.22		3,4-Dihydroxy-*trans*-cinnamate: oxygen 3,4-oxidoreductase (decyclizing)		2994
1.13.11.23		2,3-Dihydroxyindole:oxygen 2,3-oxidoreductase (decyclizing)		967
1.13.11.24		Quercetin:oxygen 2,3-oxidoreductase (decyclizing)	A copper-protein	2488
1.13.11.25	Steroid 4,5-dioxygenase, 3-Alkylcatechol 2,3-dioxygenase	3,4-Dihydroxy-9,10-secoandrosta-1,3,5(10)-triene-9,17-dione:oxygen 4,5-oxidoreductase (decyclizing)	Requires ferrous ion. Also acts on 3-isopropylcatechol and 3-*tert*-butyl-5-methylcatechol	1032
1.13.11.26	Pyrrolooxygenase	Peptidyltryptophan:oxygen 2,3-oxidoreductase (decyclizing)	Also acts on tryptophan	954
1.13.11.27		4-Hydroxyphenylpyruvate: oxygen oxidoreductase (hydroxylating, decarboxylating)	The 2-keto-acid side-chain acts instead of a second donor. Formerly EC 1.14.2.2 and 1.99.1.14	1163, 1851, 1975
1.13.11.28		2,3-Dihydroxybenzoate:oxygen 2,3-oxidoreductase (decyclizing)	Also acts more slowly with 2,3-dihydroxy-4-toluate and 2,3-dihydroxy-4-cumate	3024
1.13.11.29		3,4-Dihydroxyphenylalanine: oxygen 4,5-oxidoreductase (recyclizing)	The intermediate product undergoes ring closure and oxidation, with NAD(P)$^+$ as acceptor, to stizolobic acid. The enzyme requires Zn^{++}	2896, 2897
1.13.11.30		3,4-Dihydroxyphenylalanine: oxygen 3,4-oxidoreductase (recyclizing)	The intermediate product undergoes ring closure and oxidation, with NAD(P)$^+$ as acceptor, to stizolobinic acid. Requires Zn^{++}	2896, 2897

Number	Recommended Name	Reaction

1.13.12 WITH INCORPORATION OF ONE ATOM OF OXYGEN (INTERNAL MONOOXYGENASES OR INTERNAL MIXED FUNCTION OXIDASES)

1.13.12.1	Arginine 2-monooxygenase	L-Arginine + O_2 = 4-guanidinobutyramide + CO_2 + H_2O
1.13.12.2	Lysine 2-monooxygenase	L-Lysine + O_2 = 5-aminovaleramide + CO_2 + H_2O
1.13.12.3	Tryptophan 2-monooxygenase	L-Tryptophan + O_2 = indole-3-acetamide + CO_2 + H_2O
1.13.12.4	Lactate 2-monooxygenase	L-Lactate + O_2 = acetate + CO_2 + H_2O
1.13.12.5	*Renilla* luciferin 2-monooxygenase	*Renilla* luciferin + O_2 = oxidized *Renilla* luciferin + CO_2 + *hv*
1.13.12.6	*Cypridina* luciferin 2-monooxygenase	*Cypridina* luciferin + O_2 = oxidized *Cypridina* luciferin + CO_2 + *hv*
1.13.12.7	*Photinus* luciferin 4-monooxygenase (ATP-hydrolysing)	*Photinus* luciferin + O_2 + ATP = oxidized *Photinus* luciferin + CO_2 + H_2O + AMP + pyrophosphate + *hv*

1.13.99 MISCELLANEOUS (REQUIRES FURTHER CHARACTERIZATION)

1.13.99.1	*myo*-Inositol oxygenase	*myo*-Inositol + O_2 = D-glucuronate + H_2O
1.13.99.2	Benzoate 1,2-dioxygenase	Benzoate + O_2 = catechol + CO_2

Number	Other Names	Basis for classification (Systematic Name)	Comments	Reference
1.13.12.1		L-Arginine:oxygen 2-oxidoreductase (decarboxylating)	A flavoprotein. Also acts on canavanine and homoarginine	2496, 3405, 3406
1.13.12.2		L-Lysine:oxygen 2-oxidoreductase (decarboxylating)	A flavoprotein (FAD). Also acts on other diamino-acids	3349, 3350, 2369
1.13.12.3		L-Tryptophan:oxygen 2-oxidoreductase (decarboxylating)		1785, 1839
1.13.12.4	Lactate oxidative decarboxylase	L-Lactate:oxygen 2-oxidoreductase (decarboxylating)	A flavoprotein (FMN). Formerly EC 1.1.3.2	1243, 3302
1.13.12.5		*Renilla* luciferin:oxygen 2-oxidoreductase (decarboxylating)	The 'alkyl' group, unidentified, has a M.W. of $\sim$ 200; a benzyl group in the same position yields a fully active compound. Luciferases of other coelenterates (*Ptilosarcus, Stylatula)* also react with *Renilla* luciferin	589, 3049, 1372
1.13.12.6		*Cypridina* luciferin:oxygen 2-oxidoreductase (decarboxylating)	The luciferins (and presumably the luciferases, since they cross-react) of some luminous fish (*e.g Apogon, Parapriacanthus, Porichthys)* are apparently identical or closely similar	588, 1699, 1597, 3475
1.13.12.7		*Photinus* luciferin:oxygen 4-oxidoreductase (decarboxylating, ATP-hydrolysing)	The first step in the reaction is the formation of an acid anhydride between the carboxylic group and AMP, with the release of pyrophosphate	3669, 1360, 3670
1.13.99.1		*myo*-Inositol:oxygen oxidoreductase	Formerly EC 1.13.1.11 and 1.99.2.6	498
1.13.99.2	Benzoate hydroxylase	Benzoate:oxygen oxidoreductase	Requires NADH	1419

Number	Recommended Name	Reaction

1.14 ACTING ON PAIRED DONORS WITH INCORPORATION OF MOLECULAR OXYGEN

The enzymes in this sub-group all act on two hydrogen-donors, and oxygen from O_2 is incorporated into one or both of them. In sub-sub-class 1.14.11, 2-oxoglutarate is one donor and one atom of oxygen goes into each donor; in 1.14.12, NADH or NADPH is one donor, and two atoms of oxygen go into the other donor; in 1.14.13, NADH or NADPH is again one donor, but only one atom of oxygen goes into the other donor; in sub-sub-classes, 1.14.14-18, one atom of oxygen is incorporated into one donor, the other donor being respectively a reduced flavin or flavoprotein, a reduced iron-sulphur protein, a reduced pteridine, ascorbate, or some other compound. Sub-sub-group 1.14.99 is for cases where information about the second donor is incomplete.

It should be noted that this arrangement of sub-groups is different from that in Enzyme Nomenclature (1964).

[1.14.1.1 Transferred entry: now EC 1.14.14.1, Aryl 4-monooxygenase]

[1.14.1.2 Transferred entry: now EC 1.14.13.9, Kynurenine 3-monooxygenase]

[1.14.1.3 Deleted entry: Squalene hydroxylase. Reaction due to a mixture of EC 1.14.99.7 and EC 5.4.99.7]

[1.14.1.4 Transferred entry: now EC 1.14.99.2, Kynurenate 7,8-hydroxylase]

[1.14.1.5 Transferred entry: now EC 1.14.13.5, Imidazoleacetate 4-monooxygenase]

[1.14.1.6 Transferred entry: now EC 1.14.15.4, Steroid 11β-monooxygenase]

[1.14.1.7 Transferred entry: now EC 1.14.99.9, Steroid 17α-monooxygenase]

[1.14.1.8 Transferred entry: now EC 1.14.99.10, Steroid 21-monooxygenase]

[1.14.1.9 Deleted entry: Cholesterol 20-hydroxylase]

[1.14.1.10 Transferred entry: now EC 1.14.99.11, Oestradiol 6β-monooxygenase]

[1.14.1.11 Deleted entry: Oestriol 2-hydroxylase]

[1.14.2.1 Transferred entry: now EC 1.14.17.1, Dopamine β-monooxygenase]

[1.14.2.2 Transferred entry: now EC 1.13.11.27, 4-Hydroxyphenylpyruvate dioxygenase]

[1.14.3.1 Transferred entry: now EC 1.14.16.1, Phenylalanine 4-monooxygenase]

1.14.11 WITH 2-OXOGLUTARATE AS ONE DONOR AND INCORPORATION OF ONE ATOM EACH OF OXYGEN INTO BOTH DONORS

1.14.11.1	γ-Butyrobetaine,2-oxoglutarate dioxygenase	4-Trimethylaminobutyrate + 2-oxoglutarate + O_2 = 3-hydroxy-4-trimethylaminobutyrate + succinate + CO_2
1.14.11.2	Proline,2-oxoglutarate dioxygenase	Prolyl-glycyl-containing-peptide + 2-oxoglutarate + O_2 = hydroxyprolylglycyl-containing-peptide + succinate + CO_2

Number	Other Names	Basis for classification (Systematic Name)	Comments	Reference
1.14.11.1		4-Trimethylaminobutyrate, 2-oxoglutarate:oxygen oxidoreductase (3-hydroxylating)	Requires ferrous ion and ascorbate	1977
1.14.11.2	Protocollagen hydroxylase, Proline hydroxylase	Prolyl-glycyl-peptide, 2-oxoglutarate:oxygen oxidoreductase	Requires ferrous ion and ascorbate	1414, 1708, 274, 1706

Number	Recommended Name	Reaction

1.14.11.3   Thymidine,2-oxoglutarate dioxygenase   Thymidine + 2-oxoglutarate + O_2 = thymine ribonucleoside + succinate + CO_2

1.14.11.4   Lysine,2-oxoglutarate dioxygenase   Peptidyllysine + 2-oxoglutarate + O_2 = peptidylhydroxylysine + succinate + CO_2

[1.14.11.5 Deleted entry: 5-Hydroxymethyluracil,2-oxoglutarate dioxygenase. Now included with EC 1.14.11.6]

1.14.11.6   Thymine,2-oxoglutarate dioxygenase   Thymine + 2-oxoglutarate + O_2 = 5-hydroxymethyluracil + succinate + CO_2

1.14.12   WITH NADH OR NADPH AS ONE DONOR, AND INCORPORATION OF TWO ATOMS OF OXYGEN INTO ONE DONOR

1.14.12.1   Anthranilate 1,2-dioxygenase (deaminating, decarboxylating)   Anthranilate + NAD(P)H + O_2 + 2 H_2O = catechol + CO_2 + NAD(P)$^+$ + NH_3

1.14.12.2   Anthranilate 2,3-dioxygenase (deaminating)   Anthranilate + NADPH + O_2 = 2,3-dihydroxybenzoate + NADP$^+$ + NH_3

1.14.12.3   Benzene 1,2-dioxygenase   Benzene + NADH + O_2 = *cis*-dihydrobenzenediol + NAD$^+$

1.14.12.4   Methylhydroxypyridine-carboxylate dioxygenase   2-Methyl-3-hydroxypyridine-5-carboxylate + NAD(P)H + O_2 = 2-(*N*-acetamidomethylene)succinate + NAD(P)$^+$

1.14.12.5   5-Pyridoxate dioxygenase   5-Pyridoxate + NADPH + O_2 = 2-hydroxymethyl-3-(*N*-acetamidomethylene)succinate + NADP$^+$

1.14.12.6   2-Hydroxycyclohexanone 2-monooxygenase   2-Hydroxycyclohexan-1-one + NADPH + O_2 = 7-hydroxy-1-oxa-2-oxocycloheptane + NADP + H_2O

1.14.13   WITH NADH OR NADPH AS ONE DONOR, AND INCORPORATION OF ONE ATOM OF OXYGEN

1.14.13.1   Salicylate 1-monooxygenase   Salicylate + NADH + O_2 = catechol + NAD$^+$ + H_2O + CO_2

Number	Other Names	Basis for classification (Systematic Name)	Comments	Reference
1.14.11.3	Thymidine 2′-hydroxylase, Pyrimidine deoxyribonucleoside 2′-hydroxylase	Thymidine, 2-oxoglutarate:oxygen oxidoreductase (2′-hydroxylating)	Requires ferrous iron and ascorbate	3017, 197
1.14.11.4	Lysine hydroxylase	Peptidyllysine, 2-oxoglutarate:oxygen 5-oxidoreductase	Requires ferrous ion and ascorbate	1707, 1229
1.14.11.6	Thymine 7-hydroxylase	Thymine, 2-oxoglutarate:oxygen oxidoreductase (7-hydroxylating)	Requires ferrous iron and ascorbate. Also acts on 5-hydroxymethyluracil	2, 196, 1992
1.14.12.1	Anthranilate hydroxylase	Anthranilate,NAD(P)H:oxygen oxidoreductase (1,2-hydroxylating, deaminating, decarboxylating)		3369
1.14.12.2	Anthranilate hydroxylase	Anthranilate,NADPH:oxygen oxidoreductase (2,3-hydroxylating, deaminating)		1169, 1851, 3171
1.14.12.3	Benzene hydroxylase	Benzene,NADH:oxygen 1,2-oxidoreductase		1031
1.14.12.4	Methylhydroxypyridine-carboxylate oxidase	2-Methyl-3-hydroxypyridine-5-carboxylate,NAD(P)H:oxygen oxidoreductase (decyclizing)	A flavoprotein (FAD)	3160
1.14.12.5	5-Pyridoxate oxidase	5-Pyridoxate,NADPH:oxygen oxidoreductase (decyclizing)	A flavoprotein	3160
1.14.12.6		2-Hydroxycyclohexan-1-one, NADPH:oxygen 2-oxidoreductase (1,2-lactonizing)		644
1.14.13.1	Salicylate hydroxylase	Salicylate,NADH:oxygen oxidoreductase (1-hydroxylating, decarboxylating)	A flavoprotein (FAD)	3308, 3352, 3351, 3761

120

Number	Recommended Name	Reaction
1.14.13.2	4-Hydroxybenzoate 3-monooxygenase	4-Hydroxybenzoate + NADPH + O_2 = protocatechuate + $NADP^+$ + H_2O
1.14.13.3	4-Hydroxyphenylacetate 3-monooxygenase	4-Hydroxyphenylacetate + NADH + O_2 = 3,4-dihydroxyphenylacetate + NAD^+ + H_2O
1.14.13.4	Melilotate 3-monooxygenase	3-(2-Hydroxyphenyl) propionate + NADH + O_2 = 3-(2,3-dihydroxyphenyl) propionate + NAD^+ + H_2O
1.14.13.5	Imidazoleacetate 4-monooxygenase	Imidazoleacetate + NADH + O_2 = imidazoloneacetate + NAD^+ + H_2O
1.14.13.6	Orcinol 2-monooxygenase	Orcinol + NADH + O_2 = 2,3,5-trihydroxytoluene + NAD^+ + H_2O
1.14.13.7	Phenol 2-monooxygenase	Phenol + NADPH + O_2 = catechol + $NADP^+$ + H_2O
1.14.13.8	Dimethylaniline monooxygenase (*N*-oxide-forming)	*N,N*-Dimethylaniline + NADPH + O_2 = *N,N*-dimethylaniline *N*-oxide + $NADP^+$ + H_2O
1.14.13.9	Kynurenine 3-monooxygenase	L-Kynurenine + NADPH + O_2 = 3-hydroxy-L-kynurenine + $NADP^+$ + H_2O
1.14.13.10	2,6-Dihydroxypyridine 3-monooxygenase	2,6-Dihydroxypyridine + NADH + O_2 = 2,3,6-trihydroxypyridine + NAD^+ + H_2O
1.14.13.11	*trans*-Cinnamate 4-monooxygenase	*trans*-Cinnamate + NADPH + O_2 = 4-hydroxycinnamate + $NADP^+$ + H_2O
1.14.13.12	Benzoate 4-monooxygenase	Benzoate + NADPH + O_2 = 4-hydroxybenzoate + $NADP^+$ + H_2O
1.14.13.13	25-Hydroxycholecalciferol 1-monooxygenase	25-Hydroxycholecalciferol + NADPH + O_2 = 1,25-dihydroxycholecalciferol + $NADP^+$ + H_2O
1.14.13.14	*trans*-Cinnamate 2-monooxygenase	*trans*-Cinnamate + NADPH + O_2 = 2-hydroxycinnamate + $NADP^+$ + H_2O

Number	Other Names	Basis for classification (Systematic Name)	Comments	Reference
1.14.13.2	p-Hydroxybenzoate hydroxylase	4-Hydroxybenzoate,NADPH: oxygen oxidoreductase (3-hydroxylating)	A flavoprotein (FAD)	1380, 3161, 1386
1.14.13.3	p-Hydroxyphenyl acetate 3-hydroxylase	4-Hydroxyphenylacetate, NADH:oxygen oxidoreductase (3-hydroxylating)	A flavoprotein	14
1.14.13.4	2-Hydroxyphenyl propionate hydroxylase, Melilotate hydroxylase	3-(2-Hydroxyphenyl)- propionate,NADH:oxygen oxidoreductase (3-hydroxylating)	A flavoprotein (FAD)	1939, 1940, 3261
1.14.13.5		Imidazoleacetate,NADH:oxygen oxidoreductase (hydroxylating)	A flavoprotein (FAD). Formerly EC 1.14.1.5	2059
1.14.13.6	Orcinol hydroxylase	Orcinol,NADH: oxygen oxidoreductase (2-hydroxylating)	A flavoprotein (FAD)	2524
1.14.13.7	Phenol hydroxylase	Phenol,NADPH:oxygen oxidoreductase (2-hydroxylating)	Also active with resorcinol and o-cresol. A flavoprotein (FAD)	2350, 2397, 2398
1.14.13.8	Dimethylaniline oxidase	N,N-Dimethylaniline, NADPH:oxygen oxidoreductase (N-oxide-forming)	A flavoprotein. Acts on various dialkylarylamines	3845
1.14.13.9	Kynurenine 3-hydroxylase	L-Kynurenine,NADPH: oxygen oxidoreductase (3-hydroxylating)	A flavoprotein (FAD). Formerly EC 1.14.1.2 and EC 1.99.1.5	670, 2489, 2899
1.14.13.10		2,6-Dihydroxypyridine, NADH:oxygen oxidoreductase (3-hydroxylating)	A flavoprotein	1349, 1350
1.14.13.11		trans-Cinnamate,NADPH: oxygen oxidoreductase (4-hydroxylating)	NADH also acts, more slowly. Involves cytochrome P-450	2875, 2640
1.14.13.12		Benzoate,NADPH:oxygen oxidoreductase (4-hydroxylating)	Requires ferrous ion and tetrahydropteridine	2737
1.14.13.13	25-Hydroxy- cholecalciferol 1-hydroxylase	25-Hydroxycholecalciferol, NADPH:oxygen oxidoreductase (1-hydroxylating)		1100
1.14.13.14	Cinnamic acid 2-hydroxylase	trans-Cinnamate,NADPH: oxygen oxidoreductase (2-hydroxylating)		1010

Number	Recommended Name	Reaction
1.14.13.15	Cholestanetriol 26-monooxygenase	5β-Cholestane-$3\alpha,7\alpha,12\alpha$-triol $+$ NADPH $+$ O_2 = 5β-cholestane-$3\alpha,7\alpha,12\alpha,26$-tetraol $+$ $NADP^+$ $+$ H_2O
1.14.13.16	Cyclopentanone monooxygenase	Cyclopentanone $+$ NADPH $+$ O_2 = 5-valerolactone $+$ $NADP^+$ $+$ H_2O
1.14.13.17	Cholesterol 7α-monooxygenase	Cholesterol $+$ NADPH $+$ O_2 = 7α-hydroxycholesterol $+$ $NADP^+$ $+$ H_2O
1.14.13.18	4-Hydroxyphenylacetate 1-monooxygenase	4-Hydroxyphenylacetate $+$ NAD(P)H $+$ O_2 = homogentisate $+$ $NAD(P)^+$ $+$ H_2O
1.14.13.19	Taxifolin 8-monooxygenase	Taxifolin $+$ NAD(P)H $+$ O_2 = 2,3-dihydrogossypetin $+$ $NAD(P)^+$ $+$ H_2O

1.14.14 WITH REDUCED FLAVIN OR FLAVOPROTEIN AS ONE DONOR, AND INCORPORATION OF ONE ATOM OF OXYGEN

1.14.14.1	Flavoprotein-linked monooxygenase	RH $+$ reduced flavoprotein $+$ O_2 = ROH $+$ oxidized flavoprotein $+$ H_2O

[1.14.14.2 Deleted entry: Benzopyrene 3-monooxygenase. Now included with EC 1.14.14.1]

1.14.15 WITH A REDUCED IRON-SULPHUR PROTEIN AS ONE DONOR, AND INCORPORATION OF ONE ATOM OF OXYGEN

1.14.15.1	Camphor 5-monooxygenase	Camphor $+$ reduced putida ferredoxin $+$ O_2 = 5-exo-hydroxycamphor $+$ oxidized putida ferredoxin $+$ H_2O

Number	Other Names	Basis for classification (Systematic Name)	Comments	Reference
1.14.13.15		5β-Cholestane-3α,7α,12α-triol, NADPH:oxygen oxidoreductase (26-hydroxylating)		2492
1.14.13.16		Cyclopentanone,NADPH:oxygen oxidoreductase (5-hydroxylating, ·lactonizing)		1120
1.14.13.17		Cholesterol,NADPH:oxygen oxidoreductase (7α-hydroxylating)	A cytochrome *P*-450 enzyme	2250, 358
1.14.13.18	4-Hydroxyphenyl acetate 1-hydroxylase	4-Hydroxyphenylacetate, NAD(P)H:oxygen oxidoreductase (1-hydroxylating)	A flavoprotein (FAD). Also acts on 4-hydroxyhydratropate (forming 2-methylhomogentisate) and on 4-hydroxyphenoxy acetate (forming hydroquinone and glycollate)	1194
1.14.13.19		Taxifolin,NAD(P)H:oxygen oxidoreductase (8-hydroxylating)	A flavoprotein. Also acts on fustin, but not on catechin, quercetin or mollisacacidin	1509
1.14.14.1	Aryl 4-hydroxylase, Aryl 4-monooxygenase, Benzopyrene 3-monooxygenase, Mixed-function oxidase, RH hydroxylase	RH,reduced-flavoprotein:oxygen oxidoreductase (RH-hydroxylating)	Reactions catalysed by a family of spectrally similar cytochromes *P*-450 with different or overlapping substrate specificities. Types of substrates include steroids, fatty acids, alkanes, drugs and many xenobiotics. Formerly EC 1.14.1.1 and 1.99.1.1	341, 2011, 2245, 2246, 972, 2382
1.14.15.1	Methylene hydroxylase	Camphor,reduced-putida-ferredoxin:oxygen oxidoreductase (5-hydroxylating)	A cytochrome *P*-450 enzyme	462, 912, 1155

Number	Recommended Name	Reaction
1.14.15.2	Camphor 1,2-monooxygenase	2,5-Diketocamphane + reduced rubredoxin + O_2 = 5-keto-1,2-campholide + oxidized rubredoxin + H_2O
1.14.15.3	Alkane 1-monooxygenase	Octane + reduced rubredoxin + O_2 = 1-octanol + oxidized rubredoxin + H_2O
1.14.15.4	Steroid 11β-monooxygenase	A steroid + reduced adrenal ferredoxin + O_2 = an 11β-hydroxysteroid + oxidized adrenal ferredoxin + H_2O
1.14.15.5	Corticosterone 18-monooxygenase	Corticosterone + reduced adrenal ferredoxin + O_2 = 18-hydroxycorticosterone + oxidized adrenal ferredoxin + H_2O

1.14.16 WITH REDUCED PTERIDINE AS ONE DONOR, AND INCORPORATION OF ONE ATOM OF OXYGEN

1.14.16.1	Phenylalanine 4-monooxygenase	L-Phenylalanine + tetrahydropteridine + O_2 = L-tyrosine + dihydropteridine + H_2O
1.14.16.2	Tyrosine 3-monooxygenase	L-Tyrosine + tetrahydropteridine + O_2 = 3,4-dihydroxy-L-phenylalanine + dihydropteridine + H_2O
1.14.16.3	Anthranilate 3-monooxygenase	Anthranilate + tetrahydropteridine + O_2 = 3-hydroxyanthranilate + dihydropteridine + H_2O
1.14.16.4	Tryptophan 5-monooxygenase	L-Tryptophan + tetrahydropteridine + O_2 = 5-hydroxy-L-tryptophan + dihydropteridine + H_2O
1.14.16.5	Glyceryl-ether monooxygenase	1-Alkyl-*sn*-glycerol + tetrahydropteridine + O_2 = 1-hydroxyalkyl-*sn*-glycerol + dihydropteridine + H_2O

1.14.17 WITH ASCORBATE AS ONE DONOR, AND INCORPORATION OF ONE ATOM OF OXYGEN

1.14.17.1	Dopamine β-monooxygenase	3,4-Dihydroxyphenylethylamine + ascorbate + O_2 = noradrenaline + dehydroascorbate + H_2O

Number	Other Names	Basis for classification (Systematic Name)	Comments	Reference
1.14.15.2	2,5-Diketocamphane lactonizing enzyme	Camphor,reduced-rubredoxin: oxygen oxidoreductase (1,2-lactonizing)	Requires ferrous ion	573, 3816
1.14.15.3	Alkane 1-hydroxylase, ω-Hydroxylase, Fatty acid ω-hydroxylase	Alkane,reduced-rubredoxin: oxygen 1-oxidoreductase	Some enzymes in this group are P-450 enzymes. Also hydroxylates fatty acids in the ω-position	473, 2168, 2589
1.14.15.4	Steroid 11β-hydroxylase	Steroid,reduced-adrenal-ferredoxin:oxygen oxidoreductase (11β-hydroxylating)	A cytochrome P-450 enzyme. Formerly EC 1.14.1.6 and 1.99.1.7	1098, 1246, 3443, 3856
1.14.15.5	Corticosterone 18-hydroxylase	Corticosterone,reduced-adrenal-ferredoxin:oxygen oxidoreductase (18-hydroxylating)		2708
1.14.16.1	Phenylalaninase, Phenylalanine 4-hydroxylase	L-Phenylalanine,tetrahydro-pteridine:oxygen oxidoreductase (4-hydroxylating)	The mammalian enzyme is an iron-protein. Formerly EC 1.14.3.1 and EC 1.99.1.2	1150, 1615, 2244, 3501
1.14.16.2	Tyrosine 3-hydroxylase	L-Tyrosine, tetrahydropteridine:oxygen oxidoreductase (3-hydroxylating)	Requires ferrous ion	1434, 2332
1.14.16.3	Anthranilate 3 hydroxylase	Anthranilate,tetrahydropteridine: oxygen oxidoreductase (3-hydroxylating)	Requires ferric ion	1513
1.14.16.4	Tryptophan 5-hydroxylase	L-Tryptophan, tetrahydropteridine:oxygen oxidoreductase (5-hydroxylating)	Requires ferrous ion	1429, 1513, 935
1.14.16.5	Glyceryl ether cleaving enzyme	1-Alkyl-sn-glycerol, tetrahydropteridine:oxygen oxidoreductase	The product spontaneously breaks down to form a fatty aldehyde and glycerol. Formerly EC 1.14.99.17	3420, 2595, 3153, 3132
1.14.17.1	Dopamine β-hydroxylase	3,4-Dihydroxyphenylethyl-amine,ascorbate:oxygen oxidoreductase (β-hydroxylating)	A copper-protein. Stimulated by fumarate. Formerly EC 1.14.2.1	937, 1935

Number	Recommended Name	Reaction

1.14.17.2   p-Coumarate 3-monooxygenase

4-Hydroxycinnamate + ascorbate + O_2 = 3,4-dihydroxycinnamate + dehydroascorbate + H_2O

1.14.18 WITH ANOTHER COMPOUND AS ONE DONOR, AND INCORPORATION OF ONE ATOM OF OXYGEN

1.14.18.1   Monophenol monooxygenase

Tyrosine + dihydroxyphenylalanine + O_2 = dihydroxyphenylalanine + DOPA—quinone + H_2O

1.14.99 MISCELLANEOUS (REQUIRES FURTHER CHARACTERIZATION)

1.14.99.1   Prostaglandin synthase

8,11,14-Eicosatrienoate + AH_2 + 2 O_2 = prostaglandin E_1 + A + H_2O

1.14.99.2   Kynurenate 7,8-hydroxylase

Kynurenate + AH_2 + O_2 = 7,8-dihydro-7,8-dihydroxykynurenate + A

1.14.99.3   Haem oxygenase (decyclizing)

Haem + 3 AH_2 + 3 O_2 = biliverdin + Fe^{2+} + CO + 3A + 3 H_2O

1.14.99.4   Progesterone monooxygenase

Progesterone + AH_2 + O_2 = testosterone acetate + A + H_2O

1.14.99.5   Acyl-CoA desaturase

Stearyl-CoA + AH_2 + O_2 = oleyl-CoA + A + 2 H_2O

1.14.99.6   Acyl-[(acyl-carrier-protein)] desaturase

Stearyl-[(acyl-carrier-protein)] + AH_2 + O_2 = oleyl-[(acyl-carrier protein)] + A + 2 H_2O

1.14.99.7   Squalene monooxygenase (2,3-epoxidizing)

Squalene + AH_2 + O_2 = 2,3-oxidosqualene + A + H_2O

Number	Other Names	Basis for classification (Systematic Name)	Comments	Reference
1.14.17.2	*p*-Coumarate hydroxylase	4-Hydroxycinnamate, ascorbate:oxygen oxidoreductase (3-hydroxylating)		3542
1.14.18.1	Tyrosinase, Phenolase, Monophenol oxidase, Cresolase	Monophenol,dihydroxyphenyl alanine:oxygen oxidoreductase	A group of copper proteins that also catalyze the reaction of EC 1.10.3.1 if only 1,2-benzenediols are available as substrate	663, 900, 2636?, 2584, 2068
1.14.99.1		8,11,14- Eicosatrienoate, hydrogen-donor:oxygen oxidoreductase	Possibly two oxygenases are involved, a monooxygenase and a dioxygenase	2904, 2459
1.14.99.2		Kynurenate,hydrogen-donor: oxygen oxidoreductase (hydroxylating)	Formerly EC 1.14.1.4	3370
1.14.99.3		Haem,hydrogen-donor:oxygen oxidoreductase (α-methene-oxidizing, hydroxylating)		3390, 3805
1.14.99.4	Progesterone hydroxylase	Progesterone,hydrogen-donor:oxygen oxidoreductase (hydroxylating)	Has a wide specificity	2699
1.14.99.5	Fatty acid desaturase	Acyl-CoA, hydrogen-donor:oxygen oxidoreductase	Rat liver enzyme is an enzyme system involving a flavoprotein, cytochrome b_5 and a cyanide-sensitive factor. *Mycobacterium* enzyme is a flavoprotein requiring ferrous ion	2519, 2520, 977
1.14.99.6		Acyl-[(acyl-carrier-protein)], hydrogen-donor:oxygen oxidoreductase	An *Euglena* enzyme system involving a flavoprotein and a non-haem-iron-protein	2327
1.14.99.7	Squalene epoxidase	Squalene,hydrogen-donor: oxygen oxidoreductase (2,3-epoxidizing)	This enzyme, together with EC 5.4.99.7, was formerly known as squalene oxydocyclase. A flavoprotein (FAD)	585, 3387, 3536, 3760

Number	Recommended Name	Reaction
1.14.99.8	Arene monooxygenase (epoxidizing)	Naphthalene + AH_2 + O_2 = 1,2-epoxy-dihydronaphthalene + A + H_2O
1.14.99.9	Steroid 17α-monooxygenase	A steroid + AH_2 + O_2 = a 17α-hydroxysteroid + A + H_2O
1.14.99.10	Steroid 21-monooxygenase	A steroid + AH_2 + O_2 = a 21-hydroxysteroid + A + H_2O
1.14.99.11	Oestradiol 6β-monooxygenase	Oestradiol-17β + AH_2 + O_2 = 6β-hydroxy-oestradiol + A + H_2O
1.14.99.12	4-Androstene-3,17-dione monooxygenase	Androst-4-ene-3,17-dione + AH_2 + O_2 = 13-hydroxy-3-oxo-13,17-secoandrost-4-en-17-oic (17→13)-lactone + A + H_2O
1.14.99.13	3-Hydroxybenzoate 4-monooxygenase	3-Hydroxybenzoate + AH_2 + O_2 = 3,4-dihydroxybenzoate + A + H_2O
1.14.99.14	Progesterone 11α-monooxygenase	Progesterone + AH_2 + O_2 = 11α-hydroxyprogesterone + A + H_2O
1.14.99.15	4-Methoxybenzoate monooxygenase (O-demethylating)	4-Methoxybenzoate + AH_2 + O_2 = 4-hydroxybenzoate + formaldehyde + A + H_2O
1.14.99.16	Methylsterol monooxygenase	4,4-Dimethyl-5α-cholest-7-en-3β-ol + AH_2 + O_2 = 4α-methyl-5α-cholest-7-en-3β-ol + A + H_2O

[1.14.99.17 Transferred entry: now EC 1.14.16.5, Glyceryl-ether monooxygenase]

1.14.99.18	*N*-Acetylneuraminate monooxygenase	*N*-Acetylneuraminate + AH_2 + O_2 = *N*-glycoloylneuraminate + A + H_2O
1.14.99.19	Alkylacylglycerophosphoethanolamine desaturase	*O*-1-Alkyl-2-acyl-*sn*-glycero-3-phosphoethanolamine + AH_2 + O_2 = *O*-1-alk-1-enyl-2-acyl-*sn*-glycero-3-phosphoethanolamine + A + 2 H_2O

Number	Other Names	Basis for classification (Systematic Name)	Comments	Reference
1.14.99.8		Naphthalene,hydrogen-donor: oxygen oxidoreductase (1,2-epoxidizing)	A cytochrome P-448 or P-450 enzyme	1515, 1492?
1.14.99.9	Steroid 17α-hydroxylase	Steroid,hydrogen-donor: oxygen oxidoreductase (17α-hydroxylating)	Formerly EC 1.14.1.7 and 1.99.1.9	2022, 2191
1.14.99.10	Steroid 21-hydroxylase	Steroid,hydrogen-donor: oxygen oxidoreductase (21-hydroxylating)	An enzyme system involving cytochrome P-450 and flavoprotein. Formerly EC 1.14.1.8 and EC 1.99.1.11	1245, 2614, 2879
1.14.99.11	Oestradiol 6β-hydroxylase	Oestradiol-17β,hydrogen-donor:oxygen oxidoreductase (6β-hydroxylating)	Formerly EC 1.14.1.10 and EC 1.99.1.8	1168, 2312
1.14.99.12	Androstene-3,17-dione hydroxylase	Androst-4-ene-3,17-dione,hydrogen-donor:oxygen (13-oxidoreductase, lactonizing)		2647
1.14.99.13	3-Hydroxybenzoate 4-hydroxylase	3-Hydroxybenzoate,hydrogen-donor:oxygen oxidoreductase (4-hydroxylating)	A flavoprotein	2659
1.14.99.14	Progesterone 11α-hydroxylase	Progesterone,hydrogen-donor:oxygen oxidoreductase (11α-hydroxylating)		3044
1.14.99.15		4-Methoxybenzoate,hydrogen-donor:oxygen oxidoreductase (O-demethylating)	Also acts on 4-ethoxybenzoate, N-methyl-4-aminobenzoate and toluate	281
1.14.99.16	Methylsterol hydroxylase	4,4-Dimethyl-5α-cholest-7-en-3β-ol,hydrogen-donor:oxygen oxidoreductase		2228, 1001
1.14.99.18		N-Acetylneuraminate, hydrogen-donor:oxygen oxidoreductase (N-acetyl-hydroxylating)	Requires ferrous ion. Either NADPH or ascorbate can act as AH_2	2947, 2948
1.14.99.19		O-1-Alkyl-2-acyl-sn-glycero-3-phosphoethanolamine,hydrogen-donor:oxygen oxidoreductase	Requires NADPH or NADH. May involve cytochrome b_5. Activated by Mg^{2+} and ATP	2551, 3738

Number	Recommended Name	Reaction
1.14.99.20	Phylloquinone monooxygenase (2,3-epoxidizing)	Phylloquinone + AH_2 + O_2 = 2,3-epoxyphylloquinone + A + H_2O
1.14.99.21	*Latia* luciferin monooxygenase (demethylating)	*Latia* luciferin + AH_2 + 2 O_2 = oxidized *Latia* luciferin + CO_2 + formate + A + H_2O + hv
1.14.99.22	Ecdysone 20-monooxygenase	Ecdysone + AH_2 + O_2 = 20-hydroxy-ecdysone + A + H_2O

1.15 ACTING ON SUPEROXIDE RADICALS AS ACCEPTOR

Sub-sub-group 1.15.1 contains a single enzyme that brings about the dismutation of superoxide radicals.

1.15.1.1	Superoxide dismutase	$O_2^- + O_2^- + 2 H^+ = O_2 + H_2O_2$

1.16 OXIDIZING METAL IONS

Metal ions act as 'donor', being oxidized to a higher valency state. One sub-sub-group. is known: 1.16.3, with oxygen as acceptor.

1.16.3.1	Ferroxidase	4 Iron (II) + 4 H^+ + O_2 = 4 iron (III) + 2 H_2O

1.17 ACTING ON -CH₂- GROUPS

The -CH$_2$- group of donors is oxidized by these enzymes to -CHOH-; in the reverse direction, they are involved in the formation of deoxy-sugars. Two sub-sub-groups are known: 1.17.1 with NAD(P)$^+$ as acceptor; and 1.17.4 with a disulphide compound as acceptor.

1.17.1 WITH NAD⁺ or NADP⁺ AS ACCEPTOR

1.17.1.1	CDP-4-keto-6-deoxy-D-glucose reductase	CDP-4-keto-3,6-dideoxy-D-glucose + NAD(P)$^+$ + H_2O = CDP-4-keto-6-deoxy-D-glucose + NAD(P)H

Number	Other Names	Basis for classification (Systematic Name)	Comments	Reference
1.14.99.20	Phylloquinone epoxidase	Phylloquinone,hydrogen-donor:oxygen oxidoreductase (2,3-epoxidizing)		3695
1.14.99.21		*Latia* luciferin,hydrogen-donor: oxygen oxidoreductase (demethylating)	A flavoprotein. The reaction possibly involves two enzymes, an oxygenase followed by a monooxygenase for the actual light-emitting step	3048, 3050
1.14.99.22		Ecdysone,hydrogen-donor:oxygen oxidoreductase (20-hydroxylating)	An enzyme from insect fat body or malpighian tubules involving cytochrome *P*-450. NADPH can act as ultimate hydrogen donor	2409, 336, 1524
1.15.1.1		Superoxide:superoxide oxidoreductase	A metalloprotein; also known as erythrocuprein, haemocuprein or cytocuprein. Some enzymes (e.g. *Neurospora*, yeast, erythrocytes, heart, peas) contain copper and zinc; other (e.g. *E. coli, Streptomyces mutans*) contain manganese	2147, 2928, 2241, 3539, 1086
1.16.3.1	Coeruloplasmin	Iron (II):oxygen oxidoreductase	A cuproprotein	2512, 2513
1.17.1.1		CDP-4-keto-3,6-dideoxy-D-glucose:NAD(P)$^+$ 3-oxidoreductase	Two proteins are involved but no partial reaction has been observed in the presence of either alone	2556

Number	Recommended Name	Reaction

1.17.4 WITH A DISULPHIDE COMPOUND AS ACCEPTOR

1.17.4.1   Ribonucleoside-diphosphate reductase   2′-Deoxyribonucleoside diphosphate + oxidized thioredoxin + H_2O = ribonucleoside diphosphate + reduced thioredoxin

1.17.4.2   Ribonucleoside-triphosphate reductase   2′-Deoxyribonucleoside triphosphate + oxidized thioredoxin + H_2O = ribonucleoside triphosphate + reduced thioredoxin

1.18 ACTING ON REDUCED FERREDOXIN AS DONOR

1.18.1 WITH NAD+ or NADP+ AS ACCEPTOR

1.18.1.1   Rubredoxin-NAD+ reductase   Reduced rubredoxin + NAD^+ = oxidized rubredoxin + NADH

1.18.1.2   Ferredoxin-NADP+ reductase   Reduced ferredoxin + $NADP^+$ = oxidized ferredoxin + NADPH

1.18.1.3   Ferredoxin-NAD+ reductase   Reduced ferredoxin + NAD^+ = oxidized ferredoxin + NADH

1.18.2 WITH DINITROGEN AS ACCEPTOR

1.18.2.1   Nitrogenase   3 Reduced ferredoxin + $6H^+$ + N_2 + nATP = 3 oxidized ferredoxin + 2 NH_3 + nADP + n orthophosphate

1.18.3 WITH H+ AS ACCEPTOR

1.18.3.1   Hydrogenase   2 Reduced ferredoxin + 2 H^+ = 2 oxidized ferredoxin + H_2

1.19 ACTING ON REDUCED FLAVODOXIN AS DONOR

1.19.2 WITH DINITROGEN AS ACCEPTOR

1.19.2.1   Nitrogenase (flavodoxin)   6 reduced flavodoxin + 6 H^+ + N_2 + nATP = 6 oxidized flavodoxin + 2 NH_3 + nADP + n orthophosphate

Number	Other Names	Basis for classification (Systematic Name)	Comments	Reference
1.17.4.1		2′-Deoxyribonucleoside-diphosphate:oxidized-thioredoxin 2′-oxidoreductase	An iron-sulphur protein. Requires ATP	1882, 1883, 2270, 1881, 3395
1.17.4.2		2′-Deoxyriboncleoside-triphosphate:oxidized thioredoxin 2′-oxidoreductase	Requires a cobamide enzyme and ATP	318, 1095
1.18.1.1	Rubredoxin reductase	Rubredoxin:NAD^+ oxidoreductase	Formerly EC 1.6.7.2	2589
1.18.1.2	Adrenodoxin reductase	Ferredoxin:$NADP^+$ oxidoreductase	A flavoprotein. Formerly EC 1.6.7.1	2500, 3052
1.18.1.3		Ferredoxin:NAD^+ oxidoreductase	Formerly EC 1.6.7.3	1546
1.18.2.1		Reduced ferredoxin:dinitrogen oxidoreductase (ATP-hydrolysing)	Acetylene can also act as acceptor; in the absence of other acceptors H^+ is reduced to H_2. n is about 12-18	3857
1.18.3.1	Hydrogenase, Hydrogenlyase	Ferredoxin:H^+ oxidoreductase	Contains Fe_4S_4 centers. Can use molecular hydrogen for the reduction of a variety of substances. Formerly EC 1.98.1.1, EC 1.12.1.1 and EC 1.12.7.1	3857, 3061, 3335
1.19.2.1		Reduced flavodoxin:dinitrogen oxidoreductase (ATP-hydrolysing)		3857

Number	Recommended Name	Reaction

1.97 OTHER OXIDOREDUCTASES

1.97.1.1   Chlorate reductase

Chlorate $+ AH_2 =$ chlorite $+ A + H_2O$

[1.98 Enzymes using hydrogen as reductant: now listed as sub-group 1.12 and 1.81]

[1.98.1.1 Transferred entry: now EC 1.18.3.1, Hydrogenase]

[1.99.1 Transferred entries:Hydroxylases]

[1.99.2 Transferred entries:Oxygenases]

Entries previously in sub-sub-groups 1.99.1 and 1.99.2 are now in sub-groups 1.13 and 1.14.

2. TRANSFERASES

Transferases are enzymes transferring a group, for example, the methyl group or a glycosyl group, from one compound (generally regarded as donor) to another compound (generally regarded as acceptor). The classification is based on the scheme 'donor:acceptor grouptransferase'. The recommended names are normally formed as 'acceptor grouptransferase' or 'donor grouptransferase'. In many cases, the donor is a cofactor (coenzyme), carrying the group to be transferred. Special cases are the aminotransferases (see below, 2.6).

2.1 TRANSFERRING ONE-CARBON GROUPS

This sub-class contains the methyltransferases (2.1.1), the hydroxymethyl-, formyl and related transferases (2.1.2), the carboxyl- and carbamoyl-transferases (2.1.3), and the amidinotransferases (2.1.4).

2.1.1   METHYLTRANSFERASES

('Methyltransferase' may be replaced by 'transmethylase')

2.1.1.1   Nicotinamide methyltransferase

S-Adenosyl-L-methionine + nicotinamide = *S*-adenosyl-L-homocysteine + 1-methylnicotinamide

2.1.1.2   Guanidinoacetate methyltransferase

S-Adenosyl-L-methionine + guanidinoacetate = *S*-adenosyl-L-homocysteine + creatine

2.1.1.3   Dimethylthetin—homocysteine methyltransferase

Dimethylthetin + L-homocysteine = *S*-methylthioglycollate + L-methionine

2.1.1.4   Acetylserotonin methyltransferase

S-Adenosyl-L-methionine + *N*-acetylserotonin = *S*-adenosyl-L-homocysteine + *N*-acetyl-5-methoxytryptamine

2.1.1.5   Betaine—homocysteine methyltransferase

Betaine + L-homocysteine = dimethylglycine + L-methionine

2.1.1.6   Catechol methyltransferase

S-Adenosyl-L-methionine + catechol = *S*-adenosyl-L-homocysteine + guaiacol

Number	Other Names	Basis for classification (Systematic Name)	Comments	Reference
1.97.1.1		Chlorite:acceptor oxidoreductase	Flavins or benzylviologen can act as acceptor	158
2.1.1.1		*S*-Adenosyl-L-methionine:nicotinamide *N*-methyltransferase		461
2.1.1.2		*S*-Adenosyl-L-methionine:guanidinoacetate *N*-methyltransferase		465, 466
2.1.1.3		Dimethylthetin:L-homocysteine *S*-methyltransferase		1710, 2133, 2134
2.1.1.4		*S*-Adenosyl-L-methionine: *N*-acetylserotonin *O*-methyltransferase	Some other hydroxyindoles also act as acceptor, more slowly	155
2.1.1.5		Betaine:L-homocysteine *S*-methyltransferase		1710
2.1.1.6		*S*-Adenosyl-L-methionine: catechol *O*-methyltransferase		154

Number	Recommended Name	Reaction
2.1.1.7	Nicotinate methyltransferase	S-Adenosyl-L-methionine + nicotinate = S-adenosyl-L-homocysteine + 1-methylnicotinate
2.1.1.8	Histamine methyltransferase	S-Adenosyl-L-methionine + histamine = S-adenosyl-L-homocysteine + *tele*methylhistamine
2.1.1.9	Thiol methyltransferase	S-Adenosyl-L-methionine + a thiol = S-adenosyl-L-homocysteine + a thioether
2.1.1.10	Homocysteine methyltransferase	S-Adenosyl-L-methionine + L-homocysteine = S-adenosyl-L-homocysteine + L-methionine
2.1.1.11	Magnesium-protoporphyrin methyltransferase	S-Adenosyl-L-methionine + magnesium protoporphyrin = S-adenosyl-L-homocysteine + magnesium protoporphyrin monomethyl ester
2.1.1.12	Methionine S-methyltransferase	S-Adenosyl-L-methionine + L-methionine = S-adenosyl-L-homocysteine + S-methyl-L-methionine
2.1.1.13	Tetrahydropteroylglutamate methyltransferase	5-Methyltetrahydropteroyl-L-glutamate + L-homocysteine = tetrahydropteroylglutamate + L-methionine
2.1.1.14	Tetrahydropteroyltriglutamate methyltransferase	5-Methyltetrahydropteroyltri-L-glutamate + L-homocysteine = tetrahydropteroyltriglutamate + L-methionine
2.1.1.15	Fatty acid methyltransferase	S-Adenosyl-L-methionine + a fatty acid = S-adenosyl-L-homocysteine + a fatty acid methyl ester
2.1.1.16	Unsaturated-phospholipid methyltransferase	S-Adenosyl-L-methionine + (olefinic fatty acid)-phospholipid = S-adenosyl-L-homocysteine + (methylene-acyl)-phospholipid

Number	Other Names	Basis for classification (Systematic Name)	Comments	Reference
2.1.1.7		*S*-Adenosyl-L-methionine: nicotinate *N*-methyltransferase		1537
2.1.1.8		*S*-Adenosyl-L-methionine: histamine *N*-methyltransferase		396
2.1.1.9		*S*-Adenosyl-L-methionine:thiol *S*-methyltransferase	A variety of thiols and hydroxythiols can act as acceptor	379
2.1.1.10		*S*-Adenosyl-L-methionine: L-homocysteine *S*-methyltransferase	The bacterial enzyme uses *S*-methylmethionine as donor more actively than *S*-adenosyl-L-methionine	187, 3022, 3023
2.1.1.11		*S*-Adenosyl-L-methionine:magnesium-protoporphyrin *O*-methyltransferase		1035
2.1.1.12		*S*-Adenosyl-L-methionine: L-methionine *S*-methyltransferase	Requires Zn^{2+} or Mn^{2+}	1598
2.1.1.13	Methionine synthase	5-Methyltetrahydropteroyl-L-glutamate:L-homocysteine *S*-methyltransferase	A cobamide-protein. The bacterial enzyme requires *S*-adenosyl-L-methionine and reduced FAD. Acts on mono- or tri-glutamate derivatives	436, 918, 1139, 2007, 3379
2.1.1.14		5-Methyltetrahydropteroyl-tri-L-glutamate:L-homocysteine *S*-methyltransferase	Requires *Pi* the enzyme from *E. coli* also requires a reducing system	1139, 3675
2.1.1.15		*S*-Adenosyl-L-methionine:fatty acid *O*-methyltransferase	Oleic acid is the most effective fatty acid acceptor	32
2.1.1.16		*S*-Adenosyl-L-methionine: unsaturated-phospholipid methyltransferase	The enzyme transfers a methyl group to the 10-position of a Δ^9-olefinic acyl chain in phosphatidylglycerol, phosphatidylinositol, or phosphatidylethanolamine; subsequent proton transfer produces a methylene group	31

Number	Recommended Name	Reaction
2.1.1.17	Phosphatidylethanolamine methyltransferase	*S*-Adenosyl-L-methionine + phosphatidylethanolamine = *S*-adenosyl-L-homocysteine + phosphatidyl-*N*-methylethanolamine
2.1.1.18	Polysaccharide methyltransferase	*S*-Adenosyl-L-methionine + 1,4-α-D-gluco-oligosaccharide = *S*-adenosyl-L-homocysteine + oligosaccharide containing 6-*O*-methyl-D-glucose units
2.1.1.19	Trimethylsulphonium—tetrahydrofolate methyltransferase	Trimethylsulphonium chloride + tetrahydrofolate = dimethylsulphide + 5-methyltetrahydrofolate
2.1.1.20	Glycine methyltransferase	*S*-Adenosyl-L-methionine + glycine = *S*-adenosyl-L-homocysteine + sarcosine
2.1.1.21	Methylamine—glutamate methyltransferase	Methylamine + L-glutamate = ammonia + *N*-methyl-L-glutamate
2.1.1.22	Carnosine *N*-methyltransferase	*S*-Adenosyl-L-methionine + carnosine = *S*-adenosyl-L-homocysteine + anserine
2.1.1.23	Protein (arginine) methyltransferase	*S*-Adenosyl-L-methionine + protein = *S*-adenosyl-L-homocysteine + protein containing ω-*N*-methylarginine
2.1.1.24	Protein *O*-methyltransferase	*S*-Adenosyl-L-methionine + protein = *S*-adenosyl-L-homocysteine + *O*-methylprotein
2.1.1.25	Phenol *O*-methyltransferase	*S*-Adenosyl-L-methionine + phenol = *S*-adenosyl-L-homocysteine + anisole
2.1.1.26	Iodophenol methyltransferase	*S*-Adenosyl-L-methionine + 2-iodophenol = *S*-adenosyl-L-homocysteine + 2-iodophenol methyl ester
2.1.1.27	Tyramine *N*-methyltransferase	*S*-Adenosyl-L-methionine + tyramine = *S*-adenosyl-L-homocysteine + *N*-methyltyramine
2.1.1.28	Noradrenalin *N*-methyltransferase	*S*-Adenosyl-L-methionine + noradrenalin = *S*-adenosyl-L-homocysteine + adrenalin
2.1.1.29	tRNA (cytosine-5-)-methyltransferase	*S*-Adenosyl-L-methionine + tRNA = *S*-adenosyl-L-homocysteine + tRNA containing 5-methylcytosine

Number	Other Names	Basis for classification (Systematic Name)	Comments	Reference
2.1.1.17		S-Adenosyl-L-methionine:phosphatidyl-ethanolamine N-methyltransferase		1587
2.1.1.18		S-Adenosyl-L-methionine: 1,4-α-D-glucan 6-O-methyltransferase		872
2.1.1.19		Trimethylsulphonium-chloride:tetrahydrofolate N-methyltransferase		3576
2.1.1.20		S-Adenosyl-L-methionine:glycine methyltransferase		329
2.1.1.21	N-Methylglutamate synthase	Methylamine:L-glutamate N-methyltransferase		3035
2.1.1.22		S-Adensyl-L-methionine:carnosine N-methyltransferase		2171
2.1.1.23	Protein methylase I	S-Adenosyl-L-methionine:protein (arginine) N-methyltransferase		2542
2.1.1.24	Protein methylase II	S-Adenosyl-L-methionine:protein O-methyltransferase	Forms ester groups in a number of proteins	1687
2.1.1.25		S-Adenosyl-L-methionine:phenol O-methyltransferase	Acts on a wide variety of simple alkyl-, methoxy- and halophenols	152
2.1.1.26		S-Adenosyl-L-methionine: 2-iodophenol methyltransferase		3439
2.1.1.27		S-Adenosyl-L-methionine:tyramine N-methyltransferase	Has some activity on phenylethylamine analogues	2071
2.1.1.28		S-Adenosyl-L-methionine:phenylethanolamine N-methyltransferase	Acts on various phenylethanolamines	151, 571
2.1.1.29		S-Adenosyl-L-methionine:tRNA (cytosine-5-)-methyltransferase		303

140

Number	Recommended Name	Reaction
2.1.1.30	tRNA (purine-2- or -6-)-methyltransferase	S-Adenosyl-L-methionine + tRNA = S-adenosyl-L-homocysteine + tRNA containing 2- or 6-methylaminopurine
2.1.1.31	tRNA (guanine-1-)-methyltransferase	S-Adenosyl-L-methionine + tRNA = S-adenosyl-L-homocysteine + tRNA containing 1-methylguanine
2.1.1.32	tRNA (guanine-2-)-methyltransferase	S-Adenosyl-L-methionine + tRNA = S-adenosyl-L-homocysteine + tRNA containing N^2-methylguanine
2.1.1.33	tRNA (guanine-7-)-methyltransferase	S-Adenosyl-L-methionine + tRNA = S-adenosyl-L-homocysteine + tRNA containing 7-methylguanine
2.1.1.34	tRNA (guanosine-2′-)-methyltransferase	Transfers the methyl group from S-adenosyl-L-methionine to the 2′-hydroxyl group of a guanosine residue present in a GG sequence in tRNA^{Tyr}
2.1.1.35	tRNA (uracil-5-)-methyltransferase	S-Adenosyl-L-methionine + tRNA = S-adenosyl-L-homocysteine + tRNA containing thymine
2.1.1.36	tRNA (adenine-1-)-methyltransferase	S-Adenosyl-L-methionine + tRNA = S-adenosyl-L-homocysteine + tRNA containing 1-methyladenine
2.1.1.37	DNA (cytosine-5-)-methyltransferase	S-Adenosyl-L-methionine + DNA = S-adenosyl-L-homocysteine + DNA containing 5-methylcytosine (and 6-methylaminopurine)
2.1.1.38	O-Demethylpuromycin methyltransferase	S-Adenosyl-L-methionine + O-demethylpuromycin = S-adenosyl-L-homocysteine + puromycin
2.1.1.39	myo-Inositol 1-methyltransferase	S-Adenosyl-L-methionine + myo-inositol = S-adenosyl-L-homocysteine + 1-methyl-myo-inositol
2.1.1.40	myo-Inositol 3-methyltransferase	S-Adenosyl-L-methionine + myo-inositol = S-adenosyl-L-homocysteine + 3-methyl-myo-inositol
2.1.1.41	Δ^{24}-Sterol methyltransferase	S-Adenosyl-L-methionine + 5α-cholesta-8,24-dien-3β-ol = S-adenosyl-L-homocysteine + 24-methylene-5α-cholest-8-en-3β-ol

Number	Other Names	Basis for classification (Systematic Name)	Comments	Reference
2.1.1.30		*S*-Adenosyl-L-methionine:tRNA (purine-2- or -6-)-methyltransferase		1408
2.1.1.31		*S*-Adenosyl-L-methionine:tRNA (guanine-1-)-methyltransferase		1408
2.1.1.32		*S*-Adenosyl-L-methionine:tRNA (guanine-2-)-methyltransferase		167
2.1.1.33		*S*-Adenosyl-L-methionine:tRNA (guanine-7-)-methyltransferase		1408
2.1.1.34		*S*-Adenosyl-L-methionine:tRNA (guanosine-2′-)-methyltransferase		1002
2.1.1.35		*S*-Adenosyl-L-methionine:tRNA (uracil-5-)-methyltransferase		1408
2.1.1.36		*S*-Adenosyl-L-methionine:tRNA (adenine-1-)-methyltransferase	The enzymes from different sources are specific for different adenine residues in tRNA	167, 761
2.1.1.37		*S*-Adenosyl-L-methionine:DNA (cytosine-5-)-methyltransferase		1065, 1890, 1578
2.1.1.38		*S*-Adenosyl-L-methionine: *O*-demethylpuromycin *O*-methyltransferase		2720
2.1.1.39		*S*-Adenosyl-L-methionine:*myo*-inositol 1-methyltransferase		1336
2.1.1.40		*S*-Adenosyl-L-methionine:*myo*-inositol 3-methyltransferase		3577
2.1.1.41		*S*-Adenosyl-L-methionine:zymosterol methyltransferase	Requires glutathione	2273

Number	Recommended Name	Reaction
2.1.1.42	Luteolin methyltransferase	S-Adenosyl-L-methionine + 5,7,3',4'-tetrahydroxyflavone = S-adenosyl-L-homocysteine + 5,7,4'-trihydroxy-3'-methoxyflavone
2.1.1.43	Protein (lysine) methyltransferase	S-Adenosyl-L-methionine + protein = S-adenosyl-L-homocysteine + protein containing ϵ-N-methyllysine
2.1.1.44	Dimethylhistidine methyltransferase	S-Adenosyl-L-methionine + N^α,N^α-dimethyl-L-histidine = S-adenosyl-L-homocysteine + N^α,N^α,N^α-trimethyl-L-histidine
2.1.1.45	Thymidylate synthase	5,10-Methylenetetrahydrofolate + dUMP = dihydrofolate + dTMP
2.1.1.46	Isoflavone methyltransferase	S-Adenosyl-L-methionine + isoflavone = S-adenosyl-L-homocysteine + 4'-O-methylisoflavone
2.1.1.47	Indolepyruvate methyltransferase	S-Adenosyl-L-methionine + indolepyruvate = S-adenosyl-L-homocysteine + β-methylindolepyruvate
2.1.1.48	rRNA (adenine-6-)-methyltransferase	S-Adenosyl-L-methionine + rRNA = S-adenosyl-L-homocysteine + rRNA containing N^6-methyladenine
2.1.1.49	Tryptamine N-methyltransferase	S-Adenosyl-L-methionine + tryptamine = S-adenosyl-L-homocysteine + N-methyltryptamine
2.1.1.50	Loganate methyltransferase	S-Adenosyl-L-methionine + loganate = S-adenosyl-L-homocysteine + loganin
2.1.1.51	rRNA (guanine-1-)-methyltransferase	S-Adenosyl-L-methionine + rRNA = S-adenosyl-L-homocysteine + rRNA containing 1-methylguanine
2.1.1.52	rRNA (guanine-2-)-methyltransferase	S-Adenosyl-L-methionine + rRNA = S-adenosyl-L-homocysteine + rRNA containing N^2-methylguanine
2.1.1.53	Putrescine methyltransferase	S-Adenosyl-L-methionine + putrescine = S-adenosyl-L-homocysteine + N-methylputrescine

Number	Other Names	Basis for classification (Systematic Name)	Comments	Reference
2.1.1.42	*o*-Dihydric phenol methyltransferase	*S*-Adenosyl-L-methionine: 5,7,3′,4′-tetrahydroxyflavone 3′-*O*-methyltransferase	Also acts on luteolin-7-*O*-β-D-glucoside	783
2.1.1.43	Protein methylase III	*S*-Adenosyl-L-methionine: protein (lysine) *N*-methyltransferase		2543
2.1.1.44		*S*-Adenosyl-L-methionine: N^{α},N^{α}-dimethyl-L-histidine N^{α}-methyltransferase	Methylhistidine and histidine can also act as methyl acceptors, trimethylhistidine being formed in both cases	1458
2.1.1.45		5,10-Methylene-tetrahydrofolate: dUMP *C*-methyltransferase		319, 3578
2.1.1.46		*S*-Adenosyl-L-methionine:isoflavone 4′-*O*-methyltransferase		3658
2.1.1.47		*S*-Adenosyl-L-methionine:indolepyruvate *C*-methyltransferase		1373
2.1.1.48		*S*-Adenosyl-L-methionine:rRNA (adenine-6-)-methyltransferase	Also methylates 2-amino-adenosine to 2-methylamino-adenosine	3091
2.1.1.49		*S*-Adenosyl-L-methionine:tryptamine *N*-methyltransferase	5-Methyltetrahydrofolate can also act as donor	1388
2.1.1.50		*S*-Adenosyl-L-methionine:loganate 11-*O*-methyltransferase	Also acts on secologanate. Methylates the 11-carboxyl group of loganate	2038
2.1.1.51		*S*-Adenosyl-L-methionine:rRNA (guanine-1-)-methyltransferase		1452
2.1.1.52		*S*-Adenosyl-L-methionine:rRNA (guanine-2-)-methyltransferase		1452
2.1.1.53		*S*-Adenosyl-L-methionine:putrescine *N*-methyltransferase		2257

Number	Recommended Name	Reaction
2.1.1.54	Deoxycytidylate methyltransferase	5,10-Methylenetetrahydrofolate + dCMP = dihydrofolate + deoxy-5-methyl-cytidylate

2.1.2 HYDROXYMETHYL-, FORMYL- AND RELATED TRANSFERASES

('*Hydroxymethyltransferase*', '*formyltransferase*', '*formiminotransferase*' may be replaced by '*transhydroxymethylase*', '*transformylase*' and '*transformiminase*', respectively).

2.1.2.1	Serine hydroxymethyltransferase	5,10-Methylenetetrahydrofolate + glycine + H_2O = tetrahydrofolate + L-serine
2.1.2.2	Phosphoribosylglycinamide formyltransferase	5,10-Methenyltetrahydrofolate + 5'-phosphoribosyl-glycinamide + H_2O = tetrahydrofolate + 5'-phosphoribosyl-*N*-formylglycinamide
2.1.2.3	Phosphoribosylaminoimidazolecarboxamide formyltransferase	10-Formyltetrahydrofolate + 5'-phosphoribosyl-5-amino-4-imidazolecarboxamide = tetrahydrofolate + 5'-phosphoribosyl-5-formamido-4-imidazolecarboxamide
2.1.2.4	Glycine formiminotransferase	5-Formiminotetrahydrofolate + glycine = tetrahydrofolate + *N*-formiminoglycine
2.1.2.5	Glutamate formiminotransferase	5-Formiminotetrahydrofolate + L-glutamate = tetrahydrofolate + *N*-formimino-L-glutamate
2.1.2.6	Glutamate formyltransferase	5-Formyltetrahydrofolate + L-glutamate = tetrahydrofolate + *N*-formyl-L-glutamate
2.1.2.7	2-Methylserine hydroxymethyltransferase	5,10-Methylenetetrahydrofolate + D-alanine + H_2O = tetrahydrofolate + 2-methylserine
2.1.2.8	Deoxycytidylate hydroxymethyltransferase	5,10-Methylenetetrahydrofolate + H_2O + deoxycytidylate = tetrahydrofolate + 5-hydroxymethyldeoxycytidylate
2.1.2.9	Methionyl-tRNA formyltransferase	10-Formyltetrahydrofolate + L-methionyl-tRNA = tetrahydrofolate + *N*-formylmethionyl-tRNA

Number	Other Names	Basis for classification (Systematic Name)	Comments	Reference
2.1.1.54		5,10-Methylene-tetrahydrofolate:dCMP *C*-methyltransferase	dCMP is methylated by formaldehyde in the presence of tetrahydrofolate. CMP, dCMP and CTP can act as acceptor, more slowly	1840
2.1.2.1	Serine aldolase, Threonine aldolase, Serine hydroxymethylase	5,10-Methylene-tetrahydrofolate:glycine hydroxymethyltransferase	Also catalyses the reaction of glycine with acetaldehyde to form L-threonine. A pyridoxal-phosphate-protein	34, 317, 963, 1818, 2952
2.1.2.2		5,10-Methenyl-tetrahydrofolate: 5′-phosphoribosylglycinamide formyltransferase		1205, 3618
2.1.2.3		10-Formyltetrahydrofolate:5′-phosphoribosyl-5-amino-4-imidazolecarboxamide formyltransferase		1205
2.1.2.4		5-Formiminotetrahydrofolate: glycine *N*-formiminotransferase		2677, 2681, 2892
2.1.2.5		5-Formiminotetrahydrofolate: L-glutamate *N*-formiminotransferase		2222, 3331
2.1.2.6		5-Formyltetrahydrofolate: L-glutamate *N*-formyltransferase	A pyridoxal-phosphate-protein	3073
2.1.2.7		5,10-Methylene-tetrahydrofolate:D-alanine hydroxymethyltransferase	Also acts on 2-hydroxymethylserine	3701
2.1.2.8		5,10-Methylenetetra-hydrofolate:deoxycytidylate 5-hydroxymethyltransferase		2115
2.1.2.9		10-Formyltetrahydrofolate: L-methionyl-tRNA *N*-formyltransferase		718

Number	Recommended Name	Reaction

2.1.2.10   Glycine synthase

5,10-Methylenetetrahydrofolate + CO_2 + NH_3 + reduced hydrogen-carrier protein + H_2O = tetrahydrofolate + glycine + oxidized hydrogen-carrier protein

2.1.3   CARBOXYL- AND CARBAMOYLTRANSFERASES

(*'Carboxyltransferase' and 'carbamoyltransferase' may be replaced by 'transcarboxylase' and 'transcarbamoylase' respectively*).

2.1.3.1   Methylmalonyl-CoA carboxyltransferase

Methylmalonyl-CoA + pyruvate = propionyl-CoA + oxaloacetate

2.1.3.2   Aspartate carbamoyltransferase

Carbamoylphosphate + L-aspartate = orthophosphate + N-carbamoyl-L-aspartate

2.1.3.3   Ornithine carbamoyltransferase

Carbamoylphosphate + L-ornithine = orthophosphate + L-citrulline

[*2.1.3.4   Deleted entry: Malonyl-CoA carboxyltransferase*]

2.1.3.5   Oxamate carbamoyltransferase

Carbamoylphosphate + oxamate = orthophosphate + oxalureate

2.1.3.6   Putrescine carbamoyltransferase

Carbamoylphosphate + putrescine = orthophosphate + N-carbamoylputrescine

2.1.4   AMIDINOTRANSFERASES

(*'Amidinotransferase' may be replaced by 'transamidinase'.*)

2.1.4.1   Glycine amidinotransferase

L-Arginine + glycine = L-ornithine + guanidinoacetate

2.1.4.2   Inosamine-phosphate amidinotransferase

L-Arginine + 1-amino-1-deoxy-*scyllo*-inositol 4-phosphate = L-ornithine + 1-guanidino-1-deoxy-*scyllo*-inositol 4-phosphate

2.2   TRANSFERRING ALDEHYDE OR KETONIC RESIDUES

The single sub-sub-class (2.2.1) contains transketolase and transaldolase

2.2.1.1   Transketolase

Sedoheptulose 7-phosphate + D-glyceraldehyde 3-phosphate = D-ribose 5-phosphate + D-xylulose 5-phosphate

Number	Other Names	Basis for classification (Systematic Name)	Comments	Reference
2.1.2.10		5,10-Methylene-tetrahydrofolate:ammonia hydroxymethyltransferase (carboxylating, reducing)	A pyridoxal-phosphate-protein needing NAD^+	2301, 2818
2.1.3.1	Transcarboxylase	Methylmalonyl-CoA:pyruvate carboxyltransferase	A biotinyl-protein. Contains cobalt and zinc	3319
2.1.3.2	Carbamylasparto-transkinase, Aspartate transcarbamylase	Carbamoylphosphate: L-aspartate carbamoyltransferase		549, 2009, 2762, 3040
2.1.3.3	Citrulline phosphorylase, Ornithine transcarbamylase	Carbamoylphosphate: L-ornithine carbamoyltransferase		302, 433, 549
2.1.3.5	Oxamic transcarbamylase	Carbamoylphosphate:oxamate carbamoyltransferase		334
2.1.3.6		Carbamoylphosphate:putrescine carbamoyltransferase		2835
2.1.4.1		L-Arginine:glycine amidinotransferase	Canavanine can act instead of arginine. Formerly EC 2.6.2.1	346, 568, 2726, 2727, 3595, 3596
2.1.4.2		L-Arginine:1-amino-1-deoxy-*scyllo*-inositol-4-phosphate amidinotransferase	1D-1-Guanidino-3-amino-1,3-dideoxy-*scyllo*-inositol 6-phosphate, streptamine phosphate and 2-deoxystreptamine phosphate can also act as acceptor; canavanine can act as donor	3603
2.2.1.1	Glycolaldehydetransferase	Sedoheptulose-7-phosphate: D-glyceraldehyde-3-phosphate glycolaldehydetransferase	A thiamin-diphosphate-protein. Wide specificity for both reactants, *e.g.*	674, 742, 1369, 2691

148

2.2.1.2	Transaldolase	Sedoheptulose 7-phosphate + D-glyceraldehyde 3-phosphate = D-erythrose 4-phosphate + D-fructose 6-phosphate

2.3   ACYLTRANSFERASES

These enzymes transfer acyl groups, forming either esters or amides. The donor is in most cases the corresponding acyl-coenzyme A derivative. Aminoacyltransferases form a separate sub-sub-group (2.3.2).

2.3.1   ACYLTRANSFERASES

(*'Acetyltransferase'*, etc. *may be replaced by* *'transacetylase'*, etc.)

2.3.1.1	Amino-acid acetyltransferase	Acetyl-CoA + L-glutamate = CoA + N-acetyl-L-glutamate
2.3.1.2	Imidazole acetyltransferase	Acetyl-CoA + imidazole = CoA + N-acetylimidazole
2.3.1.3	Glucosamine acetyltransferase	Acetyl-CoA + 2-amino-2-deoxy-D-glucose = CoA + 2-acetamido-2-deoxy-D-glucose
2.3.1.4	Glucosamine-phosphate acetyltransferase	Acetyl-CoA + 2-amino-2-deoxy-D-glucose 6-phosphate = CoA + 2-acetamido-2-deoxy-D-glucose 6-phosphate
2.3.1.5	Arylamine acetyltransferase	Acetyl-CoA + arylamine = CoA + N-acetylarylamine
2.3.1.6	Choline acetyltransferase	Acetyl-CoA + choline = CoA + O-acetylcholine
2.3.1.7	Carnitine acetyltransferase	Acetyl-CoA + carnitine = CoA + O-acetylcarnitine
2.3.1.8	Phosphate acetyltransferase	Acetyl-CoA + orthophosphate = CoA + acetylphosphate

Number	Other Names	Basis for classification (Systematic Name)	Comments	Reference
			converts hydroxypyruvate and R-CHO into CO_2 and R-CHOH-CO-CH$_2$OH. Transketolase from *Alkaligenes faecalis* shows high activity with D-erythrose as acceptor	
2.2.1.2	Dihydroxyacetone-transferase	Sedoheptulose-7-phosphate: D-glyceraldehyde-3-phosphate dihydroxyacetonetransferase		1368, 2692, 3546
2.3.1.1		Acetyl-CoA:L-glutamate *N*-acetyltransferase	Also acts with L-aspartate and, more slowly, with some other amino acids	2026
2.3.1.2	Imidazole acetylase	Acetyl-CoA:imidazole *N*-acetyltransferase	Also acts with propionyl-CoA	1695
2.3.1.3	Glucosamine acetylase	Acetyl-CoA:2-amino-2-deoxy-D-glucose *N*-acetyltransferase		531
2.3.1.4	Phosphoglucosamine transacetylase, Phosphoglucosamine acetylase	Acetyl-CoA:2-amino-2-deoxy-D-glucose-6-phosphate *N*-acetyltransferase		397, 646, 2567
2.3.1.5	Arylamine acetylase	Acetyl-CoA:arylamine *N*-acetyltransferase	Wide specificity for aromatic amines, including serotonin; also catalyses acetyl-transfer between arylamines without CoA	530, 3328, 3647, 2574
2.3.1.6	Choline acetylase	Acetyl-CoA:choline *O*-acetyltransferase	Propionyl-CoA can act, more slowly, in place of acetyl-CoA	276, 284, 2976, 945
2.3.1.7		Acetyl-CoA:carnitine *O*-acetyltransferase		502, 936
2.3.1.8	Phosphotransacetylase, Phosphoacylase	Acetyl-CoA:orthophosphate acetyltransferase	Also acts with other short-chain acyl-CoA's	272, 3183, 3184

Number	Recommended Name	Reaction
2.3.1.9	Acetyl-CoA acetyltransferase	Acetyl-CoA + acetyl-CoA = CoA + acetoacetyl-CoA
2.3.1.10	Hydrogen-sulphide acetyltransferase	Acetyl-CoA + hydrogen sulphide = CoA + thioacetate
2.3.1.11	Thioethanolamine acetyltransferase	Acetyl-CoA + thioethanolamine = CoA + S-acetylthioethanolamine
2.3.1.12	Dihydrolipoamide acetyltransferase	Acetyl-CoA + dihydrolipoamide = CoA + S^6-acetyldihydrolipoamide
2.3.1.13	Glycine acyltransferase	Acyl-CoA + glycine = CoA + N-acylglycine
2.3.1.14	Glutamine phenylacetyltransferase	Phenylacetyl-CoA + L-glutamine = CoA + α-N-phenylacetyl-L-glutamine
2.3.1.15	Glycerophosphate acyltransferase	Acyl-CoA + sn-glycerol 3-phosphate = CoA + 1-acylglycerol 3-phosphate
2.3.1.16	Acetyl-CoA acyltransferase	Acyl-CoA + acetyl-CoA = CoA + 3-oxoacyl-CoA
2.3.1.17	Aspartate acetyltransferase	Acetyl-CoA + L-aspartate = CoA + N-acetyl-L-aspartate
2.3.1.18	Galactoside acetyltransferase	Acetyl-CoA + a β-D-galactoside = CoA + 6-O-acetyl-β-D-galactoside
2.3.1.19	Phosphate butyryltransferase	Butyryl-CoA + orthophosphate = CoA + butyrylphosphate
2.3.1.20	Diacylglycerol acyltransferase	Acyl-CoA + 1,2-diacylglycerol = CoA + triacylglycerol
2.3.1.21	Carnitine palmitoyltransferase	Palmitoyl-CoA + L-carnitine = CoA + L-palmitoylcarnitine
2.3.1.22	Acylglycerol palmitoyltransferase	Palmitoyl-CoA + acylglycerol = CoA + diacylglycerol
2.3.1.23	Lysolecithin acyltransferase	Acyl-CoA + 1-acylglycero-3-phosphocholine = CoA + 1,2-diacylglycero-3-phosphocholine

Number	Other Names	Basis for classification (Systematic Name)	Comments	Reference
2.3.1.9	Thiolase, Acetoacetyl-CoA thiolase	Acetyl-CoA:acetyl-CoA *C*-acetyltransferase		2021, 3215
2.3.1.10		Acetyl-CoA:hydrogen-sulphide *S*-acetyltransferase		371
2.3.1.11	Thioltransacetylase B	Acetyl-CoA:thioethanolamine *S*-acetyltransferase		371, 1145
2.3.1.12	Lipoate acetyltransferase, Thioltransacetylase A	Acetyl-CoA:dihydrolipoamide *S*-acetyltransferase	A lipoyl-protein; component of multienzyme pyruvate dehydrogenase complex	371, 1145, 1146
2.3.1.13		Acyl-CoA:glycine *N*-acyltransferase	Acts with the CoA derivatives of a number of aliphatic and aromatic acids	2941
2.3.1.14		Phenylacetyl-CoA:L-glutamine *α-N*-phenylacetyltransferase		2262
2.3.1.15		Acyl-CoA:*sn*-glycerol-3-phosphate *O*-acyltransferase	Acts only with CoA derivatives of fatty acids of chain length above C-10	1775
2.3.1.16	*β*-Ketothiolase, 3-Ketoacyl-CoA thiolase	Acyl-CoA:acetyl-CoA *C*-acyltransferase		247, 1067, 3211
2.3.1.17		Acetyl-CoA:L-aspartate *N*-acetyltransferase		1076, 1722
2.3.1.18	Thiogalactoside acetyltransferase	Acetyl-CoA:galactoside 6-*O*-acetyltransferase	Acts on thiogalactosides and phenylgalactoside	3827, 3828
2.3.1.19		Butyryl-CoA:orthophosphate butyryltransferase		3524
2.3.1.20	Diglyceride acyltransferase	Acyl-CoA:1,2-diacylglycerol *O*-acyltransferase		1073, 3644
2.3.1.21		Palmitoyl-CoA:L-carnitine *O*-palmitoyltransferase		2449
2.3.1.22	Monoglyceride acyltransferase	Palmitoyl-CoA:acylglycerol *O*-palmitoyltransferase	Various monoglycerides can act as acceptor	542
2.3.1.23		Acyl-CoA:1-acylglycero-3-phosphocholine *O*-acyltransferase	Acts preferentially with unsaturated acyl-CoA derivatives	3528

Number	Recommended Name	Reaction
2.3.1.24	Sphingosine acyltransferase	Acyl-CoA + sphingosine = CoA + N-acylsphingosine
2.3.1.25	Plasmalogen synthase	Acyl-CoA + O-1-alk-1-enyl-glycero-3-phosphocholine = CoA + O-1-alk-1-enyl-2-acylglycero-3-phosphocholine
2.3.1.26	Cholesterol acyltransferase	Acyl-CoA + cholesterol = CoA + cholesterol ester
2.3.1.27	Cortisol acetyltransferase	Acetyl-CoA + cortisol = CoA + cortisol 21-acetate
2.3.1.28	Chloramphenicol acetyltransferase	Acetyl-CoA + chloramphenicol = CoA + chloramphenicol 3-acetate
2.3.1.29	Glycine acetyltransferase	Acetyl-CoA + glycine = CoA + 2-amino-3-oxobutyrate
2.3.1.30	Serine acetyltransferase	Acetyl-CoA + L-serine = CoA + O-acetyl-L-serine
2.3.1.31	Homoserine acetyltransferase	Acetyl-CoA + L-homoserine = CoA + O-acetyl-L-homoserine
2.3.1.32	Lysine acetyltransferase	Acetyl phosphate + L-lysine = orthophosphate + N^6-acetyl-L-lysine
2.3.1.33	Histidine acetyltransferase	Acetyl-CoA + L-histidine = CoA + N-acetyl-L-histidine
2.3.1.34	D-Tryptophan acetyltransferase	Acetyl-CoA + D-tryptophan = CoA + N-acetyl-D-tryptophan
2.3.1.35	Glutamate acetyltransferase	N^2-Acetyl-L-ornithine + L-glutamate = L-ornithine + N-acetyl-L-glutamate
2.3.1.36	D-Amino-acid acetyltransferase	Acetyl-CoA + D-amino acid = CoA + N-acetyl-D-amino acid
2.3.1.37	δ-Aminolaevulinate synthase	Succinyl-CoA + glycine = δ -aminolaevulinate + CoA + CO_2
2.3.1.38	[Acyl-carrier-protein]acetyltransferase	Acetyl-CoA + [acyl-carrier protein] = CoA + acetyl-[acyl-carrier protein]

Number	Other Names	Basis for classification (Systematic Name)	Comments	Reference
2.3.1.24		Acyl-CoA:sphingosine N-acyltransferase	Acts on either *threo*- or *erythro*-sphingosine	3174
2.3.1.25		Acyl-CoA:O-1-alk-1-enyl-glycero-3-phosphocholine O-acyltransferase		3587
2.3.1.26		Acyl-CoA:cholesterol O-acyltransferase	The enzyme is highly specific for transfer of acyl groups with a single *cis* double bond 9 carbon atoms distant from the carboxyl group	3015
2.3.1.27		Acetyl-CoA:cortisol O-acetyltransferase		3412
2.3.1.28		Acetyl-CoA:chloramphenicol 3-O-acetyltransferase		3033, 3034
2.3.1.29		Acetyl-CoA:glycine C-acetyltransferase		2164
2.3.1.30		Acetyl-CoA:L-serine O-acetyltransferase		1797, 3118
2.3.1.31		Acetyl-CoA:L-homoserine O-acetyltransferase		2329
2.3.1.32		Acetyl-phosphate:L-lysine N^6-acetyltransferase		2541
2.3.1.33		Acetyl-CoA:L-histidine N-acetyltransferase		220
2.3.1.34		Acetyl-CoA:D-tryptophan N-acetyltransferase		3842
2.3.1.35		N^2-Acetyl-L-ornithine: L-glutamate N-acetyltransferase	Also has some hydrolytic activity on acetyl-L-ornithine, but rate is 1% of that of transferase activity	3202
2.3.1.36		Acetyl-CoA:D-amino-acid N-acetyltransferase		3843
2.3.1.37		Succinyl-CoA:glycine C-succinyltransferase (decarboxylating)	A pyridoxal-phosphate-protein	1680, 3340, 3617, 2969, 2711
2.3.1.38		Acetyl-CoA:[acyl-carrier-protein] S-acetyltransferase		3694, 3538, 2661

Number	Recommended Name	Reaction
2.3.1.39	[Acyl-carrier-protein]malonyltransferase	Malonyl-CoA + [acyl-carrier protein] = CoA + malonyl-[acyl-carrier protein]
2.3.1.40	Acyl-[acyl-carrier-protein]-phospholipid acyltransferase	Acyl-[acyl-carrier protein] + O-[2-acyl-sn-glycero-3-phospho]-ethanolamine = [acyl-carrier protein] + O-(1-β-acyl-2-acyl-sn-glycero-3-phospho)-ethanolamine
2.3.1.41	3-Oxoacyl-[acyl-carrier-protein]synthase	Acyl-[acyl-carrier protein] + malonyl-[acyl-carrier protein] = 3-oxoacyl-[acyl-carrier protein] + CO_2 + acyl-carrier protein
2.3.1.42	Dihydroxyacetone-phosphate acyltransferase	Acyl-CoA + dihydroxyacetone phosphate = CoA + acyldihydroxyacetone phosphate
2.3.1.43	Lecithin—cholesterol acyltransferase	A lecithin + cholesterol = 1-acylglycerophosphocholine + cholesterol ester
2.3.1.44	N-Acetylneuraminate 4-O-acetyltransferase	Acetyl-CoA + N-acetylneuraminate = CoA + N-acetyl-4-O-acetylneuraminate
2.3.1.45	N-Acetylneuraminate 7 (or 8)-O-acetyltransferase	Acetyl-CoA + N-acetylneuraminate = CoA + N-acetyl-7 (or 8)-O-acetylneuraminate
2.3.1.46	Homoserine succinyltransferase	Succinyl-CoA + L-homoserine = CoA + O-succinyl-L-homoserine
2.3.1.47	7-Oxo-8-aminononanoate synthase	Pimeloyl-CoA + L-alanine = 7-oxo-8-aminononanoate + CoA + CO_2
2.3.1.48	Histone acetyltransferase	Acetyl-CoA + histone = CoA + acetylhistone
2.3.1.49	Deacetyl-[citrate-(pro-3S)-lyase] acetyltransferase	S-Acetylphosphopantetheine + deacetyl-[citrate-oxaloacetate-lyase (pro-3S-CH$_2$COO→acetate)] = phosphopantetheine + [citrate oxaloacetate-lyase (pro-3S-CH$_2$COO→acetate)]
2.3.1.50	Serine palmitoyltransferase	Palmitoyl-CoA + L-serine = CoA + 3-oxo-D-dihydrosphingosine + CO_2

Number	Other Names	Basis for classification (Systematic Name)	Comments	Reference
2.3.1.39		Malonyl-CoA:[acyl-carrier-protein] S-malonyltransferase		3694, 2661, 41
2.3.1.40		Acyl-[acyl-carrier-protein]: O-(2-acyl-sn-glycero-3-phospho)-ethanolamine O-acyltransferase		3383
2.3.1.41		Acyl-[acyl-carrier-protein]:malonyl-[acyl-carrier-protein] C-acyltransferase (decarboxylating)		3448, 2661, 41
2.3.1.42		Acyl-CoA:dihydroxyacetone-phosphate O-acyltransferase	Uses CoA derivatives of palmitate, stearate, and oleate, with highest activity on palmitoyl-CoA	1170
2.3.1.43	Lecithin acyltransferase	Lecithin:cholesterol acyltransferase		1064
2.3.1.44		Acetyl-CoA: N-acetylneuraminate 4-O-acetyltransferase	N-Acetylneuraminate 9-phosphate and the CMP derivative of N-acetylneuraminate, can also act as acceptor	2946
2.3.1.45		Acetyl-CoA: N-acetylneuraminate 7 (or 8)-O-acetyltransferase	N-Acetylneuraminate 9-phosphate and the CMP derivative of N-acetylneuraminate, can also act as acceptor	2946
2.3.1.46	Homoserine O-transsuccinylase	Succinyl-CoA:L-homoserine O-succinyltransferase		2860
2.3.1.47		Pimeloyl-CoA:L-alanine C-pimeloyltransferase (decarboxylating)	A pyridoxal-phosphate-protein	805
2.3.1.48		Acetyl-CoA:histone acetyltransferase	A group of enzymes with differing specificity towards histone acceptors	988
2.3.1.49		S-Acetylphosphopantetheine: deacetyl-[citrate-oxaloacetate-lyase (pro-3S-CH$_2$COO→acetate)] S-acetyltransferase		3086
2.3.1.50		Palmitoyl-CoA:L-serine C-palmitoyltransferase (decarboxylating)	A pyridoxal-phosphate-protein	3236, 367

Number	Recommended Name	Reaction
2.3.1.51	1-Acylglycerophosphate acyltransferase	Acyl-CoA + 1-acyl-*sn*-glycerol 3-phosphate = CoA + 1,2-diacyl-*sn*-glycerol 3-phosphate
2.3.1.52	2-Acylglycerophosphate acyltransferase	Acyl-CoA + 2-acyl-*sn*-glycerol 3-phosphate = CoA + 1,2-diacyl-*sn*-glycerol 3-phosphate
2.3.1.53	Phenylalanine acetyltransferase	Acetyl-CoA + L-phenylalanine = CoA + *N*-acetyl-L-phenylalanine
2.3.1.54	Formate acetyltransferase	Acetyl-CoA + formate = CoA + pyruvate
2.3.1.55	Kanamycin 6'-acetyltransferase	Acetyl-CoA + kanamycin = CoA + N^6-acetylkanamycin
2.3.1.56	Aromatic-hydroxylamine acetyltransferase	*N*-Hydroxy-4-acetylaminobiphenyl + *N*-hydroxy-4-aminobiphenyl = *N*-hydroxy-4-aminobiphenyl + *N*-acetoxy-4-aminobiphenyl
2.3.1.57	Putrescine acetyltransferase	Acetyl-CoA + putrescine = CoA + monoacetylputrescine
2.3.1.58	2,3-Diaminopropionate oxalyltransferase	Oxalyl-CoA + L-2,3-diaminopropionate = CoA + N^3-oxalyl-L-2,3-diaminopropionate
2.3.1.59	Gentamicin 2'-acetyltransferase	Acetyl-CoA + gentamicin C_{1a} = CoA + $N^{2'}$-acetylgentamicin C_{1a}
2.3.1.60	Gentamicin 3-acetyltransferase	Acetyl-CoA + gentamicin C = CoA + N^3-acetylgentamicin C
2.3.1.61	Dihydrolipoamide succinyltransferase	Succinyl-CoA + dihydrolipoamide = CoA + S^6-succinylhydrolipoamide
2.3.1.62	2-Acylglycerophosphocholine acyltransferase	Acyl-CoA + 2-acyl-*sn*-glycero-3-phosphocholine = CoA + phosphatidylcholine

Number	Other Names	Basis for classification (Systematic Name)	Comments	Reference
2.3.1.51		Acyl-CoA:1-acyl-*sn*-glycerol-3-phosphate *O*-acyltransferase	Specific for the transfer of mono- and dienoic fatty-acyl-CoA thioesters	3770
2.3.1.52		Acyl-CoA:2-acyl-*sn*-glycerol-3-phosphate *O*-acyltransferase	Saturated acyl-CoA thioesters are the most effective acyl donors	3770
2.3.1.53		Acetyl-CoA:L-phenylalanine *N*-acetyltransferase	Also acts, more slowly, on L-histidine and L-alanine	1806
2.3.1.54	Pyruvate formate-lyase	Acetyl-CoA:formate *C*-acetyltransferase		3185, 1086
2.3.1.55		Acetyl-CoA:kanamycin $N^{6'}$-acetyltransferase	Kanamycin A, kanamycin B, neomycin, gentamicin C_{1a}, gentamicin C_2 and sisomicin are substrates	260, 1906
2.3.1.56		*N*-Hydroxy-4-acetylaminobiphenyl: *N*-hydroxy-4-aminobiphenyl *O*-acetyltransferase	Transfers the *N*-acetyl group of some aromatic acethydroxamates to the *O*-position of some aromatic hydroxylamines	218
2.3.1.57		Acetyl-CoA:putrescine *N*-acetyltransferase		2998
2.3.1.58	Oxalyldiaminopropionate synthase	Oxalyl-CoA:L-2,3-diaminopropionate N^3-oxalyltransferase		2062
2.3.1.59	Gentamicin acetyltransferase II	Acetyl-CoA:gentamicin-C_{1a} $N^{2'}$-acetyltransferase	Gentamicin A, sisomicin, tobramycin, paromomycin, neomycin B, kanamycin B and kanamycin C can also act as acceptors	262
2.3.1.60	Gentamicin acetyltransferase I	Acetyl-CoA:gentamicin-C N^3-acetyltransferase	Also acetylates sisomicin	413
2.3.1.61		Succinyl-CoA:dihydrolipoamide *S*-succinyltransferase	A lipoyl-protein; component of the multienzyme 2-oxo-glutarate dehydrogenase complex	2739, 699
2.3.1.62		Acyl-CoA:2-acyl-*sn*-glycero-3-phosphocholine *O*-acyltransferase		1310

Number	Recommended Name	Reaction
2.3.1.63	1-Alkylglycerophosphate acyltransferase	Acyl-CoA + 1-alkyl-*sn*-glycerol 3-phosphate = CoA + 1-alkyl-2-acyl-*sn*-glycerol 3-phosphate

2.3.2 AMINOACYLTRANSFERASES

2.3.2.1	D-Glutamyltransferase	L(or D)-Glutamine + D-glutamyl-peptide = NH$_3$ + 5-glutamyl-D-glutamyl-peptide
2.3.2.2	γ-Glutamyltransferase	(5-L-Glutamyl)-peptide + an amino acid = peptide + 5-L-glutamyl-amino acid
2.3.2.3	Lysyltransferase	L-Lysyl-tRNA + phosphatidylglycerol = tRNA + 3-phosphatidyl-1'-(3'-*O*-L-lysyl)-glycerol
2.3.2.4	γ-Glutamylcyclotransferase	(5-L-Glutamyl)-L-amino acid = pyroglutamate + L-amino acid
2.3.2.5	Glutaminyl-tRNA cyclotransferase	L-Glutaminyl-tRNA = pyroglutamyl-tRNA + NH$_3$
2.3.2.6	Leucyltransferase	L-Leucyl-tRNA + protein = tRNA + L-leucyl-protein
2.3.2.7	Aspartyltransferase	L-Asparagine + hydroxylamine = aspartylhydroxamate + NH$_3$
2.3.2.8	Arginyltransferase	L-Arginyl-tRNA + protein = tRNA + arginyl-protein
2.3.2.9	Agaritine γ-glutamyltransferase	N^3-(γ-L-Glutamyl)-4-hydroxymethylphenylhydrazine + acceptor = 4-hydroxymethylphenylhydrazine + γ-L-glutamyl-acceptor

Number	Other Names	Basis for classification (Systematic Name)	Comments	Reference
2.3.1.63		Acyl-CoA:1-alkyl-*sn*-glycero-3-phosphate *O*-acyltransferase		1310
2.3.2.1	D-Glutamyl transpeptidase	Glutamine:D-glutamyl-peptide glutamyltransferase		3693
2.3.2.2	Glutamyl transpeptidase	(5-Glutamyl)-peptide:amino-acid 5-glutamyltransferase		1083, 1913
2.3.2.3		L-Lysyl-tRNA:phosphatidylglycerol 3'-lysyltransferase		1924
2.3.2.4		(5-L-Glutamyl)-L-amino-acid 5-glutamyltransferase (cyclizing)	The enzyme acts on derivatives of L-glutamate, L-2-aminobutyrate, L-alanine and glycine	2507, 332
2.3.2.5		L-Glutaminyl-tRNA γ-glutamyltransferase (cyclizing)	Also acts on glutaminyl-peptides	280, 2147
2.3.2.6		L-Leucyl-tRNA:protein leucyltransferase	Also transfers phenylalanyl groups. Requires a monovalent cation. Peptides and proteins containing an *N*-terminal arginine, lysine or histidine residue can act as acceptors	1914, 1915, 3142
2.3.2.7	Asparagine→hydroxyl-amine transaspartase	L-Asparagine:hydroxylamine γ-aspartyltransferase		1505
2.3.2.8		L-Arginyl-tRNA:protein arginyltransferase	Requires mercaptoethanol and a monovalent cation. Peptides and proteins containing an *N*-terminal glutamate, aspartate or cystine residue can act as acceptors	3140, 3144, 3141
2.3.2.9		N^3-(γ-L-Glutamyl)-4-hydroxymethylphenylhydrazine:(acceptor) γ-glutamyltransferase	4-Hydroxyaniline, cyclohexylamine, 1-naphthylhydrazine and similar compounds can act as acceptors; the enzyme also catalyses the hydrolysis of agaritine	1037

Number	Recommended Name	Reaction
2.3.2.10	UDPacetylmuramoylpentapeptide lysine N^6-alanyltransferase	L-Alanyl-tRNA + UDP-N-acetylmuramoyl-L-alanyl-D-glutamyl-L-lysyl-D-alanyl-D-alanine = tRNA + UDP-N-acetylmuramoyl-L-alanyl-D-glutamyl-(N^6-L-alanyl)-L-lysyl-D-alanyl-D-alanine
2.3.2.11	O-Alanylphosphatidylglycerol synthase	Alanyl-tRNA + phosphatidylglycerol = tRNA + O-L-alanylphosphatidylglycerol
2.3.2.12	Peptidyltransferase	Peptidyl-tRNA^1 + aminoacyl-tRNA^2 = tRNA^1 + peptidyl-aminoacyl-tRNA^2
2.3.2.13	Glutaminyl-peptide γ-glutamyltransferase	N^2-R-glutaminyl-peptide + R'NH$_2$ = NH$_3$ + N^2-R-N^5-R'-glutaminyl-peptide

2.4 GLYCOSYLTRANSFERASES

To this class belong all enzymes transferring glycosyl groups. Some of these enzymes also have hydrolytic activities, which can be regarded as transfer of a glycosyl group from the donor to water. Also, inorganic phosphate can act as acceptor in the case of phosphorylases; phosphorolysis of glycogen is regarded as transfer of one sugar residue from glycogen to phosphate. However, the more general case is the transfer of a sugar from oligosaccharide or a high-energy compound to another carbohydrate molecule as acceptor.

The sub-class is further sub-divided, according to the nature of the sugar residue being transferred, into hexosyltransferases (2.4.1), pentosyltransferases (2.4.2) and those transferring other glycosyl groups (2.4.99).

(*'Glucosyltransferase', 'fructosyltransferase', etc. may be replaced by 'transglucosylase', 'transfructosylase', etc.*)

2.4.1 HEXOSYLTRANSFERASES

2.4.1.1	Phosphorylase	(1,4-α-D-Glucosyl)$_n$ + orthophosphate = (1,4-α-D-glucosyl)$_{n-1}$ + α-D-glucose 1-phosphate

Number	Other Names	Basis for classification (Systematic Name)	Comments	Reference
2.3.2.10		L-Alanyl-tRNA:UDP-*N*-acetylmuramoyl-L-alanyl-D-glutamyl-L-lysyl-D-alanyl-D-alanine N^6-alanytransferase	Also acts on L-seryl-tRNA	2617
2.3.2.11		Alanyl-tRNA:phosphatidylglycerol alanyltransferase		1094
2.3.2.12		Peptidyl-tRNA:aminoacyl-tRNA *N*-peptidyltransferase		2881, 3457, 2880
2.3.2.13	Transglutaminase	R-Glutaminyl-peptide:amine γ-glutamyl-yltransferase	The γ-carboxamide groups of peptide-bound glutamine residues act as acyl donors, and the 6-amino-groups of protein- and peptide-bound lysine residues act as acceptors, to give intra- and inter-molecular 6-(5-glutamyl)-lysine cross-links	906, 905, 908
2.4.1.1	P-enzyme (only for the plant enzyme), Muscle phosphorylase *a* and *b*, Amylophosphorylase, Polyphosphorylase	1,4-α-D-Glucan:orthophosphate α-D-glucosyltransferase	The recommended name should be qualified in each instance by adding the name of the natural substrate, *e.g.* maltodextrin phosphorylase, starch phosphorylase, glycogen phosphorylase	231, 511, 600, 882, 1102, 1180

Number	Recommended Name	Reaction
2.4.1.2	Dextrin dextranase	$(1,4\text{-}\alpha\text{-}D\text{-Glucosyl})_n + (1,6\text{-}\alpha\text{-}D\text{-glucosyl})_m = (1,4\text{-}\alpha\text{-}D\text{-glucosyl})_{n-1} + (1,6\text{-}\alpha\text{-}D\text{-glucosyl})_{m+1}$
[2.4.1.3	*Deleted entry: Amibomaltase, now included with EC 2.4.1.25, 4-α-D-Glucanotransferase]*	
2.4.1.4	Amylosucrase	Sucrose + $(1,4\text{-}\alpha\text{-}D\text{-glucosyl})_n$ = D-fructose + $(1,4\text{-}\alpha\text{-}D\text{-glucosyl})_{n+1}$
2.4.1.5	Dextransucrase	Sucrose + $(1,6\text{-}\alpha\text{-}D\text{-glucosyl})_n$ = D-fructose + $(1,6\text{-}\alpha\text{-}D\text{-glucosyl})_{n+1}$
[2.4.1.6	*Deleted entry:Maltose 3-glycosyltransferase]*	
2.4.1.7	Sucrose phosphorylase	Sucrose + orthophosphate = D-fructose + α-D-glucose 1-phosphate
2.4.1.8	Maltose phosphorylase	Maltose + orthophosphate = D-glucose + β-D-glucose 1-phosphate
2.4.1.9	Inulosucrase	Sucrose + $(2,1\text{-}\beta\text{-}D\text{-fructosyl})_n$ = glucose + $(2,1\text{-}\beta\text{-}D\text{-fructosyl})_{n+1}$
2.4.1.10	Levansucrase	Sucrose + $(2,6\text{-}\beta\text{-}D\text{-fructosyl})_n$ = glucose + $(2,6\text{-}\beta\text{-}D\text{-fructosyl})_{n+1}$
2.4.1.11	Glycogen (starch) synthase	UDPglucose + $(1,4\text{-}\alpha\text{-}D\text{-glucosyl})_n$ = UDP + $(1,4\text{-}\alpha\text{-}D\text{-glucosyl})_{n+1}$
2.4.1.12	Cellulose synthase (UDP-forming)	UDPglucose + $(1,4\text{-}\beta\text{-}D\text{-glucosyl})_n$ = UDP + $(1,4\text{-}\beta\text{-}D\text{-glucosyl})_{n+1}$
2.4.1.13	Sucrose synthase	UDPglucose + D-fructose = UDP + sucrose
2.4.1.14	Sucrose-phosphate synthase	UDPglucose + D-fructose 6-phosphate = UDP + sucrose 6-phosphate

Number	Other Names	Basis for classification (Systematic Name)	Comments	Reference
2.4.1.2	Dextrin 6-glucosyltransferase	1,4-α-D-Glucan:1,6-α-D-glucan 6-α-D-glucosyltransferase		1260, 1261, 1262
2.4.1.4	Sucrose-glucan glucosyltransferase	Sucrose:1,4-α-D-glucan 4-α-D-glucosyltransferase		867, 1260, 1263
2.4.1.5	Sucrose 6-glucosyltransferase	Sucrose:1,6-α-D-glucan 6-α-D-glucosyltransferase		176, 177, 1260
2.4.1.7	Sucrose glucosyltransferase	Sucrose:orthophosphate α-D-glucosyltransferase	In the forward reaction, arsenate may replace phosphate. In the reverse reaction various ketoses and L-arabinose may replace D-fructose	750, 1216, 3076
2.4.1.8		Maltose:orthophosphate 1-β-D-glucosyltransferase		750, 890, 2667, 3722
2.4.1.9	Sucrose 1-fructosyltransferase	Sucrose:2,1-β-D-fructan-β-D-fructosyltransferase	Converts sucrose into inulin and D-glucose	292, 684, 787
2.4.1.10	Sucrose 6-fructosyltransferase	Sucrose:2,6-β-D-fructan 6-β-D-fructosyltransferase	Some other sugars can act as D-fructosyl acceptors	1260, 1298, 2746
2.4.1.11	UDPglucose—glycogen glucosyltransferase	UDPglucose:glycogen 4-α-D-glucosyltransferase	The recommended name will vary according to the source of the enzyme and the nature of its synthetic product (cf phosphorylase EC 2.4.1.1). A similar enzyme utilizes ADPglucose (EC 2.4.1.21)	50, 222, 1920, 1922
2.4.1.12	UDPglucose—β-glucan glucosyltransferase, UDPglucose—cellulose glucosyltransferase	UDPglucose:1,4-β-D-glucan 4-β-D-glucosyltransferase	Involved in the synthesis of cellulose. A similar enzyme utilizes GDPglucose (EC 2.4.1.29)	1054
2.4.1.13	UDPglucose—fructose glucosyltransferase, Sucrose—UDP glucosyltransferase	UDPglucose:D-fructose 2-α-D-glucosyltransferase		472, 144
2.4.1.14	UDPglucose—fructosephosphate glucosyltransferase,	UDPglucose:D-fructose-6-phosphate 2-α-D-glucosyltransferase		2198

Number	Recommended Name	Reaction
2.4.1.15	α,α-Trehalose-phosphate synthase (UDP-forming)	UDPglucose + D-glucose 6-phosphate = UDP + α,α-trehalose 6-phosphate
2.4.1.16	Chitin synthase	UDP-2-acetamido-2-deoxy-D-glucose + [1,4-(2-acetamido-2-deoxy-β-D-glucosyl)]$_n$ = UDP + [1,4-(2-acetamido-2-deoxy-β-D-glucosyl)]$_{n+1}$
2.4.1.17	UDPglucuronosyltransferase	UDPglucuronate + acceptor = UDP + acceptor β-D-glucuronide
2.4.1.18	1,4-α-Glucan branching enzyme	Transfers a segment of a 1,4-α-D-glucan chain to a primary hydroxyl group in a similar glucan chain
2.4.1.19	Cyclomaltodextrin glucanotranferase	Cyclizes part of a 1,6-α-3-glucan chain by formation of a 1,4-α-D-glucosidic bond
2.4.1.20	Cellobiose phosphorylase	Cellobiose + orthophosphate = α-D-glucose 1-phosphate + D-glucose
2.4.1.21	Starch (bacterial glycogen) synthase	ADPglucose + (1,4-α-D-glucosyl)$_n$ = ADP + (1,4-α-D-glucosyl)$_{n+1}$

Number	Other Names	Basis for classification (Systematic Name)	Comments	Reference
	Sucrosephosphate— UDP glucosyltransferase			
2.4.1.15	UDPglucose— glucosephosphate glucosyltransferase, Trehalosephosphate— UDP glucosyltransferase	UDPglucose:D-glucose-6-phosphate 1-α-D-glucosyltransferase	See also EC 2.4.1.36	445, 454, 2316, 2006
2.4.1.16	Chitin—UDP acetylglucosaminyl-transferase, trans-N-Acetylglucosaminosylase	UDP-2-acetamido-2-deoxy-D-glucose:chitin 4-β-acetamidodeoxy-D-glucosyltransferase	Converts UDP-2-acetamido-2-deoxy-D-glucose into chitin and UDP	1058
2.4.1.17	UDPglucuronate→phenol transglucuronidase	UDPglucuronate β-D-glucuronosyltransferase (acceptor-unspecific)	A wide range of phenols, alcohols, amines and fatty acids can act as acceptor	153, 775
2.4.1.18	Q-enzyme, Branching enzyme	1,4-α-D-Glucan:1,4-α-D-glucan 6-α-D-(1,4-α-D-glucano)-transferase	Converts amylose into amylopectin. The recommended name requires qualification depending on the product, glycogen or amylopectin, i.e. glycogen branching enzyme, amylopectin branching enzyme. The latter has frequently been termed Q-enzyme	207, 231, 1260, 1437
2.4.1.19	*Bacillus macerans* amylase, Cyclodextrin glucanotransferase	1,4-α-D-Glucan 4-α-D-(1,4-α-D-glucano)-transferase (cyclizing)	Cyclomaltodextrins (Schardinger dextrins) of various sizes (6,7,8,etc. glucose units) are formed reversibly from starch and similar substrates. Will also disproportionate linear maltodextrans without cyclizing (cf. EC 2.4.1.25)	928, 679, 1260, 2982
2.4.1.20		Cellobiose:orthophosphate α-D-glucosyltransferase		48, 157
2.4.1.21	ADPglucose—starch glucosyltransferase	ADPglucose:1,4-α-D-glucan 4-α-D-glucosyltransferase	Similar to EC 2.4.1.11 but the preferred or mandatory nucleoside diphosphate sugar substrate is ADPglucose. This entry covers starch	491, 952, 1112, 1921, 2656

Number	Recommended Name	Reaction
2.4.1.22	Lactose synthase	UDPgalactose + D-glucose = UDP + lactose
2.4.1.23	UDPgalactose—sphingosine β-D-galactosyltransferase	UDPgalactose + sphingosine = UDP + psychosine
2.4.1.24	1,4-α-D-Glucan 6-α-D-glucosyltransferase	Transfers an α-glucosyl residue in a 1,4-α-D-glucan to the primary hydroxyl group of glucose, free or combined in a 1,4-α-D-glucan
2.4.1.25	4-α-D-Glucanotransferase	Transfers a segment of a 1,4-α-D-glucan to a new 4-position in an acceptor, which may be glucose or a 1,4-α-D-glucan
2.4.1.26	UDPglucose—DNA α-D-glucosyltransferase	Transfers an α-D-glucosyl residue from UDPglucose to a hydroxymethylcytosine residue in DNA
2.4.1.27	UDPglucose—DNA β-D-glucosyltransferase	Transfers a β-D-glucosyl residue from UDPglucose to a hydroxymethylcytosine residue in DNA
2.4.1.28	UDPglucose—glucosyl-DNA β-D-glucosyltransferase	Transfers a β-D-glucosyl residue from UDPglucose to a glucosyl-hydroxymethylcytosine residue in DNA
2.4.1.29	Cellulose synthase (GDP-forming)	GDPglucose + (1,4-β-D-glucosyl)$_n$ = GDP + (1,4-β-D-glucosyl)$_{n+1}$
2.4.1.30	1,3-β-D-Oligoglucan phosphorylase	(1,3-β-D-Glucosyl)$_n$ + orthophosphate = (1,3-β-D-glucosyl)$_{n-1}$ + α-D-glucose 1-phosphate

Number	Other Names	Basis for classification (Systematic Name)	Comments	Reference
			and glycogen synthases utilizing ADPglucose	
2.4.1.22	UDPgalactose—glucose galactosyltransferase	UDPgalactose:D-glucose 4-β-D-galactosyltransferase	The enzyme is a complex of two proteins A and B. In the absence of the B protein (α-lactalbumin) the enzyme catalyses the transfer of galactose from UDPgalactose to N-acetylglucosamine	891, 3621
2.4.1.23	Psychosine—UDP galactosyltransferase	UDPgalactose:sphingosine β-D-galactosyltransferase		543
2.4.1.24	T-enzyme, Oligoglucan-branching glycosyltransferase	1,4-α-D-Glucan:1,4-α-D-glucan (D-glucose) 6-α-D-glucosyltransferase		3, 208, 2914
2.4.1.25	D-enzyme, Disproportionating enzyme, Dextrin glycosyltransferase	1,4-α-D-Glucan:1,4-α-D-glucan 4-α-D-glycosyltransferase	This entry covers the former separate entry for EC 2.4.1.3 (amylomaltase). The plant enzyme has been termed D-enzyme. An enzymic activity of this nature forms part of the mammalian and yeast glycogen debranching system (See EC 3.2.1.33)	1260, 1898, 2016, 2579, 3594
2.4.1.26		UDPglucose:DNA α-D-glucosyltransferase		1778
2.4.1.27		UDPglucose:DNA β-D-glucosyltransferase		1778
2.4.1.28		UDPglucose:D-glucosyl-DNA β-D-glucosyltransferase		1778
2.4.1.29		GDPglucose:1,4-β-D-glucan 4-β-glucosyltransferase	Involved in the synthesis of cellulose. A similar enzyme (EC 2.4.1.12) utilizes UDPglucose	491, 902
2.4.1.30	β-1,3-Oligoglucan: orthophosphate glucosyltransferase II	1,3-β-D-Oligoglucan: orthophosphate glucosyltransferase	Does not act on laminarin. Differs in specificity from EC 2.4.1.31 and 2.4.1.97	2081, 2074

Number	Recommended Name	Reaction
2.4.1.31	Laminaribiose phosphorylase	3-O-β-D-Glucosylglucose + orthophosphate = D-glucose + α-D-glucose 1-phosphate
2.4.1.32	Glucomannan 4-β-D-mannosyltransferase	GDPmannose + (glucomannan)$_n$ = GDP + (glucomannan)$_{n+1}$
2.4.1.33	Alginate synthase	GDPmannuronate + (alginate)$_n$ = GDP + (alginate)$_{n+1}$
2.4.1.34	1,3-β-D-Glucan synthase	UDPglucose + (1,3-β-D-glucosyl)$_n$ = UDP + (1,3-β-D-glucosyl)$_{n+1}$
2.4.1.35	UDPglucosyltransferase	UDPglucose + a phenol = UDP + aryl β-D-glucoside
2.4.1.36	α,α-Trehalose-phosphate synthase (GDP-forming)	GDPglucose + glucose 6-phosphate = GDP + α,α-trehalose 6-phosphate
2.4.1.37	Blood-group-substance α-D-galactosyltransferase	UDPgalactose + O-α-L-fucosyl-(1,2)-D-galactose = UDP + α-D-galactosyl-(1,3)-[O-α-L-fucosyl-(1,2)]-D-galactose
2.4.1.38	Glycoprotein β-D-galactosyltransferase	UDPgalactose + 2-acetamido-2-deoxy-D-glucosyl-glycopeptide = UDP + 4-O-β-D-galactosyl-2-acetamido-2-deoxy-D-glucosyl-glycopeptide
2.4.1.39	UDPacetylglucosamine—steroid acetylglucosaminyltransferase	UDP-2-acetamido-2-deoxy-D-glucose + 17α-hydroxysteroid 3-D-glucuronoside = UDP + 17α-(2-acetamido-2-deoxy-D-glucosyloxy)-steroid 3-D-glucuronoside
2.4.1.40	Fucosyl-galactose acetylgalactosaminyltransferase	UDP-2-acetamido-2-deoxy-D-galactose + O-α-L-fucosyl-(1,2)-D-galactose = UDP + O-α-D-2-acetamido-2-deoxygalactosyl-(1,3)-[O-α-L-fucosyl-(1,2)]-D-galactose
2.4.1.41	UDPacetylgalactosamine—protein acetylgalactosaminyltransferase	UDP-2-acetamido-2-deoxy-D-galactose + protein = UDP + 2-acetamido-2-deoxy-D-galactose-protein

Number	Other Names	Basis for classification (Systematic Name)	Comments	Reference
2.4.1.31		3-O-β-D-Glucosylglucose: orthophosphate glucosyltransferase	Also acts on 1,3-β-D-oligoglucans. Differs in specificity from EC 2.4.1.30 and 2.4.1.97	1066, 2054
2.4.1.32		GDPmannose:glucomannan 1,4-β-D-mannosyltransferase		812
2.4.1.33		GDPmannuronate:alginate D-mannuronyltransferase		1970
2.4.1.34	1,3-β-D-Glucan—UDPglucosyltransferase, UDPglucose—1,3-β-D-glucan glucosyltransferase, Callose synthetase	UDPglucose:1,3-β-D-glucan 3-β-D-glucosyltransferase		2082
2.4.1.35		UDPglucose:phenol β-D-glucosyltransferase	Acts on a wide range of phenols	774
2.4.1.36	GDPglucose—glucosephosphate glucosyltransferase	GDPglucose:D-glucose-6-phosphate α-D-glucosyltransferase	See also EC 2.4.1.15	811
2.4.1.37		UDPgalactose:O-α-L-fucosyl-(1,2)-D-galactose α-D-galactosyltransferase		2683
2.4.1.38	Thyroid galactosyltransferase, UDPgalactose—glycoprotein galactosyltransferase	UDPgalactose:2-acetamido-2-deoxy-D-glucosyl-glycopeptide galactosyltransferase	This enzyme may be a component of EC 2.4.1.22	3167
2.4.1.39		UDP-2-acetamido-2-deoxy-D-glucose:17α-hydroxysteroid-3-D-glucuronoside 17α-acetamido-deoxyglucosyltransferase		560
2.4.1.40		UDP-2-acetamido-2-deoxy-D-galactose:O-α-L-fucosyl-(1,2)-D-galactose acetamido-deoxygalactosyltransferase	Can use a number of 2-fucosyl-galactosides as acceptor	1253
2.4.1.41	Protein—UDP acetylgalactosaminyl-transferase	UDP-2-acetamido-2-deoxy-D-galactose:protein acetamidodeoxygalactosyl-transferase	The enzyme is specific for the polypeptide core of ovine submaxillary mucin	2165, 1166

Number	Recommended Name	Reaction
2.4.1.42	UDPglucuronate—oestriol 17β-D-glucuronosyltransferase	UDPglucuronate + 17β-hydroxysteroid = UDP + steroid 17β-D-glucuronoside
2.4.1.43	UDPgalacturonate—polygalacturonate α-D-galacturonosyltransferase	UDPgalacturonate + (1,4-α-D-galacturonosyl)$_n$ = UDP + (1,4-α-D-galacturonosyl)$_{n+1}$
2.4.1.44	UDPgalactose—lipopolysaccharide galactosyltransferase	UDPgalactose + lipopolysaccharide = UDP + D-galactosyl-lipopolysaccharide
2.4.1.45	UDPgalactose—2-hydroxyacylsphingosine galactosyltransferase	UDPgalactose + 2-(2-hydroxyacyl)-sphingosine = UDP + 1-(β-D-galactosyl)-2-(2-hydroxyacyl)sphingosine
2.4.1.46	UDPgalactose—1,2-diacylglycerol galactosyltransferase	UDPgalactose + 1,2-diacylglycerol = UDP + 3-O-β-D-galactosyl-1,2-diacylglycerol
2.4.1.47	UDPgalactose—N-acylsphingosine galactosyltransferase	UDPgalactose + N-acylsphingosine = UDP + D-galactosylceramide
2.4.1.48	GDPmannose α-D-mannosyltransferase	GDPmannose + heteroglycan = GDP + 1,2 (1,3)-α-D-mannosylheteroglycan
2.4.1.49	Cellodextrin phosphorylase	(1,4-β-D-Glucosyl)$_n$ + orthophosphate = (1,4-β-D-glucosyl)$_{n-1}$ + α-D-glucose 1-phosphate
2.4.1.50	UDPgalactose—collagen galactosyltransferase	UDPgalactose + 5-hydroxylysine-collagen = UDP + O-D-galactosyl-5-hydroxylysine-collagen
2.4.1.51	UDP-N-acetylglucosamine—glycoprotein N-acetylglucosaminyltransferase	UDP-2-acetamido-2-deoxy-D-glucose + glycoprotein = UDP + 2-acetamido-2-deoxy-D-glucosylglycoprotein
2.4.1.52	UDPglucose—poly(glycerol phosphate) α-D-glucosyltransferase	UDPglucose + poly(glycerol phosphate) = UDP + α-D-glucosylpoly(glycerol phosphate)
2.4.1.53	UDPglucose—poly(ribitol phosphate) β-D-glucosyltransferase	UDPglucose + poly(ribitol phosphate) = UDP + β-D-glucosylpoly(ribitol phosphate)

Number	Other Names	Basis for classification (Systematic Name)	Comments	Reference
2.4.1.42	Oestriol—UDP 17β-glucuronyltransferase	UDPglucuronate: 17β-hydroxysteroid 17β-D-glucuronosyltransferase		623
2.4.1.43		UDPgalacturonate:1,4-α-poly-D-galacturonate 4-α-D-galacturonosyltransferase		3553
2.4.1.44		UDPgalactose:lipopolysaccharide galactosyltransferase	Transfers galactosyl residues to D-glucose in the partially completed core of lipopolysaccharide	826
2.4.1.45		UDPgalactose: 2-(2-hydroxyacyl)sphingosine galactosyltransferase	Highly specific	2281
2.4.1.46		UDPgalactose: 1,2-diacylglycerol 3-O-galactosyltransferase		3660, 3543
2.4.1.47		UDPgalactose: N-acylsphingosine galactosyltransferase		961
2.4.1.48		GDPmannose:heteroglycan 2,3-α-D-mannosyltransferase	The acceptor is a heteroglycan primer containing mannose, galactose and xylose. 1,2- and 1,3-mannosyl-mannose bonds are formed	88
2.4.1.49		1,4-β-D-Oligoglucan: orthophosphate α-D-glucosyltransferase		3043
2.4.1.50		UDPgalactose: 5-hydroxylysine-collagen galactosyltransferase		349
2.4.1.51		UDP-2-acetamido-2-deoxy-D-glucose:glycoprotein 2-acetamido-2-deoxy-D-glucosyltransferase		693
2.4.1.52		UDPglucose:poly (glycerol—phosphate) α-D-glucosyltransferase		1059
2.4.1.53		UDPglucose:poly(ribitol—phosphate) β-D-glucosyltransferase		526

Number	Recommended Name	Reaction
2.4.1.54	GDPmannose—undecaprenyl-phosphate mannosyltransferase	GDPmannose + undecaprenyl phosphate = GDP + D-mannosyl-1-phosphoundecaprenol
2.4.1.55	Teichoic-acid synthase	CDPribitol + teichoic acid = CMP + phospho-D-ribitol teichoic acid
2.4.1.56	UDP-*N*-acetylglucosamine—lipopolysaccharide *N*-acetylglucosaminyltransferase	UDP-2-acetamido-2-deoxy-D-glucose + lipopolysaccharide = UDP + 2-acetamido-2-deoxy-D-glucosyl-lipopolysaccharide
2.4.1.57	GDPmannose—phosphatidyl-*myo*-inositol α-D-mannosyltransferase	Transfers one or more α-D-mannose units from GDPmannose to positions 2,6-and others in 1-phosphatidyl-*myo*-inositol
2.4.1.58	UDPglucose—lipopolysaccharide glucosyltransferase I	UDPglucose + lipopolysaccharide = UDP + D-glucosyl-lipopolysaccharide
2.4.1.59	UDPglucuronate—oestradiol glucuronosyltransferase	UDPglucuronate + 17β-oestradiol = UDP + 17β-oestradiol 3-D-glucuronoside
2.4.1.60	Abequosyltransferase	CDPabequose + D-mannosyl-rhamnosyl-galactose-1-diphospholipid = CDP + abequosyl-D-mannosyl-rhamnosyl-galactose-1-diphospholipid
2.4.1.61	UDPglucuronate—oestriol 16α-glucuronosyltransferase	UDPglucuronate + oestriol = UDP + oestriol 16α-mono-D-glucuronoside
2.4.1.62	UDPgalactose—ceramide galactosyltransferase	UDPgalactose + 2-acetamido-2-deoxy-D-galactosyl-(*N*-acetylneuraminyl)-D-galactosyl-D-glucosyl-*N*-acylsphingosine = UDP + D-galactosyl-2-acetamido-2-deoxy-D-galactosyl-(*N*-acetylneuraminyl)-D-galactosyl-D-glucosyl-*N*-acylsphingosine
2.4.1.63	Linamarin synthase	UDPglucose + 2-hydroxyisobutyronitrile = UDP + linamarin
2.4.1.64	α,α-Trehalose phosphorylase	α,α-Trehalose + orthophosphate = D-glucose + β-D-glucose 1-phosphate
2.4.1.65	β-*N*-Acetylglucosaminylsaccharide fucosyltransferase	GDPfucose + 2-acetamido-2-deoxy-β-D-glucosaccharide = GDP + (1,4)-α-L-fucosyl-2-acetamido-2-deoxy-D-glucosaccharide

Number	Other Names	Basis for classification (Systematic Name)	Comments	Reference
2.4.1.54		GDPmannose:undecaprenyl-phosphate mannosyltransferase	Requires phosphatidylglycerol	1861
2.4.1.55		CDPribitol:teichoic-acid phosphoribitoltransferase		1459
2.4.1.56		UDP-2-acetamido-2-deoxy-D-glucose:lipopolysaccharide 2-acetamido-2-deoxy-D-glucosyltransferase	Transfers N-acetylglucosaminyl residues to a D-galactose residue in the partially completed lipopolysaccharide core (cf EC 2.4.1.44, 2.4.1.58 and 2.4.1.73)	2514
2.4.1.57		GDPmannose: 1-phosphatidyl-myo-inositol α-D-mannosyltransferase		380
2.4.1.58		UDPglucose:lipopolysaccharide glucosyltransferase	Transfers glucosyl residues to the backbone portion of lipopolysaccharide (cf EC 2.4.1.73)	2856
2.4.1.59	17β-Oestradiol—UDP glucuronyltransferase	UDPglucuronate:17β-oestradiol 3-glucuronosyltransferase	Oestrone can also act as acceptor	2718
2.4.1.60		CDPabequose:D-mannosyl-rhamnosyl-galactose-1-diphospholipid: abequosyltransferase		2516
2.4.1.61	Oestriol—UDP 16α-glucuronyltransferase	UDPglucuronate:oestriol 16α-D-glucuronosyltransferase	Stimulated by Mg^{2+}, Mn^{2+} and Fe^{2+}	2719
2.4.1.62		UDPgalactose:2-acetamido-2-deoxy-D-galactosyl-(N-acetylneuraminyl)-D-galactosyl-D-glucosyl-N-acylsphingosine galactosyltransferase		3787, 3789
2.4.1.63		UDPglucose: 2-hydroxyisobutyronitrile β-D-glucosyltransferase	The enzyme is specific for UDPglucose, acetone, butanone and 3-pentanone cyanohydrins	1167
2.4.1.64		α,α-Trehalose:orthophosphate β-D-glucosyltransferase		252
2.4.1.65		GDPfucose:β-2-acetamido-2-deoxy-D-glucosaccharide 4-α-L-fucosyltransferase	The product possesses Le[a] activity	1504

Number	Recommended Name	Reaction
2.4.1.66	UDPglucose—collagen glucosyltransferase	UDPglucose + 5-hydroxylysine-collagen = UDP + O-D-glucosyl-5-hydroxylysine-collagen
2.4.1.67	Galactinol—raffinose galactosyltransferase	1-O-α-D-Galactosyl-myo-inositol + raffinose = myo-inositol + stachyose
2.4.1.68	GDPfucose—glycoprotein fucosyltransferase	GDPfucose + glycoprotein = GDP + fucosyl-glycoprotein
2.4.1.69	GDPfucose—lactose fucosyltransferase	GDPfucose + lactose = GDP + fucosyllactose
2.4.1.70	UDPacetylglucosamine—poly (ribitol-phosphate) acetylglucosaminyltransferase	UDP-2-acetamido-2-deoxy-D-glucose + poly-(ribitol phosphate) = UDP + 2-acetamido-2-deoxy-D-glucosyl-poly (ribitol phosphate)
2.4.1.71	UDPglucose—arylamine glucosyltransferase	UDPglucose + an arylamine = UDP + an N-D-glucosylarylamine
[2.4.1.72	*Transferred entry: now EC 2.4.2.24, 1,4-β-D-Xylan synthase]*	
2.4.1.73	UDPglucose—lipopolysaccharide glucosyltransferase II	UDPglucose + lipopolysaccharide = UDP + D-glucosyl-lipopolysaccharide
2.4.1.74	UDPgalactose—mucopolysaccharide galactosyltransferase	UDPgalactose + mucopolysaccharide = UDP + D-galactosyl-mucopolysaccharide
2.4.1.75	UDPgalacturonosyltransferase	UDPgalacturonate + acceptor = UDP + acceptor β-galacturonide
2.4.1.76	UDPglucuronate—bilirubin glucuronosyltransferase	UDPglucuronate + bilirubin = UDP + bilirubin-glucuronoside
2.4.1.77	UDPglucuronate bilirubin-glucuronoside glucuronosyltransferase	UDPglucuronate + bilirubin-glucuronoside = UDP + bilirubin bisglucuronoside
2.4.1.78	UDPglucose phosphopolyprenol glucosyltransferase	UDPglucose + polyprenol phosphate = UDP + polyprenolphosphate-glucose

Number	Other Names	Basis for classification (Systematic Name)	Comments	Reference
2.4.1.66		UDPglucose:5-hydroxylysine-collagen glucosyltransferase	Requires Mn^{2+}	250, 252, 321 348, 350, 440
2.4.1.67		1-O-α-D-Galactosyl-*myo*-inositol:raffinose galactosyltransferase		3371
2.4.1.68		GDPfucose:glycoprotein fucosyltransferase		351
2.4.1.69		GDPfucose:lactose fucosyltransferase		1128
2.4.1.70		UDP-2-acetamido-2-deoxy-D-glucose:poly(ribitol-phosphate) 2-acetamido-2-deoxy-glucosyltransferase		2379
2.4.1.71		UDPglucose:arylamine N-glucosyltransferase		925
2.4.1.73		UDPglucose:galactosyl-lipopolysaccharide glucosyltransferase	Transfers glucosyl residues to the D-galactosyl-D-glucosyl side-chains in the partially completed core of lipopolysaccharide (*cf* EC 2.4.1.44 and 2.4.1.58)	791
2.4.1.74		UDPgalactose:muco-polysaccharide galactosyltransferase		3296
2.4.1.75	*p*-Nitrophenol conjugating enzyme	UDPgalacturonate β-galacturonosyltransferase (acceptor unspecific)		3549
2.4.1.76		UDPglucuronate:bilirubin-glucuronosyltransferase		1500
2.4.1.77		UDPglucuronate:bilirubin-glucuronoside glucuronosyltransferase		1500
2.4.1.78		UDPglucose:phosphopolyprenol glucosyltransferase	Ficaprenol is the best substrate; other polyprenols act, more slowly	1496

Number	Recommended Name	Reaction
2.4.1.79	UDPacetylgalactosamine—galactosyl-galactosyl-glucosylceramide β-N-acetyl-D-galactosaminyltransferase	UDP-2-acetamido-2-deoxy-D-galactose + D-galactosyl-(1,4)-D-galactosyl-(1,4)-D-glucosylceramide = UDP + 2-acetamido-2-deoxy-D-galactosyl-(1,3)-D-galactosyl-(1,4)-D-galactosyl-(1,4)-D-glucosylceramide
2.4.1.80	UDPglucose—ceramide glucosyltransferase	UDPglucose + N-acylsphingosine = UDP + D-glucosyl-N-acylsphingosine
2.4.1.81	UDPglucose—luteolin β-D-glucosyltransferase	UDPglucose + 5,7,3',4'-tetrahydroxyflavone = UDP + 7-O-β-D-glucosyl-5,7,3',4'-tetrahydroxyflavone
2.4.1.82	Galactinol—sucrose galactosyltransferase	1-O-α-D-Galactosyl-*myo*-inositol + sucrose = *myo*-inositol + raffinose
2.4.1.83	GDPmannose dolicholphosphate mannosyltransferase	GDPmannose + dolichol phosphate = GDP + dolichol phosphate mannose
2.4.1.84	UDPglucuronate—1,2-diacylglycerol glucuronosyltransferase	UDPglucuronate + 1,2-diacylglycerol = UDP + 1,2-diacylglycerol 3-D-glucuronoside
2.4.1.85	Cyanohydrin glucosyltransferase	UDPglucose + (S)-4-hydroxymandelonitrile = UDP + (S)-4-hydroxymandelonitrile β-D-glucoside
2.4.1.86	UDPgalactose—glucosaminyl-galactosyl-glucosylceramide β-D-galactosyltransferase	UDPgalactose + 2-acetamido-2-deoxy-D-glucosyl-(1,3)-D-galactosyl-(1,4)-D-glucosylceramide = UDP + D-galactosyl-2-acetamido-2-deoxy-D-glucosyl-(1,3)-D-galactosyl-(1,4)-D-glucosylceramide
2.4.1.87	UDPgalactose—galactosyl-glucosaminyl-galactosyl-glucosylceramide α-D-galactosyltransferase	UDPgalactose + D-galactosyl-(1,4)-2-acetamido-2-deoxy-D-glucosyl-(1,3)-D-galactosyl-(1,4)-D-glucosylceramide = UDP + D-galactosyl-D-galactosyl-(1,4)-2-acetamido-2-deoxy-D-glucosyl-(1,3)-D-galactosyl-(1,4)-D-glucosylceramide
2.4.1.88	UDPacetylgalactosamine—globoside α-N-acetyl-D-galactosaminyltransferase	UDP-2-acetamido-2-deoxy-D-galactose + 2-acetamido-2-deoxy-D-galactosyl-(1,3)-D-galactosyl-(1,4)-D-galactosyl-(1,4)-D-glucosylceramide = UDP + 2-acetamido-2-deoxy-D-galactosyl-2-acetamido-2-deoxy-D-

Number	Other Names	Basis for classification (Systematic Name)	Comments	Reference
2.4.1.79		UDP-2-acetamido-2-deoxy-D-galactose:D-galactosyl-(1,4)-D-galactosyl-(1,4)-D-glucosylceramide β-N-acetamidodeoxy-D-galactosyltransferase		1455, 523
2.4.1.80		UDPglucose:N-acylsphingosine glucosyltransferase	Sphingosine and dihydrosphingosine can also act as acceptor; CDPglucose can act as donor	226
2.4.1.81	UDPglucose—apigenin β-glucosyltransferase	UDPglucose:5,7,3',4'-tetrahydroxyflavone β-D-glucosyltransferase	A number of flavones, flavanones and flavonols can function as acceptors	3300
2.4.1.82		1-O-α-D-Galactosyl-myo-inositol:sucrose 6-galactosyltransferase	4-Nitrophenyl-α-D-galactopyranoside can also act as donor. Enzyme also catalyses an exchange reaction between raffinose and sucrose	1862
2.4.1.83		GDPmannose:dolichol-phosphate mannosyltransferase		383
2.4.1.84		UDPglucuronate:1,2-diacylglycerol 3-glucuronosyltransferase		3218
2.4.1.85		UDPglucose:(S)-4-hydroxymandelonitrile β-D-glucosyltransferase	Also acts on (S)-mandelonitrile	2733
2.4.1.86		UDPgalactose:2-acetamido-2-deoxy-D-glucosyl-(1,3)-D-galactosyl-(1,4)-D-glucosylceramide β-D-galactosyltransferase		223
2.4.1.87		UDPgalactose:D-galactosyl-(1,4)-2-acetamido-2-deoxy-D-glucosyl-(1,3)-D-galactosyl-galactosyl-(1,4)-D-glucosylceramide α-D-galactosyltransferase		224
2.4.1.88		UDP-2-acetamido-2-deoxy-D-galactose:2-acetamido-2-deoxy-D-galactosyl-(1,3)-D-galactosyl-(1,4)-D-galactosyl-(1,4)-D-glucosylceramide α-N-		1679

Number	Recommended Name	Reaction
		galactosyl-(1,3)-D-galactosyl-(1,4)-D-galactosyl-(1,4)-D-glucosylceramide
2.4.1.89	GDPfucose—galactosyl-glucosaminyl-galactosyl-glucosylceramide α-L-fucosyltransferase	GDPfucose + D-galactosyl-(1,4)-2-acetamido-2-deoxy-D-glucosyl-(1,3)-D-galactosyl-(1,4)-D-glucosylceramide = GDP + fucosyl-D-galactosyl-(1,4)-2-acetamido-2-deoxy-D-glucosyl-(1,3)-D-galactosyl-(1,4)-D-glucosylceramide
2.4.1.90	*N*-Acetyllactosamine synthase	UDPgalactose + 2-acetamido-2-deoxy-D-glucose = UDP + 4-*O*-β-D-galactosyl-2-acetamido-2-deoxy-D-glucose
2.4.1.91	UDPglucose—flavonol glucosyltransferase	UDPglucose + a flavonol = UDP + flavonol 3-*O*-glucoside
2.4.1.92	UDPacetylgalactosamine—(*N*-acetylneuraminyl)-D-galactosyl-D-glucosylceramide acetylgalactosaminyltransferase	UDP-2-acetamido-2-deoxy-D-galactose + (*N*-acetylneuraminyl)-D-galactosyl-D-glucosylceramide = UDP + 2-acetamido-2-deoxy-D-galactosyl-(*N*-acetylneuraminyl)-D-galactosyl-D-glucosylceramide
2.4.1.93	Inulin fructotransferase (depolymerizing)	Transfers a terminal fructosyl-fructofuranosyl group to the terminal 3-position, forming a cyclic anhydride
2.4.1.94	UDPacetylglucosamine-protein acetylglucosaminyltransferase	UDP-2-acetamido-2-deoxy-D-glucose + protein = UDP + 4-*N*-(2-acetamido-2-deoxy-β-D-glucosyl)-protein
2.4.1.95	Bilirubin-glucuronoside glucuronosyltransferase	2 Bilirubin glucuronoside = bilirubin + bilirubin bisglucuronoside
2.4.1.96	UDPgalactose-*sn*-glycerol-3-phosphate galactosyltransferase	UDPgalactose + *sn*-glycerol 3-phosphate = UDP + α-D-galactosyl-(1,1′)-*sn*-glycerol 3-phosphate
2.4.1.97	1,3-β-D-Glucan phosphorylase	$(1,3\text{-}\beta\text{-D-Glucosyl})_n$ + orthophosphate = $(1,3\text{-}\beta\text{-D-glucosyl})_{n-1}$ + α-D-glucose 1-phosphate

Number	Other Names	Basis for classification (Systematic Name)	Comments	Reference
		acetamidodeoxy-D-galactosyltransferase		
2.4.1.89		GDPfucose:D-galactosyl-(1,4)-2-acetamido-2-deoxy-D-glucosyl-(1,3)-D-galactosyl-(1,4)-D-glucosylceramide α-L-fucosyltransferase		225
2.4.1.90		UDPgalactose:2-acetamido-2-deoxy-D-glucose 4-β-D-galactosyltransferase		1271
2.4.1.91		UDPglucose:flavonol 3-O-glucosyltransferase	Acts on a variety of flavonols, including quercetin and quercetin 7-O-glucoside	3299
2.4.1.92		UDP-2-acetamido-2-deoxy-D-galactose: (N-acetylneuraminyl)-D-galactosyl-D-glucosylceramide acetamidodeoxygalactosyltransferase		713
2.4.1.93	Inulase II	Inulin fructosyl-β-1,2-fructofuranosyltransferase (cyclizing)	Inulin is depolymerized to di-D-fructofuranose 1,2′-2,3′-dianhydride	3500
2.4.1.94		UDP-2-acetamido-2-deoxy-D-glucose:protein β-N-acetamidodeoxy-D-glucosyltransferase	The acceptor is the asparagine residue in a sequence of the form Asn-X-Thr or Asn-X-Ser	1675, 1673, 1674
2.4.1.95	Bilirubin monoglucuronide transglucuronidase	Bilirubin-glucuronoside:bilirubin-glucuronoside glucuronosyltransferase		1501
2.4.1.96	FP synthase	UDPgalactose:sn-glycerol-3-phosphate α-D-galactosyltransferase	The product is dephosphorylated to isofloridoside, which is involved in osmoregulation	1621, 1620
2.4.1.97	Laminarin phosphorylase	1,3-β-D-Glucan: orthophosphate glucosyltransferase	Acts on a range of β-1,3-oligoglucans, and on glucans of laminarin type. Different from EC 2.4.1.30 and EC 2.4.1.31	46

Number	Recommended Name	Reaction

2.4.2 PENTOSYL TRANSFERASES

2.4.2.1	Purine-nucleoside phosphorylase	Purine nucleoside + orthophosphate = purine + α-D-ribose 1-phosphate
2.4.2.2	Pyrimidine-nucleoside phosphorylase	Pyrimidine nucleoside + orthophosphate = pyrimidine + α-D-ribose 1-phosphate
2.4.2.3	Uridine phosphorylase	Uridine + orthophosphate = uracil + α-D-ribose 1-phosphate
2.4.2.4	Thymidine phosphorylase	Thymidine + orthophosphate = thymine + 2-deoxy-D-ribose 1-phosphate
2.4.2.5	Nucleoside ribosyltransferase	D-Ribosyl-R + R' = D-ribosyl-R' + R
2.4.2.6	Nucleoside deoxyribosyltransferase	2-Deoxy-D-ribosyl-R + R' = 2-deoxy-D-ribosyl-R' + R
2.4.2.7	Adenine phosphoribosyltransferase	AMP + pyrophosphate = adenine + 5-phospho-α-D-ribose 1-diphosphate
2.4.2.8	Hypoxanthine phosphoribosyltransferase	IMP + pyrophosphate = hypoxanthine + 5-phospho-α-D-ribose 1-diphosphate
2.4.2.9	Uracil phosphoribosyltransferase	UMP + pyrophosphate = uracil + 5-phospho-α-D-ribose 1-diphosphate
2.4.2.10	Orotate phosphoribosyltransferase	Orotidine 5'-phosphate + pyrophosphate = orotate + 5-phospho-α-D-ribose 1-diphosphate
2.4.2.11	Nicotinate phosphoribosyltransferase	Nicotinate D-ribonucleotide + pyrophosphate = nicotinate + 5-phospho-α-D-ribose 1-diphosphate
2.4.2.12	Nicotinamide phosphoribosyltransferase	Nicotinamide D-ribonucleotide + pyrophosphate = nicotinamide + 5-phospho-α-D-ribose 1-diphosphate

[2.4.2.13 Transferred entry: now EC 2.5.1.6, Methionine adenosyltransferase]

Number	Other Names	Basis for classification (Systematic Name)	Comments	Reference
2.4.2.1	Inosine phosphorylase	Purine-nucleoside:orthophosphate ribosyltransferase	Specificity not completely determined. Can also catalyse ribosyltransferase reactions of the type catalysed by EC 2.4.2.5	26, 933, 1282, 1570, 3471
2.4.2.2		Pyrimidine-nucleoside:orthophosphate ribosyltransferase		933, 2923
2.4.2.3	Pyrimidine phosphorylase	Uridine:orthophosphate ribosyltransferase		455, 2538, 2633
2.4.2.4	Pyrimidine phosphorylase	Thymidine:orthophosphate deoxyribosyltransferase	The enzyme in some tissues also catalyses deoxyribosyltransferase reactions of the type catalysed by EC 2.4.2.6	934, 3848
2.4.2.5		Nucleoside:purine (pyrimidine) ribosyltransferase	R and R' represent various purines and pyrimidines	1729
2.4.2.6	*trans-N*-Glucosidase	Nucleoside:purine (pyrimidine) deoxyribosyltransferase	R and R' represent various purines and pyrimidines	1573, 2035, 2859
2.4.2.7	AMP pyrophosphorylase, Transphosphoribosidase	AMP:pyrophosphate phosphoribosyltransferase	5-Amino-4-imidazolecarboxamide can replace adenine	893, 1769, 2015
2.4.2.8	IMP pyrophosphorylase, Transphosphoribosidase	IMP:pyrophosphate phosphoribosyltransferase	Guanine and 6-mercaptopurine can replace hypoxanthine	1769, 2015, 2771, 892
2.4.2.9	UMP pyrophosphorylase	UMP:pyrophosphate phosphoribosyltransferase		605, 892
2.4.2.10	Orotidylic acid phosphorylase, Orotidine-5'-phosphate pyrophosphorylase	Orotidine-5'-phosphate:pyrophosphate phosphoribosyltransferase		1768, 1964, 892
2.4.2.11		Nicotinatenucleotide:pyrophosphate phosphoribosyltransferase		1446, 1447, 1782
2.4.2.12	NMN pyrophosphorylase	Nicotinamidenucleotide:pyrophosphate phosphoribosyltransferase		2657

Number	Recommended Name	Reaction
2.4.2.14	Amidophosphoribosyltransferase	5-Phospho-β-D-ribosylamine + pyrophosphate + L-glutamate = L-glutamine + 5-phospho-α-D-ribose 1-diphosphate + H_2O
2.4.2.15	Guanosine phosphorylase	Guanosine + orthophosphate = guanine + D-ribose 1-phosphate
2.4.2.16	Urateribonucleotide phosphorylase	Urate D-ribonucleotide + orthophosphate = urate + D-ribose 1-phosphate
2.4.2.17	ATP phosphoribosyltransferase	1-(5'-Phospho-D-ribosyl)-ATP + pyrophosphate = ATP + 5-phospho-α-D-ribose 1-diphosphate
2.4.2.18	Anthranilate phosphoribosyltransferase	N-(5'-Phospho-D-ribosyl)-anthranilate + pyrophosphate = anthranilate + 5-phospho-α-D-ribose 1-diphosphate
2.4.2.19	Nicotinatemononucleotide pyrophosphorylase (carboxylating)	Nicotinate D-ribonucleotide + pyrophosphate + CO_2 = quinolinate + 5-phospho-α-D-ribose 1-diphosphate
2.4.2.20	Dioxotetrahydropyrimidine phosphoribosyltransferase	A 2,4-dioxotetrahydropyrimidine D-ribonucleotide + pyrophosphate = a 2,4-dioxotetrahydropyrimidine + 5-phospho-α-D-ribose 1-diphosphate
2.4.2.21	Nicotinatenucleotide—dimethylbenzimidazole phosphoribosyltransferase	β-Nicotinate D-ribonucleotide + dimethylbenzimidazole = nicotinate + 1-α-D-ribosyl-5,6-dimethylbenzimidazole 5'-phosphate
2.4.2.22	Xanthine phosphoribosyltransferase	5-Phospho-α-D-ribose 1-diphosphate + xanthine = (9-D-ribosylxanthine)-5'-phosphate + pyrophosphate
2.4.2.23	Deoxyuridine phosphorylase	Deoxyuridine + orthophosphate = uracil + deoxy-D-ribose 1-phosphate
2.4.2.24	1,4-β-D-Xylan synthase	UDPxylose + (1,4-β-D-xylan)$_n$ = UDP + (1,4-β-D-xylan)$_{n+1}$
2.4.2.25	UDPapiose—flavone apiosyltransferase	UDPapiose + 7-O-β-D-glucosyl-5,7,4'-trihydroxyflavone = UDP + 7-O-(β-D-apiofuranosyl-1,2-β-D-glucosyl)-5,7,4'-trihydroxyflavone

Number	Other Names	Basis for classification (Systematic Name)	Comments	Reference
2.4.2.14	Phosphoribosyl-diphosphate 5-amidotransferase, Glutamine phosphoribosylpyrophosphate amidotransferase	5-Phosphoribosylamine:pyrophosphate phosphoribosyltransferase (glutamate-amidating)		482, 1077, 1203
2.4.2.15		Guanosine:orthophosphate ribosyltransferase	Also acts on deoxyguanosine	3742
2.4.2.16		Urateribonucleotide:orthophosphate ribosyltransferase		1888
2.4.2.17	Phosphoribosyl-ATP pyrophosphorylase	1-(5′-Phosphoribosyl)-ATP:pyrophosphate phosphoribosyltransferase		70, 2098, 3567
2.4.2.18	Phosphoribosyl-anthranilate pyrophosphorylase	N-(5′-Phosphoribosyl)-anthranilate:pyrophosphate phosphoribosyltransferase	The native enzyme in the enteric bacteria exists as a complex with anthranilate synthase (EC 4.1.3.27)	1464, 3637
2.4.2.19		Nicotinatenucleotide:pyrophosphate phosphoribosyltransferase (carboxylating)		1014, 2537
2.4.2.20	Dioxotetrahydro-pyrimidine ribonucleotide pyrophosphorylase	2,4-Dioxotetrahydropyrimidine nucleotide:pyrophosphate phosphoribosyltransferase	Acts (in the reverse direction) on uracil and other pyrimidines and pteridines containing a 2,4-diketo structure	1225
2.4.2.21		Nicotinatenucleotide:dimethylbenzimidazole phosphoribosyltransferase	Also acts on benzimidazole and the clostridial enzyme acts on adenine to form 7-α-D-ribosyladenine 5′-phosphate	938, 939, 981
2.4.2.22		5-Phospho-α-D-ribose-1-diphosphate:xanthine phosphoribosyltransferase		1801
2.4.2.23		Deoxyuridine:orthophosphate deoxyribosyltransferase		539, 3741
2.4.2.24		UDPxylose:1,4-β-D-xylan 4-β-D-xylosyltransferase	Formerly EC 2.4.1.72	179
2.4.2.25		UDPapiose:7-O-β-D-glucosyl-5,7,4′-trihydroxyflavone apiofuranosyltransferase	7-O-β-D-Glucosides of a number of flavonoids and of 4-substituted phenols can act as acceptor	2511

Number	Recommended Name	Reaction
2.4.2.26	UDPxylose—protein xylosyltransferase	Transfers a D-xylosyl residue from UDPxylose to the serine hydroxyl group of an acceptor protein substrate

2.4.99   TRANSFERRING OTHER GLYCOSYL GROUPS

2.4.99.1	CMP-N-acetylneuraminate—galactosyl-glycoprotein sialyltransferase	CMP-N-acetylneuraminate + D-galactosyl-glycoprotein = CMP + N-acetylneuraminyl-D-galactosyl-glycoprotein
2.4.99.2	CMP-N-acetylneuraminate—monosialoganglioside sialyltransferase	CMP-N-acetylneuraminate + D-galactosyl-2-acetamido-2-deoxy-D-galactosyl-(N-acetylneuraminyl)-D-galactosyl-D-glucosylceramide = CMP + N-acetylneuraminyl-D-galactosyl-2-acetamido-2-deoxy-D-galactosyl-(N-acetylneuraminyl)-D-galactosyl-D-glucosylceramide

2.5   TRANSFERRING ALKYL OR ARYL GROUPS, OTHER THAN METHYL GROUPS

This is a somewhat heterogeneous group of enzymes transferring alkyl or related groups either substituted or unsubstituted. There is no subdivision as yet in this sub-class.

2.5.1.1	Dimethylallyltransferase	Dimethylallyl diphosphate + isopentenyl diphosphate = pyrophosphate + geranyl diphosphate
2.5.1.2	Thiamin pyridinylase	Thiamin + pyridine = heteropyrithiamin + 4-methyl-5-(2′-hydroxyethyl)-thiazole
2.5.1.3	Thiamin phosphate pyrophosphorylase	2-Methyl-4-amino-5-hydroxymethyl-pyrimidine diphosphate + 4-methyl-5-(2′-phosphoethyl)-thiazole = pyrophosphate + thiamin monophosphate
2.5.1.4	Adenosylmethionine cyclotransferase	S-Adenosyl-L-methionine = 5′-(methylthio)-adenosine + 2-amino-γ-butyrolactone
2.5.1.5	Galactose-6-sulphurylase	Eliminates sulphur from the galactose 6-sulphate residues of porphyran producing 3,6-anhydrogalactose residues
2.5.1.6	Methionine adenosyltransferase	ATP + L-methionine + H_2O = orthophosphate + pyrophosphate + S-adenosyl-L-methionine
2.5.1.7	Enoylpyruvate transferase	Phosphoenolpyruvate + UDP-2-acetamido-2-deoxy-D-glucose = orthophosphate + UDP-2-acetamido-2-deoxy-3-enoylpyruvoylglucose

Number	Other Names	Basis for classification (Systematic Name)	Comments	Reference
2.4.2.26		UDPxylose:protein xylosyltransferase	Involved in the biosynthesis of chondromucoprotein	3240
2.4.99.1	Sialyltransferase	CMP-*N*-acetylneuraminate:D-galactosyl-glycoprotein *N*-acetylneuraminyltransferase		1305, 3166, 1688
2.4.99.2		CMP-*N*-acetylneuraminate:D-galactosyl-2-acetamido-2-deoxy-D-galactosyl-(*N*-acetylneuraminyl)-D-galactosyl-D-glucosylceramide *N*-acetylneuraminyltransferase		3783
2.5.1.1	Farnesylpyrophosphate synthetase, Prenyltransferase	Dimethylallyldiphosphate: isopentenyldiphosphate dimethylallyltransferase	Also transfers geranyl and farnesyl residues	1348, 2018
2.5.1.2	Pyrimidine transferase, Thiaminase I	Thiamin:base 2-methyl-4-aminopyrimidine-5-methenyltransferase	Various bases and thiol compounds can act instead of pyridine	970, 1659, 3713
2.5.1.3		2-Methyl-4-amino-5-hydroxymethyl-pyrimidinediphosphate:4-methyl-5-(2′-phosphocthyl)-thiazole 2-methyl-4-aminopyrimidine-5-methenyltransferase		449, 1895
2.5.1.4		*S*-Adenosyl-L-methionine alkyltransferase (cyclizing)		2307, 2308
2.5.1.5	Porphyran sulphatase, Galactose-6-sulphatase	Galactose-6-sulphate alkyltransferase (cyclizing)		2741, 2742
2.5.1.6		ATP:L-methionine *S*-adenosyltransferase	Formerly EC 2.4.2.13	462, 464, 2309
2.5.1.7		Phosphoenolpyruvate:UDP-2-acetamido-2-deoxy-D-glucose 2-enoyl-1-carboxyethyltransferase		1143

Number	Recommended Name	Reaction
2.5.1.8	tRNA isopentenyltransferase	2-Isopentenyl diphosphate + tRNA = pyrophosphate + tRNA containing 6-(2-isopentenyl) adenosine
2.5.1.9	Riboflavin synthase	2 6,7-Dimethyl-8-(1'-D-ribityl) lumazine = riboflavin + 4 (1'-D-ribitylamino)-5-amino-2,6-dihydroxypyrimidine
2.5.1.10	Geranyltransferase	Geranyl diphosphate + isopentenyl diphosphate = pyrophosphate + farnesyl diphosphate
2.5.1.11	Terpenoid-allyltransferase	Allylic terpene diphosphate (n isoprene units) + isopentenyl diphosphate = pyrophosphate + terpene diphosphate (n + 1 isoprene units)

[2.5.1.12 Deleted entry: Glutathione S-alkyltransferase. Now included with EC 2.5.1.18]

[2.5.1.13 Deleted entry: Glutathione S-aryltransferase. Now included with EC 2.5.1.18]

[2.5.1.14 Deleted entry: Glutathione S-aralkyltransferase. Now included with EC 2.5.1.18]

Number	Recommended Name	Reaction
2.5.1.15	Dihydropteroate synthase	2-Amino-4-hydroxy-6-hydroxymethyl-7,8-dihydropteridine diphosphate + 4-aminobenzoate = pyrophosphate + dihydropteroate
2.5.1.16	Aminopropyltransferase	(5'-Deoxy-5'-adenosyl) (3-aminopropyl) methylsulphonium salt + putrescine = 5'-methylthioadenosine + spermidine
2.5.1.17	Cob (I) alamin adenosyltransferase	ATP + cob (I) alamin + H_2O = orthophosphate + pyrophosphate + adenosylcobalamin
2.5.1.18	Glutathione transferase	RX + glutathione = HX + R-S-G

Number	Other Names	Basis for classification (Systematic Name)	Comments	Reference
2.5.1.8		2-Isopentenyldiphosphate: tRNA 2-isopentenyltransferase		1713, 2845
2.5.1.9		6,7-Dimethyl-8-(1′-D-ribityl) lumazine:6,7-dimethyl-8-(1′-D-ribityl) lumazine 2,3-butanediyltransferase	Contains bound riboflavin	2619, 3572, 2621
2.5.1.10		Geranyldiphosphate: isopentenydiphosphate geranyltransferase		255
2.5.1.11		Allylic-terpene-diphosphate:isopentenyldiphosphate terpenoid-allyltransferase	Catalyses the elongation of terpenoid allyl diphosphates by the addition of isoprene units; the main products are C-35 and C-40 terpene diphosphates	55
2.5.1.15	Dihydropteroate pyrophosphorylase	2-Amino-4-hydroxy-6-hydroxymethyl-7,8-dihydropteridine-diphosphate:4-aminobenzoate 2-amino-4-hydroxydihydropteridine-6-methenyltransferase		2786, 3055
2.5.1.16		(5′-Deoxy-5′-adenosyl) (3-aminopropyl) methylsulphonium-salt:putrescine 3-aminopropyltransferase		3326, 3330, 357, 1183
2.5.1.17	Aquacob (I) alamin adenosyltransferase	ATP:cob (I) alamin Co-β-adenosyltransferase	Requires Mn^{2+}	3557
2.5.1.18	Glutathione S-alkyltransferase, Glutathione S-aryltransferase, S-(Hydroxyalkyl) glutathione lyase, Glutathione S-aralkyltransferase	RX:glutathione R-transferase	A group of enzymes of broad specificity. R may be an aliphatic, aromatic or heterocyclic radical; X may be a sulphate, nitrite or halide radical. Also catalyses the addition of aliphatic epoxides and arene oxides to glutathione; the reduction	2534, 1156, 1155

Number	Recommended Name	Reaction

2.5.1.19   3-*enol*Pyruvoylshikimate-5-phosphate synthase   Phosphoenolpyruvate + shikimate 5-phosphate = orthophosphate + 3-enolpyruvoylshikimate 5-phosphate

2.5.1.20   Rubber allyltransferase $(cis\text{-}1,4\text{-Isoprene})_n$ diphosphate + isopentenyl diphosphate = pyrophosphate + $(cis\text{-}1,4\text{-isoprene})_{n+1}$ diphosphate

2.5.1.21   Farnesyltransferase   2 Farnesyl diphosphate = pyrophosphate + presqualene diphosphate

2.6   TRANSFERRING NITROGENOUS GROUPS

These are mainly the enzymes transferring amino groups (sub-sub-group 2.6.1) from a donor, generally an amino acid, to an acceptor, generally a 2-oxo acid. It should be kept in mind that transamination by this reaction also involves an oxidoreduction; the donor is oxidized to a ketone, while the acceptor is reduced. Nevertheless, since the transfer of the amino group is the most prominent feature of this reaction, these enzymes have been classified as aminotransferases rather than oxidoreductases, ('transaminating'). Most of these enzymes are pyridoxal-phosphate-proteins.

2.6.1   AMINOTRANSFERASES

'Aminotransferase' may be replaced by 'transaminase'.

2.6.1.1   Aspartate aminotransferase   L-Aspartate + 2-oxoglutarate = oxaloacetate + L-glutamate

2.6.1.2   Alanine aminotransferase   L-Alanine + 2-oxoglutarate = pyruvate + L-glutamate

2.6.1.3   Cysteine aminotransferase   L-Cysteine + 2-oxoglutarate = mercaptopyruvate + L-glutamate

2.6.1.4   Glycine aminotransferase   Glycine + 2-oxoglutarate = glyoxylate + L-glutamate

Number	Other Names	Basis for classification (Systematic Name)	Comments	Reference
			of polyol nitrate by glutathione to polyol and nitrite; certain isomerization reactions and disulphide interchange	
2.5.1.19		Phosphoenolpyruvate:shikimate-5-phosphate enolpyruvoyltransferase		2203
2.5.1.20	Rubber transferase	$(cis$-1,4-Isoprene$)_n$-diphosphate:isopentenyl-diphosphate cis-1,4-isoprenyltransferase	Rubber particles act as acceptor	99, 2172
2.5.1.21	Presqualene synthase	Farnesyl-diphosphate:farnesyl-diphosphate farnesyltransferase	The polymeric form of the enzyme also catalyses the reduction of presqualene diphosphate by NADPH to squalene	2673
2.6.1.1	Glutamic-oxaloacetic transaminase, Glutamic-aspartic transaminase, Transaminase A	L-Aspartate:2-oxoglutarate aminotransferase	A pyridoxal-phosphate-protein. Acts also on L-tyrosine, L-phenylalanine and L-tryptophan	198, 287, 1512, 2972, 1278, 2132, 3060, 916
2.6.1.2	Glutamic-pyruvic transaminase, Glutamic-alanine transaminase	L-Alanine:2-oxoglutarate aminotransferase	A pyridoxal-phosphate-protein. 2-Aminobutyrate acts slowly instead of alanine	766, 1103, 3698
2.6.1.3		L-Cysteine:2-oxoglutarate aminotransferase	A pyridoxal-phosphate-protein	506
2.6.1.4		Glycine:2-oxoglutarate aminotransferase	A pyridoxal-phosphate-protein	2918, 3698, 2345

Number	Recommended Name	Reaction
2.6.1.5	Tyrosine aminotransferase	L-Tyrosine + 2-oxoglutarate = 4-hydroxyphenylpyruvate + L-glutamate
2.6.1.6	Leucine aminotransferase	L-Leucine + 2-oxoglutarate = 2-oxoisocaproate + L-glutamate
2.6.1.7	Kynurenine aminotransferase	L-Kynurenine + 2-oxoglutarate = 2-aminobenzoylpyruvate + L-glutamate
2.6.1.8	Diamino-acid aminotransferase	2,5-Diaminovalerate + .2-oxoglutarate = 5-amino-2-oxovalerate + L-glutamate
2.6.1.9	Histidinol-phosphate aminotransferase	L-Histidinol phosphate + 2-oxoglutarate = imidazoleacetol phosphate + L-glutamate

[2.6.1.10 Deleted entry: now included with EC 2.6.1.21, D-Alanine aminotransferase]

2.6.1.11	Acetylornithine aminotransferase	N^2-Acetyl-L-ornithine + 2-oxoglutarate = N-acetyl-L-glutamate γ-semialdehyde + L-glutamate
2.6.1.12	Alanine—oxo-acid aminotransferase	L-Alanine + a 2-oxo acid = pyruvate + an L-amino acid
2.6.1.13	Ornithine—oxo-acid aminotransferase	L-Ornithine + a 2-oxo acid = L-glutamate γ-semialdehyde + an L-amino acid
2.6.1.14	Asparagine—oxo-acid aminotransferase	L-Asparagine + a 2-oxo acid = 2-oxosuccinamate + an amino acid
2.6.1.15	Glutamine—oxo-acid aminotransferase	L-Glutamine + a 2-oxo acid = 2-oxoglutaramate + an amino acid

[2.6.1.16 Transferred entry: now EC 5.3.1.19, Glucosaminephosphate isomerase (glutamine-forming)]

2.6.1.17	Succinyl-diaminopimelate aminotransferase	N-Succinyl-L-diaminopimelate + 2-oxoglutarate = N-succinyl-2-L-amino-6-oxopimelate + L-glutamate

Number	Other Names	Basis for classification (Systematic Name)	Comments	Reference
2.6.1.5		L-Tyrosine:2-oxoglutarate aminotransferase	A pyridoxal-phosphate-protein. Phenylalanine can act instead of tyrosine. The mitochondrial enzyme may be identical with EC 2.6.1.1	456, 1658, 2863, 3006, 1481, 2223
2.6.1.6		L-Leucine:2-oxoglutarate aminotransferase	A pyridoxal-phosphate-protein	35, 548, 2863, 3363
2.6.1.7		L-Kynurenine:2-oxoglutarate aminotransferase (cyclizing)	A pyridoxal-phosphate-protein. Also acts on 3-hydroxykynurenine	1489, 2106
2.6.1.8		2,5-Diaminovalerate:2-oxoglutarate aminotransferase	A pyridoxal-phosphate-protein. 2,5-Diaminoglutarate can act instead of diaminovalerate	2806
2.6.1.9	Imidazolylacetol-phosphate aminotransferase	L-Histidinol-phosphate:2-oxoglutarate aminotransferase	A pyridoxal-phosphate-protein	69, 2099
2.6.1.11		N^2-Acetyl-L-ornithine:2-oxoglutarate aminotransferase	A pyridoxal-phosphate-protein	3559, 43
2.6.1.12		L-Alanine:2-oxo-acid aminotransferase	A pyridoxal-phosphate-protein	62, 2864, 2902, 3698, 3414
2.6.1.13		L-Ornithine:2-oxo-acid aminotransferase	A pyridoxal-phosphate-protein	880, 2182, 2585, 2668, 1605, 3256
2.6.1.14		L-Asparagine:2-oxo-acid aminotransferase	A pyridoxal-phosphate-protein	2186
2.6.1.15	Glutaminase II	L-Glutamine:2-oxo-acid aminotransferase	A pyridoxal-phosphate-protein. L-Glutamine can be replaced by a few closely related compounds	2183
2.6.1.17		N-Succinyl-L-2,6-diaminopimelate:2-oxoglutarate aminotransferase	A pyridoxal-phosphate-protein	2587

Number	Recommended Name	Reaction
2.6.1.18	β-Alanine—pyruvate aminotransferase	L-Alanine + malonate semialdehyde = pyruvate + β-alanine
2.6.1.19	Aminobutyrate aminotransferase	4-Aminobutyrate + 2-oxoglutarate = succinate semialdehyde + L-glutamate

[2.6.1.20 Deleted entry: Tyrosine-pyruvate aminotransferase]

2.6.1.21	D-Alanine aminotransferase	D-Alanine + 2-oxoglutarate = pyruvate + D-glutamate
2.6.1.22	L-3-Aminoisobutyrate aminotransferase	L-3-Aminoisobutyrate + 2-oxoglutarate = methylmalonate semialdehyde + L-glutamate
2.6.1.23	4-Hydroxyglutamate transaminase	4-Hydroxy-L-glutamate + 2-oxoglutarate = 4-hydroxy-2-oxoglutarate + L-glutamate
2.6.1.24	Diiodotyrosine aminotransferase	3,5-Diiodo-L-tyrosine + 2-oxoglutarate = 3,5-diiodo-4-hydroxyphenylpyruvate + L-glutamate
2.6.1.25	Thyroxine aminotransferase	Thyroxine + 2-oxoglutarate = 3,5,3′,5′-tetraiodothyropyruvate + L-glutamate
2.6.1.26	Thyroid hormone aminotransferase	L-3,5,3′-Triiodothyronine + 2-oxoglutarate = 3,5,3′-triiodopyruvate + L-glutamate
2.6.1.27	Tryptophan aminotransferase	L-Tryptophan + 2-oxoglutarate = indolepyruvate + L-glutamate

Number	Other Names	Basis for classification (Systematic Name)	Comments	Reference
2.6.1.18		L-Alanine:malonate-semialdehyde aminotransferase	A pyridoxal-phosphate-protein	1237, 3228
2.6.1.19	β-Alanine—oxoglutarate aminotransferase	4-Aminobutyrate:2-oxoglutarate aminotransferase	Also acts on β-alanine	138, 2988
2.6.1.21	D-Aspartate aminotransferase	D-Alanine:2-oxoglutarate aminotransferase	A pyridoxal-phosphate-protein. Acts on the D-isomers of leucine, aspartate, glutamate, aminobutyrate, norvaline, and asparagine	2101, 3417, 3418, 2533
2.6.1.22		L-3-Aminoisobutyrate:2-oxoglutarate aminotransferase	Also acts on β-alanine and other ω-amino acids having carbon chains between 2 and 5	1566, 1841
2.6.1.23		4-Hydroxy-L-glutamate:2-oxoglutarate aminotransferase	Oxaloacetate can replace 2-oxoglutarate	1845
2.6.1.24		3,5-Diiodo-L-tyrosine:2-oxoglutarate aminotransferase	A pyridoxal-phosphate-protein. Also acts on 3,5-dichloro-, 3,5-dibromo- and 3-iodo-L-tyrosine	2362
2.6.1.25		Thyroxine:2-oxoglutarate aminotransferase	A pyridoxal-phosphate-protein. Also acts on triiodothyronine and 3,5-diiodo-L-tyrosine	2363
2.6.1.26	3,5-Dinitrotyrosine aminotransferase	L-3,5,3'-Triiodothyronine:2-oxoglutarate aminotransferase	A pyridoxal-phosphate-protein. Acts on monoiodotyrosine, diiodotyrosine, triiodothyronine, thyroxine and dinitrotyrosine (unlike EC 2.6.1.24 and 2.6.1.25, which do not act on dinitrotyrosine). Pyruvate or oxaloacetate can act as acceptors	809
2.6.1.27		L-Tryptophan:2-oxoglutarate aminotransferase	Also acts on 5-hydroxytryptophan and, to a lesser extent, on the phenyl amino acids. A pyridoxal-phosphate-protein	1007, 2502, 3365

Number	Recommended Name	Reaction
2.6.1.28	Tryptophan—phenylpyruvate aminotransferase	L-Tryptophan + phenylpyruvate = indolepyruvate + L-phenylalanine
2.6.1.29	Diamine aminotransferase	An α,ω-diamine + 2-oxoglutarate = an ω-aminoaldehyde + L-glutamate
2.6.1.30	Pyridoxamine—pyruvate transaminase	Pyridoxamine + pyruvate = pyridoxal + L-alanine
2.6.1.31	Pyridoxamine—oxaloacetate transaminase	Pyridoxamine + oxaloacetate = pyridoxal + L-aspartate
2.6.1.32	Valine—3-methyl-2-oxovalerate aminotransferase	L-Valine + 3-methyl-2-oxovalerate = 2-oxoisovalerate + L-isoleucine
2.6.1.33	dTDP-4-amino-4,6-dideoxy-D-glucose aminotransferase	dTDP-4-amino-4,6-dideoxy-D-glucose + 2-oxoglutarate = dTDP-4-keto-6-deoxy-D-glucose + L-glutamate
2.6.1.34	UDP-4-amino-2-acetamido-2,4,6-trideoxyglucose aminotransferase	UDP-4-amino-2-acetamido-2,4,6-trideoxyglucose + 2-oxoglutarate = UDP-4-keto-2-acetamido-2,6-dideoxyglucose + L-glutamate
2.6.1.35	Glycine—oxaloacetate aminotransferase	Glycine + oxaloacetate = glyoxylate + L-aspartate
2.6.1.36	L-Lysine 6-aminotransferase	L-Lysine + 2-oxoglutarate = 2-aminoadipate δ-semialdehyde + L-glutamate
2.6.1.37	(2-Aminoethyl) phosphonate aminotransferase	(2-Aminoethyl) phosphonate + pyruvate = 2-phosphonoacetaldehyde + L-alanine
2.6.1.38	Histidine aminotransferase	L-Histidine + 2-oxoglutarate = imidazol-5-yl-pyruvate + L-glutamate
2.6.1.39	2-Aminoadipate aminotransferase	L-2-Aminoadipate + 2-oxoglutarate = 2-oxoadipate + L-glutamate
2.6.1.40	(R)-3-Amino-2-methylpropionate—pyruvate aminotransferase	(R)-3-Amino-2-methylpropionate + pyruvate = methylmalonate semialdehyde + L-alanine
2.6.1.41	D-Methionine aminotransferase	D-Methionine + pyruvate = 4-methylthio-2-oxobutyrate + L-alanine

Number	Other Names	Basis for classification (Systematic Name)	Comments	Reference
2.6.1.28		L-Tryptophan:phenylpyruvate aminotransferase	Valine, leucine and isoleucine can replace tryptophan as amino donor	3284
2.6.1.29		Diamine:2-oxoglutarate aminotransferase		1685
2.6.1.30		Pyridoxamine:pyruvate aminotransferase		3575
2.6.1.31		Pyridoxamine:oxaloacetate aminotransferase		3574, 3736
2.6.1.32	Valine—isoleucine aminotransfcrase	L-Valine:3-methyl-2-oxovalerate aminotransferase		1554
2.6.1.33		dTDP-4-amino-4,6-dideoxy-D-glucose:2-oxoglutarate aminotransferase	A pyridoxal-phosphate-protein	2123
2.6.1.34		UDP-4-amino-2-acetamido-2,4,6-trideoxyglucose:2-oxoglutarate aminotransferase	A pyridoxal-phosphate-protein	725
2.6.1.35		Glycine:oxaloacetate aminotransferase	A pyridoxal-phosphate-protein	1025
2.6.1.36		L-Lysine:2-oxoglutarate 6-aminotransferase	A pyridoxal-phosphate-protein. The product is converted into the intramolecularly dehydrated form, 1-piperideine 6-carboxylate	3134, 3135
2.6.1.37		(2-Aminoethyl) phosphonate:pyruvate aminotransferase	A pyridoxal-phosphate-protein	1852
2.6.1.38		L-Histidine:2-oxoglutarate aminotransferase		584, 3678
2.6.1.39		L-2-Aminoadipate:2-oxoglutarate aminotransferase	A pyridoxal-phosphate-protein	2120
2.6.1.40	D-3-Aminoiso-butyrate—pyruvate aminotransferase	(R)-3-Amino-2-methylpropionate:pyruvate aminotransferase		1567
2.6.1.41		D-Methionine:pyruvate aminotransferase	Oxaloacetate can replace pyruvate	2079

Number	Recommended Name	Reaction
2.6.1.42	Branched-chain-amino-acid aminotransferase	L-Leucine + 2-oxoglutarate = 2-oxoisocaproate + L-glutamate
2.6.1.43	Aminolaevulinate aminotransferase	L-Alanine + 4,5-dioxovalerate = pyruvate + 5-aminolaevulinate
2.6.1.44	Alanine—glyoxylate aminotransferase	L-Alanine + glyoxylate = pyruvate + glycine
2.6.1.45	Serine—glyoxylate aminotransferase	L-Serine + glyoxylate = 3-hydroxypyruvate + glycine
2.6.1.46	Diaminobutyrate—pyruvate aminotransferase	L-2,4-Diaminobutyrate + pyruvate = L-aspartate β-semialdehyde + L-alanine
2.6.1.47	Alanine—oxomalonate aminotransferase	L-Alanine + oxomalonate = pyruvate + aminomalonate
2.6.1.48	5-Aminovalerate aminotransferase	5-Aminovalerate + 2-oxoglutarate = glutarate 5-semialdehyde + L-glutamate
2.6.1.49	Dihydroxyphenylalanine aminotransferase	3,4-Dihydroxy-L-phenylalanine + 2-oxoglutarate = dihydroxyphenylpyruvate + L-glutamate
2.6.1.50	Glutamine—*scyllo*-inosose aminotransferase	L-Glutamine + 2,4,6/3,5-pentahydroxycyclohexanone = 2-oxoglutaramate + 1-amino-1-deoxy-*scyllo*-inositol
2.6.1.51	Serine—pyruvate aminotransferase	L-Serine + pyruvate = 3-hydroxypyruvate + L-alanine
2.6.1.52	Phosphoserine aminotransferase	*O*-Phospho-L-serine + 2-oxoglutarate = 3-*O*-phosphohydroxypyruvate + L-glutamate

[*2.6.1.53 Transferred entry: now EC 1.4.1.13, Glutamate synthase (NADPH)*]

2.6.1.54	Pyridoxamine-phosphate aminotransferase	Pyridoxamine 5′-phosphate + 2-oxoglutarate = pyridoxal 5′-phosphate + D-glutamate
2.6.1.55	Taurine aminotransferase	Taurine + 2-oxoglutarate = sulphoacetaldehyde + L-glutamate
2.6.1.56	1D-1-Guanidino-3-amino-1,3-dideoxy-*scyllo*-inositol aminotransferase	1D-1-Guanidino-3-amino-1,3-dideoxy-*scyllo*-inositol + pyruvate = 1D-1-guanidino-1-deoxy-3-keto-*scyllo*-inositol + L-alanine

Number	Other Names	Basis for classification (Systematic Name)	Comments	Reference
2.6.1.42		Branched-chain-amino-acid: 2-oxoglutarate aminotransferase	Also acts on L-isoleucine and L-valine	35, 36, 1424, 3380
2.6.1.43		L-Alanine:4,5-dioxovalerate aminotransferase	A pyridoxal-phosphate-protein	1033, 2390
2.6.1.44		L-Alanine:glyoxylate aminotransferase	A pyridoxal-phosphate-protein	1745
2.6.1.45		L-Serine:glyoxylate aminotransferase	A pyridoxal-phosphate-protein	1692, 3117
2.6.1.46		L-2,4-Diaminobutyrate: pyruvate aminotransferase	A pyridoxal-phosphate-protein	2716
2.6.1.47		L-Alanine:oxomalonate aminotransferase	A pyridoxal-phosphate-protein	2333
2.6.1.48		5-Aminovalerate: 2-oxoglutarate aminotransferase	A pyridoxal-phosphate-protein	1423
2.6.1.49		3,4-Dihydroxy-L-phenylalanine:2-oxoglutarate aminotransferase	A pyridoxal-phosphate-protein	914
2.6.1.50		L-Glutamine:2,4,6/3,5-pentahydroxycyclohexanone aminotransferase	A pyridoxal-phosphate-protein	3601
2.6.1.51		L-Serine:pyruvate aminotransferase	A pyridoxal-phosphate-protein	520, 1802, 2902
2.6.1.52		O-Phospho-L-serine: 2-oxoglutarate aminotransferase	A pyridoxal-phosphate-protein	1322
2.6.1.54		Pyridoxamine-5'-phosphate: 2-oxoglutarate aminotransferase (D-glutamate-forming)	Also acts, more slowly, on pyridoxamine	3366
2.6.1.55		Taurine:2-oxoglutarate aminotransferase	A pyridoxal-phosphate-protein. Also acts on D L-3-aminoisobutyrate, β-alanine and 3-aminopropanesulphonate	3453
2.6.1.56		1D-1-Guanidino-3-amino-1,3-dideoxy-scyllo-inositol:pyruvate aminotransferase	L-Glutamate and L-glutamine can also act as amino donor	3597, 3601

Number	Recommended Name	Reaction
2.6.1.57	Aromatic-amino-acid aminotransferase	An aromatic amino acid + 2-oxoglutarate = an aromatic keto acid + L-glutamate
2.6.1.58	Phenylalanine (histidine) aminotransferase	L-Phenylalanine + pyruvate = phenylpyruvate + L-alanine
2.6.1.59	dTDP-4-amino-4,6-dideoxy-D-galactose aminotransferase	dTDP-4-amino-4,6-dideoxy-D-galactose + 2-oxoglutarate = dTDP-4-keto-6-deoxy-D-galactose + L-glutamate
2.6.1.60	Aromatic-amino-acid-glyoxylate aminotransferase	An aromatic amino acid + glyoxylate = an aromatic keto acid + glycine

[2.6.2.1   Transferred entry: now EC 2.1.4.1, Glycine amidinotransferase]

2.6.3   OXIMINOTRANSFERASES

2.6.3.1	Oximinotransferase	Pyruvate oxime + acetone = pyruvate + acetone oxime

2.7   TRANSFERRING PHOSPHORUS-CONTAINING GROUPS

This is a rather large group of enzymes comprising not only those transferring phosphate but also diphosphate, nucleotidyl residues and others. The most numerous section, that of phosphotransferases, is subdivided according to the acceptor group which may be an alcohol group (2.7.1), a carboxyl group (2.7.2), a nitrogenous group such as that of creatine (2.7.3), a phosphate group as in the case of adenylate kinase (2.7.4) or such that the overall reaction appears to be intramolecular transfer of a phosphate group (2.7.5). The pyrophosphotransferases are in sub-sub-group 2.7.6, the nucleotidyltransferases in 2.7.7 and those with other substituted phosphate groups in 2.7.8. With the enzymes of sub-sub-group 2.7.9, two phosphate groups are transferred from a donor such as ATP to two different acceptors.

2.7.1   PHOSPHOTRANSFERASES WITH AN ALCOHOL GROUP AS ACCEPTOR

2.7.1.1	Hexokinase	ATP + D-hexose = ADP + D-hexose 6-phosphate

Number	Other Names	Basis for classification (Systematic Name)	Comments	Reference
2.6.1.57		Aromatic-amino-acid: 2-oxoglutarate aminotransferase	A pyridoxal-phosphate-protein. L-Methionine can also act as donor, more slowly; oxaloacetate can act as acceptor. Controlled proteolysis converts the enzyme to EC 2.6.1.1	2132
2.6.1.58		L-Phenylalanine (L-histidine): pyruvate aminotransferase	L-Histidine and L-tyrosine can act instead of L-phenylalanine; L-methionine, L-serine and L-glutamine can replace L-alanine	2238
2.6.1.59		dTDP-4-amino-4,6-dideoxy-D-galactose:2-oxoglutarate aminotransferase	A pyridoxal-phosphate-protein	2480
2.6.1.60		Aromatic-amino-acid:glyoxylate aminotransferase	Phenylalanine, kynurenine, tyrosine and histidine can act as amino donors; glyoxylate, pyruvate and hydroxypyruvate can act as amino acceptors	1190
2.6.3.1	Transoximinase	Pyruvate-oxime:acetone oximinotransferase	Acetaldehyde can act instead of acetone; D-glucose oxime can act instead of pyruvate oxime	3753, 3754, 3755
2.7.1.1	Heterophosphatase	ATP:D-hexose 6-phosphotransferase	D-Glucose, D-mannose and D-glucosamine can act as acceptor; ITP and dATP can act as donor	175, 265, 1836, 2445

Number	Recommended Name	Reaction
2.7.1.2	Glucokinase	ATP + D-glucose = ADP + D-glucose 6-phosphate
2.7.1.3	Ketohexokinase	ATP + D-fructose = ADP + D-fructose 1-phosphate
2.7.1.4	Fructokinase	ATP + D-fructose = ADP + D-fructose 6-phosphate
2.7.1.5	Rhamnulokinase	ATP + L-rhamnulose = ADP + L-rhamnulose 1-phosphate
2.7.1.6	Galactokinase	ATP + D-galactose = ADP + α-D-galactose 1-phosphate
2.7.1.7	Mannokinase	ATP + D-mannose = ADP + D-mannose 6-phosphate
2.7.1.8	Glucosamine kinase	ATP + 2-amino-2-deoxy-D-glucose = ADP + 2-amino-2-deoxy-D-glucose phosphate
[2.7.1.9	Deleted entry: Acetylaminodeoxyglucose kinase]	
2.7.1.10	Phosphoglucokinase	ATP + D-glucose 1-phosphate = ADP + D-glucose 1,6-bisphosphate
2.7.1.11	6-Phosphofructokinase	ATP + D-fructose 6-phosphate = ADP + D-fructose 1,6-bisphosphate
2.7.1.12	Gluconokinase	ATP + D-gluconate = ADP + 6-phospho-D-gluconate
2.7.1.13	Ketogluconokinase	ATP + 2-keto-D-gluconate = ADP + 6-phospho-2-keto-D-gluconate
2.7.1.14	Sedoheptulokinase	ATP + sedoheptulose = ADP + sedoheptulose 7-phosphate
2.7.1.15	Ribokinase	ATP + D-ribose = ADP + D-ribose 5-phosphate
2.7.1.16	Ribulokinase	ATP + L(or D)-ribulose = ADP + L(or D)-ribulose 5-phosphate
2.7.1.17	Xylulokinase	ATP + D-xylulose = ADP + D-xylulose 5-phosphate
2.7.1.18	Phosphoribokinase	ATP + D-ribose 5-phosphate = ADP + D-ribose

Number	Other Names	Basis for classification (Systematic Name)	Comments	Reference
2.7.1.2		ATP:D-glucose 6-phosphotransferase	The enzyme from liver also acts on D-mannose	232, 422, 470
2.7.1.3		ATP:D-fructose 1-phosphotransferase	D-Sorbose, D-agatose and 5-keto-D-fructose can also act as acceptor	586, 1291, 1927, 2558
2.7.1.4		ATP: D-fructose 6-phosphotransferase		422, 2176
2.7.1.5		ATP:L-rhamnulose 1-phosphotransferase		3699
2.7.1.6		ATP:D-galactose 1-phosphotransferase	D-Galactosamine can also act as acceptor	471, 2392, 3687
2.7.1.7		ATP:D-mannose 6-phosphotransferase		422
2.7.1.8		ATP:2-amino-2-deoxy-D-glucose phosphotransferase		422
2.7.1.10		ATP:D-glucose-1-phosphate 6-phosphotransferase		2544
2.7.1.11	Phosphohexokinase	ATP:D-fructose-6-phosphate 1- phosphotransfcrasc	D-Tagatose 6-phosphate and sedoheptulose 7-phosphate can act as acceptor; UTP, CTP, and ITP can act as donor	150, 2472, 2559, 2685, 3519, 2076, 3147
2.7.1.12		ATP:D-gluconate 6-phosphotransferase		552, 1894, 2374, 2886
2.7.1.13		ATP:2-keto-D-gluconate 6-phosphotransferase		921
2.7.1.14	Heptulokinase	ATP:sedoheptulose 7-phosphotransferase		781
2.7.1.15		ATP:D-ribose 5-phosphotransferase	2-Deoxy-D-ribose can also act as acceptor	28, 1042
2.7.1.16		ATP:L(or D)-ribulose 5-phosphotransferase	Ribitol and L-arabinitol can also act as acceptor	431, 1901, 3079
2.7.1.17		ATP:D-xylulose 5-phosphotransferase		1302, 3097, 3272, 3078
2.7.1.18		ATP:D-ribose-5-phosphate 1-phosphotransferase		2935, 1793

Number	Recommended Name	Reaction
2.7.1.19	Phosphoribulokinase	ATP + D-ribulose 5-phosphate = ADP + D-ribulose 1,5-bisphosphate
2.7.1.20	Adenosine kinase	ATP + adenosine = ADP + AMP
2.7.1.21	Thymidine kinase	ATP + thymidine = ADP + thymidine 5'-phosphate
2.7.1.22	Ribosylnicotinamide kinase	ATP + N-ribosylnicotinamide = ADP + nicotinamide ribonucleotide
2.7.1.23	NAD$^+$ kinase	ATP + NAD$^+$ = ADP + NADP$^+$
2.7.1.24	Dephospho-CoA kinase	ATP + dephospho-CoA = ADP + CoA
2.7.1.25	Adenylylsulphate kinase	ATP + adenylylsulphate = ADP + 3'-phospho-adenylylsulphate
2.7.1.26	Riboflavin kinase	ATP + riboflavin = ADP + FMN
2.7.1.27	Erythritol kinase	ATP + erythritol = ADP + D-erythritol 4-phosphate
2.7.1.28	Triokinase	ATP + D-glyceraldehyde = ADP + D-glyceraldehyde 3-phosphate
2.7.1.29	Acetol kinase	ATP + hydroxyacetone = ADP + hydroxyacetone phosphate
2.7.1.30	Glycerol kinase	ATP + glycerol = ADP + sn-glycerol 3-phosphate
2.7.1.31	Glycerate kinase	ATP + D-glycerate = ADP + 3-phospho-D-glycerate
2.7.1.32	Choline kinase	ATP + choline = ADP + O-phosphocholine
2.7.1.33	Pantothenate kinase	ATP + pantothenate = ADP + D-4'-phosphopantothenate
2.7.1.34	Pantetheine kinase	ATP + pantetheine = ADP + pantetheine 4'-phosphate

Number	Other Names	Basis for classification (Systematic Name)	Comments	Reference
2.7.1.19	Phosphopentokinase	ATP:D-ribulose-5-phosphate 1- phosphotransferase		1411, 1490
2.7.1.20		ATP:adenosine 5′-phosphotransferase	2-Aminoadenosine can also act as acceptor	467, 1773, 1974
2.7.1.21		ATP:thymidine 5′-phosphotransferase	Deoxyuridine is also an acceptor, and dGTP is a donor	2491, 1709
2.7.1.22		ATP:N-ribosylnicotinamide 5′- phosphotransferase		2861
2.7.1.23	DPN kinase	ATP:NAD$^+$ 2′-phosphotransferase		1606, 1766, 3611, 535
2.7.1.24		ATP:dephospho-CoA 3′-phosphotransferase		1326, 3611, 6
2.7.1.25	APS-kinase	ATP:adenylylsulphate 3′-phosphotransferase		193, 2802
2.7.1.26	Flavokinase	ATP:riboflavin 5′-phosphotransferase		1046, 1630, 2149, 505
2.7.1.27		ATP:erythritol 4-phosphotransferase		1353
2.7.1.28		ATP:D-glyceraldehyde 3-phosphotransferase		1293, 3071
2.7.1.29		ATP:hydroxyacetone phosphotransferase		3002
2.7.1.30		ATP:glycerol 3-phosphotransferase	Dihydroxyacetone and L-glyceraldehyde can act as acceptors; UTP (and, in the case of the yeast enzyme, ITP and GTP) can act as donors	271, 414, 3685
2.7.1.31		ATP:D-glycerate 3-phosphotransferase		310, 751, 1421
2.7.1.32		ATP:choline phosphotransferase	Ethanolamine and its methyl and ethyl derivatives can also act as acceptor	3712, 1251
2.7.1.33		ATP:pantothenate 4′-phosphotransferase		402, 2607, 7
2.7.1.34		ATP:pantetheine 4′-phosphotransferase		2452

Number	Recommended Name	Reaction
2.7.1.35	Pyridoxal kinase	ATP + pyridoxal = ADP + pyridoxal 5-phosphate
2.7.1.36	Mevalonate kinase	ATP + mevalonate = ADP + (—)-5-phospho-mevalonate
2.7.1.37	Protein kinase	ATP + a protein = ADP + a phosphoprotein
2.7.1.38	Phosphorylase kinase	4 ATP + 2 phosphorylase b = 4 ADP + phosphorylase a
2.7.1.39	Homoserine kinase	ATP + L-homoserine = ADP + O-phospho-L-homoserine
2.7.1.40	Pyruvate kinase	ATP + pyruvate = ADP + phosphoenolpyruvate
2.7.1.41	Glucose-1-phosphate phosphodismutase	2 D-Glucose 1-phosphate = D-glucose + D-glucose 1,6-bisphosphate
2.7.1.42	Riboflavin phosphotransferase	D-Glucose 1-phosphate + riboflavin = D-glucose + FMN
2.7.1.43	Glucuronokinase	ATP + D-glucuronate = ADP + 1-phospho-α-D-glucuronate
2.7.1.44	Galacturonokinase	ATP + D-galacturonate = ADP + 1-phospho-α-D-galacturonate
2.7.1.45	2-Keto-3-deoxygluconokinase	ATP + 2-keto-3-deoxy-D-gluconate = ADP + 6-phospho-2-keto-3-deoxy-D-gluconate

Number	Other Names	Basis for classification (Systematic Name)	Comments	Reference
2.7.1.35		ATP:pyridoxal 5-phosphotransferase	Pyridoxine, pyridoxamine and various derivatives can also act as acceptor	1405, 2150, 3466
2.7.1.36		ATP:mevalonate 5-phosphotransferase	CTP, GTP or UTP can also act as donor	1269, 1947, 2087, 3386
2.7.1.37	Phosphorylase *b* kinase kinase, Glycogen synthase *a* kinase	ATP:protein phosphotransferase	An enzyme from rat tissues is stimulated by cyclic AMP and will activate phosphorylase kinase; one from skeletal muscle will, in addition, activate the *a* form of EC 2.4.1.11 to the *b* form. Will also phosphorylate certain other proteins. Some enzymes are activated by cyclic GMP, but not by cyclic AMP, and some enzymes by neither	2682, 2955, 3607, 2389, 164, 3806
2.7.1.38	Dephosphophosphorylase kinase	ATP:phosphorylase-*b* phosphotransferase		1794, 1795, 2705
2.7.1.39		ATP:L-homoserine *O*-phosphotransferase		897, 3620
2.7.1.40	Phosphoenolpyruvate kinase, Phospho*enol* transphosphorylase	ATP:pyruvate 2-*O*-phosphotransferase	UTP, GTP, CTP, ITP and dATP can also act as donor. Also phosphorylates hydroxylamine and fluoride in the presence of CO_2	360, 1773, 1809, 3266, 3421
2.7.1.41		D-Glucose-1-phosphate:D-glucose-1-phosphate 6-phosphotransferase		1923, 3064
2.7.1.42		D-Glucose-1-phosphate:riboflavin 5′-phosphotransferase		1602
2.7.1.43		ATP:D-glucuronate 1-phosphotransferase		2391
2.7.1.44		ATP:D-galacturonate 1-phosphotransferase		2393
2.7.1.45		ATP:2-keto-3-deoxy-D-gluconate 6-phosphotransferase		618

Number	Recommended Name	Reaction
2.7.1.46	L-Arabinokinase	ATP + L-arabinose = ADP + β-L-arabinose 1-phosphate
2.7.1.47	D-Ribulokinase	ATP + D-ribulose = ADP + D-ribulose 5-phosphate
2.7.1.48	Uridine kinase	ATP + uridine = ADP + UMP
2.7.1.49	Hydroxymethylpyrimidine kinase	ATP + 2-methyl-4-amino-5-hydroxymethyl-pyrimidine = ADP + 2-methyl-4-amino-5-phosphomethylpyrimidine
2.7.1.50	Hydroxyethylthiazole kinase	ATP + 4-methyl-5-(2'-hydroxyethyl) thiazole = ADP + 4-methyl-5-(2'-phosphoethyl)-thiazole
2.7.1.51	L-Fuculokinase	ATP + L-fuculose = ADP + L-fuculose 1-phosphate
2.7.1.52	Fucokinase	ATP + 6-deoxy-L-galactose = ADP + 6-deoxy-L-galactose 1-phosphate
2.7.1.53	L-Xylulokinase	ATP + L-xylulose = ADP + L-xylulose 5-phosphate
2.7.1.54	D-Arabinokinase	ATP + D-arabinose = ADP + D-arabinose 5-phosphate
2.7.1.55	Allose kinase	ATP + D-allose = ADP + D-allose 6-phosphate
2.7.1.56	1-Phosphofructokinase	ATP + D-fructose 1-phosphate = ADP + D-fructose 1,6-bisphosphate
2.7.1.57	Mannitol kinase	ATP + mannitol = ADP + mannitol 1-phosphate
2.7.1.58	2-Keto-3-deoxygalactonate kinase	ATP + 2-keto-3-deoxy-D-galactonate = ADP + 2-keto-3-deoxy-D-galactonate phosphate
2.7.1.59	N-Acetyl-D-glucosamine kinase	ATP + 2-acetamido-2-deoxy-D-glucose = ADP + 2-acetamido-2-deoxy-D-glucose 6-phosphate
2.7.1.60	N-Acyl-D-mannosamine kinase	ATP + 2-acylamino-2-deoxy-D-mannose = ADP + 2-acylamino-2-deoxy-D-mannose 6-phosphate
2.7.1.61	Acyl-phosphate—hexose phosphotransferase	Acyl-phosphate + D-hexose = an acid + D-hexose phosphate

Number	Other Names	Basis for classification (Systematic Name)	Comments	Reference
2.7.1.46		ATP:L-arabinose 1-phosphotransferase		2392
2.7.1.47		ATP:D-ribulose 5-phosphotransferase		948
2.7.1.48		ATP:uridine 5′-phosphotransferase	Cytidine can act as acceptor; GTP or ITP can act as donor	2506, 3096
2.7.1.49		ATP:2-methyl-4-amino-5-hydroxymethyl-pyrimidine 5-phosphotransferase	CTP, UTP and GTP can act as donor	1951
2.7.1.50		ATP:4-methyl-5-(2′-hydroxyethyl)-thiazole 2′-phosphotransferase		449
2.7.1.51		ATP:L-fuculose 1-phosphotransferase		1254
2.7.1.52		ATP:6-deoxy-L-galactose 1-phosphotransferase		1457
2.7.1.53		ATP:L-xylulose 5-phosphotransferase		78
2.7.1.54		ATP:D-arabinose 5-phosphotransferase		3565
2.7.1.55	Allokinase	ATP:D-allose 6-phosphotransferase		1021
2.7.1.56	Fructose-1-phosphate kinase	ATP:D-fructose-1-phosphate 6- phosphotransferase	ITP, GTP or UTP can replace ATP	2759, 2910
2.7.1.57		ATP:mannitol 1-phosphotransferase		1716
2.7.1.58		ATP:2-keto-3-deoxy-D-galactonate phosphotransferase		3243
2.7.1.59		ATP:2-acetamido-2-deoxy-D-glucose 6-phosphotransferase	Also acts on D-glucose	119, 209, 639
2.7.1.60		ATP:2-acylamino-2-deoxy-D-mannose 6-phosphotransferase	Acts on the acetyl and glycolyl derivatives	194, 1019, 1825
2.7.1.61		Acyl-phosphate:D-hexose phosphotransferase	Phosphorylates D-glucose and D-mannose at carbon atom 6, and D-fructose at carbon atom 1 or 6	77, 1581

Number	Recommended Name	Reaction
2.7.1.62	Phosphoramidate—hexose phosphotransferase	Phosphoramidate + hexose = NH_3 + hexose 1-phosphate
2.7.1.63	Polyphosphate—glucose phosphotransferase	$(Phosphate)_n$ + D-glucose = $(phosphate)_{n-1}$ + D-glucose 6-phosphate
2.7.1.64	*myo*-Inositol 1-kinase	ATP + *myo*-inositol = ADP + L-*myo*-inositol 1-phosphate
2.7.1.65	*scyllo*-Inosamine kinase	ATP + 1-amino-1-deoxy-*scyllo*-inositol = ADP + 1-amino-1-deoxy-*scyllo*-inositol 4-phosphate
2.7.1.66	Isoprenoid-alcohol kinase	ATP + C-55-isoprenoid-alcohol = ADP + C-55-isoprenoid-alcohol phosphate
2.7.1.67	Phosphatidylinositol kinase	ATP + phosphatidylinositol = ADP + phosphatidylinositol 4-phosphate
2.7.1.68	Diphosphoinositide kinase	ATP + phosphatidylinositol 4-phosphate = ADP + phosphoinositol 4,5-bisphosphate
2.7.1.69	Phosphohistidinoprotein—hexose phosphotransferase	Phosphohistidinoprotein + hexose = protein + hexose 6-phosphate
2.7.1.70	Protamine kinase	ATP + protamine = ADP + *O*-phosphoprotamine
2.7.1.71	Shikimate kinase	ATP + shikimate = ADP + shikimate 5-phosphate
2.7.1.72	Streptomycin 6-kinase	ATP + streptomycin = ADP + streptomycin 6-phosphate
2.7.1.73	Inosine kinase	ATP + inosine = ADP + IMP
2.7.1.74	Deoxycytidine kinase	NTP + deoxycytidine = NDP + dCMP

[2.7.1.75 Deleted entry: Thymidine kinase. Now EC 2.7.1.21]

Number	Other Names	Basis for classification (Systematic Name)	Comments	Reference
2.7.1.62		Phosphoramidate:hexose 1-phosphotransferase	May be identical with EC 3.1.3.9	3123
2.7.1.63		Polyphosphate:D-glucose 6-phosphotransferase	Requires a neutral salt; *e.g.* KCl for maximum activity. Also acts on glucosamine	3323, 3324
2.7.1.64		ATP:*myo*-inositol 1-phosphotransferase		832, 2263
2.7.1.65		ATP:1-amino-1-deoxy-*scyllo*-inositol 4-phosphotransferase	Also acts on streptamine, 2-deoxystreptamine and 1D-1-guanidino-3-amino-1,3-dideoxy-*scyllo*-inositol	3600, 3597
2.7.1.66		ATP:C-55-isoprenoid-alcohol phosphotransferase		1306
2.7.1.67		ATP:phosphatidylinositol 4-phosphotransferase		561, 1560
2.7.1.68		ATP:phosphatidylinositol 4-phosphate 5-phosphotransferase		1558, 1559
2.7.1.69		Phosphohistidinoprotein:hexose phosphotransferase		1826
2.7.1.70	Histone kinase	ATP:protamine *O*-phosphotransferase	Phosphorylates serine groups in protamines and histones. Requires cyclic AMP	1514, 1873, 165
2.7.1.71		ATP:shikimate 5-phosphotransferase		2280
2.7.1.72	Streptidine kinase	ATP:streptomycin 6-phosphotransferase	dATP can replace ATP; and dihydrostreptomycin, streptidine and 2-deoxystreptidine can act as acceptors	3602, 3598
2.7.1.73		ATP:inosine 5′-phosphotransferase		2608
2.7.1.74		NTP:deoxycytidine 5′-phosphotransferase	Cytosine arabinoside can act as acceptor; all natural nucleoside triphosphates (except dCTP) can act as donor	1667, 2264, 771

Number	Recommended Name	Reaction
2.7.1.76	Deoxyadenosine kinase	ATP + deoxyadenosine = ADP + dAMP
2.7.1.77	Nucleoside phosphotransferase	A nucleotide + 3′-deoxynucleoside = a nucleoside + 3′-deoxynucleoside 5′-monophosphate
2.7.1.78	Polynucleotide 5′-hydroxyl-kinase	ATP + 5′-dephospho-DNA = ADP + 5′-phospho-DNA
2.7.1.79	Pyrophosphate—glycerol phosphotransferase	Pyrophosphate + glycerol = orthophosphate + glycerol 1-phosphate
2.7.1.80	Pyrophosphate—serine phosphotransferase	Pyrophosphate + L-serine = orthophosphate + O-phospho-L-serine
2.7.1.81	Hydroxylysine kinase	GTP + 5-hydroxy-L-lysine = GDP + O-phosphohydroxy-L-lysine
2.7.1.82	Ethanolamine kinase	ATP + ethanolamine = ADP + O-phosphoethanolamine
2.7.1.83	Pseudouridine kinase	ATP + pseudouridine = ADP + pseudouridine 5′-phosphate
2.7.1.84	Alkyldihydroxyacetone kinase	ATP + O-alkyldihydroxyacetone = ADP + O-alkyldihydroxyacetone phosphate
2.7.1.85	β-D-Glucoside kinase	ATP + cellobiose = ADP + 6-phospho-β-D-glucosyl-(1,4)-D-glucose
2.7.1.86	NADH kinase	ATP + NADH = ADP + NADPH
2.7.1.87	Streptomycin 3″-kinase	ATP + streptomycin = ADP + streptomycin 3″-phosphate
2.7.1.88	Dihydrostreptomycin-6-phosphate 3′α-kinase	ATP + dihydrostreptomycin 6-phosphate = ADP + dihydrostreptomycin 3′α,6-bisphosphate

Number	Other Names	Basis for classification (Systematic Name)	Comments	Reference
2.7.1.76		ATP:deoxyadenosine 5′-phosphotransferase	Deoxyguanosine can also act as acceptor	1807
2.7.1.77		Nucleotide:3′-deoxynucleoside 5′-phosphotransferase	Phenylphosphate as well as 3′- and 5′-nucleotides can act as phosphate donor	411
2.7.1.78		ATP: 5′-dephosphopolynucleotide 5′-phosphotransferase	Also acts on 5′-dephospho-RNA and 3′-mononucleotides	2453, 2455
2.7.1.79		Pyrophosphate:glycerol · 1-phosphotransferase	May be identical with EC 3.1.3.9	3220
2.7.1.80		Pyrophosphate:L-serine O-phosphotransferase		446
2.7.1.81		GTP:5-hydroxy-L-lysine O-phosphotransferase	Allohydroxy-L-lysine can also act as acceptor	1308
2.7.1.82		ATP:ethanolamine O-phosphotransferase		3642, 3295, 864
2.7.1.83		ATP:pseudouridine 5′-phosphotransferase		3146
2.7.1.84		ATP:O-alkyldihydroxyacetone phosphotransferase		490
2.7.1.85		ATP:cellobiose 6-phosphotransferase	Phosphorylates a number of β-D-glucosides; GTP, CTP, ITP and UTP can also act as donor	2549
2.7.1.86		ATP:NADH 2′-phosphotransferase	CTP, ITP, UTP and GTP can also act as phosphate donors (in decreasing order of activity). The enzyme is specific for NADH. Activated by acetate	1123
2.7.1.87		ATP:streptomycin 3′′-phosphotransferase	Also phosphorylates dihydrostreptomycin, 3′-deoxydihydro-streptomycin and their 6-phosphates	3598
2.7.1.88		ATP:dihydrostreptomycin-6-phosphate 3′α-phosphotransferase	3′-Deoxydihydro-streptomycin 6-phosphate can also act as acceptor	3598

Number	Recommended Name	Reaction
2.7.1.89	Thiamin kinase	ATP + thiamin = ADP + thiamin monophosphate
2.7.1.90	Pyrophosphate—fructose-6-phosphate 1-phosphotransferase	Pyrophosphate + D-fructose 6-phosphate = orthophosphate + D-fructose 1,6-bisphosphate
2.7.1.91	Dihydrosphingosine kinase	ATP + D-*erythro*-dihydrosphingosine = ADP + dihydrosphingosine 1-phosphate
2.7.1.92	5-Keto-2-deoxygluconokinase	ATP + 5-keto-2-deoxy-D-gluconate = ADP + 6-phospho-5-keto-2-deoxy-D-gluconate
2.7.1.93	Alkylglycerol kinase	ATP + 1-*O*-alkyl-*sn*-glycerol = ADP + 1-*O*-alkyl-*sn*-glycerol 3-phosphate
2.7.1.94	Monoacylglycerol kinase	ATP + monoacylglycerol = ADP + monoacylglycerol 3-phosphate
2.7.1.95	Kanamycin kinase	ATP + kanamycin = ADP + kanamycin 3'-phosphate

[2.7.1.96 Deleted entry: now included with EC 2.7.1.86]

2.7.1.97	Opsin kinase	ATP + photo-bleached rhodopsin = ADP + phospho-rhodopsin
2.7.1.98	Phosphoenolpyruvate—fructose phosphotransferase	Phosphoenolpyruvate + D-fructose = pyruvate + D-fructose 1-phosphate
2.7.1.99	[Pyruvate dehydrogenase (lipoamide)] kinase	ATP + [pyruvate dehydrogenase (lipoamide)] = [pyruvate dehydrogenase (lipoamide)]phosphate

2.7.2   PHOSPHOTRANSFERASES WITH A CARBOXYL GROUP AS ACCEPTOR

2.7.2.1	Acetate kinase	ATP + acetate = ADP + acetyl phosphate
2.7.2.2	Carbamate kinase	ATP + NH_3 + CO_2 = ADP + carbamoyl phosphate
2.7.2.3	Phosphoglycerate kinase	ATP + 3-phospho-D-glycerate = ADP + 3-phospho-D-glyceroyl phosphate

Number	Other Names	Basis for classification (Systematic Name)	Comments	Reference
2.7.1.89		ATP:thiamin phosphotransferase		1469
2.7.1.90	6-Phosphofructokinase (pyrophosphate)	Pyrophosphate:D-fructose-6-phosphate 1-phosphotransferase		2758
2.7.1.91		ATP:D-*erythro*-dihydrosphingosine 1-phosphotransferase		3231, 3234
2.7.1.92		ATP:5-keto-2-deoxy-D-gluconate 6-phosphotransferase		79
2.7.1.93		ATP:1-*O*-alkyl-*sn*-glycerol 3-phosphotransferase		2825
2.7.1.94		ATP:monoacylglycerol 3-phosphotransferase	Acts on both 1- and 2-monoacylglycerol	2605, 2606
2.7.1.95	Neomycin-kanamycin phosphotransferase	ATP:kanamycin 3′-*O*-phosphotransferase	Also acts on neomycin, paromomycin, neamine, paromamine, vistamycin, and gentamicin A. An enzyme from *Pseudomonas aeruginosa* also acts on butirosin	740, 739
2.7.1.97		ATP:photo-bleached-rhodopsin phosphotransferase	Does not act on unbleached rhodopsin, histones or phosphvitin	3654
2.7.1.98		Phosphoenolpyruvate:D-fructose 1-phosphotransferase		2893
2.7.1.99		ATP:[pyruvate dehydrogenase (lipoamide)] phosphokinase	A mitochondrial enzyme associated with the pyruvate dehydrogenase complex. Phosphorylation inactivates EC 1.2.4.1	1983
2.7.2.1	Acetokinase	ATP:acetate phosphotransferase	Propionate also acts as acceptor, but more slowly	2838, 2839, 3217
2.7.2.2		ATP:carbamate phosphotransferase		301, 1060, 1532, 657
2.7.2.3		ATP:3-phospho-D-glycerate 1-phosphotransferase		149, 420, 1211, 2717

Number	Recommended Name	Reaction
2.7.2.4	Aspartate kinase	ATP + L-aspartate = ADP + 4-phospho-L-aspartate

[2.7.2.5 Transferred entry: now EC 6.3.4.16, Carbamoyl-phosphate synthetase (ammonia)]

2.7.2.6	Formate kinase	ATP + formate = ADP + formyl phosphate
2.7.2.7	Butyrate kinase	ATP + butyrate = ADP + butyryl phosphate
2.7.2.8	Acetylglutamate kinase	ATP + *N*-acetyl-L-glutamate = ADP + *N*-acetyl-L-glutamate 5-phosphate

[2.7.2.9 Transferred entry: now EC 6.3.5.5, Carbamoyl-phosphate synthetase (glutamine-hydrolysing)]

2.7.2.10	Phosphoglycerate kinase (GTP)	GTP + 3-phospho-D-glycerate = GDP + 3-phospho-D-glyceroyl phosphate
2.7.2.11	Glutamate kinase	ATP + L-glutamate = ADP + γ-L-glutamyl phosphate
2.7.2.12	Acetate kinase (pyrophosphate)	Pyrophosphate + acetate = orthophosphate + acetyl phosphate

2.7.3 PHOSPHOTRANSFERASES WITH A NITROGENOUS GROUP AS ACCEPTOR

2.7.3.1	Guanidinoacetate kinase	ATP + guanidinoacetate = ADP + phosphoguanidinoacetate
2.7.3.2	Creatine kinase	ATP + creatine = ADP + phosphocreatine
2.7.3.3	Arginine kinase	ATP + L-arginine = ADP + N^ω-phospho-L-arginine
2.7.3.4	Taurocyamine kinase	ATP + taurocyamine = ADP + N^ω-phosphotaurocyamine
2.7.3.5	Lombricine kinase	ATP + lombricine = ADP + N^ω-phospholombricine
2.7.3.6	Hypotaurocyamine kinase	ATP + hypotaurocyamine = ADP + N^ω-phosphohypotaurocyamine

Number	Other Names	Basis for classification (Systematic Name)	Comments	Reference
2.7.2.4	Aspartokinase	ATP:L-aspartate 4-phosphotransferase	The enzyme from *E. coli* is a multifunctional protein, which also catalyses the reaction of EC 1.1.1.3	307, 306, 2575
2.7.2.6		ATP:formate phosphotransferase		3101
2.7.2.7		ATP:butyrate phosphotransferase		3494
2.7.2.8		ATP:N-acetyl-L-glutamate 5-phosphotransferase		171, 858, 3561
2.7.2.10		GTP:3-phospho-D-glycerate 1-phosphotransferase		2757
2.7.2.11		ATP:L-glutamate γ-phosphotransferase	Product rapidly cyclizes to 2-pyrrolidone 5-carboxylate and orthophosphate	169
2.7.2.12		Pyrophosphate:acetate phosphotransferase		2753
2.7.3.1		ATP:guanidinoacetate N-phosphotransferase		1327, 3402, 2644, 2645
2.7.3.2	Lohmann's enzyme	ATP:creatine N-phosphotransferase	N-Ethylglycocyamine can also act as acceptor	834, 1671, 1811, 1812
2.7.3.3		ATP:L-arginine N^{ω}-phosphotransferase		822, 2296, 3321, 3554
2.7.3.4		ATP:taurocyamine N^{ω}-phosphotransferase		1327, 3402, 1601, 3408
2.7.3.5		ATP:lombricine N^{ω}-phosphotransferase	Also acts on methylated lombricines such as thalassemine; the specificity varies with the species	2555, 984, 1601, 3537
2.7.3.6		ATP:hypotaurocyamine N^{ω}-phosphotransferase	Also acts, more slowly, on taurocyamine	3408

Number	Recommended Name	Reaction
2.7.3.7	Opheline kinase	ATP + guanidinoethyl methyl phosphate = ADP + N'-phosphoguanidinoethyl methyl phosphate
2.7.3.8	Ammonia kinase	ATP + NH_3 = ADP + phosphoramide
2.7.3.9	Phosphoenolpyruvate—protein phosphotransferase	Phosphoenolpyruvate + a specific protein = pyruvate + a phosphohistidinoprotein

2.7.4 PHOSPHOTRANSFERASES WITH A PHOSPHATE GROUP AS ACCEPTOR

Number	Recommended Name	Reaction
2.7.4.1	Polyphosphate kinase	ATP + (phosphate)$_n$ = ADP + (phosphate)$_{n+1}$
2.7.4.2	Phosphomevalonate kinase	ATP + 5-phosphomevalonate = ADP + 5-diphosphomevalonate
2.7.4.3	Adenylate kinase	ATP + AMP = ADP + ADP
2.7.4.4	Nucleosidemonophosphate kinase	ATP + nucleoside monophosphate = ADP + nucleoside diphosphate

[2.7.4.5   Deleted entry:Deoxycytidylate kinase. Now included with EC 2.7.4.14]

Number	Recommended Name	Reaction
2.7.4.6	Nucleosidediphosphate kinase	ATP + nucleoside diphosphate = ADP + nucleoside triphosphate
2.7.4.7	Phosphomethylpyrimidine kinase	ATP + 2-methyl-4-amino-5-phosphomethyl-pyrimidine = ADP + 2-methyl-4-amino-5-diphosphomethyl-pyrimidine
2.7.4.8	Guanylate kinase	ATP + GMP = ADP + GDP
2.7.4.9	dTMP kinase	ATP + dTMP = ADP + dTDP

Number	Other Names	Basis for classification (Systematic Name)	Comments	Reference
2.7.3.7		ATP:guanidinoethyl-methyl-phosphate phosphotransferase	Has a little activity on taurocyamine, lombricine and phosphotaurocyamine	3403
2.7.3.8		ATP:ammonia phosphotransferase	Has a wide specificity. In the reverse direction N-phosphoglycine and N^ω-phosphohistidine can also act as phosphate donors, and ADP, dADP, GDP, CDP, dTDP, dCDP, IDP and UDP can act as phosphate acceptors (in decreasing order of activity)	752
2.7.3.9		Phosphoenolpyruvate:protein phosphotransferase		1826
2.7.4.1		ATP:polyphosphate phosphotransferase		1334, 1767, 2313
2.7.4.2		ATP:5-phosphomevalonate phosphotransferase		327, 1277, 1947
2.7.4.3	Myokinase	ATP:AMP phosphotransferase	Inorganic triphosphate can also act as donor	524, 2426, 2495, 2427, 2428
2.7.4.4		ATP:nucleosidemonophosphate phosphotransferase	Many nucleotides can act as acceptor; other nucleoside triphosphates can act instead of ATP	1029, 1286, 1965, 2427
2.7.4.6		ATP:nucleosidediphosphate phosphotransferase	Many nucleoside diphosphates can act as acceptor, while many ribo- and deoxyribo-nucleoside triphosphates can act as donor	264, 1029, 1697, 1796, 2355, 2724
2.7.4.7		ATP:2-methyl-4-amino-5-phosphomethyl-pyrimidine phosphotransferase		1951
2.7.4.8	Deoxyguanylate kinase	ATP:(d)GMP phosphotransferase	dGMP can also act as acceptor, and dATP can act as donor	416, 835, 1122, 2478, 3051
2.7.4.9		ATP:dTMP phosphotransferase		1407, 1645, 2386

Number	Recommended Name	Reaction
2.7.4.10	Nucleosidetriphosphate—adenylate kinase	Nucleoside triphosphate + AMP = nucleoside diphosphate + ADP
2.7.4.11	(Deoxy)adenylate kinase	ATP + dAMP = ADP + dADP
2.7.4.12	T_2-induced deoxynucleotide kinase	ATP + dGMP (or dTMP) = ADP + dGDP (or dTDP)
2.7.4.13	(Deoxy)nucleosidemonophosphate kinase	ATP + deoxynucleoside monophosphate = ADP + deoxynucleoside diphosphate
2.7.4.14	Cytidylate kinase	ATP + (d)CDP = ADP + (d)CMP
2.7.4.15	Thiamindiphosphate kinase	ATP + thiamin diphosphate = ADP + thiamin triphosphate
2.7.4.16	Thiamin-monophosphate kinase	ATP + thiamin monophosphate = ADP + thiamin diphosphate
2.7.4.17	3-Phosphoglyceroyl-phosphate—polyphosphate phosphotransferase	3-Phospho-D-glyceroyl phosphate + (phosphate)$_n$ = 3-phosphoglycerate + (phosphate)$_{n+1}$
2.7.4.18	Farnesyl-diphosphate kinase	ATP + farnesyl diphosphate = ADP + farnesyl triphosphate

2.7.5 PHOSPHOTRANSFERASES WITH REGENERATION OF DONORS (APPARENTLY CATALYSING INTRAMOLECULAR TRANSFERS)

Number	Recommended Name	Reaction
2.7.5.1	Phosphoglucomutase	α-D-Glucose 1,6-bisphosphate + α-D-glucose 1-phosphate = α-D-glucose 6-phosphate + α-D-glucose 1,6-bisphosphate
2.7.5.2	Acetylglucosamine phosphomutase	2-Acetamido-2-deoxy-D-glucose 1,6-bisphosphate + 2-acetamido-2-deoxy-D-glucose 1-phosphate = 2-acetamido-2-deoxy-D-glucose 6-phosphate + 2-acetamido-2-deoxy-D-glucose 1,6-bisphosphate
2.7.5.3	Phosphoglyceromutase	2,3-Bisphospho-D-glycerate + 2-phospho-D-glycerate = 3-phospho-D-glycerate + 2,3-bisphospho-D-glycerate
2.7.5.4	Bisphosphoglyceromutase	3-Phospho-D-glyceroyl phosphate + 3-phospho-D-glycerate = 3-phospho-D-glycerate + 2,3-bisphospho-D-glycerate

Number	Other Names	Basis for classification (Systematic Name)	Comments	Reference
2.7.4.10		Nucleosidetriphosphate:AMP phosphotransferase	Many nucleoside triphosphates can act as donor	44, 525
2.7.4.11		ATP:(d)AMP phosphotransferase	AMP can also act as acceptor	1122
2.7.4.12		ATP:(d)NMP phosphotransferase	dTMP and dAMP can act as acceptor; dATP can act as donor	251
2.7.4.13		ATP:deoxynucleosidemono-phosphate phosphotransferase	dATP can substitute for ATP	290
2.7.4.14	Deoxycytidylate kinase	ATP:CMP phosphotransferase	UMP and dCMP can also act as acceptors	1407, 2427, 2871
2.7.4.15		ATP:thiamin-diphosphate phosphotransferase		1467
2.7.4.16		ATP:thiamin-monophosphate phosphotransferase		2419
2.7.4.17		3-Phospho-D-glyceroyl-phosphate:polyphosphate phosphotransferase		1814, 1815
2.7.4.18		ATP:farnesyl-diphosphate phosphotransferase	ADP can also act as donor	2949
2.7.5.1	Glucose phosphomutase	α-D-Glucose-1,6-bisphosphate:α-D-glucose-1-phosphate phosphotransferase	The yeast enzyme is a zinc-protein	2152, 2343, 2344, 3297, 1536, 2732
2.7.5.2	Phosphoacetylglucosamine mutase	2-Acetamido-2-deoxy-D-glucose-1,6-bisphosphate: 2-acetamido-2-deoxy-D-glucose-1-phosphate phosphotransferase		1919, 2766, 2767
2.7.5.3	Glycerate phosphomutase	2,3-Bisphospho-D-glycerate: 2-phospho-D-glycerate phosphotransferase		601, 2613, 2827, 3298
2.7.5.4	Glycerate phosphomutase	3-Phospho-D-glyceroyl-phosphate:3-phospho-D-glycerate phosphotransferase		1542, 2722

Number	Recommended Name	Reaction
2.7.5.5	Phosphoglucomutase (glucose-cofactor)	D-Glucose 1-phosphate + D-glucose = D-glucose + D-glucose 6-phosphate
2.7.5.6	Phosphopentomutase	α-D-Glucose 1,6-bisphosphate + deoxy-D-ribose 1-phosphate = α-D-glucose 1,6-bisphosphate + deoxy-D-ribose 5-phosphate

2.7.6   DIPHOSPHOTRANSFERASES

Number	Recommended Name	Reaction
2.7.6.1	Ribosephosphate pyrophosphokinase	ATP + D-ribose 5-phosphate = AMP + 5-phospho-α-D-ribose 1-diphosphate
2.7.6.2	Thiamin pyrophosphokinase	ATP + thiamin = AMP + thiamin diphosphate
2.7.6.3	2-Amino-4-hydroxy-6-hydroxymethyldihydropteridine pyrophosphokinase	ATP + 2-amino-4-hydroxy-6-hydroxymethyl-7,8-dihydropteridine = AMP + 2-amino-4-hydroxy-6-hydroxymethyl-7,8-dihydropteridine 6'-diphosphate
2.7.6.4	Nucleotide pyrophosphokinase	ATP + nucleoside 5'-monophosphate = AMP + 5'-phosphonucleoside 3'-diphosphate

2.7.7   NUCLEOTIDYLTRANSFERASES

Number	Recommended Name	Reaction
2.7.7.1	NMN adenylyltransferase	ATP + nicotinamide ribonucleotide = pyrophosphate + NAD^+
2.7.7.2	FMN adenylyltransferase	ATP + FMN = pyrophosphate + FAD
2.7.7.3	Pantetheinephosphate adenylyltransferase	ATP + pantetheine 4'-phosphate = pyrophosphate + dephospho-CoA
2.7.7.4	Sulphate adenylyltransferase	ATP + sulphate = pyrophosphate + adenylylsulphate
2.7.7.5	Sulphate adenylyltransferase (ADP)	ADP + sulphate = orthophosphate + adenylylsulphate
2.7.7.6	RNA nucleotidyltransferase	n Nucleoside triphosphate = n pyrophosphate + RNA_n
2.7.7.7	DNA nucleotidyltransferase	n Deoxynucleoside triphosphate = n pyrophosphate + DNA_n

Number	Other Names	Basis for classification (Systematic Name)	Comments	Reference
2.7.5.5	Glucose phosphomutase	D-Glucose-1-phosphate: D-glucose 6-phosphotransferase	Fructose 1-phosphate may act as a donor while other hexoses may act as acceptors	959
2.7.5.6	Phosphodeoxyribomutase	α-D-Glucose-1,6-bisphosphate:deoxy-D-ribose-1-phosphate phosphotransferase		1583
2.7.6.1		ATP:D-ribose-5-phosphate pyrophosphotransferase	dATP can also act as donor	1399, 1404, 1768, 1947, 2771, 3320
2.7.6.2	Thiamin kinase	ATP:thiamin pyrophosphotransferase		1926, 3047, 3225
2.7.6.3		ATP:2-amino-4-hydroxy-6-hydroxymethyl-7,8-dihydropteridine 6′-pyrophosphotransferase		2786, 3055, 2785
2.7.6.4		ATP:nucleoside-5′-monophosphate pyrophosphotransferase	Enzyme acts on the 5′-mono-, di- and triphosphate derivatives of purine nucleosides	2315, 2420, 2421
2.7.7.1	NAD⁺ pyrophosphorylase	ATP:NMN adenylyltransferase	Nicotinate nucleotide can also act as acceptor. See also EC 2.7.7.18	128, 627, 1772
2.7.7.2	FAD pyrophosphorylase	ATP:FMN adenylyltransferase		1047, 2973
2.7.7.3	Dephospho-CoA pyrophosphorylase	ATP:pantetheine-4′-phosphate adenylyltransferase		2452, 1326
2.7.7.4	ATP-sulphurylase, Sulphurylase	ATP:sulphate adenylyltransferase		193, 1315, 2803
2.7.7.5	ADP-sulphurylase	ADP:sulphate adenylyltransferase		1137, 2802, 2803
2.7.7.6	RNA polymerase	Nucleosidetriphosphate:RNA nucleotidyltransferase	Needs DNA as template. See also EC 2.7.7.19	2075, 3038, 3624
2.7.7.7	DNA polymerase	Deoxynucleosidetriphosphate:DNA deoxynucleotidyl-transferase	A DNA chain acts as a template, and the enzyme forms a complimentary	337, 1908, 2940

Number	Recommended Name	Reaction
2.7.7.8	Polyribonucleotide nucleotidyltransferase	RNA_{n+1} + orthophosphate = RNA_n + a nucleoside diphosphate
2.7.7.9	Glucose-1-phosphate uridylyltransferase	UTP + α-D-glucose 1-phosphate = pyrophosphate + UDPglucose
2.7.7.10	Galactose-1-phosphate uridylyltransferase	UTP + α-D-galactose 1-phosphate = pyrophosphate + UDPgalactose
2.7.7.11	Xylose-1-phosphate uridylyltransferase	UTP + α-D-xylose 1-phosphate = pyrophosphate + UDPxylose
2.7.7.12	UDPglucose—hexose-1-phosphate uridylyltransferase	UDPglucose + α-D-galactose 1-phosphate = α-D-glucose 1-phosphate + UDPgalactose
2.7.7.13	Mannose-1-phosphate guanylyltransferase	GTP + α-D-mannose 1-phosphate = pyrophosphate + GDPmannose
2.7.7.14	Ethanolaminephosphate cytidylyltransferase	CTP + ethanolamine phosphate = pyrophosphate + CDPethanolamine
2.7.7.15	Cholinephosphate cytidylyltransferase	CTP + choline phosphate = pyrophosphate + CDPcholine

[2.7.7.16 Transferred entry: now EC 3.1.4.22, Ribonuclease I]

[2.7.7.17 Transferred entry: now EC 3.1.4.23, Ribonuclease II]

2.7.7.18	Nicotinatemononucleotide adenylyltransferase	ATP + nicotinate ribonucleotide = pyrophosphate + deamido-NAD^+
2.7.7.19	Polynucleotide adenylyltransferase	n ATP + (nucleotide)$_m$ = n pyrophosphate + (nucleotide)$_{m+n}$

[2.7.7.20 Deleted entry: this entry is identical with EC 2.7.7.25, tRNA adenylyltransferase]

2.7.7.21	tRNA cytidylyltransferase	CTP + tRNAn = pyrophosphate + tRNA $_{n+1}$
2.7.7.22	Mannose-1-phosphate guanylyltransferase (GDP)	GDP + D-mannose 1-phosphate = orthophosphate + GDPmannose

Number	Other Names	Basis for classification (Systematic Name)	Comments	Reference
			chain. Also acts on DNA-RNA hybrids	
2.7.7.8	Polynucleotide phosphorylase	Polyribonucleotide:ortho-phosphate nucleotidyltransferase	ADP, IDP, GDP, UDP and CDP can act as donor	1172, 1987, 2469
2.7.7.9	UDPglucose pyrophosphorylase	UTP:α-D-glucose-1-phosphate uridylyltransferase		45, 1043, 1568, 1723, 3109
2.7.7.10	Galactose-1-phosphate uridylyltransferase	UTP:α-D-galactose-1-phosphate uridylyltransferase		1461, 1568
2.7.7.11		UTP:α-D-xylose-1-phosphate uridylyltransferase		1044
2.7.7.12	Uridyl transferase, Hexose-1-phosphate uridylyltransferase	UDPglucose:α-D-galactose-1-phosphate uridylyltransferase		1572, 1843, 3109, 2898, 2138
2.7.7.13	GTP—mannose-1-phosphate guanylyltransferase	GTP:α-D-mannose-1-phosphate guanylyltransferase	Bacterial enzyme can also use ITP and dGTP as donor	2314, 2658
2.7.7.14	Phosphorylethanolamine transferase	CTP:ethanolaminephosphate cytidylyltransferase		1657
2.7.7.15	Phosphorylcholine transferase	CTP:cholinephosphate cytidylyltransferase		344, 1657, 3689
2.7.7.18	Deamido-NAD$^+$ pyrophosphorylase	ATP:nicotinatemononucleotide adenylyltransferase		1446
2.7.7.19	NTP polymerase, RNA adenylating enzyme	ATP:polynucleotide adenylyltransferase	Also acts slowly with CTP. The primer, depending on the source of the enzyme, may be an RNA or DNA fragment or oligo(A) bearing a 3'-OH terminal group. See also EC 2.7.7.6	132, 789, 1087, 1789, 2075, 3038
2.7.7.21	tRNA CCA-pyrophosphorylase	CTP:tRNA cytidylyltransferase	May be identical with EC 2.7.7.25	291, 711, 788
2.7.7.22	GDPmannose phosphorylase	GDP:D-mannose-1-phosphate guanylyltransferase		474

Number	Recommended Name	Reaction
2.7.7.23	UDPacetylglucosamine pyrophosphorylase	UTP + 2-acetamido-2-deoxy-D-glucose 1-phosphate = pyrophosphate + UDP-2-acetamido-2-deoxy-D-glucose
2.7.7.24	Glucose-1-phosphate thymidylyltransferase	dTTP + α-D-glucose 1-phosphate = pyrophosphate + dTDPglucose
2.7.7.25	tRNA adenylyltransferase	ATP + $tRNA_n$ = pyrophosphate + $tRNA_{n+1}$
2.7.7.26	*Transferred entry: now EC 3.1.4.8, Guanyloribonuclease]*	
2.7.7.27	Glucose-1-phosphate adenylyltransferase	ATP + α-D-glucose 1-phosphate = pyrophosphate + ADPglucose
2.7.7.28	Nucleosidetriphosphate-hexose-1-phosphate nucleotidyltransferase	Nucleoside triphosphate + hexose 1-phosphate = pyrophosphate + NDPhexose
2.7.7.29	Hexose-1-phosphate guanylyltransferase	GTP + α-D-hexose 1-phosphate = pyrophosphate + GDPhexose
2.7.7.30	Fucose-1-phosphate guanylyltransferase	GTP + L-fucose 1-phosphate = pyrophosphate + GDP-L-fucose
2.7.7.31	DNA nucleotidylexotransferase	n Deoxynucleoside triphosphate + $(deoxynucleotide)_m$ = n pyrophosphate + $(deoxynucleotide)_{m+n}$
2.7.7.32	Galactose-1-phosphate thymidylyltransferase	dTTP + α-D-galactose 1-phosphate = pyrophosphate + dTDPgalactose
2.7.7.33	Glucose-1-phosphate cytidylyltransferase	CTP + D-glucose 1-phosphate = pyrophosphate + CDPglucose
2.7.7.34	Glucose-1-phosphate guanylyltransferase	GTP + α-D-glucose 1-phosphate = pyrophosphate + GDPglucose
2.7.7.35	Ribose-5-phosphate adenylyltransferase	ADP + D-ribose 5-phosphate = orthophosphate + ADPribose
2.7.7.36	Sugar-1-phosphate adenylyltransferase	ADP + sugar 1-phosphate = orthophosphate + ADPsugar
2.7.7.37	Sugar-1-phosphate nucleotidyltransferase	NDP + sugar 1-phosphate = orthophosphate + NDPsugar

Number	Other Names	Basis for classification (Systematic Name)	Comments	Reference
2.7.7.23		UTP:2-acetamido-2-deoxy-α-D-glucose-1-phosphate uridylyltransferase		2568, 3269
2.7.7.24		dTTP:α-D-glucose-1-phosphate thymidylyltransferase		1779, 2580
2.7.7.25	tRNA CCA-pyrophosphorylase	ATP:tRNA adenylyltransferase	May be identical with EC 2.7.7.21	291, 711, 3201
2.7.7.27	ADPglucose pyrophosphorylase	ATP:α-D-glucose-1-phosphate adenylyltransferase		1015, 3039
2.7.7.28	NDPhexose pyrophosphorylase	NTP:hexose-1-phosphate nucleotidyltransferase	In decreasing order of activity, guanosine, inosine and adenosine diphosphate hexoses are substrates in the reverse reaction with either glucose or mannose as the sugar	3547
2.7.7.29	GDPhexose pyrophosphorylase	GTP:α-D-hexose-1-phosphate guanylyltransferase		1184
2.7.7.30	GDPfucose pyrophosphorylase	GTP:fucose-1-phosphate guanylyltransferase		1456
2.7.7.31	Terminal deoxyribonucleotidyl transferase, Terminal addition enzyme	Nucleosidetriphosphate:DNA deoxynucleotidylexotransferase	Nucleoside may be ribo- or deoxyribo-, n must be greater than 3 and a 3′-OH is required	338, 1087, 1789
2.7.7.32	dTDPgalactose pyrophosphorylase	dTTP:α-D-galactose-1-phosphate thymidylyltransferase		2578
2.7.7.33	CDPglucose pyrophosphorylase	CTP:D-glucose-1-phosphate cytidylyltransferase		2137
2.7.7.34	GDPglucose pyrophosphorylase	GTP:α-D-glucose-1-phosphate guanylyltransferase	Also acts to a lesser degree on D-mannose 1-phosphate	634
2.7.7.35	ADPribose phosphorylase	ADP:D-ribose-5-phosphate adenylyltransferase		849, 3207
2.7.7.36	ADPsugar phosphorylase	ADP:sugar-1-phosphate adenylyltransferase		635
2.7.7.37	NDPsugar phosphorylase	NDP:sugar-1-phosphate nucleotidyltransferase		444

Number	Recommended Name	Reaction
2.7.7.38	3-Deoxy-*manno*-octulosonate cytidylyltransferase	CTP + 3-deoxy-D-*manno*-octulosonate = pyrophosphate + CMP-3-deoxy-D-*manno*-octulosonate
2.7.7.39	Glycerol-3-phosphate cytidylyltransferase	CTP + *sn*-glycerol 3-phosphate = pyrophosphate + CDPglycerol
2.7.7.40	D-Ribitol-5-phosphate cytidylyltransferase	CTP + D-ribitol 5-phosphate = pyrophosphate + CDPribitol
2.7.7.41	Phosphatidate cytidylyltransferase	CTP + phosphatidate = pyrophosphate + CDPdiacylglycerol
2.7.7.42	Glutamine-synthetase adenylyltransferase	ATP + [(L-glutamate:ammonia ligase(ADP-forming)]= pyrophosphate + adenylyl-[(L-glutamate:ammonia ligase (ADP-forming)]
2.7.7.43	Acylneuraminate cytidylyltransferase	CTP + *N*-acylneuraminate = pyrophosphate + CMP-*N*-acylneuraminate
2.7.7.44	Glucuronate-1-phosphate uridylyltransferase	UTP + 1-phospho-α-D-glucuronate = pyrophosphate + UDP-D-glucuronate
2.7.7.45	Guanosinetriphosphate guanylyltransferase	2 GTP = pyrophosphate + P^1,P^4-bis(5'-guanosyl) tetraphosphate
2.7.7.46	Gentamicin 2″-nucleotidyltransferase	Nucleoside triphosphate + gentamicin = pyrophosphate + 2″-nucleotidylgentamicin
2.7.7.47	Streptomycin 3″-adenylyltransferase	ATP + streptomycin = pyrophosphate + 3″-adenylylstreptomycin

2.7.8 TRANSFERASES FOR OTHER SUBSTITUTED PHOSPHATE GROUPS

2.7.8.1	Ethanolaminephosphotransferase	CDPethanolamine + 1,2-diacylglycerol = CMP + a phosphatidylethanolamine
2.7.8.2	Cholinephosphotransferase	CDPcholine + 1,2-diacylglycerol = CMP + a phosphatidylcholine
2.7.8.3	Ceramide cholinephosphotransferase	CDPcholine + *N*-acylsphingosine = CMP + sphingomyelin
2.7.8.4	Serine-phosphoethanolamine synthase	CDPethanolamine + L-serine = CMP + L-serine-phosphoethanolamine

Number	Other Names	Basis for classification (Systematic Name)	Comments	Reference
2.7.7.38	CMP-3-deoxy-D-*manno*-octulosonate pyrophosphorylase	CTP:3-deoxy-D-*manno*-octulosonate cytidylyltransferase		1012
2.7.7.39	CDPglycerol pyrophosphorylase	CTP:*sn*-glycerol-3-phosphate cytidylyltransferase		3029
2.7.7.40	CDPribitol pyrophosphorylase	CTP:D-ribitol-5-phosphate cytidylyltransferase		3029
2.7.7.41	CDPdiglyceride pyrophosphorylase	CTP:phosphatide cytidylyltransferase		479, 2144, 2594
2.7.7.42		ATP:[(L-glutamate:ammonia ligase (ADP-forming)]adenylyltransferase		1694, 2173, 3714, 2174, 3021, 786
2.7.7.43	CMPsialate pyrophosphorylase, CMPsialate synthase	CTP:*N*-acylneuraminate cytidylyltransfcrase	Acts on *N*-acetyl- and *N*-glycolyl-derivatives	1629
2.7.7.44		UTP:1-phospho-α-D-glucuronate uridylyltransferase	Also acts slowly with CTP	2815
2.7.7.45		GTP:GTP guanylyltransferase	Also acts to a lesser degree on GDP to form P^1,P^3-bis(5′-guanosyl) triphosphate	3614
2.7.7.46		NTP:gentamicin 2″-nucleotidyltransferase	ATP, dATP, CTP, ITP and GTP can act as donors; kanamycin, tobramycin and sisomicin can also act as acceptors	261, 2331
2.7.7.47		ATP:streptomycin 3″-adenylyltransferase	Also acts on spectinomycin	1207
2.7.8.1		CDPethanolamine: 1,2-diacylglycerol ethanolaminephosphotransferase		1657
2.7.8.2	Phosphorylcholine-glyceride transferase	CDPcholine:1,2-diacylglycerol cholinephosphotransferase		1657, 3645, 1656
2.7.8.3		CDPcholine: *N*-acylsphingosine cholinephosphotransferase		3175, 1656
2.7.8.4		CDPethanolamine:L-serine ethanolaminephosphotransferase		54

Number	Recommended Name	Reaction
2.7.8.5	Glycerophosphate phosphatidyltransferase	CDPdiacylglycerol + sn-glycerol 3-phosphate = CMP + 3-phosphatidyl-phospho-sn-glycerol
2.7.8.6	Poly (isoprenol)-phosphate galactosephosphotransferase	UDPgalactose + C-55-poly(isoprenol) phosphate = UMP + D-galactose 1-diphospho-C-55-poly(isoprenol)
2.7.8.7	Holo-[acyl-carrier-protein]synthase	CoA + apo-[acyl-carrier protein] = 3′, 5′-ADP + holo-[acyl-carrier protein]
2.7.8.8	Phosphatidylserine synthase	CDPdiacylglycerol + L-serine = CMP + phosphatidylserine
2.7.8.9	Phosphomannan mannosephosphotransferase	GDPmannose + $(\text{phosphomannan})_n$ = GMP + $(\text{phosphomannan})_{n+1}$
2.7.8.10	Sphingosine cholinephosphotransferase	CDPcholine + sphingosine = CMP + sphingosyl-phosphocholine
2.7.8.11	CDPdiacylglycerol—inositol 3-phosphatidyltransferase	CDPdiacylglycerol + myo-inositol = CMP + 1D-myo-1- inositol 1-phosphatide
2.7.8.12	CDPglycerol glycerophosphotransferase	CDPglycerol + $(\text{glycerophosphate})_n$ = CMP + $(\text{glycerophosphate})_{n+1}$
2.7.8.13	Phospho-N-acetylmuramoyl-pentapeptide-transferase	UDP-N-acetylmuramoyl-L-alanyl-D-γ-glutamyl-L-lysyl-D-alanyl-D-alanine + undecaprenoid-1-ol phosphate = UMP + undecaprenoid-1-ol-diphospho-N-acetylmuramoyl-L-alanyl-D-γ-glutamyl-L-lysyl-D-alanyl-D-alanine

2.7.9 PHOSPHOTRANSFERASES WITH PAIRED ACCEPTORS

Number	Recommended Name	Reaction
2.7.9.1	Pyruvate,orthophosphate dikinase	ATP + pyruvate + orthophosphate = AMP + phosphoenolpyruvate + pyrophosphate
2.7.9.2	Pyruvate,water dikinase	ATP + pyruvate + H_2O = AMP + phosphoenolpyruvate + orthophosphate

2.8 TRANSFERRING SULPHUR-CONTAINING GROUPS

These are enzymes transferring sulphur atoms (2.8.1), sulphate groups (2.8.2) or coenzyme A (2.8.3).

2.8.1 SULPHURTRANSFERASES

Number	Recommended Name	Reaction
2.8.1.1	Thiosulphate sulphurtransferase	Thiosulphate + cyanide = sulphite + thiocyanate

Number	Other Names	Basis for classification (Systematic Name)	Comments	Reference
2.7.8.5		CDPdiacylglycerol:*sn*-glycerol-3-phosphate phosphatidyltransferase		495
2.7.8.6		UDPgalactose:C-55-poly-(isoprenol)-phosphate galactosephosphotransferase		2517, *3734*
2.7.8.7		CoA:apo-[acyl-carrier-protein] pantetheinephosphotransferase		823, 2663
2.7.8.8	CDPdiglyceride—serine *O*-phosphatidyl-transferase	CDPdiacylglycerol:L-serine *O*-phosphatidyltransferase		1588
2.7.8.9		GDPmannose:phosphomannan mannosephosphotransferase		382
2.7.8.10		CDPcholine:sphingosine cholinephosphotransferase		962
2.7.8.11	CDP-diglyceride—inositol phosphatidyltransferase	CDPdiacylglycerol:*myo*-inositol 3-phosphatidyltransferase		2576, 2665, 2903
2.7.8.12		CDPglycerol:poly-(glycerophosphate) glycerophosphotransferase		428
2.7.8.13		UDP-*N*-acetylmuramoyl-L-alanyl-D-γ-glutamyl-L-lysyl-D-alanyl-D-alanine: undecaprenoid-1-ol-phosphate phospho-*N*-acetylmuramoyl-pentapeptide-transferase		1300, 1307, 3270
2.7.9.1		ATP:pyruvate,orthophosphate phosphotransferase		1221, 2751, 2754, 2752
2.7.9.2	Phosphoenolpyruvate synthase	ATP:pyruvate,water phosphotransferase	Contains Mn	581, 580, 275, 582
2.8.1.1	Rhodanese, Thiosulphate cyanide transsulphurase	Thiosulphate:cyanide sulphurtransferase	A few other sulphur compounds can act as donor	3155, 3664

Number	Recommended Name	Reaction
2.8.1.2	3-Mercaptopyruvate sulphurtransferase	3-Mercaptopyruvate + cyanide = pyruvate + thiocyanate

2.8.2 SULPHOTRANSFERASES

2.8.2.1	Aryl sulphotransferase	3′-Phosphoadenylylsulphate + a phenol = adenosine 3′,5′-bisphosphate + an aryl sulphate
2.8.2.2	3β-Hydroxysteroid sulphotransferase	3′-Phosphoadenylylsulphate + a 3β-hydroxysteroid = adenosine 3′,5′-bisphosphate + a steroid 3β-sulphate
2.8.2.3	Arylamine sulphotransferase	3′-Phosphoadenylylsulphate + an arylamine = adenosine 3′,5′-bisphosphate + an aryl-sulphamate
2.8.2.4	Oestrone sulphotransferase	3′-Phosphoadenylylsulphate + oestrone = adenosine 3′,5′-bisphosphate + oestrone 3-sulphate
2.8.2.5	Chondroitin sulphotransferase	3′-Phosphoadenylylsulphate + chondroitin = adenosine 3′,5′-bisphosphate + chondroitin 4′-sulphate
2.8.2.6	Choline sulphotransferase	3′-Phosphoadenylylsulphate + choline = adenosine 3′,5′-bisphosphate + choline sulphate
2.8.2.7	UDP-N-acetylgalactosamine-4-sulphate sulphotransferase	3′-Phosphoadenylylsulphate + UDP-2-acetamido-2-deoxygalactose 4-sulphate = adenosine 3′,5′-bisphosphate + UDP-2-acetamido-2-deoxygalactose 4,6-bissulphate
2.8.2.8	Desulphoheparin sulphotransferase	3′-Phosphoadenylylsulphate + N-desulphoheparin = adenosine 3′,5′-bisphosphate + heparin
2.8.2.9	Tyrosine-ester sulphotransferase	3′-Phosphoadenylylsulphate + L-tyrosine methyl ester = adenosine 3′,5′-bisphosphate + L-tyrosine methyl ester 4-sulphate
2.8.2.10	Luciferin sulphotransferase	3′-Phosphoadenylylsulphate + luciferin = adenosine 3′,5′-bisphosphate + luciferyl sulphate

Number	Other Names	Basis for classification (Systematic Name)	Comments	Reference
2.8.1.2		3-Mercaptopyruvate:cyanide sulphurtransferase	Sulphite, sulphinates, mercaptoethanol and mercaptopyruvate can also act as acceptor. The bacterial enzyme is a zinc-protein	877, 1418, 3154, 3529, 3522
2.8.2.1	Phenol sulphotransferase, Sulfokinase	3'-Phosphoadenylylsulphate: phenol sulphotransferase		409, 2801, 2803, 904
2.8.2.2		3'-Phosphoadenylylsulphate: 3β-hydroxysteroid sulphotransferase		2450, 2801, 21, 1140
2.8.2.3		3'-Phosphoadenylylsulphate: arylamine sulphotransferase		2801, 2866
2.8.2.4		3'-Phosphoadenylylsulphate: oestrone 3-sulphotransferase		2450, 22
2.8.2.5		3'-Phosphoadenylylsulphate: chondroitin 4'-sulphotransferase	Oligo- and polysaccharides containing 2-acetyl-D-galactosamine can act as acceptor	647, 2801, 3309
2.8.2.6		3'-Phosphoadenylylsulphate: choline sulphotransferase		2510
2.8.2.7		3'-Phosphoadenylylsulphate: UDP-2-acetamido-2-deoxy-D-galactose-4-sulphate 6-sulphotransferase		1191
2.8.2.8		3'-Phosphoadenylylsulphate: N-desulphoheparin N-sulphotransferase	The enzyme also catalyses the sulphation of heparin sulphate, and to a much more limited extent, of chondroitin 4-sulphate and dermatan sulphate	807, 1518
2.8.2.9		3'-Phosphoadenylylsulphate: L-tyrosine-methyl-ester sulphotransferase	Only carboxyl-substituted tyrosine can act as acceptor	2117
2.8.2.10		3'-Phosphoadenylylsulphate: luciferin sulphotransferase		590

Number	Recommended Name	Reaction
2.8.2.11	Galactosylceramide sulphotransferase	3'-Phosphoadenylylsulphate + a galactosylceramide = adenosine 3',5'-bisphosphate + galactosylceramidesulphate
2.8.2.12	Heparitin sulphotransferase	3'-Phosphoadenylylsulphate + heparitin = adenosine 3',5'-bisphosphate + N-sulphoheparitin
2.8.2.13	Psychosine sulphotransferase	3'-Phosphoadenylylsulphate + galactosylsphingosine = adenosine 3',5'-bisphosphate + psychosine sulphate
2.8.2.14	Bile-salt sulphotransferase	3'-Phosphoadenylylsulphate + taurolithocholate = adenosine 3',5'-bisphosphate + taurolithocholate sulphate

2.8.3   CoA-TRANSFERASES

Number	Recommended Name	Reaction
2.8.3.1	Propionate CoA-transferase	Acetyl-CoA + propionate = acetate + propionyl-CoA
2.8.3.2	Oxalate CoA-transferase	Succinyl-CoA + oxalate = succinate + oxalyl-CoA
2.8.3.3	Malonate CoA-transferase	Acetyl-CoA + malonate = acetate + malonyl-CoA
[2.8.3.4	*Deleted entry: Butyrate CoA-transferase]*	
2.8.3.5	3-Ketoacid CoA-transferase	Succinyl-CoA + 3-oxo acid = succinate + a 3-oxo-acyl-CoA
2.8.3.6	3-Oxoadipate CoA-transferase	Succinyl-CoA + 3-oxoadipate = succinate + 3-oxoadipyl-CoA
2.8.3.7	Citramalate CoA-transferase	Succinyl-CoA + citramalate = succinate + citramalyl-CoA
2.8.3.8	Acetate CoA-transferase	Acyl-CoA + acetate = a fatty acid anion + acetyl-CoA

3.   HYDROLASES

These enzymes catalyse the hydrolysis of various bonds. Some of these enzymes pose problems because they have a very wide specificity, and it is not easy to decide if two preparations described by different authors are the same, or if they should be listed under different entries.

Number	Other Names	Basis for classification (Systematic Name)	Comments	Reference
2.8.2.11		3′-Phosphoadenylylsulphate: galactosylceramide 3′- sulphotransferase		2170
2.8.2.12		3′-Phosphoadenylylsulphate: heparitin N-sulphotransferase		3310
2.8.2.13		3′-Phosphoadenylylsulphate: galactosylsphingosine sulphotransferase		2463
2.8.2.14		3′-Phosphoadenylylsulphate: taurolithocholate sulphotransferase	Both conjugated and unconjugated bile salts can act as acceptor	514
2.8.3.1		Acetyl-CoA:propionate CoA- transferase	Butyrate and lactate can also act as acceptor	3181
2.8.3.2	Succinyl—β-ketoacyl-CoA transferase	Succinyl-CoA:oxalate CoA- transferase		2671
2.8.3.3		Acetyl-CoA:malonate CoA- transferase		1232
2.8.3.5		Succinyl-CoA:3-oxo-acid CoA- transferase	Acetoacetate or, more slowly, malonate semialdehyde, 3-oxovalerate, 3-oxoisocaproate or 3-oxocaproate, can act as acceptor; malonyl-CoA can act instead of succinyl-CoA	2021, 2201, 3212, 1294
2.8.3.6		Succinyl-CoA:3-oxoadipate CoA-transferase		1604
2.8.3.7		Succinyl-CoA:citramalate CoA- transferase		583
2.8.3.8		Acyl-CoA:acetate CoA- transferase	Acts on butyryl-CoA and valeryl-CoA	3540

Number	Recommended Name	Reaction

While the systematic name always includes 'hydrolase', the recommended name is, in most cases, formed by the name of the substrate with the suffix *-ase.* It is understood that the name of the substrate with this suffix means a hydrolytic enzyme.

3.1 ACTING ON ESTER BONDS

The esterases are subdivided into those acting on carboxylic esters (3.1.1), thiolesterases (3.1.2), phosphoric monoester hydrolases, the phosphatases (3.1.3), phosphodiester hydrolases (3.1.4), triphosphoric monoester hydrolases (3.1.5), sulphatases (3.1.6), and diphosphoric monoesterases (3.1.7). The nucleases, previously included under 3.1.4, are now placed in a number of new sub-groups: the exonucleases (3.1.11-16) and the endonucleases (3.1.21-31).

3.1.1 CARBOXYLIC ESTER HYDROLASES

3.1.1.1	Carboxylesterase	A carboxylic ester + H_2O = an alcohol + a carboxylic acid anion
3.1.1.2	Arylesterase	A phenyl acetate + H_2O = a phenol + acetate
3.1.1.3	Triacylglycerol lipase	Triacylglycerol + H_2O = diacylglycerol + a fatty acid anion
3.1.1.4	Phospholipase A_2	A lecithin + H_2O = 1-acylglycerophosphocholine + an unsaturated fatty acid anion
3.1.1.5	Lysophospholipase	A lysolecithin + H_2O = glycerophosphocholine + a fatty acid anion
3.1.1.6	Acetylesterase	An acetic ester + H_2O = an alcohol + acetate
3.1.1.7	Acetylcholinesterase	Acetylcholine + H_2O = choline + acetate
3.1.1.8	Cholinesterase	An acylcholine + H_2O = choline + a carboxylic acid anion

Number	Other Names	Basis for classification (Systematic Name)	Comments	Reference
3.1.1.1	Ali-esterase, B-esterase, Methylbutyrase, Monobutyrase, Cocain esterase, Procaine esterase	Carboxylic-ester hydrolase	Wide specificity. Also hydrolyses vitamin A esters	133, 202, 288, 425, 1371, 2874
3.1.1.2	A-esterase, Paraoxonase	Aryl-ester hydrolase	Acts on many phenolic esters; the enzyme from sheep serum also hydrolyses paraoxon	47, 136, 2056, 347
3.1.1.3	Steapsin, Tributyrase, Triglyceride lipase, Lipase	Triacylglycerol acylhydrolase	The pancreatic enzyme acts only on an ester-water interface; the outer ester links are preferentially hydrolysed	1764, 2023, 2911, 3081
3.1.1.4	Lecithinase A, Phosphatidase, Phosphatidolipase	Phosphatide 2-acylhydrolase	Also acts on phosphatidylethanolamine, choline plasmalogen and phosphatides, removing the fatty acid attached to the 2-position	672, 736, 919, 1175, 2272, 2895
3.1.1.5	Lecithinase B, Lysolecithinase, Phospholipase B	Lysolecithin acylhydrolase		574, 665, 855, 3527
3.1.1.6	C-esterase (in animal tissues)	Acetic-ester acetylhydrolase		47, 268, 1499
3.1.1.7	True cholinesterase, Choline esterase I, Cholinesterase	Acetylcholine acetylhydrolase	Acts on a variety of acetic esters; also catalyses transacetylations.	134, 269, 1928, 2324, 3852, 540
3.1.1.8	Pseudocholinesterase, Butyrylcholine esterase, Choline esterase II (unspecific), Benzoylcholinesterase	Acylcholine acylhydrolase	Acts on a variety of choline esters and a few other compounds	134, 136, 1732, 2324, 2931, 3259

236

Number	Recommended Name	Reaction

[3.1.1.9   Deleted entry: Benzoylcholinesterase: a side reaction of EC 3.1.1.8]

3.1.1.10   Tropinesterase

Atropine + H_2O = tropine + tropate

3.1.1.11   Pectinesterase

Pectin + n H_2O = n methanol + pectate

[3.1.1.12   Deleted entry: previously Vitamin A esterase, now believed to be identical with EC 3.1.1.1]

3.1.1.13   Cholesterol esterase

A cholesterol ester + H_2O = cholesterol + a fatty acid anion

3.1.1.14   Chlorophyllase

Chlorophyll + H_2O = phytol + chlorophyllide

3.1.1.15   Arabinonolactonase

L-Arabinono-γ-lactone + H_2O = L-arabinonate

[3.1.1.16   Deleted entry: This reaction was due to a mixture of muconolactone isomerase (EC 5.3.3.4) and 3-oxoadipate-enol-lactone hydrolase (EC 3.1.1.24)]

3.1.1.17   Gluconolactonase

D-Glucono-δ-lactone + H_2O = D-gluconate

3.1.1.18   Aldonolactonase

L-Gulono-γ-lactone + H_2O = gulonate

3.1.1.19   Uronolactonase

D-Glucurono-δ-lactone + H_2O= D-glucuronate

3.1.1.20   Tannase

Digallate + H_2O = 2 gallate

3.1.1.21   Retinol-palmitate esterase

Retinol palmitate + H_2O = retinol + palmitate

3.1.1.22   Hydroxybutyrate-dimer hydrolase

3-D-(3-D-Hydroxybutyryloxy)-butyrate + H_2O = 2 3-D-hydroxybutyrate

3.1.1.23   Monoacylglycerol lipase

Hydrolyses glycerol monoesters of long-chain fatty acids

3.1.1.24   3-Oxoadipate enol-lactonase

4-Carboxymethylbut-3-enolide(1,4) + H_2O = 3-oxoadipate

3.1.1.25   γ-Lactonase

γ-Lactone + H_2O = 4-hydroxyacid

Number	Other Names	Basis for classification (Systematic Name)	Comments	Reference
3.1.1.10		Atropine acylhydrolase	Also acts on cocaine and other tropine esters	1062, 2269
3.1.1.11	Pectin demethoxylase, Pectin methoxylase, Pectin methylesterase	Pectin pectylhydrolase		708, 1979, 2234
3.1.1.13		Sterol-ester acylhydrolase	Also acts on esters of some other sterols	442, 1290, 1781, 3318
3.1.1.14		Chlorophyll chlorophyllidohydrolase	Also catalyses chlorophyllide transfer, *e.g.* converts chlorophyll to methylchlorophyllide	1343, 1712
3.1.1.15		L-Arabinono-γ-lactone lactonohydrolase		3641
3.1.1.17	Lactonase	D-Glucono-δ-lactone lactonohydrolase		391, 803
3.1.1.18		L-Gulono-γ-lactone lactonohydrolase		415, 507
3.1.1.19		D-Glucurono-δ-lactone lactonohydrolase		3706
3.1.1.20		Tannin acylhydrolase	Also hydrolyses ester links in other tannins	776
3.1.1.21		Retinol-palmitate palmitohydrolase		2045
3.1.1.22		3-D-(3-D-Hydroxybutyryloxy)-butyrate hydroxybutyrohydrolase		688
3.1.1.23		Glycerol-monoester acylhydrolase		2636
3.1.1.24		4-Carboxymethylbut-3-enolide (1,4) enol-lactonohydrolase		1819
3.1.1.25		γ-Lactone hydroxyacylhydrolase	The enzyme is specific for 4-8 carbon γ-lactones. It does not hydrolyse simple aliphatic esters,	887, 888

Number	Recommended Name	Reaction
3.1.1.26	Galactolipase	2,3-Di-O-acyl-1-O-(β-D-galactosyl)-D-glycerol + 2 H_2O = 1-O-(β-D-galactosyl)-D-glycerol + 2 fatty acid anions
3.1.1.27	4-Pyridoxolactonase	4-Pyridoxolactone + H_2O = 4-pyridoxate
3.1.1.28	Acylcarnitine hydrolase	O-Acylcarnitine + H_2O = a fatty acid + L-carnitine
3.1.1.29	Aminoacyl-tRNA hydrolase	N-Substituted aminoacyl-tRNA + H_2O = N-substituted amino acid + tRNA
3.1.1.30	D-Arabinonolactonase	D-Arabinono-γ-lactone + H_2O = D-arabinonate
3.1.1.31	6-Phosphogluconolactonase	6-Phospho-D-gluconate δ-lactone + H_2O = 6-phospho-D-gluconate
3.1.1.32	Phospholipase A_1	A lecithin + H_2O = 2-acylglycerophosphocholine + a fatty acid anion
3.1.1.33	6-O-Acetylglucose deacetylase	6-O-Acetyl-D-glucose + H_2O = glucose + acetate
3.1.1.34	Lipoprotein lipase	Triacylglycerol + H_2O = diacylglycerol + a fatty acid anion
3.1.1.35	Dihydrocoumarin hydrolase	Dihydrocoumarin + H_2O = melilotate
3.1.1.36	Limonin-D-ring-lactonase	Limonoate D-ring-lactone + H_2O = limonoate
3.1.1.37	Steroid-lactonase	Testololactone + H_2O = testolate
3.1.1.38	Triacetate-lactonase	Triacetate lactone + H_2O = triacetate

Number	Other Names	Basis for classification (Systematic Name)	Comments	Reference
			acetylcholine, sugar lactones or substituted aliphatic lactones, *e.g.* 3-hydroxy-4-butyrolactone; requires Ca^{2+}	
3.1.1.26		2,3-Di-*O*-acyl-1-*O*(β-D-galactosyl)-D-glycerol acylhydrolase	Also acts on 2,3-di-*O*-acyl-1-*O*-(6-*O*-α-D-galactosyl-β-D-galactosyl)-D-glycerol, and phosphatidylcholine and other phospholipids	1270, 1321
3.1.1.27		4-Pyridoxolactone lactonohydrolase		427
3.1.1.28		*O*-Acylcarnitine acylhydrolase	Acts on higher fatty acid (C-6 to C-18) esters of L-carnitine; highest activity is with *O*-decanoyl-L-carnitine	2046
3.1.1.29		Aminoacyl-tRNA aminoacylhydrolase		1538
3.1.1.30		D-Arabinono-γ-lactone lactonohydrolase		2546
3.1.1.31		6-Phospho-D-gluconate-δ-lactone lactonohydrolase		1624
3.1.1.32		Phosphatidate 1-acylhydrolase		997
3.1.1.33		6-*O*-Acetyl-D-glucose acetylhydrolase		765
3.1.1.34	Clearing factor lipase, Diglyceride lipase, Diacylglycerol lipase	Triacylglycero-protein acylhydrolase	Hydrolyses triacylglycerols in chylomicrons and low-density lipoproteins. Also hydrolyses diacylglycerol	878, 1118, 796, 2293, 2411
3.1.1.35		Dihydrocoumarin lactonohydrolase	Also hydrolyses some other benzenoid γ-lactones	1784
3.1.1.36		Limonoate-D-ring-lactone lactonohydrolase		2055
3.1.1.37		Testololactone lactonohydrolase		1351
3.1.1.38		Triacetolactone lactonohydrolase		1608

Number	Recommended Name	Reaction
3.1.1.39	Actinomycin lactonase	Actinomycin + H_2O = actinomycinic monolactone
3.1.1.40	Orsellinate-depside hydrolase	Orsellinate depside + H_2O = 2 orsellinate
3.1.1.41	Cephalosporin-C deacetylase	Cephalosporin C + H_2O = deacetylcephalosporin C + acetate

3.1.2   THIOLESTER HYDROLASES

3.1.2.1	Acetyl-CoA hydrolase	Acetyl-CoA + H_2O = CoA + acetate
3.1.2.2	Palmitoyl-CoA hydrolase	Palmitoyl-CoA + H_2O = CoA + palmitate
3.1.2.3	Succinyl-CoA hydrolase	Succinyl-CoA + H_2O = CoA + succinate
3.1.2.4	3-Hydroxyisobutyryl-CoA hydrolase	3-Hydroxyisobutyryl-CoA + H_2O = CoA + 3-hydroxyisobutyrate
3.1.2.5	Hydroxymethylglutaryl-CoA hydrolase	3-Hydroxy-3-methylglutaryl-CoA + H_2O = CoA + 3-hydroxy-3-methylglutarate
3.1.2.6	Hydroxyacylglutathione hydrolase	S-2-Hydroxyacylglutathione + H_2O = glutathione + a 2-hydroxyacid anion
3.1.2.7	Glutathione thiolesterase	S-Acylglutathione + H_2O = glutathione + a carboxylic acid anion

[3.1.2.8   *Deleted entry: now included with EC 3.1.2.6, Hydroxyacylglutathione hydrolase*]

[3.1.2.9   *Deleted entry: S-Acetoacetylhydrolipoate hydrolase*]

3.1.2.10	Formyl-CoA hydrolase	Formyl-CoA + H_2O = CoA + formate
3.1.2.11	Acetoacetyl-CoA hydrolase	Acetoacetyl-CoA + H_2O = CoA + acetoacetate
3.1.2.12	S-Formylglutathione hydrolase	S-Formylglutathione + H_2O = glutathione + formate
3.1.2.13	S-Succinylglutathione hydrolase	S-Succinylglutathione + H_2O = glutathione + succinate

Number	Other Names	Basis for classification (Systematic Name)	Comments	Reference
3.1.1.39		Actinomycin lactonohydrolase		1382
3.1.1.40	Lecanorate hydrolase	Orsellinate-depside hydrolase	Enzyme will only hydrolyse those substrates based on the 2,4-dihydroxy-6-methylbenzoate structure that also have a free hydroxyl *ortho* to the depside linkage	2978
3.1.1.41		Cephalosporin-C acetylhydrolase	Hydrolyses the acetyl ester bond on the 10-position	969
3.1.2.1	Acetyl-CoA deacylase, Acetyl-CoA acylase	Acetyl-CoA hydrolase		1008
3.1.2.2		Palmitoyl-CoA hydrolase		210, 2638, 3173
3.1.2.3	Succinyl-CoA acylase	Succinyl-CoA hydrolase		1008
3.1.2.4		3-Hydroxyisobutyryl-CoA hydrolase	Also hydrolyses 3-hydroxypropionyl-CoA	2773
3.1.2.5		3-Hydroxy-3-methylglutaryl-CoA hydrolase		685
3.1.2.6	Glyoxalase II	S-2-Hydroxyacylglutathione hydrolase	Also hydrolyses S-acetylglutathione, more slowly	2684, 3512, 3516
3.1.2.7		S-Acylglutathione hydrolase		1678
3.1.2.10		Formyl-CoA hydrolase		3100
3.1.2.11		Acetoacetyl-CoA hydrolase		757
3.1.2.12		S-Formylglutathione hydrolase	Also hydrolyses S-acetylglutathione, more slowly	3516, 3513
3.1.2.13		S-Succinylglutathione hydrolase		3513, 3514

Number	Recommended Name	Reaction

3.1.3 PHOSPHORIC MONOESTER HYDROLASES

3.1.3.1 Alkaline phosphatase   ·An orthophosphoric monoester + H_2O = an alcohol + orthophosphate

3.1.3.2 Acid phosphatase   An orthophosphoric monoester + H_2O = an alcohol + orthophosphate

3.1.3.3 Phosphoserine phosphatase   L(or D)-O-Phosphoserine + H_2O = L(or D)-serine + orthophosphate

3.1.3.4 Phosphatidate phosphatase   An L-α-phosphatidate + H_2O = a D-2,3-(or L-1,2)-diglyceride + orthophosphate

3.1.3.5 5′-Nucleotidase   A 5′-ribonucleotide + H_2O = a ribonucleoside + orthophosphate

3.1.3.6 3′-Nucleotidase   A 3′-ribonucleotide + H_2O = a ribonucleoside + orthophosphate

3.1.3.7 Phosphoadenylate 3′-nucleotidase   Adenosine 3′,5′-bisphosphate + H_2O = 5′-AMP + orthophosphate

3.1.3.8 3-Phytase   myo-Inositol hexakisphosphate + H_2O = D-myo-inositol 1,2,4,5,6-pentakisphosphate + orthophosphate

3.1.3.9 Glucose-6-phosphatase   D-Glucose 6-phosphate + H_2O = D-glucose + orthophosphate

3.1.3.10 Glucose-1-phosphatase   D-Glucose 1-phosphate + H_2O = D-glucose + orthophosphate

Number	Other Names	Basis for classification (Systematic Name)	Comments	Reference
3.1.3.1	Alkaline phosphomonoesterase, Phosphomonoesterase, Glycerophosphatase	Orthophosphoric-monoester phosphohydrolase (alkaline optimum)	Wide specificity. Also catalyses transphosphorylations. The human placental enzyme is a zinc-protein. Some enzymes hydrolyse pyrophosphate	833, 1195, 2061, 2299, 3187
3.1.3.2	Acid phosphomonoesterase, Phosphomonoesterase, Glycerophosphatase	Orthophosphoric-monoester phosphohydrolase (acid optimum)	Wide specificity. Also catalyses transphosphorylations	1543, 1838, 3470, 3472
3.1.3.3		O-Phosphoserine phosphohydrolase		345, 441, 2396
3.1.3.4		L-α-Phosphatidate phosphohydrolase		3126
3.1.3.5		5′-Ribonucleotide phosphohydrolase	Wide specificity for 5′-nucleotides	1142, 1281, 2993
3.1.3.6		3′-Ribonucleotide phosphohydrolase	Wide specificity for 3′-nucleotides	3063
3.1.3.7		Adenosine-3′,5′-bisphosphate 3′-phosphohydrolase	Also acts on 3′-phosphoadenylylsulphate	409
3.1.3.8		*myo*-Inositol-hexakisphosphate 3-phosphohydrolase		593, 1521
3.1.3.9		D-Glucose-6-phosphate phosphohydrolase	Wide distribution in animal tissues. Also catalyses potent transphosphorylations from carbamoyl phosphate, hexose phosphates, phosphoenolpyruvate, and nucleoside di- and triphosphates, to D-glucose, D-mannose, 3-O-methyl-D-glucose, or 2-deoxy-D-glucose.	2434, 2435, 2436, 3010, 2013, 1189, 73, 559
3.1.3.10		D-Glucose-1-phosphate phosphohydrolase	Also acts, more slowly, on D-galactose 1-phosphate	865, 3489

Number	Recommended Name	Reaction
3.1.3.11	Fructose-bisphosphatase	D-Frucose 1,6-bisphosphate + H_2O = D-fructose 6-phosphate + orthophosphate
3.1.3.12	Trehalose-phosphatase	Trehalose 6-phosphate + H_2O = trehalose + orthophosphate
3.1.3.13	Bisphosphoglycerate phosphatase	2,3-Bisphospho-D-glycerate + H_2O = 3-phospho-D-glycerate + orthophosphate
3.1.3.14	Methylthiophosphoglycerate phosphatase	1-Methylthio-3-phospho-D-glycerate + H_2O = methylthio-D-glycerate + orthophosphate
3.1.3.15	Histidinol phosphatase	L-Histidinol phosphate + H_2O = L-histidinol + orthophosphate
3.1.3.16	Phosphoprotein phosphatase	A phosphoprotein + n H_2O = a protein + n orthophosphate
3.1.3.17	Phosphorylase phosphatase	Phosphorylase a + 4 H_2O = 2 phosphorylase b + 4 orthophosphate
3.1.3.18	Phosphoglycollate phosphatase	2-Phosphoglycollate + H_2O = glycollate + orthophosphate
3.1.3.19	Glycerol-2-phosphatase	Glycerol 2-phosphate + H_2O = glycerol + orthophosphate
3.1.3.20	Phosphoglycerate phosphatase	D-Glycerate 2-phosphate + H_2O = D-glycerate + orthophosphate
3.1.3.21	Glycerol-1-phosphatase	Glycerol 1-phosphate + H_2O = glycerol + orthophosphate
3.1.3.22	Mannitol-1-phosphatase	D-Mannitol 1-phosphate + H_2O = D-mannitol + orthophosphate
3.1.3.23	Sugar-phosphatase	Sugar phosphate + H_2O = sugar + orthophosphate

Number	Other Names	Basis for classification (Systematic Name)	Comments	Reference
3.1.3.11	Hexosediphosphatase	D-Fructose-1,6-bisphosphate 1- phosphohydrolase	The animal enzyme also acts on sedoheptulose 1,7-bisphosphate	1078, 2261, 2635, 810
3.1.3.12		Trehalose-6-phosphate phosphohydrolase		445, 454
3.1.3.13		2,3-Bisphospho-D-glycerate 2-phosphohydrolase		1541, 2721
3.1.3.14		1-Methylthio-3-phospho-D-glycerate phosphohydrolase		310
3.1.3.15		Histidinol-phosphate phosphohydrolase		67
3.1.3.16	Protein phosphatase	Phosphoprotein phosphohydrolase	Acts on casein and other phosphoproteins; the spleen enzyme also acts on phenolic phosphates and phosphoamides	2539, 2840, 3293
3.1.3.17	PR-enzyme	Phosphorylase *a* phosphohydrolase		1099, 2705
3.1.3.18		2-Phosphoglycollate phosphohydrolase		2784
3.1.3.19		Glycerol-2-phosphate phosphohydrolase		2962, 3472
3.1.3.20		D-Glycerate-2-phosphate phosphohydrolase		856
3.1.3.21		Glycerol-1-phosphate phosphohydrolase	Also acts on 1-propanediol phosphate; but not on a variety of other phosphate esters	3470
3.1.3.22		D-Mannitol-1-phosphate phosphohydrolase		3748
3.1.3.23		Sugar-phosphate phosphohydrolase	Has a wide specificity, acting on aldohexose 1-phosphates, ketohexose 1-phosphates, aldohexose 6-phosphates, ketohexose 6-phosphates, both phosphate ester bonds of fructose 1,6-bisphosphate, phosphoric esters of disaccharides and pentose and triose phosphates, but at a slower rate	1904

Number	Recommended Name	Reaction
3.1.3.24	Sucrose-phosphatase	Sucrose 6^F-phosphate + H_2O = sucrose + orthophosphate
3.1.3.25	1L-*myo*-Inositol-1-phosphatase	1L-*myo*-Inositol 1-phosphate + H_2O = *myo*-inositol + orthophosphate
3.1.3.26	6-Phytase	*myo*-Inositol hexakisphosphate + H_2O = 1L-*myo*-inositol 1,2,3,4,5-pentakisphosphate + orthophosphate
3.1.3.27	Phosphatidylglycerophosphatase	Phosphatidylglycerophosphate + H_2O = phosphatidylglycerol + orthophosphate
3.1.3.28	ADPphosphoglycerate phosphatase	ADPphosphoglycerate + H_2O = ADPglycerate + orthophosphate
3.1.3.29	*N*-Acylneuraminate-9-phosphatase	*N*-Acylneuraminate 9-phosphate + H_2O = *N*-acylneuraminate + orthophosphate
3.1.3.30	3'-Phosphoadenylylsulphate 3'-phosphatase	3'-Phosphoadenylylsulphate + H_2O = adenylylsulphate + orthophosphate
3.1.3.31	Nucleotidase	Nucleotide + H_2O = nucleoside + orthophosphate
3.1.3.32	Polynucleotide 3'-phosphatase	3'-Phosphopolynucleotide + H_2O = polynucleotide + orthophosphate
3.1.3.33	Polynucleotide 5'-phosphatase	A 5'-phosphopolynucleotide + H_2O = polynucleotide + orthophosphate
3.1.3.34	Deoxynucleotide 3'-phosphatase	Deoxynucleoside 3'-phosphate + H_2O = deoxynucleoside + orthophosphate
3.1.3.35	Thymidylate 5'-phosphatase	Thymidylate + H_2O = thymidine + orthophosphate

Number	Other Names	Basis for classification (Systematic Name)	Comments	Reference
3.1.3.24		Sucrose-6^F-phosphate phosphohydrolase		1231
3.1.3.25		1L-*myo*-Inositol-1-phosphate phosphohydrolase	Acts primarily on phosphoric esters of secondary equatorial hydroxyl groups	512, 806
3.1.3.26	Phytase, Phytate 6-phosphatase	*myo*-Inositol-hexakisphosphate 6-phosphohydrolase		1521
3.1.3.27		Phosphatidylglycerophosphate phosphohydrolase		496
3.1.3.28		ADPphosphoglycerate phosphohydrolase	Also acts on 2,3-bisphosphoglycerate	3832
3.1.3.29		*N*-Acylneuraminate-9-phosphate phosphohydrolase		1539
3.1.3.30		3′-Phosphoadenylylsulphate 3′-phosphohydrolase		862
3.1.3.31		Nucleotide phosphohydrolase	A wide specificity for 2′-, 3′- and 5′-nucleotides; also hydrolyses glycerol phosphate and 4-nitrophenyl phosphate	111
3.1.3.32	2′(3′)-Polynucleotidase	Polynucleotide 3′-phosphohydrolase	Also hydrolyses nucleoside 2′-, 3′- and 5′-monophosphates, but only 2′- and 3′-phosphopolynucleotides	238
3.1.3.33	5′-Polynucleotidase	Polynucleotide 5′-phosphohydrolase	Does not act on nucleoside monophosphates. Induced in *Escherichia coli* by T-even phages	238
3.1.3.34	3′-Deoxynucleotidase	Deoxyribonucleotide 3′-phosphohydrolase	Also catalyses the selective removal of 3′-phosphate groups from DNA and oligodeoxynucleotides. Induced in *Escherichia coli* by T-even phages	238
3.1.3.35	Thymidylate 5′-nucleotidase	Thymidylate 5′-phosphohydrolase	Also acts on 5′-methyl-dCMP and 5′-ribothymidylate but at lower rates	94

Number	Recommended Name	Reaction
3.1.3.36	Phosphatidyl-inositol-bisphosphate phosphatase	Phosphatidyl-*myo*-inositol 4,5-bisphosphate + H_2O = phosphatidyl-inositol monophosphate + orthophosphate
3.1.3.37	Sedoheptulose-bisphosphatase	Sedoheptulose 1,7-bisphosphate + H_2O = sedoheptulose 7-phosphate + orthophosphate
3.1.3.38	3-Phosphoglycerate phosphatase	D-Glycerate 3-phosphate + H_2O = D-glycerate + orthophosphate
3.1.3.39	Streptomycin-6-phosphatase	Streptomycin 6-phosphate + H_2O = streptomycin + orthophosphate
3.1.3.40	Guanidinodeoxy-*scyllo*-inositol-4-phosphatase	1-Guanidino-1-deoxy-*scyllo*-inositol 4-phosphate + H_2O = 1-guanidino-1-deoxy-*scyllo*-inositol + orthophosphate
3.1.3.41	4-Nitrophenylphosphatase	4-Nitrophenyl phosphate + H_2O = 4-nitrophenol + orthophosphate
3.1.3.42	Glycogen-synthase-D phosphatase	Glycogen-synthase D = glycogen-synthase 1 + orthophosphate
3.1.3.43	[Pyruvate dehydrogenase (lipoamide)]-phosphatase	[Pyruvate dehydrogenase (lipoamide)] phosphate + H_2O = [pyruvate dehydrogenase (lipoamide)] + orthophosphate

3.1.4   PHOSPHORIC DIESTER HYDROLASES

3.1.4.1	Phosphodiesterase I	Hydrolytically removes 5'-nucleotides successively from the 3'-hydroxy termini of 3'-hydroxy-terminated oligonucleotides
3.1.4.2	Glycerophosphocholine phosphodiesterase	L-3-Glycerophosphocholine + H_2O = choline + glycerol 1-phosphate
3.1.4.3	Phospholipase C	A phosphatidylcholine + H_2O = 1,2-diacylglycerol + choline phosphate

Number	Other Names	Basis for classification (Systematic Name)	Comments	Reference
3.1.3.36	Triphosphoinositide phosphatase	Phosphatidyl-*myo*-inositol-4,5-bisphosphate phosphohydrolase		668
3.1.3.37		Sedoheptulose-1,7-bisphosphate 1-phosphohydrolase		3455, 2693
3.1.3.38		D-Glycerate-3-phosphate phosphohydrolase	Wide specificity, but 3-phosphoglycerate is the best substrate	2713
3.1.3.39		Streptomycin-6-phosphate phosphohydrolase	Also acts on dihydrostreptomycin 3′α,6-bisphosphate and streptidine 6-phosphate	3604, 3599
3.1.3.40		1-Guanidino-1-deoxy-*scyllo*-inositol-4-phosphate 4-phosphohydrolase		3604
3.1.3.41		4-Nitrophenylphosphate phosphohydrolase	A number of other substrates including phenyl phosphate, 4-nitrophenyl sulphate, acetyl phosphate and glycerol phosphate are not substrates	129, 130
3.1.3.42		[UDPglucose—glycogen glucosyltransferase-D] phosphohydrolase		4
3.1.3.43		[Pyruvate dehydrogenase (lipoamide)]-phosphate phosphohydrolase	A mitochondrial enzyme associated with the pyruvate dehydrogenase complex	1983
3.1.4.1	5′-Exonuclease	Oligonucleate 5′-nucleotidohydrolase	Low activity towards polynucleotides. A 3′-phosphate terminus on the substrate inhibits hydrolysis	1676
3.1.4.2		L-3-Glycerophosphocholine glycerophosphohydrolase	Also acts on L-3-glycerophospho-ethanolamine	664, 1236, 3629
3.1.4.3	Lipophosphodiesterase I, Lecithinase C, *Clostridium welchii* α-toxin, *Clostridium*	Phosphatidylcholine cholinephosphohydrolase	Also acts on sphingomyelin. The enzyme from *B. cereus* is a zinc-protein	758, 1177, 2000, 1989

Number	Recommended Name	Reaction

3.1.4.4   Phospholipase D   A phosphatidylcholine + H_2O = choline + a phosphatidate

[3.1.4.5   Transferred entry: now EC 3.1.21.1, Deoxyribonuclease I]

[3.1.4.6   Transferred entry: now EC 3.1.22.1, Deoxyribonuclease II]

[3.1.4.7   Transferred entry: now EC 3.1.31.1, Micrococcal endonuclease]

[3.1.4.8   Transferred entry: now EC 3.1.27.3, Ribonuclease T_1]

[3.1.4.9   Transferred entry: now EC 3.1.30.2, Endonuclease (Serratia marcescens)]

3.1.4.10   Monophosphatidylinositol phosphodiesterase   Monophosphatidylinositol + H_2O = D-*myo*-1:2-cyclic-inositol phosphate + diacylglycerol

3.1.4.11   Triphosphoinositide phosphodiesterase   Triphosphoinositide + H_2O = inositol trisphosphate + diacylglycerol

3.1.4.12   Sphingomyelin phosphodiesterase   Sphingomyelin + H_2O = *N*-acylsphingosine + choline phosphate

3.1.4.13   Serine-ethanolaminephosphate phosphodiesterase   Serine-phospho-ethanolamine + H_2O = serine + ethanolamine phosphate

3.1.4.14   [Acyl-carrier-protein]phosphodiesterase   [Acyl-carrier protein]+ H_2O = 4′-phosphopantetheine + apoprotein

3.1.4.15   Adenylyl-[glutamine-synthetase]hydrolase   Adenylyl-[L-glutamate:ammonia ligase (ADP-forming)] + H_2O = adenylate + [L-glutamate: ammonia ligase (ADP-forming)]

3.1.4.16   2′:3′-Cyclic-nucleotide 2′-phosphodiesterase   Nucleoside 2′:3′-cyclic phosphate + H_2O = nucleoside 3′-phosphate

Number	Other Names	Basis for classification (Systematic Name)	Comments	Reference
	oedematiens β- and γ-toxins			
3.1.4.4	Lipophosphodiesterase II, Lecithinase D, Choline phosphatase	Phosphatidylcholine phosphatidohydrolase	Also acts on other phosphatidyl esters	800, 1176, 3446, 125
3.1.4.10		Monophosphatidylinositol inositolphosphohydrolase		931
3.1.4.11		Triphosphoinositide inositol-trisphosphohydrolase	Also hydrolyses diphosphoinositides	3416
3.1.4.12		Sphingomyelin cholinephosphohydrolase	Has very little activity on phosphatidylcholine	211, 1268, 1589
3.1.4.13		Serine-phospho-ethanolamine ethanolaminephosphohydrolase	Acts only on those phosphodiesters that have ethanolamine as a component part of the molecule	1164
3.1.4.14		[Acyl-carrier-protein] 4′-pantetheine-phosphohydrolase		3523
3.1.4.15		Adenylyl-[L-glutamate:ammonia ligase (ADP-forming)] adenylylhydrolase		1264, 3020, 3021
3.1.4.16		Nucleoside-2′:3′-cyclic-phosphate 3′-nucleotidohydrolase	Also hydrolyses 3′-nucleoside monophosphates and bis-*p*-nitrophenyl phosphate, but not 3′-deoxynucleotides. Similar reactions are carried out by EC 3.1.4.8 and 3′:5′-cyclic AMP, 3.1.4.22	91, 488, 2493, 3510

Number	Recommended Name	Reaction
3.1.4.17	3′:5′-Cyclic-nucleotide phosphodiesterase	Nucleoside 3′:5′-cyclic phosphate + H_2O = nucleoside 5′-phosphate

[*3.1.4.18 Transferred entry: now EC 3.1.16.1, Spleen exonuclease*]

[*3.1.4.19 Transferred entry: now EC 3.1.13.3, Oligonucleotidase*]

[*3.1.4.20 Transferred entry: now EC 3.1.13.1, Ribonuclease II*]

[*3.1.4.21 Transferred entry: now EC 3.1.30.1, Endonuclease S₁(Aspergillus)*]

[*3.1.4.22 Transferred entry: now EC 3.1.27.5, Ribonuclease (pancreatic)*]

[*3.1.4.23 Transferred entry: now EC 3.1.27.1, Ribonuclease T₂*]

[*3.1.4.24 Deleted entry: Endoribonuclease III*]

[*3.1.4.25 Transferred entry: now EC 3.1.11.1, Exodeoxyribonuclease I*]

[*3.1.4.26 Deleted entry: Exodeoxyribonuclease II*]

[*3.1.4.27 Transferred entry: now EC 3.1.11.2, Exodeoxyribonuclease III*]

[*3.1.4.28 Transferred entry: now EC 3.1.11.3, Exodeoxyribonuclease (Lambda-induced)*]

[*3.1.4.29 Deleted entry: Oligodeoxyribonucleate exonuclease*]

[*3.1.4.30 Transferred entry: now EC 3.1.21.2, Endodeoxyribonuclease IV (Phage T4-induced)*]

[*3.1.4.31 Transferred entry: now EC 3.1.11.4, Exodeoxyribonuclease (Phage SP3-induced)*]

[*3.1.4.32 Deleted entry: Endodeoxyribonuclease(ATP- and S-adenosyl-methionine-dependent). See subgroup 3.1.24*]

[*3.1.4.33 Deleted entry: Endodeoxyribonuclease(ATP-hydrolysing). See subgroup 3.1.24*]

[*3.1.4.34 Deleted entry: Hybrid nuclease. See subgroups 3.1.15, 3.1.16, 3.1.30 and 3.1.31*]

3.1.4.35	3′:5′-Cyclic-GMP phosphodiesterase	Guanosine 3′:5′-cyclic phosphate + H_2O = guanosine 5′-phosphate
3.1.4.36	1:2-Cyclic-inositol-monophosphate phosphodiesterase	D-*myo*-Inositol 1:2-cyclic phosphate + H_2O = D-*myo*-inositol 1-phosphate
3.1.4.37	2′:3′-Cyclic-nucleotide 3′-phosphodiesterase	Nucleoside 2′:3′-cyclic phosphate + H_2O = nucleoside 2′-phosphate

Number	Other Names	Basis for classification (Systematic Name)	Comments	Reference
3.1.4.17		3':5'-Cyclic-nucleotide 5'-nucleotidohydrolase	Acts on 3':5'-cyclic AMP, 3':5'-cyclic dAMP, 3':5'-cyclic IMP, 3':5'-cyclic GMP and 3':5'-cyclic CMP	2335
3.1.4.35		3':5'-Cyclic-GMP 5'-nucleotidohydrolase		2089
3.1.4.36		D-*myo*-Inositol-1:2-cyclic-phosphate 2-inositolphosphohydrolase		666, 667
3.1.4.37		Nucleoside-2':3'-cyclic-phosphate 2'-nucleotidohydrolase		756, 3279

Number	Recommended Name	Reaction
3.1.4.38	Glycerophosphocholine cholinephosphodiesterase	L-3-Glycerophosphocholine + H_2O = glycerol + choline phosphate
3.1.4.39	Alkylglycerophosphoethanolamine phosphodiesterase	1-Alkyl-*sn*-glycero-3-phosphoethanolamine + H_2O = 1-alkyl-*sn*-glycerol 3-phosphate + ethanolamine
3.1.4.40	CMP-*N*-acylneuraminate phosphodiesterase	CMP-*N*-acylneuraminate + H_2O = CMP + *N*-acylneuraminate
3.1.4.41	Sphingomyelin phosphodiesterase D	Spingomyelin = ceramide phosphate + choline

3.1.5   TRIPHOSPHORIC MONOESTER HYDROLASES

3.1.5.1	dGTPase	dGTP + H_2O = deoxyguanosine + triphosphate

3.1.6   SULPHURIC ESTER HYDROLASES

3.1.6.1	Arylsulphatase	A phenol sulphate + H_2O = a phenol + sulphate
3.1.6.2	Sterol-sulphatase	3β-Hydroxyandrost-5-en-17-one 3-sulphate + H_2O = 3β-hydroxyandrost-5-en-17-one + sulphate
3.1.6.3	Glycosulphatase	D-Glucose 6-sulphate + H_2O = D-glucose + sulphate
3.1.6.4	Chondroitinsulphatase	Hydrolysis of the 6-sulphate groups of the 2-acetamido-2-deoxy-D-galactose 6-sulphate units of chondroitin sulphate
[3.1.6.5	*Deleted entry: Sinigrin sulphohydrolase, myrosulphatase]*	
3.1.6.6	Cholinesulphatase	Choline sulphate + H_2O = choline + sulphate
3.1.6.7	Cellulose polysulphatase	Hydrolysis of the 2- and 3-sulphate groups of the polysulphates of cellulose and charonin
3.1.6.8	Cerebroside-sulphatase	A cerebroside 3-sulphate + H_2O = a cerebroside + sulphate

Number	Other Names	Basis for classification (Systematic Name)	Comments	Reference
3.1.4.38		L-3-Glycerophosphocholine cholinephosphohydrolase	No activity on L-3-glycerophosphoethanol-amine	10
3.1.4.39	Lysophospholipase D	1-Alkyl-*sn*-glycero-3-phosphoethanolamine ethanolaminehydrolase	Also acts on acyl and choline analogues	3739
3.1.4.40	CMP-sialate hydrolase	CMP-*N*-acylneuraminate *N*-acylneuraminohydrolase		1628
3.1.4.41		Sphingomyelin ceramide-phosphohydrolase	Does not act on phosphatidylcholine, but hydrolyses 2-lysophosphatidylcholine to choline and 2-lysophosphatidate	3156, 475
3.1.5.1	Deoxy-GTPase	dGTP triphosphohydrolase	Also acts on GTP	1777
3.1.6.1	Sulphatase	Aryl-sulphate sulphohydrolase	A group of enzymes with rather similar specificities	735, 2865, 3626
3.1.6.2		Sterol-sulphate sulphohydrolase	Also acts on some related sterol sulphates	2865, 2867, 3229
3.1.6.3	Glucosulphatase	Sugar-sulphate sulphohydrolase	Also acts on other sulphates of mono- and disaccharides and on adenosine 5′-sulphate	732, 794, 2865
3.1.6.4	Chondroitinase	Chondroitin-sulphate sulphohydrolase		733, 734, 2610
3.1.6.6		Choline-sulphate sulphohydrolase		3348
3.1.6.7		Cellulose-sulphate sulphohydrolase		3344
3.1.6.8		Cerebroside-3-sulphate 3-sulphohydrolase		2177

Number	Recommended Name	Reaction
3.1.6.9	Chondro-4-sulphatase	Δ4,5-β-D-Glucuronosyl-(1,4)-2-acetamido-2-deoxy-D-galactose 4-sulphate + H_2O = Δ4,5-β-D-glucuronosyl-(1,4)-2-acetamido-2-deoxy-D-galactose + sulphate
3.1.6.10	Chondro-6-sulphatase	Δ4,5-β-D-Glucuronosyl-(1,4)-2-acetamido-2-deoxy-D-galactose 6-sulphate + H_2O = Δ4,5-β-D-glucuronosyl-(1,4)-2-acetamido-2-deoxy-D-galactose + sulphate
3.1.6.11	Disulphoglucosamine-6-sulphatase	2-Sulphamido-2-deoxy-6-O-sulpho-D-glucose + H_2O = 2-sulphamido-2-deoxy-D-glucose + sulphate

3.1.7 DIPHOSPHORIC MONOESTER HYDROLASES

Number	Recommended Name	Reaction
3.1.7.1	Prenol pyrophosphatase	Prenol diphosphate + H_2O = prenol + pyrophosphate

3.1.11 EXODEOXYRIBONUCLEASES PRODUCING 5'-PHOSPHOMONOESTERS

Number	Recommended Name	Reaction
3.1.11.1	Exodeoxyribonuclease I	Exonucleolytic cleavage in the 3' to 5' direction to yield 5'-phosphomononucleotides
3.1.11.2	Exodeoxyribonuclease III	Exonucleolytic cleavage in the 3' to 5' direction to yield 5'-phosphomononucleotides
3.1.11.3	Exodeoxyribonuclease (Lambda-induced)	Exonucleolytic cleavage in the 5' to 3' direction to yield 5'-phosphomononucleotides
3.1.11.4	Exodeoxyribonuclease (Phage SP3-induced)	Exonucleolytic cleavage in the 5' to 3' direction to yield 5'-phosphodinucleotides
3.1.11.5	Exodeoxyribonuclease V	Exonucleolytic cleavage (in presence of ATP) in either 5' to 3' or 3' to 5' direction to yield 5'-phosphooligonucleotides

Number	Other Names	Basis for classification (Systematic Name)	Comments	Reference
3.1.6.9		Δ4,5-β-D-Glucuronosyl-(1,4)-2-acetamido-2-deoxy-D-galactose-4-sulphate 4-sulphohydrolase	Also acts on the saturated analogue, but not on higher oligosaccharides or any 6-sulphates	1266, 3758
3.1.6.10		Δ4,5-β-D-Glucuronosyl-(1,4)-2-acetamido-2-deoxy-D-galactose-6-sulphate 6-sulphohydrolase	Also acts on the saturated analogue and acetamidodeoxygalactose 4,6-disulphate, but not higher oligosaccharides, or any 4-sulphate	3758
3.1.6.11		2-Sulphamido-2-deoxy-6-O-sulpho-D-glucose 6-sulphohydrolase		721
3.1.7.1		Prenol-diphosphate pyrophosphohydrolase		3467
3.1.11.1	*E. coli* exonuclease I. Similar enzymes: Mammalian DNase III, Exonuclease IV, T2 and T4 induced exodeoxyribonucleases		Preference for single stranded DNA. The *E. coli* enzyme hydrolyses glucosylated DNA. Formerly EC 3.1.4.25	1909, 1653, 316
3.1.11.2	*E. coli* exonuclease III. Similar enzymes: *Haemophilus influenzae* exonuclease		Preference for double-stranded DNA. Has endonucleolytic activity near apurinic sites on DNA. Formerly EC 3.1.4.27	1972, 2781, 2782
3.1.11.3	Lambda exonuclease. Similar enzymes: T4, T5 and T7 exonucleases, Mammalian DNase IV		Preference for double-stranded DNA. Does not attack single-strand breaks. Formerly EC 3.1.4.28	1973, 1991
3.1.11.4	Phage SP3 DNase, DNA 5′-dinucleotidohydrolase		Preference for single-stranded DNA. Formerly EC 3.1.4.31	3462
3.1.11.5	*E. coli* exonuclease V. Similar enzyme: *H. influenzae* ATP-dependent DNase		Preference for double-stranded DNA. Possesses DNA-dependent ATPase activity. Acts endonucleolytically on	2487, 1075, 3735, 799

Number	Recommended Name	Reaction
3.1.11.6	Exodeoxyribonuclease VII	Exonucleolytic cleavage in either 5′ to 3′ or 3′ to 5′ direction to yield 5′-phosphomononucleotides

3.1.13 EXORIBONUCLEASES PRODUCING 5′-PHOSPHOMONOESTERS

3.1.13.1	Ribonuclease II	Exonucleolytic cleavage in the 3′ to 5′ direction to yield 5′-phosphomononucleotides
3.1.13.2	Exoribonuclease H	Exonucleolytic cleavage to 5′-phosphomonoester oligonucleotides in both 5′ to 3′ and 3′ to 5′ directions
3.1.13.3	Oligonucleotidase	Exonucleolytic cleavage of oligonucleotides to yield 5′-phosphomononucleotides

3.1.14 EXORIBONUCLEASES PRODUCING OTHER THAN 5′-PHOSPHOMONOESTERS

3.1.14.1	Yeast ribonuclease	Exonucleolytic cleavage to 3′-phosphomononucleotides

3.1.15 EXONUCLEASES ACTIVE WITH EITHER RIBO- OR DEOXYRIBONUCLEIC ACIDS AND PRODUCING 5′-PHOSPHOMONOESTERS

3.1.15.1	Venom exonuclease	Exonucleolytic cleavage in the 3′ to 5′ direction to yield 5′-phosphomononucleotides

3.1.16 EXONUCLEASES ACTIVE WITH EITHER RIBO- OR DEOXYRIBONUCLEIC ACIDS AND PRODUCING OTHER THAN 5′-PHOSPHOMONOESTERS

3.1.16.1	Spleen exonuclease	Exonucleolytic cleavage in the 5′ to 3′ direction to yield 3′-phosphomononucleotides

Number	Other Names	Basis for classification (Systematic Name)	Comments	Reference
			single-stranded circular DNA	
3.1.11.6	*E. coli* exonuclease VII. Similar to: *Micrococcus luteus* exonuclease		Preference for single-stranded DNA	503, 504
3.1.13.1	Exoribonuclease. Similar enzymes: *Lactobacillus plantarum* RNase, Mouse nuclear RNase, Oligoribonuclease of *E. coli*		Preference for single-stranded RNA. Formerly EC 3.1.4.20	1650, 2451, 3158, 3168
3.1.13.2			Attacks RNA in duplex with DNA strand. Found in certain oncorna viruses and animal cells	3548
3.1.13.3			Also hydrolyzes NAD^+ to NMN and AMP. Formerly EC 3.1.4.19	980
3.1.14.1	Similar enzyme: RNase U_4			2485
3.1.15.1	Similar enzymes: Hog kidney phosphodiesterase, *Lactobacillus* exonuclease		Preference for single-stranded substrate. The venom enzyme is also active with superhelical turns	1886
3.1.16.1	3′-Exonuclease, Spleen phosphodiesterase. Similar enzymes: *Lactobacillus acidophilus* nuclease, *B. subtilis* nuclease, Salmon testis nuclease		Preference for single-stranded substrate. Formerly EC 3.1.4.18	278

Number	Recommended Name	Reaction

3.1.21   ENDODEOXYRIBONUCLEASES PRODUCING 5′-PHOSPHOMONOESTERS

3.1.21.1   Deoxyribonuclease I

Endonucleolytic cleavage to 5′-phospho-di-
and -oligonucleotide end products

3.1.21.2   Endodeoxyribonuclease IV
(Phage T4-induced)

Endonucleolytic cleavage to 5′-phospho-
oligonucleotide end products

3.1.22   ENDODEOXYRIBONUCLEASES PRODUCING OTHER THAN 5′-PHOSPHOMONOESTERS

3.1.22.1   Deoxyribonuclease II

Endonucleolytic cleavage to 3′-phospho-mono-
and -oligonucleotides

3.1.22.2   *Aspergillus* Deoxyribonuclease K₁

Endonucleolytic cleavage to 3′-phospho-mono-
and -oligonucleotides

3.1.22.3   Endodeoxyribonuclease V

Endonucleolytic cleavage at apurinic or
apyrimidinic sites to products with a
3′-phosphate

Number	Other Names	Basis for classification (Systematic Name)	Comments	Reference
3.1.21.1	Pancreatic DNase, DNase, thymonuclease. Similar enzymes: Streptococcal DNase (Streptodornase), T4 Endonuclease II, T7 Endonuclease II, *E.coli* endonuclease I "Nicking" nuclease of calf thymus, Colicin E2 and E3		Preference for double-stranded DNA. Formerly EC 3.1.4.5	671, 1835, 1887
3.1.21.2	*E. coli* endonuclease IV Similar enzymes: DNase V (mammalian), *Aspergillus sojae* DNase, *B. subtilis* endonuclease , T4 Endonuclease III, T7 Endonuclease I, *Aspergillus* DNase K2, Vaccinia virus DNase VI, Yeast DNase, *Chlorella* DNase		Preference for single-stranded DNA. Formerly EC 3.1.4.30	929, 930, 1157, 2891
3.1.22.1	DNase II, Pancreatic DNase II. Similar enzymes: Crab testes DNase, Snail DNase, Salmon testis DNase, Liver acid DNase, Human acid DNase of gastric mucosa and cervix		Preference for double-stranded DNA. Formerly EC 3.1.4.6	279
3.1.22.2	*Aspergillus* DNase K₁		Preference for single-stranded DNA	1607, 3057
3.1.22.3	Similar enzymes from: Thymus endonuclease, *E. coli* endonuclease II, Human placenta endonuclease			996

Number	Recommended Name	Reaction

3.1.23   SITE-SPECIFIC ENDODEOXYRIBONUCLEASES: CLEAVAGE IS SEQUENCE-SPECIFIC

These enzymes recognize a specific sequence of DNA and carry out endonucleolytic cleavage at the sequence indicated by an arrow to yield double-stranded fragments; both strands are cleaved. This sub-group includes the "restriction enzymes". For some of these enzymes the recognition sequence has not yet been determined and is recorded as unknown. For reasons of space, the hyphens representing the phosphodiester bond have been omitted as has the "d" for deoxyribonucleotide. The column headed "Reaction" for this group should be considered as "Reaction or Recognition Sequence".

3.1.23.1	Endodeoxyribonuclease *Alu*I	AG↓CT
3.1.23.2	Endodeoxyribonuclease *Asu*I	G↓GNCC
3.1.23.3	Endodeoxyribonuclease *Ava*I	C↓YCGRG
3.1.23.4	Endodeoxyribonuclease *Ava*II	G↓G(A or T)CC
3.1.23.5	Endodeoxyribonuclease *Bal*I	TGG↓CCA
3.1.23.6	Endodeoxyribonuclease *Bam*HI	G↓GATCC
3.1.23.7	Endodeoxyribonuclease *Bbv*I	GC(A or T)GC
3.1.23.8	Endodeoxyribonuclease *Bcl*I	T↓GATCA
3.1.23.9	Endodeoxyribonuclease *Bgl*I	Unknown
3.1.23.10	Endodeoxyribonuclease *Bgl*II	A↓GATCT
3.1.23.11	Endodeoxyribonuclease *Bpu*I	Unknown
3.1.23.12	Endodeoxyribonuclease *Dpn*I	Gm6ATC

Number	Other Names	Basis for classification (Systematic Name)	Comments	Reference
3.1.23.1			Isolated from *Arthrobacter luteus*	2811
3.1.23.2			Isolated from *Anabaena subcylindrica*	1398
3.1.23.3			Isolated from *Anabaena variabilis*	2317
3.1.23.4			Isolated from *Anabaena variabilis*	2317
3.1.23.5			Isolated from *Brevibacterium albidum*	1005
3.1.23.6	Similar enzymes from: *Bacillus amyloliquefaciens* F (*Bam*FI), (*Bam*KI), *Bacillus amyloliquefaciens* N (*Bam*NI), *Bacillus stearothermophilus* (*Bst*I)		Isolated from *Bacillus amyloliquefaciens* H	2814, 3403
3.1.23.7			Isolated from *Bacillus brevis*	1040
3.1.23.8			Isolated from *Bacillus caldolyticus*	2984
3.1.23.9			Isolated from *Bacillus globigii*	3703
3.1.23.10			Isolated from *Bacillus globigii*	2612, 3703
3.1.23.11			Isolated from *Bacillus pumilus*	1433
3.1.23.12			Requires m^6A at the site shown and will not cleave unmethylated sequence; cleavage site is unknown.	1858

Number	Recommended Name	Reaction
3.1.23.13	Endodeoxyribonuclease *Eco*RI	G↓AATTC
3.1.23.14	Endodeoxyribonuclease *Eco*RII	↓CC(A or T)GG
3.1.23.15	Endodeoxyribonuclease *Hae*I	(A or T)GG↓CC(A or T)
3.1.23.16	Endodeoxyribonuclease *Hae*II	RGCGC↓Y
3.1.23.17	Endodeoxyribonuclease *Hae*III	GG↓CC
3.1.23.18	Endodeoxyribonuclease *Hga*I	(5′-3′) GACGCNNNNN↓ (3′-5′) CTGCGNNNNNNNNNN↓
3.1.23.19	Endodeoxyribonuclease *Hha* I	GCG↓C
3.1.23.20	Endodeoxyribonuclease *Hind*II	GTY↓RAC
3.1.23.21	Endodeoxyribonuclease *Hind*III	A↓AGCTT

Number	Other Names	Basis for classification (Systematic Name)	Comments	Reference
			Isolated from pneumococcus	
3.1.23.13			Isolated from *E. coli* carrying a fi⁺ plasmid	1258, 3807
3.1.23.14			Isolated from *E. coli* carrying a fi⁻ plasmid	3807
3.1.23.15			Isolated from *Haemophilus aegyptius*	2318
3.1.23.16	Similar enzymes from: *Haemophilus influenzae* HI (*Hin* HI) and *Neisseria gonorrhoea* (*Ngo*I)		Isolated from *Haemophilus aegyptius*	2810, 3079, 2813
3.1.23.17	Endonuclease Z. Similar enzymes from: *Bacillus sphaericus* (*Bsp*RI), *Bacillus subtilis* X5 (*Bsu*I), *Brevibacterium luteum* (*Blu*II), *Providencia alcalifaciens* (*Pal*I), *Haemophilus haemoglobinophilus* (*Hhg*I) and *Streptococcus faecalis* (*Sfa*I)		Isolated from *Haemophilus aegyptius*. Cleaves duplex regions in single-stranded DNA	392, 2217
3.1.23.18			Isolated from *Haemophilus gallinarum*	405, 3346
3.1.23.19			Isolated from *Haemophilus haemolyticus*. Cleaves duplex regions in single-stranded DNA	1318, 2713
3.1.23.20	Endonuclease R. Similar enzymes from: *Corynebacterium humiferum* (*Chu*II), *Haemophilus influenzae* serotype c (*Hinc*II), *Moraxella nonliquefaciens* (*Mnn*I)		Isolated from *Haemophilus influenzae*, serotype d	1655, 3116
3.1.23.21	Similar enzymes from: *Bordetella bronchiseptica* (*Bbr*I), *Corynebacterium humiferum* (*Chu*I),		Isolated from *Haemophilus influenzae*, serotype d	2494

Number	Recommended Name	Reaction
3.1.23.22	Endodeoxyribonuclease *Hinf*I	G↓ANTC
3.1.23.23	Endodeoxyribonuclease *Hpa*I	GTT↓AAC
3.1.23.24	Endodeoxyribonuclease *Hpa*II	C↓CGG
3.1.23.25	Endodeoxyribonuclease *Hph*I	(5′-3′) GGTGANNNNNNNN↓ (3′-5′) CCACTNNNNNNN↓
3.1.23.26	Endodeoxyribonuclease *Kpn*I	GGTAC↓C
3.1.23.27	Endodeoxyribonuclease *Mbo*I	↓GATC
3.1.23.28	Endodeoxyribonuclease *Mbo*II	(5′-3′) GAAGANNNNNNN↓ (3′-5′) CTTCTNNNNNNN↓
3.1.23.29	Endodeoxyribonuclease *Mnl*I	CCTC
3.1.23.30	Endodeoxyribonuclease *Pfa*I	Unknown
3.1.23.31	Endodeoxyribonuclease *Pst*I	CTGCA↓G
3.1.23.32	Endodeoxyribonuclease *Pvu*I	Unknown

Number	Other Names	Basis for classification (Systematic Name)	Comments	Reference
	Haemophilus influenzae serotype b (*Hinb*III), *Haemophilus influenzae* serotype f (*Hinf*II), *Haemophilus suis* (*Hsu*I)			
3.1.23.22	Similar enzyme from: *Haemophilus haemolyticus (Hha*II)		Isolated from *Haemophilus influenzae,* serotype f	1413, 2218
3.1.23.23			Isolated from *Haemophilus parainfluenzae*	993, 3025
3.1.23.24	Similar enzymes from: *Haemophilus aphrophilus (Hap*II), *Moraxella nonliquefaciens*(*Mno*I)		Isolated from *Haemophilus parainfluenzae*	636, 3025
3.1.23.25			Isolated from *Haemophilus parahaemolyticus*	1711, 2218
3.1.23.26			Isolated from *Klebsiella pneumoniae*	3108
3.1.23.27	Similar enzymes from: *Diplococcus pneumoniae (Dpn*II), *Moraxella osloensis (Mos*I), *Staphylococcus aureus* (*Sau*3AI)		Isolated from *Moraxella bovis*	1004
3.1.23.28			Isolated from *Moraxella bovis*	547, 1004
3.1.23.29			Cleavage site unknown but is a few nucleotides away from recognition sequence. Isolated from *Moraxella nonliquefaciens*	3826
3.1.23.30			Isolated from *Pseudomonas facilis*	
3.1.23.31	Similar enzymes from: *Bacillus subtilis* (*Bsu*1247), *Streptomyces albus (Sal*PI), *Xanthomonas malvacearum (Xma*II)		Isolated from *Providencia stuartii*	406, 3108
3.1.23.32			Isolated from *Proteus vulgaris*	1040

Number	Recommended Name	Reaction
3.1.23.33	Endodeoxyribonuclease *Pvu*II	CAG↓CTG
3.1.23.34	Endodeoxyribonuclease *Sac*I	G↓AGCTC
3.1.23.35	Endodeoxyribonuclease *Sac*II	CCGC↓GG
3.1.23.36	Endodeoxyribonuclease *Sac*III	Unknown
3.1.23.37	Endodeoxyribonuclease *Sal*I	G↓TCGAC
3.1.23.38	Endodeoxyribonuclease *Sgr*I	Unknown
3.1.23.39	Endodeoxyribonuclease *Taq*I	T↓CGA
3.1.23.40	Endodeoxyribonuclease *Taq*II	Unknown
3.1.23.41	Endodeoxyribonuclease *Xba*I	T↓CTAGA
3.1.23.42	Endodeoxyribonuclease *Xho*I	C↓TCGAG
3.1.23.43	Endodeoxyribonuclease *Xho*II	Unknown
3.1.23.44	Endodeoxyribonuclease *Xma*I	C↓CCGGG
3.1.23.45	Endodeoxyribonuclease *Xni*I	Unknown

3.1.24 SITE-SPECIFIC ENDODEOXYRIBONUCLEASES: CLEAVAGE IS NOT SEQUENCE-SPECIFIC

These enzymes recognize a specific sequence of DNA although cleavage does not occur at that sequence.

Number	Other Names	Basis for classification (Systematic Name)	Comments	Reference
3.1.23.33			Isolated from *Proteus vulgaris*	1040
3.1.23.34	Similar enzymes from: *Streptomyces stanford* (*Sst*I)		Isolated from *Streptomyces achromogenes*	110
3.1.23.35	Similar enzymes from: *Streptomyces stanford* (*Sst*II), *Thermopolyspora glauca* (*Tgl*I)		Isolated from *Streptomyces achromogenes*	110
3.1.23.36	Similar enzymes from: *Streptomyces stanford* (*Sst*III)		Isolated from *Streptomyces achromogenes*	110
3.1.23.37	Similar enzymes from: *Xanthomonas amaranthicola* (*Xam*I)		Isolated from *Streptomyces albus* G	110
3.1.23.38			Isolated from *Streptomyces griseus*	109
3.1.23.39			Isolated from *Thermus aquaticus*	2920
3.1.23.40			Isolated from *Thermus aquaticus*	2813
3.1.23.41			Isolated from *Xanthomonas badrii*	3829
3.1.23.42	Similar enzymes: *Brevibacterium luteum* (*Blu*I), *Xanthomonas papavericola* (*Xpa*I)		Isolated from *Xanthomonas holicola*	1039
3.1.23.43			Isolated from *Xanthomonas holicola*	1039
3.1.23.44	Similar enzyme from: *Serratia marcescens* (*Sma*I)		Isolated from *Xanthomonas malvacearum*	827
3.1.23.45			Isolated from *Xanthomonas nigromaculans*	1178

Number	Recommended Name	Reaction
3.1.24.1	Endodeoxyribonuclease *Eco*B	Unknown
3.1.24.2	Endodeoxyribonuclease *Eco*K	Unknown
3.1.24.3	Endodeoxyribonuclease *Eco*Pl	Unknown
3.1.24.4	Endodeoxyribonuclease *Eco*P15	Unknown

3.1.25   SITE-SPECIFIC ENDODEOXYRIBONUCLEASES; SPECIFIC FOR ALTERED BASES

3.1.25.1	Endodeoxyribonuclease (pyrimidine dimer)	Endonucleolytic cleavage near pyrimidine dimers to products with 5'-phosphate
3.1.25.2	Endodeoxyribonuclease (apurinic or apyrimidinic)	Endonucleolytic cleavage near apurinic or apyrimidinic sites to products with 5'-phosphate

3.1.26   ENDORIBONUCLEASES PRODUCING 5'-PHOSPHOMONOESTERS

3.1.26.1	Ribonuclease (*Physarum polycephalum*)	Endonucleolytic cleavage to 5'-phosphomonoester
3.1.26.2	Ribonuclease alpha	Endonucleolytic cleavage to 5'-phosphomonoester
3.1.26.3	Ribonuclease III	Endonucleolytic cleavage to 5'-phosphomonoester

Number	Other Names	Basis for classification (Systematic Name)	Comments	Reference
3.1.24.1			Absolute requirement for ATP (or dATP) and S-adenosyl-L-methionine. Cleaves randomly. ATPase activity in presence of DNA. Isolated from *E. coli* B	843
3.1.24.2			Absolute requirement for ATP (or dATP) and S-adenosyl-L-methionine. Cleaves randomly. ATPase activity in presence of DNA. Isolated from *E. coli* K	1153, 2207
3.1.24.3			Requirement for S-adenosyl-L-methionine. Isolated from *E. coli* lysogenic for P10	1154, 2207
3.1.24.4			Requirement for S-adenosyl-L-methionine. Isolated from *E. coli* lysogenic for P15	2765
3.1.25.1	Similar enzymes from: T4 endonuclease V, *E. coli* endonucleases III and V, Correndonuclease II		Acts on the damaged strand, 5' from the damaged site	394, 2778
3.1.25.2			Acts on damaged strand, 5' from the damaged site	1996
3.1.26.1	Similar enzymes from: Pig liver nuclease, HeLa cell RNase, *E. coli* RNase, Bovine adrenal cortex RNase			1318
3.1.26.2			Specific for O-methylated RNA	2446
3.1.26.3	Similar enzyme from: Calf thymus RNase		Specific for double-stranded RNA	2817, 611

Number	Recommended Name	Reaction
3.1.26.4	Endoribonuclease H (calf thymus)	Endonucleolytic cleavage to 5′-phosphomonoester
3.1.26.5	Ribonuclease P	Endonucleolytic cleavage to 5′-phosphomonoester

3.1.27 ENDORIBONUCLEASES PRODUCING OTHER THAN 5′-PHOSPHOMONOESTERS

3.1.27.1	Ribonuclease T$_2$	Two-stage endonucleolytic cleavage to 3′-phosphomononucleotides and oligonucleotides with 2′,3′-cyclic phosphate intermediates
3.1.27.2	Ribonuclease (*Bacillus subtilis*)	Endonucleolytic cleavage to 2′,3′-cyclic nucleotides
3.1.27.3	Ribonuclease T$_1$	Two-stage endonucleolytic cleavage to 3′-phospho-mono- and oligonucleotides ending in Gp with 2′,3′-cyclic phosphate intermediates
3.1.27.4	Ribonuclease U$_2$	Two-stage endonucleolytic cleavage to 3′-phospho-mono- and oligonucleotides ending in Ap or Gp with 2,3′-cyclic phosphate intermediates
3.1.27.5	Ribonuclease (pancreatic)	Endonucleolytic cleavage to 3′-phosphomono- and oligonucleotides ending in Cp or Up with 2′,3′-cyclic phosphate intermediates

Number	Other Names	Basis for classification (Systematic Name)	Comments	Reference
3.1.26.4	Similar enzymes from: *E. coli,* chicken embryo, human KB cells, rat liver, *Ustilago maydis,* human leukaemic cells, *Saccharomyces cerevisiae* (H_2), and *Tetrahymena pyriformis*		Acts on RNA · DNA hybrids	3204, 1152
3.1.26.5	Similar enzyme from: RNase NU from KB cells		Specificity for tRNA precursors	2816
3.1.27.1	Ribonuclease II. Similar enzymes from: Plant RNase, *E. coli* RNase I, RNase N_2, Microbial RNase II		Formerly EC 2.7.7.17 and 3.1.4.23	1279, 2736, 3496
3.1.27.2	Similar enzymes from: *Azotobacter agilis* RNase, *Proteus mirabilis* RNase			2417, 3767, 3768
3.1.27.3	Guanyloribonuclease, *Aspergillus oryzae* ribonuclease, RNase N_1 and N_2 . Similar enzymes: *N. crassa* RNase N_1 and N_2, *Ustilago sphaerogena* RNase, *Chalaropsis* RNase, *B. subtilis* RNase, Microbial RNase I		Formerly EC 3.1.4.8	1599, 3343
3.1.27.4	Similar enzymes: RNase U_3, *Pleospora* RNase, *Trichoderma koningi* RNase III			1063, 3497
3.1.27.5	RNase, RNase I, Pancreatic RNase, Ribonuclease I Similar enzymes: Venom RNase, *Thiobacillus thioparus* RNase, *Xenopus laevis* RNase, *Rhizopus oligosporus* RNase		Formerly EC 2.7.7.16 and 3.1.4.22	84, 236

Number	Recommended Name	Reaction
3.1.27.6	Ribonuclease (*Enterobacter*)	Endonucleolytic cleavage to 3'-phosphomono- and oligonucleotides ending with 2',3'-cyclic phosphate intermediates

3.1.30 ENDONUCLEASES ACTIVE WITH EITHER RIBO- OR DEOXYRIBONUCLEIC ACIDS AND PRODUCING 5'-PHOSPHOMONOESTERS

3.1.30.1	Endonuclease S_1 (*Aspergillus*)	Endonucleolytic cleavage to 5'-phosphomono- and oligonucleotide end products
3.1.30.2	Endonuclease (*Serratia marcescens*)	Endonucleolytic cleavage to 5'-phosphomono and oligonucleotide end products

3.1.31 ENDONUCLEASES ACTIVE WITH EITHER RIBO- OR DEOXYRIBONUCLEIC ACIDS AND PRODUCING OTHER THAN 5'-PHOSPHOMONOESTERS

3.1.31.1	Micrococcal endonuclease	Endonucleolytic cleavage to 3'-phosphomono- and oligonucleotide end products

3.2 GLYCOSIDASES

Glycosidases are classified under hydrolases, though some of them can also transfer glycosyl residues to oligosaccharides, polysaccharides and other alcoholic acceptors. The glycosidases are subdivided into those hydrolysing *O*-glycosyl-, *N*-glycosyl- and *S*-glycosyl-compounds (3.2.1, 3.2.2 and 3.2.3, respectively).

3.2.1 HYDROLYSING *O*-GLYCOSYL COMPOUNDS

3.2.1.1	α-Amylase	Endohydrolysis of 1,4-α-D-glucosidic linkages in polysaccharides containing three or more 1,4-α-linked D-glucose units
3.2.1.2	β-Amylase	Hydrolysis of 1,4-α-D-glycosidic linkages in polysaccharides so as to remove successive maltose units from the non-reducing ends of the chains

Number	Other Names	Basis for classification (Systematic Name)	Comments	Reference
3.1.27.6			Preference for cleavage at CpA. Homopolymers of A, U or G are not hydrolyzed	1941, 2091
3.1.30.1	*Aspergillus* nuclease S₁; Single-stranded-nucleate endonuclease, Deoxyribonuclease S₁. Similar enzymes: *N. crassa* nuclease 1423, Mung bean nuclease, *Penicillium citrium* nuclease P₁		Preference for single-stranded substrate. Formerly EC 3.1.4.21	80, 3303, 3563
3.1.30.2	Similar enzymes: Silkworm nuclease, Potato nuclease, *Azotobacter* nuclease		Hydrolyses double - or single-stranded substrate. Formerly EC 3.1.4.9	2220, 3221, 3635
3.1.31.1	Similar enzymes: *Chlamydomonas* nuclease, Spleen phosphodiesterase, Spleen endonuclease		Hydrolyses double - or single-stranded substrate. Formerly EC 3.1.4.7	85
3.2.1.1	Diastase, Ptyalin, Glycogenase	1,4-α-D-Glucan glucanohydrolase	Acts on starch, glycogen, and related polysaccharides and oligosaccharides in a random manner; reducing groups are liberated in the α-configuration	884, 2072, 2983
3.2.1.2	Diastase, Saccharogen amylase, Glycogenase	1,4-α-D-Glucan maltohydrolase	Acts on starch, glycogen and related polysaccharides and oligosaccharides producing β-maltose by an inversion	191, 926, 2072

Number	Recommended Name	Reaction
3.2.1.3	Exo-1,4-α-D-glucosidase	Hydrolysis of terminal 1,4-linked α-D-glucose residues successively from non-reducing ends of the chains with release of β-D-glucose
3.2.1.4	Cellulase	Endohydrolysis of 1,4-β-glucosidic linkages in cellulose, lichenin and cereal β-D-glucans
[3.2.1.5	Deleted entry: Licheninase]	
3.2.1.6	Endo-1,3(4)-β-D-glucanase	Endohydrolysis of 1,3- or 1,4-linkages in β-D-glucans when the glucose residue whose reducing group is involved in the linkage to be hydrolysed is itself substituted at C-3
3.2.1.7	Inulinase	Endohydrolysis of 2,1-β-D-fructosidic linkages in inulin
3.2.1.8	Endo-1,4-β-D-xylanase	Endohydrolysis of 1,4-β-D-xylosidic linkages in xylans
[3.2.1.9	Deleted entry: Amylopectin-1,6-glucosidase]	
3.2.1.10	Oligo-1,6-glucosidase	Hydrolysis of 1,6-α-D-glucosidic linkages in isomaltose and dextrins produced from starch and glycogen by α-amylase
3.2.1.11	Dextranase	Endohydrolysis of 1,6-α-D-glucosidic linkages in dextran
[3.2.1.12	Deleted entry: Cycloheptaglucanase. Now included with EC 3.2.1.54]	
[3.2.1.13	Deleted entry: Cyclohexaglucanase. Now included with EC 3.2.1.54]	
3.2.1.14	Chitinase	Random hydrolysis of 1,4-β-acetamido-2-deoxy-D-glucoside linkages in chitin and chitodextrin

Number	Other Names	Basis for classification (Systematic Name)	Comments	Reference
3.2.1.3	Glucoamylase, Amyloglucosidase, γ-Amylase, Lysosomal α-glucosidase, Acid maltase	1,4-α-D-Glucan glucohydrolase	Most forms of the enzyme are able also rapidly to hydrolyse 1,6-α-D-glucosidic bonds when the next bond in sequence is 1,4, and some preparations of this enzyme hydrolyse 1,6- and 1,3-α-D-glucosidic bonds in other polysaccharides. This entry covers all such enzymes acting on polysaccharides more rapidly than on oligosaccharides	927, 1438, 1511, 2224, 3476, 1654
3.2.1.4	Endo-1,4-β-glucanase	1,4-(1,3;1,4)-β-D-Glucan 4-glucanohydrolase	Will also hydrolyse 1,4-linkages in β-D-glucans also containing 1,3-linkages	641, 1877, 2319, 2422, 3666
3.2.1.6	Endo-1,3-β-D-glucanase, Laminarinase	1,3-(1,3;1,4)-β-D-Glucan 3 (4)-glucanohydrolase	Substrates include laminarin, lichenin and cereal β-D-glucans; different from EC 3.2.1.39	213, 214, 616, 2747, 3157
3.2.1.7	Inulase	2,1-β-D-Fructan fructanohydrolase		23
3.2.1.8		1,4-β-D-Xylan xylanohydrolase		1385, 1449, 3665
3.2.1.10	Limit dextrinase, Isomaltase	Dextrin 6-α-D-glucanohydrolase		1877, 1878
3.2.1.11		1,6-α-D-Glucan 6-glucanohydrolase		178, 883, 2868
3.2.1.14	Chitodextrinase, 1,4-β-Poly-N-acetylglucosaminidase, Poly-β-glucosaminidase	Poly (1,4-β-(2-acetamido-2-deoxy-D-glucoside)) glycanohydrolase	Some chitinases also display the activity defined in EC 3.2.1.17	883, 3454, 3837

Number	Recommended Name	Reaction
3.2.1.15	Polygalacturonase	Random hydrolysis of 1,4-α-D-galactosiduronic linkages in pectate and other galacturonans
[3.2.1.16	Deleted entry: Alginase]	
3.2.1.17	Lysozyme	Hydrolysis of 1,4-β-linkages between N-acetylmuramic acid and 2-acetamido-2-deoxy- D-glucose residues in a mucopolysaccharide or muropeptide
3.2.1.18	Neuraminidase	Hydrolysis of 2,3-, 2,6- and 2,8-glucosidic linkages joining terminal non-reducing N- or O-acylneuraminyl residues to galactose, N-acetylhexosamine, or N- or O-acylated neuraminyl residues in oligosaccharides, glycoproteins, glycolipids or colominic acid
[3.2.1.19	Deleted entry: Heparinase]	
3.2.1.20	α-D-Glucosidase	Hydrolysis of terminal, non-reducing 1,4-linked α-D-glucose residues with release of α-glucose
3.2.1.21	β-D-Glucosidase	Hydrolysis of terminal non-reducing β-D-glucose residues with release of β-D-glucose
3.2.1.22	α-D-Galactosidase	Hydrolysis of terminal, non-reducing α-D-galactose residues in α-D-galactosides, including galactose oligosaccharides, galactomannans and galactolipids
3.2.1.23	β-D-Galactosidase	Hydrolysis of terminal non-reducing β-D-galactose residues in β-D-galactosides
3.2.1.24	α-D-Mannosidase	Hydrolysis of terminal, non-reducing α-D-mannose residues in α-D-mannosides

Number	Other Names	Basis for classification (Systematic Name)	Comments	Reference
3.2.1.15	Pectin depolymerase, Pectinase	Poly(1,4-α-D-galacturonide) glycanohydrolase		708, 1979, 2153, 2221, 2601
3.2.1.17	Muramidase, Mucopeptide glycohydrolase	Mucopeptide N-acetylmuramoylhydrolase	See also EC 3.2.1.14	313, 314, 1526
3.2.1.18	Sialidase	Acylneuraminyl hydrolase		1089, 678
3.2.1.20	Maltase, Glucoinvertase, Glucosidosucrase	α-D-Glucoside glucohydrolase	This single entry covers a group of enzymes whose specificity is directed mainly towards the exohydrolysis of 1,4-α-glucosidic linkages, and that hydrolyse oligosaccharides rapidly, relative to polysaccharides, which are hydrolysed relatively slowly, or not at all	1877, 1960, 410
3.2.1.21	Gentiobiase, Cellobiase, Amygdalase	β-D-Glucoside glucohydrolase	Wide capability of hydrolysing β-D-glucosides. Some examples also hydrolyse one or more of the following: β-D-galactosides, α-L-arabinosides, β-D-xylosides	567, 622, 1301, 1877, 2908
3.2.1.22	Melibiase	α-D-Galactoside galactohydrolase	Also hydrolyses α-D-fucosides	1340, 3304, 3544, 3683
3.2.1.23	Lactase	β-D-Galactoside galactohydrolase	Some enzymes in this group also hydrolyse α-L-arabinosides	315, 1810, 1869, 2268, 3314, 3606
3.2.1.24		α-D-Mannoside mannohydrolase	Also hydrolyses α-D-lyxosides and heptopyranosides with the	1955

Number	Recommended Name	Reaction

3.2.1.25   β-D-Mannosidase

Hydrolysis of terminal, non-reducing β-D-mannose residues in β-D-mannosides

3.2.1.26   β-D-Fructofuranosidase

Hydrolysis of terminal non-reducing β-D-fructofuranoside residues in β-D-fructofuranosides

[3.2.1.27   Deleted entry: α-1,3-Glucosidase]

3.2.1.28   α,α-Trehalase

α,α-Trehalose + H_2O = 2 D-glucose

[3.2.1.29   Deleted entry: Chitobiase. Now included with EC 3.2.1.30]

3.2.1.30   β-N-Acetyl-D-glucosaminidase

Hydrolysis of terminal, non-reducing 2-acetamido-2-deoxy-β-D-glucose residues in chitobiose and higher analogues and in glycoproteins

3.2.1.31   β-D-Glucuronidase

A β-D-glucuronide + H_2O = an alcohol + D-glucuronate

3.2.1.32   Endo-1,3-β-D-xylanase

Random hydrolysis of 1,3-β-D-xylosidic linkages in 1,3-β-D-xylans

3.2.1.33   Amylo-1,6-glucosidase

Endohydrolysis of 1,6-α-D-glucoside linkages at points of branching in chains of 1,4-linked α-D-glucose residues

[3.2.1.34   Deleted entry: Chondroitinase. Now included with EC 3.2.1.35]

3.2.1.35   Hyaluronoglucosaminidase

Random hydrolysis of 1,4-linkages between 2-acetamido-2-deoxy-β-D-glucose and D-glucuronate residues in hyaluronate

3.2.1.36   Hyaluronoglucuronidase

Random hydrolysis of 1,3-linkages between 2-acetamido-2-deoxy-D-glucose and β-D-glucuronate residues in hyaluronate

Number	Other Names	Basis for classification (Systematic Name)	Comments	Reference
			same configuration at C-2, C-3 and C-4 as mannose	
3.2.1.25	Mannanase, Mannase	β-D-Mannoside mannohydrolase		23, 707, 1417, 217
3.2.1.26	Invertase, Invertin, Saccharase, β-h-Fructosidase	β-D-Fructofuranoside fructohydrolase	Substrates include sucrose; also catalyses fructotransferase reactions	2322, 2399
3.2.1.28		α,α-Trehalose glucohydrolase		1574, 2323
3.2.1.30		2-Acetamido-2-deoxy-β-D-glucoside acetamidodeoxyglucohydrolase	Various examples of the enzyme act on N-acetyl-β-D-galactosaminides and chitobiose	881, 1954, 2666, 3730, 3837
3.2.1.31		β-D-Glucuronide glucuronosohydrolase		753, 889, 1937, 3582
3.2.1.32	Xylanase	1,3-β-D-Xylan xylanohydrolase		973
3.2.1.33		Dextrin 6-α-D-glucosidase	In mammals and yeast this enzyme is linked to a glycosyltransferase similar to EC 2.4.1.25; together these two activities constitute the glycogen debranching system	398, 1897, 2387
3.2.1.35	Mucinase, Spreading factor, Hyaluronidase, Hyaluronoglucosidase	Hyaluronate 4-glycanohydrolase	Also hydrolyses 1,4-β-glycosidic linkages between N-acetylgalactosamine or N-acetylgalactosamine sulphate and glucuronic acid in chondroitin, chondroitin 4- and 6-sulphates and dermatan	2212, 2213, 2723, 3649
3.2.1.36	Mucinase, Spreading factor, Hyaluronidase	Hyaluronate 3-glycanohydrolase		1981, 2212

Number	Recommended Name	Reaction
3.2.1.37	Exo-1,4-β-D-xylosidase	Hydrolysis of 1,4-β-D-xylans so as to remove successive D-xylose residues from the non-reducing termini
3.2.1.38	β-D-Fucosidase	Hydrolysis of terminal non-reducing β-D-fucose residues in β-D-fucosides
3.2.1.39	Endo-1,3-β-D-glucanase	Hydrolysis of 1,3-β-D-glucosidic linkages in 1,3-β-D-glucans
3.2.1.40	α-L-Rhamnosidase	Hydrolysis of terminal, non-reducing α-L-rhamnose residues in α-L-rhamnosides
3.2.1.41	Pullulanase	Hydrolysis of 1,6-α-D-glucosidic linkages in pullulan, amylopectin and glycogen, and in the α- and β-amylase limit dextrins of amylopectin and glycogen
3.2.1.42	GDPglucosidase	GDPglucose + H_2O = GDP + D-glucose
3.2.1.43	β-L-Rhamnosidase	Hydrolysis of terminal, non-reducing β-L-rhamnose residues in β-L-rhamnosides
3.2.1.44	Fucoidanase	Endohydrolysis of 1,2-α-L-fucoside linkages in fucoidan without release of sulphate
3.2.1.45	Glucosylceramidase	D-Glucosyl-N-acylsphingosine + H_2O = D-glucose + N-acylsphingosine
3.2.1.46	Galactosylceramidase	D-Galactosyl-N-acylsphingosine + H_2O = D-galactose + N-acylsphingosine
3.2.1.47	Galactosylgalactosylglucosylceramidase	Galactosylgalactosylglucosylceramide + H_2O = lactosylceramide + galactose
3.2.1.48	Sucrose α-D-glucohydrolase	Hydrolysis of sucrose and maltose by an α-D-glucosidase-type action
3.2.1.49	α-N-Acetyl-D-galactosaminidase	R-2-acetamido-2-deoxy-α-D-galactoside + H_2O = ROH + 2-acetamido-2-deoxy-D-galactose

Number	Other Names	Basis for classification (Systematic Name)	Comments	Reference
3.2.1.37	Xylobiase, β-Xylosidase	1,4-β-D-Xylan xylohydrolase	Also hydrolyses xylobiose	1385
3.2.1.38		β-D-Fucoside fucohydrolase	Also hydrolyses β-D-galactosides	3681, 3684
3.2.1.39	Laminarinase	1,3-β-D-Glucan glucanohydrolase	Different from EC 3.2.1.6. Very limited action on mixed-link (1,3-, 1,4-) β-D-glucans. Hydrolyses laminarin, paranylon and pachyman	519, 2747
3.2.1.40		α-L-Rhamnoside rhamnohydrolase		2848
3.2.1.41	R-enzyme, Limit dextrinase, Debranching enzyme, Amylopectin 6-glucanohydrolase	Pullulan 6-glucanohydrolase		1899, 767
3.2.1.42		GDPglucose glucohydrolase		3152
3.2.1.43		β-L-Rhamnoside rhamnohydrolase		206
3.2.1.44		Poly(1,2-α-L-fucoside-4-sulphate) glycanohydrolase		3393
3.2.1.45		D-Glucosyl-N-acylsphingosine glucohydrolase		370
3.2.1.46		D-Galactosyl-N-acylsphingosine galactohydrolase		369
3.2.1.47		D-Galactosyl-D-galactosyl-D-glucosylceramide galactohydrolase		368
3.2.1.48	Sucrase	Sucrose α-D-glucohydrolase	This enzyme is isolated from intestinal mucosa as a complex also displaying activity towards isomaltose (oligo-1,6-glucosidase, EC 3.2.1.10)	1753
3.2.1.49		2-Acetamido-2-deoxy-α-D-galactoside acetamidodeoxygalactohydrolase	Also acts on the seryl and threonyl derivatives. Does not act on N-acetyl-α-D-glucosaminides	3487, 3650, 3661

Number	Recommended Name	Reaction
3.2.1.50	α-*N*-Acetyl-D-glucosaminidase	R-2-acetamido-2-deoxy-α-D-glucoside + H_2O = ROH + 2-acetamido-2-deoxy-D-glucose
3.2.1.51	α-L-Fucosidase	An α-L-fucoside + H_2O = an alcohol + L-fucose
3.2.1.52	β-*N*-Acetyl-D-hexosaminidase	R-2-acetamido-2-deoxy-β-D-hexoside + H_2O = ROH + 2-acetamido-2-deoxy-hexose
3.2.1.53	β-*N*-Acetyl-D-galactosaminidase	R-2-acetamido-2-deoxy-β-D-galactoside + H_2O = ROH + 2-acetamido-2-deoxy-D-galactose
3.2.1.54	Cyclomaltodextrinase	Cyclomaltodextrin + H_2O = linear maltodextrin
3.2.1.55	α-L-Arabinofuranosidase	Hydrolysis of terminal non-reducing α-L-arabinofuranoside residues in α-L-arabinosides
3.2.1.56	Glucuronosyl-disulphoglucosamine glucuronidase	1,3-D-Glucuronosyl-2-sulphamido-2-deoxy-6-*O*-sulpho-β-D-glucose + H_2O = 2-sulphamido-2-deoxy-6-*O*-sulpho-D-glucose + glucuronate
3.2.1.57	Isopullulanase	Hydrolysis of pullulan to isopanose (6-α-maltosylglucose)
3.2.1.58	Exo-1,3-β-D-glucosidase	Successive hydrolysis of β-D-glucose units from the non-reducing ends of 1,3-β-D-glucans, releasing α-glucose
3.2.1.59	Endo-1,3-α-D-glucanase	Endohydrolysis of 1,3-α-D-glucosidic linkages in isolichenin, pseudonigeran and nigeran
3.2.1.60	Exo-maltotetraohydrolase	Hydrolysis of 1,4-α-D-glucosidic linkages in amylaceous polysaccharides so as to remove successive maltotetraose residues from the non-reducing chain ends

Number	Other Names	Basis for classification (Systematic Name)	Comments	Reference
3.2.1.50		2-Acetamido-2-deoxy-α-D-glucoside acetamidodeoxyglucohydrolase	Hydrolyses UDP-*N*-acetylglucosamine. Does not act on *N*-acetyl-α-D-galactosaminides	3651, 3661
3.2.1.51		α-L-Fucoside fucohydrolase		1938, 3360, 3379
3.2.1.52		2-Acetamido-2-deoxy-β-D-hexoside acetamidodeoxyhexohydrolase	Acts on glucosides and galactosides and also hydrolyses several oligosaccharides	947, 1953
3.2.1.53		2-Acetamido-2-deoxy-β-D-galactoside acetamidodeoxygalactohydrolase		947
3.2.1.54		Cyclomaltodextrin dextrin-hydrolase (decyclizing)	Also hydrolyses linear maltodextrin	698
3.2.1.55	Arabinosidase	α-L-Arabinofuranoside arabinofuranohydrolase	The enzyme acts on α-L-arabinofuranosides, α-L-arabinans containing (1,3)- and/or (1,5)-linkages, arabinoxylans and arabinogalactans	1564, 3336, 1557
3.2.1.56		1,3-D-Glucuronosyl-2-sulphamido-2-deoxy-6-*O*-sulpho-β-D-glucose glucuronohydrolase		722
3.2.1.57		Pullulan 4-glucanohydrolase	The enzyme has practically no action on starch. Panose (4 α-isomaltosylglucose) is hydrolysed to isomaltose and glucose	2901
3.2.1.58		1,3-β-D-Glucan glucohydrolase	Acts on oligosaccharides but very slowly on laminaribiose	213, 214
3.2.1.59		1,3-(1,3;1,4)-α-D-Glucan 3-glucanohydrolase	Products from pseudonigeran (1,3-α-D-glucan) are nigerose and α-D-glucose	1209
3.2.1.60		1,4-α-D-Glucan maltotetraohydrolase	Compare EC 3.2.1.2, β-amylase, which removes successive maltose residues	2820

Number	Recommended Name	Reaction
3.2.1.61	Mycodextranase	Endohydrolysis of 1,4-α-D-glucosidic linkages in α-D-glucans containing both 1,3- and 1,4-bonds
3.2.1.62	Glycosylceramidase	Glycosyl-N-acylsphingosine + H_2O = a sugar + N-acylsphingosine
3.2.1.63	1,2-α-L-Fucosidase	2-O-α-L-Fucopyranosyl-β-D-galactoside + H_2O = L-fucose + D-galactose
3.2.1.64	2,6-β-D-Fructan 6-levanbiohydrolase	Hydrolysis of 2,6-β-D-fructan so as to remove successive levanbiose residues from the end of the chain
3.2.1.65	Levanase	Random hydrolysis of 2,6-β-D-fructofuranosidic linkages in 2,6-β-D-fructans (levans) containing more than 3 fructose units
3.2.1.66	Quercitrinase	Quercitrin + H_2O = rhamnose + quercitrin aglycone
3.2.1.67	Exopolygalacturonase	$(1,4$-α-D-Galacturonide$)_n$ + H_2O = $(1,4$-α-D-galacturonide$)_{n-1}$ + D-galacturonate
3.2.1.68	Isoamylase	Hydrolysis of 1,6-α-D-glucosidic branch linkages in glycogen, amylopectin and their β-limit dextrins

[3.2.1.69 Deleted entry: Amylopectin 6-glucanohydrolase. Now included with EC 3.2.1.41]

3.2.1.70	Exo-1,6-α-D-glucosidase	Hydrolysis of successive glucose residues from 1,6-α-D-glucans and derived oligosaccharides

Number	Other Names	Basis for classification (Systematic Name)	Comments	Reference
3.2.1.61		1,3-1,4-α-D-Glucan 4-glucanohydrolase	Products are nigerose and 4-α-D-nigerosylglucose. No hydrolysis of α-D-glucans containing only 1,3- or 1,4-bonds	3486
3.2.1.62	Phlorizin hydrolase, Phloretin-glucosidase	Glycosyl-N-acylsphingosine glycohydrolase	Broad specificity (*cf* EC 3.2.1.45 and EC 3.2.1.46). Also hydrolyses phlorizin to phloretin and glucose	2063, 1905
3.2.1.63		2-O-α-L-Fucopyranosyl-β-D-galactoside fucohydrolase	Highly specific for non-reducing terminal L-fucose residues linked to D-galactose residues by 1,2 α-linkage	168
3.2.1.64		2,6-β-D-Fructan 6-β-D-fructofuranosylfructohydrolase		145
3.2.1.65		2,6-β-D-Fructan fructanohydrolase		142
3.2.1.66		Quercitrin 3-rhamnohydrolase		3663
3.2.1.67	Poly(galacturonate) hydrolase	Poly(1,4-α-D-galacturonide) galacturonohydrolase		1208
3.2.1.68	Debranching enzyme	Glycogen 6-glucanohydrolase	Distinguished from pullulanase (EC 3.2.1.41) by the inability of isoamylase to attack pullulan, and by limited action on α-limit dextrins. Action of bacterial enzyme on glycogen however is complete in contrast to limited action by pullulanase. 1,6-Linkage hydrolysed only if at a branch point	3792
3.2.1.70	Glucodextranase	1,6-α-D-Glucan glucohydrolase	β-D-Glucose is formed by an inversion. Dextrans and isomaltosaccharides are attacked and, very slowly, isomaltose. Some	3593, 2930, 2486

Number	Recommended Name	Reaction
3.2.1.71	Endo-1,2-β-D-glucanase	Random hydrolysis of 1,2-glucosidic linkages in 1,2-β-D-glucans
3.2.1.72	Exo-1,3-β-D-xylosidase	Hydrolysis of successive xylose residues from the non-reducing termini of 1,3-β-D-xylans
3.2.1.73	Lichenase	Hydrolysis of 1,4-β-D-glucosidic linkages in β-D-glucans containing 1,3- and 1,4-bonds
3.2.1.74	Exo-1,4-β-D-glucosidase	Hydrolysis of 1,4-linkages in 1,4-β-D-glucans so as to remove successive glucose units
3.2.1.75	Endo-1,6-β-D-glucanase	Random hydrolysis of 1,6-linkages in 1,6-β-D-glucans
3.2.1.76	L-Iduronidase	Hydrolysis of α-L-iduronosidic linkages in desulphated dermatan
3.2.1.77	Exo-1,2-1,3-α-D-mannosidase	Hydrolysis of 1,2- and 1,3-linkages in yeast mannan, releasing mannose
3.2.1.78	Endo-1,4-β-D-mannanase	Random hydrolysis of 1,4-β-D-mannosidic linkages in mannans, galactomannans and glucomannans

[3.2.1.79 Deleted entry: α-L-Arabinofuranoside hydrolase. Now included with EC 3.2.1.55]

Number	Recommended Name	Reaction
3.2.1.80	Exo-β-D-fructosidase	Hydrolysis of terminal, non-reducing 2,1- and 2,6-linked β-D-fructofuranose residues in fructans
3.2.1.81	Agarase	Hydrolysis of 1,3-β-D-galactosidic linkages in agarose, giving the tetramer as the predominant product
3.2.1.82	Exo-poly-α-D-galacturonosidase	Hydrolysis of pectic acid from the non-reducing end, releasing digalacturonate

Number	Other Names	Basis for classification (Systematic Name)	Comments	Reference
			members of this group attack 1,3-α-D-glucosidic bonds in dextrans	
3.2.1.71		1,2-β-D-Glucan glucanohydrolase		2748
3.2.1.72		1,3-β-D-Xylan xylohydrolase		973
3.2.1.73		1,3-1,4-β-D-Glucan 4-glucanohydrolase	Acts on lichenin and cereal β-D-glucans, but not on β-D-glucans containing only 1,3- or 1,4-bonds	212
3.2.1.74		1,4-β-D-Glucan glucohydrolase	Acts on 1,4-β-D-glucans and related oligosaccharides. Cellobiose is hydrolysed very slowly	212
3.2.1.75		1,6-β-D-Glucan glucanohydrolase	Acts on lutean, pustulan and 1,6-oligo-β-D-glucosides	2749
3.2.1.76		Mucopolysaccharide α-L-iduronohydrolase		2114
3.2.1.77		1,2-1,3-α-D-Mannan mannohydrolase	A 1,6-α-D-mannan backbone remains after action on yeast mannan. This is further attacked, but slowly	1530
3.2.1.78		1,4-β-D-Mannan mannanohydrolase		840, 2745
3.2.1.80		β-D-Fructan fructohydrolase	Hydrolyses inulin and levan; also hydrolyses sucrose	619
3.2.1.81		Agarose 3-glycanohydrolase	Also acts on porphyran	762
3.2.1.82		Poly (1,4-α-D-galactosiduronate) digalacturonohydrolase		1208, 1219

Number	Recommended Name	Reaction
3.2.1.83	κ-Carrageenanase	Hydrolysis of 1,4-β-linkages between D-galactose 4-sulphate and 3,6-anhydro-D-galactose in various carrageenans
3.2.1.84	Exo-1,3-α-glucanase	Hydrolysis of terminal 1,3-α-D-glucosidic links in 1,3-α-D-glucans
3.2.1.85	6-Phospho-β-D-galactosidase	A 6-phospho-β-D-galactoside + H_2O = an alcohol + 6-phospho-D-galactose
3.2.1.86	6-Phospho-β-D-glucosidase	6-Phospho-β-D-glucosyl-(1,4)-D-glucose + H_2O = D-glucose 6-phosphate + D-glucose
3.2.1.87	Capsular-polysaccharide galactohydrolase	Random hydrolysis of 1,3-α-D-galactosidic linkages in *Aerobacter aerogenes* capsular polysaccharide
3.2.1.88	β-L-Arabinosidase	β-L-Arabinoside + H_2O = an alcohol + L-arabinose
3.2.1.89	Endo-1,4-β-D-galactanase	Endohydrolysis of 1,4-β-D-galactosidic linkages in arabinogalactans
3.2.1.90	Endo-1,3-β-D-galactanase	Endohydrolysis of 1,3-β-D-galactosidic linkages in arabinogalactans
3.2.1.91	Exo-cellobiohydrolase	Hydrolysis of 1,4-β-D-glucosidic linkages in cellulose and cellotetraose, releasing cellobiose from the non-reducing ends of the chains
3.2.1.92	Exo-β-N-acetylmuramidase	Hydrolysis of terminal, non-reducing N-acetylmuramic residues
3.2.1.93	α,α-Phosphotrehalase	α,α-Trehalose 6-phosphate + H_2O = D-glucose + D-glucose 6-phosphate
3.2.1.94	Exo-isomaltohydrolase	Hydrolysis of 1,6-α-D-glucosidic linkages in polysaccharides so as to remove successive isomaltose units from the non-reducing ends of the chains

Number	Other Names	Basis for classification (Systematic Name)	Comments	Reference
3.2.1.83		κ-Carrageenan 4-β-D-glycanohydrolase	Dominant products are the oligomers [3-O-(3,6-anhydro-α-D-galactopyranosyl)-D-galactose 4-O-sulphate]$_{1-4}$	3638
3.2.1.84		1,3-α-D-Glucan 3-glucohydrolase	Does not act on nigeran	3853
3.2.1.85		6-Phospho-β-D-galactoside 6-phosphogalactohydrolase		1274
3.2.1.86		6-Phospho-β-D-glucosyl-(1,4)-D-glucose glucohydrolase	Also hydrolyses several other phospho-β-D-glucosides, but not their non-phosphorylated forms	2550
3.2.1.87	Polysaccharide depolymerase	$Aerobacter$-capsular-polysaccharide galactohydrolase	Hydrolyses the galactosyl-α-1,3-D-galactose linkages only in the complex substrate, bringing about depolymerization	3824, 3825
3.2.1.88		β-L-Arabinoside arabinohydrolase		712
3.2.1.89	Galactanase, Arabinogalactanase	Arabinogalactan 4-β-D-galactanohydrolase		824
3.2.1.90	Galactanase, Arabinogalactanase	Arabinogalactan 3-β-D-galactanohydrolase		1212
3.2.1.91	C$_1$ enzyme	1,4-β-D-Glucan cellobiohydrolase		266, 841
3.2.1.92		Mucopolysaccharide β-N-acetylmuramoylexo-hydrolase		687
3.2.1.93		α,α-Trehalose-6-phosphate phosphoglucohydrolase		294
3.2.1.94	Isomalto-dextranase	1,6-α-D-Glucan isomaltohydrolase	Optimum activity is on those 1,6-α-D-glucans containing 6,7, and 8 glucose units; those containing 3,4, and 5 glucose units are hydrolysed at slower rates	2929, 3312

Number	Recommended Name	Reaction
3.2.1.95	Exo-isomaltotriohydrolase	Hydrolysis of 1,6-α-D-glucosidic linkages in dextrans, so as to remove successive isomaltotriose units from the non-reducing ends of the chains
3.2.1.96	Endo-β-N-acetylglucosaminidase	Endohydrolysis of 1,4-N-acetamidodeoxy-β-D-glucosidic linkages in mannosyl-glycoproteins
3.2.1.97	Endo-α-N-acetylgalactosaminidase	Hydrolysis of terminal D-galactosyl-N-acetamidodeoxy-α-D-galactosidic residue from a variety of glycopeptides and glycoproteins
3.2.1.98	Exo-maltohexaohydrolase	Hydrolysis of 1,4-α-D-glucosidic linkages in amylaceous polysaccharides so as to remove successive maltohexaose residues from the non-reducing chain ends

3.2.2 HYDROLYSING N-GLYCOSYL COMPOUNDS

Number	Recommended Name	Reaction
3.2.2.1	Nucleosidase	An N-ribosyl-purine + H_2O = a purine + D-ribose
3.2.2.2	Inosine nucleosidase	Inosine + H_2O = hypoxanthine + D-ribose
3.2.2.3	Uridine nucleosidase	Uridine + H_2O = uracil + D-ribose
3.2.2.4	AMP nucleosidase	AMP + H_2O = adenine + D-ribose 5-phosphate
3.2.2.5	NAD$^+$ nucleosidase	NAD$^+$ + H_2O = nicotinamide + ADPribose
3.2.2.6	NAD(P)$^+$ nucleosidase	NAD(P)$^+$ + H_2O = nicotinamide + ADPribose(P)
3.2.2.7	Adenosine nucleosidase	Adenosine + H_2O = adenine + D-ribose
3.2.2.8	N-Ribosylpyrimidine nucleosidase	An N-ribosylpyrimidine + H_2O = a pyrimidine + D-ribose

Number	Other Names	Basis for classification (Systematic Name)	Comments	Reference
3.2.1.95		1,6-α-D-Glucan isomaltotriohydrolase		3281
3.2.1.96		Mannosyl-glycoprotein 1,4-N-acetamidodeoxy-β-D-glycohydrolase		1746, 3339
3.2.1.97		D-Galactosyl-N-acetamidodeoxy-α-D-galactoside D-galactosyl-N-acetamidodeoxy-D-galactohydrolase	The aglycone may be serine or threonine	293
3.2.1.98		1,4-α-D-Glucan maltohexaohydrolase	Compare EC 3.2.1.3, glucohydrolase which removes successive glucose residues; EC 3.2.1.2, β-amylase which removes successive maltose residues; and EC 3.2.1.60, exo-maltotetraohydrolase which removes successive maltotetraose residues	1562
3.2.2.1	Purine nucleosidase	N-Ribosyl-purine ribohydrolase		1282, 1571, 3342, 3373
3.2.2.2	Inosinase	Inosine ribohydrolase		1729, 3373
3.2.2.3		Uridine ribohydrolase		477
3.2.2.4		AMP phosphoribohydrolase		1409
3.2.2.5	NADase, DPNase, DPN hydrolase	NAD$^+$ glycohydrolase	Also catalyses transfer of ADPribose residues	1335, 2368
3.2.2.6		NAD(P)$^+$ glycohydrolase	Also catalyses transfer of ADPribose(P) residues	52, 3835, 3836
3.2.2.7	Adenosinase	Adenosine ribohydrolase	Also acts on adenosine N-oxide	2139
3.2.2.8		Nucleoside ribohydrolase	Also hydrolyses purine ribonucleosides but at a slower rate	3391

Number	Recommended Name	Reaction
3.2.2.9	Adenosylhomocysteine nucleosidase	S-Adenosyl-L-homocysteine + H_2O = adenine + S-ribosyl-L-homocysteine
3.2.2.10	Pyrimidine-5′-nucleotide nucleosidase	A pyrimidine 5′-nucleotide + H_2O = a pyrimidine + D-ribose 5-phosphate
3.2.2.11	β-Aspartylacetylglucosaminidase	1-β-Aspartyl-2-acetamido-1,2-dideoxy-D-glucosylamine + H_2O = 2-acetamido-2-deoxy-D-glucose + L-asparagine
3.2.2.12	Inosinate nucleosidase	5′-Inosinate + H_2O = hypoxanthine + D-ribose 5-phosphate
3.2.2.13	1-Methyladenosine nucleosidase	1-Methyladenosine + H_2O = 1-methyladenine + D-ribose
3.2.2.14	NMN nucleosidase	Nicotinamide D-ribonucleotide + H_2O = nicotinamide + D-ribose 5-phosphate

3.2.3 HYDROLYSING S-GLYCOSYL COMPOUNDS

3.2.3.1	Thioglucosidase	A thioglucoside + H_2O = a thiol + a sugar

3.3 ACTING ON ETHER BONDS

3.3.1 THIOETHER HYDROLASES

3.3.1.1	Adenosylhomocysteinase	S-Adenosyl-L-homocysteine + H_2O = adenosine + L-homocysteine
3.3.1.2	Adenosylmethionine hydrolase	S-Adenosyl-L-methionine + H_2O = methylthioadenosine + L-homoserine
3.3.1.3	Ribosylhomocysteinase	S-Ribosyl-L-homocysteine + H_2O = ribose + L-homocysteine

3.3.2 ETHER HYDROLASES

3.3.2.1	Isochorismatase	Isochorismate + H_2O = 2,3-dihydroxy-2,3-dihydrobenzoate + pyruvate
3.3.2.2	Alkenyl-glycerophosphocholine hydrolase	1-(1-Alkenyl)-glycero-3-phosphocholine + H_2O = an aldehyde + glycero-3-phosphocholine

Number	Other Names	Basis for classification (Systematic Name)	Comments	Reference
3.2.2.9		*S*-Adenosyl-L-homocysteine homocysteinylribohydrolase	Also acts on 5′-methylthioadenosine to give adenine and 5-methylthioribose	763
3.2.2.10		Pyrimidine-5′-nucleotide phosphoribo(deoxyribo)-hydrolase	Also acts on dUMP, dTMP and dCMP	1439, 1440
3.2.2.11		1-*β*-Aspartyl-2-acetamido-1,2-dideoxy-D-glucosylamine-L-asparagino-hydrolase		853
3.2.2.12		5′-Inosinate phosphoribohydrolase		1831
3.2.2.13		1-Methyladenosine ribohydrolase		3374
3.2.2.14	NMNase	Nicotinamidenucleotide phosphoribohydrolase		83
3.2.3.1	Myrosinase, Sinigrinase, Sinigrase	Thioglucoside glucohydrolase	Has a wide specificity for thioglycosides	594, 1082, 2609
3.3.1.1		*S*-Adenosyl-L-homocysteine hydrolase		673
3.3.1.2	*S*-Adenosylmethionine cleaving enzyme, Methylmethionine-sulphonium-salt hydrolase	*S*-Adenosyl-L-methionine hydrolase	Also hydrolyses methylmethionine sulphonium salt to dimethyl sulphide and homoserine	2140
3.3.1.3		*S*-Ribosyl-L-homocysteine ribohydrolase		764
3.3.2.1	2,3-Dihydro-2,3-dihydroxy-benzoate synthase	Isochorismate pyruvate-hydrolase		3813
3.3.2.2		1-(1-Alkenyl)-glycero-3-phosphocholine aldehydrohydrolase		816, 3616

Number	Recommended Name	Reaction
3.3.2.3	Epoxide hydrolase	An epoxide + H_2O = a glycol

3.4   ACTING ON PEPTIDE BONDS (PEPTIDE HYDROLASE)

This is a large group of enzymes, the nomenclature of which is rather difficult. This is mainly due to the fact that the overall reaction is essentially the same in all peptidases and proteinases, and they do not have a substrate specificity in the ordinary sense; most of the enzymes in groups 3.4.21-3.4.24 attack all denatured and many native proteins.

The classification used here is based on a number of different criteria. As it differs substantially from that used in Enzyme Nomenclature (1964), the original sub-sub-group numbers have been abandoned and the new sub-sub-groups are numbered from 3.4.11 onwards.

There are two sets of sub-sub-groups; the peptidases (exopeptidases: 3.4.11-17) and the proteinases (3.4.21-24). The peptidases are divided according to their specificity into those hydrolysing single amino acids from the *N*-terminus of the peptide chain (3.4.11), those hydrolysing single residues from the *C*-terminus (3.4.16-17), those specific for dipeptide substrates (3.4.13), and those splitting off dipeptide units either from the *N*-terminus (3.4.14) or the *C*-terminus (3.4.15). The group hydrolysing single residues from the *C*-terminus is subdivided into two distinct classes. One class displays maximum activity in the acid range and is inhibited by the substitution of a serine residue by organic fluorophosphates (serine carboxypeptidases 3.4.16), the second class requires for activity divalent cations (metallocarboxypeptidases, 3.4.17).

The proteinases (proteolytic enzymes, endopeptidases, peptidyl-peptide hydrolases) are divided into sub-sub-groups on the basis of the catalytic mechanism, as shown by active centre studies or the effect of pH, and specificity is only used to identify individual enzymes within the groups. The enzymes of sub-sub-groups 3.4.21 (usually called "serine-proteinases") have an active centre histidine and serine involved in the catalytic process; those of 3.4.22 ("SH-proteinases") have a cysteine in the active centre; those of 3.4.23 have a pH optimum below 5, due to the involvement of an acidic residue in the catalytic process ("carboxyl proteinases") and those of 3.4.24 are metalloproteins using a metal ion in the catalytic mechanism. Many proteinases have a very close but not entirely identical specificity. The products resulting from their action on the same protein as substrate differ quantitatively and qualitatively. In those cases a separate entry is allocated. On the other hand differences in origin or in structure alone do not substantiate a separate entry.

A number of proteolytic enzymes have been isolated, sometimes in a high degree of purity, without obtaining the information about their catalytic mechanism that would allow them to be allocated to the sub-sub-groups so far described. As an interim measure, they are listed in sub-sub-group 3.4.99.

It should be noted that many enzymes in this group give different products from the same substrate and are classified as separate enzymes, even though in the table the reaction is written in the same general way, e.g. "hydrolyses protein". In some cases the entry under Reaction indicates the particular kind of peptide bond that is acted on.

[3.4.1.1   Transferred entry: now EC 3.4.11.1, Aminopeptidase(cytosol)]

[3.4.1.2   Transferred entry: now EC 3.4.11.2, Aminopeptidase(microsomal)]

[3.4.1.3   Transferred entry: now EC 3.4.11.4, Tripeptide aminopeptidase]

[3.4.1.4   Transferred entry: now EC 3.4.11.5, Proline iminopeptidase]

Number	Other Names	Basis for classification (Systematic Name)	Comments	Reference
3.3.2.3	Epoxide hydratase, Arene-oxide hydratase	Epoxide hydrolase	Acts on a variety of epoxides and arene oxides. Formerly EC 4.2.1.63, 4.2.1.64	2474, 2012, 1491, 2473

Number	Recommended Name	Reaction

[3.4.2.1   Transferred entry: now EC 3.4.17.1, Carboxypeptidase A]

[3.4.2.2   Transferred entry: now EC 3.4.17.2, Carboxypeptidase B]

[3.4.2.3   Transferred entry: now EC 3.4.17.4, Glycine carboxypeptidase]

[3.4.3.1   Transferred entry: now EC 3.4.13.11, Dipeptidase]

[3.4.3.2   Transferred entry: now EC 3.4.13.11, Dipeptidase]

[3.4.3.3   Transferred entry: now EC 3.4.13.3, Aminoacyl-histidine dipeptidase]

[3.4.3.4   Transferred entry: now EC 3.4.13.5, Aminoacyl-methylhistidine dipeptidase]

[3.4.3.5   Transferred entry: now EC 3.4.13.6, Cysteinyl-glycine dipeptidase]

[3.4.3.6   Transferred entry: now EC 3.4.13.8, Prolyl dipeptidase]

[3.4.3.7   Transferred entry: now EC 3.4.13.9, Proline dipeptidase]

[3.4.4.1   Transferred entry: now EC 3.4.23.1, Pepsin A]

[3.4.4.2   Transferred entry: now EC 3.4.23.2, Pepsin B]

[3.4.4.3   Transferred entry: now EC 3.4.23.4, Chymosin]

[3.4.4.4   Transferred entry: now EC 3.4.21.4, Trypsin]

[3.4.4.5   Transferred entry: now EC 3.4.21.1, Chymotrypsin]

[3.4.4.6   Transferred entry: now EC 3.4.21.1, Chymotrypsin]

[3.4.4.7   Transferred entry: now EC 3.4.21.11, Elastase]

[3.4.4.8   Transferred entry: now EC 3.4.21.9, Enteropeptidase]

[3.4.4.9   Transferred entry: now EC 3.4.14.1, Dipeptidyl peptidase I]

[3.4.4.10 Transferred entry: now EC 3.4.22.2, Papain]

[3.4.4.11 Transferred entry: now EC 3.4.22.6, Chymopapain]

[3.4.4.12 Transferred entry: now EC 3.4.22.3, Ficin]

[3.4.4.13 Transferred entry: now EC 3.4.21.5, Thrombin]

[3.4.4.14 Transferred entry: now EC 3.4.21.7, Plasmin]

[3.4.4.15 Transferred entry: now EC 3.4.99.19, Renin]

[3.4.4.16 Transferred entry: now EC 3.4.21.14, Microbial serine proteinases]

[3.4.4.17 Transferred entry: now EC 3.4.23.6, Microbial carboxyl proteinases]

Number	Other Names	Basis for classification (Systematic Name)	Comments	Reference

Number	Recommended Name	Reaction

[3.4.4.18 Transferred entry: now EC 3.4.22.10, Streptococcal proteinase]

[3.4.4.19 Transferred entry: now EC 3.4.24.3, Clostridium histolyticum *collagenase]*

[3.4.4.20 Transferred entry: now EC 3.4.22.8, Clostripain]

[3.4.4.21 Transferred entry: now EC 3.4.21.8, Kallikrein]

[3.4.4.22 Transferred entry: now EC 3.4.23.3, Pepsin C]

[3.4.4.23 Transferred entry: now EC 3.4.23.5, Cathepsin D]

[3.4.4.24 Transferred entry: now EC 3.4.22.4, Bromelain]

[3.4.4.25 Transferred entry: now EC 3.4.99.11, Streptomyces *alkalophilic keratinase]*

3.4.11 α-AMINOACYLPEPTIDE HYDROLASES

3.4.11.1 Aminopeptidase (cytosol) Aminoacyl-peptide + H_2O = amino acid + peptide

3.4.11.2 Aminopeptidase (microsomal) Aminoacyl-peptide + H_2O = amino acid + oligopeptide

3.4.11.3 Cystyl aminopeptidase Cystyl peptide + H_2O = amino acid + peptide

3.4.11.4 Tripeptide aminopeptidase Aminoacyl-dipeptide + H_2O = amino acid + dipeptide

Number	Other Names	Basis for classification (Systematic Name)	Comments	Reference
3.4.11.1	Leucine amino-peptidase ('LAP')	α-Aminoacyl-peptide hydrolase (cytosol)	Zn-enzyme of broad specificity isolated from hog kidney, bovine lens. Hydrolyses most L-peptides, splitting off an N-terminal residue with a free amino group. Does not act on lysyl- and arginyl peptides. Also has esterase activity. Is activated by heavy metal ions. Formerly EC 3.4.1.1. Another aminopeptidase, close but not identical, has been isolated from pig kidney; it does not cleave Arg- and Lys-bonds	1185, 689 1186 1185 1316, 1317
3.4.11.2	Particle-bound aminopeptidase	α-Aminoacyl-peptide hydrolase (microsomal)	Splits α-amino acids (preferentially alanine, not proline) from peptides, amides and p-nitroanilides. A Zn-enzyme. Is not activated by heavy metal ions. Formerly EC 3.4.1.2	2596, 2597, 3621, 3571, 3570, 1187
3.4.11.3	Oxytocinase	α-Aminoacyl-peptide hydrolase	Degrades cystine peptides such as oxytocin and vasopressin	3094, 3791
3.4.11.4		α-Aminoacyl-dipeptide hydrolase	Cleaves exclusively unsubstituted tripeptides. Formerly EC 3.4.1.3	3112

Number	Recommended Name	Reaction
3.4.11.5	Proline iminopeptidase	L-Prolyl-peptide + H_2O = L-proline + peptide
3.4.11.6	Arginine aminopeptidase	L-Arginyl-peptide + H_2O = L-arginine + peptide
3.4.11.7	Aspartate aminopeptidase	L-α-Aspartyl-peptide + H_2O = L-aspartate + peptide
3.4.11.8	Pyroglutamyl aminopeptidase	Pyroglutamyl-peptide + H_2O = pyroglutamate + peptide
3.4.11.9	Aminopeptidase P	Aminoacylprolyl-peptide + H_2O = amino acid + prolyl-peptide
3.4.11.10	*Aeromonas proteolytica* aminopeptidase	Aminoacyl-peptide + H_2O = amino acid + peptide
3.4.11.11	Aminopeptidase	Neutral or aromatic aminoacyl-peptide + H_2O = neutral or aromatic amino acid + peptide
3.4.11.12	Thermophilic aminopeptidase	Aminoacyl-peptide + H_2O = amino acid + peptide

Number	Other Names	Basis for classification (Systematic Name)	Comments	Reference
3.4.11.5		L-Prolyl-peptide hydrolase	An Mn^{2+} -requiring enzyme from *E. coli*, removes *N*-terminal proline only. Formerly EC 3.4.1.4	2912
3.4.11.6		L-Arginyl (L-lysyl)-peptide hydrolase	Removes *N*-terminal L-arginine or L-lysine only. Activated by Cl$^-$	1900, 1364, 1365
3.4.11.7		L-α-Aspartyl (L-α-glutamyl)-peptide hydrolase	Also acts on L-α-glutamyl-peptides	1061
3.4.11.8	Pyrrolidone-carboxylate peptidase	L-Pyroglutamyl-peptide hydrolase	Removes pyroglutamate from various penultimate amino acid residues except L-proline. Occurs in *Pseudomonas*, *Bacillus subtilis*, rat liver	107, 747, 3288
3.4.11.9	Aminoacylproline aminopeptidase	Aminoacylprolyl-peptide hydrolase	Releases any *N*-terminal amino acid only if adjacent to proline residue; requires Mn^{2+} Isolated from *E. coli*	3782, 3781
3.4.11.10		α-Aminoacyl-peptide hydrolase (*Aeromonas proteolytica*)	A zinc enzyme. Acts most rapidly on L-leucyl-peptides, amide and β-naphthylamide. Does not cleave Glu- and Asp-bonds. Similar aminopeptidases were isolated from *E. coli* and *Staphylococcus thermophilus*	2662 715 2674
3.4.11.11	Peptidase a	α-Aminoacyl-peptide hydrolase	An Mn^{2+} -requiring enzyme from guinea pig ileum	745
3.4.11.12	AP I, II, III	α-Aminoacyl-peptide hydrolase	Metalloenzymes of high temperature stability and of broad specificity, releasing all *N*-terminal amino acids, including arginine and lysine. Isolated from *Bacillus stearothermophilus*, *Talaromyces duponti*, *Mucor*	2834, 2833

Number	Recommended Name	Reaction
3.4.11.13	*Clostridium histolyticum* aminopeptidase	Aminoacyl-peptide + H₂O = amino acid + peptide
3.4.11.14	Aminopeptidase (human liver)	Aminoacyl-peptide + H₂O = amino acid + peptide. Preferential cleavage: Ala⁻

3.4.12 PEPTIDYLAMINO–ACID HYDROLASES OR ACYLAMINO–ACID HYDROLASES

[3.4.12.1 Transferred entry: now EC 3.4.16.1, Serine carboxypeptidase]

[3.4.12.2 Transferred entry: now EC 3.4.17.1, Carboxypeptidase A]

[3.4.12.3 Transferred entry: now EC 3.4.17.2, Carboxypeptidase B]

[3.4.12.4 Transferred entry: now EC 3.4.16.2, Proline acid carboxypeptidase]

[3.4.12.5 Transferred entry: now EC 3.4.17.7, Acylmuramoyl-alanine carboxypeptidase]

[3.4.12.6 Transferred entry: now EC 3.4.17.8, Muramoyl-pentapeptide carboxypeptidase]

[3.4.12.7 Transferred entry: now EC 3.4.17.3, Arginine carboxypeptidase]

[3.4.12.8 Transferred entry: now EC 3.4.17.4, Glycine carboxypeptidase]

[3.4.12.9 Transferred entry: now EC 3.4.17.5, Aspartate carboxypeptidase]

[3.4.12.10 Transferred entry: now EC 3.4.22.12, γ-Glutamyl hydrolase]

[3.4.12.11 Transferred entry: now EC 3.4.17.6, Alanine carboxypeptidase]

[3.4.12.12 Transferred entry: now EC 3.4.16.3, Tyrosine carboxypeptidase]

[3.4.12.13 Deleted entry: formerly γ-Glutamylglutamate carboxypeptidase]

3.4.13 DIPEPTIDE HYDROLASES

[3.4.13.1 Transferred entry: now EC 3.4.13.11, Dipeptidase]

[3.4.13.2 Transferred entry: now EC 3.4.13.11, Dipeptidase]

| 3.4.13.3 | Aminoacyl-histidine dipeptidase | Aminoacyl-L-histidine + H₂O = amino acid + L-histidine |

Number	Other Names	Basis for classification (Systematic Name)	Comments	Reference
3.4.11.13	CAP	α-Aminoacyl-peptide hydrolase	Extracellular; releases any *N*-terminal amino acid including proline and hydroxyproline. Does not cleave an amino acid-proline bond. Mn^{2+} or Co^{2+} dependent	1670
3.4.11.14	HLA	α-Aminoacyl-peptide hydrolase	Zn-metalloenzyme from human liver, pancreas, kidney and duodenum. Zn^{2+} can be replaced by Co^{2+}. Also cleaves amino acid amides	3200, 244
3.4.13.3	Carnosinase	Aminoacyl-L-histidine hydrolase	Acts on many aminoacyl-L-histidine dipeptides and their amides. Activated by Zn^{2+} or Mn^{2+}. Formerly EC 3.4.3.3	1188, 2846, 3110, 3112

Number	Recommended Name	Reaction
3.4.13.4	Aminoacyl-lysine dipeptidase	Aminoacyl-L-lysine + H_2O = amino acid + L-lysine
3.4.13.5	Aminoacyl-methylhistidine dipeptidase	Anserine + H_2O = β-alanine + *pros*methyl-L-histidine
3.4.13.6	Cysteinyl-glycine dipeptidase	L-Cysteinyl-glycine + H_2O = L-cysteine + glycine
3.4.13.7	α-Glutamyl-glutamate dipeptidase	2-L-Glutamyl-L-glutamate + H_2O = 2 L-glutamate
3.4.13.8	Prolyl dipeptidase	L-Prolyl-amino acid + H_2O = L-proline + amino acid
3.4.13.9	Proline dipeptidase	Aminoacyl-L-proline + H_2O = amino acid + L-proline
3.4.13.10	β-Aspartyldipeptidase	β-L-Aspartyl-L-leucine + H_2O = L-aspartate + L-leucine
3.4.13.11	Dipeptidase	Dipeptide + H_2O = 2 amino acid
3.4.13.12	Methionyl dipeptidase	L-Methionyl-amino acid + H_2O = L-methionine + amino acid

Number	Other Names	Basis for classification (Systematic Name)	Comments	Reference
3.4.13.4	N^2-(4-Aminobutyryl)-L-lysine hydrolase	Aminoacyl-L-lysine (-L-arginine) hydrolase	Hydrolyses dipeptides having lysine, arginine or ornithine as the C-terminal amino acid	1823
3.4.13.5	Anserinase	Aminoacyl-prosmethyl-L-histidine hydrolase	Formerly EC 3.4.3.4	1533
3.4.13.6		L-Cysteinyl-glycine hydrolase	Formerly EC 3.4.3.5	298, 3005
3.4.13.7		2-L-Glutamyl-L-glutamate hydrolase		2649
3.4.13.8	Iminodipeptidase, Prolinase, L-Prolylglycine dipeptidase	L-Prolyl-amino acid hydrolase	Also acts on hydroxyproline derivatives and on amides. Formerly EC 3.4.3.6	2912
3.4.13.9	Prolidase, Iminodipeptidase	Aminoacyl-L-proline hydrolase	Also acts on hydroxyproline dipeptides and amides. Formerly EC 3.4.3.7	656
3.4.13.10	β-Aspartyl peptidase	β-L-Aspartyl-amino acid hydrolase	Specific for β-aspartyl dipeptides	1173
3.4.13.11		Dipeptide hydrolase	Many dipeptidases from various sources have comparable specificities, hydrophobic dipeptides being cleaved preferentially. They have been found in bacteria (*Mycobacterium phlei*, *E. coli*, *Streptococcus thermophilus*), yeast, mouse ascites tumor, rat liver and swine kidney. Formerly EC 3.4.3.1, 3.4.3.2, 3.4.13.1 and 3.4.13.2	2615 2571, 2570 2674 2829 2571, 1252 2644 451, 450, 1114, 3111, 3456
3.4.13.12	Dipeptidase M	L-Methionyl-amino acid hydrolase		404

Number	Recommended Name	Reaction

3.4.14 DIPEPTIDYLPEPTIDE HYDROLASES

| 3.4.14.1 | Dipeptidyl peptidase I | Dipeptidyl-polypeptide + H_2O = dipeptide + polypeptide |

| 3.4.14.2 | Dipeptidyl peptidase II | Dipeptidyl-polypeptide + H_2O = dipeptide + polypeptide |

| 3.4.14.3 | Acylamino-acid-releasing enzyme | Acylaminoacyl-peptide + H_2O = acylamino acid + peptide |

3.4.15 PEPTIDYLDIPEPTIDE HYDROLASES

| 3.4.15.1 | Dipeptidyl carboxypeptidase | Polypeptidyl-dipeptide + H_2O = polypeptide + dipeptide
Also: Angiotensin I + H_2O = angiotensin II + His-Leu |

| 3.4.15.2 | Peptidyl carboxyamidase | Peptidyl-glycinamide + H_2O = peptide + glycinamide |

3.4.16 SERINE CARBOXYPEPTIDASES

Carboxypeptidases with optimum activity at acidic pH contain a diisopropyl fluorophosphate-sensitive serine in their catalytic site, in analogy to serine proteinases. Therefore the group name 'serine carboxypeptidases' is preferred to that of 'acid carboxypeptidases'.

Number	Other Names	Basis for classification (Systematic Name)	Comments	Reference
3.4.14.1	Cathepsin C, Dipeptidyl-amino-peptidase I, Dipeptidyl transferase	Dipeptidylpeptide hydrolase	Also actively polymerizes dipeptide amides and transfers dipeptide residues; SH-group forms part of the active centre. Formerly EC᾿ 3.4.4.9	2159, 1721, 2209, 2616, 3568
3.4.14.2	Dipeptidyl-aminopeptidase II	Dipeptidylpeptide hydrolase	Analogous to EC 3.4.14.1, but with complementary specificity. Cleaves Lys-Ala-naphthylamide.	2158
			Other less defined dipeptidylpeptidases (III and IV) have been described	2156
3.4.14.3		*N*-Acylaminoacyl-peptide hydrolase	Group of similar enzymes liberating	3478, 3801
			N-acetyl	3478
			or *N*-formyl	3801
			amino acid from proteins and peptides	3478, 3801
3.4.15.1	Angiotensin converting enzyme. Peptidase P, Kinase II, Carboxycathepsin	Peptidyldipeptide hydrolase	Cleaves *C*-terminal dipeptides from a variety of substrates including bradykinin and angiotensin I. Enzyme isolated from *E. coli* cleaves N-blocked tripeptides, free tetra- and higher peptides, except those with a penultimate proline residue or those consisting of a chain of several glycine residues	814, 839, 2505, 2611, 3774, 3777, 87, 3780, 3222
3.4.15.2	Carboxyamidase	Peptidylaminoacylamide hydrolase	A group of enzymes inactivating vasopressin or oxytocin by splitting off glycinamide	949, 2372, 2624

1248, 1430, 3027

Number	Recommended Name	Reaction
3.4.16.1	Serine carboxypeptidase	Peptidyl-L-amino acid + H_2O = peptide + L-amino acid (broad specificity)
3.4.16.2	Proline carboxypeptidase	Peptidylprolyl-L-amino acid + H_2O = peptidyl-proline + amino acid
3.4.16.3	Tyrosine carboxypeptidase	Peptidyl-L-tyrosine + H_2O = peptide + L-tyrosine

3.4.17   METALLO-CARBOXYPEPTIDASES

3.4.17.1	Carboxypeptidase A	Peptidyl-L-amino acid + H_2O = peptide + L-amino acid

Number	Other Names	Basis for classification (Systematic Name)	Comments	Reference
3.4.16.1	See comments	Peptidyl-L-amino-acid hydrolase	Carboxypeptidases of broad specificity with optimum pH 4.5-6.0 and sensitive to the action of diisopropyl fluorophosphate have been isolated from microorganisms:	
			Penicillium (S$_1$, S$_2$, P)	3794, 3793, 1534, 1400
			yeast (carboxypeptidase Y)	1249, 1218, 1813, 1250
			Aspergillus	1425, 97
			higher plants: citrus (carboxypeptidase C)	3854, 3855, 3169, 3468
			germinating barley	3556
			germinating cotton	1430
			beans (phaseolin) higher animals:	3656, 3026
			lysosomal carboxypeptidase A (cathepsin A)	1451, 355, 738, 3382, 2157
			lysosomal carboxypeptidase B (cathepsin B$_2$, cathepsin IV). Formerly EC 3.4.12.1 and EC 3.4.21.13	2530, 2413, 2157
3.4.16.2	Angiotensinase C, Lysosomal carboxypeptidase C	Peptidylprolyl-amino acid hydrolase	Releases the *C*-terminal amino acid only if proline is in the penultimate position. Inactivates angiotensin II by this action. Formerly EC 3.4.12.4	3775, 3776, 3381
3.4.16.3	Thyroid peptide carboxypeptidase, Thyroid peptidase	Peptidyl-L-tyrosine hydrolase	Acts particularly on peptides with at least one aromatic or hydrophobic amino acid. Formerly EC 3.4.12.12	749, 2202
3.4.17.1	Carboxypolypeptidase	Peptidyl-L-amino-acid hydrolase	Zn-metalloenzyme formed from procarboxypeptidase A. Releases *C*-terminal amino acids, with the exception of *C*-terminal arginine, lysine and proline. Isolated from bovine, porcine, and	615, 909, 1857, 2400, 3525

Number	Recommended Name	Reaction
3.4.17.2	Carboxypeptidase B	Peptidyl-L-lysine (-L-arginine) + H_2O = peptide + L-lysine (or L-arginine)
3.4.17.3	Arginine carboxypeptidase	Peptidyl-L-arginine + H_2O = peptide + L-arginine
3.4.17.4	Glycine carboxypeptidase	Peptidyl-glycine + H_2O = peptide + glycine
3.4.17.5	Aspartate carboxypeptidase	Peptidyl-L-aspartate + H_2O = peptide + L-aspartate
3.4.17.6	Alanine carboxypeptidase	Peptidyl-L-alanine + H_2O = peptide + L-alanine
3.4.17.7	Acylmuramoyl-alanine carboxypeptidase	N-acyl-muramoyl-L-alanine + H_2O = N-acyl-muramate + L-alanine
3.4.17.8	Muramoyl-pentapeptide carboxypeptidase	UDP-N-acetylmuramoyl-L-alanyl-D-glutamyl-2,6-diaminopimelyl-D-alanyl-D-alanine + H_2O = UDP-N-acetylmuramoyl-L-alanyl-D-glutamyl-2,6-diaminopimelyl-D-alanine + D-alanine

Number	Other Names	Basis for classification (Systematic Name)	Comments	Reference
			dogfish pancreas. Formerly EC 3.4.2.1 and EC 3.4.12.2	
3.4.17.2	Protaminase	Peptidyl-L-lysine (L-arginine) hydrolase	Zn-metalloenzyme formed from procarboxypeptidase B. Releases *C*-terminal L-lysine or L-arginine preferentially. Isolated from bovine, porcine, and dogfish pancreas. Formerly EC 3.4.2.2 and 3.4.12.3	911, 2400, 2646, 3708
3.4.17.3	Carboxypeptidase N	Peptidyl-L-arginine hydrolase	Inactivates bradykinin in serum. Formerly EC 3.4.12.7	837, 2563, 838
3.4.17.4	Yeast carboxypeptidase	Peptidyl-glycine hydrolase	Also acts on peptides having a *C*-terminal L-leucine residue. Formerly EC 3.4.2.3 and EC 3.4.12.8	869, 877
3.4.17.5		Peptidyl-L-aspartate hydrolase	Metallocarboxypeptidase (Zn or Co). Peptide can be replaced by a pteroyl group or variety of acyl groups. Formerly EC 3.4.12.9	1943
3.4.17.6		Peptidyl-L-alanine hydrolase	Metallocarboxypeptidase from soil bacteria. Peptide can be replaced by a variety of pteroyl or acyl groups. Formerly EC 3.4.12.11	1944
3.4.17.7	Acylmuramoylalaninase	*N*-Acylmuramoyl-L-alanine hydrolase	Formerly EC 3.4.12.5	3812
3.4.17.8		UDP-*N*-acetylmuramoyl-tetrapeptidyl-D-alanine alanine-hydrolase	Formerly EC 3.4.12.6	1470

Number	Recommended Name	Reaction

3.4.21 SERINE PROTEINASES

3.4.21.1	Chymotrypsin	Preferential cleavage* :Tyr-, Trp-, Phe-, Leu-
3.4.21.2	Chymotrypsin C	Preferential cleavage: Leu-, Tyr-, Phe-, Met-,Trp-, Gln-, Asn-
3.4.21.3	*Metridium* proteinase A	Preferential cleavage: Tyr-, Phe-, Leu-
3.4.21.4	Trypsin	Preferential cleavage: Arg-, Lys-

---

*In this reaction, "preferential cleavage", means cleavage at the carbonyl end of the amino acid whose symbol is given."

Number	Other Names	Basis for classification (Systematic Name)	Comments	Reference
3.4.21.1	Chymotrypsin A and B		Chymotrypsin A is formed from bovine and porcine chymotrypsinogen A; a number of chymotrypsins are formed according to the number of bonds hydrolysed in the precursor. Hydrolyses peptides, amides, esters, preferentially at carboxyl groups of hydrophobic amino acids. Inhibited by tosylphenylalanine chlormethylketone.	706, 1832, 3107
			Formerly EC 3.4.4.5. Chymotrypsin B, formerly EC 3.4.4.6, formed from chymotrypsinogen B, is homologous with chymotrypsin A. Enzymes with specificity similar to that of chymotrypsin A and B have been isolated from many species:	1447
			ox, pig, sheep,	1744
			fin whale,	2127
			chicken,	2877
			salmon,	610
			hornet	3150
3.4.21.2			Formed from porcine chymotrypsinogen C, and from bovine subunit II of procarboxypeptidase A. Reacts more readily with N-tosyl-L-leucine chlormethylketone than with the phenylalanine derivative	907, 910, 2581
3.4.21.3	Sea anemone protease A		Does not attack the tryptophyl bond in glucagon	1027
3.4.21.4	α-and β-Trypsin		β-Trypsin is formed from trypsinogen by cleavage of one peptide bond; single polypeptide chain	2445

Number	Recommended Name	Reaction

3.4.21.5 Thrombin   Preferential cleavage: Arg-; activates
                                                       fibrinogen to fibrin

Number	Other Names	Basis for classification (Systematic Name)	Comments	Reference
			α-Trypsin is derived from β-trypsin by additional cleavage of a peptide bond.	
			When two bonds are cleaved, ψ-trypsin is formed	2974, 3124
			Formerly EC 3.4.4.4. Hydrolyses peptides, amides, esters at bonds involving the carboxyl group of L-arginine or L-lysine.	
			Inhibited by substituted lysine	3031
			and arginine,	3030, 3068
			chlormethylketones and many natural polypeptidic inhibitors.	1884
			Isolated from many species:	
			pancreas of ox, man,	3459
			pig,	501, 1289, 3536
			sheep,	384, 3458
			turkey,	2877
			dogfish,	1857, 3428
			lungfish,	2738
			shrimp,	995
			silk moth ('cocoonase'),	1551, 773, 1011
			sea urchin eggs,	50
			two strains of *Streptomyces* (component of 'Pronase');	1549, 3580, 3465, 2285
			paromotrypsin from *S. paromomycinus*	509
			E. coli ('Protease II').	196
			Crayfish	2599, 3436, 3859
			and bee	1036
			trypsin-like proteinase have no enzyme precursors	
3.4.21.5	Fibrinogenase		Formed from prothrombin. Is more selective than trypsin and plasmin. Formerly EC 3.4.4.13.	1196, 828, 1864, 2042, 2225
			A proteinase from Malayan pit viper (*Ancistrodom rhodastoma*) has a similar specificity	846

Number	Recommended Name	Reaction
3.4.21.6	Coagulation Factor Xa	Preferential cleavage: Arg-Ile, Arg-Gly; activates prothrombin to thrombin
3.4.21.7	Plasmin	Preferential cleavage: Lys>Arg-; higher selectivity than trypsin
3.4.21.8	Kallikrein	Preferential cleavage: Arg-, Lys-; produces kinin from kininogen by limited proteolysis
3.4.21.9	Enteropeptidase	Selective cleavage of Lys^6-Ile^7 bond in trypsinogen

Number	Other Names	Basis for classification (Systematic Name)	Comments	Reference
3.4.21.6	Stuart factor, Thrombokinase		Formed from zymogen X by limited proteolysis. Homologous with trypsin and thrombin. A proteinase from *Oxyuranus sentellatus* venom (3.4.99) has a similar specificity	1473, 1474, 2042, 2236
3.4.21.7	Fibrinase, Fibrinolysin		Formed from plasminogen by proteolysis which results in multiple forms of the active plasmin. Forms activated complex with streptokinase. Converts fibrin into soluble products. Formerly EC 3.4.4.14	9, 607, 3464
3.4.21.8	Kininogenin, Kininogenase		At least three kallikrein types with different functions are known. Plasma kallikrein, releases bradykinin and activates the Hageman factor. Pancreatic and submandibular kallikreins release lysyl-bradykinin (kallidin) from kininogens, activate procollagenase and liberate other biologically active polypeptides other than kinins from high molecular weight plasma proteins. Kidney (urinary) kallikrein acts on kininogen and has an antagonistic function to the renin-angiotensin system. Kallikreins also hydrolyse acyl-arginine esters. Formerly EC 3.4.4.21	2822, 3430, 3638?, 2944
3.4.21.9	Enterokinase		Activates trypsinogen. Also acts on benzoyl-arginine ethyl ester. Is not inhibited by natural peptidic inhibitors of trypsin. Formerly EC 3.4.4.8	1833, 2092, 2445, 3769

Number	Recommended Name	Reaction
3.4.21.10	Acrosin	Hydrolysis of Arg- and Lys-bonds; preferential cleavage Arg-X>> Lys-Lys>>Lys-X
3.4.21.11	Elastase	Preferentially cleaves bonds involving the carbonyl groups of amino acids bearing uncharged non-aromatic side chains. Hydrolysis of elastin
3.4.21.12	*Myxobacter* α-lytic proteinase	Hydrolysis of proteins, especially bonds adjacent to L-alanine residues

[3.4.21.13 Transferred entry: now EC 3.4.16.1, Serine carboxypeptidase]

| 3.4.21.14 | Microbial serine proteinases | |
| | Subtilisin | Hydrolysis of proteins; and peptide amides |

Number	Other Names	Basis for classification (Systematic Name)	Comments	Reference
3.4.21.10			Occurs in spermatozoa; formed from proacrosin by limited proteolysis. Inhibited by naturally occurring trypsin inhibitors. Esterolytic activity is competitively inhibited by L-arginine but not by L-lysine	2627, 3196, 2626, 395, 2957, 1392, 3194
3.4.21.11			Formed from proelastase from mammalian pancreas. Formerly EC 3.4.4.7. Other elastolytic enzymes with different composition and kinetic parameters have been isolated from	195, 1202
			hog pancrease (elastase II),	101, 102
			from lysosomes of polymorphonuclear leucocytes (lysosomal elastase)	1498, 2482, 3200
			and from blood platelets	1908
3.4.21.12			A myxobacterium proteinase	1592, 2499
3.4.21.14				
	See comments		Converts ovalbumin into plakalbumin. Two major types differing in primary structure are known: 'Carlsberg' (also subtilisin A, subtilopeptidase A, alcalase *Novo*) and BNP (also subtilisin B, subtilopeptidase A, Nagarse proteinase, bacterial proteinase *Novo*). Formerly EC 3.4.4.16. Similar enzymes are produced by	1147, 2118, 2528
			B. pumulis,	1632
			B. licheniformis,	
			B. amyloliquefaciens (BPN)	3653, 2975

Number	Recommended Name	Reaction
	Escherichia coli periplasmic proteinase	Preferentially cleaves bonds between hydrophobic residues
	Aspergillus alkaline proteinase	Hydrolysis of proteins; does not split peptide amides
	Tritirachium alkaline proteinase	Hydrolysis of keratin and of native proteins at the carboxyl of the aromatic or hydrophobic amino acid residues
	Arthrobacter serine proteinase	Hydrolysis of proteins; no clear specificity
	Pseudomonas serine proteinase	Hydrolysis of proteins; no clear specificity
	Thermomycolin	Preferential cleavage: Ala-, Tyr-, Phe-
	Thermophilic *Streptomyces* serine proteinase	Hydrolyses proteins in nonpolar sequences
	Candida lipolytica serine proteinase	

Number	Other Names	Basis for classification (Systematic Name)	Comments	Reference
			and *B. subtilis-amylosacchariticus*	2085
	E. coli protease I		Does not cleave Ac-Phe-ethyl ester	2536
	Aspergillus proteinase B		From *Aspergillus oryzae* and *Aspergillus flavus*, with subtilisin specificity. Alkaline proteinases with similar properties were isolated from	3276, 3488
			Aspergillus sojae,	1247
			Aspergillus sulphureus,	638
			Aspergillus sydowi,	478
			and *Aspergillus melleus*. *Bacillus amyloliquefaciens* serine proteinase also belongs here. Formerly EC 3.4.21.15	1465
	Proteinase K		From the mold *Tritirachium album* Limber. In specificity on synthetic substrates resembles bacterial rather than mold serine proteinases	784, 2284, 1791
				1338, 3579
			Extracellular proteinase from *Pseudomonas aeruginosa*. Cleaves the Pro-Leu bond in PZ-Pro-Leu-Gly-Pro-D-Arg. Is also inhibited by EDTA	188
	Thermomycolase		From the fungus *Malbranchea pulchella* var. *sulfurea*. Is inhibited by Z-Phe-chloromethyl ketone	3223, 2503
			Thermophilic proteinase from *Streptomyces rectus* is inhibited by both fluorophosphates and mercurials	2258
			Inhibited by diisopropyl-fluorophosphate. Hydrolyses salmine and acetyltyrosine-ethylester	3432

Number	Recommended Name	Reaction

[3.4.21.15 Transferred entry: now included with EC 3.4.21.14]

3.4.21.16   *Alternaria* serine proteinase   Hydrolysis of proteins; no clear specificity

[3.4.21.17 Transferred entry: now included with EC 3.4.21.14]

3.4.21.18   *Tenebrio* α-proteinase   Hydrolysis of proteins; no clear specificity

3.4.21.19   Staphylococcal serine proteinase   Preferential specificity: Glu-, Asp-

3.4.21.20   Cathepsin G   Specificity similar to chymotrypsin

3.4.21.21   Coagulation factor VIIa (bovine)   Hydrolyses one Arg-Ile bond in factor X to form factor Xa

3.4.21.22   Coagulation Factor IXa   Cleaves Arg-Ile bond to convert Factor X to Factor Xa

Number	Other Names	Basis for classification (Systematic Name)	Comments	Reference
3.4.21.16			Fungal proteinase	1535
3.4.21.18			From the insect *Tenebrio molitor*. Hydrolyses native proteins. NH_2-substituted esters of arginine and tyrosine are not cleaved. The enzyme is inhibited by sulphofluorides. A trypsin-like β-proteinase was isolated from the same source. Formerly EC 3.4.99.24	3859
3.4.21.19			Isolated from *Staphylococcus aureus*; in appropriate buffer the specificity is restricted to Glu-; bonds involving bulky side chains of hydrophobic amino acids are cleaved at a lower rate	754, 1383, 304, 114
3.4.21.20			In lysosomes of polymorphonuclear leucocytes	3199, 2795, 2963
3.4.21.21			Inhibited by diisopropyl fluorophosphate in contrast to the factor VII from	1701, 1516, 2695, 1049, 957
			human plasma which is not readily inhibited by the reagent.	2388
			Single-chain factor VII is activated to a two-chain form by factor X_a and thrombin	2695
3.4.21.22	Activated Christmas Factor		Homologous with trypsin, thrombin and Factor Xa; a member of the class of vitamin K-related blood coagulation factors.	958, 957
			The zymogen IX is activated by Factor XIa	957

Number	Recommended Name	Reaction
3.4.21.23	*Vipera russelli* proteinase	Converts coagulation Factor V and X to their active forms
3.4.21.24	Red cell neutral endopeptidase	Preferential cleavage of hydrophobic region in low molecular weight peptides. Does not degrade large proteins
3.4.21.25	Cucumisin	Hydrolysis of proteins, preferentially at the carboxyl of acidic amino acid residues
3.4.21.26	Post-proline endopeptidase	Specific hydrolysis of prolyl bonds
3.4.21.27	Coagulation Factor XIa	Converts Factor IX to Factor IXa
3.4.21.28	*Agkistrodon* serine proteinase	Preferential cleavage: Arg- in limited number of polypeptideic substrates
3.4.21.29	*Bothrops atrox* serine proteinase	Preferential cleavage: Arg-Gly bonds in fibrinogen α(A) chains
3.4.21.30	*Crotalus adamanteus* serine proteinase	Preferential cleavage: Arg-Gly bonds in fibrinogen β (B) chains
3.4.21.31	Urokinase	Preferential cleavage: Arg-Val in plasminogen
3.4.21.32	*Uca pugilator* collagenolytic proteinase	Broad specificity; degrades native collagen

Number	Other Names	Basis for classification (Systematic Name)	Comments	Reference
3.4.21.23	Coagulant protein of Russell's viper venom		An arginine esterase. Extent of activation of Factor V the same as obtained with thrombin. In contrast to thrombin, *Vipera russelli* enzyme neither clots fibrinogen nor catalyses proteolysis of prothrombin	844, 2951, 979
3.4.21.24				3711
3.4.21.25			From the sarcocarp of melon (*Cucumis melo*)	1586
3.4.21.26	Post-proline cleaving enzyme		Isolated from lamb kidney, ubiquitous at low levels in vertebrates. Inactivates oxytocin and vasopressin. Does not split Pro-D,L-Pro-bond. Specifically inhibited by substituted proline chloromethylketones	3608, 1742
3.4.21.27				957
3.4.21.28	Ancrod		From *Agkistrodon rhodostoma* (Malayan pit viper). Converts fibrinogen to fibrin. Is without effect on other clotting factors	1344, 852, 846
3.4.21.29	Batroxobin, Reptilase, *B. atrox* coagulant enzyme		Does not activate human Factor VIII or plasminogen. Does not cleave fibrinogen β-chain	3230
3.4.21.30	Crotalase		Does not activate clotting factors	2086, 633
3.4.21.31	Plasminogen activator		Converts plasminogen to plasmin. Isolated from urine and kidney cells	1749, 2627, 3042, 3673, 1870, 2800
3.4.21.32			From the fiddler crab *Uca pugilator*. Also hydrolyses trypsin and chymotrypsin substrates. Inhibited by tosyl lysine	802, 801

Number	Recommended Name	Reaction
3.4.21.33	*Entomophthora* collagenolytic proteinase	Broad specificity; degrades native collagen

3.4.22 THIOL PROTEINASES

3.4.22.1	Cathepsin B	Hydrolyses proteins, with a specificity resembling that of papain
3.4.22.2	Papain	Preferential cleavage: Arg-, Lys-, Phe-X- (the bond next-but-one to the carboxyl group of phenylalanine); limited hydrolysis of native immunoglobulins
3.4.22.3	Ficin	Preferential cleavage: Lys-, Ala-, Tyr-, Gly-, Asn-, Leu-, Val-
3.4.22.4	Bromelain	Preferential cleavage: Lys-, Ala-, Tyr-, Gly-

[3.4.22.5 Transferred entry: now included with EC 3.4.22.4]

3.4.22.6	Chymopapain	Specificity close but not identical to that of papain

Number	Other Names	Basis for classification (Systematic Name)	Comments	Reference
			chlormethylketone. Inactivated by acid pH	
3.4.21.33			From the fungus, *Entomophthora coronata*. Degrades Z-Pro-Leu-Gly-Ala-D-Arg. Does not degrade trypsin substrates	1403
3.4.22.1	Cathepsin B₁		An intracellular enzyme in vertebrates. The amino-terminal blocking is cleaved from the synthetic substrates; pH optimum of 6.0 for low molecular substrates; 4.0 for proteins	1107, 1648, 2529
3.4.22.2	Papainase		From *Papaya latex*. Also has esterases thiolesterase, transamidase and transesterase activity. Formerly EC 3.4.4.10	190, 755, 3099, 3114, 1517
3.4.22.3			From *Ficus glabrata* or *Ficus carica latex*. Wide specificity on protein substrates. Formerly EC 3.4.4.12	614, 1780, 1790, 1978
3.4.22.4			Two bromelains (A and B) with similar specificity have been isolated from pineapple stem.	1448, 527, 2521, 3075, 3074, 2985
			Highly active thiol proteinase which splits glycine esters but not lysine esters.	3075
			Bromelain A corresponds to Bromelain II,	3075 2985
			Bromelain B to Bromelain III. Formerly EC 3.4.4.24	3075 2985
3.4.22.6			Enzymes with slightly different properties from that of papain have been	481, 782, 1498, 1830, 1517

Number	Recommended Name	Reaction
3.4.22.7	Asclepain	Hydrolysis of protein; no clear specificity. Specificity similar to that of papain
3.4.22.8	Clostripain	Preferential cleavage: Arg-, also Arg-Pro bond
3.4.22.9	Yeast proteinase B	Hydrolysis of proteins; no clear specificity
3.4.22.10	Streptococcal proteinase	The amino acid adjacent, at the amino terminal end, to the residue contributing the carbonyl to the susceptible peptide bond, must contain a bulky side chain
	[3.4.22.11 Deleted entry: Insulinase]	
3.4.22.12	γ-Glutamyl hydrolase	Hydrolyses γ-glutamyl bonds in *N*-pteroyl-γ-oligoglutamate; pteroyl-γ-diglutamate is a prominent product
3.4.22.13	Staphylococcal thiol proteinase	Broad specificity on protein substrates

Number	Other Names	Basis for classification (Systematic Name)	Comments	Reference
			isolated from papaya latex:	
			chymopapain,	1498
			chymopapain A,	782
			chymopapain B	1830
			and papaya 'peptidase'	2943
			Formerly EC 3.4.4.11	
3.4.22.7	*Clostridium histolyticum* proteinase B′		From milkweed latex.	476, 1554
3.4.22.8	Clostridiopeptidase B, *Clostridium histolyicum* proteinase B		From *Clostridium histolyticum*. Activated by Ca^{2+}; inactivated by substituted arginine and lysine chloromethylketones. Formerly EC 3.4.4.20	1129, 2243, 1421 3068 2639
3.4.22.9	Baker's yeast proteinase, Brewer's yeast proteinase		A group of enzymes inhibited by diisopropyl fluorophosphate and *p*-chloromercuribenzoate. Differ in pH optima. Baker's yeast proteinase B, C, and brewer's yeast proteinase β ('peptidase β') belong here	869, 1217, 1925, 3026 1217 1217, 1925, 3026 869
3.4.22.10			Formed from inactive zymogen by proteolytic action. Formerly EC 3.4.4.18	817, 1993, 1009
3.4.22.12	Conjugase, 'Carboxypeptidase G'		Endoproteinase formerly considered as a carboxypeptidase with an acid pH optimum from liver, intestine and pancreas. Differs from the Zn-metalloenzyme with a close specificity. Formerly EC 3.4.12.10	2894, 3070, 229 1553
3.4.22.13	Staphylococcal proteinase II			304, 114

Number	Recommended Name	Reaction
3.4.22.14	Actinidin	Specificity close to that of papain
3.4.22.15	Cathepsin L	Hydrolysis of proteins; no action on acylamino acid esters

3.4.23 CARBOXYL (ACID) PROTEINASES

3.4.23.1	Pepsin A	Preferential cleavage: Phe-, Leu-
3.4.23.2	Pepsin B	More restricted specificity than pepsin A; degradation of gelatine; little activity with haemoglobin as substrate
3.4.23.3	Pepsin C	More restricted specificity than pepsin A; high activity towards haemoglobin as substrate

Number	Other Names	Basis for classification (Systematic Name)	Comments	Reference
3.4.22.14	*Actinidia* anionic protease		From *Actinidia chinensis* (Chinese gooseberry)	2160
3.4.22.15			From rat liver lysosomes. Inhibited by leupeptin	1698
3.4.23.1	Pepsin		Formed from porcine pepsinogen by cleavage of one peptide bond; single polypeptide chain. Inhibited by diazoacetyl-norleucine methyl ester. Enzymes analogous to porcine pepsin A have been characterized from many species:	1450, 2442, 2704
			ox,	2444
			man,	2994
			'Funduspepsin';	3384
			'PII';	1847
			'PI';	869
			chicken;	333, 746, 1929
			tuna;	2441
			salmon;	2440
			dogfish.	2204
			Porcine pepsin D is dephosphorylated pepsin A. Formerly EC 3.4.4.1	1896
3.4.23.2			Formed from porcine pepsinogen B. Cleaves acetyl-L-phenylalanyl-L-diiodotyrosine.	2076, 2077
			Identical with porcine gelatinase	1847
			and parapepsin I.	2882
			Human pepsin IV is related to porcine pepsin B. Formerly EC 3.4.4.2	2443
3.4.23.3			Optimum pH when acting on haemoglobin is 3, higher than for pepsin A to which it is structurally related.	2882, 2884
			Identical with porcine parapepsin II.	2882
			The human enzymes, gastricsin,	2787, 1390, 3364
			'Pyloruspepsin',	3384

Number	Recommended Name	Reaction
3.4.23.4	Chymosin	Cleaves a single bond in casein κ
3.4.23.5	Cathepsin D	Specificity similar to, but narrower than, that of pepsin A
3.4.23.6	Microbial carboxyl proteinases	
	Aspergillus oryzae carboxyl proteinase	Broad specificity. Activates trypsinogen. Does not clot milk
	Aspergillus saitoi carboxyl proteinase	Broad specificity. Activates trypsinogen. Does not clot milk
	Aspergillus niger var. *macrosporus* carboxyl proteinase	Preferential cleavage in B chain of insulin: Asn^3-Gln^4, Gly^{13}-Ala^{14}, Tyr^{26}-Thr^{27}
	Penicillium janthinellum carboxyl proteinase	Broad specificity. Activates trypsinogen
	Rhizopus carboxyl proteinase	Broad specificity. Activates trypsinogen. Milk clotting activity

Number	Other Names	Basis for classification (Systematic Name)	Comments	Reference
			'PII' and 'PIII' are related to porcine pepsin C. Formerly EC 3.4.4.22	1181 1847
3.4.23.4	Rennin, Chymase		Formed from prochymosin (prorennin). Formerly EC 3.4.4.3	702, 912
3.4.23.5			Intracellular proteinase inhibited by diazoacetylnorleucine methyl ester. Enzymes homologous with cathepsin D have been characterized, *e.g.* cathepsin E. Formerly EC 3.4.4.23	1646, 2664, 2763 1649, 1875
3.4.23.6				
	Takadiastase, Trypsinogen kinase		Slow action on low molecular weight synthetic substrates can be due to contaminating peptidases	2359, 2360, 273
	Aspergillopeptidase A		No low molecular weight synthetic substrates cleaved. A carboxyl proteinase that is inhibited by diazoacetyl-2,4-DNP-ethylenediamine has also been isolated from *A. awamori* (Awamorin). Formerly EC 3.4.4.17	8, 982, 1426, 1427, 3803, 3804 1787
			Not inhibited by pepstatin	1432
	Penicillopepsin, Peptidase A		Resembles porcine pepsin structurally. Splits the Phe-Phe bond in Z-Gly-Gly-Phe-Phe-Y. Another carboxyl proteinase has been isolated from *P. notatum,* and *P. roqueforti.* Formerly EC 3.4.23.7	3138, 3139; 3394, 1337
			From *Rhizopus chinensis.* Structurally homologous to pepsin. No low molecular weight synthetic	974, 3479, 976

Number	Recommended Name	Reaction
	Endothia carboxyl proteinase	Milk clotting activity
	Mucor pusillus carboxyl proteinase	Does not activate trypsinogen. Milk clotting activity
	Candida albicans carboxyl proteinase	Preferential cleavage at the carboxyl of hydrophobic amino acid residues
	Paecilomyces varioti carboxyl proteinase	Broad specificity
	Saccharomyces carboxyl proteinase	Broad specificity
	Rhodotorula carboxyl proteinase	
	Physarum carboxyl proteinase	Milk clotting activity
	Tetrahymena carboxyl proteinase	
	Plasmodium carboxyl proteinase	

[3.4.23.7 Transferred entry: now included in EC 3.4.23.6, Microbial carboxyl proteinases]

Number	Other Names	Basis for classification (Systematic Name)	Comments	Reference
			substrates cleaved. Inhibited by diazoacetyl norleucine methyl ester. Formerly EC 3.4.23.9 Another fungal carboxyl proteinase was isolated from	
			Trametes sanguinea (formerly EC 3.4.99.25)	3482
			and *Acrocylindrium.* (Formerly 3.4.99.1)	3498, 1453
			From *Endothia parasitica.* Does not hydrolyse L-Glu-L-Tyr. Formerly EC 3.4.23.10	1159, 1160, 1880
	Mucor rennin		Specificity similar to that of pepsin and rennin. A similar enzyme, isolated from	106
			Mucor miehei, preferentially hydrolyses bonds involving aromatic residues	2527
			Diazoacetylnorleucine methyl ester does not inhibit activity	2770
			Low molecular weight synthetic substrates are not cleaved. Formerly EC 3.4.99.15	2927, 2926
	Baker's yeast proteinase A		Intracellular enzyme. Does not act on esters of tyrosine and arginine. Formerly EC 3.4.23.8	1217
			From *Rhodotorula glutinis.* Is inhibited by diazoacetylnorleucine methyl ester	1967
			From *Physarum polycephalum*	863
			Intracellular enzyme from *Tetrahymena pyriformis*	719
			From *Plasmodium berghei*	1950

Number	Recommended Name	Reaction

[3.4.23.8 Transferred entry: now included in EC 3.4.23.6, Microbial carboxyl proteinases]

[3.4.23.9 Transferred entry: now included in EC 3.4.23.6, Microbial carboxyl proteinases]

[3.4.23.10 Transferred entry: now included in EC 3.4.23.6, Microbial carboxyl proteinases]

3.4.23.11   Thyroid carboxyl proteinase

3.4.23.12   *Nepenthes* carboxyl proteinase   Preferential cleavage: -Asp, Asp-, Lys-, Ala-

3.4.23.13   *Lotus* carboxyl proteinase

3.4.23.14   Sorghum carboxyl proteinase   Preferential cleavage: Asp-, Glu-

3.4.24   METALLOPROTEINASES

3.4.24.1   *Crotalus atrox* metalloproteinase   Preferential cleavage: Leu-, Phe-, Val-, Ile-

3.4.24.2   *Sepia* proteinase

3.4.24.3   *Clostridium histolyticum* collagenase   Degradation of helical regions of native collagen to small fragments. Preferential cleavage: -Gly in the sequence -Z-Pro-X-Gly-Pro-X

3.4.24.4   Microbial metalloproteinases

Aeromonas proteolytica neutral proteinase   Preferential cleavage: -Val, -Leu, -Ile, -Phe

Number	Other Names	Basis for classification (Systematic Name)	Comments	Reference
3.4.23.11			Inhibited by diazoacetylnorleucine methyl ester	3115
3.4.23.12			From the insectivorous plants *Nepenthes, Drosera peltata.* Formerly EC 3.4.99.4	64
3.4.23.13			From the *Lotus* seeds	3053
3.4.23.14			From *Sorghum vulgare.* Does not cleave the Asp-Lys bond	994
3.4.24.1			From snake venom	2192, 2598, 2599
3.4.24.2			Not inhibited by chloromethylketones, but by EDTA and *o*-phenanthroline	2855
3.4.24.3	Clostridiopeptidase A, Collagenase A, Collagenase I		Extracellular Zn^{2+}-enzyme	3252, 2070, 1757, 3802, 2996, 1199, 1198, 1634
			Other forms with a broader specificity were isolated from the same source:	1198
			collagenase B, collagenase II, pseudocollagenase. Many bacterial collagenases of similar specificity have been isolated:	1757, 2070, 2242
			Streptomyces madurae,	2797
			Bacteroides melaninogenicus. Formerly EC 3.4.4.19. Collagenase B; formerly EC 3.4.99.5	1022
3.4.24.4				
			Zn^{2+}-protein.	1121

Number	Recommended Name	Reaction
	Pseudomonas aeruginosa neutral proteinase	Preferential cleavage of bonds adjacent to a hydrophobic amino acid residue
	Pseudomonas aeruginosa alkaline proteinase	Preferential cleavage of bonds adjacent to a hydrophobic amino acid residue
	Escherichia freundii proteinase	Preferential cleavage of bonds adjacent to a hydrophobic amino acid residue
	Bacillus thermoproteolyticus neutral proteinase	Preferential cleavage: -Leu$>$-Phe
	Bacillus subtilis neutral proteinase	Preferential cleavage:-Leu$>$-Phl
	Staphylococcus aureus neutral proteinase	Converts plasminogen to plasmin
	Micrococcus caseolyticus neutral proteinase	Preferential cleavage of bonds adjacent to a hydrophobic amino acid residue
	Sarcina neutral proteinase	Preferential cleavage of bonds adjacent to a hydrophobic amino acid residue
	Streptococcus thermophilus intracellular proteinase	Preferential cleavage: -Leu, -Phe, -Tyr, -Ala
	Streptomyces griseus neutral proteinase	Preferential cleavage of bonds adjacent to a hydrophobic amino acid residue
	Aspergillus oryzae neutral proteinase	Preferential cleavage of bonds adjacent to a hydrophobic amino acid residue

Number	Other Names	Basis for classification (Systematic Name)	Comments	Reference
			The rate of hydrolysis of X-Phe bond is 40-fold greater than that for the X-Leu bond	1631
			Degrades elastin, clots milk probably contains Zn^{2+}	2289
			Ca^{2+}-protein. Does not degrade elastin	2283, 2286, 2287, 2288
			Contains one Zn atom	2352
	Thermolysin		Contains one Zn atom. Retains activity at 80	2287, 3206, 2523, 365, 3430
			Contains one essetial Zn atom. Homologous with thermolysin.	866, 2025, 2287, 2553, 2554, 2230
			Similar enzymes have been isolated from *B. amyloliquefaciens*	3481
			and *B. megaterium* (megateriopeptidase)	2231, 2232
	Protease III, Staphylokinase		Ca^{2+} is essential to stabilize the enzyme. Separated in forms A, B, C. Formerly EC 3.4.99.22	112, 648, 1051, 1855, 3550
			Ca^{2+} protein	705
	Coccus P (proteinase)		Ca^{2+} protein	113
			Does not hydrolyse disubstituted dipeptides. Optimum pH 6.5. Inhibited by EDTA, reactivated by Mn^{2+}.	2287
			Similar enzyme is in *Streptococcus diacetilactis*	704
	'Pronase'(component)		Similar enzymes are in other *Streptomyces*	3206, 2287 1319
				3206

Number	Recommended Name	Reaction
	Penicillium roqueforti neutral proteinase	Preferential cleavage of bonds adjacent to a hydrophobic amino acid residue
	Myxobacter β-lytic proteinase	Preferential cleavage of bonds adjacent to a hydrophobic amino acid residue
	Serratia marcescens extracellular proteinase	Preferential cleavage at the carboxyl of proline residues
3.4.24.5	Lens neutral proteinase	Degradation of α_2-cristallin
3.4.24.6	*Leucostoma* neutral proteinase	Preferential cleavage: -Leu, -Phe, -Val
3.4.24.7	Vertebrate collagenase	Cleaves preferentially one bond in native collagen leaving a *N*-terminal (75%) and *C*-terminal (25%) fragment
3.4.24.8	*Achromobacter iophagus* collagenase	Degradation of helical regions of native collagen to large fragments

Number	Other Names	Basis for classification (Systematic Name)	Comments	Reference
	P. roqueforti protease II		Inactivated by metal chelators. Does not hydrolyse disubstituted dipeptides.	1125, 3206
			Similar enzyme is in *Aspergillus sojae*	3000
			Formerly EC 3.4.99.13	3667, 3668, 57
			Inactivated by metal chelators; reactivated by Fe^{2+}	30
3.4.24.5			Activated by Ca^{2+}, Mg^{2+}	328
3.4.24.6			From the venom of *Agkistrodon piscivorus leucostoma*. Ca^{2+}:Zn^{2+} ratio 2:1.	
			A similar neutral proteinase is in the venom of	
			A. halys blomhoffii	2484, 3345
			and other snake venoms	2518
3.4.24.7			Group of Zn^{2+}-enzymes of similar specificity and ubiquitous occurrence in animals.	2995
			Isolated from tadpole tail,	2330
			cornea,	1356
			granulocytes,	2483
			uterus,	1510
			rheumatoid synovium,	1200
			bones,	978
			tumors.	2154
			Enzyme from human gastric mucosa hydrolyses collagen in the same chain segment as the tadpole enzyme but between different amino acid residues	3731
3.4.24.8			Extracellular Zn^{2+}-enzyme. In contrast to EC 3.4.24.3, degrades Z-Pro-Leu-Gly-Ala-D-Arg, X-Gly-Pro and X-Gly-Ala bonds in other proteins. Partially degraded forms of the enzyme are still active	1634, 3657, 1893, 1635

Number	Recommended Name	Reaction
3.4.24.9	*Trichophyton schoenleinii* collagenase	Degradation of native collagen in acid pH range
3.4.24.10	*Trichophyton mentagrophytes* keratinase	Hydrolyses keratin; preferential cleavage at hydrophobic residues
3.4.24.11	Kidney brush border neutral proteinase	Preferential cleavage at the amino group of hydrophobic residues of B-chain of insulin
3.4.24.12	Sea urchin hatching proteinase	Preferential cleavage: Glu-, Asp-

3.4.99 PROTEINASES OF UNKNOWN CATALYTIC MECHANISM

[3.4.99.1 Transferred entry: now EC 3.4.23.6, Microbial carboxyl proteinases]

3.4.99.2	Agavain	
3.4.99.3	Angiotensinase	Inactivates angiotensin II by preferential cleavage of the bond Tyr-Ile

[3.4.99.4 Transferred entry: now EC 3.4.23.12, Nepenthes carboxyl proteinase]

[3.4.99.5 Transferred entry: now EC 3.4.24.3, Clostridium histolyticum collagenase]

3.4.99.6	Crayfish low-molecular-weight proteinase
3.4.99.7	Euphorbain
3.4.99.8	*Gliocladium* proteinase
3.4.99.9	Hurain

[3.4.99.10 Deleted entry: Insulinase]

3.4.99.11	*Streptomyces* alkalophilic keratinase	Hydrolyses keratin and poly-L-lysine; preferential cleavage: Ser-His, Leu-Val, Phe-Tyr, Lys-Ala

[3.4.99.12 Transferred entry: now EC 3.4.24.10, Trichophython mentagrophytes keratinase]

[3.4.99.13 Transferred entry: now EC 3.4.24.4, Microbial metallo proteinases]

Number	Other Names	Basis for classification (Systematic Name)	Comments	Reference
3.4.24.9			Acid-stable fungal metalloenzyme	2798
3.4.24.10			Keratinase I is extracellular; II and III cell-bound. Formerly EC 3.4.99.12	669, 3818, 3819, 3817
3.4.24.11			In rabbit kidney. Inhibited by EDTA, reactivated by Zn^{2+}	1664
3.4.24.12		Hatching enzyme	Extracellular Ca^{2+}-dependent enzyme. Dissolves fertilization envelope in embryos of sea urchin *Strongylocentrotus purpuratus.*Is inhibited by Z-Glu. Enzymes of similar specificity exist in other *Echinoidea*	216, 215, 2003
3.4.99.2				3425
3.4.99.3				1750
3.4.99.6				2599, 3151
3.4.99.7				821
3.4.99.8				1700
3.4.99.9				1483
3.4.99.11			Optimum activity at pH 13. Formerly EC 3.4.4.25	2407, 2326

Number	Recommended Name	Reaction
3.4.99.14	Mexicanain	
[3.4.99.15 Transferred entry: now EC 3.4.23.6, Microbial carboxy proteinases]		
3.4.99.16	*Penicillium notatum* extracellular proteinase	
3.4.99.17	Peptidoglycan endopeptidase	Hydrolyses pentaglycine cross-bridges at D-alanylglycine and glycyl-glycine linkages in the cell wall peptidoglycan
3.4.99.18	Pinguinain	
3.4.99.19	Renin	
3.4.99.20	Scopulariopsis proteinase	
3.4.99.21	Solanain	
[3.4.99.22 Transferred entry: now EC 3.4.24.4, Microbial metallo proteinases]		
3.4.99.23	Tabernamontanain	
[3.4.99.24 Transferred entry: now EC 3.4.21.18, Tenebrio α-proteinase]		
[3.4.99.25 Transferred entry: now EC 3.4.23.6, Microbial carboxyl proteinases]		
[3.4.99.26 Deleted entry: Urokinase]		
3.4.99.27	*Echis carnatus* prothrombin activating proteinase	Cleaves specifically Arg-Ile bond in prothrombin
3.4.99.28	*Oxyuranus scutellatus* prothrombin-activating proteinase	Cleaves prothrombin to thrombin and two inactive fragments
3.4.99.29	*Myxobacter* AL-1 proteinase I	Preferential cleavage of bonds adjacent to a hydrophobic amino acid residue
3.4.99.30	*Myxobacter* AL-1 proteinase II	Preferential cleavage: -Lys

Number	Other Names	Basis for classification (Systematic Name)	Comments	Reference
3.4.99.14				483
3.4.99.16			Broad specificity for protein substrates	249, 2097, 248
3.4.99.17		Glycyl-glycine endopeptidase, Endo-β-N-acetyl-glucosaminidase	From *Staphylococcus staphylolyticus*	1854, 1135
3.4.99.18				118, 3451
3.4.99.19			Converts angiotensinogen to angiotensin I. Inhibited by a phospholipid similar to bovine phosphatidylserine. Formerly EC 3.4.4.15	3103
3.4.99.20				3811
3.4.99.21				1110
3.4.99.23				1484
3.4.99.27	Echarin		Not inhibited by diisopropyl fluorophosphate	1765, 923
3.4.99.28			Not inhibited by diisopropyl fluorophosphate, mercurials, iodoacetate. Does not hydrolyse arginine esters	2532, 1867, 697
3.4.99.29			Inhibition studies suggest that mechanism of action is different from the four groups 3.4.21-3.4.24	1475, 1476
3.4.99.30				3705

348

Number	Recommended Name	Reaction
3.4.99.31	Tissue endopeptidase degrading collagenase synthetic substrate	Degrades specifically Z-Pro-Leu-Gly-Pro-D-Arg. Does not degrade native collagen
3.4.99.32	*Armillaria mellea* neutral proteinase	Preferential cleavage: -Lys

3.5 Acting On Carbon-Nitrogen Bonds, Other Than Peptide Bonds

To this sub-class belong those enzymes hydrolysing amides, amidines and other C-N bonds. The sub-subgroups are separated on the basis of the substrate: linear amides (3.5.1), cyclic amides (3.5.2), linear amidines (3.5.3), cyclic amidines (3.5.4) and nitriles(3.5.5).

3.5.1 IN LINEAR AMIDES

3.5.1.1	Asparaginase	L-Asparagine + H_2O = L-aspartate + NH_3
3.5.1.2	Glutaminase	L-Glutamine + H_2O = L-glutamate + NH_3
3.5.1.3	ω-Amidase	An ω-amido-dicarboxylic acid + H_2O = a dicarboxylate + NH_3
3.5.1.4	Amidase	A monocarboxylic acid amide + H_2O = a monocarboxylate + NH_3
3.5.1.5	Urease	Urea + H_2O = CO_2 + 2 NH_3
3.5.1.6	β-Ureidopropionase	N-Carbamoyl-β-alanine + H_2O = β-alanine + CO_2 + NH_3
3.5.1.7	Ureidosuccinase	N-Carbamoyl-L-aspartate + H_2O = L-aspartate + CO_2 + NH_3
3.5.1.8	Formylaspartate deformylase	N-Formyl-L-aspartate + H_2O = formate + L-aspartate
3.5.1.9	Formamidase	N-Formyl-L-kynurenine + H_2O = formate + L-kynurenine
3.5.1.10	Formyltetrahydrofolate deformylase	10-Formyltetrahydrofolate + H_2O = formate + tetrahydrofolate

Number	Other Names	Basis for classification (Systematic Name)	Comments	Reference
3.4.99.31			Enzymes completing the degradation of collagen made by collagenase in vertebrate tissues.	
			Isolated from kidney	127
			granuloma tissue	126
			rheumatoid synovium,	1201
			tadpole tissues,	1197
			spermatozoa	1760, 1759
3.4.99.32			Also cleaves peptide bonds at the amino group of 2-aminoethylcystine	3056
3.5.1.1	Asparaginase II	L-Asparagine amidohydrolase		1174, 1325, 3285
3.5.1.2		L-Glutamine amidohydrolase		1022, 1233, 1829, 2064
3.5.1.3		ω Amidodicarboxylate amidohydrolase	Acts on glutaramate, succinamate and the corresponding α-keto-ω-amido-acids	2188
3.5.1.4	Acylamidase, Acylase	Acylamide amidohydrolase		376, 377
3.5.1.5		Urea amidohydrolase	A nickeloprotein	3290, 3541, 731
3.5.1.6		N-Carbamoyl-β-alanine amidohydrolase	The animal enzyme also acts on β-ureidoisobutyrate	453, 469, 1126
3.5.1.7		N-Carbamoyl-L-aspartate amidohydrolase		1963
3.5.1.8		N-Formyl-L-aspartate amidohydrolase		2483
3.5.1.9	Kynurenine formamidase, Formylase, Formylkynureninase	Aryl-formylamine amidohydrolase	Also acts on other aromatic formylamines	1242, 1486, 2178
3.5.1.10		10-Formyltetrahydrofolate		1396

Number	Recommended Name	Reaction
3.5.1.11	Penicillin amidase	Penicillin + H_2O = penicin + a carboxylic acid anion
3.5.1.12	Biotinidase	Biotin amide + H_2O = biotin + NH_3
3.5.1.13	Aryl acylamidase	An anilide + H_2O = a fatty acid anion + aniline
3.5.1.14	Aminoacylase	An N-acyl-amino acid + H_2O = a fatty acid anion + an amino acid
3.5.1.15	Aspartoacylase	N-Acyl-L-aspartate + H_2O = a fatty acid anion + aspartate
3.5.1.16	Acetylornithine deacetylase	N^2-Acetyl-L-ornithine + H_2O = acetate + L-ornithine
3.5.1.17	Acyl-lysine deacylase	N^6-Acyl-L-lysine + H_2O = a fatty acid anion + L-lysine
3.5.1.18	Succinyl-diaminopimelate desuccinylase	N-Succinyl-LL-2,6-diaminopimelate + H_2O = succinate + LL-2,6-diaminopimelate
3.5.1.19	Nicotinamidase	Nicotinamide + H_2O = nicotinate + NH_3
3.5.1.20	Citrullinase	L-Citrulline + 2 H_2O = L-ornithine + CO_2 + NH_3
3.5.1.21	N-Acetyl-β-alanine deacetylase	N-Acetyl-β-alanine + H_2O = acetate + β-alanine
3.5.1.22	Pantothenase	Pantothenate + H_2O = pantoate + β-alanine
3.5.1.23	Acylsphingosine deacylase	N-Acylsphingosine + H_2O = sphingosine + a fatty acid anion
3.5.1.24	Choloylglycine hydrolase	$3\alpha,7\alpha,12\alpha$-Trihydroxy-5β-cholan-24-oylglycine + H_2O = $3\alpha,7\alpha,12\alpha$-trihydroxy-5β-cholanate + glycine
3.5.1.25	N-Acetylglucosamine-6-phosphate deacetylase	2-Acetamido-2-deoxy-D-glucose 6-phosphate + H_2O = 2-amino-2-deoxy-D-glucose 6-phosphate + acetate
3.5.1.26	4-N-(2-β-D-Glucosaminyl)-L-asparaginase	N^4-(2-acetamido-2-deoxy-β-D-glucopyranosyl)-L-asparagine + H_2O = 2-acetamido-2-deoxy-β-D-glucopyranosylamine + L-asparate

Number	Other Names	Basis for classification (Systematic Name)	Comments	Reference
3.5.1.11		Penicillin amidohydrolase		2900
3.5.1.12		Biotin-amide amidohydrolase	Also acts on other biotinides	1718, 3410
3.5.1.13		Aryl-acylamide amidohydrolase	Also acts on 1,4-substituted acyl-anilides	2412
3.5.1.14	Dehydropeptidase II, Histozyme, Hippuricase, Benzamidase	N-Acylamino-acid amidohydrolase	Wide specificity; also hydrolyses dehydro-peptides	300, 913, 2557
3.5.1.15	Aminoacylase II	N-Acyl-L-aspartate amidohydrolase		299, 300
3.5.1.16		N^2-Acetyl-L-ornithine amidohydrolase	Also hydrolyses N-acetylmethionine	3559, 3560
3.5.1.17		N^6-Acyl-L-lysine amidohydrolase		2540
3.5.1.18		N-Succinyl-LL-2,6-diaminopimelate amidohydrolase		1690
3.5.1.19		Nicotinamide amidohydrolase		2591, 2913
3.5.1.20		L-Citrulline N^5-carbamoyldihydrolase		1309
3.5.1.21		N-Acetyl-β-alanine amidohydrolase		960
3.5.1.22		Pantothenate amidohydrolase		2460
3.5.1.23		N-Acylsphingosine amidohydrolase	Does not act on N-acetylsphingosine	998
3.5.1.24	Glycocholase	$3\alpha,7\alpha,12\alpha$-Trihydroxy-5β-cholan-24-oylglycine amidohydrolase	Also acts on the $3\alpha,12\alpha$-dihydroxy derivative	2342
3.5.1.25		2-Acetamido-2-deoxy-D-glucose-6-phosphate amidohydrolase		3672
3.5.1.26	Aspartylglucosylamine deaspartylase, Aspartylglucosyl-aminase	4-N-(2-Acetamido-2-deoxy-β-D-glucopyranosyl)-L-asparagine amidohydrolase		2047, 3372, 1741

Number	Recommended Name	Reaction
3.5.1.27	*N*-Formylmethionylaminoacyl-tRNA deformylase	*N*-Formyl-L-methionylaminoacyl-tRNA + H_2O = formate + L-methionylaminoacyl-tRNA
3.5.1.28	*N*-Acetylmuramoyl-L-alanine amidase	Hydrolyses the link between *N*-acetylmuramoyl residues and L-amino acid residues in certain cell wall glycopeptides
3.5.1.29	α-(Acetamidomethylene) succinate hydrolase	α-(Acetamidomethylene) succinate + 2 H_2O = acetate + succinate semialdehyde + NH_3 + CO_2
3.5.1.30	5-Aminovaleramidase	5-Amino-*n*-valeramide + H_2O = 5-amino-*n*-valerate + NH_3
3.5.1.31	Formylmethionine deformylase	*N*-Formyl-L-methionine + H_2O = formate + L-methionine
3.5.1.32	Hippurate hydrolase	Hippurate + H_2O = benzoate + glycine
3.5.1.33	*N*-Acetylglucosamine deacetylase	2-Acetamido-2-deoxy-D-glucose + H_2O = 2-amino-2-deoxy-D-glucose + acetate
3.5.1.34	Acetylhistidine deacetylase	*N*-Acetyl-L-histidine + H_2O = L-histidine + acetate
3.5.1.35	D-Glutaminase	D-Glutamine + H_2O = D-glutamate + NH_3
3.5.1.36	*N*-Methyl-2-oxoglutaramate hydrolase	*N*-Methyl-2-oxoglutaramate + H_2O = 2-oxoglutarate + methylamine

[*3.5.1.37 Deleted entry: 4-L-Aspartylglycosylamine amidohydrolase. Identical with EC 3.5.1.26*]

3.5.1.38	Glutamin-(asparagin-)ase	L-Glutamine + H_2O = L-glutamate + NH_3
3.5.1.39	Alkylamidase	*N*-Methylhexanoamide + H_2O = hexanoate + methylamine

Number	Other Names	Basis for classification (Systematic Name)	Comments	Reference
3.5.1.27		*N*-Formyl-L-methionylaminoacyl-tRNA amidohydrolase		1994
3.5.1.28		Mucopeptide amidohydrolase		1020, 492
3.5.1.29		α-(Acetamidomethylene)succinate amidohydrolase (deaminating, decarboxylating)		2466
3.5.1.30		5-Amino-*n*-valeramide amidohydrolase	The enzyme from *Pseudomonas putida* also acts on 4-aminobutyramide and, more slowly, on 6-aminocaproamide	3350, 2768
3.5.1.31		*N*-Formyl-L-methionine amidohydrolase		108
3.5.1.32		*N*-Benzoylamino-acid amidohydrolase	Acts on various *N*-benzoylamino acids	2830, 2831
3.5.1.33		2-Acetamido-2-deoxy-D-glucose amidohydrolase		2842
3.5.1.34		*N*-Acetyl-L-histidine amidohydrolase		221
3.5.1.35		D-Glutamine amidohydrolase		744
3.5.1.36	5-Hydroxy-*N*-methylpyroglutamate synthase	*N*-Methyl-2-oxoglutaramate methylamidohydrolase	In the reverse reaction, the product cyclizes non-enzymically to 5-hydroxy-*N*-methylpyroglutamate	1295, 1297
3.5.1.38		L-Glutamine (L-asparagine) amidohydrolase	L-Asparagine is hydrolysed at 0.8 of the rate of L-glutamine; the D-isomers are also hydrolysed, more slowly	2809
3.5.1.39		*N*-Methylhexanoamide amidohydrolase	The enzyme hydrolyses *N*-monosubstituted and *N,N*-disubstituted amides, and there is some activity towards primary amides. It has little or no activity	515

Number	Recommended Name	Reaction
3.5.1.40	Acylagmatine amidase	Benzoylagmatine + H_2O = benzoate + agmatine
3.5.1.41	Chitin deacetylase	Chitin + H_2O = chitosan + acetate
3.5.1.42	Nicotinamidenucleotide amidase	β-Nicotinamide D-ribonucleotide + H_2O = β-nicotinate D-ribonucleotide + NH_3
3.5.1.43	Peptidyl-glutaminase	α-N-Peptidyl-L-glutamine + H_2O = α-N-peptidyl-L-glutamate + NH_3
3.5.1.44	Glutaminyl-peptide glutaminase	L-Glutaminyl-peptide + H_2O = L-glutamyl-peptide + NH_3
3.5.1.45	Urease (ATP-hydrolysing)	ATP + urea + H_2O = ADP + orthophosphate + CO_2 + 2 NH_3

3.5.2 IN CYCLIC AMIDES

Number	Recommended Name	Reaction
3.5.2.1	Barbiturase	Barbiturate + H_2O = malonate + urea
3.5.2.2	Dihydropyrimidinase	5,6-Dihydrouracil + H_2O = 3-ureidopropionate
3.5.2.3	Dihydro-orotase	L-5,6-Dihydro-orotate + H_2O = N-carbamoyl-L-aspartate

Number	Other Names	Basis for classification (Systematic Name)	Comments	Reference
			towards short chain substrates	
3.5.1.40		Benzoylagmatine amidohydrolase	Also acts on acetylagmatine, propionylagmatine and bleomycin B$_2$	3508
3.5.1.41		Chitin amidohydrolase	Hydrolyses the *N*-acetamido groups of 2-acetamido-2-deoxy-D-glucose residues in chitin	98
3.5.1.42		Nicotinamidenucleotide amidohydrolase	Also acts more slowly on β-nicotinamide D-ribonucleoside	1443
3.5.1.43	Peptidoglutaminase I	Peptidyl-L-glutamine amidohydrolase	Specific for the hydrolysis of the γ-amide of glutamine substituted at the α-amino group, *e.g.* glycyl-L-glutamine, *N*-acetyl-L-glutamine and L-leucylglycyl-L-glutamine	1682
3.5.1.44	Peptidoglutaminase II	L-Glutaminyl-peptide amidohydrolase	Specific for the hydrolysis of the γ-amide of glutamine substituted at the carboxyl position or both the α-amino and carboxyl positions, *e.g.* L-glutaminylglycine and L-phenylalanyl-L-glutaminylglycine	1682
3.5.1.45		Urea amidohydrolase (ATP-hydrolysing)	Requires catalytic amounts of bicarbonate. Also acts, more slowly on propionamide, acetamide, formamide, *N*-methylurea, cyanamide and biuret	2836
3.5.2.1		Barbiturate amidohydrolase		1235
3.5.2.2	Hydantoinase	5,6-Dihydropyrimidine amidohydrolase	Also acts on dihydrothymine and hydantoin	778, 1126, 3605
3.5.2.3	Carbamoylaspartic dehydrase	L-5,6-Dihydro-orotate amidohydrolase		578, 1962

Number	Recommended Name	Reaction
3.5.2.4	Carboxymethylhydantoinase	L-5-Carboxymethylhydantoin + H_2O = N-carbamoyl-L-aspartate
3.5.2.5	Allantoinase	Allantoin + H_2O = allantoate
3.5.2.6	Penicillinase	Penicillin + H_2O = penicilloate
3.5.2.7	Imidazolonepropionase	4-Imidazolone-5-propionate + H_2O = N-formimino-L-glutamate

[3.5.2.8   Deleted entry: now included with EC 3.5.2.6, Penicillinase]

Number	Recommended Name	Reaction
3.5.2.9	5-Oxoprolinase (ATP-hydrolysing)	ATP + 5-oxo-L-proline + 2 H_2O = ADP + orthophosphate + L-glutamate
3.5.2.10	Creatininase	Creatinine + H_2O = creatine

3.5.3   IN LINEAR AMIDINES

Number	Recommended Name	Reaction
3.5.3.1	Arginase	L-Arginine + H_2O = L-ornithine + urea
3.5.3.2	Glycocyaminase	Guanidinoacetate + H_2O = glycine + urea
3.5.3.3	Creatinase	Creatine + H_2O = sarcosine + urea
3.5.3.4	Allantoicase	Allantoate + H_2O = (—)-ureidoglycollate + urea
3.5.3.5	Formiminoaspartate deiminase	N-Formimino-L-aspartate + H_2O = N-formyl-L-aspartate + NH_3
3.5.3.6	Arginine deiminase	L-Arginine + H_2O = L-citrulline + NH_3
3.5.3.7	Guanidinobutyrase	4-Guanidinobutyrate + H_2O = 4-aminobutyrate + urea
3.5.3.8	Formiminoglutamase	N-Formimino-L-glutamate + H_2O = L-glutamate + formamide

Number	Other Names	Basis for classification (Systematic Name)	Comments	Reference
3.5.2.4		L-5-Carboxymethylhydantoin amidohydrolase		1962
3.5.2.5		Allantoin amidohydrolase		408, 901
3.5.2.6	β-Lactamase I, β-Lactamase II, Cephalosporinase	Penicillin amido-β-lactamhydrolase	A group of enzymes hydrolysing β-lactams of varying specificity; some act more rapidly on penicillin antibiotics, some more rapidly on cephalosporin and cephaloridine. The latter were formerly listed as EC 3.2.5.8	1850, 2629, 2630, 2850, 1275, 541, 3016
3.5.2.7		4-Imidazolone-5-propionate amidohydrolase		2715, 3133
3.5.2.9	Pyroglutamase (ATP-hydrolysing)	5-Oxo-L-proline amidohydrolase (ATP-hydrolysing)		3532
3.5.2.10		Creatinine amidohydrolase		3480
3.5.3.1	Arginine amidinase, Canavanase	L-Arginine amidinohydrolase	Also hydrolyses α-N-substituted L-arginines and canavanine	161, 443, 1108, 1109
3.5.3.2		Guanidinoacetate amidinohydrolase		2824
3.5.3.3		Creatine amidinohydrolase		2824, 3810
3.5.3.4		Allantoate amidinohydrolase	Also hydrolyses (+)-ureidoglycollate to glyoxylate and urea	901, 3460
3.5.3.5		N-Formimino-L-aspartate iminohydrolase		822
3.5.3.6	Arginine dihydrolase	L-Arginine iminohydrolase	Also acts on canavanine	2479, 2725
3.5.3.7		4-Guanidinobutyrate amidinohydrolase	Also acts on 3-guanidinopropionate and D-arginine, but not on L-arginine	2276, 3409
3.5.3.8		N-Formimino-L-glutamate formiminohydrolase		2017

Number	Recommended Name	Reaction
3.5.3.9	Allantoate deiminase	Allantoate + H_2O = ureidoglycine + NH_3 + CO_2
3.5.3.10	D-Arginase	D-Arginine + H_2O = D-ornithine + urea
3.5.3.11	Agmatinase	Agmatine + H_2O = putrescine + urea
3.5.3.12	Agmatine deiminase	Agmatine + H_2O = N-carbamoylputrescine + NH_3
3.5.3.13	Formiminoglutamate deiminase	N-Formimino-L-glutamate + H_2O = N-formyl-L-glutamate + NH_3
3.5.3.14	Amidinoaspartase	N-Amidino-L-aspartate + H_2O = L-aspartate + urea

3.5.4   IN CYCLIC AMIDINES

3.5.4.1	Cytosine deaminase	Cytosine + H_2O = uracil + NH_3
3.5.4.2	Adenine deaminase	Adenine + H_2O = hypoxanthine + NH_3
3.5.4.3	Guanine deaminase	Guanine + H_2O = xanthine + NH_3
3.5.4.4	Adenosine deaminase	Adenosine + H_2O = inosine + NH_3
3.5.4.5	Cytidine deaminase	Cytidine + H_2O = uridine + NH_3
3.5.4.6	AMP deaminase	AMP + H_2O = IMP + NH_3
3.5.4.7	ADP deaminase	ADP + H_2O = IDP + NH_3
3.5.4.8	Aminoimidazolase	4-Aminoimidazole + H_2O = unidentified product + NH_3
3.5.4.9	Methenyltetrahydrofolate cyclohydrolase	5,10-Methenyltetrahydrofolate + H_2O = 10-formyltetrahydrofolate
3.5.4.10	IMP cyclohydrolase	IMP + H_2O = 5′-phosphoribosyl-5-formamido-4-imidazolecarboxamide
3.5.4.11	Pterin deaminase	2-Amino-4-hydroxypteridine + H_2O = 2,4-dihydroxypteridine + NH_3
3.5.4.12	dCMP deaminase	dCMP + H_2O = dUMP + NH_3

Number	Other Names	Basis for classification (Systematic Name)	Comments	Reference
3.5.3.9		Allantoate amidinohydrolase (decarboxylating)		3562
3.5.3.10		D-Arginine amidinohydrolase		2325
3.5.3.11		Agmatine amidinohydrolase		1323
3.5.3.12		Agmatine iminohydrolase		3127
3.5.3.13		N-Formimino-L-glutamate iminohydrolase		3677
3.5.3.14		N-Amidino-L-aspartate amidinohydrolase	Also acts slowly on N-amidino-L-glutamate	2235
3.5.4.1		Cytosine aminohydrolase	Also acts on 5-methylcytosine	554, 1235, 1792
3.5.4.2	Adenase, Adenine aminase	Adenine aminohydrolase		325, 1284
3.5.4.3	Guanase, Guanine aminase	Guanine aminohydrolase		1324, 1569, 2676
3.5.4.4		Adenosine aminohydrolase		1595, 2643
3.5.4.5		Cytidine aminohydrolase		2805, 3612
3.5.4.6	Adenylic acid deaminase, AMP aminase	AMP aminohydrolase	*cf.* EC 3.5.4.17	1569, 1903, 2199, 3490, 3640
3.5.4.7		ADP aminohydrolase		710
3.5.4.8		4-Aminoimidazole aminohydrolase		2679
3.5.4.9		5,10-Methenyltetrahydrofolate 5-hydrolase (decyclizing)		2678, 3331
3.5.4.10		IMP 1,2-hydrolase (decyclizing)		894
3.5.4.11		2-Amino-4-hydroxypteridine aminohydrolase	The animal enzyme is specific for pterin, isoxanthopterin and tetrahydropterin	1931, 2769
3.5.4.12		dCMP aminohydrolase	Also acts on some 5-substituted dCMP's	2933, 2934, 3007

Number	Recommended Name	Reaction
3.5.4.13	dCTP deaminase	$dCTP + H_2O = dUTP + NH_3$
3.5.4.14	Deoxycytidine deaminase	Deoxycytidine $+ H_2O =$ deoxyuridine $+ NH_3$
3.5.4.15	Guanosine deaminase	Guanosine $+ H_2O =$ xanthosine $+ NH_3$
3.5.4.16	GTP cyclohydrolase	$GTP + H_2O =$ formate $+$ 2-amino-4-hydroxy-6-(*erythro*-1′,2′,3′-trihydroxypropyl)-dihydropteridine triphosphate
3.5.4.17	Adenosine (phosphate) deaminase	$5'$-AMP $+ H_2O = 5'$-IMP $+ NH_3$
3.5.4.18	ATP deaminase	$ATP + H_2O = ITP + NH_3$
3.5.4.19	Phosphoribosyl-AMP cyclohydrolase	1-N-($5'$-phospho-D-ribosyl)-AMP $+ H_2O =$ 5-($5'$-phospho-D-ribosyl-aminoformimino)-1-($5''$-phosphoribosyl)-imidazole-4-carboxamide
3.5.4.20	Pyrithiamin deaminase	1-(4-Amino-2-methylpyrimid-5-ylmethyl)-3-(β-hydroxyethyl)-2-methylpyridinium bromide $+ H_2O =$ 1-(4-hydroxy-2-methylpyrimid-5-ylmethyl)-3-(β-hydroxyethyl)-2-methylpyridinium bromide $+ NH_3$
3.5.4.21	Creatinine deiminase	Creatinine $+ H_2O = N$-methylhydantoin $+ NH_3$
3.5.4.22	1-Pyrroline-4-hydroxy-2-carboxylate deaminase	1-Pyrroline-4-hydroxy-2-carboxylate $+ H_2O =$ 2,5-dioxovalerate $+ NH_3$
3.5.4.23	Blasticidin-S deaminase	Blasticidin S $+ H_2O =$ deaminohydroxyblasticidin S $+ NH_3$
3.5.4.24	Sepiapterin deaminase	Sepiapterin $+ H_2O =$ xanthopterin-B$_2$ $+ NH_3$

Number	Other Names	Basis for classification (Systematic Name)	Comments	Reference
3.5.4.13		dCTP aminohydrolase		3438
3.5.4.14		Deoxycytidine aminohydrolase		551
3.5.4.15	Guanosine aminase	Guanosine aminohydrolase		1460
3.5.4.16		GTP 7,8-8,9-dihydrolase	The reaction involves hydrolysis of two C-N bonds and isomerization of the pentose unit; the recyclization may be non-enzymic	426, 3715
3.5.4.17		Adenosine (phosphate) aminohydrolase	Acts on 5′-AMP, ADP, ATP, NAD$^+$ and adenosine, in decreasing order of activity. *cf.* EC 3.5.4.6	3273, 3783
3.5.4.18		ATP aminohydrolase		538
3.5.4.19	Phosphoribosyl-ATP pyrophosphohydrolase, Phosphoribosyl-AMP pyrophosphorylase	1-*N*-(5′-Phospho-D-ribosyl)-AMP 1,6-hydrolase	The *Neurospora crassa* enzyme also catalyses the reaction of EC 1.1.1.23, and cleaves pyrophosphate from 1-*N*-(5′-phosphoribosyl)-ATP	2239
3.5.4.20		1-(4-Amino-2-methylpyrimid-5-ylmethyl)-3-(β-hydroxyethyl)-2-methylpyridinium-bromide aminohydrolase		3090
3.5.4.21		Creatinine iminohydrolase		3322
3.5.4.22		1-Pyrroline-4-hydroxy-2-carboxylate aminohydrolase (decyclizing)		3088
3.5.4.23		Blasticidin-S aminohydrolase	Catalyses the deamination of the cytosine moiety of blasticidin S, cytomycin and acetylblasticidin S	3759
3.5.4.24		Sepiapterin aminohydrolase	Also acts on isosepiapterin, more slowly	3482

Number	Recommended Name	Reaction

3.5.5   IN NITRILES

3.5.5.1   Nitrilase

A nitrile + H_2O = a carboxylate + NH_3

3.5.5.2   Ricinine nitrilase

Ricinine + H_2O = N-methyl-3-carboxy-4-methoxy-2-pyridone + NH_3

3.5.5.3   Cyanate hydrolase

Cyanate + H_2O = CO_2 + NH_3

3.5.99   IN OTHER COMPOUNDS

3.5.99.1   Riboflavinase

Riboflavin + H_2O = ribitol + lumichrome

3.5.99.2   Thiaminase

Thiamin + H_2O = 2-methyl-4-amino-5-hydroxymethyl-pyrimidine + 4-methyl-5-(2′-hydroxyethyl)-thiazole

3.6   ACTING ON ACID ANHYDRIDES

To this group belong mainly the enzymes acting on diphosphate bonds in compounds such as nucleoside di- and tri- phosphates (3.6.1) and on sulphonyl-containing anhydrides such as adenylylsulphate (3.6.2)

3.6.1   IN PHOSPHORYL-CONTAINING ANHYDRIDES

3.6.1.1   Inorganic pyrophosphatase

Pyrophosphate + H_2O = 2 orthophosphate

3.6.1.2   Trimetaphosphatase

Trimetaphosphate + H_2O = triphosphate

3.6.1.3   Adenosinetriphosphatase

ATP + H_2O = ADP + orthophosphate

Number	Other Names	Basis for classification (Systematic Name)	Comments	Reference
3.5.5.1		Nitrile aminohydrolase	Acts on a wide range of aromatic nitriles including 3-indoleacetonitrile and also on some aliphatic nitriles	3401
3.5.5.2		Ricinine aminohydrolase		1357
3.5.5.3	Cyanase	Cyanate aminohydrolase		2858, 3375, 3376
3.5.99.1		Riboflavin hydrolase		3772
3.5.99.2	Thiaminase II	Thiamin hydrolase		971, 1435
3.6.1.1		Pyrophosphate phosphohydrolase	Specificity varies with the source and with the activating metal ion. The enzyme from some sources may be identical with EC 3.1.3.1 or 3.1.3.9	174, 1834, 2697
3.6.1.2		Trimetaphosphate hydrolase		1776, 2215
3.6.1.3	Adenylpyrophosphatase, ATP monophosphatase, Triphosphatase, ATPase, Adenosinetriphosphatase (Na$^+$,K$^+$-activated)	ATP phosphohydrolase	This activity occurs in myosin and actomyosin, mitochondria, microsomes and cell membranes. In some cases the ATPase activity is activated by Mg^{2+}, in some by Ca^{2+} and in other cases by both Ca^{2+} and Mg^{2+}. Another form of ATPase is stimulated by Na$^+$ + K$^+$ and is inhibited by ouabain. Some ATPases also hydrolyse ITP and other nucleoside 5'-triphosphates, as well as triphosphate and	942, 1677, 2383, 2586

Number	Recommended Name	Reaction

[3.6.1.4   Deleted entry: adenosinetriphosphatase (Mg-activated)]

3.6.1.5	Apyrase	$ATP + 2 H_2O = AMP + 2$ orthophosphate
3.6.1.6	Nucleosidediphosphatase	A nucleoside diphosphate $+ H_2O =$ a nucleotide $+$ orthophosphate
3.6.1.7	Acylphosphatase	An acylphosphate $+ H_2O =$ a fatty acid anion $+$ orthophosphate
3.6.1.8	ATP pyrophosphatase	$ATP + H_2O = AMP +$ pyrophosphate
3.6.1.9	Nucleotide pyrophosphatase	A dinucleotide $+ H_2O = 2$ mononucleotides
3.6.1.10	Endopolyphosphatase	Polyphosphate $+ n H_2O = n$ pentaphosphate
3.6.1.11	Exopolyphosphatase	$(Polyphosphate)_n + H_2O = (polyphosphate)_{n-1} +$ orthophosphate
3.6.1.12	Deoxycytidinetriphosphatase	$dCTP + H_2O = dCMP +$ pyrophosphate
3.6.1.13	ADPribose pyrophosphatase	$ADPribose + H_2O = AMP +$ D-ribose 5-phosphate
3.6.1.14	Adenosinetetraphosphatase	Adenosine 5′-tetraphosphate $+ H_2O = ATP +$ orthophosphate

Number	Other Names	Basis for classification (Systematic Name)	Comments	Reference
			adenosine tetraphosphate. The substrate specificity may depend on the activating divalent cation and on the presence of monovalent cations	
3.6.1.5	ATP-diphosphatase, Adenosine diphosphatase, ADPase	ATP diphosphohydrolase	Activated by Ca^{++}. Also acts on ADP. Acts on other nucleoside triphosphates and diphosphates	1805, 1957, 2618
3.6.1.6		Nucleosidediphosphate phosphohydrolase	Acts on IDP, GDP, UDP and also on D-ribose 5-diphosphate	1028, 1367, 2623
3.6.1.7		Acylphosphate phosphohydrolase		1985, 2701, 3054
3.6.1.8	ATPase	ATP pyrophosphohydrolase	Also acts on ITP, GTP, CTP and UTP	1283, 1523
3.6.1.9		Dinucleotide nucleotidohydrolase	Substrates include NAD$^+$, NADP$^+$, FAD, CoA and also ATP and ADP	1480, 1770, 1822, 3315
3.6.1.10	Polyphosphate depolymerase, Metaphosphatase, Polyphosphatase	Polyphosphate polyphosphohydrolase		2066, 2130
3.6.1.11	Metaphosphatase	Polyphosphate phosphohydrolase		1134, 1805, 2066
3.6.1.12	Deoxy-CTPase, dCTPase	dCTP nucleotidohydrolase	Also hydrolyses dCDP to dCMP and orthophosphate	3849
3.6.1.13		ADPribose ribophosphohydrolase		737
3.6.1.14		Adenosinetetraphosphate phosphohydrolase	Also acts on inosine tetraphosphate and tripolyphosphate but shows little or no activity with other nucleotides or polyphosphates	3102

Number	Recommended Name	Reaction
3.6.1.15	Nucleoside triphosphatase	NTP + H_2O = NDP + orthophosphate
3.6.1.16	CDPglycerol pyrophosphatase	CDPglycerol + H_2O = CMP + sn-glycerol 3-phosphate
3.6.1.17	Bis(5'-guanosyl) tetraphosphatase	P^1,P^4-Bis(5'-guanosyl) tetraphosphate + H_2O = GTP + GMP
3.6.1.18	FAD pyrophosphatase	FAD + H_2O = AMP + FMN
3.6.1.19	Nucleosidetriphosphate pyrophosphatase	A nucleoside triphosphate + H_2O = a nucleotide + pyrophosphate
3.6.1.20	5'-Acylphosphoadenosine hydrolase	5'-Acylphosphoadenosine + H_2O = AMP + a fatty acid anion
3.6.1.21	ADPsugar pyrophosphatase	ADPsugar + H_2O = AMP + sugar 1-phosphate
3.6.1.22	NAD$^+$ pyrophosphatase	NAD$^+$ + H_2O = AMP + NMN
3.6.1.23	Deoxyuridinetriphosphatase	dUTP + H_2O = dUMP + pyrophosphate
3.6.1.24	Nucleoside phosphoacylhydrolase	Hydrolyses mixed phospho-anhydride bonds
3.6.1.25	Triphosphatase	Triphosphate + H_2O = pyrophosphate + orthophosphate
3.6.1.26	CDPdiacylglycerol pyrophosphatase	CDPdiacylglycerol + H_2O = CMP + phosphatidate
3.6.1.27	C-55-Isoprenyldiphosphatase	C-55-Isoprenyl diphosphate + H_2O = C-55-isoprenyl phosphate + orthophosphate
3.6.1.28	Thiamin-triphosphatase	Thiamin triphosphate + H_2O = thiamin diphosphate + orthophosphate
3.6.1.29	Bis(5'-adenosyl) triphosphatase	P^1,P^3-Bis(5'-adenosyl) triphosphate + H_2O = ADP + AMP

Number	Other Names	Basis for classification (Systematic Name)	Comments	Reference
3.6.1.15		Unspecific diphosphate phosphohydrolase	Also hydrolyses other nucleoside triphosphates, diphosphate, thiamin diphosphate and FAD	386, 1952, 2128
3.6.1.16		CDPglycerol phosphoglycerylhydrolase		1053
3.6.1.17	Diguanosinetetra-phosphatase	P^1,P^4-Bis(5'-guanosyl)-tetraphosphate guanylohydrolase		3615
3.6.1.18		FAD nucleotidohydrolase	Also hydrolyses NAD^+ and NADH	2731
3.6.1.19		Nucleosidetriphosphate pyrophosphohydrolase		518
3.6.1.20		5'-Acylphosphoadenosine acylhydrolase	Also acts on inosine and uridine compounds	1651
3.6.1.21		ADPsugar sugarphosphohydrolase		2826
3.6.1.22		NAD^+ phosphohydrolase	Also acts on $NADP^+$, 3-acetylpyridine and the thionicotinamide analogues of NAD^+ and $NADP^+$	74, 2353
3.6.1.23	dUTPase	dUTP nucleotidohydrolase		285, 1113, 1124
3.6.1.24		Nucleoside-5'-phosphoacylate acylhydrolase	Attacks ribonucleoside 5'-nitrophenylphosphates, but is inactive against phosphodiester	3159
3.6.1.25		Triphosphate phosphohydrolase		1816, 3509
3.6.1.26		CDPdiacylglycerol phosphatidylhydrolase		2696
3.6.1.27		C-55-Isoprenyl-diphosphate phosphohydrolase		1074
3.6.1.28		Thiamin-triphosphate phosphohydrolase		1213
3.6.1.29	Dinucleoside-triphosphatase	P^1,P^3-Bis(5'-adenosyl)-triphosphate adenylohydrolase		3072

Number	Recommended Name	Reaction
3.6.1.30	m⁷G (5′) pppN pyrophosphatase	7-Methylguanosine 5′-triphosphoryl-5′-polynucleotide + H_2O = 7-methylguanosine 5′-phosphate + polynucleotide

3.6.2   IN SULPHONYL-CONTAINING ANHYDRIDES

3.6.2.1	Adenylylsulphatase	Adenylylsulphate + H_2O = AMP + sulphate
3.6.2.2	Phosphoadenylylsulphatase	3′-Phosphoadenylylsulphate + H_2O = adenosine 3′,5′-bisphosphate + sulphate

3.7   ACTING ON CARBON-CARBON BONDS

There are very few carbon-carbon hydrolases; they mostly catalyse the hydrolysis of 3-oxo-carboxylic acids.

3.7.1   IN KETONIC SUBSTANCES

3.7.1.1	Oxaloacetase	Oxaloacetate + H_2O = oxalate + acetate
3.7.1.2	Fumarylacetoacetase	4-Fumarylacetoacetate + H_2O = acetoacetate + fumarate
3.7.1.3	Kynureninase	L-Kynurenine + H_2O = anthranilate + L-alanine
3.7.1.4	Phloretin hydrolase	Phloretin + H_2O = phloretate + phloroglucinol
3.7.1.5	Acylpyruvate hydrolase	An acylpyruvate + H_2O = a fatty acid + pyruvate

3.8   ACTING ON HALIDE BONDS

These enzymes hydrolyse organic halides.

3.8.1   IN C-HALIDE COMPOUNDS

3.8.1.1	Alkylhalidase	CH_2BrCl + H_2O = HCHO + bromide + chloride
3.8.1.2	2-Haloacid dehalogenase	L-2-Haloacid + OH^- = D-2-hydroxyacid + halide
3.8.1.3	Haloacetate dehalogenase	Haloacetate + OH^- = glycollate + halide

Number	Other Names	Basis for classification (Systematic Name)	Comments	Reference
3.6.1.30	Decapase	7-Methylguanosine 5'-triphosphoryl-5'-polynucleotide 7-methylguanosine-5'-phosphohydrolase		2462, 2074
3.6.2.1		Adenylylsulphate sulphohydrolase		180
3.6.2.2		3'-Phosphoadenylylsulphate sulphohydrolase	Activated by Mn^{2+}	184
3.7.1.1		Oxaloacetate acetylhydrolase		1241
3.7.1.2	β-Diketonase	4-Fumarylacetoacetate fumarylhydrolase	Also acts on other 3,5-and 2,4-diketoacids	572, 793, 2187
3.7.1.3		L-Kynurenine hydrolase	A pyridoxal-phosphate-protein. Also acts on 3'-hydroxykynurenine and some other (3-arylcarbonyl)-alanines	1488, 1724, 3710
3.7.1.4		2',4,4',6'-Tetrahydroxy-dehydrochalcone 1,3,5-trihydroxybenzene-hydrolase	Also hydrolyses other C-acylated phenols related to phloretin	2237
3.7.1.5		Acylpyruvate acylhydrolase	Acts on formylpyruvate, 2,4-dioxovalerate, 2,4-dioxohexanoate and 2,4-dioxoheptanoate	3623
3.8.1.1	Halogenase	Alkyl-halide halidohydrolase		1285
3.8.1.2		2-Haloacid halidohydrolase	Acts on acids of short chain lengths, C-2 to C-4	1072
3.8.1.3		Haloacetate halidohydrolase		1069, 1071

Number	Recommended Name	Reaction

3.8.2 IN P-HALIDE COMPOUNDS

3.8.2.1 Di-isopropyl-fluorophosphatase

Di-isopropyl fluorophosphate + H_2O = di-isopropyl phosphate + fluoride

3.9 ACTING ON PHOSPHORUS-NITROGEN BONDS

This is the phospho-amidase group.

3.9.1.1 Phosphoamidase

N-Phosphocreatine + H_2O = creatine + orthophosphate

3.10 ACTING ON SULPHUR-NITROGEN BONDS

3.10.1.1 Sulphoglucosamine sulphamidase

2-Sulphamido-2-deoxy-D-glucose + H_2O = 2-amino-2-deoxy-D-glucose + sulphate

3.10.1.2 Cyclamate sulphamidase

Cyclohexylsulphamate + H_2O = cyclohexylamine + sulphate

3.11 ACTING ON CARBON-PHOSPHORUS BONDS

This enzyme hydrolyses C-phosphono-groups.

3.11.1.1 Phosphonoacetaldehyde hydrolase

Phosphonoacetaldehyde + H_2O = acetaldehyde + orthophosphate

4. LYASES

Lyases are enzymes cleaving C-C, C-O, C-N, and other bonds by other means than by hydrolysis or oxidation. They differ from other enzymes in that two substrates are involved in one reaction direction, but only one in the other direction. When acting on the single substrate, a molecule is eliminated leaving an unsaturated residue. The systematic name is formed according to 'substrate group-lyase'. In recommended names, expressions like decarboxylase, aldolase, *etc.* are used. 'Dehydratase' is used for those enzymes eliminating water. In cases where the reverse reaction is the more important, or the only one to be demonstrated, 'synthase' may be used in the name.

4.1 CARBON-CARBON LYASES

This sub-class contains the decarboxylases (4.1.1), the aldehyde-lyases catalysing the reversal of an aldol condensation (4.1.2), and the oxo-acid-lyases, catalysing the cleavage of a 3-hydroxy-acid (4.1.3), or the reverse reactions.

Number	Other Names	Basis for classification (Systematic Name)	Comments	Reference
3.8.2.1	Diisopropylfluorophos-phonate halogenase, Tabunase, DFPase, Di-isopropyl phosphorofluoridase	Di-isopropyl-fluorophosphate fluorohydrolase	Acts on other organophosphorus compounds and 'nerve' gases	135, 547, 2302
3.9.1.1		Phosphoamide hydrolase	Also acts on *N*-phospho-arginine and other phosphoamides. Possibly identical with EC 3.1.3.16 or 3.1.3.9	2561, 3080, 3293
3.10.1.1		2-Sulphamido-2-deoxy-D-glucose sulphamidase		721
3.10.1.2	Cyclamate sulphamatase	Cyclohexylsulphamate sulphamidase	Also readily hydrolyses aliphatic sulphamates with 3 to 8 carbons	2410
3.11.1.1		(2-Oxoethyl) phosphonate phosphonohydrolase		1866, 1852

Number	Recommended Name	Reaction

4.1.1 CARBOXY-LYASES

4.1.1.1   Pyruvate decarboxylase   A 2-oxo acid = an aldehyde + CO_2

4.1.1.2   Oxalate decarboxylase   Oxalate = formate + CO_2

4.1.1.3   Oxaloacetate decarboxylase   Oxaloacetate = pyruvate + CO_2

4.1.1.4   Acetoacetate decarboxylase   Acetoacetate = acetone + CO_2

4.1.1.5   Acetolactate decarboxylase   (+)-2-Hydroxy-2-methyl-3-oxobutyrate = (—)-2-acetoin + CO_2

4.1.1.6   Aconitate decarboxylase   *cis*-Aconitate = itaconate + CO_2

4.1.1.7   Benzoylformate decarboxylase   Benzoylformate = benzaldehyde + CO_2

4.1.1.8   Oxalyl-CoA decarboxylase   Oxalyl-CoA = formyl-CoA + CO_2

4.1.1.9   Malonyl-CoA decarboxylase   Malonyl-CoA = Acetyl-CoA + CO_2

[4.1.1.10 Deleted entry: formerly Aminomalonate decarboxylase. Now included with EC 4.1.1.12]

[4.1.1.11 Deleted entry: formerly Aspartate 1-decarboxylase. Now included with EC 4.1.1.15]

4.1.1.12   Aspartate 4-decarboxylase   L-Aspartate = L-alanine + CO_2

[4.1.1.13 Deleted entry: Carbamoylaspartate decarboxylase]

4.1.1.14   Valine decarboxylase   L-Valine = isobutylamine + CO_2

4.1.1.15   Glutamate decarboxylase   L-Glutamate = 4-aminobutyrate + CO_2

Number	Other Names	Basis for classification (Systematic Name)	Comments	Reference
4.1.1.1	α-Carboxylase, Pyruvic decarboxylase, α-Ketoacid carboxylase	2-Oxo-acid carboxy-lyase	A thiamin-diphosphate-protein. Also catalyses acyloin formation	3084
4.1.1.2		Oxalate carboxy-lyase		1494
4.1.1.3	Oxalacetate β-decarboxylase	Oxaloacetate carboxy-lyase	The enzyme from *Aerobacter aerogenes* is a biotinyl-protein and requires Na$^+$. Some animal enzymes require Mn^{2+}. The enzyme from *Pseudomonas ovalis* is inhibited by acetyl-CoA	592, 1287, 1374, 1783, 2965
4.1.1.4		Acetoacetate carboxy-lyase		654, 3844
4.1.1.5		2-Hydroxy-2-methyl-3-oxobutyrate carboxy-lyase		161, 3241
4.1.1.6		*cis*-Aconitate carboxy-lyase		258
4.1.1.7		Benzoylformate carboxy-lyase	A thiamin-diphosphate-protein	1144
4.1.1.8		Oxalyl-CoA carboxy-lyase	A thiamin-diphosphate-protein	2669
4.1.1.9		Malonyl-CoA carboxy-lyase	Specific for malonyl-CoA	421
4.1.1.12	Desulphinase	L-Aspartate 4-carboxy-lyase	A pyridoxal-phosphate-protein. Also catalyses the decarboxylation of aminomalonate (formerly listed as EC 4.1.1.10) and the desulphination of L-cysteine-sulphinate to sulphite and alanine	1565, 2454, 2545, 3700
4.1.1.14		L-Valine carboxy-lyase	A pyridoxal-phosphate-protein. Also acts on L-leucine	3301
4.1.1.15		L-Glutamate 1-carboxy-lyase	A pyridoxal-phosphate-protein. The brain enzyme also acts on	66, 2808, 3691

Number	Recommended Name	Reaction
4.1.1.16	Hydroxyglutamate decarboxylase	3-Hydroxy-L-glutamate = 4-amino-3-hydroxybutyrate + CO_2
4.1.1.17	Ornithine decarboxylase	L-Ornithine = putrescine + CO_2
4.1.1.18	Lysine decarboxylase	L-Lysine = cadaverine + CO_2
4.1.1.19	Arginine decarboxylase	L-Arginine = agmatine + CO_2
4.1.1.20	Diaminopimelate decarboxylase	*meso*-2,6-Diaminopimelate = L-lysine + CO_2
4.1.1.21	Phosphoribosylaminoimidazole carboxylase	5'-Phosphoribosyl-5-amino-4-imidazolecarboxylate = 5'-phosphoribosyl-5-aminoimidazole + CO_2
4.1.1.22	Histidine decarboxylase	L-Histidine = histamine + CO_2
4.1.1.23	Orotidine-5'-phosphate decarboxylase	Orotidine 5'-phosphate = UMP + CO_2
4.1.1.24	Aminobenzoate decarboxylase	4-(or 2-) Aminobenzoate = aniline + CO_2
4.1.1.25	Tyrosine decarboxylase	L-Tyrosine = tyramine + CO_2

[4.1.1.26 Deleted entry: DOPA decarboxylase. Now included with EC 4.1.1.28]

[4.1.1.27 Deleted entry: Tryptophan decarboxylase. Now included with EC 4.1.1.28]

4.1.1.28	Aromatic-L-amino-acid decarboxylase	L-Tryptophan = tryptamine + CO_2

Number	Other Names	Basis for classification (Systematic Name)	Comments	Reference
			L-cysteate, L-cysteine sulphinate and L-aspartate	
4.1.1.16		3-Hydroxy-L-glutamate 1-carboxy-lyase	A pyridoxal-phosphate-protein	3507
4.1.1.17		L-Ornithine carboxy-lyase	A pyridoxal-phosphate-protein	3377, 2504
4.1.1.18		L-Lysine carboxy-lyase	A pyridoxal-phosphate-protein. Also acts on 5-hydroxy-L-lysine	986, 3136
4.1.1.19		L-Arginine carboxy-lyase	A pyridoxal-phosphate-protein	326, 3377
4.1.1.20		meso-2,6-Diaminopimelate carboxy-lyase	A pyridoxal-phosphate-protein	695
4.1.1.21		5′-Phosphoribosyl-5-amino-4-imidazolecarboxylate carboxy-lyase		2014
4.1.1.22		L-Histidine carboxy-lyase	A pyridoxal-phosphate-protein (in animal tissues). The bacterial enzyme does not require pyridoxal phosphate. (In Lactobacillus the prosthetic group is a pyruvoyl residue)	836, 2792, 2849
4.1.1.23		Orotidine-5′-phosphate carboxy-lyase		1964
4.1.1.24		Aminobenzoate carboxy-lyase	A pyridoxal-phosphate-protein	2155
4.1.1.25		L-Tyrosine carboxy-lyase	A pyridoxal-phosphate-protein. The bacterial enzyme also acts on 3-hydroxytyrosine and, more slowly, on 3-hydroxyphenylalanine	2163
4.1.1.28	DOPA decarboxylase, Tryptophan decarboxylase, Hydroxytryptophan	Aromatic-L-amino-acid carboxy-lyase	A pyridoxal-phosphate enzyme. The enzyme acts on L-tryptophan, 5-hydroxy-L-tryptophan	532, 2008, 2163, 2999, 3646

Number	Recommended Name	Reaction
4.1.1.29	Cysteine sulphinate decarboxylase	L-Cysteine sulphinate = hypotaurine + CO_2
4.1.1.30	Pantothenoylcysteine decarboxylase	N-(L-Pantothenoyl)-L-cysteine = pantetheine + CO_2
4.1.1.31	Phosphoenolpyruvate carboxylase	Orthophosphate + oxaloacetate = H_2O + phosphoenolpyruvate + CO_2
4.1.1.32	Phosphoenolpyruvate carboxykinase (GTP)	GTP + oxaloacetate = GDP + phosphoenolpyruvate + CO_2
4.1.1.33	Pyrophosphomevalonate decarboxylase	ATP + 5-diphosphomevalonate = ADP + orthophosphate + isopentenyl diphosphate + CO_2
4.1.1.34	Keto-L-gulonate decarboxylase	3-Keto-L-gulonate = L-xylulose + CO_2
4.1.1.35	UDPglucuronate decarboxylase	UDPglucuronate = UDPxylose + CO_2
4.1.1.36	Phosphopantothenoyl-cysteine decarboxylase	4'-Phospho-N-(L-pantothenoyl)-L-cysteine = pantotheine 4'-phosphate + CO_2
4.1.1.37	Uroporphyrinogen decarboxylase	Uroporphyrinogen III = coproporphyrinogen + 4 CO_2
4.1.1.38	Phosphoenolpyruvate carboxykinase (pyrophosphate)	Pyrophosphate + oxaloacetate = orthophosphate + phosphoenolpyruvate + CO_2
4.1.1.39	Ribulosebisphosphate carboxylase	D-Ribulose 1,5-bisphosphate + CO_2 = 2,3-D-phospho-D-glycerate
4.1.1.40	Hydroxypyruvate decarboxylase	Hydroxypyruvate = glycolaldehyde + CO_2
4.1.1.41	Propionyl-CoA carboxylase	(R)-Methylmalonyl-CoA = propionyl-CoA + CO_2
4.1.1.42	Carnitine decarboxylase	Carnitine = 2-methylcholine + CO_2

Number	Other Names	Basis for classification (Systematic Name)	Comments	Reference
	decarboxylase		and dihydroxyphenylalanine	
4.1.1.29		L-Cysteine-sulphinate carboxy-lyase	A pyridoxal-phosphate-protein. Also acts on L-cysteate	1477, 1141
4.1.1.30		N-(L-Pantothenoyl)-L-cysteine carboxy-lyase		401
4.1.1.31		Orthophosphate:oxaloacetate carboxy-lyase (phosphorylating)		2141, 513
4.1.1.32	Phosphoenolpyruvate carboxylase, Phosphopyruvate carboxylase	GTP:oxaloacetate carboxy-lyase (transphosphorylating)	ITP can act as phosphate donor	497, 1842
4.1.1.33		ATP:5-diphosphomevalonate carboxy-lyase (dehydrating)		327
4.1.1.34		3-Keto-L-gulonate carboxy-lyase		3105
4.1.1.35		UDPglucuronate carboxy-lyase	Requires NAD^+	89
4.1.1.36		4'-Phospho-N-(L-pantothenoyl)-L-cysteine carboxy-lyase		400
4.1.1.37		Uroporphyrinogen III carboxy-lyase	Acts on a number of porphyrinogens	2131, 3437
4.1.1.38	Phosphopyruvate carboxylase, Phosphoenolpyruvate carboxylase, PEP carboxyphosphotransferase	Pyrophosphate:oxaloacetate carboxy-lyase (transphosphorylating)	Also catalyses the reaction: phosphoenolpyruvate + orthophosphate = pyruvate + pyrophosphate	1997
4.1.1.39	Carboxydismutase	3-Phospho-D-glycerate carboxy-lyase (dimerizing)	A copper-protein. Will utilize O_2 instead of CO_2, forming 3-phospho-D-glycerate and 2-phosphoglycollate	356, 3709
4.1.1.40		Hydroxypyruvate carboxy-lyase		1259
4.1.1.41	Methylmalonyl-CoA decarboxylase	(R)-Methylmalonyl-CoA carboxy-lyase	A biotinyl-protein	987
4.1.1.42		Carnitine carboxy-lyase	Requires ATP	1672

Number	Recommended Name	Reaction
4.1.1.43	Phenylpyruvate decarboxylase	Phenylpyruvate = phenylacetaldehyde + CO_2
4.1.1.44	4-Carboxymuconolactone decarboxylase	4-Carboxymuconolactone = 3-oxoadipate enol-lactone + CO_2
4.1.1.45	Aminocarboxymuconate-semialdehyde decarboxylase	(3'-Oxo-prop-2'-enyl)-2-amino-but-2-ene-dioate = 2-aminomuconate semialdehyde + CO_2
4.1.1.46	o-Pyrocatechuate decarboxylase	2,3-Dihydroxybenzoate = catechol + CO_2
4.1.1.47	Tartronate-semialdehyde synthase	2-Glyoxylate = tartronate semialdehyde + CO_2
4.1.1.48	Indole-3-glycerol-phosphate synthase	1-(2'-Carboxyphenylamino)-1-deoxyribulose 5-phosphate = 1-C-(3'-indolyl)-glycerol 3-phosphate + CO_2 + H_2O
4.1.1.49	Phosphoenolpyruvate carboxykinase (ATP)	ATP + oxaloacetate = ADP + phosphoenolpyruvate + CO_2
4.1.1.50	Adenosylmethionine decarboxylase	S-Adenosyl-L-methionine = (5'-deoxy-5'-adenosyl) (3-aminopropyl) methyl sulfonium salt + CO_2
4.1.1.51	3-Hydroxy-2-methyl-pyridine-4,5-dicarboxylate 4-decarboxylase	3-Hydroxy-2-methyl-pyridine-4,5-dicarboxylate = 3-hydroxy-2-methyl-pyridine-5-carboxylate + CO_2
4.1.1.52	6-Methylsalicylate decarboxylase	6-Methylsalicylate = 3-cresol + CO_2
4.1.1.53	Phenylalanine decarboxylase	L-Phenylalanine = phenylethylamine + CO_2
4.1.1.54	Dihydroxyfumarate decarboxylase	Dihydroxyfumarate = tartronate semialdehyde + CO_2
4.1.1.55	4,5-Dihydroxyphthalate decarboxylase	4,5-Dihydroxyphthalate = protocatechuate + CO_2
4.1.1.56	3-Oxolaurate decarboxylase	3-Oxolaurate = 2-undecanone + CO_2

Number	Other Names	Basis for classification (Systematic Name)	Comments	Reference
4.1.1.43		Phenylpyruvate carboxy-lyase	Also acts on indolepyruvate	117
4.1.1.44		4-Carboxymuconolactone carboxy-lyase		2509
4.1.1.45		(3'-Oxo-prop-2'-enyl)-2-amino-but-2-ene-dioate carboxy-lyase	Product rearranges non-enzymically to picolinate	1428
4.1.1.46		2,3-Dihydroxybenzoate carboxy-lyase		3274
4.1.1.47	Tartronate-semialdehyde carboxylase, Glyoxylate carbo-ligase	Glyoxylate carboxy-lyase (dimerizing)	A flavoprotein	1149, 1788
4.1.1.48		1-(2'-Carboxyphenylamino)-1-deoxyribulose-5-phosphate carboxy-lyase (cyclizing)		608
4.1.1.49	Phosphopyruvate carboxylase (ATP), Phosphoenolpyruvate carboxylase, Phosphoenolpyruvate carboxykinase	ATP:oxaloacetate carboxy-lyase (transphosphorylating)		457, 459, 460
4.1.1.50		S-Adenosyl-L-methionine carboxy-lyase	The *Escherichia coli* enzyme contains pyruvate	3325
4.1.1.51		3-Hydroxy-2-methyl-pyridine-4,5-dicarboxylate 4-carboxy-lyase		3129
4.1.1.52		6-Methylsalicylate carboxy-lyase		1966, 3558
4.1.1.53		L-Phenylalanine carboxy-lyase	A pyridoxal-phosphate-protein. Also acts on tyrosine and other aromatic amino acids	2008, 2979
4.1.1.54		Dihydroxyfumarate carboxy-lyase		974
4.1.1.55		4,5-Dihydroxyphthalate carboxy-lyase		2780
4.1.1.56		3-Oxolaurate carboxy-lyase	Also decarboxylates other C-14 to C-16 3-oxo acids	922

Number	Recommended Name	Reaction
4.1.1.57	Methionine decarboxylase	L-Methionine = 3-methylthiopropylamine + CO_2
4.1.1.58	Orsellinate decarboxylase	2,4-Dihydroxy-6-methylbenzoate = orcinol CO_2
4.1.1.59	Gallate decarboxylase	3,4,5-Trihydroxybenzoate = pyrogallol + CO_2
4.1.1.60	Stipitatonate decarboxylase	Stipitatonate = stipitatate + CO_2
4.1.1.61	4-Hydroxybenzoate decarboxylase	4-Hydroxybenzoate = phenol + CO_2
4.1.1.62	Gentisate decarboxylase	2,5-Dihydroxybenzoate = hydroquinone + CO_2
4.1.1.63	Protocatechuate decarboxylase	3,4-Dihydroxybenzoate = catechol + CO_2
4.1.1.64	Dialkylamino-acid decarboxylase (pyruvate)	2,2-Dialkyl-L-amino acid + pyruvate = dialkyl-ketone + CO_2 + L-alanine
4.1.1.65	Phosphatidylserine decarboxylase	Phosphatidylserine = phosphatidylethanolamine + CO_2
4.1.1.66	Uracil-5-carboxylate decarboxylase	Uracil 5-carboxylate = uracil + CO_2

4.1.2 ALDEHYDE-LYASES

[4.1.2.1 *Deleted entry: now included with EC 4.1.2.31]*

4.1.2.2	Ketotetrose-phosphate aldolase	Erythrulose 1-phosphate = dihydroxyacetone phosphate + formaldehyde

[4.1.2.3 *Deleted entry: Pentosealdolase]*

4.1.2.4	Deoxyribose-phosphate aldolase	2-Deoxy-D-ribose 5-phosphate = D-glyceraldehyde 3-phosphate + acetaldehyde
4.1.2.5	Threonine aldolase	L-Threonine = glycine + acetaldehyde

[4.1.2.6 *Deleted entry: Allothreonine aldolase (reaction due to EC 2.1.2.1)]*

[4.1.2.7 *Deleted entry: Ketose-1-phosphate aldolase. Now included with EC 4.1.2.13]*

[4.1.2.8 *Deleted entry: Indoleglycerolphosphate aldolase (a side-reaction of EC 4.2.1.20)]*

Number	Other Names	Basis for classification (Systematic Name)	Comments	Reference
4.1.1.57		L-Methionine carboxy-lyase		1165
4.1.1.58		Orsellinate carboxy-lyase		2593
4.1.1.59		Gallate carboxy-lyase		1097
4.1.1.60		Stipitatonate carboxy-lyase (decyclizing)		259
4.1.1.61		4-Hydroxybenzoate carboxy-lyase		1097
4.1.1.62		Gentisate carboxy-lyase		1097
4.1.1.63		Protocatechuate carboxy-lyase		1097
4.1.1.64		2,2-Dialkyl-L-amino-acid carboxy-lyase (amino-transferring)	A pyridoxal-phosphate-protein. Acts on isovaline, 2-aminoisobutyrate and cycloleucine	173
4.1.1.65		Phosphatidylserine carboxy-lyase	A pyridoxal-phosphate-protein	1588
4.1.1.66		Uracil-5-carboxylate carboxy-lyase		2548
4.1.2.2	Phosphoketotetrose aldolase, Erythrulose-1-phosphate synthetase	Erythrulose-1-phosphate formaldehyde-lyase		499
4.1.2.4	Phosphodeoxyriboaldolase, Deoxyriboaldolase	2-Deoxy-D-ribose-5-phosphate acetaldehyde-lyase		1333, 2688, 2844, 1507
4.1.2.5		L-Threonine acetaldehyde-lyase	A pyridoxal-phosphate-protein	1596, 250, 1818

Number	Recommended Name	Reaction
4.1.2.9	Phosphoketolase	D-Xylulose 5-phosphate + orthophosphate = acetylphosphate + D-glyceraldehyde 3-phosphate + H_2O
4.1.2.10	Mandelonitrile lyase	Mandelonitrile = HCN + benzaldehyde
4.1.2.11	Hydroxymandelonitrile lyase	4-Hydroxymandelonitrile = HCN + 4-hydroxybenzaldehyde
4.1.2.12	Ketopantoaldolase	2-Oxopantoate = 2-oxoisovalerate + formaldehyde
4.1.2.13	Fructose-bisphosphate aldolase	D-Fructose 1,6-bisphosphate = dihydroxyacetone phosphate + D-glyceraldehyde 3-phosphate
4.1.2.14	Phospho-2-keto-3-deoxy-gluconate aldolase	6-Phospho-2-keto-3-deoxy-D-gluconate = pyruvate + D-glyceraldehyde 3-phosphate
4.1.2.15	Phospho-2-keto-3-deoxy-heptonate aldolase	7-Phospho-2-keto-3-deoxy-D-*arabino*-heptonate + orthophosphate = phosphoenolpyruvate + D-erythrose 4-phosphate + H_2O
4.1.2.16	Phospho-2-keto-3-deoxy-octonate aldolase	8-Phospho-2-keto-3-deoxy-D-octonate + orthophosphate = phosphoenolpyruvate + D-arabinose 5-phosphate + H_2O
4.1.2.17	L-Fuculosephosphate aldolase	L-Fuculose 1-phosphate = dihydroxyacetone phosphate + L-lactaldehyde
4.1.2.18	2-Keto-3-deoxy-L-pentonate aldolase	2-Keto-3-deoxy-L-pentonate = pyruvate + glycollaldehyde
4.1.2.19	Rhamnulosephosphate aldolase	L-Rhamnulose 1-phosphate = dihydroxyacetone phosphate + L-lactaldehyde
4.1.2.20	2-Keto-3-deoxy-D-glucarate aldolase	2-Keto-3-deoxy-D-glucarate = pyruvate + tartronate semialdehyde
4.1.2.21	6-Phospho-2-keto-3-deoxy-galactonate aldolase	2-Keto-3-deoxy-D-galactonate 6-phosphate = pyruvate + D-glyceraldehyde 3-phosphate
4.1.2.22	Fructose-6-phosphate phosphoketolase	D-Fructose 6-phosphate + orthophosphate = acetyl phosphate + D-erythrose 4-phosphate + H_2O
4.1.2.23	3-Deoxy-D-*manno*-octulosonate aldolase	3-Deoxy-D-*manno*-octulosonate = pyruvate + D-arabinose

Number	Other Names	Basis for classification (Systematic Name)	Comments	Reference
4.1.2.9		D-Xylulose-5-phosphate D-glyceraldehyde-3-phosphate-lyase (phosphate-acetylating)	A thiamin-diphosphate-protein	1256, 2971
4.1.2.10		Mandelonitrile benzaldehyde-lyase	A flavoprotein	241, 242, 1804, 2169
4.1.2.11		4-Hydroxymandelonitrile hydroxybenzaldehyde-lyase		353, 2992
4.1.2.12		2-Oxopantoate formaldehyde-lyase		2167
4.1.2.13	Zymohexase, Aldolase, Fructose-1,6-bisphosphate triosephosphate-lyase	D-Fructose-1,6-bisphosphate D-glyceraldehyde-3-phosphate-lyase	Also acts on (3S,4R)-ketose 1-phosphates. The yeast and bacterial enzymes are zinc-proteins	2298
4.1.2.14		6-Phospho-2-keto-3-deoxy-D-gluconate D-glyceraldehyde-3-phosphate-lyase	Also acts on 2-oxobutyrate	2197
4.1.2.15	DHAP synthase, KDHP synthetase	7-Phospho-2-keto-3-deoxy-D-*arabino*heptonate D-erythrose-4-phosphate-lyase (pyruvate-phosphorylating)		3178
4.1.2.16		8-Phospho-2-keto-3-deoxy-D-octonate D-arabinose-5-phosphate-lyase (pyruvate-phosphorylating)		1933
4.1.2.17		L-Fuculose-1-phosphate L-lactaldehyde-lyase		1011
4.1.2.18	2-Keto-3-deoxy-L-arabonate aldolase	2-Keto-3-deoxy-L-pentonate glycollaldehyde-lyase		630
4.1.2.19		L-Rhamnulose-1-phosphate L-lactaldehyde-lyase		528, 2925
4.1.2.20		2-Keto-3-deoxy-D-glucarate tartronate-semialdehyde-lyase		886
4.1.2.21		2-Keto-3-deoxy-D-galactonate-6-phosphate D-glyceraldehyde-3-phosphate-lyase		3062
4.1.2.22		D-Fructose-6-phosphate D-erythrose-4-phosphate-lyase (phosphate-acetylating)	Also acts on D-xylulose 5-phosphate	2971
4.1.2.23		3-Deoxy-D-*manno*-octulosonate D-arabinose-lyase		1013

Number	Recommended Name	Reaction
4.1.2.24	Dimethylaniline-*N*-oxide aldolase	*N*,*N*-Dimethylaniline *N*-oxide = *N*-methylaniline + formaldehyde
4.1.2.25	Dihydroneopterin aldolase	2-Amino-4-hydroxy-6-(D-*erythro*-1′,2′,3′-trihydroxypropyl)-7,8-dihydropteridine = 2-amino-4-hydroxy-6-hydroxymethyl-7,8-dihydropteridine + glycolaldehyde
4.1.2.26	Phenylserine aldolase	L-*threo*-3-Phenylserine = glycine + benzaldehyde
4.1.2.27	Dihydrosphingosine-1-phosphate aldolase	Dihydrosphingosine 1-phosphate = phosphoethanolamine + palmitaldehyde
4.1.2.28	2-Keto-3-deoxy-D-pentonate aldolase	2-Keto-3-deoxy-D-pentonate = pyruvate + glycolaldehyde
4.1.2.29	Phospho-5-keto-2-deoxy-gluconate aldolase	6-Phospho-5-keto-2-deoxy-D-gluconate = dihydroxyacetone phosphate + malonate semialdehyde
4.1.2.30	17α-Hydroxyprogesterone aldolase	17α-Hydroxyprogesterone = 4-androstene-3,17-dione + acetaldehyde
4.1.2.31	2-Oxo-4-hydroxyglutarate aldolase	2-Oxo-4-hydroxyglutarate = pyruvate + glyoxylate
4.1.2.32	Trimethylamine-oxide aldolase	$(CH_3)_3NO = (CH_3)_2NH$ + formaldehyde

4.1.3 OXO-ACID-LYASES

4.1.3.1	Isocitrate lyase	*threo*-D$_s$-Isocitrate = succinate + glyoxylate
4.1.3.2	Malate synthase	L-Malate + CoA = acetyl-CoA + H_2O + glyoxylate
4.1.3.3	*N*-Acetylneuraminate lyase	*N*-Acetylneuraminate = 2-acetamido-2-deoxy-D-mannose + pyruvate
4.1.3.4	Hydroxymethylglutaryl-CoA lyase	3-Hydroxy-3-methylglutaryl-CoA = acetyl-CoA + acetoacetate
4.1.3.5	Hydroxymethylglutaryl-CoA synthase	3-Hydroxy-3-methylglutaryl-CoA + CoA = acetyl-CoA + H_2O + acetoacetyl-CoA

Number	Other Names	Basis for classification (Systematic Name)	Comments	Reference
4.1.2.24		*N,N*-Dimethylaniline-*N*-oxide formaldehyde-lyase	Acts on various *N,N*-dialkylarylamides	2030
4.1.2.25		2-Amino-4-hydroxy-6-(D-*erythro*-1′,2′,3′-trihydroxypropyl)-7,8-dihydropteridine glycolaldehyde-lyase		2116
4.1.2.26		L-*threo*-3-Phenylserine benzaldehyde-lyase	A pyridoxal-phosphate-protein	233
4.1.2.27		Dihydrosphingosine-1-phosphate palmitaldehyde-lyase	A pyridoxal-phosphate-protein	3235
4.1.2.28		2-Keto-3-deoxy-D-pentonate glycollaldehyde-lyase		628
4.1.2.29		6-Phospho-5-keto-2-deoxy-D-gluconate malonate-semialdehyde-lyase		79
4.1.2.30		17α-Hydroxyprogesterone acetaldehyde-lyase		2456
4.1.2.31		2-Oxo-4-hydroxyglutarate glyoxylate-lyase	Also acts, slowly, on some other hydroxy-ketoacids, including 2-keto-4-hydroxybutyrate	2416, 2415
4.1.2.32		Trimethylamine-*N*-oxide formaldehyde-lyase		2321, 1876
4.1.3.1	Isocitrase, Isocitritase, Isocitratase	*threo*-D$_s$-Isocitrate glyoxylate-lyase		2497, 3045, 3122
4.1.3.2	Malate condensing enzyme, Glyoxylate transacetylase, Malate synthetase	L-Malate glyoxylate-lyase (CoA-acetylating)		728
4.1.3.3	*N*-Acetylneuraminic acid aldolase	*N*-Acetylneuraminate pyruvate-lyase	Also acts on *N*-glycolylneuraminate	566
4.1.3.4		3-Hydroxy-3-methylglutaryl-CoA acetoacetate-lyase		162
4.1.3.5		3-Hydroxy-3-methylglutaryl-CoA acetoacetyl-CoA-lyase (CoA-acetylating)		2869

Number	Recommended Name	Reaction
4.1.3.6	Citrate (*pro*-3*S*)-lyase	Citrate = acetate + oxaloacetate
4.1.3.7	Citrate (*si*)-synthase	Citrate + CoA = acetyl-CoA + H_2O + oxaloacetate
4.1.3.8	ATP citrate (*pro*-3*S*)-lyase	ATP + citrate + CoA = ADP + orthophosphate + acetyl-CoA + oxaloacetate
4.1.3.9	2-Hydroxyglutarate synthase	2-Hydroxyglutarate + CoA = propionyl-CoA + H_2O + glyoxylate
4.1.3.10	2-Ethylmalate synthase	2-Ethylmalate + CoA = butyryl-CoA + H_2O + glyoxylate
4.1.3.11	3-Propylmalate synthase	3-Propylmalate + CoA = valeryl-CoA + H_2O + glyoxylate
4.1.3.12	2-Isopropylmalate synthase	3-Hydroxy-4-methyl-3-carboxyvalerate + CoA = acetyl-CoA + 2-oxo-3-methylbutyrate + H_2O
4.1.3.13	Oxalomalate lyase	3-Oxalomalate = oxaloacetate + glyoxylate
4.1.3.14	3-Hydroxyaspartate aldolase	*erythro*-3-Hydroxy-L$_s$-aspartate = glycine + glyoxylate
4.1.3.15	2-Hydroxy-3-oxoadipate carboxylase	2-Hydroxy-3-oxoadipate + CO_2 = 2-oxoglutarate + glyoxylate
4.1.3.16	4-Hydroxy-2-oxoglutarate aldolase	4-Hydroxy-2-oxoglutarate = pyruvate + glyoxylate
4.1.3.17	4-Hydroxy-4-methyl-2-oxoglutarate aldolase	4-Hydroxy-4-methyl-2-oxoglutarate = 2 pyruvate

Number	Other Names	Basis for classification (Systematic Name)	Comments	Reference
4.1.3.6	Citrase, Citratase, Citritase, Citridesmolase, Citrate aldolase	Citrate oxaloacetate-lyase (*pro*-3S-CH₂COO →acetate)		621
4.1.3.7	Condensing enzyme, Citrate condensing enzyme, Citrogenase, Oxaloacetate transacetase	Citrate oxaloacetate-lyase (*pro*-3S-CH₂COO⁻→acetyl-CoA)		1092, 2470, 3210
4.1.3.8	Citrate cleavage enzyme	ATP:citrate oxaloacetate-lyase (*pro*-3S-CH₂COO⁻→acetyl-CoA; ATP-dephosphorylating)		3172
4.1.3.9		2-Hydroxyglutarate glyoxylate-lyase (CoA-propionylating)		2750
4.1.3.10		2-Ethylmalate glyoxylate-lyase (CoA-butyrylating)	Also acts on 2-(*n*-propyl)-malate	2675, 3245
4.1.3.11		3-Propylmalate glyoxylate-lyase (CoA-valerylating)		1442
4.1.3.12		3-Hydroxy-4-methyl-3-carboxyvalerate 2-oxo-3-methylbutyrate-lyase (CoA-acetylating)	Requires K⁺	1734, 3634
4.1.3.13		3-Oxalomalate glyoxylate-lyase		3001
4.1.3.14		*erythro*-3-Hydroxy-Lₛ-aspartate glyoxylate-lyase		1026
4.1.3.15		2-Hydroxy-3-oxoadipate glyoxylate-lyase (carboxylating)	The bacterial enzyme requires thiamin diphosphate. Reaction goes in the reverse direction; the product decarboxylates to 4-oxo-5-hydroxyvalerate. The enzyme can decarboxylate 2-oxoglutarate. Acetaldehyde can replace glyoxylate	2958, 2959, 3224
4.1.3.16		4-Hydroxy-2-oxoglutarate glyoxylate-lyase	The enzyme from *E. coli* is specific for the L-substrate, whereas the liver enzyme acts on both stereoisomers	1844, 2057, 2416, 3725
4.1.3.17		4-Hydroxy-4-methyl-2-oxoglutarate pyruvate-lyase	Also acts on 4-hydroxy-4-methyl-2-oxoadipate	3018, 3334, 3725

Number	Recommended Name	Reaction
4.1.3.18	Acetolactate synthase	2-Acetolactate + CO_2 = 2 pyruvate
4.1.3.19	N-Acetylneuraminate synthase	N-Acetylneuraminate + orthophosphate = 2-acetamido-2-deoxy-D-mannose + phosphoenolpyruvate + H_2O
4.1.3.20	N-Acylneuraminate-9-phosphate synthase	N-Acylneuraminate 9-phosphate + orthophosphate = 2-acetamido-2-deoxy-D-mannose 6-phosphate + phosphoenolpyruvate + H_2O
4.1.3.21	Homocitrate synthase	3-Hydroxy-3-carboxyadipate + CoA = acetyl-CoA + H_2O + 2-oxoglutarate
4.1.3.22	Citramalate lyase	Citramalate = acetate + pyruvate
4.1.3.23	Decylcitrate synthase	(—)-Dicarboxy-3-hydroxytetradecanoate + CoA = lauroyl-CoA + H_2O + oxaloacetate
4.1.3.24	Malyl-CoA lyase	Malyl-CoA = acetyl-CoA + glyoxylate
4.1.3.25	Citramalyl-CoA lyase	Citramalyl-CoA = acetyl-CoA + pyruvate
4.1.3.26	3-Hydroxy-3-isohexenylglutaryl-CoA lyase	3-Hydroxy-3-isohex-3-enylglutaryl-CoA = isopent-2-enyl-2-acetoacetyl-CoA + acetate
4.1.3.27	Anthranilate synthase	Chorismate + L-glutamine = anthranilate + pyruvate + L-glutamate
4.1.3.28	Citrate (re)-synthase	Citrate + CoA = acetyl-CoA + H_2O + oxaloacetate
4.1.3.29	Decylhomocitrate synthase	4,5-Dicarboxy-4-hydroxypentadecanoate + CoA = lauroyl-CoA + H_2O + 2-oxoglutarate

Number	Other Names	Basis for classification (Systematic Name)	Comments	Reference
4.1.3.18		Acetolactate pyruvate-lyase (carboxylating)	A flavoprotein. Requires thiamin diphosphate; also catalyses the formation of 2-aceto-2-hydroxybutyrate	228, 1412, 3242
4.1.3.19		N-Acetylneuraminate pyruvate-lyase (pyruvate-phosphorylating)		311
4.1.3.20		N-Acylneuraminate-9-phosphate pyruvate-lyase (pyruvate-phosphorylating)	Acts on N-glycolyl and N-acetyl-derivatives	2843, 3622
4.1.3.21		3-Hydroxy-3-carboxyadipate 2-oxoglutarate-lyase (CoA-acetylating)		3246
4.1.3.22		Citramalate pyruvate-lyase		203
4.1.3.23		3,4-Dicarboxy-3-hydroxytetradecanoate oxaloacetate-lyase (CoA-acylating)		2048
4.1.3.24		Malyl-CoA glyoxylate-lyase		3485
4.1.3.25		Citramalyl-CoA pyruvate-lyase		583
4.1.3.26		3-Hydroxy-3-isohex-3-enylglutaryl-CoA isopentenylacetoacetyl-CoA-lyase	Also acts on the hydroxy derivative of farnesoyl-CoA	3011
4.1.3.27		Chorismate pyruvate-lyase (amino-accepting)	The native enzyme in the enteric bacteria exists as a complex with anthranilate phosphoribosyltransferase (EC 2.4.2.18) which can use either glutamine or (less efficiently) NH_3. The enzyme separated from the complex uses NH_3 only	183, 1464, 3831
4.1.3.28		Citrate oxaloacetate-lyase (*pro*-3R-CH_2COO^-→acetyl-CoA)	The stereospecificity of this enzyme is opposite to that of EC 4.1.3.7	726, 1092, 1093
4.1.3.29		4,5-Dicarboxy-4-hydroxypentadecanoate 2-oxoglutarate-lyase (CoA-acylating)	In the reverse reaction decanoyl-CoA can act instead of lauroyl-CoA, but 2-oxoglutarate cannot	2064

Number	Recommended Name	Reaction
4.1.3.30	Methylisocitrate lyase	*threo*-D$_s$-2-Methylisocitrate = pyruvate + succinate
4.1.3.31	Methylcitrate synthase	Methylcitrate + CoA = propionyl-CoA + H$_2$O + oxaloacetate

4.1.99   OTHER CARBON-CARBON LYASES

4.1.99.1	Tryptophanase	L-Tryptophan + H$_2$O = indole + pyruvate + NH$_3$
4.1.99.2	Tyrosine phenol-lyase	L-Tyrosine + H$_2$O = phenol + pyruvate + NH$_3$
4.1.99.3	Deoxyribodipyrimidine photolyase	Cyclobutadipyrimidine (in DNA) = 2 pyrimidine residues (in DNA)

Number	Other Names	Basis for classification (Systematic Name)	Comments	Reference
			be replaced by oxaloacetate or pyruvate	
4.1.3.30		*threo*-D_s-2-Methylisocitrate pyruvatelyase	The enzyme does not act on *threo*-D_s-isocitrate, *threo*-DL-isocitrate or *erthyro*-L_s-isocitrate	3332, 3333
4.1.3.31		Methylcitrate oxaloacetate-lyase	The enzyme acts on acetyl-CoA, propionyl-CoA, butyryl-CoA and pentanoyl-CoA. The relative rate of condensation of acetyl-CoA and oxaloacetate is 140% of that of propionyl-CoA and oxaloacetate but the enzyme has been separated from EC 4.1.3.7. Oxaloacetate cannot be replaced by glyoxylate, pyruvate or 2-oxoglutarate	3499
4.1.99.1		L-Tryptophan indole-lyase (deaminating)	A pyridoxal-phosphate-protein. Also catalyses 2,3-elimination and β-replacement reactions of some indole-substituted tryptophan analogues of L-cysteine, L-serine and other 3-substituted amino acids	434, 2402, 599
4.1.99.2	β-Tyrosinase	L-Tyrosine phenol-lyase (deaminating)	A pyridoxal-phosphate-protein. The enzyme also slowly catalyses pyruvate formation from D-tyrosine, S-methyl-L-cysteine, L-cysteine, L-serine and D-serine	1819, 1820
4.1.99.3	Photoreactivating enzyme, PR-enzyme	Deoxyribocyclobutadipyrimidine pyrimidine-lyase	The enzyme catalyses the reactivation by light of irradiated DNA. A similar reactivation of irradiated RNA is probably due to a separate enzyme	575, 3009, 808

4.2 CARBON-OXYGEN LYASES

These enzymes catalyse the breakage of a carbon-oxygen bond leading to unsaturated products. In the case of hydro-lyases (4.2.1), this is by elimination of water; in sub-sub-group 4.2.2, it is by the elimination of an alcohol from a polysaccharide. A few other cases are grouped in 4.2.99.

4.2.1 HYDRO-LYASES

Number	Recommended Name	Reaction
4.2.1.1	Carbonate dehydratase	H_2CO_3 (or $H^+ + HCO_3^-$) $= CO_2 + H_2O$
4.2.1.2	Fumarate hydratase	L-Malate $=$ fumarate $+ H_2O$
4.2.1.3	Aconitate hydratase	Citrate $= cis$-aconitate $+ H_2O$
4.2.1.4	Citrate dehydratase	Citrate $= cis$-aconitate $+ H_2O$
4.2.1.5	Arabinonate dehydratase	D-Arabinonate $=$ 2-keto-3-deoxy-D-arabinonate $+ H_2O$
4.2.1.6	Galactonate dehydratase	D-Galactonate $=$ 2-keto-3-deoxy-D-galactonate $+ H_2O$
4.2.1.7	Altronate dehydratase	D-Altronate $=$ 2-keto-3-deoxy-D-gluconate $+ H_2O$
4.2.1.8	Mannonate dehydratase	D-Mannonate $=$ 2-keto-3-deoxy-D-gluconate $+ H_2O$
4.2.1.9	Dihydroxyacid dehydratase	2,3-Dihydroxyisovalerate $=$ 2-oxoisovalerate $+ H_2O$
4.2.1.10	3-Dehydroquinate dehydratase	3-Dehydroquinate $=$ 3-dehydroshikimate $+ H_2O$
4.2.1.11	Enolase	2-Phospho-D-glycerate $=$ phosphoenolpyruvate $+ H_2O$
4.2.1.12	Phosphogluconate dehydratase	6-Phospho-D-gluconate $=$ 6-phospho-2-keto-3-deoxy-D-gluconate $+ H_2O$
4.2.1.13	L-Serine dehydratase	L-Serine $+ H_2O =$ pyruvate $+ NH_3 + H_2O$
4.2.1.14	D-Serine dehydratase	D-Serine $+ H_2O =$ pyruvate $+ NH_3 + H_2O$

[4.2.1.15 Deleted entry: Homoserine dehydratase. Identical with EC 4.4.1.1]

Number	Other Names	Basis for classification (Systematic Name)	Comments	Reference
4.2.1.1	Carbonic anhydrase	Carbonate hydro-lyase	A zinc-protein	658, 1643
4.2.1.2	Fumarase	L-Malate hydro-lyase		42, 1585
4.2.1.3	Aconitase	Citrate (isocitrate) hydro-lyase	Also converts isocitrate into *cis*-aconitate	720, 2295
4.2.1.4		Citrate hydro-lyase	Does not act on isocitrate	2385
4.2.1.5		D-Arabinonate hydro-lyase		3641
4.2.1.6		D-Galactonate hydro-lyase		675
4.2.1.7		D-Altronate hydro-lyase		3104
4.2.1.8		D-Mannonate hydro-lyase		121, 2804
4.2.1.9		2,3-Dihydroxyacid hydro-lyase		1584, 2320
4.2.1.10		3-Dehydroquinate hydro-lyase		2248, 2249
4.2.1.11	Phosphopyruvate hydratase, 2-Phosphoglycerate dehydratase	2-Phospho-D-glycerate hydro-lyase	Also acts on 3-phospho-D-erythronate	1352, 2067, 3662
4.2.1.12		6-Phospho-D-gluconate hydro-lyase		2196
4.2.1.13	Serine deaminase, L-Hydroxyaminoacid dehydratase	L-Serine hydro-lyase (deaminating)	A pyridoxal-phosphate-protein. This reaction is also carried out by EC 4.2.1.16 from a number of sources	2349, 3077
4.2.1.14	D-Hydroxyaminoacid dehydratase	D-Serine hydro-lyase (deaminating)	A pyridoxal-phosphate-protein. Also acts, slowly, on D-threonine	769, 2210

Number	Recommended Name	Reaction
4.2.1.16	Threonine dehydratase	L-Threonine + H_2O = 2-oxobutyrate + NH_3 + H_2O
4.2.1.17	Enoyl-CoA hydratase	An L-3-hydroxyacyl-CoA = a 2,3-(or 3,4-)trans-enoyl-CoA + H_2O
4.2.1.18	Methylglutaconyl-CoA hydratase	3-Hydroxy-3-methylglutaryl-CoA = trans-3-methylglutaconyl-CoA + H_2O
4.2.1.19	Imidazoleglycerol-phosphate dehydratase	D-erythro-Imidazoleglycerol phosphate = imidazoleacetol phosphate + H_2O
4.2.1.20	Tryptophan synthase	L-Serine + indoleglycerol phosphate = L-tryptophan + glyceraldehyde phosphate

[4.2.1.21 Deleted entry: Cystathionine β-synthase. Now EC 4.2.1.22]

4.2.1.22	Cystathionine β-synthase	L-Serine + L-homocysteine = cystathionine + H_2O

[4.2.1.23 Deleted entry: Methylcysteine synthase (a side-reaction of EC 4.2.1.22)]

4.2.1.24	Porphobilinogen synthase	2 5-Aminolaevulinate = porphobilinogen + 2 H_2O
4.2.1.25	L-Arabinonate dehydratase	L-Arabinonate = 2-keto-3-deoxy-L-arabinonate + H_2O

Number	Other Names	Basis for classification (Systematic Name)	Comments	Reference
4.2.1.16	Threonine deaminase, L-Serine dehydratase, Serine deaminase	L-Threonine hydro-lyase (deaminating)	The enzyme from many sources is a pyridoxal-phosphate-protein; that from *P. putida* is not. The enzyme from a number of sources also acts on L-serine (*cf* EC 4.2.1.13)	2418, 2603, 3058, 557
4.2.1.17	Crotonase, Enoyl hydrase, Unsaturated acyl-CoA hydratase	L-3-Hydroxyacyl-CoA hydro-lyase	Also acts (in the reverse reaction) on the *cis*-compounds, yielding D-3-hydroxyacyl-CoA	2300, 3209
4.2.1.18		3-Hydroxy-3-methylglutaryl-CoA hydro-lyase		1314
4.2.1.19		D-*erythro*-Imidazoleglycerol-phosphate hydro-lyase		68
4.2.1.20	Tryptophan desmolase	L-Serine hydro-lyase (adding indoleglycerol-phosphate)	A pyridoxal-phosphate-protein. Also catalyses the conversion of serine and indole to tryptophan and water and of indoleglycerol phosphate to indole and glyceraldehyde phosphate (the latter reaction formerly listed as EC 4.1.2.8)	606
4.2.1.22	Serine sulphhydrase, β-Thionase, Methylcysteine synthase	L-Serine hydro-lyase (adding homocysteine)	A pyridoxal-phosphate-protein. A multifunctional enzyme: catalyses β-replacement reactions between L-serine, L-cysteine, cysteine thioethers or some other β-substituted α-L-amino acids and a variety of mercaptans	375, 2348, 2960
4.2.1.24	Aminolaevulinate dehydratase	5-Aminolaevulinate hydro-lyase (adding 5-aminolaevulinate and cyclizing)	Fungal enzyme is a metalloprotein	1034, 1754, 3766
4.2.1.25		L-Arabinonate hydro-lyase		3641

Number	Recommended Name	Reaction
4.2.1.26	Aminodeoxygluconate dehydratase	2-Amino-2-deoxy-D-gluconate + H_2O = 2-keto-3-deoxy-D-gluconate + NH_3 + H_2O
4.2.1.27	Malonate-semialdehyde dehydratase	Malonate semialdehyde = acetylene monocarboxylate + H_2O
4.2.1.28	Propanediol dehydratase	1,2-Propanediol = propionaldehyde + H_2O
4.2.1.29	Indoleacetaldoxime dehydratase	3-Indoleacetaldoxime = 3-indoleacetonitrile + H_2O
4.2.1.30	Glycerol dehydratase	Glycerol = 3-hydroxy-propionaldehyde + H_2O
4.2.1.31	Maleate hydratase	D-Malate = maleate + H_2O
4.2.1.32	Tartrate dehydratase	L(+)-Tartrate = oxaloacetate + H_2O
4.2.1.33	3-Isopropylmalate dehydratase	2-Hydroxy-4-methyl-3-carboxyvalerate = dimethylcitraconate + H_2O
4.2.1.34	Mesaconate hydratase	(+)-Citramalate = mesaconate + H_2O
4.2.1.35	Citraconate hydratase	(—)-Citramalate = citraconate + H_2O
4.2.1.36	Homoaconitate hydratase	2-Hydroxy-3-carboxyadipate = 3-carboxy-hex-2-ene-dioate + H_2O
4.2.1.37	*trans*-Epoxysuccinate hydratase	*meso*-Tartrate = *trans*-2,3-epoxysuccinate + H_2O
4.2.1.38	*erythro*-3-Hydroxyaspartate dehydratase	*erythro*-3-Hydroxy-Lₛ-aspartate + H_2O = oxaloacetate + NH_3 + H_2O
4.2.1.39	Gluconate dehydratase	D-Gluconate = 2-keto-3-deoxy-D-gluconate + H_2O
4.2.1.40	Glucarate dehydratase	D-Glucarate = 5-keto-4-deoxy-D-glucarate + H_2O
4.2.1.41	5-Keto-4-deoxy-D-glucarate dehydratase	5-Keto-4-deoxy-D-glucarate = 2,5-dioxovalerate + H_2O + CO_2
4.2.1.42	Galactarate dehydratase	D-Galactarate = 5-keto-4-deoxy-D-glucarate + H_2O

Number	Other Names	Basis for classification (Systematic Name)	Comments	Reference
4.2.1.26		2-Amino-2-deoxy-D-gluconate hydro-lyase (deaminating)	A pyridoxal-phosphate-protein	3744
4.2.1.27		Malonate-semialdehyde hydro-lyase		3740
4.2.1.28		1,2-Propanediol hydro-lyase	Requires a cobamide coenzyme. Also dehydrates ethylene glycol to acetaldehyde	5, 1900
4.2.1.29		3-Indoleacetaldoxime hydro-lyase		1821, 2044
4.2.1.30		Glycerol hydro-lyase	Requires a cobamide coenzyme	2967, 2968, 3106
4.2.1.31		D-Malate hydro-lyase		388, 2888
4.2.1.32		L(+)-Tartrate hydro-lyase		1405
4.2.1.33		2-Hydroxy-4-methyl-3-carboxyvalerate hydro-lyase	The enzyme also hydrates the product to 3-hydroxy-4-methyl-3-carboxyvalerate thus bringing about an interconversion between the two isomers	1132
4.2.1.34		(+)-Citramalate hydro-lyase	Also hydrates fumarate to L-malate	312, 3609
4.2.1.35		(−)-Citramalate hydro-lyase	Requires Fe^{2+}	3277
4.2.1.36		2-Hydroxy-3-carboxyadipate hydro-lyase		3248
4.2.1.37		*meso*-Tartrate hydro-lyase	The enzyme acts on both optical isomers of *trans*-2,3-epoxysuccinate	58
4.2.1.38		*erythro*-3-Hydroxy-Ls-aspartate hydro-lyase (deaminating)	A pyridoxal-phosphate-protein	1024
4.2.1.39		D-Gluconate hydro-lyase		81
4.2.1.40		D-Glucarate hydro-lyase		330
4.2.1.41		5-Keto-4-deoxy-D-glucarate hydro-lyase (decarboxylating)		1508
4.2.1.42		D-Galactarate hydro-lyase		331

Number	Recommended Name	Reaction
4.2.1.43	2-Keto-3-deoxy-L-arabonate dehydratase	2-Keto-3-deoxy-L-arabonate = 2-oxoglutarate semialdehyde + H_2O
4.2.1.44	*myo*-Inosose-2 dehydratase	2,4,6/3,5-Pentahydroxycyclohexanone = 4,6/5-trihydroxycyclohexa-1,2-dione + H_2O
4.2.1.45	CDPglucose 4,6-dehydratase	CDPglucose = CDP-4-keto-6-deoxy-D-glucose + H_2O
4.2.1.46	dTDPglucose 4,6-dehydratase	dTDPglucose = dTDP-4-keto-6-deoxy-D-glucose + H_2O
4.2.1.47	GDPmannose 4,6-dehydratase	GDPmannose = GDP-4-keto-6-deoxy-D-mannose + H_2O
4.2.1.48	D-Glutamate cyclase	D-Glutamate = D-pyrrolidone carboxylate + H_2O
4.2.1.49	Urocanate hydratase	4-Imidazolone-5-propionate = urocanate + H_2O
4.2.1.50	Pyrazolylalanine synthase	L-Serine + pyrazole = 3-pyrazol-1-ylalanine + H_2O
4.2.1.51	Prephenate dehydratase	Prephenate = phenylpyruvate + H_2O + CO_2
4.2.1.52	Dihydrodipicolinate synthase	L-Aspartate β-semialdehyde + pyruvate = dihydrodipicolinate + 2 H_2O
4.2.1.53	Oleate hydratase	10-D-Hydroxystearate = oleate + H_2O
4.2.1.54	Lactoyl-CoA dehydratase	Lactoyl-CoA = acryloyl-CoA + H_2O
4.2.1.55	D-3-Hydroxybutyryl-CoA dehydratase	D-3-Hydroxybutyryl-CoA = crotonoyl-CoA + H_2O
4.2.1.56	Itaconyl-CoA hydratase	Citramalyl-CoA = itaconyl-CoA + H_2O
4.2.1.57	Isohexenylglutaconyl-CoA hydratase	3-Hydroxy-3-isohexenylglutaryl-CoA = 3-isohexenylglutaconyl-CoA + H_2O
4.2.1.58	Crotonoyl-[acyl-carrier-protein] hydratase	D-3-Hydroxybutyryl-[acyl-carrier protein] = crotonoyl-[acyl-carrier protein] + H_2O

Number	Other Names	Basis for classification (Systematic Name)	Comments	Reference
4.2.1.43		2-Keto-3-deoxy-L-arabonate hydro-lyase		3239
4.2.1.44		2,4,6/3,5-Pentahydroxycyclo-hexanone hydro-lyase	Requires Co^{2+} or Mn^{2+}	277
4.2.1.45		CDPglucose 4,6-hydro-lyase	Requires bound NAD^+	1299, 2125, 2194
4.2.1.46		dTDPglucose 4,6-hydro-lyase	Requires bound NAD^+	1038, 2194, 3610
4.2.1.47		GDPmannose 4,6-hydro-lyase	The bacterial enzyme requires bound NAD^+	813, 2194, 1956
4.2.1.48		D-Glutamate hydro-lyase (cyclizing)	Also acts on various derivatives of D-glutamate	2185
4.2.1.49	Urocanase	4-Imidazolone-5-propionate hydro-lyase	Contain bound 2-oxobutyrate	1006, 3313
4.2.1.50		L-Serine hydro-lyase (adding pyrazole)	A pyridoxal-phosphate-protein	768
4.2.1.51		Prephenate hydro-lyase (decarboxylating)	This enzyme in the enteric bacteria also possesses chorismate mutase activity and converts chorismate into prephenate	489, 596, 2964
4.2.1.52		L-Aspartate-β-semialdehyde hydro-lyase (adding pyruvate and cyclizing)		3036, 3821
4.2.1.53		10-D-Hydroxystearate 10-hydro-lyase	Acts on a number of 10-hydroxy-acids	655, 2408
4.2.1.54		Lactoyl-CoA hydro-lyase		185
4.2.1.55		D-3-Hydroxybutyryl-CoA hydro-lyase	Also acts on crotonoyl thioesters of pantetheine and acyl-carrier protein	2300
4.2.1.56		Citramalyl-CoA hydro-lyase		583
4.2.1.57		3-Hydroxy-3-isohexenylglutaryl-CoA hydro-lyase	Also acts on dimethylacryloyl-CoA and farnesoyl-CoA	3011
4.2.1.58		D-3-Hydroxybutyryl-[acyl-carrier-protein]hydro-lyase	Is specific for short chain length 3-hydroxyacyl-[acyl-carrier protein] derivatives (C-4 to C-8)	2058, 2255

Number	Recommended Name	Reaction
4.2.1.59	3-Hydroxyoctanoyl-[acyl-carrier-protein] dehydratase	3-Hydroxyoctanoyl-[acyl-carrier protein] = 2-octenoyl-[acyl-carrier protein] + H_2O
4.2.1.60	D-3-Hydroxydecanoyl-[acyl-carrier-protein]dehydratase	D-3-Hydroxydecanoyl-[acyl-carrier protein] = 2,3-decenoyl-[acyl-carrier protein] or 3,4-decenoyl-[acyl-carrier protein] + H_2O
4.2.1.61	3-Hydroxypalmitoyl-[acyl-carrier-protein] dehydratase	3-Hydroxypalmitoyl-[acyl-carrier protein] = 2-hexadecenoyl-[acyl-carrier protein] + H_2O
4.2.1.62	5α-Hydroxysterol dehydratase	5α-Ergosta-7,22-diene-3β,5-diol = ergosterol + H_2O

[*4.2.1.63 Transferred entry: now EC 3.3.2.3, Epoxide hydrolase*]

[*4.2.1.64 Transferred entry: now EC 3.3.2.3, Epoxide hydrolase*]

4.2.1.65	3-Cyanoalanine hydratase	L-Asparagine = 3-cyanoalanine + H_2O
4.2.1.66	Cyanide hydratase	Formamide = hydrogen cyanide + H_2O
4.2.1.67	D-Fuconate dehydratase	D-Fuconate = 2-keto-3-deoxy-D-fuconate + H_2O
4.2.1.68	L-Fuconate dehydratase	L-Fuconate = 2-keto-3-deoxy-L-fuconate + H_2O
4.2.1.69	Cyanamide hydratase	Urea = cyanamide + H_2O
4.2.1.70	Pseudouridylate synthase	Uracil + D-ribose 5-phosphate = pseudouridine 5'-phosphate + H_2O
4.2.1.71	Acetylenemonocarboxylate hydrase	3-Hydroxyacrylate = acetylenemonocarboxylate + H_2O
4.2.1.72	Acetylenedicarboxylate hydrase	2-Hydroxyethylenedicarboxylate = acetylenedicarboxylate + H_2O
4.2.1.73	Protoaphin-aglucone dehydratase (cyclizing)	Protoaphin-aglucone = xanthoaphin + H_2O

Number	Other Names	Basis for classification (Systematic Name)	Comments	Reference
4.2.1.59		3-Hydroxyoctanoyl-[acyl-carrier-protein] hydro-lyase	Is specific for 3-hydroxyacyl-[acyl-carrier protein]derivatives (C-6 to C-12)	2254
4.2.1.60		D-3-Hydroxydecanoyl-[acyl-carrier-protein] hydro-lyase	Specific for C-10 chain length	390, 2439
4.2.1.61		3-Hydroxypalmitoyl-[acyl-carrier-protein] hydro-lyase	Is specific for 3-hydroxyacyl-[acyl-carrier protein]derivatives (C-12 to C-16) and has the highest activity on the C-16 derivative	2254
4.2.1.62		5α-Ergosta-7,22-diene-3β,5-diol 5,6-hydro-lyase		3449
4.2.1.65		L-Asparagine hydro-lyase		485
4.2.1.66		Formamide hydro-lyase		951
4.2.1.67		D-Fuconate hydro-lyase	Also acts on L-arabonate	629
4.2.1.68		L-Fuconate hydro-lyase	Also acts, slowly, on D-arabonate	3820
4.2.1.69		Urea hydro-lyase		3244
4.2.1.70		Uracil hydro-lyase (adding D-ribose 5-phosphate)		3311, 1265, 2129, 2775
4.2.1.71		3-Hydroxyacrylate hydro-lyase	Hydrates acetylenemonocarboxylate to malonic semialdehyde	3744
4.2.1.72		2-Hydroxyethylene-dicarboxylate hydro-lyase	Hydrates acetylenedicarboxylate to pyruvate + CO_2	3743
4.2.1.73		Protoaphin-aglucone hydro-lyase (cyclizing)	The product is converted non-enzymically to erythroaphin. A glycoprotein	448

Number	Recommended Name	Reaction

4.2.2 ACTING ON POLYSACCHARIDES

4.2.2.1 Hyaluronate lyase

Hyaluronate = n 3(Δ-4,5-β-D-glucuronosyl)-2-acetamido-2-deoxy-D-glucose

4.2.2.2 Pectate lyase

Elimination of Δ-4,5-D-galacturonate residues from pectate, thus bringing about depolymerization

4.2.2.3 Alginate lyase

Elimination of Δ-4,5-D-mannuronate residues from alginate, thus bringing about depolymerization

4.2.2.4 Chondroitin ABC lyase

Elimination of Δ-4,5-D-glucuronate residues from polysaccharides containing 1,4-β-D-hexosaminyl and 1,3-β-D-glucuronosyl or 1,3-α-L-iduronosyl linkages, thus bringing about depolymerization

4.2.2.5 Chondroitin AC lyase

Elimination of Δ-4,5-D-glucuronate residues from polysaccharides containing 1,4-β-D-hexosaminyl and 1,3-β-D-glucuronosyl linkages, thus bringing about depolymerization

4.2.2.6 Oligogalacturonide lyase

Elimination of 4-deoxy-5-hexoseulose-uronate residues from O-(4-deoxy-β-L-5-*threo*-hexopyranos-4-enyl-uronate)-(1,4)-D-galacturonate, and related oligosaccharides, thus bringing about depolymerization

4.2.2.7 Heparin lyase

Elimination of Δ-4,5-D-glucuronate residues from polysaccharides containing 1,4-linked glucuronate or iduronate residues and 1,4-α-linked 2-sulphamido-2-deoxy-6-O-sulpho-D-glucose residues

4.2.2.8 Heparitinsulphate lyase

Elimination of sulphate; appears to act on linkages between N-acetamido-2-deoxy-D-glucose and uronate. Product is an unsaturated sugar

4.2.2.9 Exopolygalacturonate lyase

Elimination of Δ-4,5-D-galacturonoso-D-galacturonate residues from the reducing end of de-esterified pectin

Number	Other Names	Basis for classification (Systematic Name)	Comments	Reference
4.2.2.1	Hyaluronidase (but *cf* EC 3.2.1.35 and 3.2.1.36), Mucinase, Spreading factor	Hyaluronate lyase	Also acts on chondroitin Formerly EC 4.2.99.1.	1980, 2213, 2277
4.2.2.2	Pectate transeliminase	Poly(1,4-α-D-galacturonide) lyase	Also acts on other polygalacturonides. Does not act on pectin. Formerly EC 4.2.99.3	39, 792, 2334, 2378
4.2.2.3		Poly(1,4-β-D-mannuronide) lyase	Possibly also acts on the guluronide residues in alginate. Formerly EC 4.2.99.4	2346, 2653
4.2.2.4	Chondroitinase (but *cf* EC 3.1.6.4), Chondroitin ABC eliminase	Chondroitin ABC lyase	Acts on chondroitin 4-sulphate, chondroitin 6-sulphate and dermatan sulphate and slowly on hyaluronate	3758
4.2.2.5	Chondroitinase (but *cf* EC 3.1.6.4), Chondroitin sulphate lyase, Chondroitin AC eliminase	Chondroitin AC lyase	Acts on chondroitin 4-sulphate and chondroitin 6-sulphate, but less well on hyaluronate. Formerly EC 4.2.99.6	2347
4.2.2.6		Oligogalacturonide lyase		2278
4.2.2.7	Heparin eliminase, Heparinase	Heparin lyase		1384
4.2.2.8	Heparin-sulphate eliminase	Heparin-sulphate lyase	Does not act on *N,O*-desulphated glucosamine or *N*-acetyl-*O*-sulphated glucosamine linkages	1384
4.2.2.9		Poly(1,4-α-D-galacturonide) exo-lyase		2034

Number	Recommended Name	Reaction

4.2.2.10   Pectin lyase

Elimination of 6-methyl-Δ-4,5-D-galacturonate residues from pectin, thus bringing about depolymerization

4.2.99   OTHER CARBON-OXYGEN LYASES

[4.2.99.1 Transferred entry: now EC 4.2.2.1, Hyaluronate lyase]

4.2.99.2   Threonine synthase

O-Phosphohomoserine + H_2O = threonine + orthophosphate

[4.2.99.3 Transferred entry: now EC 4.2.2.2, Pectate lyase]

[4.2.99.4 Transferred entry: now EC 4.2.2.3, Alginate lyase]

[4.2.99.5 Deleted entry: Polyglucuronide lyase]

[4.2.99.6 Deleted entry: Chondroitin sulphate lyase. Now included in EC 4.2.2.4 and 4.2.2.5]

4.2.99.7   Ethanolaminephosphate phospho-lyase

Ethanolamine phosphate + H_2O = acetaldehyde + NH_3 + orthophosphate

4.2.99.8   D-Acetylserine (thiol)-Lyase

O-Acetyl-L-serine + hydrogen sulphide = L-cysteine + acetate

4.2.99.9 O-Succinylhomoserine (thiol)-lyase

O-Succinyl-L-homoserine + L-cysteine = cystathionine + succinate

4.2.99.10 O-Acetylhomoserine (thiol)-lyase

O-Acetyl-L-homoserine + methanethiol = L-methionine + acetate

Number	Other Names	Basis for classification (Systematic Name)	Comments	Reference
4.2.2.10		Poly(methoxygalacturonide) lyase	Does not act on de-esterified pectin	40
4.2.99.2		*O*-Phosphohomoserine phospho-lyase (adding water)	A pyridoxal-phosphate-protein	897
4.2.99.7		Ethanolaminephosphate phospho-lyase (deaminating)	A pyridoxal-phosphate-protein. Also acts on D(or L)-1-aminopropan-2-ol-*O*-phosphate	899, 1527
4.2.99.8	Cysteine synthase	*O*-Acetyl-L-serine acetate-lyase (adding hydrogen sulphide)	A pyridoxal-phosphate-protein. Also uses some alkyl thiols as replacing agents	240, 3413
4.2.99.9	Cystathionine γ-synthase	*O*-Succinyl-L-homoserine succinate-lyase (adding cysteine)	A pyridoxal-phosphate-protein. Also reacts with hydrogen sulphide and methanethiol as replacing agents, producing homocysteine and methionine respectively. In the absence of thiol, can also catalyse β-γ-elimination to form 2-oxobutyrate, succinate and ammonia	896, 1593, 1594, 3679, 3680
4.2.99.10	Methionine synthase	*O*-Acetyl-L-homoserine acetate-lyase (adding methanethiol)	Also reacts with other thiols or H_2S, producing homocysteine or its thio-ethers. The name methionine synthase is more commonly applied to EC 2.1.1.13 An enzyme from baker's yeast also catalyses the reaction of EC 4.2.99.8, but more slowly	1662, 3119, 3757, 3756

Number	Recommended Name	Reaction
4.2.99.11	Methylglyoxal synthase	Dihydroxyacetone phosphate = methylglyoxal + orthophosphate
4.2.99.12	Carboxymethyloxysuccinate lyase	Carboxymethyloxysuccinate = fumarate + glycollate

4.3 CARBON-NITROGEN LYASES

This group contains the enzymes eliminating ammonia with the formation of a double bond (in contrast to dehydrogenation and hydrolysis); sub-sub-groups are the ammonia-lyases (4.3.1) and the amidine-lyases (4.3.2).

4.3.1 AMMONIA-LYASES

4.3.1.1	Aspartate ammonia-lyase	L-Aspartate = fumarate + NH$_3$
4.3.1.2	Methylaspartate ammonia-lyase	L-*threo*-3-Methylaspartate = mesaconate + NH$_3$
4.3.1.3	Histidine ammonia-lyase	L-Histidine = urocanate + NH$_3$
4.3.1.4	Formiminotetrahydrofolate cyclodeaminase	5-Formiminotetrahydrofolate = 5,10-methenyltetrahydrofolate + NH$_3$
4.3.1.5	Phenylalanine ammonia-lyase	L-Phenylalanine = *trans*-cinnamate + NH$_3$
4.3.1.6	β-Alanyl-CoA ammonia-lyase	β-Alanyl-CoA = acrylyl-CoA + NH$_3$
4.3.1.7	Ethanolamine ammonia-lyase	Ethanolamine = acetaldehyde + NH$_3$
4.3.1.8	Uroporphyrinogen I synthase	4 Porphobilinogen = uroporphyrinogen I + 4 NH$_3$
4.3.1.9	Glucosaminate ammonia-lyase	2-Amino-2-deoxy-D-gluconate = 2-keto-3-deoxy-D-gluconate + NH$_3$
4.3.1.10	Serinesulphate ammonia-lyase	L-Serine *O*-sulphate + H$_2$O = pyruvate + NH$_3$ + sulphate
4.3.1.11	Dihydroxyphenylalanine ammonia-lyase	3,4-Dihydroxy-L-phenylalanine = *trans*-caffeate + NH$_3$
4.3.1.12	Ornithine cyclodeaminase	L-Ornithine = L-proline + NH$_3$

Number	Other Names	Basis for classification (Systematic Name)	Comments	Reference
4.2.99.11		Dihydroxyacetone-phosphate phospho-lyase	Does not act on D-glyceraldehyde 3-phosphate	579, 1362
4.2.99.12		Carboxymethyloxysuccinate glycollate-lyase		2588
4.3.1.1	Aspartase, Fumaric aminase	L-Aspartate ammonia-lyase		815
4.3.1.2	β-Methylaspartase	L-threo-3-Methylaspartate ammonia-lyase	A cobalamin-protein	205, 385
4.3.1.3	Histidase, Histidinase, Histidine α-deaminase	L-Histidine ammonia-lyase		2180
4.3.1.4		5-Formiminotetrahydrofolate ammonia-lyase (cyclizing)		2681
4.3.1.5		L-Phenylalanine ammonia-lyase	This enzyme may also act on L-tyrosine	1230, 1786, 3814
4.3.1.6		β-Alanyl-CoA ammonia-lyase		3182
4.3.1.7		Ethanolamine ammonia-lyase	Requires a cobamide coenzyme	363, 364, 1591
4.3.1.8		Porphobilinogen ammonia-lyase (polymerizing)	In the presence of a second protein, often called co-synthase, uroporphyrinogen III instead of uroporphyrinogen I is formed	1934, 953
4.3.1.9		2-Amino-2-deoxy-D-gluconate ammonia-lyase (isomerizing)	A pyridoxal-phosphate-protein	2205
4.3.1.10		L-Serine-O-sulphate ammonia-lyase (pyruvate-forming)		3411
4.3.1.11		3,4-Dihydroxy-L-phenylalanine ammonia-lyase		2032
4.3.1.12		L-Ornithine ammonia-lyase (cyclizing)	Requires NAD$^+$	595

Number	Recommended Name	Reaction
4.3.1.13	Carbamoylserine ammonia-lyase	O-Carbamoyl-L-serine + H_2O = pyruvate + 2 NH_3 + CO_2

4.3.2 AMIDINE-LYASES

4.3.2.1	Argininosuccinate lyase	L-Argininosuccinate = fumarate + L-arginine
4.3.2.2	Adenylosuccinate lyase	Adenylosuccinate = fumarate + AMP
4.3.2.3	Ureidoglycollate lyase	(—)-Ureidoglycollate = glyoxylate + urea

4.4 CARBON-SULPHUR LYASES

Enzymes eliminating H_2S or substituted H_2S; a single sub-sub-group 4.4.1.

4.4.1.1	Cystathionine γ-lyase	L-Cystathionine + H_2O = L-cysteine + NH_3 + 2-oxobutyrate
4.4.1.2	Homocysteine desulphhydrase	L-Homocysteine + H_2O = 2-oxobutyrate + NH_3 + hydrogen sulphide
4.4.1.3	Dimethylpropiothetin dethiomethylase	S-Dimethyl-β-propiothetin = acrylate + dimethyl sulphide
4.4.1.4	Alliin lyase	An S-alkyl-L-cysteine sulphoxide = 2-aminoacrylate + an alkyl sulphenate
4.4.1.5	Lactoyl-glutathione lyase	S-D-Lactoyl-glutathione = glutathione + methylglyoxal
4.4.1.6	S-Alkylcysteine lyase	An S-alkyl-L-cysteine + H_2O = pyruvate + NH_3 + an alkyl thiol

Number	Other Names	Basis for classification (Systematic Name)	Comments	Reference
4.3.1.13	*O*-Carbamoyl-L-serine deaminase	*O*-Carbamoyl-L-serine ammonia-lyase (pyruvate-forming)	A pyridoxal-phosphate-protein	577
4.3.2.1	Argininosuccinase	L-Argininosuccinate arginine-lyase		660
4.3.2.2	Adenylosuccinase	Adenylosuccinate AMP-lyase	Also acts on 5′-phosphoribosyl-4-(*N*-succinocarboxamide)-5-aminoimidazole	478
4.3.2.3		(—)-Ureidoglycollate urea-lyase		3461
4.4.1.1	Homoserine deaminase, Homoserine dehydratase, Cystine desulphhydrase, Cysteine desulphhydrase, γ-Cystathionase, Cystathionase	L-Cystathionine cysteine-lyase (deaminating)	A multifunctional pyridoxal-phosphate-protein. Also catalyses elimination reactions of homoserine to form H_2O, NH_3 and 2-oxobutyrate and of L-cystine, producing thiocysteine, pyruvate and NH_3 and of cysteine producing pyruvate, NH_3 and H_2S	373, 374, 895, 2126
4.4.1.2		L-Homocysteine hydrogen-sulphide-lyase (deaminating)	A pyridoxal-phosphate-protein	1575
4.4.1.3		*S*-Dimethyl-β-propiothetin dimethyl-sulphide-lyase		463
4.4.1.4	Alliinase	Alliin alkyl-sulphenate-lyase	A pyridoxal-phosphate-protein	1085, 1478, 770
4.4.1.5	Glyoxalase I, Methylglyoxalase, Aldoketomutase, Ketone-aldehyde mutase	*S*-D-Lactoyl-glutathione methylglyoxal-lyase (isomerizing)	Also acts on 3-phosphoglycerol-glutathione	2687, 809
4.4.1.6		*S*-Alkyl-L-cysteine alkylthiol-lyase (deaminating)	A pyridoxal-phosphate-protein. Decomposes *S*-alkyl-L-cysteines by α,β-elimination. Possibly identical, in yeast, with EC 4.4.1.8	2432

Number	Recommended Name	Reaction

[4.4.1.7 Deleted entry: S-(Hydroxyalkyl)glutathione lyase. Now included with EC 2.5.1.18]

4.4.1.8　　Cystathionine β-lyase　　　　Cystathionine + H_2O = pyruvate + NH_3 + L-homocysteine

4.4.1.9　　β-Cyanoalanine synthase　　　L-Cysteine + HCN = 3-cyanoalanine + hydrogen sulphide

4.4.1.10　　Cysteine lyase　　　　　　L-Cysteine + sulphite = L-cysteate + hydrogen sulphide

4.4.1.11　　L-Methionine γ-lyase　　　L-Methionine = methanethiol + NH_2 + 2-oxobutyrate

4.4.1.12　　Sulphoacetaldehyde lyase　　Sulphoacetaldehyde + H_2O = sulphite + acetate

4.5　CARBON-HALIDE LYASES

This group was set up on the basis of the enzyme eliminating HCl from DDT.

4.5.1.1　　DDT-dehydrochlorinase　　1,1,1-Trichloro-2,2-bis(4-chlorophenyl)-ethane = 1,1-dichloro-2,2-bis(4-chlorophenyl)-ethylene + HCl

4.6　PHOSPHORUS-OXYGEN LYASES

The so-called 'nucleotidyl-cyclases' are included here, on the basis that pyrophosphate is eliminated from the nucleoside triphosphate.

4.6.1.1　　Adenylate cyclase　　ATP = 3':5'-cyclic AMP + pyrophosphate

4.6.1.2　　Guanylate cyclase　　GTP = 3':5'-cyclic GMP + pyrophosphate

4.6.1.3　　3-Dehydroquinate synthase　　7-Phospho-3-deoxy-D-*arabino*-heptulosonate = 3-dehydroquinate + orthophosphate

4.6.1.4　　Chorismate synthase　　3-Phospho-5-enolpyruvoylshikimate = chorismate + orthophosphate

4.99　OTHER LYASES

A miscellaneous sub-group.

Number	Other Names	Basis for classification (Systematic Name)	Comments	Reference
4.4.1.8	β-Cystathionase	Cystathionine L-homocysteine-lyase (deaminating)	A pyridoxal-phosphate-protein	898
4.4.1.9		L-Cysteine hydrogen-sulphide-lyase (adding HCN)		1272, 1273, 484, 37
4.4.1.10		L-Cysteine hydrogen-sulphide-lyase (adding sulphite)	A pyridoxal-phosphate-protein. Can use a second molecule of cysteine (producing lanthionine) or other alkylthiols as replacing agents	3435
4.4.1.11	L-Methioninase	L-Methionine methanethiol-lyase (deaminating)	A pyridoxal-phosphate-protein	1799
4.4.1.12		Sulphoacetaldehyde sulpho-lyase	Requires thiamin diphosphate	1755
4.5.1.1		1,1,1-Trichloro-2,2-bis-(4-chlorophenyl)-ethane hydrogen-chloride-lyase		1984, 2275
4.6.1.1	Adenylyl cyclase, Adenyl cyclase	ATP pyrophosphate-lyase (cyclizing)	Also acts on dATP to form 3':5'-cyclic dAMP. Requires pyruvate	1320
4.6.1.2	Guanylyl cyclase, Guanyl cyclase	GTP pyrophosphate-lyase (cyclizing)	Also acts on ITP and deoxy-GTP	1193, 992
4.6.1.3		7-Phospho-3-deoxy-D-*arabino*-heptulosonate phosphate-lyase (cyclizing)	A Co^{++}-requiring enzyme. The hydrogen atoms on C-7 of the substrate are retained on C-2 of the product.	3177, 2853
4.6.1.4		3-Phospho-5-enolpyruvoyl-shikimate phosphate-lyase		2279, 983, 3652

Number	Recommended Name	Reaction
4.99.1.1	Ferrochelatase	Protoporphyrin + Fe^{2+} = protohaem + 2 H^+
4.99.1.2	Alkylmercury lyase	$RH + Hg^{2+} = R Hg^+ + H^+$

5. ISOMERASES

These enzymes catalyse changes within one molecule.

5.1 RACEMASES AND EPIMERASES

These enzymes, catalysing either racemization or epimerization of a centre of chirality, are sub-divided according to their substrates: amino acids (5.1.1), hydroxy acids (5.1.2), carbohydrates and derivatives (5.1.3), or other compounds (5.1.99).

5.1.1 ACTING ON AMINO ACIDS AND DERIVATIVES

5.1.1.1	Alanine racemase	L-Alanine = D-alanine
5.1.1.2	Methionine racemase	L-Methionine = D-methionine
5.1.1.3	Glutamate racemase	L-Glutamate = D-glutamate
5.1.1.4	Proline racemase	L-Proline = D-proline
5.1.1.5	Lysine racemase	L-Lysine = D-lysine
5.1.1.6	Threonine racemase	L-Threonine = D-threonine
5.1.1.7	Diaminopimelate epimerase	2,6-LL-Diaminopimelate = *meso*-diaminopimelate
5.1.1.8	Hydroxyproline epimerase	L-Hydroxyproline = D-allohydroxyproline
5.1.1.9	Arginine racemase	L-Arginine = D-arginine
5.1.1.10	Amino-acid racemase	L-Amino acid = D-amino acid
5.1.1.11	Phenylalanine racemase (ATP-hydrolysing)	ATP + L-phenylalanine + H_2O = AMP + pyrophosphate + D-phenylalanine

Number	Other Names	Basis for classification (Systematic Name)	Comments	Reference
4.99.1.1		Protohaem ferro-lyase		2637, 2789, 2794
4.99.1.2		Alkylmercury mercuric-lyase	Acts on CH_3Hg^+ and a number of other alkylmercury and arylmercury compounds in the presence of cysteine or other thiols, liberating mercury as a mercaptide	3392
5.1.1.1		Alanine racemase	A pyridoxal-phosphate-protein	2093, 3724, 3726
5.1.1.2		Methionine racemase	A pyridoxal-phosphate-protein	1576
5.1.1.3		Glutamate racemase	A pyridoxal-phosphate-protein	1056
5.1.1.4		Proline racemase		3190
5.1.1.5		Lysine racemase		1390
5.1.1.6		Threonine racemase		71
5.1.1.7		2,6-LL-Diaminopimelate 2-epimerase		93
5.1.1.8		Hydroxyproline 2-epimerase	Also interconverts D-hydroxyproline and L-allohydroxyproline	19
5.1.1.9		Arginine racemase	A pyridoxal-phosphate-protein	3798
5.1.1.10		Amino-acid racemase	A pyridoxal-phosphate-protein	3137
5.1.1.11		Phenylalanine racemase (ATP-hydrolysing)		3750

Number	Recommended Name	Reaction
5.1.1.12	Ornithine racemase	L-Ornithine = D-ornithine
5.1.1.13	Aspartate racemase	L-Aspartate = D-aspartate

5.1.2 ACTING ON HYDROXY ACIDS AND DERIVATIVES

5.1.2.1	Lactate racemase	L-Lactate = D-lactate
5.1.2.2	Mandelate racemase	L-Mandelate = D-mandelate
5.1.2.3	3-Hydroxybutyryl-CoA epimerase	L-3-Hydroxybutyryl-CoA = D-3-hydroxybutyryl-CoA
5.1.2.4	Acetoin racemase	L-Acetoin = D-acetoin
5.1.2.5	Tartrate epimerase	L-Tartate = *meso*-tartrate

5.1.3 ACTING ON CARBOHYDRATES AND DERIVATIVES

5.1.3.1	Ribulosephosphate 3-epimerase	D-Ribulose 5-phosphate = D-xylulose 5-phosphate
5.1.3.2	UDPglucose 4-epimerase	UDPglucose = UDPgalactose
5.1.3.3	Aldose 1-epimerase	α-D-Glucose = β-D-glucose
5.1.3.4	L-Ribulosephosphate 4-epimerase	L-Ribulose 5-phosphate = D-xylulose 5-phosphate
5.1.3.5	UDParabinose 4-epimerase	UDP-L-arabinose = UDP-D-xylose
5.1.3.6	UDPglucuronate 4-epimerase	UDP-D-glucuronate = UDP-D-galacturonate
5.1.3.7	UDPacetylglucosamine 4-epimerase	UDP-2-acetamido-2-deoxy-D-glucose = UDP-2-acetamido-2-deoxy-D-galactose
5.1.3.8	Acylglucosamine 2-epimerase	2-Acylamido-2-deoxy-D-glucose = 2-acylamido-2-deoxy-D-mannose
5.1.3.9	Acylglucosamine-6-phosphate 2-epimerase	2-Acylamido-2-deoxy-D-glucose 6-phosphate = 2-acylamido-2-deoxy-D-mannose 6-phosphate
5.1.3.10	CDPabequose epimerase	CDP-3,6-dideoxy-D-glucose = CDP-3,6-dideoxy-D-mannose

Number	Other Names	Basis for classification (Systematic Name)	Comments	Reference
5.1.1.12		Ornithine racemase		3148
5.1.1.13		Aspartate racemase	Also acts, at half the rate, on L-alanine	1865
5.1.2.1	Hydroxyacid racemase, Lacticoracemase	Lactate racemase		1397, 1703
5.1.2.2		Mandelate racemase		1144
5.1.2.3		3-Hydroxybutyryl-CoA 3-epimerase		3213, 3583
5.1.2.4	Acetylmethylcarbinol racemase	Acetoin racemase		3378
5.1.2.5		Tartrate epimerase		2714
5.1.3.1	Phosphoribulose epimerase	D-Ribulose-5-phosphate 3-epimerase		120, 717, 1410, 3272
5.1.3.2	Galactowaldenase	UDPglucose 4-epimerase	NAD^+ acts as cofactor. Also acts on UDP-2-deoxyglucose	1917, 2135, 3697
5.1.3.3	Mutarotase, Aldose mutarotase	Aldose 1-epimerase	Also acts on L-arabinose D-xylose, D-galactose, maltose and lactose	257, 1638, 1946
5.1.3.4		L-Ribulose-5-phosphate 4-epimerase		432, 709, 1902, 3720
5.1.3.5		UDP-L-arabinose 4-epimerase		868
5.1.3.6		UDPglucuronate 4-epimerase		868
5.1.3.7		UDP-2-acetamido-2-deoxy-D-glucose 4-epimerase		1055
5.1.3.8		2-Acylamido-2-deoxy-D-glucose 2-epimerase	Requires catalytic amounts of ATP	1018
5.1.3.9		2-Acylamido-2-deoxy-D-glucose-6-phosphate 2-epimerase		1017
5.1.3.10		CDPabequose 2-epimerase, CDPparatose epimerase	Requires NAD^+	2124

Number	Recommended Name	Reaction
5.1.3.11	Cellobiose epimerase	Cellobiose = glucosylmannose
5.1.3.12	UDPglucuronate 5′-epimerase	UDP-D-glucuronate = UDP-L-iduronate
5.1.3.13	dTDP-4-ketorhamnose 3,5-epimerase	dTDP-4-keto-6-deoxy-D-glucose = dTDP-4-keto-6-deoxy-L-mannose
5.1.3.14	UDPacetylglucosamine 2-epimerase	UDP-2-acetamido-2-deoxy-D-glucose = UDP-2-acetamido-2-deoxy-D-mannose
5.1.3.15	Glucose-6-phosphate 1-epimerase	α-D-Glucose 6-phosphate = β-D-glucose 6-phosphate

5.1.99 ACTING ON OTHER COMPOUNDS

5.1.99.1	Methylmalonyl-CoA racemase	R-Methylmalonyl-CoA = S-methylmalonyl-CoA
5.1.99.2	16-Hydroxysteroid epimerase	16α-Hydroxysteroid = 16β-hydroxysteroid
5.1.99.3	Allantoin racemase	$(S)(+)$-Allantoin = $(R)(-)$-allantoin

5.2 *Cis-trans*-ISOMERASES

These enzymes rearrange the geometry at double bonds.

5.2.1.1	Maleate isomerase	Maleate = fumarate
5.2.1.2	Maleylacetoacetate isomerase	4-Maleylacetoacetate = 4-fumarylacetoacetate
5.2.1.3	Retinal isomerase	all-*trans*-Retinal = 11-*cis*-retinal
5.2.1.4	Maleylpyruvate isomerase	3-Maleylpyruvate = 3-fumarylpyruvate
5.2.1.5	Linoleate isomerase	9-*cis*-12-*cis*-Octadecadienoate = 9-*cis*-11-*trans*-octadecadienoate
5.2.1.6	Furylfuramide isomerase	2-(2-Furyl)-3-*cis*-(5-nitro-2-furyl) acrylamide = 2-(2-furyl)-3-*trans*-(5-nitro-2-furyl) acrylamide

5.3 INTRAMOLECULAR OXIDOREDUCTASES

These enzymes bring about the oxidation of one part of a molecule with a corresponding reduction of another part. They include the enzymes converting, in the sugar series, aldoses to ketoses and *vice-versa* (sugar isomerases, 5.3.1), enzymes catalysing a keto-enol equilibrium (tautomerases, 5.3.2), enzymes shifting a carbon-carbon double bond from one position to another (5.3.3) and enzymes transposing S-S bonds (5.3.4).

Number	Other Names	Basis for classification (Systematic Name)	Comments	Reference
5.1.3.11		Cellobiose 2-epimerase		3495
5.1.3.12		UDPglucuronate 5'-epimerase	Requires NAD+	1479
5.1.3.13		dTDP-4-keto-6-deoxy-D-glucose 3,5-epimerase		999, 2195
5.1.3.14		UDP-2-acetamido-2-deoxy-D-glucose 2-epimerase	The enzyme hydrolyses the product to UDP and 2-acetamido-2-deoxy-D-mannose	1681
5.1.3.15		Glucose-6-phosphate 1-epimerase		3737
5.1.99.1		Methylmalonyl-CoA racemase		2143, 2531
5.1.99.2		16-Hydroxysteroid 16-epimerase		626
5.1.99.3		Allantoin racemase		3531
5.2.1.1		Maleate *cis-trans*-isomerase		245
5.2.1.2		4-Maleylacetoacetate *cis-trans*-isomerase	Also acts on maleylpyruvate	793, 1856, 3004
5.2.1.3	Retinene isomerase	all-*trans*-Retinal 11-*cis-trans*-isomerase	Light shifts the equilibrium towards the *cis*-isomer	1393
5.2.1.4		3-Maleylpyruvate *cis-trans*-isomerase		1856
5.2.1.5		Linoleate Δ^{12}-*cis*-Δ^{11}-*trans*-isomerase		1661
5.2.1.6		2-(2-Furyl)-3-(5-nitro-2-furyl) acrylamide *cis-trans*-isomerase	Requires NADH	3445

Number	Recommended Name	Reaction

5.3.1   INTERCONVERTING ALDOSES AND KETOSES

| 5.3.1.1 | Triosephosphate isomerase | D-Glyceraldehyde 3-phosphate = dihydroxyacetone phosphate |

[5.3.1.2   Deleted entry: Erythrose isomerase]

5.3.1.3	Arabinose isomerase	D-Arabinose = D-ribulose
5.3.1.4	L-Arabinose isomerase	L-Arabinose = L-ribulose
5.3.1.5	Xylose isomerase	D-Xylose = D-xylulose
5.3.1.6	Ribosephosphate isomerase	D-Ribose 5-phosphate = D-ribulose 5-phosphate
5.3.1.7	Mannose isomerase	D-Mannose = D-fructose
5.3.1.8	Mannosephosphate isomerase	D-Mannose 6-phosphate = D-fructose 6-phosphate
5.3.1.9	Glucosephosphate isomerase	D-Glucose 6-phosphate = D-fructose 6-phosphate
5.3.1.10	Glucosaminephosphate isomerase	2-Amino-2-deoxy-D-glucose 6-phosphate + H_2O = D-fructose 6-phosphate + NH_3

[5.3.1.11   Deleted entry: Acetylglucosaminephosphate isomerase]

Number	Other Names	Basis for classification (Systematic Name)	Comments	Reference
5.3.1.1	Phosphotriose isomerase, Triosephosphate mutase	D-Glyceraldehyde-3-phosphate ketol-isomerase		2211, 2214
5.3.1.3		D-Arabinose ketol-isomerase	Also acts on L-fucose and, more slowly, on L-galactose and D-altrose	553, 1105
5.3.1.4		L-Arabinose ketol-isomerase		1255, 2354
5.3.1.5		D-Xylose ketol-isomerase	Same enzymes also convert D-glucose to D-fructose	1330, 3097, 3762
5.3.1.6	Phosphopentosisomerase, Phosphoriboisomerase	D-Ribose-5-phosphate ketol-isomerase	Also acts on D-ribose 5-diphosphate and D-ribose 5-triphosphate	717, 1370, 1411
5.3.1.7		D-Mannose ketol-isomerase	Also acts on D-lyxose and rhamnose	2547
5.3.1.8	Mannose-6-phosphate isomerase, Phosphomannose isomerase, Phosphohexoisomerase, Phosphohexomutase	D-Mannose-6-phosphate ketol-isomerase	A zinc-protein	412, 1096, 3098
5.3.1.9	Phosphohexose isomerase, Phosphohexomutase, Oxoisomerase, Hexosephosphate isomerase, Glucose-6-phosphate isomerase, Phosphosaccharomutase, Phosphoglucoisomerase, Phosphohexoisomerase	D-Glucose-6-phosphate ketol-isomerase	Also catalyses the anomerization of D-glucose-6-phosphate	172, 2351, 2431, 2709, 3469, 2429
5.3.1.10		2-Amino-2-deoxy-D-glucose-6-phosphate ketol-isomerase (deaminating)	Acetylglucosamine 6-phosphate, which is not broken down, activates the enzyme	565, 2569, 3717

Number	Recommended Name	Reaction
5.3.1.12	Glucuronate isomerase	D-Glucuronate = D-fructuronate
5.3.1.13	Arabinosephosphate isomerase	D-Arabinose 5-phosphate = D-ribulose 5-phosphate
5.3.1.14	L-Rhamnose isomerase	D-Rhamnose = L-rhamnulose
5.3.1.15	D-Lyxose ketol-isomerase	D-Lyxose = D-xylulose
5.3.1.16	N-(5′-Phospho-D-ribosylformimino)-5-amino-1-(5″-phosphoribosyl)-4-imidazolecarboxamide isomerase	N-(5′-phospho-D-ribosylformimino)-5-amino-1-(5″-phosphoribosyl)-4-imidazolecarboxamide = N-(5′-phospho-D-1′-ribulosylformimino)-5-amino-1-(5″-phosphoribosyl)-4-imidazolecarboxamide
5.3.1.17	4-Deoxy-L-*threo*-5-hexosulose-uronate ketol-isomerase	4-Deoxy-L-*threo*-5-hexosulose uronate = 3-deoxy-D-*glycero*-2,5-hexodiulosonate

[5.3.1.18 *Deleted entry: Glucose isomerase (reaction due to EC 5.3.1.9, in presence of arsenate, or EC 5.3.1.5)*]

Number	Recommended Name	Reaction
5.3.1.19	Glucosaminephosphate isomerase (glutamine-forming)	2-Amino-2-deoxy-D-glucose 6-phosphate + L-glutamate = D-fructose 6-phosphate + L-glutamine
5.3.1.20	Ribose isomerase	D-Ribose = D-ribulose

5.3.2   INTERCONVERTING KETO- AND ENOL-GROUPS

Number	Recommended Name	Reaction
5.3.2.1	Phenylpyruvate tautomerase	*keto*-Phenylpyruvate = *enol*-phenylpyruvate
5.3.2.2	Oxaloacetate tautomerase	*keto*-Oxaloacetate = *enol*-oxaloacetate

5.3.3   TRANSPOSING C=C BONDS

Number	Recommended Name	Reaction
5.3.3.1	Steroid Δ-isomerase	A 3-oxo-Δ^5-steroid = a 3-oxo-Δ^4-steroid
5.3.3.2	Isopentenyldiphosphate Δ-isomerase	Isopentenyl diphosphate = dimethylallyl diphosphate
5.3.3.3	Vinylacetyl-CoA Δ-isomerase	Vinylacetyl-CoA = crotonoyl-CoA
5.3.3.4	Muconolactone Δ-isomerase	(+)-4-Hydroxy-4-carboxymethylisocrotonolactone = 3-oxoadipate *enol*-lactone

Number	Other Names	Basis for classification (Systematic Name)	Comments	Reference
5.3.1.12	Uronic isomerase	D-Glucuronate ketol-isomerase	Also converts D-galacturonate to D-tagaturonate	122, 1683
5.3.1.13		D-Arabinose-5-phosphate ketol-isomerase		3564
5.3.1.14		L-Rhamnose ketol-isomerase		743
5.3.1.15		D-Lyxose ketol-isomerase		76
5.3.1.16		N-(5′-Phospho-D-ribosylformimino)-5-amino-1-(5″-phosphoribosyl)-4-imidazolecarboxamide ketol-isomerase		2083
5.3.1.17		4-Deoxy-L-*threo*-5-hexosulose-uronate ketol-isomerase		2652
5.3.1.19	Hexosephosphate aminotransferase, Glutamine-fructose-6-phosphate aminotransferase	2-Amino-2-deoxy-D-glucose-6-phosphate ketol-isomerase (amino-transferring)	Formerly EC 2.6.1.16	1016, 1138, 1918
5.3.1.20		D-Ribose ketol-isomerase	Also acts on L-lyxose and L-rhamnose	1471, 1472
5.3.2.1		Phenylpyruvate keto—enol-isomerase	Also acts on other acyl pyruvates	324, 1726, 1728
5.3.2.2		Oxaloacetate keto—enol-isomerase		90
5.3.3.1		3-Oxosteroid Δ^5—Δ^4-isomerase	May be at least three distinct enzymes	851, 1625, 3357
5.3.3.2		Isopentenyldiphosphate Δ^3— Δ^2-isomerase		29
5.3.3.3		Vinylacetyl-CoA Δ^3—Δ^2-isomerase	Also acts on 3-methyl-vinylacetyl-CoA	2019, 2793
5.3.3.4		(+)-4-Hydroxy-4-carboxymethylisocrotonolactone Δ^2—Δ^3-isomerase		2509

Number	Recommended Name	Reaction
5.3.3.5	Cholestenol Δ-isomerase	5α-Cholest-7-en-3β-ol = 5α-cholest-8-en-3β-ol
5.3.3.6	Methylitaconate Δ-isomerase	Methylitaconate = dimethylmaleate
5.3.3.7	Aconitate Δ-isomerase	*trans*-Aconitate = *cis*-aconitate
5.3.3.8	Dodecenoyl-CoA Δ-isomerase	3-*cis*-Dodecenoyl-CoA = 2-*trans*-dodecenoyl-CoA
5.3.3.9	Prostaglandin-A$_1$ Δ-isomerase	Prostaglandin A$_1$ = prostaglandin C$_1$

5.3.4 TRANSPOSING S-S BONDS

5.3.4.1	Protein disulphide-isomerase	Catalyses the rearrangement of -S-S-bonds in proteins

5.3.99 OTHER INTRAMOLECULAR OXIDOREDUCTASES

5.3.99.1	Hydroperoxide isomerase	13-Hydroperoxy-octadeca-9,11-dienoate = 12-oxo-13-hydroxyoctadeca-9-enoate
5.3.99.2	Prostaglandin R$_2$ D-isomerase	Prostaglandin R$_2$ = prostaglandin D$_2$
5.3.99.3	Prostaglandin R$_2$ E-isomerase	Prostaglandin R$_2$ = prostaglandin E$_2$

5.4 INTRAMOLECULAR TRANSFERASES

These enzymes transfer acyl-, phospho-, amino- or other groups from one position to another.

5.4.1 TRANSFERRING ACYL-GROUPS

5.4.1.1	Lysolecithin acylmutase	2-Lysolecithin = 3-lysolecithin

5.4.2 TRANSFERRING PHOSPHORYL GROUPS

5.4.2.1	Phosphoglycerate phosphomutase	2-Phospho-D-glycerate = 3-phospho-D-glycerate

Number	Other Names	Basis for classification (Systematic Name)	Comments	Reference
5.3.3.5		Δ^7-Cholestenol Δ^7—Δ^8-isomerase		3704
5.3.3.6		Methylitaconate Δ^2—Δ^3-isomerase		1828
5.3.3.7	Δ^3-*cis*-Δ^2-*trans*-enoyl-CoA isomerase	Aconitate Δ^2—Δ^3-isomerase	This isomerization could take place either in a direct *cis-trans* interconversion or by an allelic rearrangement; the enzyme has been shown to catalyse the latter change	1715
5.3.3.8		Dodecenoyl-CoA Δ^3-*cis*—Δ^2-*trans*-isomerase		3233
5.3.3.9		Prostaglandin- A_1 Δ^{10}—Δ^{11}-isomerase		2628
5.3.4.1	'S-S rearrangase'	Protein disulphide-isomerase	Needs reducing agents or partly-reduced enzyme; the reaction depends on sulphhydryl-disulphide interchange	677, 955
5.3.99.1		Fatty-acid-hydroperoxide isomerase	Acts on the products of EC 1.13.11.12	3847
5.3.99.2		Prostaglandin R_2 D-isomerase	Brings about the opening of the peroxide ring	2458
5.3.99.3		Prostaglandin R_2 D-isomerase	Brings about the opening of the peroxide ring	2458
5.4.1.1	Lysolecithin migratase	Lysolecithin 2,3-acylmutase		3521
5.4.2.1	Bisphosphoglyceromutase	D-Phosphoglycerate 2,3-phosphomutase	See also EC 2.7.5.3	873, 1466

Number	Recommended Name	Reaction

5.4.3   TRANSFERRING AMINO GROUPS

[5.4.3.1   Deleted entry: Ornithine 4,5-aminomutase. This reaction was due to a mixture of EC 5.1.1.12 and 5.4.3.5]

5.4.3.2	Lysine 2,3-aminomutase	L-Lysine = L-3,6-diaminohexanoate
5.4.3.3	β-Lysine 5,6-aminomutase	L-3,6-Diaminohexanoate = 3,5-diaminohexanoate
5.4.3.4	D-Lysine 5,6-aminomutase	D-Lysine = 2,5-diaminohexanoate
5.4.3.5	D-Ornithine 4,5-aminomutase	D-Ornithine = D-*threo*-2,4-diaminopentanoate
5.4.3.6	Tyrosine 2,3-aminomutase	L-Tyrosine = 3-amino-3-(4-hydroxyphenyl)-propionate

5.4.99   TRANSFERRING OTHER GROUPS

5.4.99.1	Methylaspartate mutase	L-*threo*-3-Methylaspartate = L-glutamate
5.4.99.2	S-Methylmalonyl-CoA mutase	S-Methylmalonyl-CoA = succinyl-CoA
5.4.99.3	2-Acetolactate mutase	2-Acetolactate = 3-hydroxy-2-oxo-isovalerate
5.4.99.4	2-Methylene-glutarate mutase	2-Methylene-glutarate = 2-methylene-3-methyl-succinate
5.4.99.5	Chorismate mutase	Chorismate = prephenate
5.4.99.6	Isochorismate synthase	Chorismate = isochorismate
5.4.99.7	2,3-Oxidosqualene lanosterol-cyclase	2,3-Oxidosqualene = lanosterol
5.4.99.8	2,3-Oxidosqualene cycloartenol-cyclase	2,3-Oxidosqualene = cycloartenol

Number	Other Names	Basis for classification (Systematic Name)	Comments	Reference
5.4.3.2		L-Lysine 2,3-aminomutase	Activity is stimulated by S-adenosyl-L-methionine and pyridoxal phosphate	3834
5.4.3.3	β-Lysine mutase	L-3,6-Diaminohexanoate aminomutase	Requires a cobamide coenzyme	3191
5.4.3.4	D-α-Lysine mutase	D-2,6-Diaminohexanoate aminomutase	Requires a cobamide coenzyme	2292, 3192
5.4.3.5		D-Ornithine 4,5-aminomutase	A pyridoxal-phosphate-protein, which requires a cobamide coenzyme and dithiothreitol for activity	3148
5.4.3.6		L-Tyrosine 2,3-aminomutase	Requires ATP	1846
5.4.99.1	Glutamate mutase	L-threo-3-Methylaspartate carboxy-aminomethylmutase	Requires a cobamide coenzyme	204, 3648
5.4.99.2		Methylmalonyl-CoA CoA-carbonylmutase	Requires a cobamide coenzyme	237, 458, 797, 1652, 2531
5.4.99.3		2-Acetolactate methylmutase	Requires ascorbic acid; also converts 2-hydroxy-2-acetobutyrate to 2-hydroxy-2-oxo-3-methylvalerate	53
5.4.99.4		2-Methylene-glutarate carboxy-methylenemethylmutase	Requires a cobamide coenzyme	1827, 1828
5.4.99.5		Chorismate pyruvatemutase		596, 2005, 3170, 3728
5.4.99.6		Isochorismate hydroxymutase		3813
5.4.99.7		2,3-Oxidosqualene mutase (cyclizing, lanosterol-forming)		681
5.4.99.8		2,3-Oxidosqualene mutase (cyclizing, cycloartenol-forming)		2743

Number	Recommended Name	Reaction

5.5 INTRAMOLECULAR LYASES

These catalyse reactions in which a group can be regarded as eliminated from one part of a molecule, leaving a double bond, while remaining covalently attached to the molecule.

5.5.1.1	Muconate cycloisomerase	(+)-4-Carboxymethyl-4-hydroxyisocrotonolactone = *cis-cis*-muconate
5.5.1.2	3-Carboxy-*cis-cis*-muconate cycloisomerase	4-Carboxymuconolactone = 3-carboxy-*cis-cis*-muconate
5.5.1.3	Tetrahydroxypteridine cycloisomerase	Tetrahydroxypteridine = xanthine-8-carboxylate
5.5.1.4	*myo*-Inositol-1-phosphate synthase	D-Glucose 6-phosphate = 1L-*myo*-inositol 1-phosphate
5.5.1.5	Carboxy-*cis-cis*-muconate cyclase	3-Carboxymuconolactone = 3-carboxy-*cis-cis*-muconate
5.5.1.6	Chalcone isomerase	A chalcone = a flavanone

5.99 OTHER ISOMERASES

A miscellaneous sub-group.

5.99.1.1	Thiocyanate isomerase	Benzyl isothiocyanate = benzyl thiocyanate

6. LIGASES (SYNTHETASES)

Ligases are enzymes catalysing the joining of two molecules with concomitant hydrolysis of the pyrophosphate bond in ATP or a similar triphosphate. The bonds formed are often 'high-energy' bonds. 'Synthetase' is commonly used for the recommended name.

6.1 FORMING CARBON-OXYGEN BONDS

To this sub-class belong the enzymes acylating a transfer RNA with the corresponding amino acid (6.1.1).

6.1.1 LIGASES FORMING AMINOACYL-tRNA AND RELATED COMPOUNDS

6.1.1.1	Tyrosyl-tRNA synthetase	ATP + L-tyrosine + tRNA^{Tyr} = AMP + pyrophosphate + L-tyrosyl-tRNA^{Tyr}
6.1.1.2	Tryptophanyl-tRNA synthetase	ATP + L-tryptophan + tRNA^{Trp} = AMP + pyrophosphate + L-tryptophanyl-tRNA^{Trp}
6.1.1.3	Threonyl-tRNA synthetase	ATP + L-threonine + tRNA^{Thr} = AMP + pyrophosphate + L-threonyl-tRNA^{Thr}
6.1.1.4	Leucyl-tRNA synthetase	ATP + L-leucine + tRNA^{Leu} = AMP + pyrophosphate + L-leucyl-tRNA^{Leu}

Number	Other Names	Basis for classification (Systematic Name)	Comments	Reference
5.5.1.1		4-Carboxymethyl-4-hydroxyisocrotonolactone lyase (decyclizing)	Also acts (in the reverse reaction) very slowly, on *cis-trans*-muconate	2509, 3092
5.5.1.2		4-Carboxymuconolactone lyase (decyclizing)		2508
5.5.1.3		Tetrahydroxypteridine lyase (isomerizing)		2036
5.5.1.4		1L-*myo*-Inositol-1-phosphate lyase (isomerizing)	Requires NAD^+	806, 3041
5.5.1.5		3-Carboxymuconolactone lyase (decyclizing)		1133
5.5.1.6		Flavanone lyase (decyclizing)		2303
5.99.1.1		Benzyl-thiocyanate isomerase		3555
6.1.1.1		L-Tyrosine:tRNA^{Tyr} ligase (AMP-forming)		56, 1345, 2981, 602
6.1.1.2		L-Tryptophan:tRNA^{Trp} ligase (AMP-forming)		650, 2650, 3721
6.1.1.3		L-Threonine:tRNA^{Thr} ligase (AMP-forming)		56, 1345
6.1.1.4		L-Leucine:tRNA^{Leu} ligase (AMP-forming)		56, 263, 267

Number	Recommended Name	Reaction
6.1.1.5	Isoleucyl-tRNA synthetase	ATP + L-isoleucine + tRNA^{Ile} = AMP + pyrophosphate + L-isoleucyl-tRNA^{Ile}
6.1.1.6	Lysyl-tRNA synthetase	ATP + L-lysine + tRNA^{Lys} = AMP + pyrophosphate + L-lysyl-tRNA^{Lys}
6.1.1.7	Alanyl-tRNA synthetase	ATP + L-alanine + tRNA^{Ala} = AMP + pyrophosphate + L-alanyl-tRNA^{Ala}
[6.1.1.8	*Deleted entry:* D-*Alanine-sRNA synthetase]*	
6.1.1.9	Valyl-tRNA synthetase	ATP + L-valine + tRNA^{Val} = AMP + pyrophosphate + L-valyl-tRNA^{Val}
6.1.1.10	Methionyl-tRNA synthetase	ATP + L-methionine + tRNA^{Met} = AMP + pyrophosphate + L-methionyl-tRNA^{Met}
6.1.1.11	Seryl-tRNA synthetase	ATP + L-serine + tRNA^{Ser} = AMP + pyrophosphate + L-seryl-tRNA^{Ser}
6.1.1.12	Aspartyl-tRNA synthetase	ATP + L-aspartate + tRNA^{Asp} = AMP + pyrophosphate + L-aspartyl-tRNA^{Asp}
6.1.1.13	D-Alanyl-poly(phosphoribitol) synthetase	ATP + D-alanine + poly(ribitol phosphate) = AMP + pyrophosphate + *O*-D-alanyl-poly(ribitol phosphate)
6.1.1.14	Glycyl-tRNA synthetase	ATP + glycine + tRNA^{Gly} = AMP + pyrophosphate + glycyl-tRNA^{Gly}
6.1.1.15	Prolyl-tRNA synthetase	ATP + L-proline + tRNA^{Pro} = AMP + pyrophosphate + L-prolyl-tRNA^{Pro}
6.1.1.16	Cysteinyl-tRNA synthetase	ATP + L-cysteine + tRNA^{Cys} = AMP + pyrophosphate + L-cysteinyl-tRNA^{Cys}
6.1.1.17	Glutamyl-tRNA synthetase	ATP + L-glutamate + tRNA^{Glu} = AMP + pyrophosphate + L-glutamyl-tRNA^{Glu}
6.1.1.18	Glutaminyl-tRNA synthetase	ATP + L-glutamine + tRNA^{Gln} = AMP + pyrophosphate + L-glutaminyl-tRNA^{Gln}
6.1.1.19	Arginyl-tRNA synthetase	ATP + L-arginine + tRNA^{Arg} = AMP + pyrophosphate + L-arginyl-tRNA^{Arg}
6.1.1.20	Phenylalanyl-tRNA synthetase	ATP + L-phenylalanine + tRNA^{Phe} = AMP + pyrophosphate + L-phenylalanyl-tRNA^{Phe}
6.1.1.21	Histidyl-tRNA synthetase	ATP + L-histidine + tRNA^{His} = AMP + pyrophosphate + L-histidyl-tRNA^{His}
6.1.1.22	Asparaginyl-tRNA synthetase	ATP + L-asparagine + tRNA = AMP + pyrophosphate + L-asparaginyl-tRNA

Number	Other Names	Basis for classification (Systematic Name)	Comments	Reference
6.1.1.5		L-Isoleucine:tRNA^{Ile} ligase (AMP-forming)		56, 263, 267
6.1.1.6		L-Lysine:tRNA^{Lys} ligase (AMP-forming)		56, 529, 3219, 1860
6.1.1.7		L-Alanine:tRNA^{Ala} ligase (AMP-forming)		1346, 3627
6.1.1.9		L-Valine:tRNA^{Val} ligase (AMP-forming)		263, 267
6.1.1.10		L-Methionine:tRNA^{Met} ligase (AMP-forming)		267
6.1.1.11		L-Serine:tRNA^{Ser} ligase (AMP-forming)		1610, 2060, 3630
6.1.1.12		L-Aspartate:tRNA^{Asp} ligase (AMP-forming)		2448, 991
6.1.1.13		D-Alanine:poly(phosphoribitol) ligase (AMP-forming)	Involved in the synthesis of teichoic acids	163
6.1.1.14		Glycine:tRNA^{Gly} ligase (AMP-forming)		924, 2425
6.1.1.15		L-Proline:tRNA^{Pro} ligase (AMP-forming)		2447, 2590
6.1.1.16		L-Cysteine-tRNA^{Cys} ligase (AMP-forming)		2151
6.1.1.17		L-Glutamate:tRNA^{Glu} ligase (AMP-forming)		2730
6.1.1.18		L-Glutamine:tRNA^{Gln} ligase (AMP-forming)		2730
6.1.1.19		L-Arginine:tRNA^{Arg} ligase (AMP-forming)		61, 2179, 2247
6.1.1.20		L-Phenylalanine:tRNA^{Phe} ligase (AMP-forming)		3271
6.1.1.21		L-Histidine:tRNA^{His} ligase (AMP-forming)		676, 3422
6.1.1.22		L-Asparagine:tRNA ligase (AMP-forming)		653

Number	Recommended Name	Reaction

6.2   FORMING CARBON-SULPHUR BONDS

To this sub-class belong the enzymes synthesizing acyl-CoA derivatives.

6.2.1   ACID-THIOL LIGASES

6.2.1.1   Acetyl-CoA synthetase

ATP + acetate + CoA = AMP + pyrophosphate + acetyl-CoA

6.2.1.2   Butyryl-CoA synthetase

ATP + an acid + CoA = AMP + pyrophosphate + an acyl-CoA

6.2.1.3   Acyl-CoA synthetase

ATP + an acid + CoA = AMP + pyrophosphate + an acyl-CoA

6.2.1.4   Succinyl-CoA synthetase (GDP-forming)

GTP + succinate + CoA = GDP + orthophosphate + succinyl-CoA

6.2.1.5   Succinyl-CoA synthetase (ADP-forming)

ATP + succinate + CoA = ADP + orthophosphate + succinyl-CoA

6.2.1.6   Glutaryl-CoA synthetase

ATP + glutarate + CoA = ADP + orthophosphate + glutaryl-CoA

6.2.1.7   Choloyl-CoA synthetase

ATP + cholate + CoA = AMP + pyrophosphate + choloyl-CoA

6.2.1.8   Oxalyl-CoA synthetase

ATP + oxalate + CoA = AMP + pyrophosphate + oxalyl-CoA

6.2.1.9   Malyl-CoA synthetase

ATP + malate + CoA = ADP + orthophosphate + malyl-CoA

6.2.1.10   Acyl-CoA synthetase (GDP-forming)

GTP + an acid + CoA = GDP + orthophosphate + acyl-CoA

6.2.1.11   Biotinyl-CoA synthetase

ATP + biotin + CoA = AMP + pyrophosphate + biotinyl-CoA

6.2.1.12   4-Coumaroyl-CoA synthetase

ATP + 4-coumarate + CoA = AMP + pyrophosphate + 4-coumaroyl-CoA

6.2.1.13   Acetyl-CoA synthetase (ADP-forming)

ATP + acetate + CoA = ADP + orthophosphate + acetyl-CoA

Number	Other Names	Basis for classification (Systematic Name)	Comments	Reference
6.2.1.1	Acetyl activating enzyme, Acetate thiokinase, Acyl-activating enzyme	Acetate:CoA ligase (AMP-forming)	Also acts on propionate and acrylate	530, 804, 1267, 2229
6.2.1.2	Fatty acid thiokinase (medium chain), Acyl-activating enzyme	Butyrate:CoA ligase (AMP-forming)	Acts on acids from C-4 to C-11 and on the corresponding 3-hydroxy- and 2,3- or 3,4-unsaturated acids	2053, 2107, 3631
6.2.1.3	Fatty acid thiokinase (long chain), Acyl-activating enzyme	Acid:CoA ligase (AMP-forming)	Acts on acids from C-6 to C-20	343, 1774
6.2.1.4	Succinic thiokinase	Succinate:CoA ligase (GDP-forming)	Itaconate can act instead of succinate and ITP instead of GTP	1161. 1618, 2142, 2905
6.2.1.5	Succinic thiokinase	Succinate:CoA ligase (ADP-forming)		1161, 1614, 1617
6.2.1.6		Glutarate:CoA ligase (ADP-forming)	GTP or ITP can act instead of ATP	2200
6.2.1.7	Cholate thiokinase	Cholate:CoA ligase (AMP-forming)		818, 819
6.2.1.8		Oxalate:CoA ligase (AMP-forming)		1045
6.2.1.9		Malate:CoA ligase (ADP-forming)		2310
6.2.1.10		Acid:CoA ligase (GDP-forming)		2851
6.2.1.11		Biotin:CoA ligase (AMP-forming)		533
6.2.1.12		4-Coumarate:CoA ligase (AMP-forming)		1976
6.2.1.13		Acetate:CoA ligase (ADP-forming)	Also acts on propionate and, very slowly, on n-butyrate	2760

Number	Recommended Name	Reaction

6.3 FORMING CARBON-NITROGEN BONDS

To this sub-class belong the amide synthetases (6.3.1), the peptide synthetases (6.3.2), enzymes forming heterocyclic rings (6.3.3), and a few others (6.3.4).

6.3.1 ACID—AMMONIA (OR AMINE) LIGASES (AMIDE SYNTHETASES)

6.3.1.1	Asparagine synthetase	$ATP + L\text{-aspartate} + NH_3 = AMP +$ pyrophosphate $+ L\text{-asparagine}$
6.3.1.2	Glutamine synthetase	$ATP + L\text{-glutamate} + NH_3 = ADP +$ orthophosphate $+ L\text{-glutamine}$

[6.3.1.3   Transferred entry: now EC 6.3.4.13, Phosphoribosyl-glycinamide synthetase]

6.3.1.4	Asparagine synthetase (ADP-forming)	$ATP + L\text{-aspartate} + NH_3 = ADP +$ orthophosphate $+ L\text{-asparagine}$
6.3.1.5	NAD$^+$ synthetase	$ATP + \text{deamido-NAD}^+ + NH_3 = AMP +$ pyrophosphate $+ NAD^+$
6.3.1.6	N^5-Ethyl-L-glutamine synthetase	$ATP + L\text{-glutamate} + \text{ethylamine} = ADP$ $+ \text{orthophosphate} + N^5\text{-ethyl-L-glutamine}$

6.3.2 ACID—AMINO-ACID LIGASES (PEPTIDE SYNTHETASES)

6.3.2.1	Pantothenate synthetase	$ATP + L\text{-pantoate} + \beta\text{-alanine} = AMP +$ pyrophosphate $+ L\text{-pantothenate}$
6.3.2.2	γ-Glutamylcysteine synthetase	$ATP + L\text{-glutamate} + L\text{-cysteine} = ADP$ $+ \text{orthophosphate} + \gamma\text{-L-glutamyl-L-cysteine}$
6.3.2.3	Glutathione synthetase	$ATP + \gamma\text{-L-glutamyl-L-cysteine} + \text{glycine} =$ $ADP + \text{orthophosphate} + \text{glutathione}$
6.3.2.4	D-Alanylalanine synthetase	$ATP + D\text{-alanine} + D\text{-alanine} = ADP +$ orthophosphate $+ D\text{-alanyl-alanine}$
6.3.2.5	Phosphopantothenoylcysteine synthetase	$CTP + 4'\text{-phospho-L-pantothenate} +$ $L\text{-cysteine} = \text{unidentified products of CTP}$ breakdown $+ 4'\text{-phospho-L-pantothenoyl-L-}$ cysteine
6.3.2.6	Phosphoribosylaminoimidazole-succinocarboxamide synthetase	$ATP + 5'\text{-phosphoribosyl-4-carboxy-5-}$ aminoimidazole $+ L\text{-aspartate} = ADP +$ orthophosphate $+ 5'\text{-phosphoribosyl-4-}(N\text{-}$ succinocarboxamide)-5-aminoimidazole
6.3.2.7	UDP-N-acetylmuramoyl-L-alanyl-D-glutamyl-L-lysine synthetase	$ATP + \text{UDP-}N\text{-acetylmuramoyl-L-alanyl-D-}$ glutamate $+ L\text{-lysine} = ADP +$ orthophosphate $+ \text{UDP-}N\text{-acetylmuramoyl-L-}$ alanyl-D-glutamyl-L-lysine

Number	Other Names	Basis for classification (Systematic Name)	Comments	Reference
6.3.1.1		L-Aspartate:ammonia ligase (AMP-forming)		2729, 3628
6.3.1.2		L-Glutamate:ammonia ligase (ADP-forming)		820, 950, 1863, 2184, 3729
6.3.1.4		L-Aspartate:ammonia ligase (ADP-forming)		2336
6.3.1.5		Deamido-NAD$^+$:ammonia ligase (AMP-forming)	L-Glutamine also acts, more slowly, as amido-donor (*cf* EC 6.3.5.1)	3163
6.3.1.6	Theanine synthetase	L-Glutamate:ethylamine ligase (ADP-forming)		2916, 2915, 2917
6.3.2.1	Pantoate activating enzyme	L-Pantoate:β-alanine ligase (AMP-forming)		1041, 2024, 2025
6.3.2.2		L-Glutamate:L-cysteine γ-ligase (ADP-forming)		2069, 3131
6.3.2.3		γ-L-Glutamyl-L-cysteine:glycine ligase (ADP-forming)		1463, 2395
6.3.2.4		D-Alanine:D-alanine ligase (ADP-forming)		402
6.3.2.5		4'-Phospho-L-pantothenate: L-cysteine ligase	Cysteine can be replaced by some of its derivatives	402
6.3.2.6		5'-Phosphoribosyl-4-carboxy-5-aminoimidazole:L-aspartate ligase (ADP-forming)		2014
6.3.2.7		UDP-*N*-acetylmuramoyl-L-alanyl-D-glutamate:L-lysine ligase (ADP-forming)		1462

Number	Recommended Name	Reaction
6.3.2.8	UDP-*N*-acetylmuramoylalanine synthetase	ATP + UDP-*N*-acetylmuramate + L-alanine = ADP + orthophosphate + UDP-*N*-acetylmuramoyl-L-alanine
6.3.2.9	UDP-*N*-acetylmuramoyl-L-alanyl-D-glutamate synthetase	ATP + UDP-*N*-acetylmuramoyl-L-alanine + D-glutamate = ADP + orthophosphate + UDP-*N*-acetylmuramoyl-L-alanyl-D-glutamate
6.3.2.10	UDP-*N*-acetylmuramoyl-L-alanyl-D-glutamyl-L-lysyl-D-alanyl-D-alanine synthetase	ATP + UDP-*N*-acetylmuramoyl-L-alanyl-D-glutamyl-L-lysine + D-alanyl-D-alanine = ADP + orthophosphate + UDP-*N*-acetylmuramoyl-L-alanyl-D-glutamyl-L-lysyl-D-alanyl-D-alanine
6.3.2.11	Carnosine synthetase	ATP + L-histidine + β-alanine = AMP + pyrophosphate + carnosine
6.3.2.12	Dihydrofolate synthetase	ATP dihydropteroate + L-glutamate = ADP + orthophosphate + dihydrofolate
6.3.2.13	UDP-*N*-acetylmuramoyl-L-alanyl-D-glutamyl-*meso*-2,6-diaminopimelate synthetase	ATP + UDP-*N*-acetylmuramoyl-L-alanyl-D-glutamate + *meso*-2,6-diaminopimelate = ADP + orthophosphate + UDP-*N*-acetylmuramoyl-L-alanyl-D-glutamyl-*meso*-2,6-diaminopimelate
6.3.2.14	2,3-Dihydroxybenzoylserine synthetase	ATP + 2,3-dihydroxybenzoate + L-serine = products of ATP breakdown + 2,3-dihydroxybenzoyl-L-serine
6.3.2.15	UDP-*N*-acetylmuramoyl-L-alanyl-D-glutamyl-*meso*-2,6-diaminopimeloyl-D-alanyl-D-alanine synthetase	ATP + UDP-*N*-acetylmuramoyl-L-alanyl-D-glutamyl-*meso*-2,6-diaminopimelate + D-alanyl-D-alanine = ADP + orthophosphate + UDP-*N*-acetylmuramoyl-L-alanyl-D-glutamyl-*meso*-2,6-diaminopimeloyl-D-alanyl-D-alanine
6.3.2.16	D-Alanyl-alanyl-poly (glycerophosphate) synthetase	ATP + D-alanine + alanyl-poly-(glycerophosphate) = ADP + orthophosphate + D-alanyl-alanyl-poly-(glycerophosphate)

6.3.3 CYCLO-LIGASES

Number	Recommended Name	Reaction
6.3.3.1	Phosphoribosylaminoimidazole synthetase	ATP + 5′-phosphoribosylformylglycinamidine = ADP + orthophosphate + 5′-phosphoribosyl-5-aminoimidazole
6.3.3.2	5,10-Methenyltetrahydrofolate synthetase	ATP + 5-formyltetrahydrofolate = ADP + orthophosphate + 5,10-methenyltetrahydrofolate

Number	Other Names	Basis for classification (Systematic Name)	Comments	Reference
6.3.2.8		UDP-*N*-acetylmuramate: L-alanine ligase (ADP-forming)		1462, 2380
6.3.2.9		UDP-*N*-acetylmuramoyl-L-alanine:D-glutamate ligase (ADP-forming)		1462
6.3.2.10		UDP-*N*-acetylmuramoyl-L-alanyl-D-glutamyl-L-lysine: D-alanyl-D-alanine ligase (ADP-forming)	Involved with enzymes EC 6.3.2.4 and 6.3.2.7-9 in the synthesis of a cell-wall peptide	1463
6.3.2.11		L-Histidine:β-alanine ligase (AMP-forming)		1579, 3206
6.3.2.12		Dihydropteroate:L-glutamate ligase (ADP-forming)		1119
6.3.2.13		UDP-*N*-acetylmuramoyl-L-alanyl-D-glutamate:*meso*-2,6-diaminopimelate ligase (ADP-forming)		2256
6.3.2.14		2,3-Dihydroxybenzoate: L-serine ligase		393
6.3.2.15		UDP-*N*-acetylmuramoyl-L-alanyl-D-glutamyl-*meso*-2,6-diaminopimelate:D-alanyl-D-alanine ligase (ADP-forming)	The enzyme from *Bacillus subtilis* also catalyses the reverse, hydrolytic, reaction in the presence of ADP, orthophosphate and Co^{2+}. No formation of ATP, however, was observed	795, 563
6.3.2.16	D-Alanine:membrane-acceptor ligase	D-Alanine:alanyl-poly-(glycerophosphate) ligase (ADP-forming)	Involved in the synthesis of teichoic acids	2777
6.3.3.1		5'-Phosphoribosylformyl-glycinamidine cyclo-ligase (ADP-forming)		1930
6.3.3.2	5-Formyltetrahydrofolate cyclo-ligase	5-Formyltetrahydrofolate cyclo-ligase (ADP-forming)		1111

Number	Recommended Name	Reaction
6.3.3.3	Dethiobiotin synthetase	ATP + 7,8-diaminononanoate + CO_2 = ADP + orthophosphate + dethiobiotin

6.3.4 OTHER CARBON-NITROGEN LIGASES

6.3.4.1	GMP synthetase	ATP + xanthosine 5′-phosphate + NH_3 = AMP + pyrophosphate + GMP
6.3.4.2	CTP synthetase	ATP + UTP + NH_3 = ADP + orthophosphate + CTP
6.3.4.3	Formyltetrahydrofolate synthetase	ATP + formate + tetrahydrofolate = ADP + orthophosphate + 10-formyltetrahydrofolate
6.3.4.4	Adenylosuccinate synthetase	GTP + IMP + L-aspartate = GDP + orthophosphate + adenylosuccinate
6.3.4.5	Argininosuccinate synthetase	ATP + L-citrulline + L-aspartate = AMP + pyrophosphate + L-argininosuccinate
6.3.4.6	Urea carboxylase (hydrolysing)	ATP + urea + CO_2 = ADP + orthophosphate + 2 NH_3 + 2 CO_2
6.3.4.7	5′-Phosphoribosylamine synthetase	ATP + ribose 5-phosphate + NH_3 = ADP + orthophosphate + 5′-phosphoribosylamine
6.3.4.8	5′-Phosphoribosylimidazoleacetate synthetase	ATP + imidazoleacetate + 5′-phosphoribosyldiphosphate = ADP + orthophosphate + 5′-phosphoribosylimidazoleacetate + pyrophosphate
6.3.4.9	Biotin—[methylmalonyl-CoA-carboxyltransferase] synthetase	ATP + biotin + apo-[methylmalonyl-CoA:pyruvate carboxyltransferase] = AMP + pyrophosphate + [methylmalonyl-CoA:pyruvate carboxyltransferase]
6.3.4.10	Biotin—[propionyl-CoA-carboxylase(ATP-hydrolysing)] synthetase	ATP + biotin + apo-[propionyl-CoA:carbon-dioxide ligase (ADP-forming)] = AMP + pyrophosphate + [propionyl-CoA:carbon-dioxide ligase (ADP-forming)]
6.3.4.11	Biotin—[methylcrotonoyl-CoA-carboxylase] synthetase	ATP + biotin + apo-[3-methylcrotonoyl-CoA:carbon-dioxide ligase (ADP-forming)] = AMP + pyrophosphate + [3-methylcrotonoyl-CoA:carbon-dioxide ligase (ADP-forming)]
6.3.4.12	γ-Glutamylmethylamide synthetase	ATP + L-glutamate + methylamine = ADP + orthophosphate + N^γ-methyl-L-glutamine

Number	Other Names	Basis for classification (Systematic Name)	Comments	Reference
6.3.3.3		7,8-Diaminononanoate:-carbon-dioxide cyclo-ligase (ADP-forming)	CTP has half the activity of ATP	1800, 3773
6.3.4.1		Xanthosine-5′-phosphate:ammonia ligase (AMP-forming)		2304
6.3.4.2		UTP:ammonia ligase (ADP-forming)	Glutamine can replace NH_3	1958, 2001
6.3.4.3		Formate:tetrahydrofolate ligase (ADP-forming)		1482, 2001, 2680, 3674
6.3.4.4	Adenylosuccinate synthase	IMP:L-aspartate ligase (GDP-forming)		643, 1959, 3785
6.3.4.5		L-Citrulline:L-aspartate ligase (AMP-forming)		2725, 2977
6.3.4.6		Urea:carbon-dioxide ligase (ADP-forming) (decarboxylating, deaminating)	A biotinyl-protein	2837
6.3.4.7		Ribose-5-phosphate:ammonia ligase (ADP-forming)		2740
6.3.4.8		Imidazoleacetate:5′-phosphoribosyldiphosphate ligase (ADP and pyrophosphate-forming)		612
6.3.4.9		Biotin:apo-[methylmalonyl-CoA:pyruvate carboxyltransferase] ligase (AMP-forming)		1872
6.3.4.10		Biotin:apo-[propionyl-CoA:carbon-dioxide ligase (ADP-forming)] ligase (AMP-forming)		3066
6.3.4.11		Biotin:apo-[3-methylcrotonoyl-CoA:carbon-dioxide ligase (ADP-forming)] ligase (AMP-forming)		1361
6.3.4.12		L-Glutamate:methylamine ligase (ADP-forming)		1829

Number	Recommended Name	Reaction
6.3.4.13	Phosphoribosylglycinamide synthetase	ATP + 5-phosphoribosylamine + glycine = ADP + orthophosphate + 5′-phosphoribosylglycinamide
6.3.4.14	Biotin carboxylase	ATP + biotin-carboxyl-carrier protein + CO_2 = ADP + orthophosphate + carboxybiotin-carboxyl-carrier protein
6.3.4.15	Biotin-[acetyl-CoA carboxylase]synthetase	ATP + biotin + apo-[acetyl-CoA:carbon-dioxide ligase (ADP-forming)]= AMP + pyrophosphate + [acetyl-CoA:carbon-dioxide ligase (ADP-forming)]
6.3.4.16	Carbamoyl-phosphate synthetase (ammonia)	2 ATP + NH_3 + CO_2 + H_2O = 2 ADP + orthophosphate + carbamoyl phosphate

6.3.5 CARBON-NITROGEN LIGASES WITH GLUTAMINE AS AMIDO-*N*-DONOR

6.3.5.1	NAD$^+$ synthetase (glutamine-hydrolysing)	ATP + deamido-NAD$^+$ + L-glutamine + H_2O = AMP + pyrophosphate + NAD$^+$ + L-glutamate
6.3.5.2	GMP synthetase (glutamine-hydrolysing)	ATP + xanthosine 5′-phosphate + L-glutamine + H_2O = AMP + pyrophosphate + GMP + L-glutamate
6.3.5.3	Phosphoribosylformylglycinamidine synthetase	ATP + 5′-phosphoribosylformylglycinamide + L-glutamine + H_2O = ADP + orthophosphate + 5′-phosphoribosylformylglycinamidine + L-glutamate
6.3.5.4	Asparagine synthetase (glutamine-hydrolysing)	ATP + L-aspartate + L-glutamine = AMP + pyrophosphate + L-asparagine + L-glutamate
6.3.5.5	Carbamoyl-phosphate synthetase (glutamine-hydrolysing)	2 ATP + glutamine + CO_2 + H_2O = 2 ADP + orthophosphate + glutamate + carbamoyl phosphate

6.4 FORMING CARBON-CARBON BONDS

These are the carboxylating enzymes, mostly biotinyl-proteins.

6.4.1.1	Pyruvate carboxylase	ATP + pyruvate + CO_2 + H_2O = ADP + orthophosphate + oxaloacetate

Number	Other Names	Basis for classification (Systematic Name)	Comments	Reference
6.3.4.13	Glycinamide ribonucleotide synthetase	5-Phosphoribosylamine:glycine ligase (ADP-forming)	Formerly EC 6.3.1.3	1077, 1204
6.3.4.14		Biotin-carboxyl-carrier-protein:carbon-dioxide ligase (ADP-forming)		724
6.3.4.15		Biotin:apo-[acetyl-CoA:carbon-dioxide ligase (ADP-forming)] ligase (AMP-forming)		1868
6.3.4.16	Carbamoyl-phosphate synthase (ammonia)	Carbon-dioxide:ammonia ligase (ADP-forming, carbamate-phosphorylating)	Formerly EC 2.7.2.5	1531, 2095, 2096, 854
6.3.5.1		Deamido-NAD$^+$:L-glutamine amido-ligase (AMP-forming)	NH$_3$ can act instead of glutamine	1446, 1447
6.3.5.2		Xanthosine-5'-phosphate: L-glutamine amido-ligase (AMP-forming)		11, 1859
6.3.5.3		5'-Phosphoribosylformyl-glycinamide:L-glutamine amido-ligase (ADP-forming)		2193
6.3.5.4		L-Aspartate:L-glutamine amido-ligase (AMP-forming)		2572
6.3.5.5	Carbamoyl-phosphate synthase (glutamine)	Carbon-dioxide:L-glutamine amido-ligase (ADP-forming, carbamate-phosphorylating)	Formerly EC 2.7.2.9	75, 3790
6.4.1.1	Pyruvic carboxylase	Pyruvate:carbon-dioxide ligase (ADP-forming)	A biotinyl enzyme containing Mn (animal tissues) or Zn (yeast). The animal enzyme requires acetyl-CoA	3014, 3518, 2145, 2990

Number	Recommended Name	Reaction
6.4.1.2	Acetyl-CoA carboxylase	$ATP + acetyl\text{-}CoA + CO_2 + H_2O = ADP + orthophosphate + malonyl\text{-}CoA$
6.4.1.3	Propionyl-CoA carboxylase (ATP-hydrolysing)	$ATP + propionyl\text{-}CoA + CO_2 + H_2O = ADP + orthophosphate + methylmalonyl\text{-}CoA$
6.4.1.4	Methylcrotonoyl-CoA carboxylase	$ATP + 3\text{-}methylcrotonoyl\text{-}CoA + CO_2 + H_2O = ADP + orthophosphate + 3\text{-}methylglutaconyl\text{-}CoA$
6.4.1.5	Geranoyl-CoA carboxylase	$ATP + geranoyl\text{-}CoA + CO_2 + H_2O = ADP + orthophosphate + 3\text{-}isohexenylglutaconyl\text{-}CoA$

6.5 FORMING PHOSPHORIC ESTER BONDS

These are the enzymes restoring broken phosphodiester bonds in nucleic acids (often, improperly, called repair enzymes).

Number	Recommended Name	Reaction
6.5.1.1	Polydeoxyribonucleotide synthetase(ATP)	$ATP + (deoxyribonucleotide)_n + (deoxyribonucleotide)_m = AMP + pyrophosphate + (deoxyribonucleotide)_{n+m}$
6.5.1.2	Polydeoxyribonucleotide synthetase (NAD$^+$)	$NAD^+ + (deoxyribonucleotide)_n + (deoxyribonucleotide)_m = AMP + NMN + (deoxyribonucleotide)_{n+m}$
6.5.1.3	Polyribonucleotide synthetase(ATP)	$ATP + (ribonucleotide)_n + (ribonucleotide)_m = AMP + pyrophosphate + (ribonucleotide)_{n+m}$

Number	Other Names	Basis for classification (Systematic Name)	Comments	Reference
6.4.1.2		Acetyl-CoA:carbon-dioxide ligase (ADP-forming)	A biotinyl-protein. Also catalyses transcarboxylation; the plant enzyme also carboxylates propionyl-CoA and butyryl-CoA	1222, 2121, 2122, 3584
6.4.1.3		Propionyl-CoA:carbon-dioxide ligase (ADP-forming)	A biotinyl-protein. Also carboxylates butyryl-CoA and catalyses transcarboxylation	1627, 1871
6.4.1.4		3-Methylcrotonoyl-CoA:carbon-dioxide ligase (ADP-forming)	A biotinyl-protein	1719, 2020, 2793
6.4.1.5		Geranoyl-CoA:carbon-dioxide ligase (ADP-forming)	A biotinyl-protein. Also carboxylates dimethylacryloyl-CoA and farnesoyl-CoA	3012
6.5.1.1	Polynucleotide ligase, Sealase, DNA repair enzyme, DNA joinase, DNA ligase	Poly(deoxyribonucleotide): poly(deoxyribonucleotide) ligase(AMP-forming)	Catalyses the formation of a phosphodiester at the site of a single-strand break in duplex DNA. RNA can also act as substrate to some extent	239, 3643, 286
6.5.1.2	Polynucleotide ligase (NAD$^+$), DNA repair enzyme, DNA joinase, DNA ligase	Poly(deoxyribonucleotide): poly(deoxyribonucleotide) ligase (AMP-forming, NMN-forming)	Catalyses the formation of a phosphodiester at the site of a single-strand break in duplex DNA. RNA can also act as substrate to some extent	3850
6.5.1.3	RNA ligase	Poly(ribonucleotide): poly(ribonucleotide) ligase (AMP-forming)		

REFERENCES

The EC number follows the reference.

1. Aarts, E.M. & Hinnen-Bouwmans, C. (1972) *Biochim. Biophys. Acta* **268**, 21
2. Abbott, M.T., Kadner, R.J. & Fink, R.M. (1964) *J. Biol. Chem.* **239**, 156 [1.14.11.6]
3. Abdullah, M. & Whelan, W.J. (1960) *Biochem. J.* **75**, 12P [2.4.1.24]
4. Abe, N. & Tsuiki, S. (1974) *Biochim. Biophys. Acta* **350**, 383 [3.1.3.42]
5. Abeles, R.H. & Lee, H.A. Jr. (1961) *J. Biol. Chem.* **236**, 2347 [4.2.1.28]
6. Abiko, Y. (1970) *Methods Enzymol.* **18A**, 358 [2.7.1.24]
7. Abiko, Y., Ashida, S.-I. & Shimizu, M. (1972) *Biochim. Biophys. Acta* **268**, 364 [2.7.1.33]
8. Abita J.P., Delaage M., Lazdunski M. & Savrda J. (1969) *Eur. J. Biochem.* **8**, 314 [3.4.23.6]
9. Ablondi, F.B. & Hagan, J.J. (1960) in *The Enzymes,* 2nd edn (Boyer, P.D., Lardy, H. & Myrbäck, K., ed.) Vol. 4, p. 176, Academic Press, New York [3.4.21.7]
10. Abra, R.M. & Quinn, P.J. (1975) *Biochim. Biophys. Acta* **380**, 436 [3.1.4.38]
11. Abrams, R. & Bentley, M. (1959) *Arch. Biochem. Biophys.* **79**, 91 [6.3.5.2]
12. Adachi, K., Iwayama, Y., Tanioka, H. & Takeda, Y. (1966) *Biochim. Biophys. Acta* **118**, 88 [1.13.11.5]
13. Adachi, K. & Okuyama, T. (1972) *Biochim. Biophys. Acta* **268**, 629 [1.6.99.3]
14. Adachi, K., Takeda, Y., Senoh, S. & Kita, H. (1964) *Biochim. Biophys. Acta* **93**, 483 [1.13.11.15, 1.14.13.3]
15. Adams, E. (1954) *J. Biol. Chem.* **209**, 829 [1.1.1.23]
16. Adams, E. (1955) *J. Biol. Chem.* **217**, 325 [1.1.1.23]
17. Adams, E. & Goldstone, A. (1960) *J. Biol. Chem.* **235**, 3499 [1.5.1.2]
18. Adams, E. & Goldstone, A. (1960) *J. Biol. Chem.* **235**, 3504 [1.5.1.12]
19. Adams, E. & Norton, I.L. (1964) *J. Biol. Chem.* **239**, 1525 [5.1.1.8]
20. Adams, E. & Rosso, G. (1967) *J. Biol. Chem.* **242**, 1803 [1.2.1.26]
21. Adams, J.B. & Edwards, A.M. (1968) *Biochim. Biophys. Acta* **167**, 122 [2.8.2.2]
22. Adams, J.B., Ellyard, R.K. & Law, J. (1974) *Biochim. Biophys. Acta* **370**, 160 [2.8.2.4]
23. Adams, M., Richtmyer, N.K. & Hudson, C.S. (1943) *J. Amer. Chem. Soc.* **65**, 1369 [3.2.1.25, 3.2.1.7]
24. Adler, E., Günther, G. & Everett, J.E. (1938) *Hoppe-Seyler's Z. Physiol. Chem.* **225**, 27 [1.4.1.4]
25. Adler, E., Hellström, V., Günther, G. & Euler, H. von. (1938) *Hoppe-Seyler's Z. Physiol. Chem.* **225**, 14 [1.4.1.4]
26. Agarwal, R.P. & Parks, R.E. (1969) *J. Biol. Chem.* **244**, 644 [2.4.2.1]
27. Agosin, M. & Weinbach, E.C. (1956) *Biochim. Biophys. Acta* **21**, 117 [1.1.1.42]
28. Agranoff, B.W. & Brady, R.O. (1956) *J. Biol. Chem.* **219**, 221 [2.7.1.15]
29. Agranoff, B.W., Eggerer, H., Henning, U. & Lynen, F. (1960) *J. Biol. Chem.* **235**, 326 [5.3.3.2]
30. Aiyappa P.S. & Harris J.O. (1976) *Mol. Cell. Biochem.* **13**, 131 [3.4.24.4]
31. Akamatsu, Y. & Law, J.H. (1970) *J. Biol. Chem.* **245**, 701 [2.1.1.16]
32. Akamatsu, Y. & Law, J.H. (1970) *J. Biol. Chem.* **245**, 709 [2.1.1.15]
33. Åkesson, Ã, Ehrenberg, A. & Theorell, H. (1963) in *The Enzymes,* 2nd edn (Boyer, P.D., Lardy, H. & Myrbäck, K., ed.) Vol. 7, p. 477, Academic Press, New York [1.6.99.1]
34. Akhtar, M. & El-Obeid, H.A. (1972) *Biochim. Biophys. Acta* **258**, 791 [2.1.2.1]
35. Aki, K., Ogawa, K. & Ichihara, A. (1968) *Biochim. Biophys. Acta* **159**, 276 [2.6.1.42, 2.6.1.6]
36. Aki, K., Yokojima, A. & Ichihara, A. (1969) *J. Biochem. (Tokyo)* **65**, 539 [2.6.1.42]
37. Akopyan, T.N., Braunstein, A.E. & Goryachenkova, E.V. (1975) *Proc. Nat. Acad. Sci. USA* **72**, 1617 [4.4.1.9]
38. Albers, R.W. & Koval, G.J. (1961) *Biochim. Biophys. Acta* **52**, 29 [1.2.1.24]
39. Albersheim, P. & Killias, U. (1962) *Arch. Biochem. Biophys.* **97**, 107 [4.2.2.2]
40. Albersheim, P., Neukom, H. & Deuel, H. (1960) *Helv. Chim. Acta* **43**, 1422 [4.2.2.10]
41. Alberts, A.W., Majerus, P.W. & Vagelos, P.R. (1969) *Methods Enzymol.* **14**, 50 [2.3.1.39, 2.3.1.41]
42. Alberty, R.A. (1961) in *The Enzymes,* 2nd edn (Boyer, P.D., Lardy, H. & Myrbäck, K.,ed.) Vol. 5, p. 531, Academic Press, New York [4.2.1.2]
43. Albrecht, A. & Vogel, H.J. (1964) *J. Biol. Chem.* **239**, 1872 [2.6.1.11]
44. Albrecht, G.J. (1970) *Biochemistry* **9**, 2462 [2.7.4.10]

45. Albrecht, G.J., Bass, S.T., Seifert, L.L. & Hansen, R.G. (1966) *J. Biol. Chem.* **241**, 2968 [2.7.7.9]
46. Albrecht, G.J. & Kauss, H. (1971) *Phytochemistry* **10**, 1293 [2.4.1.97]
47. Aldridge, W.N. (1953) *Biochem. J.* **53**, 110 [3.1.1.2, 3.1.1.6]
48. Alexander, J.K. (1968) *J. Biol. Chem.* **243**, 2899 [2.4.1.20]
49. Alexander, M., Heppel, L.A. & Hurwitz, J. (1961) *J. Biol. Chem.* **236**, 3014
50. Algranati, I.D. & Cabib, E. (1960) *Biochim. Biophys. Acta* **43**, 141 [2.4.1.11, 3.4.21.4]
51. Ali, S.Y. & Evans, L. (1968) *Biochem. J.* **107**, 293
52. Alivasatos, S.G.A. & Woolley, D.W. (1956) *J. Biol. Chem.* **219**, 823 [3.2.2.6]
53. Allaudeen, H.S. & Ramakrishnan, T. (1968) *Arch. Biochem. Biophys.* **125**, 199 [5.4.99.3]
54. Allen, A.K. & Rosenberg, H. (1968) *Biochim. Biophys. Acta* **151**, 504 [2.7.8.4]
55. Allen, C.M. Alworth, W., MacRae, A. & Bloch, K. (1967) *J. Biol. Chem.* **242**, 1895 [2.5.1.11]
56. Allen, E.H., Glassman, E. & Schweet, R.S. (1960) *J. Biol. Chem.* **235**, 1061 [6.1.1.1, 6.1.1.3, 6.1.1.4, 6.1.1.5, 6.1.1.6]
57. Allen L.C., Ph. D Thesis, Univ. of Ottawa, 1973 [3.4.24.4]
58. Allen, R.H. & Jakoby, W.B. (1969) *J. Biol. Chem.* **244**, 2078 [4.2.1.37]
59. Allen, S.H.G. (1966) *J. Biol. Chem.* **241**, 5266 [1.1.99.7]
60. Allen, S.H.G. & Patil, J.R. (1972) *J.Biol. Chem.* **247**, 909 [1.1.99.7]
61. Allende, C.C. & Allende, J.E. (1964) *J. Biol. Chem.* **239**, 1102 [6.1.1.19]
62. Altenbern, R.A. & Housewright, R.D. (1953) *J. Biol. Chem.* **204**, 159 [2.6.1.12]
63. Altschul, A.M., Abrams, R. & Hogness, T.R. (1940) *J. Biol. Chem.* **136**, 777 [1.11.1.5]
64. Amagase S. (1972) *J. Biochem. (Tokyo)* **72**, 73 [3.4.23.12]
65. Amagase, S., Nakayama, S. & Tsugita, A. (1969) *J. Biochem. (Tokyo)* **66**, 431
66. Ambe, L. & Sohonie, K. (1963) *Enzymologia* **26**, 98 [4.1.1.15]
67. Ames, B.N. (1957) *J. Biol. Chem.* **226**, 583 [3.1.3.15]
68. Ames, B.N. (1957) *J. Biol. Chem.* **228**, 131 [4.2.1.19]
69. Ames, B.N. & Horecker, B.L. (1956) *J. Biol. Chem.* **220**, 113 [2.6.1.9]
70. Ames, B.N., Martin, R.G. & Garry, B.J. (1961) *J. Biol. Chem.* **236**, 2019 [2.4.2.17]
71. Amos, H. (1954) *J. Amer. Chem. Soc.* **76**, 3858 [5.1.1.6]
72. Anai, M., Hirahashi, T. & Takagi, Y. (1970) *J. Biol. Chem.* **245**, 767
73. Anchors, J.M. & Karnovsky, M.L. (1975) *J. Biol. Chem.* **250**, 6048 [3.1.3.9]
74. Anderson, B.M. & Lang, C.A. (1966) *Biochem. J.* **101**, 392 [3.6.1.22]
75. Anderson, P.M. & Meister, A. (1965) *Biochemistry* **4**, 2803 [6.3.5.5]
76. Anderson, R.L. & Allison, D.P. (1965) *J. Biol. Chem.* **240**, 2367 [5.3.1.15]
77. Anderson, R.L. & Kamel, M.Y. (1966) *Methods Enzymol.* **9**, 392 [2.7.1.61]
78. Anderson, R.L. & Wood, W.A. (1962) *J. Biol. Chem.* **237**, 1029 [2.7.1.53]
79. Anderson, W.A. & Magasanik, B. (1971) *J. Biol. Chem.* **246**, 5662 [2.7.1.92, 4.1.2.29]
80. Ando, T. (1966) *Biochim. Biophys. Acta* **114**, 158 [3.1.30.1]
81. Andreesen, J.R. & Gottschalk, G. (1969) *Arch. Mikrobiol.* **69**, 160 [4.2.1.39]
82. Andreesen, J.R. & Ljungdahl, L.G. (1974) *J. Bacteriol.* **120**, 6 [1.2.1.43]
83. Andreoli, A.J., Okita, T.W., Bloom, R. & Grover, T.A. (1972 *Biochem. Biophys. Res. Commun.* **49**, 264 [3.2.2.14]
84. Anfinsen, C.B. & White, F.H. Jr. (1961) in *The Enzymes*, 2nd edn (Boyer, P.D., Lardy, H. & Myrbäck, K., ed.) Vol. 5, p. 95, Academic Press, New York [3.1.27.5]
85. Anfinsen, C., Cuatrecasas, P. & Taniuchi, H. (1971) in *The Enzymes*. 3rd edn (Boyer, P.D. ed.) Vol. 4, p177, Academic Press, New York. [3.1.31.1]
86. Anggard, E. & Samuelsson, B. (1966) *Ark. Kem.* **25**, 293 [1.1.1.141]
87. Angus C.W., Lee H.J. & Wilson I.B. (1973) *Biochim. Biophys. Acta* **309**, 169 [3.4.15.1]
88. Ankel, H., Ankel, E., Schutzbach, J. & Garancis, J.C. (1970) *J. Biol. Chem.* **245**, 3945 [2.4.1.48]
89. Ankel, H. & Feingold, D.S. (1965) *Biochemistry* **4**, 2468 [4.1.1.35]
90. Annett, R.G. & Kosicki, G.W. (1969) *J. Biol. Chem.* **244**, 2059 [5.3.2.2]
91. Anraku, Y. (1964) *J. Biol. Chem.* **239**, 3412 & 3420 [3.1.4.16]
92. Anthony, C. & Zatman, L.J. (1967) *Biochem. J.* **104**, 953 [1.1.99.8]
93. Antia, M., Hoare, D.S. & Work, E. (1957) *Biochem. J.* **65**, 448 [5.1.1.7]
94. Aposhian, H.V. & Tremblay, G.Y. (1966) *J. Biol. Chem.* **241**, 5095 [3.1.3.35]
95. Appleby, C.A. & Morton, R.K. (1959) *Biochem. J.* **71**, 492 [1.1.2.3]

96.   Appleby, C.A. & Morton, R.K. (1959) *Biochem. J.* **73**, 539   [1.1.2.3]
97.   Arai T.& Ichishima E. (1974) *J. Biochem. (Tokyo)* **76**, 765   [3.4.16.1]
98.   Araki, Y. & Ito, E. (1974) *Biochem. Biophys. Res. Commun.* **56**, 669   [3.5.1.41]
99.   Archer, B.L. & Cockbain, E.G. (1969) *Methods Enzymol.* **15**, 476   [2.5.1.20]
100.   Arcus, A.C. & Edson, N.L. (1956) *Biochem. J.* **64**, 385   [1.1.2.2.]
101.   Ardelt W. (1975) *Biochim. Biophys. Acta* **393**, 267   [3.4.21.11]
102.   Ardelt W. (1974) *Biochim. Biophys. Acta* **341**, 318   [3.4.21.11]
103.   Arfin, S.M. & Umbarger, H.E. (1969) *J. Biol. Chem.* **244**, 1118   [1.1.1.86]
104.   Arima, K., Iwasaki, S. & Tamura, G. (1967) *Agr. Biol. Chem. (Tokyo)* **31**, 540
105.   Arima, K., Yu, J. & Iwasaki, S. (1970) *Methods Enzymol.* **19**, 447
106.   Arima, K., Yu J. & Iwasaki S. (1970) *Methods Enzymol.* **19, 446**   [3.4.23.6]
107.   Armentrout, R.W. & Doolittle, R.F. (1969) *Arch. Biochem. Biophys.* **132**, 80   [3.4.11.8]
108.   Aronson, J.N. & Lugay, J.C. (1969) *Biochem. Biophys. Res. Commun.* **34**, 311
       [3.5.1.31]
109.   Arrand, J.R., Myers, P.A. & Roberts, R.J. (1978) *J. Mol. Biol.* **118**, 127   [3.1.23.38]
110.   Arrand, J.R., Myers. P.A. & Roberts, R.J., unpublished results.
       [3.1.23.34, 3.1.23.35, 3.1.23.36, 3.1.23.37]
111.   Arsenis, C. & Touster, O. (1968) *J. Biol. Chem.* **243**, 5702   [3.1.3.31]
112.   Arvidson S. (1973) *Biochim. Biophys. Acta* **302**, 149   [3.4.24.4]
113.   Arvidson S. & Holme T. (1973) *Biochim. Biophys. Acta* **302**, 135   [3.4.24.4]
114.   Arvidson S., Holme T. & Lindholm B. (1973) *Biochim. Biophys. Acta* **302**, 135
       [3.4.21.19, 3.4.22.13]
115.   Asada, K. (1967) *J. Biol. Chem.* **242**, 3646   [1.8.99.1]
116.   Asada, K., Tamura, G. & Bandurski, R.S. (1969) *J. Biol. Chem.* **244**, 4904   [1.8.99.1]
117.   Asakawa, T., Wada, H. & Yamano, T. (1968) *Biochim. Biophys. Acta* **170**, 375
       [4.1.1.43]
118.   Asenjo, C.F. & Capella de Fernandez, M. (1942) *Science* **95**, 48   [3.4.99.18]
119.   Asensio, C. & Ruiz-Amil, M. (1966) *Methods Enzymol.* **9**, 421   [2.7.1.59]
120.   Ashwell, G. & Hickman, J. (1957) *J. Biol. Chem.* **226**, 65   [5.1.3.1]
121.   Ashwell, G., Wahba, A.J. & Hickman, J. (1958) *Biochim. Biophys. Acta* **30**, 186
       [4.2.1.8]
122.   Ashwell, G., Wahba, A.J. & Hickman, J. (1960) *J. Biol. Chem.* **235**, 1559   [5.3.1.12]
123.   Asnis, R.E. & Brodie, A.F. (1953) *J. Biol. Chem.* **203**, 153   [1.1.1.6]
124.   Aspen, A.J. & Jakoby, W.B. (1964) *J. Biol. Chem.* **239**, 710   [1.1.1.129]
125.   Astrachan, L. (1973) Biochim. Biophys. Acta **296**, 79   [3.1.4.4]
126.   Aswanikumar S. & Radhakrishnan A.N. (1972) *Biochim. Biophys. Acta* **276**, 241
       [3.4.99.31]
127.   Aswanikumar S. & Radhakrisnan A.N. (1975) *Biochim. Biophys. Acta* **384**,194
       [3.4.99.31]
128.   Atkinson, M.R., Jackson, J.F. & Morton, R.K. (1961) *Biochem. J.* **80**, 318   [2.7.7.1]
129.   Attias, J. & Bonnet, J.L. (1972) *Biochim. Biophys. Acta* **268**, 422   [3.1.3.41]
130.   Attias, J. & Durand, H. (1973) *Biochim. Biophys. Acta* **321**, 561   [3.1.3.41]
131.   Attwood, M.A. & Doughty, C.C. (1974) *Biochim. Biophys. Acta* **370**, 358   [1.1.1.21]
132.   August, J.T., Ortiz, P.J. & Hurwitz, J. (1962) *J. Biol. Chem.* **237**, 3786   [2.7.7.19]
133.   Augusteyn, R.C., de Jersey, J., Webb, E.C. & Zerner, B. (1969) *Biochim. Biophys. Acta*
       171, 128   [3.1.1.1]
134.   Augustinsson, K.-B. (1948) *Acta Physiol. Scand.* **15**, Suppl. 52   [3.1.1.7, 3.1.1.8]
135.   Augustinsson, K.-B. & Heimbürger, G. (1954) *Acta Chem. Scand.* **8**, 753, 762 & 1533
       [3.8.2.1]
136.   Augustinsson, K.-B. & Olsson, B. (1959) *Biochem. J.* **71**, 477   [3.1.1.2, 3.1.1.8]
137.   Aurbach, G.D. & Jakoby, W.B. (1962) *J. Biol. Chem.* **237**, 565   [1.8.3.2]
138.   Aurich, H. (1961) *Hoppe-Seyler's Z. Physiol. Chem.* **326**, 25   [2.6.1.19]
139.   Aurich, H., Kleber, H.-P., Sorger, H. & Tauchert, H. (1968) *Eur. J. Biochem.* **6**, 196
       [1.1.1.108]
140.   Avigad, G., Alroy, Y. & England, S. (1968) *J. Biol. Chem.* **243**, 1936   [1.1.1.119]
141.   Avigad, G., Amaral, D., Asensio, C. & Horecker, B.L. (1962) *J. Biol. Chem.* **237**, 2736
       [1.1.3.9]
142.   Avigad, G. & Bauer, S. (1966) *Methods Enzymol.* **8**, 621   [3.2.1.65]
143.   Avigad, G., England, S. & Pifco, S. (1966) *J. Biol. Chem.* **241**, 373   [1.1.1.124]
144.   Avigad, G. & Milner, Y. (1966) *Methods Enzymol.* **8**, 341   [2.4.1.13]

145. Avigad, G. & Zelikson, R. (1963) *Bull. Res. Counc. (Israel)* **11**, 253   [3.2.1.64]
146. Avis, P.G., Bergel, F. & Bray, R.C. (1955) *J. Chem. Soc.* 1100   [1.2.3.2]
147. Avron, M., & Jagendorf, A.T. (1957) *Arch. Biochem. Biophys.* **72**, 17   [1.6.99.1]
148. Axcell, B.C. & Geary, P.J. (1973) *Biochem. J.* **136**, 927   [1.3.1.19]
149. Axelrod, B. & Bandurski, R.S. (1953) *J. Biol. Chem.* **204**, 939   [2.7.2.3]
150. Axelrod, B., Saltman, P., Bandurski, R.S. & Baker, R.S. (1952) *J. Biol. Chem.* **197**, 89   [2.7.1.11]
151. Axelrod, J. (1962) *J. Biol. Chem.* **237**, 1656   [2.1.1.28]
152. Axelrod, J. & Daly, J. (1968) *Biochim. Biophys. Acta* **159**, 472   [2.1.1.25]
153. Axelrod, J., Inscoe, J.K. & Tomkins, G.M. (1958) *J. Biol. Chem.* **232**, 835   [2.4.1.17]
154. Axelrod, J. & Tomchick, R. (1958) *J. Biol. Chem.* **233**, 702   [2.1.1.6]
155. Axelrod, J. & Weissbach, H. (1961) *J. Biol. Chem.* **236**, 211   [2.1.1.4]
156. Ayengar, P.K., Hayaishi, O., Nakajima, M. & Tomida, I. (1959) *Biochim. Biophys. Acta* **33**, 111   [1.3.1.20]
157. Ayers, W.A. (1959) *J. Biol. Chem.* **234**, 2819   [2.4.1.20]
158. Azoulay, E., Mutaftshiev, S. & de Sousa, M.L. (1971) *Biochim. Biophys. Acta* **237**, 559   [1.97.1.1]
159. Bacchawat, B.K., Robinson, W.G. & Coon, M.J. (1956) *J. Biol. Chem.* **219**, 539   [1.3.99.10]
160. Bach, S.J., Dixon, M. & Zerfas, L.G. (1946) *Biochem. J.* **40**, 229   [1.1.2.3]
161. Bach, S.J. & Killip, J.D. (1961) *Biochim. Biophys. Acta* **47**, 336   [3.5.3.1, 4.1.1.5]
162. Bachhawat, B.K., Robinson, W.G. & Coon, M.J. (1955) *J. Biol. Chem.* **216**, 727   [4.1.3.4]
163. Baddiley, J. & Neuhaus, F.C. (1960) *Biochem. J.* **75**, 579   [6.1.1.13]
164. Baggio, B. & Moret, V. (1973) *Biochim. Biophys. Acta* **315**, 347   [2.7.1.37]
165. Baggio, B., Pinna, L.A., Moret, V. & Siliprandi, N. (1970) *Biochim. Biophys. Acta* **212**, 515   [2.7.1.70]
166. Baginsky, M.L. & Rodwell, V.W. (1967) *J. Bacteriol.* **94**, 1034   [1.5.99.3]
167. Baguley, B.C. & Staehelin, M. (1968) *Biochemistry* **7**, 45   [2.1.1.32, 2.1.1.36]
168. Bahl, O.P. (1970) *J. Biol. Chem.* **245**, 299   [3.2.1.63]
169. Baich, A. (1969) *Biochim. Biophys. Acta* **192**, 462   [2.7.2.11]
170. Baich, A. (1971) *Biochim. Biophys. Acta* **244**, 129   [1.2.1.41]
171. Baich, A. & Vogel, H.J. (1962) *Biochem. Biophys. Res. Commun.* **7**, 491   [1.2.1.38, 2.7.2.8]
172. Baich, A., Wolfe, R.G. & Reithel, F.J. (1960) *J. Biol. Chem.* **235**, 3130   [5.3.1.9]
173. Bailey, G.B. & Dempsey, W.B. (1967) *Biochemistry* **6**, 1526   [4.1.1.64]
174. Bailey, K. & Webb, E.C. (1944) *Biochem. J.* **38**, 394   [3.6.1.1]
175. Bailey, K. & Webb, E.C. (1948) *Biochem. J.* **42**, 60   [2.7.1.1]
176. Bailey, R.W. (1959) *Biochem. J.* **72**, 42   [2.4.1.5]
177. Bailey, R.W., Barker, S.A., Bourne, E.J. & Stacey, M. (1957) *J. Chem. Soc.* 3530   [2.4.1.5]
178. Bailey, R.W. & Clarke, R.T.J. (1959) *Biochem. J.* **72**, 49   [3.2.1.11]
179. Bailey, R.W. & Hassid, W.Z. (1966) *Proc. Nat. Acad. Sci. USA* **56**, 1586   [2.4.2.24]
180. Bailey-Wood, R., Dodgson, K.S. & Rose, F.A. (1969) *Biochem. J.* **112**, 257   [3.6.2.1]
181. Bak, T.-G. (1967) *Biochim. Biophys. Acta* **139**, 277   [1.1.99.10]
182. Baker, J.J., Jeng, I. & Barker, H.A. (1972) *J. Biol. Chem* **247**, 7724   [1.4.1.11]
183. Baker, T. & Crawford, I.P. (1966) *J. Biol. Chem.* **241**, 5577   [4.1.3.27]
184. Balasubramanian, A.S. & Bachhawat, B.K. (1962) *Biochim. Biophys. Acta* **59**, 389   [3.6.2.2]
185. Baldwin, R.L., Wood, W.A. & Emery, R.S. (1965) *Biochim. Biophys. Acta* **97**, 202   [4.2.1.54]
186. Balinsky, D. & Davies, D.D. (1961) *Biochem. J.* **80**, 292   [1.1.1.25]
187. Balish, E. & Shapiro, S.K. (1967) *Arch. Biochem. Biophys.* **119**, 62   [2.1.1.10]
188. Balke E. & Scharmann W. (1974) Hoppe-Seyler's *Z. Physiol. Chem.* **355**, 958   [3.4.21.14]
189. Ballal, N.R., Bhattacharyya, P.K. & Rangachari, P.N. (1966) *Biochem. Biophys. Res. Commun.* **23**, 473   [1.1.1.144]
190. Balls, A.K., Lineweaver, H. & Thompson, R.R. (1937) *Science* **86**, 379   [3.4.22.2]
191. Balls, A.K., Walden, M.K. & Thompson, R.R. (1948) *J. Biol. Chem.* **173**, 9   [3.2.1.2]

446

192. Banauch, D., Brümmer, W., Ebeling, W., Metz, H., Rindfrey, H., Lang, H., Leybold, K. & Rick, W. (1975) *Z. Klin. Chem. Klin. Biochem.* **13**, 101 [1.1.1.47]
193. Bandurski, R.S., Wilson, L.G. & Squires, C.L. (1956) *J. Amer. Chem. Soc.* **78**, 6408 [2.7.1.25, 2.7.7.4]
194. Banerjee, S. & Ghosh, S. (1969) *Eur. J. Biochem.* **8**, 200 [2.7.1.60]
195. Banga, J. (1952) *Acta Physiol. Acad. Sci. (Hungary)* **3**, 317 [3.4.21.11]
196. Bankel, L., Holme, E., Lindstedt, G., & Lindstedt, S. (1972) *FEBS Lett.* **21**, 135 [1.14.11.6, 3.4.21.4]
197. Bankel, L., Lindstedt, G. & Lindstedt, S. (1972) *J. Biol. Chem* **247**, 6128 [1.14.11.3]
198. Banks, B.E.C. & Vernon, C.A. (1961) *J. Chem. Soc.* 1698 [2.6.1.1]
199. Baranowski, T. (1949) *J. Biol. Chem.* **180**, 535 [1.1.1.8]
200. Baranowski, T. (1963) in *The Enzymes*, 2nd edn (Boyer, P.D., Lardy, H. & Myrbäck, K. ed.) Vol. 7, p. 85, Academic Press, New York [1.1.1.8]
201. Barban, S. (1954) *J. Bacteriol*, **68**, 493 [1.4.1.2]
202. Barker, D.L. & Jencks, W.P. (1969) *Biochemistry* **8**, 3879 [3.1.1.1]
203. Barker, H.A. (1967) *Arch. Microbiol.* **59**, 4 [4.1.3.22]
204. Barker, H.A., Rooze, V., Suzuki, F. & Iodice, A.A. (1964) *J. Biol. Chem.* **239**, 3260 [5.4.99.1]
205. Barker, H.A., Smyth, R.O., Wawszkiewicz, E.J., Lee, M.N. & Wilson, R.M. (1958) *Arch. Biochem. Biophys.* **78**, 468 [4.3.1.2]
206. Barker, S.A. (1965) *Carbohyd. Res.* **1**, 106 [3.2.1.43]
207. Barker, S.A., Bourne, E. & Peat, S. (1949) *J. Chem. Soc.* 1705 [2.4.1.18]
208. Barker, S.A. & Carrington, T.R. (1953) *J. Chem. Soc.* 3588 [2.4.1.24]
209. Barkulis, S.S. (1966) *Methods Enzymol.* **9**, 415 [2.7.1.59]
210. Barnes, E.M. Jr. & Wakil, S.J. (1968) *J. Biol. Chem.* **243**, 2955 [3.1.2.2]
211. Barnholz, Y., Roitman, A. & Gatt, S. (1966) *J. Biol. Chem.* **241**, 3731 [3.1.4.12]
212. Barras, D.R., Moore, A.E. & Stone, B.A. (1969) *Advan. Chem. Ser.* **95**, 105 [3.2.1.73, 3.2.1.74]
213. Barras, D.R. & Stone, B.A. (1969) *Biochim. Biophys. Acta* **191**, 329 [3.2.1.58, 3.2.1.6]
214. Barras, D.R. & Stone, B.A. (1969) *Biochim. Biophys. Acta* **191**, 342 [3.2.1.58, 3.2.1.6]
215. Barrett D. (1968) *Amer. Zool.* **8**, 816 [3.4.24.12]
216. Barrett D., Edwards B.F., Wood D.O. & Lane D.J. (1971) *Arch. Biochem. Biophys.* **143**, 261 [3.4.24.12]
217. Bartholomew, B.A. & Perry, A.L. (1973) *Biochim. Biophys. Acta* **315**, 123 [3.2.1.25]
218. Bartsch, H., Dworkin, C., Miller, E.C. & Miller, J.A. (1973 *Biochim. Biophys. Acta* **304**, 42 [2.3.1.56]
219. Baskin, F.K. & Kitabchi, A.E. (1973) *Eur. J. Biochem.* **37**, 48
220. Baslow, M.H. (1966) *Brain Res.* **3**, 210 [2.3.1.33]
221. Baslow, M.H. & Lenney, J.F. (1967) *Can. J. Biochem.* **45**, 337 [3.5.1.34]
222. Basu, D.K. & Bachhawat, B.K. (1961) *Biochim. Biophys. Acta* **50**, 123 [2.4.1.11]
223. Basu, M. & Basu, S. (1972) *J. Biol. Chem.* **247**, 1489 [2.4.1.86]
224. Basu, M. & Basu, S. (1973) *J. Biol. Chem.* **248**, 170 [2.4.1.87]
225. Basu, S., Basu, M. & Chien, J.L. (1975) *J. Biol. Chem.* **250**, 2956 [2.4.1.89]
226. Basu, S., Kaufman, B. & Roseman, S. (1973) *J. Biol. Chem.* **248**, 1388 [2.4.1.80]
227. Battle, A.M., Benson, A. & Rimington, C. (1965) *Biochem. J.* **97**, 731 [1.3.3.3]
228. Bauerle, R.H., Freundlich, M., Størmer, F.C. & Umbarger, H.E. (1964) *Biochim. Biophys. Acta* **92**, 142 [4.1.3.18]
229. Baugh C.M. & Krumdieck C.L. (1971) *Ann. N.Y. Acad. Sci.* **186**, 7 [3.4.22.12]
230. Baugh, C.M., Stevens, J.C. & Krumdieck, C.L. (1970) *Biochim Biophys. Acta* **212**, 116
231. Baum, H. & Gilbert, G.A. (1953) *Nature (London)* **171**, 983 [2.4.1.1, 2.4.1.18]
232. Baumann, P. (1969) *Biochemistry* **8**, 5011 [2.7.1.2]
233. Bauns, F.H. & Fiedler, L. (1958) *Nature (London)* **181**, 1553 [4.1.2.26]
234. Bean, R.C. & Hassid, W.Z. (1956) *J. Biol. Chem.* **218**, 425 [1.1.3.5]
235. Bean, R.C., Porter, G.G. & Steinberg, B.M. (1961) *J. Biol. Chem.* **236**, 1235 [1.1.3.5]
236. Beard, J.R. & Razzel, W.E. (1964) *J. Biol. Chem.* **239**, 4186 [3.1.27.5]
237. Beck, W.S. & Ochoa, S. (1958) *J. Biol. Chem.* **232**, 931 [5.4.99.2]
238. Becker, A. & Hurwitz, J. (1967) *J. Biol. Chem.* **242**, 936 [3.1.3.32, 3.1.3.33, 3.1.3.34]
239. Becker, A., Lyn, G., Gefter, M. & Hurwitz, J. (1967) *Proc. Nat. Acad. Sci. U.S.* **58**, 1996 [6.5.1.1]

240. Becker, M.A., Kredich, N.M. & Tomkins, G.M. (1969) *J. Biol. Chem.* **244**, 2418 [4.2.99.8]
241. Becker, W., Benthin, U., Eschenhof, E. & Pfeil, E. (1963) *Biochem. Z.* **337**, 156 [4.1.2.10]
242. Becker, W. & Pfeil, E. (1964) *Naturwiss.* **51**, 193 [4.1.2.10]
243. Beevers, H. & French, R.C. (1954) *Arch. Biochem. Biophys.* **50**, 427 [1.7.3.2]
244. Behal F.J., Asserson B., Dawson F. & Hardman J. (1965) *Arch. Biochem. Biophys.* **111**, 335 [3.4.11.14]
245. Behrman, E.J. & Stanier, R.Y. (1957) *J. Biol. Chem.* **228**, 923 [1.13.11.9, 5.2.1.1]
246. Beinert, H. (1963) in *The Enzymes* 2nd edn (Boyer, P.D., Lardy, H. & Myrbäck, K., ed.) Vol. 7, p. 447, Academic Press, New York [1.3.99.2, 1.3.99.3]
247. Beinert, H., Bock, R.M., Goldman, D.S., Green, D.E., Mahler, H.R., Mii, S., Stansly, P.G. & Wakil, S.J. (1953) *J. Amer. Chem. Soc.* **75**, 4111 [2.3.1.16]
248. Belew M. & Porath H.C., (1970) Methods Enzymol. **19**, 576 [3.4.99.16]
249. Belew, M. & Porath, J. (1968) *Eur. J. Biochem.* **6**, 425 [3.4.99.16]
250. Bell, S.C. & Turner, J.M. (1973) Biochem. Soc. Trans. **1**, 678 [2.4.1.66, 4.1.2.5]
251. Bello, L.J. & Bessman, M.J. (1963) *J. Biol. Chem.* **238**, 1777 [2.7.4.12]
252. Belocopitow, E. & Maréchal, L.R. (1970) *Biochim. Biophys. Acta* **198**, 151 [2.4.1.64, 2.4.1.66]
253. Ben-Amotz, A. & Avron, M. (1973) *FEBS Lett.* **29**, 153 [1.1.1.156]
254. Ben-Gershom, E. (1955) *Nature (London)* **175**, 593
255. Benedict, C.R., Kett, J. & Porter, J.W. (1965) *Arch. Biochem. Biophys.* **110**, 611 [2.5.1.10]
256. Bentley, R. (1963) in *The Enzymes*, 2nd edn (Boyer, P.D., Lardy, H. & Myrbäck, K. ed.) Vol. 7, p. 567, Academic Press, New York [1.1.3.4]
257. Bentley, R. & Bhate, D.S. (1960) *J. Biol. Chem.* **235**, 1219 and 1225 [5.1.3.3]
258. Bentley, R. & Thiessen, C.P. (1957) *J. Biol. Chem.* **226**, 703 [4.1.1.6]
259. Bentley, R. & Thiessen, C.P. (1963) *J. Biol. Chem.* **238**, 3811 [4.1.1.60]
260. Benveniste, R. & Davies, J. (1971) *Biochemistry* **10**, 1787 [2.3.1.55]
261. Benveniste, R. & Davies, J. (1971) *FEBS Lett.* **14**, 293 [2.7.7.46]
262. Benveniste, R. & Davies, J. (1973) *Proc. Nat. Acad. Sci. USA* **70**, 2276 [2.3.1.59]
263. Berg, P., Bergmann, F.H., Ofengand, E.J. & Dieckmann, M. (1961) *J. Biol. Chem.* **236**, 1726 [6.1.1.4, 6.1.1.5, 6.1.1.9]
264. Berg, P. & Joklik, W.K. (1954) *J. Biol. Chem.* **210**, 657 [2.7.4.6]
265. Berger, L., Slein, M.W., Colowick, S.P. & Cori, C.F. (1946) *J. Gen. Physiol.* **29**, 379 [2.7.1.1]
266. Berghem, L.E.R. & Pettersson, L.G (1973) *Eur. J. Biochem.* **37**, 21 [3.2.1.91]
267. Bergmann, F.H., Berg, P. & Dieckmann, M. (1961) *J. Biol. Chem.* **236**, 1735 [6.1.1.10, 6.1.1.4, 6.1.1.5, 6.1.1.9]
268. Bergmann, F. & Rimon, S. (1960) *Biochem. J.* **77**, 209 [3.1.1.6]
269. Bergmann, F., Rimon, S. & Segal, R. (1958) *Biochem. J.* **68**, 493 [3.1.1.7]
270. Bergmeyer, H.-U., Gawehn, K., Klotzsch, H., Krebs, H.A. & Williamson, D.H. (1967) *Biochem. J.* **102**, 423 [1.1.1.30]
271. Bergmeyer, H.-U., Holz, G., Kauder, E.M., Möllering, H. & Wieland, O. (1961) *Biochem. Z.* **333**, 471 [2.7.1.30]
272. Bergmeyer, H.-U., Holz, G., Klotzsch, H. & Lang, G. (1963) *Biochem. Z.* **338**, 114 [2.3.1.8]
273. Bergwist R. (1963) *Acta Chem. Scand.* **17**, 1541 [3.4.23.6]
274. Berk, R.A. & Prockop, D.J. (1973) *J. Biol. Chem.* **248**, 1175 [1.14.11.2]
275. Berman, K.M. & Cohn, M. (1970) *J. Biol. Chem.* **245**, 5309 & 5319 [2.7.9.2]
276. Berman, R., Wilson, I.B. & Nachmansohn, D. (1953) *Biochim. Biophys. Acta* **12**, 315 [2.3.1.6]
277. Berman, T. & Magasanik, B. (1966) *J. Biol. Chem.* **241**, 800 [1.1.1.18, 4.2.1.44]
278. Bernardi, A. & Bernardi, G. (1971) in *The Enzymes.* 3rd edn (Boyer, P.D., ed.) Vol. 4, p329, Academic Press, New York. [3.1.16.1]
279. Bernardi, G. (1966) in *Procedures in Nucleic Acid Research* (Cantoni, G.L. & Davies, D.R., ed.) p. 102, Harper & Row, New York [3.1.22.1]
280. Bernfield, M.R. & Nestor, L. (1968) *Biochem. Biophys. Res. Commun.* **33**, 843 [2.3.2.5]
281. Bernhardt, F.H., Staudinger, H. & Ullrich, V. (1970) *Hoppe-Seyler's Z. Physiol. Chem.* **351**, 467 [1.14.99.15]

448

282. Bernheim, M.L.C. (1969) *Arch. Biochem. Biophys.* **134**, 408   [1.6.6.11]
283. Bernheim, M.L.C. & Hochstein, P. (1968) *Arch. Biochem. Biophys.* **124**, 436   [1.6.6.11]
284. Berry, J.F. & Whittaker, V.P. (1959) *Biochem. J.* **73**, 447   [2.3.1.6]
285. Bertani, L.E., Häggmark, A. & Reichard, P. (1963) *J. Biol. Chem.* **238**, 3407   [3.6.1.23]
286. Bertazzoni, U., Mathelet, M. & Campagnari, F. (1972) *Biochim. Biophys. Acta* **287**, 404
      [6.5.1.1]
287. Bertland, L.H. & Kaplan, N.O. (1968) *Biochemistry* **7**, 134   [2.6.1.1]
288. Bertram, J. & Krisch, K. (1969) *Eur. J. Biochem.* **11**, 122   [3.1.1.1]
289. Besrat, A., Polan, C.E. & Henderson, L.M. (1969) *J. Biol. Chem.* **244**, 1461   [1.3.99.7]
290. Bessman, M.J., Herriott, S.T. & Orr, M.J.V.B. (1965) *J. Biol. Chem.* **240**, 439
      [2.7.4.13]
291. Best, A. & Novelli, G.D. (1971) *Arch. Biochem. Biophys.* **142**, 527 & 539
      [2.7.7.21, 2.7.7.25]
292. Bhatia, I.S., Satvanaravana, M.N. & Srinivasan, M. (1955) *Biochem. J.* **61**, 171
      [2.4.1.9]
293. Bhavanandan, V.P., Umemote, J. & Davidson, E.A. (1976) *Biochem. Biophys. Res.
      Commun.* **70**, 738   [3.2.1.97]
294. Bhumiratana, A., Anderson, R.L. & Costilow, R.N. (1974) J. Bacteriol. **119**, 484
      [3.2.1.93]
295. Bigger, C.H., Murray, K. & Murray, N.E. *Nature New Biology* **244**, 7 (1973).
296. Billings, R.E., Sullivan, H.R. & McMahon, R.E. (1971) *J. Biol. Chem.* **246**, 3512
      [1.1.1.112]
297. Bingham, A.H.A., Sharp, R.J. & Atkinson, A., unpublished results.
298. Binkley, F. (1951) *Nature (London)* **167**, 888   [3.4.13.6]
299. Birnbaum, S.M. (1955) *Methods Enzymol.* **2**, 115   [3.5.1.15]
300. Birnbaum, S.M., Levintow, L., Kingsley, R.B. & Greenstein, J.P. (1952) *J. Biol. Chem.*
      194, 455   [3.5.1.14, 3.5.1.15]
301. Bishop, S.H. & Grisolia, S. (1966) *Biochim. Biophys. Acta* **118**, 211   [2.7.2.2]
302. Bishop, S.H. & Grisolia, S. (1967) *Biochim. Biophys. Acta* **139**, 344   [2.1.3.3]
303. Bjork, G.R. & Svensson, I. (1969) *Eur. J. Biochem.* **9**, 207   [2.1.1.29]
304. Björklind A. & Jörnvall H. (1974) *Biochim. Biophys. Acta* **370**, 524
      [3.4.21.19, 3.4.22.13]
305. Black, S. (1951) *Arch. Biochem. Biophys.* **34**, 86   [1.2.1.5]
306. Black, S. (1962) *Methods Enzymol.* **5**, 820   [2.7.2.4]
307. Black, S. & Wright, N.G. (1955) *J. Biol. Chem.* **213**, 27   [2.7.2.4]
308. Black, S. & Wright, N.G. (1955) *J. Biol. Chem.* **213**, 39   [1.2.1.11]
309. Black, S. & Wright, N.G. (1955) *J. Biol. Chem.* **213**, 51   [1.1.1.3]
310. Black, S. & Wright, N.G. (1956) *J. Biol. Chem.* **221**, 171   [2.7.1.31, 3.1.3.14]
311. Blacklow, R.S. & Warren, L. (1962) *J. Biol. Chem.* **237**, 3520   [4.1.3.19]
312. Blair, A.H. & Barker, H.A. (1966) *J. Biol. Chem.* **241**, 400   [4.2.1.34]
313. Blake, C.C.F., Johnson, L.N., Mair, G.A., North, A.C.T., Phillips, D.C. & Sarma, V.R.
      (1967) *Proc. Roy. Soc. Ser. B* **167**, 378   [3.2.1.17]
314. Blake, C.C.F., Mair, G.A., North, A.C.T., Phillips, D.C. & Sarma, V.R. (1967) *Proc. Roy.
      Soc. Ser. B* **167**, 365   [3.2.1.17]
315. Blakely, J.A. & MacKenzie, S.L. (1969) *Can. J. Biochem.* **47** 1021   [3.2.1.23]
316. Blakesley, R.W., Dodgson, J.B., Nes, I.F. & Wells, R.D. (1977) *J. Biol. Chem.* **252**, 7300.
      [3.1.11.1]
317. Blakley, R.L. (1960) *Biochem. J.* **77**, 459   [2.1.2.1]
318. Blakley, R.L. (1965) *J. Biol. Chem.* **240**, 2173   [1.17.4.2]
319. Blakley, R.L. (1963) *J. Biol. Chem.* **238**, 2113   [2.1.1.45]
320. Blakley, R.L. & MacDougall, B.M. (1961) *J. Biol. Chem.* **236**, 1163   [1.5.1.3]
321. Blanchard, M., Green, D.E., Nocito-Carroll, V. & Ratner, S. (1946) *J. Biol. Chem.* **163**,
      137   [1.4.3.2, 2.4.1.66]
322. Blaschko, H. (1963) in *The Enzymes*, 2nd edn (Boyer, P.D., Lardy H. & Myrbäck, K.,
      ed.) Vol. 8, p. 337, Academic Press, New York   [1.4.3.4, 1.4.3.6]
323. Blaschko, H. & Buffoni, F. (1965) *Proc. Roy. Soc. Ser. B* **163**, 45   [1.4.3.6]
324. Blasi, F., Fragomele, F. & Covelli, I. (1969) *J. Biol. Chem.* **244**, 4864   [5.3.2.1]
325. Blauch, M., Koch, F.C. & Hane, M.E. (1939) *J. Biol. Chem.* **130**, 471   [3.5.4.2]
326. Blethen, S.L., Boeker, E.A. & Snell, E.E. (1968) *J. Biol. Chem.* **243**, 1671   [4.1.1.19]

327. Bloch, K., Chaykin, S., Phillips, A.H. & de Waard, A. (1959) *J. Biol. Chem.* **234**, 2595 [2.7.4.2, 4.1.1.33]
328. Blow A.M.J., Van Heyningen R. & Barrett A.J. (1975) *Biochem. J.***145**, 591 [3.4.24.5]
329. Blumenstein, J. & Williams, G.R. (1963) *Can. J. Biochem. Physiol.* **41**, 201 [2.1.1.20]
330. Blumenthal, H.J. (1966) *Methods Enzymol.* **9**, 660 [4.2.1.40]
331. Blumenthal, H.J. & Jepson, T. (1966) *Methods Enzymol.* **9**, 665 [4.2.1.42]
332. Bodnaryk, R.P. & McGirr, L. (1973) Biochim. Biophys. Acta **315**, 352 [2.3.2.4]
333. Bohak, Z. (1969) *J. Biol. Chem.* **244**, 4638 [3.4.23.1]
334. Bojanowski, R., Gaudy, E., Valentine, R.C. & Wolfe, R.S. (1964) *J. Bacteriol.* **87**, 75 [2.1.3.5]
335. Boland, M.J. & Benny, A.G (1977) *Eur. J. Biochem.* **79**, 355
336. Bollenbacher, W.E., Smith, S.L., Wielgus, J.J. & Gilbert, L.I. (1977) *Nature* **268**, 660 [1.14.99.22]
337. Bollum, F.J. (1960) *J. Biol. Chem.* **235**, 2399 [2.7.7.7]
338. Bollum, F.J. (1971) *J. Biol. Chem.* **246**, 909 [2.7.7.31]
339. Bone, D.H., Bernstein, S. & Vishniac, W. (1963) *Biochim. Biophys. Acta* **67**, 581 [1.12.1.2]
340. Bonnichsen, R.K. (1950) *Acta Chem. Scand.* **4**, 715 [1.1.1.1]
341. Booth, J. & Boyland, E. (1957) *Biochem. J.* **66**, 73 [1.14.14.1]
342. Booth, J., Boyland, E. & Sims, P. (1961) *Biochem. J.* **79**, 516
343. Borgström, B. & Wheeldon, L.W. (1961) *Biochim. Biophys. Acta* **50**, 171 [6.2.1.3]
344. Borkenhagen, L.F. & Kennedy, E.P. (1957) *J. Biol. Chem.* **227**, 951 [2.7.7.15]
345. Borkenhagen, L.F. & Kennedy, E.P. (1959) *J. Biol. Chem.* **234**, 849 [3.1.3.3]
346. Borsook, H. & Dubnoff, J.W. (1941) *J. Biol. Chem.* **138**, 389 [2.1.4.1]
347. Bosmann, H.B. (1972) *Biochim. Biophys. Acta* **276**, 180 [3.1.1.2]
348. Bosmann, H.B. & Eylar, E.H. (1968) *Biochem. Biophys. Res. Commun.* **30**, 89 [2.4.1.66]
349. Bosmann, H.B. & Eylar, E.H. (1968) *Biochem. Biophys. Res. Commun.* **33**, 340 [2.4.1.50]
350. Bosmann, H.B. & Eylar, E.H. (1968) *Nature (London)* **218**, 582 [2.4.1.66]
351. Bosmann, H.B., Hagopian, A. & Eylar, E.H. (1968) *Arch. Biochem. Biophys.* **128**, 470 [2.4.1.68]
352. Bottger, M., Fittkan, S., Niese, S. & Nanson, H. (1968) *Acta Biol. Med. Germ.* **21**, 143
353. Bové, C. & Conn, E.E. (1961) *J. Biol. Chem.* **236**, 207 [4.1.2.11]
354. Bowden, J.A. & Connelly, J.L. (1968) *J. Biol. Chem.* **243**, 3526
355. Bowen D.M. & Davidson D.M. (1973) *Biochem. J.* **131**, 417 [3.4.16.1]
356. Bowles, G., Ogren, W.L. & Hageman, R.H. (1971) *Biochem. Biophys. Res. Commun.* **45**, 716 [4.1.1.39]
357. Bowman, W.H., Tabor, C.W. & Tabor, H. (1973) *J. Biol. Chem.* **248**, 2480 [2.5.1.16]
358. Boyd, G.S., Grimwade, A.M. & Lawson, M.E. (1973) *Eur. J. Biochem.* **37**, 334 [1.14.13.17]
359. Boyer, H.W., Chow, L.T., Dugaiczyk, A., Hedgpeth, J. & Goodman, H.M. *Nature New Biology* **244**, 40 (1973).
360. Boyer, P.D. (1962) in *The Enzymes*, 2nd edn (Boyer, P.D., Lardy, H. & Myrbäck, K., ed.) Vol. 6, p. 95, Academic Press, New York [2.7.1.40]
361. Boyland, E. & Chasseaud, L.F. (1969) *Biochem. J.* **115**, 985
362. Boyland, E. & Williams, K. (1965) *Biochem. J.* **94**, 190
363. Bradbeer, C. (1965) *J. Biol. Chem.* **240**, 4669 [4.3.1.7]
364. Bradbeer, C. (1965) *J. Biol. Chem.* **240**, 4675 [4.3.1.7]
365. Bradshaw R.A. (1969) *Biochemistry* **8**, 3871 [3.4.24.4]
366. Brady, F.O., Monaco, M.E., Forman, H.J., Schutz, G. & Feigelson, P. (1972) J. Biol. Chem. **247**, 7915 [1.13.11.11]
367. Brady, R.N., DiMari, S.J. & Snell, E.E. (1969) *J. Biol. Chem.* **244**, 491 [2.3.1.50]
368. Brady, R.O., Gal, A.E., Bradley, R.M. & Matensson, E. (1967) *J. Biol. Chem.* **242**, 1021 [3.2.1.47]
369. Brady, R.O., Gal, A.E., Kanfer, J.N. & Bradley, R.M. (1965) *J. Biol. Chem.* **240**, 3766 [3.2.1.46]
370. Brady, R.O., Kanfer, J. & Shapiro, D. (1965) *J. Biol. Chem.* **240**, 39 [3.2.1.45]
371. Brady, R.O. & Stadtman, E.R. (1954) *J. Biol. Chem.* **211**, 621 [2.3.1.10, 2.3.1.11, 2.3.1.12]

372. Braunstein, A.E. (1960) in *The Enzymes*, 2nd edn (Boyer, P.D., Lardy, H. & Myrbäck, K., ed.) Vol. 2, p. 113, Academic Press, New York
373. Braunstein, A.E. & Azarkh, R.M. (1950) *Dokl. Akad. Nauk. S.S.S.R.* **71**, 93 [4.4.1.1]
374. Braunstein, A.E. & Azarkh, R.M. (1952) *Dokl. Akad. Nauk. S.S.S.R.* **85**, 385 [4.4.1.1]
375. Braunstein, A.E., Goryachinkova, E.V., Tolosa, E.A., Willhardt, I.H. & Yefremova, L.L. (1971) *Biochim. Biophys. Acta* **242**, 247 [4.2.1.22]
376. Bray, H.G., James, S.P., Raffan, I.M., Ryman, B.E. & Thorpe, W.V. (1949) *Biochem. J.* **44**, 618 [3.5.1.4]
377. Bray, H.G., James, S.P., Thorpe. W.V. & Wasdell, M.R. (1950) *Biochem. J.* **47**, 294 [3.5.1.4]
378. Bray, R.C. (1963) in *The Enzymes*, 2nd edn (Boyer, P.D., Lardy, H. & Myrbäck, K., ed.) Vol. 7, p. 533, Academic Press, New York [1.2.3.2]
379. Bremer, J. & Greenberg, D.M. (1961) *Biochim. Biophys. Acta* **46**, 217 [2.1.1.9]
380. Brennan, P. & Ballou, C.E. (1968) *Biochem. Biophys. Res. Commun.* **30**, 69 [2.4.1.57]
381. Brenneman, F.N. & Volk, W.A. (1959) *J. Biol. Chem.* **234**, 2443 [1.2.1.13]
382. Bretthauer, R.K., Kozak, L.P. & Irwin, W.E. (1969) *Biochem. Biophys. Res. Commun.* **37**, 820 [2.7.8.9]
383. Bretthauer, R.K., Wu, S. & Irwin, W.E. (1973) *Biochim. Biophys. Acta* **304**, 736 [2.4.1.83]
384. Bricteux-Gregoire, S., Schyns, R. & Florkin, M. (1966) *Biochim. Biophys. Acta* **127**, 277 [3.4.21.4]
385. Bright, H.J. & Ingraham, L.L. (1960) *Biochim. Biophys. Acta* **44**, 586 [4.3.1.2]
386. Brightwell, R. & Tappel, A.L. (1968) *Arch. Biochem. Biophys.* **124**, 333 [3.6.1.15]
387. Brink, N.G. (1953) *Acta Chem. Scand.* **7**, 1081 [1.1.1.47]
388. Britten, J.S., Morell, H. & Taggart, J.V. (1969) *Biochim. Biophys. Acta* **185**, 220 [4.2.1.31]
389. Broadbent D., Turner R.W. & Walton P.L. (1969) U.K. Pat. No. 1263956
390. Brock, D.J.H., Kass, L.R. & Bloch, K. (1967) *J. Biol. Chem.* **242**, 4432 [4.2.1.60]
391. Brodie, A.F. & Lipmann, F. (1955) *J. Biol. Chem.* **212**, 677 [3.1.1.17]
392. Bron, S., & Murray, K., (1975) *Mol. Gen. Genet.* **143**, 25 [3.1.23.17]
393. Brot, N. & Goodwin, J. (1968) *J. Biol. Chem.* **243**, 510 [6.3.2.14]
394. Brown, A.G., Radman, M. & Grossman, L. (1976) *Biochemistry* **15**, 416. [3.1.25.1]
395. Brown C.R., Andani Z. & Hartree E.F. (1975) *Biochem. J.* **149**, 133 [3.4.21.10]
396. Brown, D.D., Tomchick, R. & Axelrod, J. (1959) *J. Biol. Chem.* **234**, 2948 [2.1.1.8]
397. Brown, D.H. (1955) *Biochim. Biophys. Acta* **16**, 429 [2.3.1.4]
398. Brown, D.H. & Brown, B.I. (1966) *Methods Enzymol.* **8**, 515 [3.2.1.33]
399. Brown, F.C. & Ward, D.N. (1957) *J. Amer. Chem. Soc.* **79**, 2647 [1.10.3.1]
400. Brown, G.M. (1958) *J. Amer. Chem. Soc.* **80**, 3161 [4.1.1.36]
401. Brown, G.M. (1957) *J. Biol. Chem.* **226**, 651 [4.1.1.30]
402. Brown, G.M. (1959) *J. Biol. Chem.* **234**, 370 [2.7.1.33, 6.3.2.4, 6.3.2.5]
403. Brown-Grant, K., Forchielli, E. & Dorfman, R.I. (1960) *J. Biol. Chem.* **235**, 1317 [1.3.1.3]
404. Brown, J.L. (1973) *J. Biol. Chem.* **248**, 409 [3.4.13.12]
405. Brown, N.L., Hutchison, C.A. III & Smith, M. (1977) *Proc. Nat. Acad. Sci. USA* **74**, 3213 [3.1.23.18]
406. Brown, N.L., & Smith, M. FEBS Letters **65**, 284 (1976). [3.1.23.31]
407. Brühmüller, M., Möhler, H. & Decker, K. (1972) *Eur. J. Biochem.* **29**, 143 [1.5.3.6]
408. Brunel, A. (1936) *Thesis, University of Paris*, quoted by M. Florkin & G. Duchateau-Bosson (1940) *Enzymologia* **9**, 5 [1.6.1.1, 3.5.2.5]
409. Brungraber, E.G. (1958) *J. Biol. Chem.* **233**, 472 [2.8.2.1, 3.1.3.7]
410. Bruni, C.B., Sica, V., Auricchio, F. & Covelli, I. (1970) *Biochim. Biophys. Acta* **212**, 470 [3.2.1.20]
411. Brunngraber, E.F. & Chargaff, E. (1967) *J. Biol. Chem.* **242**, 4834 [2.7.1.77]
412. Bruns, F.H., Noltmann, E. & Willemsen, Λ. (1958) *Biochem. Z.* **330**, 411 [5.3.1.8]
413. Brzezinska, M., Benveniste, R., Davies, J., Daniels, P.J.L. & Weinstein, J. (1972) *Biochemistry* **11**, 761 [2.3.1.60]
414. Bublitz, C. & Kennedy, E.P. (1955) *J. Biol. Chem.* **211**, 951 [2.7.1.30]
415. Bublitz, C. & Lehninger, A.L. (1961) *Biochim. Biophys. Acta* **47**, 288 [3.1.1.18]
416. Buccino, R.J. Jr. & Roth, J.S. (1969) *Arch. Biochem. Biophys.* **132**, 49 [2.7.4.8]
417. Buchanan, B.B. (1969) *J. Biol. Chem.* **244**, 4218 [1.2.7.2]

418. Buchanan, B.B. & Evans, M.C.W. (1965) *Proc. Nat. Acad. Sci. USA* **54**, 1212
[1.2.7.3]

419. Bucher, N.L.R., Overath, P. & Lynen, F. (1960) *Biochim. Biophys. Acta* **40**, 491
[1.1.1.34]

420. Bucher, T. (1947) *Biochim. Biophys. Acta* **1**, 292 [2.7.2.3]

421. Buckner, J.S., Kolattudy, P.E. & Poulose, A.J. (1976) *Arch. Biochem. Biophys.* **177**, 539
[4.1.1.9]

422. Bueding, E. & MacKinnon, J.A. (1955) *J. Biol. Chem.* **215**, 495
[2.7.1.2, 2.7.1.4, 2.7.1.7, 2.7.1.8]

423. Buffoni, F. & Blaschko, H. (1964) *Proc. Roy. Soc. Ser B* **161**, 153   [1.4.3.6]

424. Bulen, W.A. (1956) *Arch. Biochem. Biophys.* **62**, 173   [1.4.1.2]

425. Burch, J. (1954) *Biochem. J.* **58**, 415   [3.1.1.1]

426. Burg, A.W. & Brown, G.M. (1968) *J. Biol. Chem.* **243**, 2349   [3.5.4.16]

427. Burg, R.W. & Snell, E.E. (1969) *J. Biol. Chem.* **244**, 2585   [1.1.1.107, 3.1.1.27]

428. Burger, M.M. & Glaser, L. (1964) *J. Biol. Chem.* **239**, 3168   [2.7.8.12]

429. Burghen, G.A., Kitabchi, A.E. & Brush, J.S. (1972) *Endocrinology* **91**, 633

430. Burgi, W., Richterich, R. & Colombo, J.P. (1966) *Nature (London)* **211**, 854   [1.4.1.15]

431. Burma, D.P. & Horecker, B.L. (1958) *J. Biol. Chem.* **231**, 1039   [2.7.1.16]

432. Burma. D.P. & Horecker, B.L. (1958) *J. Biol. Chem.* **231**, 1053   [5.1.3.4]

433. Burnett, G.H. & Cohen, P.P. (1957) *J. Biol. Chem.* **229**, 337   [2.1.3.3]

434. Burns, R.O. & DeMoss, R.D. (1962) *Biochim. Biophys. Acta* **65**, 233   [4.1.99.1]

435. Burns, R.O., Umbarger, H.E. & Gross, S.R. (1963) *Biochemistry* **2**, 1053   [1.1.1.85]

436. Burton, E.G. & Sakami, W. (1969) *Biochem. Biophys. Res. Commun.* **36**, 229   [2.1.1.13]

437. Burton, R.M. & Kaplan, N.O. (1953) *J. Amer. Chem. Soc.* **75**, 1005   [1.1.1.6]

438. Burton, R.M. & Stadtman, E.R. (1953) *J. Biol. Chem.* **202**, 873   [1.2.1.10]

439. Bush, I.E., Hunter, S.A. & Meigs, R.A. (1968) *Biochem. J.* **107**, 239   [1.1.1.146]

440. Butler, W.T. & Cunningham, L.W. (1966) *J. Biol. Chem.* **241**, 3882   [2.4.1.66]

441. Byrne, W.L. (1961) in *The Enzymes*, 2nd edn (Boyer, P.D., Lardy, H. & Myrbäck, K.,
ed.) Vol. 5, p. 73, Academic Press, New York   [3.1.3.3]

442. Byron, J.E., Wood, W.A. & Treadwell, C.R. (1953) *J. Biol. Chem.* **205**, 483   [3.1.1.13]

443. Cabello, J., Basilio, C. & Prajoux, V. (1961) *Biochim. Biophys. Acta* **48**, 148   [3.5.3.1]

444. Cabib, E., Carminatti, H. & Woyskovsky, N.M. (1965) *J. Biol. Chem.* **240**, 2114
[2.7.7.37]

445. Cabib, E. & Leloir, L.F. (1958) *J. Biol. Chem.* **231**, 259   [1.3.1.12, 2.4.1.15, 3.1.3.12]

446. Cagen, L.M. & Friedmann, H.C. (1972) *J. Biol. Chem.* **247**, 3382   [2.7.1.80]

447. Calvert, A.F. & Rodwell, V.W. (1966) *J. Biol. Chem.* **241**, 409   [1.2.1.31]

448. Cameron, D.W., Sawyer, W.H. & Trikojus, V.M. (1977) *Aust. J. Biol. Sci.* **30**, 173
[4.2.1.73]

449. Camiener, G.W. & Brown, G.M. (1960) *J. Biol. Chem.* **235**, 2411   [2.5.1.3, 2.7.1.50]

450. Campbell, B.J. (1970) *Methods Enzymol.* **19**, 722   [3.4.13.11]

451. Campbell, B., Lin, H., Davis, R. & Ballew, E. (1966) *Biochim. Biophys. Acta* **118**, 371
[3.4.13.11]

452. Campbell, L.L. (1957) *J. Biol. Chem.* **227**, 693   [1.3.1.1]

453. Campbell, L.L. (1960) *J. Biol. Chem.* **235**, 2375   [3.5.1.6]

454. Candy, D.J. & Kilby, B.A. (1961) *Biochem. J.* **78**, 531   [2.4.1.15, 3.1.3.12]

455. Canellakis, E.S. (1957) *J. Biol. Chem.* **227**, 329   [2.4.2.3]

456. Canellakis, Z.N. & Cohen, P.P. (1956) *J. Biol. Chem.* **222**, 53 & 63   [2.6.1.5]

457. Cannata, J.J.B. (1970) *J. Biol. Chem.* **245**, 792   [4.1.1.49]

458. Cannata, J.J.B., Focesi, A., Mazumder, R., Warner, R.C. & Ochoa, S. (1965) *J. Biol.
Chem.* **240**, 3249   [5.4.99.2]

459. Cannata, J.J.B. & Stoppani, A.O.M. (1963) *J. Biol. Chem.* **238**, 1196   [4.1.1.49]

460. Cannata, J.J.B. & Stoppani, A.O.M. (1963) *J. Biol. Chem.* **238**, 1208   [4.1.1.49]

461. Cantoni, G.L. (1951) *J. Biol. Chem.* **189**, 203   [2.1.1.1]

462. Cantoni, G.L. (1953) *J. Biol. Chem.* **204**, 403   [1.14.15.1, 2.5.1.6]

463. Cantoni, G.L. & Anderson, D.G. (1956) *J. Biol. Chem.* **222**, 171   [4.4.1.3]

464. Cantoni, G.L. & Durell, J. (1957) *J. Biol. Chem.* **225**, 1033   [2.5.1.6]

465. Cantoni, G.L. & Scarano, E. (1954) *J. Amer. Chem. Soc.* **76**, 4744   [2.1.1.2]

466. Cantoni, G.L. & Vignos, P.J. (1954) *J. Biol. Chem.* **209**, 647   [2.1.1.2]

467. Caputto, R. (1951) *J. Biol. Chem.* **189**, 801   [2.7.1.20]

468. Caputto, R. & Dixon, M. (1945) *Nature (London)* **156**, 630   [1.2.1.12]

469. Caravaca, J. & Grisolia, S. (1958) *J. Biol. Chem.* **231**, 357 [3.5.1.6]
470. Cardini, C.E. (1951) *Enzymologia* **14**, 362 [2.7.1.2]
471. Cardini, C.E. & Leloir, L.F. (1953) *Arch. Biochem. Biophys.* **45**, 55 [2.7.1.6]
472. Cardini, C.E., Leloir, L.F. & Chiriboga, J. (1955) *J. Biol. Chem.* **214**, 149 [2.4.1.13]
473. Cardini, G. & Jurtshuk, P. (1970) *J. Biol. Chem.* **245**, 2789 [1.14.15.3]
474. Carminatti, H. & Cabib, E. (1961) *Biochim. Biophys. Acta* **53**, 417 [2.7.7.22]
475. Carne, H.R. & Onon, E. (1978) *Nature (London)* **271**, 246 [3.1.4.41]
476. Carpenter, D.C. & Lovelace, F.E. (1943) *J. Amer. Chem. Soc.* **65**, 2364 [3.4.22.7]
477. Carter, C.E. (1951) *J. Amer. Chem. Soc.* **73**, 1508 [3.2.2.3]
478. Carter, C.E. & Cohen, L.H. (1956) *J. Biol. Chem.* **222**, 17 [3.4.21.14, 4.3.2.2]
479. Carter, J.R. & Kennedy, E.P. (1966) *J. Lipid. Res.* **7**, 678 [2.7.7.41]
480. Cartwright, L.N. & Hullin, R.P. (1966) *Biochem. J.* **101**, 781 [1.1.1.79]
481. Carty, R.P. & Kirschenbaum, D.M. (1964) *Biochim. Biophys. Acta* **85**, 446 [3.4.22.6]
482. Caskey, C.T., Ashton, D.M. & Wyngaarden, J.B. (1964) *J. Biol. Chem.* **239**, 2570 [2.4.2.14]
483. Castañeda, M., Balcazar, M.R. & Gavarrôn, F.F. (1942) *Science* **96**, 365 [3.4.99.14]
484. Castric, P.A. & Conn, E.E. (1971) *J. Bacteriol.* **108**, 132 [4.4.1.9]
485. Castric, P.A., Farnden, K.J.F. & Conn, E.E. (1972) *Arch. Biochem. Biophys.* **152**, 62 [4.2.1.65]
486. Cathou, R.E. & Buchanan, J.M. (1963) *J. Biol. Chem.* **238**, 1746
487. Cavallini, D., de Marco, C., Scandurra, R., Dupré, S. & Graziani, M.T. (1966) *J. Biol. Chem.* **241**, 3189 [1.13.11.19]
488. Center, M.S. & Behal, F.J. (1968) *J. Biol. Chem.* **243**, 138 [3.1.4.16]
489. Cerutti, P. & Guroff, G. (1965) *J. Biol. Chem.* **240**, 3034 [4.2.1.51]
490. Chae, K., Piantadosi, C. & Snyder, F. (1973) *J. Biol. Chem.* **248**, 6718 [2.7.1.84]
491. Chambers, J.C. & Elbein, A.D. (1970) *Arch. Biochem. Biophys.* **138**, 620 [2.4.1.21, 2.4.1.29]
492. Chan, L. & Glaser, L. (1972) *J. Biol. Chem.* **247**, 5391 [3.5.1.28]
493. Chang, S.H. & Wilken, D.R. (1966) *J. Biol. Chem.* **241**, 4251 [1.8.4.3]
494. Chang, Y.-F. & Adams, E. (1974) *J. Bacteriol.* **117**, 753 [1.5.1.14]
495. Chang, Y.Y. & Kennedy, E.P. (1967) *J. Lipid Res.* **8**, 447 [2.7.8.5]
496. Chang, Y.Y. & Kennedy, E.P. (1967) *J. Lipid Res.* **8**, 456 [3.1.3.27]
497. Change, H.-C. & Lane, M.D. (1966) *J. Biol. Chem.* **241**, 2413 [4.1.1.32]
498. Charalampous, F.C. (1959) *J. Biol. Chem.* **234**, 220 [1.13.99.1]
499. Charalampous, F.C. & Mueller, G.C. (1953) *J. Biol. Chem.* **201**, 161 [4.1.2.2]
500. Charles, A.M. & Suzuki, I. (1966) *Biochim. Biophys. Acta* **128**, 522 [1.8.2.1]
501. Charles, M., Rovery, M., Guidoni, A. & Desnuelle, P. (1963) *Biochim. Biophys. Acta* **69**, 115 [3.4.21.4]
502. Chase, J.F.A., Pearson, D.J. & Tubbs, P.K. (1965) *Biochim. Biophys. Acta* **96**, 162 [2.3.1.7]
503. Chase, J.W. & Richardson, C.C. (1974) *J. Biol. Chem.* **249**, 4545. [3.1.11.6, 3.4.21.12]
504. Chase, J.W. & Richardson, C.C. (1974) *J. Biol. Chem.* **249**, 4553 [3.1.11.6]
505. Chassy, B.M., Arsenis, C. & McCormick, D.B. (1965) *J. Biol. Chem.* **240**, 1338 [2.7.1.26]
506. Chatagner, F. & Sauret-Ignazi, G. (1956) *Bull. Soc. Chim. Biol. (Paris)* **38**, 415 [2.6.1.3]
507. Chatterjee, I.B., Chatterjee, G.C., Ghosh, N.C., Ghosh, J.J. & Guha, B.C. (1960) *Biochem. J.* **74**, 193 [3.1.1.18]
508. Chatterjee, I.B., Chatterjee, G.C., Ghosh, N.C., Ghosh, J.J. & Guha, B.C. (1960) *Biochem. J.* **76**, 279 [1.1.3.8]
509. Chauvet J., Dostal J.P. & Acher R. (1976) *Int. J. Peptide Protein Res.* **8**, 45 [3.4.21.4]
510. Cheatum, S.G. & Warren, J.C. (1966) *Biochim. Biophys. Acta* **122**, 1 [1.1.1.145]
511. Chen, G.S. & Segel, I.H. (1968) *Arch. Biochem. Biophys.* **127**, 175 [2.4.1.1]
512. Chen. I.-W. & Charalampous, F.C. (1966) *Arch. Biochem. Biophys.* **117**, 154 [3.1.3.25]
513. Chen, J.H. & Jones, R.F. (1970) *Biochim. Biophys. Acta* **214**, 318 [4.1.1.31]
514. Chen, L.-J., Bolt, R.J. & Admirand, W.H. (1977) *Biochim. Biophys. Acta* **480**, 219 [2.8.2.14]
515. Chen, P.R.S. & Dauterman, W.C. (1971) *Biochim. Biophys. Acta* **250**, 216 [3.5.1.39]
516. Cheng, Y.-J. & Karavolas, H.J. (1975) *Steroids* **26**, 57 [1.3.1.30]
517. Cheng, Y.-J. & Karavolas, H.J. (1975) *J. Biol. Chem.* **250**, 7997 [1.3.1.30]

518. Chern, C.J., MacDonald, A.B. & Morris, A.J. (1969) *J. Biol. Chem.* **244**, 5489
     [3.6.1.19]
519. Chesters, C.G.C. & Bull, A.T. (1963) *Biochem. J.* **86**, 31   [3.2.1.39]
520. Cheung, G.P., Rosenblum, I. & Sallach, H.J. (1968) *Plant Physiol.* **43**, 1813   [2.6.1.51]
521. Chiang, C. & Knight, S.G. (1960) *Biochem. Biophys. Res. Commun.* **3**, 554
     [1.1.1.12, 1.1.1.9]
522. Chiang, C. & Knight, S.G. (1961) *Biochim. Biophys. Acta* **46**, 271   [1.1.1.12, 1.1.1.13]
523. Chien, J.-L., Williams, T. & Basu, S. (1973) *J. Biol. Chem.* **248**, 1778   [2.4.1.79]
524. Chiga, M. & Plaut, G.W.E. (1960) *J. Biol. Chem.* **235**, 3260   [2.7.4.3]
525. Chiga, M., Rogers, A.E. & Plaut, G.W.E. (1961) *J. Biol. Chem.* **236**, 1800   [2.7.4.10]
526. Chin, T., Burger, M.M. & Glaser, L. (1966) *Arch. Biochem. Biophys.* **116**, 358
     [2.4.1.53]
527. Chittenden, R.H. (1892) *Trans. Conn. Acad. Sci.* **8**, 281   [3.4.22.4]
528. Chiu, T.-H. & Feingold, D.S. (1969) *Biochemistry* **8**, 98   [4.1.2.19]
529. Chlumecka, V., von Tigerstrom, M., D'Obrenan, P. & Smith, C.J. (1969) *J. Biol. Chem.*
     244, 5481   [6.1.1.6]
530. Chou, T.C. & Lipmann, F. (1952) *J. Biol. Chem.* **196**, 89   [2.3.1.5, 6.2.1.1]
531. Chou, T.C. & Soodak, M. (1952) *J. Biol. Chem.* **196**, 105   [2.3.1.3]
532. Christenson, J.G., Dairman, W. & Udenfriend, S. (1972) *Proc. Nat. Acad. Sci. U.S.* **69**,
     343   [4.1.1.28]
533. Christner, J.E., Schlesinger, M.J. & Coon, M.J. (1964) *J. Biol. Chem.* **239**, 3997
     [6.2.1.11]
534. Christopher, J., Pistorius, E. & Axelrod, B. (1970) *Biochim. Biophys. Acta* **198**, 12
     [1.13.11.12]
535. Chung, A.E. (1967) *J. Biol. Chem.* **242**, 1182   [2.7.1.23]
536. Chung, C.W. & Najjar, V.A. (1956) *J. Biol. Chem.* **218**, 617   [1.7.99.3]
537. Chung, C.W. & Najjar, V.A. (1956) *J. Biol. Chem.* **218**, 627   [1.7.99.2]
538. Chung, S.-T. & Aida, K. (1967) *J. Biochem. (Tokyo)* **61**, 1   [3.5.4.18]
539. Čihák, A. & Šorm, F. (1964) *Biochim. Biophys. Acta* **80**, 672   [2.4.2.23]
540. Ciliv, G. & Özand, P.T. (1972) *Biochim. Biophys. Acta* **284**, 136   [3.1.1.7]
541. Citri, N. (1971) in The Enzymes, 3rd ed (Boyer, P.D., ed.) Vol. 4, p. 23, Academic
     Press, New York   [3.5.2.6]
542. Clark, B. & Hübscher, G. (1963) *Biochim. Biophys. Acta* **70**, 43   [2.3.1.22]
543. Cleland, W.W. & Kennedy, E.P. (1960) *J. Biol. Chem.* **235**, 45   [2.4.1.23]
544. Cline, A.L. & Hu, A.S.L. (1965) *J. Biol. Chem.* **240**, 4488
     [1.1.1.117, 1.1.1.120, 1.1.1.121]
545. Cline, A.L. & Hu, A.S.L. (1965) *J. Biol. Chem.* **240**, 4493
     [1.1.1.117, 1.1.1.120, 1.1.1.121]
546. Cline, A.L. & Hu, A.S.L. (1965) *J. Biol. Chem.* **240**, 4498
     [1.1.1.117, 1.1.1.120, 1.1.1.121]
547. Cohen, J.A. & Warringa, M.G.P.J. (1957) *Biochim. Biophys. Acta* **26**, 29
     [3.1.23.28, 3.8.2.1]
548. Cohen, P.P. (1951) in *The Enzymes*, 1st edn (Sumner, J.B. & Myrbäck, K., ed.) Vol. 1,
     p. 1040, Academic Press, New York   [2.6.1.6]
549. Cohen, P.P. & Marshall, M. (1962) in *The Enzymes*, 2nd edn (Boyer, P.D., Lardy, H. &
     Myrbäck, K., ed.) Vol. 6, p. 327, Academic Press, New York   [2.1.3.2, 2.1.3.3]
550. Cohen, P.T. & Kaplan, N.O. (1970) *J. Biol. Chem.* **245**, 2825
551. Cohen, S.S. (1953) *Cold Spring Harbor Symp. Quant. Biol.* **18**, 221   [3.5.4.14]
552. Cohen, S.S. (1951) *J. Biol. Chem.* **189**, 617   [2.7.1.12]
553. Cohen, S.S. (1953) *J. Biol. Chem.* **201**, 71   [5.3.1.3]
554. Cohen, S.S. & Barner, H.D. (1957) *J. Biol. Chem.* **226**, 631   [3.5.4.1]
555. Cohn, D.V. (1958) *J. Biol. Chem.* **233**, 299   [1.1.3.3]
556. Cohn, D.V. (1958) *J. Biol. Chem.* **233**, 299   [1.4.1.14]
557. Cohn, M.S. & Phillips, A.T. (1974) *Biochemistry* **13**, 1208   [4.2.1.16]
558. Colby, J. & Zatman, L.J. (1971) *Biochem. J.* **121**, 9P   [1.5.99.7]
559. Colilla, W., Jorgenson, R.A. & Nordlie, R.C. (1975) *Biochim. Biophys. Acta* **377**, 117
     [3.1.3.9]
560. Collins, D.C., Jirku, H. & Layne, D.S. (1968) *J. Biol. Chem.* **243**, 2928   [2.4.1.39]
561. Colodzin, M. & Kennedy, E.P. (1965) *J. Biol. Chem.* **240**, 3771   [2.7.1.67]

454

562. Colowick, S.P., Kaplan, N.O., Neufeld, E.F. & Ciotti, M.M. (1952) *J. Biol. Chem.* **195**, 951
563. Comb, D.G. (1962) *J. Biol. Chem.* **237**, 1601   [6.3.2.15]
564. Comb, D.G. & Roseman, S. (1957) *Fed. Proc.* **16**, 166
565. Comb, D.G. & Roseman, S. (1958) *J. Biol. Chem.* **232**, 807   [5.3.1.10]
566. Comb, D.G. & Roseman, S. (1960) *J. Biol. Chem.* **235**, 2529   [4.1.3.3]
567. Conchie, J. (1954) *Biochem. J.* **58**, 552   [3.2.1.21]
568. Conconi, F. & Grazi, E. (1965) *J. Biol. Chem.* **240**, 2461   [2.1.4.1]
569. Conn, E.E., Kraemer, L.M., Liu, P.N. & Vennesland, B. (1952) *J. Biol. Chem.* **194**, 143   [1.11.1.2]
570. Connelly, J.L., Danner, D.J. & Bowden, J.A. (1968) *J. Biol. Chem.* **243**, 1198   [1.2.4.4]
571. Connett, R.J. & Kirschner, N. (1970) *J. Biol. Chem.* **245**, 329   [2.1.1.28]
572. Connors, W.M. & Stotz, E. (1949) *J. Biol. Chem.* **178**, 881   [3.7.1.2]
573. Conrad, H.E., DuBus, R. & Gunsalus, I.C. (1961) *Biochem. Biophys. Res. Commun.* **6**, 293   [1.14.15.2]
574. Contardi, A. & Ercoli, A. (1933) *Biochem. Z.* **261**, 275   [3.1.1.5]
575. Cook, J.S. (1967) *Photochem. Photobiol.* **6**, 97   [4.1.99.3]
576. Coon, M.J., Kupiecki, F.P., Dekker, E.E., Schlesinger, M.J. & del Campillo, A. (1959) in *CIBA Symposium on the Biosynthesis of Terpenes and Sterols* (Wolstenholme, G.E.W. & O'Connor, M., ed.) p. 62, Churchill, London   [1.1.1.33]
577. Cooper, A.J.L. & Meister, A. (1973) *Biochem. Biophys. Res. Commun.* **55**, 780   [4.3.1.13]
578. Cooper, C. & Wilson, D.W. (1954) *Fed. Proc.* **13**, 194   [3.5.2.3]
579. Cooper, R.A. & Anderson, A. (1970) *FEBS Lett.* **11**, 273   [4.2.99.11]
580. Cooper, R.A. & Kornberg, H.L. (1967) *Biochem. J.* **105**, 49C   [2.7.9.2]
581. Cooper, R.A. & Kornberg, H.L. (1965) *Biochim. Biophys. Acta* **104**, 618   [2.7.9.2]
582. Cooper, R.A. & Kornberg, H.L. (1969) *Methods Enzymol.* **13**, 309   [2.7.9.2]
583. Cooper, R.A. & Kornberg, H.L. (1964) *Biochem. J.* **91**, 82   [2.8.3.7, 4.1.3.25, 4.2.1.56]
584. Coote, J.G. & Hassal, H. (1969) *Biochem. J.* **111**, 237   [1.1.1.111, 2.6.1.38]
585. Corey, E.J., Russey, W.E. & Ortiz de Montellano, P.R. (1966) *J. Amer. Chem. Soc.* **88**, 4750   [1.14.99.7]
586. Cori, G.T., Ochoa, S., Slein, M.W. & Cori, C.F. (1951) *Biochim. Biophys. Acta* **7**, 304   [2.7.1.3]
587. Cori, G.T., Slein, M.W. & Cori, C.F. (1948) *J. Biol. Chem.* **173**, 605   [1.2.1.12]
588. Cormier, M.J., Crane, J.M. Jr. & Nakano, Y. (1967) *Biochem. Biophys. Res. Commun.* **29**, 747   [1.13.12.6]
589. Cormier, M.J., Hori, K. & Anderson, J.M. (1974) *Biochim. Biophys. Acta* **346**, 137   [1.13.12.5]
590. Cormier, M.J., Hori, K. & Karkhanis, Y.D. (1970) *Biochemistry* **9**, 1184   [2.8.2.10]
591. Cortese, R., Brevet, J., Hedegaard, J. & Roche, J. (1968) *C.R. Soc. Biol.* **162**, 390   [1.1.1.111]
592. Corwin, L.M. (1959) *J. Biol. Chem.* **234**, 1338   [4.1.1.3]
593. Cosgrove, D.J. (1969) *Ann. N.Y. Acad. Sci.* **165**, 677   [3.1.3.8]
594. Costa, O.A. (1937) *Rev. Quim. Farm. (Rio de Janiero)* **2**, 71   [3.2.3.1]
595. Costilow, R.N. & Laycock, L. (1971) *J. Biol. Chem.* **246**, 6655   [4.3.1.12]
596. Cotton, R.G.H. & Gibson, F. (1965) *Biochim. Biophys. Acta* **100**, 76   [4.2.1.51, 5.4.99.5]
597. Coulthard, C.E., Michaelis, R., Short, W.F., Sykes, G., Skrimshire, G.E.H., Standfast, A.F.B., Birkinshaw, J.H. & Raistrick, H. (1945) *Biochem. J.* **39**, 24   [1.1.3.4]
598. Coval, M.L. & Taurog, A. (1967) *J. Biol. Chem.* **242**, 5511   [1.11.1.8]
599. Cowell, J.L., Maser, K. & DeMoss, R.D. (1973) *Biochim. Biophys. Acta* **315**, 449   [4.1.99.1]
600. Cowgill, R.W. (1959) *J. Biol. Chem.* **234**, 3146   [2.4.1.1]
601. Cowgill, R.W. & Pizer, L.I. (1956) *J. Biol. Chem.* **223**, 885   [2.7.5.3]
602. Cowles, J.R. & Key, J.L. (1972) *Biochim. Biophys. Acta* **281**, 33   [6.1.1.1]
603. Crandall, D.I. & Halikis, D.N. (1954) *J. Biol. Chem.* **208**, 629   [1.13.11.5]
604. Crane, F.L., Mii, S., Hauge, J.G., Green, D.E. & Beinert, H. (1956) *J. Biol. Chem.* **218**, 701   [1.3.99.3]
605. Crawford, I., Kornberg, A. & Simms, E.S. (1957) *J. Biol. Chem.* **226**, 1093   [2.4.2.9]
606. Crawford, I.P. & Yanofsky, C. (1958) *Proc. Nat. Acad. Sci. USA* **44**, 1161   [4.2.1.20]

607. Creig, H.B.W. & Cornelius, E.M. (1963) *Biochim. Biophys. Acta* **67**, 658   [3.4.21.7]
608. Creighton, T.E. & Yanofsky, C. (1966) *J. Biol. Chem.* **241**, 4616   [4.1.1.48]
609. Crook, E.M. (1941) *Biochem. J.* **35**, 226   [1.8.5.1]
610. Croston, C.B. (1965) *Arch. Biochem. Biophys.* **112**, 219   [3.4.21.1]
611. Crouch, R.J. (1974) *J. Biol. Chem.* **249**, 1314   [3.1.26.3]
612. Crowley, G.M. (1964) *J. Biol. Chem.* **239**, 2593   [6.3.4.8]
613. Cunningham, B.A. & Kirkwood, S. (1961) *J. Biol. Chem.* **236**, 485   [1.11.1.8]
614. Cunningham, L. (1965) in *Comprehensive Biochemistry* (Florkin, M. & Stotz, E.H., ed.) Vol. 16, p. 85, Elsevier Publishing Co., Amsterdam   [3.4.22.3]
615. Cunningham, L. (1965) in *Comprehensive Biochemistry* (Florkin, M. & Stotz, E.H., ed.) Vol. 16, p. 151, Elsevier Publishing Co., Amsterdam   [3.4.17.1]
616. Cunningham, L.W. & Manners, D.J. (1961) *Biochem. J.* **80**, 42P   [3.2.1.6]
617. Cushman, D.W., Tsai, R.L. & Gunsalus, I.C. (1967) *Biochem. Biophys. Res. Commun.* **26**, 577
618. Cynkin, M.A. & Ashwell, G. (1960) *J. Biol. Chem.* **235**, 1576   [2.7.1.45]
619. DaCosta, T. & Gibbons, R.J. (1968) *Arch. Oral Biol.* **13**, 609   [3.2.1.80]
620. Dagley, S., Chapman, P.J. & Gibson, D.T. (1965) *Biochem. J.* **97**, 643   [1.13.11.16]
621. Dagley, S. & Dawes, E.A. (1955) *Biochim. Biophys. Acta* **17**, 177   [4.1.3.6]
622. Dahlqvist, A. (1961) *Biochim. Biophys. Acta* **50**, 55   [3.2.1.21]
623. Dahm, K. & Breuer, H. (1966) *Biochim. Biophys. Acta* **128**, 306   [2.4.1.42]
624. Dahm, K. & Breuer, H. (1964) *Hoppe-Seyler's Z. Physiol. Chem.* **336**, 63   [1.1.1.51]
625. Dahm, K., Breuer, H. & Lindlau, M. (1966) *Hoppe-Seyler's Z. Physiol. Chem.* **345**, 139
626. Dahm, K., Lindlau, M. & Breuer, H. (1968) *Biochim. Biophys. Acta* **159**, 377   [5.1.99.2]
627. Dahmen, W., Webb, B. & Preiss, J. (1967) *Arch. Biochem. Biophys.* **120**, 440   [2.7.7.1]
628. Dahms, A.S. (1974) *Biochem. Biophys. Res. Commun.* **60**, 1433   [4.1.2.28]
629. Dahms, A.S. & Anderson, R.L. (1972) J. Biol. Chem. **247**, 2233   [4.2.1.67]
630. Dahms, A.S. & Anderson, R.L. (1969) *Biochim. Biophys. Res. Commun.* **36**, 809   [4.1.2.18]
631. Dai, V.D., Decker, K. & Sund, H. (1968) *Eur. J. Biochem.* **4**, 95   [1.5.3.5]
632. Dalziel, K. (1961) *Biochem. J.* **80**, 440   [1.1.1.1]
633. Damus P.S., Markland F.S., Davidson T.M.& Shanley J.D. (1972) *J. Lab. Clin. Med.* **79**, 906   [3.4.21.30]
634. Danishefsky, I. & Heritier-Watkins, O. (1967) *Biochim. Biophys. Acta* **139**, 349   [2.7.7.34]
635. Dankert, M., Gonçalves, I.R.J. & Recondo, E. (1964) *Biochim. Biophys. Acta* **81**, 79   [2.7.7.36]
636. Danna, K.J., Sack, G.H., & Nathans, P. (1973) *J. Mol. Biol.* **78**, 363   [3.1.23.24]
637. Danno, G. (1970) *Agr. Biol. Chem. (Tokyo)* **34**, 264
638. Danno, G. & Yoshimura, S. (1967) *Agr. Biol. Chem. (Tokyo)* **31**, 1151   [3.4.21.14]
639. Datta, A. (1970) *Biochim. Biophys. Acta* **220**, 51   [2.7.1.59]
640. Datta, A.G. & Katznelson, H. (1956) *Arch. Biochem. Biophys* **65**, 576   [1.1.99.4]
641. Datta, P.K., Hanson, K.R. & Whitaker, D.R. (1963) *Can. J. Biochem. Physiol.* **41**, 697   [3.2.1.4]
642. Datta, P.K., Meeuse, B.J.D., Engstrom-Heg, V. & Hilal, S.H. (1955) *Biochim. Biophys. Acta* **17**, 602   [1.2.3.4]
643. Davey, C.L. (1959) *Nature (London)* **183**, 995   [6.3.4.4]
644. Davey, J.F. & Trudgill, P.W. (1977) *Eur. J. Biochem. (in press)*   *[1.1.1.174, 1.14.12.6]*
645. Davidson, E.A. (1966) *Methods Enzymol.* **9**, 704
646. Davidson, E.A., Blumenthal, H.J. & Roseman, F. (1957) *J. Biol. Chem.* **226**, 125   [2.3.1.4]
647. Davidson, E.A. & Riley, J.G. (1960) *J. Biol. Chem.* **235**, 3367   [2.8.2.5]
648. Davidson, F.M. (1960) *Biochem. J.* **76**, 56   [3.4.24.4]
649. Davidson, S.J. & Talalay, P. (1966) *J. Biol. Chem.* **241**, 906   [1.3.99.6]
650. Davie, E.W., Koningsberger, V.V. & Lipmann, F. (1956) *Arch. Biochem. Biophys.* **65**, 21   [6.1.1.2]
651. Davies, D.D. (1960) *J. Exp. Bot.* **11**, 289   [1.2.1.21]
652. Davies, D.D. & Kun, E. (1957) *Biochem. J.* **66**, 307   [1.1.1.37]
653. Davies, M.R. & Marshall, R.D. (1976) Biochim. Biophys. Acta **429**, 1   [6.1.1.22]
654. Davies, R. (1943) *Biochem. J.* **37**, 230   [4.1.1.4]

456

655. Davis, E.N., Wallen, L.L., Goodwin, J.C., Rohwedder, W.K. & Rhodes, R.A. (1969) *Lipids* **4**, 356 [4.2.1.53]
656. Davis, N.C. & Smith, E.L. (1957) *J. Biol. Chem.* **224**, 261 [3.4.13.9]
657. Davis, R.H. (1965) *Biochim. Biophys. Acta* **107**, 44 [2.7.2.2]
658. Davis, R.P. (1961) in *The Enzymes*, 2nd edn (Boyer, P.D., Lardy, H. & Myrbäck, K., ed.) Vol. 5, p. 545, Academic Press, New York [4.2.1.1]
659. Davison, D.C. (1951) *Biochem. J.* **49**, 520 [1.2.1.2]
660. Davison, D.C. & Elliott, W.H. (1952) *Nature (London)* **169**, 313 [4.3.2.1]
661. Dawson, C.R. & Magee, R.J. (1955) *Methods Enzymol.* **2**, 817
662. Dawson, C.R. & Magee, R.J. (1955) *Methods Enzymol.* **2**, 817
663. Dawson, C.R. & Tarpley, W.B. (1951) in *The Enzymes*, 1st edn (Sumner, J.B. & Myrbäck, K., ed.) Vol. 2, p. 454, Academic Press, New York [1.10.3.1, 1.10.3.2, 1.14.18.1]
664. Dawson, R.M.C. (1956) *Biochem. J.* **62**, 689 [3.1.4.2]
665. Dawson, R.M.C. (1958) *Biochem. J.* **70**, 559 [3.1.1.5]
666. Dawson, R.M.C. & Clarke, N.G. (1972) *Biochem. J.* **127**, 113 [3.1.4.36]
667. Dawson, R.M.C. & Clarke, N.G. (1973) *Biochem. J.* **134**, 59 [3.1.4.36]
668. Dawson, R.M.C. & Thompson, W. (1964) *Biochem. J.* **91**, 244 [3.1.3.36]
669. Day, W.C., Tocnic, P., Stratman, S.L., Leeman, U. & Harmon, S.R. (1968) *Biochim. Biophys. Acta* **167**, 597 [3.4.24.10]
670. de Castro, F.T., Price, J.M. & Brown, R.R. (1956) *J. Amer. Chem. Soc.* **78**, 2904 [1.14.13.9]
671. de Garilhe, M.P. & Laskowski, M. (1955) *J. Biol. Chem.* **215**, 269 [3.1.21.1]
672. de Haas, G.H., Postema, M.M., Nienwenhuizen, W. & van Deenan, L.L.M. (1968) *Bull. Soc. Chim. Biol. (Paris)* **50**, 1383 [3.1.1.4]
673. de la Haba, G. & Cantoni, G.L. (1959) *J. Biol. Chem.* **234**, 603 [3.3.1.1]
674. de la Haba, G., Leder, I.G. & Racker, E. (1955) *J. Biol. Chem.* **214**, 409 [2.2.1.1]
675. de Ley, J. & Doudoroff, M. (1957) *J. Biol. Chem.* **227**, 745 [1.1.1.48, 4.2.1.6]
676. De Lorenzo, F. & Ames, B.N. (1970) *J. Biol. Chem.* **245**, 1710 [6.1.1.21]
677. De Lorenzo, F., Goldberger, R.F., Steers, E., Givol, D. & Anfinsen, C.B. (1966) *J. Biol. Chem.* **241**, 1562 [5.3.4.1]
678. de Martinez, N.R. & Olavarria, J.M. (1973) *Biochim. Biophys. Acta* **320**, 301 [3.2.1.18]
679. De Pinto, J.A. & Campbell, L.L. (1968) *Biochemistry* **7**, 114 [2.4.1.19]
680. de Renzo, E.C. (1956) *Advan. Enzymol.* **17**, 293 [1.2.3.2]
681. Dean, P.D.G., Ortiz de Montellano, P.R., Bloch, K. & Corey, E.J. (1967) *J. Biol. Chem.* **242**, 3014 [5.4.99.7]
682. Decker, K. & Bleeg, H. (1965) *Biochim. Biophys. Acta* **105**, 313 [1.5.3.5, 1.5.3.6, 1.5.99.4]
683. Decker, R.H., Kang, H.H., Leach, F.R. & Henderson, L.M. (1961) *J. Biol. Chem.* **236**, 3076 [1.13.11.6]
684. Dedonder, R. (1952) *Bull. Soc. Chim. Biol. (Paris)* **34**, 171 [2.4.1.9]
685. Dekker, E.E., Schlesinger, M.J. & Coon, M.J. (1958) *J. Biol. Chem.* **233**, 434 [3.1.2.5]
686. Dekker, E.E. & Swain, R.R. (1968) *Biochim. Biophys. Acta* **158**, 306
687. Del Rio, L.A., Berkeley, R.C.W., Brewer, S.J. & Roberts, S.E. (1973) *FEBS Lett.* **37**, 7 [3.2.1.92]
688. Delafield, F.P., Cooksey, K.E. & Doudoroff, M. (1965) *J. Biol. Chem.* **240**, 4023 [1.1.1.30, 3.1.1.22]
689. Delange, R.J. & Smith, E.L. (1971) in *The Enzymes*, 3 edn (Boyer, P.D., ed.) Vol. 3, Academic Press, New York, p. 81 [3.4.11.1]
690. DeLey, J. (1966) *Methods Enzymol.* **9**, 200 [1.1.1.69]
691. DeMoss, R. (1953) *Bacteriol. Proc.* 81 [1.1.1.2]
692. Dempsey, M.E., Seaton, J.D., Schroepfer, G.J. & Trockman, R.W. (1964) *J. Biol. Chem.* **239**, 1381 [1.3.1.21, 1.3.3.2]
693. Den, H., Kaufman, B. & Roseman, S. (1970) *J. Biol. Chem.* **245**, 6607 [2.4.1.51]
694. Den, H., Robinson, W.G. & Coon, M.J. (1959) *J. Biol. Chem.* **234**, 1666 [1.1.1.59]
695. Denman, R.F., Hoare, D.S. & Work, E. (1955) *Biochim. Biophys. Acta* **16**, 442 [4.1.1.20]
696. Dennis, D. & Kaplan, N.O. (1960) *J. Biol. Chem.* **235**, 810 [1.1.1.27, 1.1.1.28]
697. Denson K.W.E., Barrett R. & Biggs R. (1971) *Brit. J. Haemat.* **21**, 219 [3.4.99.28]
698. DePinto, J.A. & Campbell, L.L. (1968) *Biochemistry* **7**, 121 [3.2.1.54]

699.	Derosier, D.J., Oliver, R.M. & Reed, L.J. (1971) *Proc. Nat. Acad. Sci. USA,* **68**, 1135
	[2.3.1.61]
700.	DerVartanian, D.V. & Le Gall, J. (1974) *Biochim. Biophys. Acta* **346**, 799 [1.12.2.1]
701.	Desa, R.J. (1972) *J. Biol. Chem.* **247**, 5527 [1.4.3.10]
702.	Deschamps, N. (1840) *J. Pharm.* **26**, 412 [3.4.23.4]
703.	Desmazeaud M.J. (1974) *Biochimie* **56**, 1173
704.	Desmazeaud M.J. & Zevaco C. (1976) Ann. Biol. Anim. Biochim Biophys. **16**, 851
	[3.4.24.4]
705.	Desnazeaud M. & Hermier J. (1968) *Ann. Biol. Anim. Biochim. Biophys.* **8**, 565
	[3.4.24.4]
706.	Desnuelle, P. (1960) in *The Enzymes*, 2nd edn (Boyer, P.D., Lardy, H. & Myrbäck, K.,
	ed.) Vol. 4, p. 93, Academic Press, New York [3.4.21.1]
707.	Deuel, H., Leuenberger, R. & Huber, G. (1950) *Helv. Chim. Acta* **33**, 942 [3.2.1.25]
708.	Deuel, H. & Stutz, E. (1958) *Advan. Enzymol.* **20**, 341 [3.1.1.11, 3.2.1.15]
709.	Deupree, J.D. & Wood, W.A. (1970) *J. Biol. Chem.* **245**, 3988 [5.1.3.4]
710.	Deutsch, A. & Nilsson, R. (1954) *Acta Chem. Scand.* **8**, 1898 [3.5.4.7]
711.	Deutscher, M. (1972) *Prog. Nucleic Acid Res. Mol. Biol.* **13**, 51 [2.7.7.21, 2.7.7.25]
712.	Dey, P.M. (1973) *Biochim. Biophys. Acta* **302**, 393 [3.2.1.88]
713.	Di Cesare, J.L. & Dain, J.A. (1971) *Biochim. Biophys. Acta* **231**, 385 [2.4.1.92]
714.	di Prisco, G., Casola, L. & Giuditta, A. (1967) *Biochem. J.* **105**, 455 [1.6.99.2]
715.	Dick A.J., Matheson A.T, & Wang J.H. (1970) *Canad. J. Biochem.* **48**, 1181
	[3.4.11.10]
716.	Dickens, F. & Glock, G.E. (1951) *Biochem. J.* **50**, 81 [1.1.1.44]
717.	Dickens, F. & Williamson, D.H. (1956) *Biochem. J.* **64**, 567 [5.1.3.1, 5.3.1.6]
718.	Dickerman, H.W., Steers, E. Jr., Redfield, B.G. & Weissbach, H. (1967) *J. Biol. Chem.*
	242, 1522 [2.1.2.9]
719.	Dickie N. & Liener I.E. (1962) *Biochim. Biophys. Acta* **64**, 41 [3.4.23.6]
720.	Dickman, S.R. (1961) in *The Enzymes*, 2nd edn (Boyer, P.D., Lardy, H. & Myrbäck, K.,
	ed.) Vol. 5, p. 495, Academic Press, New York [4.2.1.3]
721.	Dietrich, C.P. (1969) *Biochem. J.* **111**, 91 [3.1.6.11, 3.10.1.1]
722.	Dietrich, C.P. (1969) *Biochemistry* **8**, 2089 [3.2.1.56]
723.	Diez, V., Burgos, J. & Martin, R. (1974) *Biochim. Biophys. Acta* **350**, 253 [1.1.1.5]
724.	Dimroth, P., Guchhait, R.B., Stoll, E. & Lane, M.D. (1970) *Proc. Nat. Acad. Sci. USA*
	67, 1353 [6.3.4.14]
725.	Distler, J., Kaufman, B. & Roseman, S. (1966) *Arch. Biochem. Biophys.* **116**, 466
	[2.6.1.34]
726.	Dittbrenner, S., Chowdhury, A.A. & Gottschalk, G. (1969) *Biochem. Biophys. Res.
	Commun.* **36**, 802 [4.1.3.28]
727.	Divorsky, P. & Hoffmann-Ostenhof, O. (1964) *Acta Biochim. Pol.* **11**, 269 [1.1.1.45]
728.	Dixon, G.H., Kornberg, H.L. & Lund, P. (1960) *Biochim. Biophys. Acta* **41**, 217
	[4.1.3.2]
729.	Dixon, M. & Kenworthy, P. (1967) *Biochim. Biophys. Acta* **146**, 54 [1.4.3.1]
730.	Dixon, M. & Kleppe, K. (1965) *Biochim. Biophys. Acta* **96**, 357, 368 & 383 [1.4.3.3]
731.	Dixon, N.E., Gazzola, C., Blakeley, R.L. & Zerner, B. (1976) *Science,* **191**, 1144
	[3.5.1.5]
732.	Dodgson, K.S. (1961) *Biochem. J.* **78**, 324 [3.1.6.3]
733.	Dodgson, K.S. & Lloyd, A.G. (1958) *Biochem. J.* **68**, 88 [3.1.6.4]
734.	Dodgson, K.S. & Spencer, B. (1954) *Biochem. J.* **57**, 310 [3.1.6.4]
735.	Dodgson, K.S. & Spencer, B. & Williams, K. (1956) *Biochem. J.* **64**, 216 [3.1.6.1]
736.	Doery, H.M. & Pearson, J.E. (1961) *Biochem. J.* **78**, 820 [3.1.1.4]
737.	Doherty, M.D. & Morrison, J.F. (1962) *Biochim. Biophys. Acta* **65**, 364 [3.6.1.13]
738.	Doi E. (1974) *J. Biochem. (Tokyo)* **75**, 881 [3.4.16.1]
739.	Doi, O., Miyamoto, M., Tanaka, N. & Umezawa, H. (1968) *Appl. Microbiol.* **16**, 1282
	[2.7.1.95]
740.	Doi, O., Ogura, M., Tanaka, N. & Umezawa, H. (1968) *Appl. Microbiol.* **16**, 1276
	[2.7.1.95]
741.	Dolen, M.I. (1957) *J. Biol. Chem.* **225**, 557 [1.11.1.1]
742.	Domagk, G.F. & Horecker, B.L. (1965) *Arch. Biochem. Biophys.* **109**, 342 [2.2.1.1]
743.	Domagk, G.F. & Zeck, R. (1963) *Biochem. Z.* **339**, 145 [5.3.1.14]
744.	Domnas, A. & Catimo, E.C. (1965) *Phytochemistry* **4**, 273 [3.5.1.35]

458

745. Donlon J. & Fottrell P.F. (1973) *Biochim. Biophys. Acta* **327**, 425 [3.4.11.11]
746. Donta, S.T. & van Vunakis, H. (1970) *Biochemistry* **9**, 2791 and 2798 [3.4.23.1]
747. Doolittle, R.F. & Armentrout, R.W. (1968) *Biochemistry* **7**, 516 [3.4.11.8]
748. Doonan S., Doonan H.J., Hanford R., Vernon C.A., Walker J.M., Bossa F., Barra D., Carloni M., Fasella P., Riva F. & Walton P.L. (1974) *FEBS Letters* **38**, 229
749. Dopheide, T.A.A., Menzies, C.A., McQuillan, M.T. & Trikojus, V.M. (1969) *Biochim. Biophys. Acta* **181**, 105 [3.4.16.3]
750. Doudoroff, M. (1961) in *The Enzymes*, 2nd edn (Boyer, P.D., Lardy, H. & Myrbäck, K., ed.) Vol. 5, p. 229, Academic Press, New York [2.4.1.7, 2.4.1.8]
751. Doughty, C.C., Hayashi, J.A. & Guenther, H.L. (1966) *J. Biol. Chem.* **241**, 568 [2.7.1.31]
752. Dowler, M.J. & Nakada, H.I. (1968) *J. Biol. Chem.* **243**, 1435 [2.7.3.8]
753. Doyle, M.L., Katzman, P.A. & Doisy, E.A. (1955) *J. Biol. Chem.* **217**, 921 [3.2.1.31]
754. Drapeau G.R., Boily Y. & Houmard J. (1972) *J. Biol. Chem.* **247**, 6720 [3.4.21.19]
755. Drenth, J., Jansonius, J.N., Koekoek, R., Swan, H.M. & Wolthers, B.G. (1968) *Nature (London)* **218**, 929 [3.4.22.2]
756. Drummond, G.I., Iyer, N.T. & Keith, J. (1962) *J. Biol. Chem.* **237**, 3535 [3.1.4.37]
757. Drummond, G.I. & Stern, J.R. (1960) *J. Biol. Chem.* **235**, 318 [3.1.2.11]
758. Druzhinina, K.V. & Kritzman, M.G. (1952) *Biokhimiya* **17**, 77 [3.1.4.3]
759. Druzhinina, T.N., Kusov, Y.Y., Shibaev, V.N., Kochetkov, N.K., Bielý, P., Kučár, Š. & Bauer, Š. (1975) *Biochim. Biophys. Acta* **381**, 301 [1.1.1.22]
760. Du Toit, P.J. & Kotzé, J.P. (1970) *Biochim. Biophys. Acta* **206**, 333 [1.1.1.140]
761. Dubois, E.E., Dirheimer, G. & Weil, J.H. (1974) *Biochim. Biophys. Acta* **374**, 332 [2.1.1.36]
762. Duckworth, M. & Turvey, J.R. (1969) *Biochem. J.* **113**, 687 [3.2.1.81]
763. Duerre, J.A. (1962) *J. Biol. Chem.* **237**, 3737 [3.2.2.9]
764. Duerre, J.A. & Miller, C.H. (1966) *J. Bacteriol.* **91**, 1210 [3.3.1.3]
765. Duff, R.B. & Webley, D.M. (1958) *Biochem. J.* **70**, 520 [3.1.1.33]
766. Dumitru, I.F., Iorfăchescu, D. & Niculescu, S. (1970) *Rev. Rom. Biochim.* **7**, 31 [2.6.1.2]
767. Dunn, G. & Manners, D.J. (1975) *Carbohyd. Res. (in press)* *[3.2.1.41]*
768. Dunnill, P.M. & Fowden, L. (1963) *J. Exp. Bot.* **14**, 237 [4.2.1.50]
769. Dupourque, D., Newton, W.A. & Snell, E.E. (1966) *J. Biol. Chem.* **241**, 1233 [4.2.1.14]
770. Durbin, R.D. & Uchytil, T.F. (1971) *Biochim. Biophys. Acta* **235**, 518 [4.4.1.4]
771. Durham, J.P. & Ives, D.H. (1970) *J. Biol. Chem.* **245**, 2276 & 2285 [2.7.1.74]
772. Durr, I.F. & Rudney, H. (1960) *J. Biol. Chem.* **235**, 2572 [1.1.1.34]
773. Duspiva, F. (1950) *Z. Naturforsch.* **5b**, 273 [3.4.21.4]
774. Dutton, G.J. (1966) *Arch. Biochem. Biophys.* **116**, 399 [2.4.1.35]
775. Dutton, G.J. (1956) *Biochem. J.* **64**, 693 [2.4.1.17]
776. Dyckerhoff, H. & Armbruster, R. (1933) *Hoppe-Seyler's Z. Phy siol. Chem.* **219**, 38 [3.1.1.20]
777. Dyer, J.K. & Costilow, R.N. (1970) *J. Bacteriol.* **101**, 77
778. Eadie, G.S., Bernheim, F. & Bernheim, M.L.C. (1949) *J. Biol. Chem.* **181**, 449 [3.5.2.2]
779. Eady, R.R. & Large, P.J. (1968) *Biochem. J.* **106**, 245 [1.4.99.3]
780. Eady, R.R. & Large, P.J. (1971) *Biochem. J.* **123**, 757 [1.4.99.3]
781. Ebata, M., Sato, R. & Bak, T. (1955) *J. Biochem. (Tokyo)* **42**, 715 [2.7.1.14]
782. Ebata, M. & Yasunobu, K.T. (1962) *J. Biol. Chem.* **237**, 1086 [3.4.22.6]
783. Ebel, J., Hahlbrock, K. & Grisebach, H. (1972) *Biochim. Biophys. Acta* **268**, 313 [2.1.1.42]
784. Ebeling W., Hennrich N., Klockow M., Metz H., Orth H.D. & Lang H. (1974) *Eur. J. Biochem.* **47**, 91 [3.4.21.14]
785. Ebisuzaki, K. & Williams, J.N. (1955) *Biochem. J.* **60**, 644 [1.1.99.1]
786. Ebner, E., Wolf, D., Gancedo, C., Elsässer, S. & Holzer, H. (1970) *Eur. J. Biochem.* **14**, 535 [2.7.7.42]
787. Edelman, J. & Bacon, J.S.D: (1951) *Biochem. J.* **49**, 529 [2.4.1.9]
788. Edmonds, M. (1965) *J. Biol. Chem.* **240**, 4621 [2.7.7.21]
789. Edmonds, M. & Abrams, R. (1960) *J. Biol. Chem.* **235**, 1142 [2.7.7.19]
790. Edmundowicz, J.M. & Wriston, J.C. Jr. (1963) *J. Biol. Chem.* **238**, 3539 [1.1.1.138]

791. Edstrom. R.D. & Heath, E.C. (1967) *J. Biol. Chem.* **242**, 3581 [2.4.1.73]
792. Edstrom. R.D. & Phaff, H.J. (1964) *J. Biol. Chem.* **239**, 2403 & 2409 [4.2.2.2]
793. Edwards, S.W. & Knox, W.E. (1956) *J. Biol. Chem.* **220**, 79 [3.7.1.2, 5.2.1.2]
794. Egami, F. & Takahashi, N. (1955) *Bull. Chem. Soc. (Japan)* **25**, 666 [3.1.6.3]
795. Egan, A., Lawrence, P. & Strominger, J.L. (1973) *J. Biol. Chem.* **248**, 3122 [6.3.2.15]
796. Egelrud, T. & Olivecrona, T. (1973) *Biochim. Biophys. Acta* **306**, 115 [3.1.1.34]
797. Eggerer, H., Overath, P., Lynen, F. & Stadtman, E.R. (1960) *J. Amer. Chem. Soc.* **82**, 2643 [5.4.99.2]
798. Eichhorn, M.M. & Cynkin, M.A. (1965) *Biochemistry* **4**, 159 [1.1.1.125]
799. Eichler, D.C. & Lehman, I.R. (1977) *J. Biol. Chem.* **252**, 499. [3.1.11.5]
800. Einset, E. & Clark, W.L. (1958) *J. Biol. Chem.* **231**, 703 [3.1.4.4]
801. Eisen, A.Z., Henderson, K.O., Jeffrey, J.J.& Bradshaw, R.A. (1973) *Biochemistry* **12**, 1814 [3.4.21.32]
802. Eisen, A.Z. & Jeffrey, J.J. (1969) Biochim. Biophys. Acta **191**, 517 [3.4.21.32]
803. Eisenberg, F. & Field, J.B. (1956) *J. Biol. Chem.* **222**, 293 [3.1.1.17]
804. Eisenberg, M.A. (1955) *Biochim. Biophys. Acta* **16**, 58 [6.2.1.1]
805. Eisenberg, M.A. & Star, C. (1968) *J. Bacteriol.* **96**, 1291 [2.3.1.47]
806. Eisenberg, P. Jr. (1967) *J. Biol. Chem.* **242**, 1375 [3.1.3.25, 5.5.1.4]
807. Eisenman, R.A., Balasubramanian, A.S. & Marx, W. (1967) *Arch. Biochem. Biophys.* **119**, 387 [2.8.2.8]
808. Eker, A.P.M. & Fichtinger-Schepman, A.M.J. (1975) *Biochim. Biophys. Acta* **378**, 54 [4.1.99.3]
809. Ekwall, K. & Mannervik, B. (1973) *Biochim. Biophys. Acta* **297**, 297 [2.6.1.26, 4.4.1.5]
810. El-Badry, A.M. (1974) *Biochim. Biophys. Acta* **333**, 366 [3.1.3.11]
811. Elbein, A.D. (1967) *J. Biol. Chem.* **242**, 403 [2.4.1.36]
812. Elbein, A.D. (1969) *J. Biol. Chem.* **244**, 1608 [2.4.1.32]
813. Elbein, A.D. & Heath, E.C. (1965) *J. Biol. Chem.* **240**, 1926 [4.2.1.47]
814. Elisseeva, Y. & Orekhovich, V. (1963) *Dokl. Akad. Nauk. S.S.S.R.* **153**, 954 [3.4.15.1]
815. Ellfolk, N. (1953) *Acta Chem. Scand.* **7**, 824 [4.3.1.1]
816. Ellington, J.S. & Lands, W.E.M. (1968) *Lipids* **3**, 111 [3.3.2.2]
817. Elliott, S.D. (1945) *J. Exp. Med.* **81**, 573 [3.4.22.10]
818. Elliott, W.H. (1956) *Biochem. J.* **62**, 427 [6.2.1.7]
819. Elliott, W.H. (1957) *Biochem. J.* **65**, 315 [6.2.1.7]
820. Elliott, W.H. (1953) *J. Biol. Chem.* **201**, 661 [6.3.1.2]
821. Ellis, W.J. & Lennox, F.G. (1942) *Aust. J. Sci.* **4**, 187 [3.4.99.7]
822. Elödi, P. & Szörényi, E.T. (1956) *Acta Physiol. Acad. Sci. (Hung.)* **9**, 367 [2.7.3.3, 3.5.3.5]
823. Elovson, J. & Vagelos, P.R. (1968) *J. Biol. Chem.* **243**, 3603 [2.7.8.7]
824. Emi, S. & Yamamoto, T. (1972) *Agr. Biol. Chem. (Tokyo)* **36**, 1945 [3.2.1.89]
825. Endahl, G.L., Kochakian, C.D. & Hamm, D. (1960) *J. Biol. Chem.* **235**, 2792 [1.1.1.63, 1.1.1.64]
826. Endo, A. & Rothfield, L. (1969) *Biochemistry* **8**, 3500 [2.4.1.44]
827. Endow, S.A. & Roberts, R.J. (1977) *J. Mol. Biol.,* **112**, 521 [3.1.23.44]
828. Enfield D.L., Erucssin L.H., Fujikawa F., Titani K., Walsh K.A. & Neurath H. (1974) *FEBS-Letters* **47**, 132 [3.4.21.5]
829. Engel, H.J., Domschke, W., Alberti, M. & Domagk, G.F. (1969) *Biochim. Biophys. Acta* **191**, 509 [1.1.1.49]
830. Englard, S. & Breiger, H.H. (1962) *Biochim. Biophys. Acta* **56**, 571 [1.1.1.37]
831. Englard, S., Kaysen, G. & Avigad, G. (1970) *J. Biol. Chem.* **245**, 1311 [1.1.1.123]
832. English, P.D., Dietz, M. & Albersheim, P. (1966) *Science* **151**, 198 [2.7.1.64]
833. Engström, L. (1961) *Biochim. Biophys. Acta* **52**, 36 [3.1.3.1]
834. Ennor, A.H., Rosenberg, H. & Armstrong, M.D. (1955) *Nature (London)* **175**, 120 [2.7.3.2]
835. Entner, N. & Gonzalez, C. (1966) *Biochim. Biophys. Acta* **114**, 416 [2.7.4.8]
836. Epps, H.M.R. (1945) *Biochem. J.* **39**, 42 [4.1.1.22]
837. Erdös, E.G. & Sloane, E.M. (1962) *Biochem. Pharmacol.* **11**, 585 [3.4.17.3]
838. Erdös, E.G., Sloane, E.M. & Wohler, I.M. (1964) *Biochem. Pharmacol.* **13**, 893 [3.4.17.3]
839. Erdös, E.G. & Yang, H.Y.T. (1967) *Life Sci.* **6**, 569 [3.4.15.1]

840. Eriksson, A.F.V. (1968) *Acta Chem. Scand.* **22**, 1924   [3.2.1.78]
841. Eriksson, K.E. (1975) *Eur. J. Biochem.* **51**, 213   [3.2.1.91]
842. Erwin, V.G. & Hellerman, L. (1967) *J. Biol. Chem.* **242**, 4230   [1.4.3.4]
843. Eskin, B. & Linn, S. *J. Biol. Chem.* **247**, 6183-6191 (1972).   [3.1.24.1]
844. Esmon C.T. & Jackson C.M. (1973) *Thrombosis Research* **2**, 509   [3.4.21.23]
845. Esmon C.T., Owen W.G. & Jackson C.M. (1974) *J. Biol. Chem.* **249**, 594
846. Esnouf M.P., Tunnah G.W. (1967) Brit. J. Haematol. **13**, 581   [3.4.21.28, 3.4.21.5]
847. Euler, H. von., Adler, E., Gunther, G. & Das, N.B. (1938) *Hoppe-Seyler's Z. Physiol. Chem.* **254**, 61   [1.4.1.2]
848. Evans, M.C.W. & Buchanan, B.B. (1965) *Proc. Nat. Acad. Sci. U.S.* **53**, 1420   [1.2.7.1]
849. Evans, W.R. & San Pietro, A. (1966) *Arch. Biochem. Biophys.* **113**, 236   [2.7.7.35]
850. Everling, F.B., Weis, W. & Staudinger, H. (1969) *Hoppe-Seyler's Z. Physiol. Chem.* **350**, 1485   [1.10.2.1]
851. Ewald, W., Werbein, H. & Chaikoff, I.L. (1965) *Biochim. Biophys. Acta* **111**, 306   [5.3.3.1]
852. Ewart M.R., Hatton M.W.C., Basford J.M. & Dodge K.S., (1970) *Biochem. J.* **118**, 603   [3.4.21.28]
853. Eylar, E.H. & Murakami, M. (1966) *Methods Enzymol.* **8**, 597   [3.2.2.11]
854. Fahien, L.A. & Cohen, P.P. (1964) *J. Biol. Chem.* **239**, 1925   [6.3.4.16]
855. Fairbairn, D. (1948) *J. Biol. Chem.* **173**, 705   [3.1.1.5]
856. Fallon, H.J. & Byrne, W.L. (1965) *Biochim. Biophys. Acta* **105**, 43   [3.1.3.20]
857. Fan, D.-F., John, C.E., Zalitis, J. & Feingold, D.S. (1969) *Arch. Biochem. Biophys.* **135**,45   [1.1.1.136]
858. Faragó, A. & Dénes, G. (1967) *Biochim. Biophys. Acta* **136**, 6   [2.7.2.8]
859. Farkas, W. & Gilvarg, C. (1965) *J. Biol. Chem.* **240**, 4717   [1.3.1.26]
860. Farmer, J.J., III & Eagon, R.G. (1969) *J.Bacteriol.* **97**, 97   [1.1.1.131, 1.2.1.34]
861. Farmer, V.C., Henderson, M.E.K. & Russell, J.D. (1960) *Biochem. J.* **74**, 257   [1.1.3.7]
862. Farooqui, A.A. & Balasubramanian, A.S. (1970) *Biochim. Biophys. Acta* **198**,56   [3.1.3.30]
863. Farr D.R., Horisberger M. & Jolles P. (1974) *Biochim. Biophys. Acta* **334**, 410   [3.4.23.6]
864. Faulkner, A. & Turner, J.M. (1974) *Biochem. Soc. Trans.* **2**, 133   [2.7.1.82]
865. Faulkner, P. (1955) *Biochem. J.* **60**, 590   [3.1.3.10]
866. Feder, J. & Lewis, C. Jr. (1967) *Biochem. Biophys. Res. Commun.* **28**, 318   [3.4.24.4]
867. Feingold, D.S., Avigad, G. & Hestrin, S. (1957) *J. Biol. Chem.* **224**, 295   [2.4.1.4]
868. Feingold, D.S., Neufeld, E.F. & Hassid, W.Z. (1960) *J. Biol. Chem.* **235**, 910   [5.1.3.5, 5.1.3.6]
869. Félix, F. & Brouillet, N. (1966) *Biochim. Biophys. Acta* **122**, 127   [3.4.17.4, 3.4.22.9, 3.4.23.1]
870. Félix, F. & Labouesse-Mercouroff, J. (1956) *Biochim. Biophys. Acta* **21**, 303
871. Feraudi, M. & Schmolz, G. (1976) *Int. J. Biochem.* **7**, 461   [1.1.1.1]
872. Ferguson, J.A. & Ballou, C.E. (1970) *J. Biol. Chem.* **245**, 4213   [2.1.1.18]
873. Fernandez, M. & Grisolia, S. (1960) *J. Biol. Chem.* **235**, 2188   [5.4.2.1]
874. Fewson, C.A. & Nicholas, D.J.D. (1961) *Biochem. J.* **78**, 9P   [1.7.99.2]
875. Fewson, C.A. & Nicholas, D.J.D. (1961) *Biochim. Biophys. Acta* **49**, 335   [1.6.6.1]
876. Fidge, N.H. & Goodman, D.S. (1968) *J. Biol. Chem.* **243**, 4372   [1.1.1.71]
877. Fiedler, H. & Wood, J.L. (1956) *J. Biol. Chem.* **222**, 387   [2.8.1.2, 3.4.17.4]
878. Fielding, C.J. (1970) *Biochim. Biophys. Acta* **206**, 109   [3.1.1.34]
879. Fimognari, G.M. & Rodwell, V.W. (1965) *Biochemistry* **4**, 2086   [1.1.1.88]
880. Fincham, J.R.S. (1953) *Biochem. J.* **53**, 313   [2.6.1.13]
881. Findlay, J. & Levvy, G.A. (1960) *Biochem. J.* **77**, 170   [3.2.1.30]
882. Fischer, E.H., Pocke, A. & Saari, J.C. (1970) in *Essays in Biochemistry* (Campbell, P.N. & Greville, G.D., ed.) Vol. 6, p. 23, Academic Press, London & New York   [2.4.1.1]
883. Fischer, E.H. & Stein, E.A. (1960) in *The Enzymes*, 2nd edn (Boyer, P.D., Lardy, H. & Myrbäck, K., ed.) Vol. 4, p. 301, Academic Press, New York   [3.2.1.11, 3.2.1.14]
884. Fischer, E.H. & Stein, E.A. (1960) in *The Enzymes*, 2nd edn (Boyer, P.D., Lardy, H. & Myrbäck, K., ed.) Vol. 4, p. 313, Academic Press, New York   [3.2.1.1]
885. Fischer, U. & Amrhein, N. (1974) *Biochim. Biophys. Acta* **341**, 412
886. Fish, D.C. & Blumenthal, H.J. (1966) *Methods Enzymol.* **9**, 529   [4.1.2.20]
887. Fishbein, W.N. & Bessman, S.P. (1966) *J. Biol. Chem.* **241**, 4835   [3.1.1.25]

888. Fishbein, W.N. & Bessman, S.P. (1966) *J. Biol. Chem.* **241**, 4842 [3.1.1.25]
889. Fishman, W.H. (1955) *Advan. Enzymol.* **16**, 361 [3.2.1.31]
890. Fitting, C. & Doudoroff, M. (1952) *J. Biol. Chem.* **199**, 153 [2.4.1.8]
891. Fitzgerald, D.K., Brodbeck, U., Kiyosawa, I., Mawal, R., Colvin, B. & Ebner, K.E. (1970) *J. Biol. Chem.* **245**, 2103 [2.4.1.22]
892. Flaks, J.G. (1963) *Methods Enzymol.* **6**, 136 [2.4.2.10, 2.4.2.8, 2.4.2.9]
893. Flaks, J.G., Erwin, M.J. & Buchanan, J.M. (1957) *J. Biol. Chem.* **228**, 201 [2.4.2.7]
894. Flaks, J.G., Erwin, M.J. & Buchanan, J.M. (1957) *J. Biol. Chem.* **229**, 603 [3.5.4.10]
895. Flavin, M. & Segal, A. (1964) *J. Biol. Chem.* **239**, 2220 [4.4.1.1]
896. Flavin, M. & Slaughter, C. (1967) *Biochim. Biophys. Acta* **132**, 400 [4.2.99.9]
897. Flavin, M. & Slaughter, C. (1960) *J. Biol. Chem.* **235**, 1103 [2.7.1.39, 4.2.99.2]
898. Flavin, M. & Slaughter, C. (1964) *J. Biol. Chem.* **239**, 2212 [4.4.1.8]
899. Fleshood, H.L. & Pitot, H.C. (1970) *J. Biol. Chem.* **245**, 4414 [4.2.99.7]
900. Fling, M., Horowitz, N.H. & Heinemann, S.F. (1963) *J. Biol. Chem.* **238**, 2045 [1.10.3.1, 1.14.18.1]
901. Florkin, M. & Duchateau-Bosson, G. (1940) *Enzymologia* **9**, 5 [3.5.2.5, 3.5.3.4]
902. Flowers, H.M., Batra, K.K., Kemp, J. & Hassid, W.Z. (1969) *J. Biol. Chem.* **244**, 4969 [2.4.1.29]
903. Fodor E.J.B., Ako H. & Walsh K.A. (1975) *Biochemistry* **22**, 4923
904. Foldes, A. & Meek, J.L. (1973) *Biochim. Biophys. Acta* **327**, 365 [2.8.2.1]
905. Folk, J.E. & Chung, S.I. (1973) *Advanc. Enzymol.* **38**, 109 [2.3.2.13]
906. Folk, J.E. & Cole, P.W. (1966) *J. Biol. Chem.* **241**, 5518 [2.3.2.13]
907. Folk, J.E. & Cole, P.W. (1965) *J. Biol. Chem.* **240**, 193 [3.4.21.2]
908. Folk, J.E. & Finlayson, J.S. (1977) *Advanc. Protein Chem.* **31**, 1 [2.3.2.13]
909. Folk, J.E. & Schirmer, E.W. (1963) *J. Biol. Chem.* **238**, 3884 [3.4.17.1]
910. Folk, J.E. & Schirmer, E.W. (1965) *J. Biol. Chem.* **240**, 181 [3.4.21.2]
911. Folk, J., Piez, K., Carroll, W. & Gladner, J. (1960) *J. Biol. Chem.* **235**, 2272 [3.4.17.2]
912. Foltmann, B. (1966) *C.R. Trav. Lab. Carlsberg* **35**, 143 [1.14.15.1, 3.4.23.4]
913. Fones, W.S. & Lee, M. (1953) *J. Biol. Chem.* **201**, 847 [3.5.1.14]
914. Fonnum, F. & Larsen, K. (1965) *J. Neurochem.* **12**, 589 [2.6.1.49]
915. Forchielli, E. & Dorfman, R.I. (1956) *J. Biol. Chem.* **223**, 443 [1.3.1.23]
916. Foresst, J.C. & Wightman, F. (1973) *Can. J. Biochem.* **50**, 813 [2.6.1.1]
917. Forrester, P.I. & Gaucher, G.M. (1972) *Biochemistry* **11**, 1108 [1.1.1.97]
918. Foster, M.A., Dilworth, M.J. & Woods, D.D. (1964) *Nature (London)* **201**, 39 [2.1.1.13]
919. Fraenkel-Conrat, H. & Fraenkel-Conrat, J. (1950) *Biochim. Biophys. Acta* **5**, 98 [3.1.1.4]
920. Frampton, E.W. & Wood, W.A. (1961) *J. Biol. Chem.* **236**, 2571 [1.1.1.43]
921. Frampton, E.W. & Wood, W.A. (1961) *J. Biol. Chem.* **236**, 2578 [2.7.1.13]
922. Franke, W., Platzeck, A. & Eichhorn, G. (1961) *Arch. Mikrobiol.* **40**, 73 [4.1.1.56]
923. Franza R., Aronson D. & Finlayson J. (1975) J. Biol. Chem. **250**, 7057 [3.4.99.27]
924. Fraser, M.J. (1963) *Can. J. Biochem. Physiol.* **41**, 1123 [6.1.1.14]
925. Frear, D.S. (1968) *Phytochemistry* **7**, 381 [2.4.1.71]
926. French, D. (1960) in *The Enzymes*, 2nd edn (Boyer, P.D., Lardy, H. & Myrbäck, K., ed.) Vol. 4, p. 345, Academic Press, New York [3.2.1.2]
927. French, D. & Knapp. D.W. (1950) *J. Biol. Chem.* **187**, 463 [3.2.1.3]
928. French, D., Levine, M.L., Norberg, E., Norden, P., Pazur, J.H. & Wild, G.M. (1954) *J. Amer. Chem. Soc.* **76**, 2387 [2.4.1.19]
929. Friedberg, E.C. & Goldthwait, D.A. (1969) *Proc. Nat. Acad. Sci. U.S.* **62**, 934 [3.1.21.2]
930. Friedberg, E.C., Hadi, S.-M. & Goldthwait, D.A. (1969) *J. Biol. Chem.* **244**, 5879 [3.1.21.2]
931. Friedel, R.O., Brown, J.D. & Durell, J. (1967) *Biochim. Biophys. Acta* **144**, 684 [3.1.4.10]
932. Frieden, C. (1963) in *The Enzymes*, 2nd edn (Boyer, P.D., Lardy, H. & Myrbäck, K., ed.) Vol. 7, p. 3, Academic Press, New York [1.4.1.2]
933. Friedkin, M. & Kalckar, H. (1961) in *The Enzymes*, 2nd edn (Boyer, P.D., Lardy, H. & Myrbäck, K., ed.) Vol. 5, p. 237, Academic Press, New York [2.4.2.1, 2.4.2.2]
934. Friedkin, M. & Roberts, D. (1954) *J. Biol. Chem.* **207**, 245 [2.4.2.4]

462

935. Friedman, P.A., Kappelman, A.H. & Kaufman, S. (1972) *J. Biol. Chem.* **247**, 4165 [1.14.16.4]
936. Friedman, S. & Fraenkel, G. (1955) *Arch. Biochem. Biophys.* **59**, 491 [2.3.1.7]
937. Friedman, S. & Kaufman, S. (1965) *J. Biol. Chem.* **240**, 4763 [1.14.17.1]
938. Friedmann, H.C. (1965) *J. Biol. Chem.* **240**, 413 [2.4.2.21]
939. Friedmann, H.C. & Fyfe, J.A. (1969) *J. Biol. Chem.* **244**, 1667 [2.4.2.21]
940. Friedmann, H.C. & Vennesland, B. (1958) *J. Biol. Chem.* **233**, 1398 [1.3.1.14]
941. Friedmann, H.C. & Vennesland, B. (1960) *J. Biol. Chem.* **235**, 1526 [1.3.1.14, 1.3.3.1]
942. Friess, E.T. & Morales, M.F. (1955) *Arch. Biochem. Biophys.* **56**, 326 [3.6.1.3]
943. Frisell, W.R. & MacKenzie, C.G. (1955) *J. Biol. Chem.* **217**, 275 [1.5.3.1]
944. Frisell, W.R. & MacKenzie, C.G. (1962) *J. Biol. Chem.* **237**, 94 [1.5.99.1, 1.5.99.2]
945. Fritz, I.B. & Schultz, S.K. (1965) *J. Biol. Chem.* **240**, 2188 [2.3.1.6]
946. Fritzson, P. (1960) *J. Biol. Chem.* **235**, 719 [1.3.1.2]
947. Frohwein, Y.Z. & Gatt, S. (1967) *Biochemistry* **6**, 2775 [3.2.1.52, 3.2.1.53]
948. Fromm, H.J. (1959) *J. Biol. Chem.* **234**, 3097 [2.7.1.47]
949. Fruhanfova L., Suska-Brezezinska E., Barth T. & Rychlik I. (1973) *Coll. Czech. Chem. Commun.* **38**, 2793 [3.4.15.2]
950. Fry, B.A. (1955) *Biochem. J.* **59**, 579 [6.3.1.2]
951. Fry, W.E. & Millar, R.L. (1972) *Arch. Biochem. Biophys.* **151**, 468 [4.2.1.66]
952. Frydman, R.B. & Cardini, C.E. (1965) *Biochim. Biophys. Acta* **96**, 294 [2.4.1.21]
953. Frydman, R.B. & Feinstein, G. (1974) *Biochim. Biophys. Acta* **350**, 358 [4.3.1.8]
954. Frydman, R.B., Tomaro, M.L. & Frydman, B. (1972) *Biochim. Biophys. Acta* **284**, 63 [1.13.11.26]
955. Fuchs, S., De Lorenzo, F. & Anfinsen, C.B. (1967) *J. Biol. Chem.* **242**, 398 [5.3.4.1]
956. Fujikawa K., Legaz M.E. & Davie E.W. (1972) *Biochemistry* **11**, 4892
957. Fujikawa K., Legaz M.E., Kato H. & Davie E.W. (1974) *Biochemistry* **13**, 4508 [3.4.21.21, 3.4.21.22, 3.4.21.27]
958. Fujikawa K., Thompson A.R., Legaz M.E., Meyer R.G. & David E.W. (1973) *Biochemistry* **12**, 4938 [3.4.21.22]
959. Fujimoto, A., Ingram, P. & Smith, R.A. (1965) *Biochim. Biophys. Acta* **96**, 91 [2.7.5.5]
960. Fujimoto, D., Koyama, T. & Tamiya, N. (1968) *Biochim. Biophys. Acta* **167**, 407 [3.5.1.21]
961. Fujino, Y. & Nakano, Mo. (1969) *Biochem. J.* **113**, 573 [2.4.1.47]
962. Fujino, Y., Nigishi, T. & Ito, S. (1968) *Biochem. J.* **109**, 310 [2.7.8.10]
963. Fujioka, M. (1969) *Biochim. Biophys. Acta* **185**, 338 [2.1.2.1]
964. Fujioka, M., Asakawa, H., Wada, H. & Yamano, T. (1966) Koso Kagaku Shimpojiumu **18**, 106 [1.2.1.39]
965. Fujioka, M., Morino, Y. & Wada, H. (1970) *Methods Enzymol.* **17A**, 593 [1.2.1.39]
966. Fujioka, M. & Nakatani, Y. (1974) *Eur. J. Biochem.* **25**, 301 [1.5.1.7]
967. Fujioka, M. & Wada, H. (1968) *Biochim. Biophys. Acta* **158**,70 [1.13.11.23]
968. Fujisawa, H. & Hayaishi, O. (1968) *J. Biol. Chem.* **243**, 2673 [1.13.11.3]
969. Fujisawa, Y., Shirafuji, H., Kida, M. & Nara, K. (1973) *Nature (New Biol.)* **246**, 154 [3.1.1.41]
970. Fujita, A. (1954) *Advan. Enzymol.* **15**, 389 [2.5.1.2]
971. Fujita, A., Nose, Y. & Kuratani, K. (1954) *J. Vitaminol. (Kyoto)* **1**, 1 [3.5.99.2]
972. Fujita, T. & Mannering, G.J. (1971) *Chem.-Biol. Interactions* **3**, 264 [1.14.14.1]
973. Fukui, S., Suzuki, T., Kitahara, K. & Miwa, T. (1960) *J. Gen. Appl. Microbiol.* **6**, 270 [3.2.1.32, 3.2.1.72]
974. Fukumaga, K. (1960) *J. Biochem. (Tokyo)* **47**, 741 [3.4.23.6, 4.1.1.54]
975. Fukumoto, J., Tsuru, D. & Yamamoto, T. (1967) *Agr. Biol. Chem. (Tokyo)* **31**, 710
976. Fukumoto J., Tsuru D.& Yamamoto T. (1967) *Agr. Biol. Chem. (Tokyo)* **31**, 710 [3.4.23.6]
977. Fulco, A.J. & Bloch, K. (1964) *J. Biol. Chem.* **239**, 993 [1.14.99.5]
978. Fullmer H.M. & Lazarus G. (1967) *Isruel J. Med. Sci.* **3**, 758 [3.4.24.7]
979. Furie B.C. & Furie B. (1976) *Methods Enzymol.* **45**, 191 [3.4.21.23]
980. Futai, M. & Mizuno, D. (1967) *J. Biol. Chem.* **242**, 5301 [3.1.13.3]
981. Fyfe, J.A. & Friedmann, H.C. (1969) *J. Biol. Chem.* **244**, 1659 [2.4.2.21]
982. Gabeloteau G. & Desnuelle P. (1960) *Biochim. Biophys. Acta* **42**, 230 [3.4.23.6]
983. Gaertner, F.H. & Cole, K.W. (1973) *J. Biol. Chem.* **248**, 4602 [4.6.1.4]
984. Gaffney, T.J., Rosenberg, H. & Ennor, A.H. (1964) *Biochem. J.* **90**, 170 [2.7.3.5]

985. Gale, E.F. (1939) *Biochem. J.* **33**, 1012   [1.2.2.1]
986. Gale, E.F. & Epps, H.M.R. (1944) *Biochem. J.* **38**, 232   [4.1.1.18]
987. Galivan, J.H. & Allen, S.H.G. (1968) *J. Biol. Chem.* **243**, 1253   [4.1.1.41]
988. Gallwitz, D. & Sures, I. (1972) *Biochim. Biophys. Acta* **263**, 315   [2.3.1.48]
989. Gamborg, O.L. (1966) *Biochim. Biophys. Acta* **128**, 483   [1.1.1.24]
990. Gamborg, O.L. & Keeley, F.W. (1966) *Biochim. Biophys. Acta* **115**, 65   [1.3.1.13]
991. Gangloff, J. & Dirheimer, G. (1973) Biochim. Biophys. Acta **294**, 263   [6.1.1.12]
992. Garbers, D.L., Suddath, J.L. & Hardman, J.G. (1975) *Biochim. Biophys. Acta* **377**, 174
      [4.6.1.2]
993. Garfin, D.E. & Goodman, H.M. (1974) *Biochem. Biophys. Res. Comm.* **59**, 108
      [3.1.23.23]
994. Garg G.K. & Virupaksha T.K. (1970) *Eur. J. Biochem.* **17**, 4   [3.4.23.14]
995. Gates, B.J. & Travis, J. (1969) *Biochemistry* **8**, 4483   [3.4.21.4]
996. Gates, F.T. & Linn, S. (1977) *J. Biol. Chem.* **252**, 1647   [3.1.22.3]
997. Gatt, S. (1968) *Biochim. Biophys. Acta* **159**, 304   [3.1.1.32]
998. Gatt, S. (1966) *J. Biol. Chem.* **241**, 3724   [3.5.1.23]
999. Gaugler, R.W. & Gabriel, O. (1970) *Fed. Proc.* **29**, 337   [1.1.1.134, 5.1.3.13]
1000. Gauthier, J.J. & Rittenberg, S.C. (1971) *J. Biol. Chem.* **246**, 3737 & 3743   [1.13.11.9]
1001. Gaylor, J.L. & Mason, H.S. (1968) *J. Biol. Chem.* **243**, 4966   [1.14.99.16]
1002. Gefter, M.L. (1969) *Biochem. Biophys. Res. Commun.* **36**, 435   [2.1.1.34]
1003. Gehring, U. & Arnon, D.I. (1972) *J. Biol. Chem.* **247**, 6963   [1.2.7.1, 1.2.7.3]
1004. Gelinas, R.E., Myers, P.A. & Roberts, R.J., (1977) *J. Mol. Biol.,* **114**, 169
      [3.1.23.27, 3.1.23.28]
1005. Gelinas, R.E., Myers, P.A., Weiss, G.H., Roberts, R.J. & Murray, K. *J. Mol. Biol.* **114**,
      433   [3.1.23.5]
1006. George, D.J. & Phillips, A.T. (1970) *J. Biol. Chem.* **245**, 528   [4.2.1.49]
1007. George, H. & Gabay, S. (1968) *Biochim. Biophys. Acta* **167**, 555   [2.6.1.27]
1008. Gergely, J., Hele, P. & Ramakrishnan, C.V. (1952) *J. Biol. Chem.* **198**, 323
      [3.1.2.1, 3.1.2.3]
1009. Gerwin B.I., Stein W.H. & Moore S. (1966) *J. Biol. Chem.* **241**, 3331   [3.4.22.10]
1010. Gestetner, B. & Conn, E.E. (1974) *Arch. Biochem. Biophys.* **163**, 617   [1.14.13.14]
1011. Ghalambor, M.A. & Heath, E.C. (1962) *J. Biol. Chem.* **237**, 2427   [3.4.21.4, 4.1.2.17]
1012. Ghalambor, M.A. & Heath, E.C. (1966) *J. Biol. Chem.* **241**, 3216   [2.7.7.38]
1013. Ghalambor, M.A. & Heath, E.C. (1966) *J. Biol. Chem.* **241**, 3222   [4.1.2.23]
1014. Gholson, R.K., Ueda, I., Ogasawara, N. & Henderson, L.M. (1964) *J. Biol. Chem.* **239**,
      1208   [2.4.2.19]
1015. Ghosh, H.P. & Preiss, J. (1966) *J. Biol. Chem.* **241**, 4491   [2.7.7.27]
1016. Ghosh, S., Blumenthal, H.J., Davidson, E. & Roseman, S. (1960) *J. Biol. Chem.* **235**,
      1265   [5.3.1.19]
1017. Ghosh, S. & Roseman, S. (1965) *J. Biol. Chem.* **240**, 1525   [5.1.3.9]
1018. Ghosh, S. & Roseman, S. (1965) *J. Biol. Chem.* **240**, 1531   [5.1.3.8]
1019. Ghosh, S. & Roseman, S. (1961) *Proc. Nat. Acad. Sci. USA* **47**, 955   [2.7.1.60]
1020. Ghuysen, J.-M., Dierickx, L., Coyette, J., Leyh-Bouille, M., Guinand, M. & Campbell,
      J.N. (1969) *Biochemistry* **8**, 213   [3.5.1.28]
1021. Gibbins, L.N. & Simpson, F.J. (1963) *Can. J. Microbiol.* **9**, 769   [2.7.1.55]
1022. Gibbons, R.G. & MacDonald, J.B. (1961) *J. Bacteriol.* **81**, 614   [3.4.24.3, 3.5.1.2]
1023. Gibbs, M. (1955) *Methods Enzymol.* **1**, 411   [1.2.1.13]
1024. Gibbs, R.G. & Morris, J.G. (1965) *Biochem. J.* **97**, 547   [4.2.1.38]
1025. Gibbs, R.G. & Morris, J.G. (1966) *Biochem. J.* **99**, 27P   [2.6.1.35]
1026. Gibbs, R.G. & Morris, J.G. (1964) *Biochim. Biophys. Acta* **85**, 501   [4.1.3.14]
1027. Gibson, D. & Dixon, G.H. (1969) *Nature (London)* **222**, 753   [3.4.21.3]
1028. Gibson, D.M., Ayengar, P. & Sanadi, D.R. (1955) *Biochim. Biophys. Acta* **16**, 536
      [3.6.1.6]
1029. Gibson, D.M., Ayengar, P. & Sanadi, D.R. (1956) *Biochim. Biophys. Acta* **21**, 86
      [2.7.4.4, 2.7.4.6]
1030. Gibson, D.M., Davisson, E.O., Bachhawat, B.K., Ray, B.R. & Vestling, C.S. (1953)
      J. Biol. Chem. **203**, 397   [1.1.1.27]
1031. Gibson, D.T., Koch, J.R. & Kallio, R.E. (1968) *Biochemistry* **7**, 2653
      [1.14.12.3, 1.3.1.19]

464

1032. Gibson, D.T., Wang, K.C., Sih, C.J. & Whitlock, J.H. (1966) *J. Biol. Chem.* **241**, 551
       [1.13.11.25]
1033. Gibson, K.D., Matthew, M. & Neuberger, A. (1961) *Nature (London)* **192**, 204
       [2.6.1.43]
1034. Gibson, K.D., Neuberger, A. & Scott, J.J. (1955) *Biochem. J.* **61**, 618   [4.2.1.24]
1035. Gibson, K.D., Neuberger, A. & Tait, G.H. (1963) *Biochem. J.* **88**, 325   [2.1.1.11]
1036. Giebel, W., Zwilling, R. & Pfleiderer, G. (1971) *Comp. Biochem. Physiol.* **38**, 197
       [3.4.21.4]
1037. Gigliotti, H.J. & Levenberg, B. (1964) *J. Biol. Chem.* **239**, 2274   [2.3.2.9]
1038. Gilbert, J.M., Matsuhashi, M. & Strominger, J.L. (1965) *J. Biol. Chem.* **240**, 1305
       [4.2.1.46]
1039. Gingeras, T.R., Myers, P.A., Olsen, J.A., Hamberg, F.A. & Roberts, R.J. (1978) *J. Mol. Biol.* **118**, 113   [3.1.23.42, 3.1.23.43]
1040. Gingeras, T.R. & Roberts, R.J., unpublished results   [3.1.23.32, 3.1.23.33, 3.1.23.7]
1041. Ginoza, H.S. & Altenbern, R.A. (1955) *Arch. Biochem. Biophys.* **56**, 537   [6.3.2.1]
1042. Ginsburg, A. (1959) *J. Biol. Chem.* **234**, 481   [2.7.1.15]
1043. Ginsburg, V. (1958) *J. Biol. Chem.* **232**, 55   [2.7.7.9]
1044. Ginsburg, V., Neufeld, E.F. & Hassid, W.Z. (1956) *Proc. Nat. Acad. Sci. USA* **42**, 333
       [2.7.7.11]
1045. Giovanelli, J. (1966) *Biochim. Biophys. Acta* **118**, 124   [6.2.1.8]
1046. Giri, K.V., Krishnaswamy, P.R. & Rao, N.A. (1958) *Biochem. J.* **70**, 66   [2.7.1.26]
1047. Giri, K.V., Rao, N.A., Cama, H.R. & Kumar, S.A. (1960) *Biochem. J.* **75**, 381
       [2.7.7.2]
1048. Giuditta, A. & Strecker, H.J. (1961) *Biochim. Biophys. Acta* **48**, 10   [1.6.99.2]
1049. Gladhaug A. & Prydz H. (1970) *Biochim. Biophys. Acta* **215**, 105   [3.4.21.21]
1050. Glansdorff, N. & Sand, G. (1965) *Biochim. Biophys. Acta* **108**, 808   [1.2.1.38]
1051. Glanville, K.L.A. (1963) *Biochem. J.* **88**, 11   [3.4.24.4]
1052. Glaser, L. (1963) *Biochim. Biophys. Acta* **67**, 525   [1.1.1.137]
1053. Glaser, L. (1965) *Biochim. Biophys. Acta* **101**, 6   [3.6.1.16]
1054. Glaser, L. (1958) *J. Biol. Chem.* **232**, 627   [2.4.1.12]
1055. Glaser, L. (1959) *J. Biol. Chem.* **234**, 2801   [5.1.3.7]
1056. Glaser, L. (1960) *J. Biol. Chem.* **235**, 2095   [5.1.1.3]
1057. Glaser, L. & Brown, D.H. (1955) *J. Biol. Chem.* **216**, 67   [1.1.1.49]
1058. Glaser, L. & Brown, D.H. (1957) *J. Biol. Chem.* **228**, 729   [2.4.1.16]
1059. Glaser, L. & Burger, M.M. (1964) *J. Biol. Chem.* **239**, 3187   [2.4.1.52]
1060. Glasziou, K.T. (1956) *Aust. J. Biol. Sci.* **9**, 253   [2.7.2.2]
1061. Glenner, G.G., McMillan, P.J. & Folk, J.E. (1962) *Nature (London)* **194**, 867
       [3.4.11.7]
1062. Glick, D., Glaubach, S. & Moore, D.H. (1942) *J. Biol. Chem.* **144**, 525   [3.1.1.10]
1063. Glitz, D.G. & Decker, C.A. (1964) *Biochemistry* **3**, 1391 & 1399   [3.1.27.4]
1064. Glomset, J.A. (1968) *J. Lipid Res.* **9**, 155   [2.3.1.43]
1065. Gold, M. & Hurwitz, J. (1964) *J. Biol. Chem.* **239**, 3858   [2.1.1.37]
1066. Goldemberg, S.H., Maréchal, L.R. & De Souza, B.C. (1966) *J. Biol. Chem.* **241**, 45
       [2.4.1.31]
1067. Goldman, D.S. (1954) *J. Biol. Chem.* **208**, 345   [2.3.1.16]
1068. Goldman, D.S. & Wagner, M.J. (1962) *Biochim. Biophys. Acta* **65**, 297   [1.4.1.10]
1069. Goldman, P. (1965) *J. Biol. Chem.* **240**, 3434   [3.8.1.3]
1070. Goldman, P. & Levy, C.C. (1967) *Proc. Nat. Acad. Sci. USA* **58**, 1229
1071. Goldman, P. & Milne, G.W.A. (1966) *J. Biol. Chem.* **241**, 5557   [3.8.1.3]
1072. Goldman, P., Milne, G.W.A. & Keister, D.B. (1968) *J. Biol. Chem.* **243**, 428   [3.8.1.2]
1073. Goldman, P. & Vagelos, P.R. (1961) *J. Biol. Chem.* **236**, 2620   [2.3.1.20]
1074. Goldman, R. & Strominger, J.L. (1972) *J. Biol. Chem.* **247**, 5116   [3.6.1.27]
1075. Goldmark, P.J. & Liun, S. (1972) *J. Biol. Chem.* **247**, 1849   [3.1.11.5]
1076. Goldstein, F.B. (1959) *J. Biol. Chem.* **234**, 2702   [2.3.1.17]
1077. Goldthwait, D.A., Peabody, R.A. & Greenberg, G.R. (1956) *J. Biol. Chem.* **221**, 569
       [2.4.2.14, 6.3.4.13]
1078. Gomori, G. (1943) *J. Biol. Chem.* **148**, 139   [3.1.3.11]
1079. Goodhue, C.T. & Snell, E.E. (1966) *Biochemistry* **5**, 403   [1.1.1.106]
1080. Goodman, D.S., Huang, H.S., Kanai, M. & Shiratori, T. (1967) *J. Biol. Chem.* **242**, 3543
       [1.13.11.21]

1081. Goodman, D.S., Huang, H.S. & Shiratori, T. (1966) *J. Biol. Chem.* **241**, 1929 [1.13.11.21]

1082. Goodman, I., Fouts, J.R., Bresnick, E., Menegas, R. & Hitchings, G.H. (1959) *Science* **130**, 450 [3.2.3.1]

1083. Goore, M.Y. & Thompson, J.F. (1967) *Biochim. Biophys. Acta* **132**, 15 [2.3.2.2]

1084. Gordon, A.H., Green, D.E. & Subrahmanyan, V. (1940) *Biochem. J.* **34**, 764 [1.2.3.1]

1085. Goryachenkova, E.V. (1952) *Dokl. Akad. Nauk. S.S.S.R.* **87**, 457 [4.4.1.4]

1086. Goscin, S.A. & Fridovich, I. (1972) *Biochim. Biophys. Acta* **289**, 276 [1.15.1.1, 2.3.1.54]

1087. Gottesman, M.E. & Canellakis, E.S. (1966) *J. Biol. Chem.* **241**, 4339 [2.7.7.19, 2.7.7.31]

1088. Gotto, A.M. & Kornberg, H.L. (1961) *Biochem. J.* **81**, 273 [1.1.1.60]

1089. Gottschalk, A. (1958) *Advan, Enzymol.* **20**, 135 [3.2.1.18]

1090. Gottschalk, A. (1957) *Biochim. Biophys. Acta* **23**, 645

1091. Gottschalk, A. (1960) in *The Enzymes*, 2nd edn (Boyer, P.D., Lardy, H. & Myrbäck, K., ed.) Vol. 4, p. 461, Academic Press, New York

1092. Gottschalk, G. (1969) *Eur. J. Biochem.* **7**, 301 [4.1.3.28, 4.1.3.7]

1093. Gottschalk, G. & Barker, H.A. (1966) *Biochemistry* **5**, 1125 [4.1.3.28]

1094. Gould, R.M., Thornton, M.P., Liepkalns, V. & Lennarz, W.J. (1968) *J. Biol. Chem.* **243**, 3096 [2.3.2.11]

1095. Goulian, M. & Beck, W.S. (1966) *J. Biol. Chem.* **241**, 4233 [1.17.4.2]

1096. Gracy, R.W. & Noltmann, E.A. (1968) *J. Biol. Chem.* **243**, 4109 [5.3.1.8]

1097. Grant, D.J.W. & Patel, J.C. (1969) *J. Microbiol. Serol.* **35**, 325 [4.1.1.59, 4.1.1.61, 4.1.1.62, 4.1.1.63]

1098. Grant, J.K. & Brownie, A.C. (1955) *Biochim. Biophys. Acta* **18**, 433 [1.14.15.4]

1099. Graves, D.J., Fischer, E.H. & Krebs, E.G. (1960) *J. Biol. Chem.* **235**, 805 [3.1.3.17]

1100. Gray, R.W., Omdahl, J.L., Ghazarian, J.G. & DeLuca, H.F. (1972) *J. Biol. Chem.* **247**, 7528 [1.14.13.13]

1101. Grazi, E., Barbieri, G. & Gagliano, R. (1974) *Biochim. Biophys. Acta* **341**, 248

1102. Green, A.A. & Cori, G.T. (1943) *J. Biol. Chem.* **151**, 21 [2.4.1.1]

1103. Green, D.E., Leloir, L.F. & Nocito, W. (1945) *J. Biol. Chem.* **161**, 559 [2.6.1.2]

1104. Green, D.E., Mii, S., Mahler, H.R. & Bock, R.M. (1954) *J. Biol. Chem.* **206**, 1 [1.3.99.2]

1105. Green, M. & Cohen, S.S. (1956) *J. Biol. Chem.* **219**, 557 [5.3.1.3]

1106. Green, M.L. & Elliott, W.H. (1964) *Biochem. J.* **92**, 537 [1.1.1.103]

1107. Greenbaum, L.M. & Fruton, J.S. (1957) *J. Biol. Chem.* **226**, 173 [3.4.22.1]

1108. Greenberg, D.M. (1960) in *The Enzymes*, 2nd edn (Boyer, P.D., Lardy, H. & Myrbäck, K., ed.) Vol. 4, p. 257, Academic Press, New York [3.5.3.1]

1109. Greenberg, D.M., Bagot, A.E. & Roholt, O.A. (1956) *Arch. Biochem. Biophys.* **62**, 446 [3.5.3.1]

1110. Greenberg, D.M. & Winnick, T. (1940) *J. Biol. Chem.* **135**, 761 [3.4.99.21]

1111. Greenberg, D.M., Wynston, L.K. & Nagabhushanan, A. (1965) *Biochemistry* **4**, 1872 [6.3.3.2]

1112. Greenberg, E. & Preiss, J. (1965) *J. Biol. Chem.* **240**, 2341 [2.4.1.21]

1113. Greenberg, G.R. & Somerville, R.L. (1962) *Proc. Nat. Acad. Sci. USA* **48**, 247 [3.6.1.23]

1114. Greenstein, J.P. (1948) *Advan. Enzymol.* **8**, 117 [3.4.13.11]

1115. Gregolin, C. & Singer, T.P. (1963) *Biochim. Biophys. Acta* **67**, 201 [1.1.2.4]

1116. Gregolin, C., Singer, T.P., Kearney, E.B. & Boeri, E. (1961) *Ann. N.Y. Acad. Sci.* **94**, 780 [1.1.2.4, 1.1.99.6]

1117. Gregory, R.P.F. & Bendall, D.S. (1966) *Biochem. J.* **101**, 569 [1.10.3.1]

1118. Greten, H., Levy, R.I., Fales, H. & Fredrickson, D.S. (1970) *Biochim. Biophys. Acta* **210**, 39 [3.1.1.34]

1119. Griffin, M.J. & Brown, G.M. (1964) *J. Biol. Chem.* **239**, 310 [6.3.2.12]

1120. Griffin, M. & Trudgill, P.W. (1972) *Biochem. J.* **129**, 595 [1.1.1.163, 1.14.13.16]

1121. Griffin, T.B. & Prescott, J.M. (1970) *J. Biol. Chem.* **245**, 1348 [3.4.24.4]

1122. Griffith, T.J. & Helleiner, C.W. (1965) *Biochim. Biophys. Acta* **108**, 114 [2.7.4.11, 2.7.4.8]

1123. Griffiths, M.M. & Bernofsky, C. (1972) *J. Biol. Chem.* **247**, 1473 [2.7.1.86]

1124. Grindey, G.B. & Nichol, C.A. (1971) *Biochim. Biophys. Acta* **240**, 180 [3.6.1.23]

1125. Gripon J.C. & Hermier J. (1974) *Biochimie* **56**, 1323 [3.4.24.4]
1126. Grisolia, S. & Cardoso, S. (1957) *Biochim. Biophys. Acta* **25**, 430
 [1.3.1.2, 3.5.1.6, 3.5.2.2]
1127. Grisolia, S., Quijada, C.L. & Fernandez, M. (1964) *Biochim. Biophys. Acta* **81**, 61
 [1.4.1.4]
1128. Grollman, A.P. (1966) *Methods Enzymol.* **8**, 351 [2.4.1.69]
1129. Gros, P. & Labonesse, B. (1960) *Bull. Soc. Chim. Biol. (Paris)* **42**, 559 [3.4.22.8]
1130. Gross, G.G. & Zenk, M.H. (1969) *Eur. J. Biochem.* **8**, 413 [1.2.1.30]
1131. Gross, G.G. & Zenk, M.H. (1969) *Eur. J. Biochem.* **8**, 420 [1.1.1.91]
1132. Gross, S.R., Burns, R.O. & Umbarger, H.-E. (1963) *Biochemistry* **2**, 1046 [4.2.1.33]
1133. Gross, S.R., Gafford, R.D. & Tatum, E.L. (1956) *J. Biol. Chem.* **219**, 781
 [1.13.11.3, 5.5.1.5]
1134. Grossman, D. & Lang, K. (1962) *Biochem. Z.* **336**, 351 [3.6.1.11]
1135. Grov A., Iveresn O.-J. & Endresen C. (1974) *Eur. J. Biochem.* **48**, 193 [3.4.99.17]
1136. Grover, P.L. & Sims, P. (1964) *Biochem. J.* **90**, 603
1137. Grunberg-Manago, M., del Campillo-Campbell, A., Dondon, L. & Michelson, A.M.
 (1966) *Biochim. Biophys. Acta* **123**, 1 [2.7.7.5]
1138. Gryder, R.M. & Pogell, B.M. (1960) *J. Biol. Chem.* **235**, 558 [5.3.1.19]
1139. Guest, J.R., Friedman, S., Foster, M.A., Tejerina, G. & Woods, D.D. (1964) *Biochem. J.*
 92, 497 [2.1.1.13, 2.1.1.14]
1140. Gugler, R., Rao, G.S. & Breuer, H. (1970) *Biochim. Biophys. Acta* **220**, 69 [2.8.2.2]
1141. Guion-Rain, M.C., Portemer, C. & Chatagner, F. (1975) Biochim. Biophys. Acta **384**, 265
 [4.1.1.29]
1142. Gulland, J.M. & Jackson, E.M. (1938) *Biochem. J.* **32**, 597 [3.1.3.5]
1143. Gunetileke, K.G. & Anwar, R.A. (1968) *J. Biol. Chem.* **243**, 5770 [2.5.1.7]
1144. Gunsalus, C.F., Stanier, R.Y. & Gunsalus, I.C. (1953) *J. Bacteriol.* **66**, 548
 [1.2.1.28, 1.2.1.7, 4.1.1.7, 5.1.2.2]
1145. Gunsalus, I.C. (1954) in *A Symposium on the Mechanism of Enzyme Action,* (McElroy,
 W.D. & Glass, B., ed.) p. 545, Johns Hopkins Press, Baltimore [2.3.1.11, 2.3.1.12]
1146. Gunsalus, I.C., Barton, L.S. & Gruber, W. (1956) *J. Amer. Chem. Soc.* **78**, 1763
 [2.3.1.12]
1147. Güntelberg, A.V. & Ottesen, M. (1954) *C.R. Trav. Lab. Carlsberg* **29**, 36 [3.4.21.14]
1148. Gupta, N.K. & Robinson, W.G. (1960) *J. Biol. Chem.* **235**, 1609 [1.1.1.55]
1149. Gupta, N.K. & Vennesland, B. (1964) *J. Biol. Chem.* **239**, 3787 [4.1.1.47]
1150. Guroff, G. & Rhoads, C.A. (1969) *J. Biol. Chem.* **244**, 142 [1.14.16.1]
1151. Haas, E., Horecker, B.L. & Hogness, T.R. (1940) *J. Biol. Chem.* **136**, 747 [1.6.2.4]
1152. Haberkern, R.C. & Cantoni, G.L. (1973) *Biochemistry* **12**, 2389. [3.1.26.4]
1153. Haberman, A., Heywood, J. & Meselson, M. *Proc. Nat. Acad. Sci. USA* **69**, 3138 (1972).
 [3.1.24.2]
1154. Haberman, A. J. (1974) *Mol. Biol.* **89, 545. [3.1.24.3]**
1155. Habig, W.H., Pabst, M.J., Fleischner, G., Gatmaitan, Z., Arias, I.M. & Jakoby, W.B.
 (1974) *Proc. Nat. Acad. Sci. USA* **71**, 3879 [1.14.15.1, 2.5.1.18]
1156. Habig, W.H., Pabst, M.J. & Jakoby, W.B. (1974) *J. Biol. Chem.* **249**, 7130 [2.5.1.18]
1157. Hadi, S.M. & Goldthwait, D.A. (1971) *Biochemistry* **10**, 4986 [3.1.21.2]
1158. Hageman, R.H. & Arnon, D.I. (1955) *Arch. Biochem. Biophys.* **55**, 162 [1.2.1.12]
1159. Hagemeyer K., Fawwal I. & Whitaker J.R. (1968) *J. Dairy Sci.* **51**, 1916 [3.4.23.6]
1160. Hagemeyer, K., Fawwal, I. & Whitaker, J.R. (1968) *J. Dairy Sci.* **51**, 1916 [3.4.23.6]
1161. Hager, L.P. (1962) in *The Enzymes*, 2nd edn (Boyer, P.D., Lardy, H. & Myrbäck, K.,
 ed.) Vol. 6, p. 387, Academic Press, New York [6.2.1.4, 6.2.1.5]
1162. Hager, L.P., Geller, D.M. & Lipmann, F. (1954) *Fed. Proc.* **13**, 734 [1.2.3.3]
1163. Hager, S.E., Gregerman, R.I. & Knox, W.E. (1957) *J. Biol. Chem.* **225**, 935
 [1.13.11.27]
1164. Hagerman, D.D., Rosenberg, H., Ennor, A.H., Schiff, P. & Inove, S. (1965) *J. Biol.
 Chem.* **240**, 1108 [3.1.4.13]
1165. Hagino, H. & Nakayama, K. (1968) *Agr. Biol. Chem. (Tokyo)* **32**, 727 [4.1.1.57]
1166. Hagopian, A. & Eyler, E.M. (1969) *Arch. Biochem. Biophys.* **129**, 515 [2.4.1.41]
1167. Hahlbrock, K. & Conn, E.E. (1970) *J. Biol. Chem.* **245**, 917 [2.4.1.63]
1168. Haines, W.J. (1952) *Recent Progr. Hormone Res.* **7**, 255 [1.14.99.11]
1169. Hajer, S.E., Gregerman, R.I. & Knox, W.E. (1957) *J. Biol. Chem.* **225**, 935 [1.14.12.2]
1170. Hajra, A.K. (1968) *J. Biol. Chem.* **243**, 3458 [2.3.1.42]

1171. Hajra, A.K. & Agranoff, B.W. (1968) *J. Biol. Chem.* **243**, 3542   [1.1.1.101]
1172. Hakim, A.A. (1959) *Nature (London)* **183**, 334   [2.7.7.8]
1173. Haley, E.E. (1968) *J. Biol. Chem.* **243**, 5748   [3.4.13.10]
1174. Halpern, Y.S. & Grossowicz, N. (1957) *Biochem. J.* **65**, 716   [3.5.1.1]
1175. Hanahan, D.J., Brockerhoff, H. & Barron, E.J. (1960) *J. Biol. Chem.* **235**, 1917
   [3.1.1.4]
1176. Hanahan, D.J. & Chaikoff, I.L. (1948) *J. Biol. Chem.* **172**, 191   [3.1.4.4]
1177. Hanahan, D.J. & Vercomer, R. (1954) *J. Amer. Chem. Soc.* **76**, 1804   [3.1.4.3]
1178. Hanberg, F., Myers, P.A. & Roberts, R.J., unpublished results.   [3.1.23.45]
1179. Handler, P., Bernheim, M.L.C. & Klein, J.R. (1941) *J. Biol. Chem.* **138**, 211   [1.5.3.1]
1180. Hanes, C.S. (1940) *Proc. Roy. Soc. Ser. B.* **128**, 421   [2.4.1.1]
1181. Hanley, W.B., Boyer, S.H. & Haughton, M.A. (1966) *Nature (London)* **209**, 996
   [3.4.23.3]
1182. Hanna, R., Picken, M. & Mendicino, J. (1973) *Biochim. Biophys. Acta* **315**, 259
   [1.1.1.114]
1183. Hannonen, P., Jänne, J. & Raina, A. (1972) *Biochim. Biophys. Acta* **289**, 225
   [2.5.1.16]
1184. Hansen, R.G., Verachtert, H., Rodriguez, P. & Bass, S.T. (1966) *Methods Enzymol.* **8**, 269
   [2.7.7.29]
1185. Hanson H., Glasser D. & Kirsche H. (1965) Hoppe-Seyler's *Z. Physiol. Chem.* **340**, 107
   [3.4.11.1]
1186. Hanson H., Glässer, D., Ludewig, M., Mannsfeldt, H.G., John, M. & Nesvadba, H.
(1967) Hoppe-Seyler's *Z. Physiol. Chem.* **348**, 689   [3.4.11.1]
1187. Hanson H., Hutter H.J., Mannsfeldt H.G., Kretschmer K. & Sohr C. (1967) Hoppe-
Seyler's *Z. Physiol. Chem.* **348**, 680   [3.4.11.2]
1188. Hanson, H.T. & Smith, E.L. (1949) *J. Biol. Chem.* **179**, 789   [3.4.13.3]
1189. Hanson, T.L., Lueck, J.D., Horne, R.N. & Nordlie, R.C. (1970) *J. Biol. Chem.* **245**, 6078
   [3.1.3.9]
1190. Harada, I., Noguchi, T. & Kido, R. (1978) *Hoppe-Seyler's Z. Phsiol. Chem.* (in press)
   [2.6.1.60]
1191. Harada, T., Shimizu, S., Nakanishi, Y. & Suzuki, S. (1967) *J. Biol. Chem.* **242**, 2288
   [2.8.2.7]
1192. Harary, I., Korey, S.R. & Ochoa, S. (1953) *J. Biol. Chem.* **203**, 595   [1.1.1.40]
1193. Hardman, J.G. & Sutherland, E.W. (1969) *J. Biol. Chem.* **244**, 6363   [4.6.1.2]
1194. Hareland, W.A., Crawford, R.L., Chapman, P.J. & Dagley, S. (1975) *J. Bacteriol.* **121**,
272   [1.14.13.18]
1195. Harkness, D.R. (1968) *Arch. Biochem. Biophys.* **126**, 513   [3.1.3.1]
1196. Harmison, C.R. & Mammen, E.F. (1967) in *Blood Clotting Enzymology* (Seegers, W.H.,
ed.) p. 23, Academic Press, London   [3.4.21.5]
1197. Harper E. & Gross J. (1970) *Biochim. Biophys. Acta* **198**, 286   [3.4.99.31]
1198. Harper E. & Kang A.H. (1970) *Biochem. Biophys. Res. Commun.* **41**, 482   [3.4.24.3]
1199. Harper E., Seifter S. & Hospelhorn V. (1965) *Biochem. Biophys. Res. Commun.* **18**, 627
   [3.4.24.3]
1200. Harris E.D., Dibona D.R. & Krane S.M. (1969) *J. Clin. Invest.* **48**, 2104   [3.4.24.7]
1201. Harris E.D. & Krane S.M. (1972) *Biochim. Biophys. Acta* **258**, 566   [3.4.99.31]
1202. Hartley, B.S., Naughton, M.A. & Sanger, F. (1959) *Biochim. Biophys. Acta* **34**, 243
   [3.4.21.11]
1203. Hartman, S.C. & Buchanan, J.M. (1958) *J. Biol. Chem.* **233**, 451   [2.4.2.14]
1204. Hartman, S.C. & Buchanan, J.M. (1958) *J. Biol. Chem.* **233**, 456   [6.3.4.13]
1205. Hartman, S.C. & Buchanan, J.M. (1959) *J. Biol. Chem.* **234**, 1812   [2.1.2.2, 2.1.2.3]
1206. Hartshorne, D. & Greenberg, D.M. (1964) *Arch. Biochem. Biophys.* **105**, 173
   [1.1.1.103]
1207. Harwood, J.H. & Smith, D.H. (1969) *J. Bacteriol.* **97**, 1262   [2.7.7.47]
1208. Hasegawa, S. & Nagel, C.W. (1968) *Arch. Biochem. Biophys.* **124**, 513
   [3.2.1.67, 3.2.1.82]
1209. Hasegawa, S., Nordin, J.H. & Kirkwood, S. (1969) *J. Biol. Chem.* **244**, 5460   [3.2.1.59]
1210. Hasegawa, H. (1977) *J. Biochem. (Tokyo)* **81**, 169   [1.6.1.1, 1.6.99.10]
1211. Hashimoto, T. & Yoshikawa, H. (1962) *Biochim. Biophys. Acta* **65**, 355   [2.7.2.3]
1212. Hashimoto, Y. (1971) *J. Agr. Chem. Soc. (Japan)* **45**, 147   [3.2.1.90]
1213. Hashitani, Y. & Cooper, J.R. (1972) *J. Biol. Chem.* **247**, 2117   [3.6.1.28]

468

1214. Haslewood, E.S. & Haslewood, G.A.D. (1976) *Biochem. J.* **157**, 207
1215. Hassall, H. & Greenberg, D.M. (1971) *Methods Enzymol.* **17B** 84
1216. Hassid, W.Z. & Doudoroff, M. (1950) *Advan. Carbohyd. Chem.* **5**, 29   [2.4.1.7]
1217. Hata, T., Hayaishi, R. & Doi, E. (1967) *Agr. Biol. Chem. (Tokyo)* **31**, 357
       [3.4.22.9, 3.4.23.6]
1218. Hata T., Hayashi R. & Doi E. (1967) *Agr. Biol. Chem. (Tokyo)* **31**, 150   [3.4.16.1]
1219. Hatanaka, C. & Ozawa, J. (1968) *J. Agr. Chem. Soc. (Japan)* **43**, 764; (1971) *Ber. des
       O'Hara Inst.* **15**, 47   [3.2.1.82]
1220. Hatch, M.D. & Slack, C.R. (1969) *Biochem. Biophys. Res. Commum.* **34**, 589   [1.1.1.82]
1221. Hatch, M.D. & Slack, C.R. (1968) *Biochem. J.* **106**, 141   [2.7.9.1]
1222. Hatch, M.D. & Stumpf, P.K. (1961) *J. Biol. Chem.* **236**, 2879   [6.4.1.2]
1223. Hatch, M.D. & Turner, J.F. (1960) *Biochem. J.* **76**, 556   [1.6.4.4]
1224. Hatefi, Y., Osborn, M.J., Kay, L.D. & Huennekens, F.M. (1957) *J. Biol. Chem.* **227**, 637
       [1.5.1.5]
1225. Hatfield, D. & Wyngaarden, J.B. (1964) *J. Biol. Chem.* **239**, 2580   [2.4.2.20]
1226. Hathaway, J.A. & Atkinson, D.E. (1963) *J. Biol. Chem.* **238**, 2875   [1.1.1.41]
1227. Hauge, J.G., Crane, F.L. & Beinert, H. (1956) *J. Biol. Chem.* **219**, 727
       [1.3.99.2, 1.3.99.3]
1228. Hausen, P. & Stein, H. (1970) *Eur. J. Biochem.* **14**, 278
1229. Hausmann, E. (1967) *Biochim. Biophys. Acta* **133**, 591   [1.14.11.4]
1230. Havir, E.A. Private Communication   [4.3.1.5]
1231. Hawker, J.S. & Hatch, M.D. (1966) *Biochem. J.* **99**, 102   [3.1.3.24]
1232. Hayaishi, O. (1955) *J. Biol. Chem.* **215**, 125   [2.8.3.3]
1233. Hayaishi, O. (1963) in *The Enzymes, 2nd ed. (Boyer, P.D., Lardy, H. and Myrbäck, K.,
       ed.) Vol. 8, p. 353, Academic Press, New York*
       [1.13.11.1, 1.13.11.2, 1.13.11.4, 1.13.11.5, 1.13.11.6, 3.5.1.2]
1234. Hayaishi, O., Katagiri, M. & Rothberg, S. (1957) *J. Biol. Chem.* **229**, 905   [1.13.11.1]
1235. Hayaishi, O. & Kornberg, A. (1952) *J. Biol. Chem.* **197**, 717   [1.2.99.1, 3.5.2.1, 3.5.4.1]
1236. Hayaishi, O. & Kornberg, A. (1954) *J. Biol. Chem.* **206**, 647   [3.1.4.2]
1237. Hayaishi, O., Nishizuka, Y., Tatibana, M., Takeshita, M. & Kuno, S. (1961) *J. Biol.
       Chem.* **236**, 781   [1.2.1.18, 2.6.1.18]
1238. Hayaishi, O. Private Communication
1239. Hayaishi, O., Rothberg, S., Mehler, A.H. & Saito, Y. (1957) *J. Biol. Chem.* **229**, 889
       [1.13.11.11]
1240. Hayaishi, O., Saito, Y., Jakoby, W.B. & Stohlman, E.F. (1955) *Arch. Biochem. Biophys.*
       56, 554   [1.1.1.52]
1241. Hayaishi, O., Shimazono, H., Katagiri, M. & Saito, Y. (1956) *J. Amer. Chem. Soc.* **78**,
       5126   [3.7.1.1]
1242. Hayaishi, O. & Stanier, R.Y. (1951) *J. Bacteriol.* **62**, 691   [3.5.1.9]
1243. Hayaishi, O. & Sutton, W.B. (1957) *J. Amer. Chem. Soc.* **79**, 4809   [1.13.12.4]
1244. Hayano, K. & Fukui, S. (1967) *J. Biol. Chem.* **242**, 3665   [1.1.99.13]
1245. Hayano, M. & Dorfman, R.I. (1952) *Arch. Biochem. Biophys.* **36**, 237   [1.14.99.10]
1246. Hayano, M. & Dorfman, R.I. (1954) *J. Biol. Chem.* **211**, 227   [1.14.15.4]
1247. Hayashi, K., Fukushima, D. & Mogi, K. (1967) *Agr. Biol. Chem. (Tokyo)* **31**, 1171
       [3.4.21.14]
1248. Hayashi, R., (1976) Methods Enzymol. **45**, 568   [3.4.15.2]
1249. Hayashi R., Bai Y. & Hata T. (1975) *J. Biochem. (Tokyo)* **77**, 69   [3.4.16.1]
1250. Hayashi R., Moore S. & Stein W.H. (1973) *J. Biol. Chem.* **248**, 2296   [3.4.16.1]
1251. Hayashi, S. & Lin, E.C.C. (1967) *J. Biol. Chem.* **242**, 1030   [2.7.1.32]
1252. Hayman S. & Patterson E.K. (1971) *J. Biol. Chem.* **246**, 660   [3.4.13.11]
1253. Hearn, V.M., Smith, Z.G. & Watkins, W.M. (1968) *Biochem. J.* **109**, 315   [2.4.1.40]
1254. Heath, E.C. & Ghalambor, M.A. (1962) *J. Biol. Chem.* **237**, 2423   [2.7.1.51]
1255. Heath, E.C., Horecker, B.L., Smyrniotis, P.Z. & Takagi, Y. (1958) *J. Biol. Chem.* **231**,
       1031   [5.3.1.4]
1256. Heath, E.C., Hurwitz, J., Horecker, B.L. & Ginsburg, A. (1958) *J. Biol. Chem.* **231**, 1009
       [4.1.2.9]
1257. Hedegaard, J. & Gunsalus, I.C. (1965) *J. Biol. Chem.* **240**, 4038
1258. Hedgpeth, J., Goodman, H.M. & Boyer, H.W. (1972) *Proc. Nat. Acad. Sci. USA* **69**, 3448
       [3.1.23.13]
1259. Hedrick, J.L. & Sallach, H.J. (1964) *Arch. Biochem. Biophys.* **105**, 261   [4.1.1.40]

1260. Hehre, E.J. (1951) *Advan. Enzymol.* **11**, 297
        [2.4.1.10, 2.4.1.18, 2.4.1.19, 2.4.1.2, 2.4.1.25, 2.4.1.4, 2.4.1.5]
1261. Hehre, E.J. & Hamilton, D.M. (1953) *J. Biol. Chem.* **192**, 161   [2.4.1.2]
1262. Hehre, E.J. & Hamilton, D.M. (1949) *Proc. Soc. Exp. Biol. (New York)* **71**, 336
        [2.4.1.2]
1263. Hehre, E.J., Hamilton, D.M. & Carlson, A.S. (1949) *J. Biol. Chem.* **177**, 267   [2.4.1.4]
1264. Heilmeyer, L., Battig, F. & Holzer, H. (1968) *Eur. J. Biochem.* **9**, 259   [3.1.4.15]
1265. Heinrikson, R.L. & Goldwasser, E. (1964) *J. Biol. Chem.* **239**, 1177   [4.2.1.70]
1266. Held, V.E. & Buddecke, E. (1967) *Hoppe-Seyler's Z. Physiol. Chem.* **348**, 1047
        [3.1.6.9]
1267. Hele, P. (1954) *J. Biol. Chem.* **206**, 671   [6.2.1.1]
1268. Heller, M. & Shapiro, B. (1966) *Biochem. J.* **98**, 763   [3.1.4.12]
1269. Hellig, H. & Popjak, G. (1961) *J. Lipid Res.* **2**, 235   [2.7.1.36]
1270. Helmsing, P.J. (1969) *Biochim. Biophys. Acta* **178**, 519   [3.1.1.26]
1271. Helting, T. & Erbing, B. (1973) *Biochim. Biophys. Acta* **293**, 94   [2.4.1.90]
1272. Hendrickson, H.R. (1968) *Fed. Proc.* **27**, 593   [4.4.1.9]
1273. Hendrickson, H.R. & Conn, E.E. (1969) *J. Biol. Chem.* **244**, 2632   [4.4.1.9]
1274. Hengstenberg, W., Penberthy, W.K. & Morse, M.L. (1970) *Eur. J. Biochem.* **14**, 27
        [3.2.1.85]
1275. Hennessey, T.D. & Richmond, M.H. (1968) *Biochem. J.* **109**, 469   [3.5.2.6]
1276. Hennessey, T.D. & Richmond, M.H. (1968) *Biochem. J.* **109**, 469
1277. Henning, U., Möslein, E.M. & Lynen, F. (1959) *Arch. Biochem. Biophys.* **83**, 259
        [2.7.4.2]
1278. Henson, C.P. & Cleland W.W. (1964) *Biochemistry* **3**, 338   [2.6.1.1]
1279. Heppel, L.A. (1966) in *Procedures in Nucleic Acid Research* (Cantoni, G.L. & Davies,
        D.R., ed.) Harper & Row, New York, p. 31   [3.1.27.1]
1280. Heppel, L.A. & Hilmoe, R.J. (1950) *J. Biol. Chem.* **183**, 129
1281. Heppel, L.A. & Hilmoe, R.J. (1951) *J. Biol. Chem.* **188**, 665   [3.1.3.5]
1282. Heppel, L.A. & Hilmoe, R.J. (1952) *J. Biol. Chem.* **198**, 683   [2.4.2.1, 3.2.2.1]
1283. Heppel, L.A. & Hilmoe, R.J. (1953) *J. Biol. Chem.* **202**, 217   [3.6.1.8]
1284. Heppel, L.A., Hurwitz, J. & Horecker, B.L. (1957) *J. Amer. Chem. Soc.* **79**, 630
        [3.5.4.2]
1285. Heppel, L.A. & Porterfield, V.T. (1948) *J. Biol. Chem.* **176**, 763   [3.8.1.1]
1286. Heppel, L.A., Strominger, J.L. & Maxwell, E.S. (1959) *Biochim. Biophys. Acta* **32**, 422
        [2.7.4.4]
1287. Herbert, D. (1955) *Methods Enzymol.* **1**, 753   [4.1.1.3]
1288. Herbert, D. & Pinsent, J. (1948) *Biochem. J.* **43**, 193 & 203   [1.11.1.6]
1289. Hermodson M.A., Ericcson L.H., Neurath H. & Walsh K.A. (1973) *Biochemistry* **12**, 3146
        [3.4.21.4]
1290. Hernandez, H.H. & Chaikoff, I.L. (1957) *J. Biol. Chem.* **228** 447   [3.1.1.13]
1291. Hers, H.G. (1952) *Biochim. Biophys. Acta* **8**, 416   [2.7.1.3]
1292. Hers, H.G. (1960) *Biochim. Biophys. Acta* **37**, 120   [1.1.1.21]
1293. Hers, H.G. & Kusaka, T. (1953) *Biochim. Biophys. Acta* **11**, 427   [2.7.1.28]
1294. Hersch, L.B. & Jencks, W.P. (1967) *J. Biol. Chem.* **242**, 3468   [2.8.3.5]
1295. Hersh, L.B. (1970) *J. Biol. Chem.* **245**, 3526   [3.5.1.36]
1296. Hersh, L.B., Stark, M.J., Worthen, S. & Fiero, M.K. (1972) *Arch. Biochem. Biophys.* **150**,
        219   [1.5.99.5]
1297. Hersh, L.B., Tsai, L. & Stadtman, E.R. (1969) *J. Biol. Chem.* **244**, 4677   [3.5.1.36]
1298. Hestrin, S., Feingold, D.S. & Avigad, G. (1956) *Biochem. J.* **64**, 340   [2.4.1.10]
1299. Hey, A.E. & Elbein, A.D. (1966) *J. Biol. Chem.* **241**, 5473   [4.2.1.45]
1300. Heydanek, M.G. Jr. & Neuhaus, F.C. (1969) *Biochemistry* **8**, 1474   [2.7.8.13]
1301. Heyworth, R. & Walker, P.G. (1962) *Biochem. J.* **83**, 331   [3.2.1.21]
1302. Hickman, J. & Ashwell, G. (1958) *J. Biol. Chem.* **232**, 737   [2.7.1.17]
1303. Hickman, J. & Ashwell, G. (1959) *J. Biol. Chem.* **234**, 758   [1.1.1.10, 1.1.1.9]
1304. Hickman, J. & Ashwell, G. (1960) *J. Biol. Chem.* **235**, 1566   [1.1.1.57, 1.1.1.58]
1305. Hickman, J., Ashwell, G., Morell, A.G., van den Hamer, C.J.A. & Scheinberg, I.H.
        (1970) *J. Biol. Chem.* **245**, 759   [2.4.99.1]
1306. Higashi, Y., Siewert, G. & Strominger, J.L. (1970) *J. Biol. Chem.* **245**, 3683   [2.7.1.66]
1307. Higashi, Y., Strominger, J.L. & Sweeley, C.C. (1967) *Proc. Nat. Acad. Sci. USA* **57**, 1878
        [2.7.8.13]

470

1308. Hiles, R.A. & Henderson, L.M. (1972) *J. Biol. Chem.* **247**, 646 [2.7.1.81]
1309. Hill, D.L. & Chambers, P. (1967) *Biochim. Biophys. Acta* **148**, 435 [3.5.1.20]
1310. Hill, E.E. & Lands, W.E.M. (1968) *Biochim. Biophys. Acta* **152**, 645 [2.3.1.62, 2.3.1.63]
1311. Hill, R.L. (1965) *Advan. Prot. Chem.* **20**, 64
1312. Hillmer, P. & Gottschalk, G. (1974) *Biochim. Biophys. Acta* **334**, 12 [1.1.1.35]
1313. Hilz, H., Kittler, M. & Knape, G. (1959) *Biochem. Z.* **332**, 151 [1.8.1.2]
1314. Hilz, H., Knappe, J., Ringelmann, E. & Lynen, F. (1958) *Biochem. Z.* **329**, 476 [4.2.1.18]
1315. Hilz, H. & Lipmann, F. (1955) *Proc. Nat. Acad. Sci. USA* **41**, 880 [2.7.7.4]
1316. Himmelhoch, S.R. (1969) *Arch. Biochem. Biophys.* **134**, 597 [3.4.11.1]
1317. Himmelhoch S.R. (1970) in Methods in Enzymology **19**, (Colowick S.P. & Kaplan N.O. eds.) Acad. Press, New York & London, p. 508 [3.4.11.1]
1318. Hiramaru, M., Uchida, T. & Egami, F. (1969) *J. Biochem. (Tokyo)* **65**, 701. [3.1.23.19, 3.1.26.1]
1319. Hiramatsu, A. (1967) *J. Biochem. (Tokyo)* **62**, 353 [3.4.24.4]
1320. Hirata, M. & Hayaishi, O. (1967) *Biochim. Biophys. Acta* **149**, 1 [4.6.1.1]
1321. Hirayama, O., Matsuda, H., Takeda, H., Maenaka, K. & Takatsuka, H. (1975) *Biochim. Biophys. Acta* **384**, 127 [3.1.1.26]
1322. Hirsch, H. & Greenberg, D.M. (1967) *J. Biol. Chem.* **242**, 2283 [2.6.1.52]
1323. Hirshfeld, I.N., Rosenfeld, H.J., Leifer, Z. & Maas, W.K. (1970) *J. Bacteriol.* **101**, 725 [3.5.3.11]
1324. Hitchings, G.H. & Falco, E.A. (1944) *Proc. Nat. Acad. Sci. USA* **30**, 294 [3.5.4.3]
1325. Ho, P.P.K., Frank, B.H. & Burck, P.J. (1969) *Science* **165**, 510 [3.5.1.1]
1326. Hoagland, M.B. & Novelli, G.D. (1954) *J. Biol. Chem.* **207**, 767 [2.7.1.24, 2.7.7.3]
1327. Hobson, G.E. & Rees, K.R. (1957) *Biochem. J.* **65**, 305 [2.7.3.1, 2.7.3.4]
1328. Hochstein, L.I. & Dalton, B.P. (1973) *Biochim. Biophys. Acta* **302**, 216 [1.6.99.3]
1329. Hochstein, L.I. & Dalton, B.P. (1967) *Biochim. Biophys. Acta* **139**, 56 [1.5.99.4]
1330. Hochster, R.M. & Watson, R.W. (1954) *Arch. Biochem. Biophys.* **48**, 120 [5.3.1.5]
1331. Hodgins, D. & Abeles, R.H. (1967) *J. Biol. Chem.* **242**, 5158 [1.4.4.1]
1332. Hodgins, D.S. & Abeles, R.H. (1969) *Arch. Biochem. Biophys.* **130**, 274 [1.4.4.1]
1333. Hoffee, P.A. (1968) *Arch. Biochem. Biophys.* **126**, 795 [4.1.2.4]
1334. Hoffmann-Ostenhof, O., Kenedy, J., Keck, K., Gabriel, O. & Schönfellinger, H.W. (1954) *Biochim. Biophys. Acta* **14**, 285 [2.7.4.1]
1335. Hofmann, E.C.G. & Rapoport, S. (1955) *Biochim. Biophys. Acta* **18**, 296 [3.2.2.5]
1336. Hofmann, H., Wagner, I. & Hoffmann-Ostenhof, O. (1969) *Hoppe Seyler's Z. Physiol. Chem.* **350**, 1465 [2.1.1.39]
1337. Hofmann T. & Shaw R. (1964) *Biochim. Biophys. Acta* **92**, 543 [3.4.23.6]
1338. Hofsten, B.V. & Reinhammar, B. (1965) *Biochim. Biophys. Acta* **110**, 599 [3.4.21.14]
1339. Hofsten, B.V., van Kley, H. & Eaker, D. (1965) *Biochim. Biophys. Acta,* **110**, 585
1340. Hogness, D.S. & Battley, E.H. (1957) *Fed. Proc.* **16**, 197 [3.2.1.22]
1341. Holcenberg, J.S. & Stadtman, E.R. (1969) *J. Biol. Chem.* **244**, 1194 [1.5.1.13]
1342. Holcenberg, J.S. & Tsai, L. (1969) *J. Biol. Chem.* **244**, 1204 [1.3.7.1]
1343. Holden, M. (1961) *Biochem. J.* **78**, 359 [3.1.1.14]
1344. Holleman W.H. & Coen L.J. (1970) *Biochim. Biophys. Acta* **200**, 587 [3.4.21.28]
1345. Holley, R.W., Brunngraber, E.F., Saad, F. & Williams, H.H. (1961) *J. Biol. Chem.* **236**, 197 [6.1.1.1, 6.1.1.3]
1346. Holley, R.W. & Goldstein, J. (1959) *J. Biol. Chem.* **234**, 1765 [6.1.1.7]
1347. Hollmann, S. & Touster, O. (1957) *J. Biol. Chem.* **225**, 87 [1.1.1.10, 1.1.1.56]
1348. Holloway, P.W. & Popják, G. (1967) *Biochem. J.* **104**, 57 [2.5.1.1]
1349. Holmes, P.E. & Rittenberg, S.C. (1972) *J. Biol. Chem.* **247**, 7622 [1.14.13.10]
1350. Holmes, P.E., Rittenberg, S.C. & Knackmuss, H.J. (1972) *J. Biol. Chem.* **247**, 7628 [1.14.13.10]
1351. Holmlund, C.E. & Blank, R.H. (1965) *Arch. Biochem. Biophys.* **109**, 29 [3.1.1.37]
1352. Holt, A. & Wold, F. (1961) *J. Biol. Chem.* **236**, 3227 [4.2.1.11]
1353. Holten, D. & Fromm, H.J. (1961) *J. Biol. Chem.* **236**, 2581 [2.7.1.27]
1354. Holzer, H. & Holldorf, A. (1957) *Biochem. Z.* **329**, 292 [1.1.1.29]
1355. Holzer, H. & Schneider, S. (1961) *Biochim. Biophys. Acta* **48**, 71 [1.1.1.65]
1356. Hook C.W., Bull F.G., Iwanij V. & Brown S.I. (1972) *Invest. Ophthalmol.* **11**, 728 [3.4.24.7]
1357. Hook, R.H. & Robinson, W.G. (1964) *J. Biol. Chem.* **239**, 4257 & 4263 [3.5.5.2]

1358. Hopgood, M.F. & Walker, D.J. (1969) *Aust. J. Biol. Sci.* **22**, 1413 [1.3.1.6]
1359. Hopkins, R.P., Drummond, E.C. & Callaghan, P. (1973) *Biochem. Soc. Trans.* **1**, 989 [1.10.1.1]
1360. Hopkins, T.A., Seliger, H.H., White, E.H. & Cass, M.W. (1967) *J. Amer. Chem. Soc.* **89**, 7148 [1.13.12.7]
1361. Hopner, T. & Knappe, J. (1965) *Biochem. Z.* **342**, 190 [6.3.4.11]
1362. Hopper, D.J. & Cooper, R.A. (1971) *FEBS Lett.* **13**, 213 [4.2.99.11]
1363. Hopsu, V.K., Kantonen, U.-M. & Glenner, G.G. (1964) *Life Sci.* **3**, 1449
1364. Hopsu, V.K., Mäkinen, K.K. & Glenner, G.G. (1966) *Arch. Biochem. Biophys.* **114**, 557 [3.4.11.6]
1365. Hopsu, V.K., Mäkinen, K.K & Glenner, G.G. (1966) *Arch. Biochem. Biophys.* **114**, 567 [3.4.11.6]
1366. Horecker, B.L. (1950) *J. Biol. Chem.* **183**, 593 [1.6.2.4]
1367. Horecker, B.L., Hurwitz, J. & Heppel, L.A. (1957) *J. Amer. Chem. Soc.* **79**, 701 [3.6.1.6]
1368. Horecker, B.L. & Smyrniotis, P.Z. (1955) *J. Biol. Chem.* **212**, 811 [2.2.1.2]
1369. Horecker, B.L., Smyrniotis, P.Z. & Hurwitz, J. (1956) *J. Biol. Chem.* **223**, 1009 [2.2.1.1]
1370. Horecker, B.L., Smyrniotis, P.Z. & Seegmiller, J.E. (1951) *J. Biol. Chem.* **193**, 383 [5.3.1.6]
1371. Horgan, D.J., Stoops, J.K., Webb, E.C. & Zerner, B. (1969) *Biochemistry* **8**, 2000 [3.1.1.1]
1372. Hori, K., Anderson, J.M., Ward, W.W. & Cormier, M.J. (1975) *Biochemistry* **14**, 2371 [1.13.12.5]
1373. Hornemann, U., Speedie, M.K., Hurley, L.H. & Floss, H.G. (1970) *Biochem. Biophys. Res. Commun.* **39**, 594 [2.1.1.47]
1374. Horton, A.A. & Kornberg, H.L. (1964) *Biochim. Biophys. Acta* **89**, 381 [4.1.1.3]
1375. Horwitz, S.B. & Kaplan, N.O. (1964) *J. Biol. Chem.* **239**, 830 [1.1.1.14]
1376. Hoshino, K. (1960) *Nippon Nogei Kagaku Kaishi* **34**, 606 [1.1.1.80]
1377. Hoshino, K. & Udagawa, K. (1960) *Nippon Nogei Kagaku Kaishi* **34**, 616 [1.1.1.80]
1378. Hoskins, D.D. & Bjur, R.A. (1964) *J. Biol. Chem.* **239**, 1856 [1.5.3.1]
1379. Hoskins, D.D. & MacKenzie, C.G. (1961) *J. Biol. Chem.* **236**, 177 [1.5.99.1, 1.5.99.2]
1380. Hosokawa, K. & Stainer, R.Y. (1966) *J. Biol. Chem.* **241**, 2453 [1.14.13.2]
1381. Hosoya, T., Kondo, Y. & Ui, N. (1962) *J. Biochem. (Tokyo)* **52**, 180 [1.11.1.8]
1382. Hou, C.T. & Perlman, D. (1970) *J. Biol. Chem.* **245**, 1289 [3.1.1.39]
1383. Houmard J. & Drapeau G.R. (1972) *Proc. Nat. Acad. Sci. USA* **69**, 3506 [3.4.21.19]
1384. Hovingh, P. & Linker, A. (1970) *J. Biol. Chem.* **245**, 6170 [4.2.2.7, 4.2.2.8]
1385. Howard, B.H., Jones, G. & Purdom, M.R. (1960) *Biochem. J.* **74**, 173 [3.2.1.37, 3.2.1.8]
1386. Howell, L.G., Spector, T. & Massey, V. (1972) *J. Biol. Chem.* **247**, 4340 [1.14.13.2]
1387. Hruska, J.F., Law, J.H. & Kézdy, F.J. (1969) *Biochem. Biophys. Res. Commun.* **36**, 272
1388. Hsu, L.L. & Mandell, A.J. (1973) *Life Sci.* **13**, 847 [2.1.1.49]
1389. Hu, A.S.L. & Cline, A.L. (1964) *Biochim. Biophys. Acta* **93**, 237 [1.1.1.118, 1.1.1.48]
1390. Huang, H.T., U.S. Pat. 2944943; (1960) *Chem. Abstr.* **54**, 20073 [3.4.23.3, 5.1.1.5]
1391. Huang, W.Y. & Tang, J. (1970) *J. Biol. Chem.* **245**, 2189 [1.1.1.99]
1392. Huang-Yang Y.H.J. & Meizel S. (1975) *Biol. Reprod.* **12**, 232 [3.4.21.10]
1393. Hubbard, R. (1956) *J. Gen. Physiol.* **39**, 935 [5.2.1.3]
1394. Hübener, H.G. & Sahrholz, F.G. (1960) *Biochem. Z.* **333** 95 [1.1.1.53]
1395. Hübener, H.J., Sahrholz, F.G., Schmidt-Thomé, J., Nesemann, G. & Junk, R. (1959) *Biochim. Biophys. Acta* **35**, 270 [1.1.1.53]
1396. Huennekens, F.M. (1957) *Fed. Proc.* **16**, 199 [3.5.1.10]
1397. Huennekens, F.M., Mahler, H.R. & Nordmann, J. (1951) *Arch. Biochem.* **30**, 77 [5.1.2.1]
1398. Huges, S.G., Bruce, T. & Murray, K., unpublished observations. [3.1.23.2]
1399. Hughes, D.E. & Williamson, D.H. (1952) *Biochem. J.* **51**, 45 [2.7.6.1]
1400. Hui A., Rao L., Kurosky A., Jones S.R., Mains G., Dixon J.W., Szewezuk A. & Hofmann T. (1974) *Arch. Biochem. Biophys.* **160**, 577 [3.4.16.1]
1401. Humphrey, G.F. (1957) *Biochem. J.* **65**, 546
1402. Hurión N., Fromentin H. & Keil B., unpublished results

472

1403. Hurion, N., Fromention, H. & Keil, B. (1977) *Comp. Biochem. Physiol.* **56B**, 259 [3.4.21.33]
1404. Hurlbert, R.B. & Reichard, P. (1955) *Acta Chem. Scand.* **9**, 251 [2.7.6.1]
1405. Hurlbert, R.E. & Jakoby, W.B. (1965) *J. Biol. Chem.* **240**, 2772 [2.7.1.35, 4.2.1.32]
1406. Hurwitz, J. (1953) *J. Biol. Chem.* **205**, 935
1407. Hurwitz, J. (1959) *J. Biol. Chem.* **234**, 2351 [2.7.4.14, 2.7.4.9]
1408. Hurwitz, J., Gold, M. & Anders, M. (1964) *J. Biol. Chem.* **239**, 3462 [2.1.1.30, 2.1.1.31, 2.1.1.33, 2.1.1.35]
1409. Hurwitz, J., Heppel, L.A. & Horecker, B.L. (1957) *J. Biol. Chem.* **226**, 525 [3.2.2.4]
1410. Hurwitz, J. & Horecker, B.L. (1956) *J. Biol. Chem.* **223**, 993 [5.1.3.1]
1411. Hurwitz, J., Weissbach, A., Horecker, B.L. & Smyrniotis, P.Z. (1956) *J. Biol. Chem.* **218**, 769 [2.7.1.19, 5.3.1.6]
1412. Huseby, N.E., Christensen, T.B., Olsen, B.R. & Størmer, F.C. (1971) *Eur. J. Biochem.* **20**, 209 [4.1.3.18]
1413. Huthison, C.A. & Barrell, B.G., unpublished observations [3.1.23.22]
1414. Hutton, J.J. Jr., Tappel, A.L. & Udenfriend, S. (1967) *Arch. Biochem. Biophys.* **118**, 231 [1.14.11.2]
1415. Hutzler, J. & Dancis, J. (1968) *Biochim. Biophys. Acta* **158**, 62 [1.5.1.8]
1416. Hutzler, J. & Dancis, J. (1970) *Biochim. Biophys. Acta* **206**, 205 [1.5.1.9]
1417. Hylin, J.W. & Sawai, K. (1964) *J. Biol. Chem.* **239**, 990 [3.2.1.25]
1418. Hylin, J.W. & Wood, J.L. (1959) *J. Biol. Chem.* **234**, 2141 [2.8.1.2]
1419. I. Emöd & B. Keil (1977) *FEBS Letters* **77**, 51 [1.13.99.2]
1420. Ichihara, A., Adachi, K., Hosokawa, K. & Takeda, Y. (1962) *J. Biol. Chem.* **237**, 2296
1421. Ichihara, A. & Greenberg, D.M. (1957) *J. Biol. Chem.* **225**, 949 [2.7.1.31, 3.4.22.8]
1422. Ichihara, A. & Ichihara, E.A. (1961) *J. Biochem. (Tokyo)* **49**, 154 [1.2.1.20]
1423. Ichihara, A., Ichihara, E.A. & Suda, M. (1960) *J. Biochem. (Tokyo)* **48**, 412 [2.6.1.48]
1424. Ichihara, A. & Koyama, E. (1966) *J. Biochem. (Tokyo)* **59**, 160 [2.6.1.42]
1425. Ichishima E. (1972) *Biochim. Biophys. Acta* **258**, 274 [3.4.16.1]
1426. Ichishima, E. & Yoshida, F. (1965) *Biochim. Biophys. Acta* **99**, 360 [3.4.23.6]
1427. Ichishma E. (1970) in *Methods Enzymol.* **193**, 397 [3.4.23.6]
1428. Ichiyama, A., Nakamura, S., Kawai, H., Honjo, T., Nishizuka, Y., Hayaishi, O. & Senoh, S. (1965) *J. Biol. Chem.* **240**, 740 [1.2.1.32, 4.1.1.45]
1429. Ichiyama, A., Nakamura, S., Nisehizuka, Y. & Hayaishi, O. (1970) *J. Biol. Chem.* **245**, 1699 [1.14.16.4]
1430. Ihle J.W. & Dure L.S. (1972) *J. Biol. Chem.* **247**, 5034, & 5041 [3.4.15.2, 3.4.16.1]
1431. Ii, I. & Sakai, H. (1974) *Biochim. Biophys. Acta* **350**, 141 [1.6.4.2]
1432. Iio K. & Yamasaki M., *Biochem. Biophys. Acta (1976)* **429**, 912 [3.4.23.6]
1433. Ikawa, S., Shibata, T. & Ando, T. *J. Biochem. (Tokyo)* **80**, 1457-1460 (1976). [3.1.23.11]
1434. Ikeda, M., Levitt, M. & Udenfriend, S. (1967) *Arch. Biochem. Biophys.* **120**, 420 [1.14.16.2]
1435. Ikehata, H. (1960) *J. Gen. Appl. Microbiol.* **6**, 30 [3.5.99.2]
1436. Ikuta, S., Imamura, S., Misaki, H. & Horiuti, Y. (1977) *J. Biochem. (Tokyo)* **82**, 1741 [1.1.3.17, 1.6.1.1]
1437. Illingworth Brown, B. & Brown, D.H. (1966) *Methods Enzymol.* **8**, 395 [2.4.1.18]
1438. Illingworth Brown, B. & Brown, D.H. (1965) *Biochim. Biophys. Acta* **110**, 124 [3.2.1.3]
1439. Imada, A. (1967) *J. Gen. Appl. Microbiol.* **13**, 267 [3.2.2.10]
1440. Imada, A., Kuno, M. & Igarasi, S. (1967) *J. Gen. Appl. Microbiol.* **13**, 255 [3.2.2.10]
1441. Imai, D. & Brodie, A.F. (1973) *J. Biol. Chem.* **248**, 7487 [1.1.99.16]
1442. Imai, K., Reeves, H.C. & Ajl, S.J. (1963) *J. Biol. Chem.* **238**, 3193 [4.1.3.11]
1443. Imai, T. (1973) *J. Biochem. (Tokyo)* **73**, 139 [3.5.1.42]
1444. Imai, T. (1978) *Biochim. Biophys. Acta* (in press) [1.1.99.16]
1445. Imanaga, Y. (1958) *J. Biochem. (Tokyo)* **45**, 647
1446. Imsande, J. (1961) *J. Biol. Chem.* **236**, 1494 [2.4.2.11, 2.7.7.18, 6.3.5.1]
1447. Imsande, J. & Handler, P. (1961) *J. Biol. Chem.* **236**, 525 [2.4.2.11, 3.4.21.1, 6.3.5.1]
1448. Inagami, T. & Murachi, T. (1963) *Biochemistry* **2**, 1439 [3.4.22.4]
1449. Inaoka, M. & Soda, H. (1956) *Nature (London)* **178**, 202 [3.2.1.8]
1450. Inouye, K. & Fruton, J.S. (1967) *Biochemistry* **6**, 1765 [3.4.23.1]
1451. Iodice A.A. (1967) *Arch. Biochem. Biophys.* **121**, 241 [3.4.16.1]
1452. Isaksson, L.A. (1973) *Biochim. Biophys. Acta* **312**, 122 [2.1.1.51, 2.1.1.52]

473

1453. Ischihara S. & Uchino F. (1975) *Agr. Biol. Chem. (Tokyo)* **39**, 423   [3.4.23.6]
1454. Isherwood, F.A., Mapson, L.W. & Chen, Y.T. (1960) *Biochem. J.* **76**, 157   [1.1.3.8]
1455. Ishibashi, T., Kijimoto, S. & Makita, A. (1974) *Biochim. Biophys. Acta* **337**, 92   [2.4.1.79]
1456. Ishihara, H. & Heath, E.C. (1968) *J. Biol. Chem.* **243**, 1110   [2.7.7.30]
1457. Ishihara, H., Massaro, D.J. & Heath, E.C. (1968) *J. Biol. Chem.* **243**, 1103   [2.7.1.52]
1458. Ishikawa, Y. & Melville, D.B. (1970) *J. Biol. Chem.* **245**, 5967   [2.1.1.44]
1459. Ishimoto, N. & Strominger, J.L. (1966) *J. Biol. Chem.* **241**, 639   [2.4.1.55]
1460. Isihida, Y., Shirafiji, H., Kida, M. & Yoneda, M. (1969) *Agr. Biol. Chem. (Tokyo)* **33**, 384   [3.5.4.15]
1461. Isselbacher, K.J. (1958) *J. Biol. Chem.* **232**, 429   [2.7.7.10]
1462. Ito, E. & Strominger, J.L. (1962) *J. Biol. Chem.* **237**, 2689   [6.3.2.7, 6.3.2.8, 6.3.2.9]
1463. Ito, E. & Strominger, J.L. (1962) *J. Biol. Chem.* **237**, 2696   [6.3.2.10, 6.3.2.3]
1464. Ito, J. & Yanofsky, C. (1969) *J. Bacteriol.* **97**, 734   [2.4.2.18, 4.1.3.27]
1465. Ito, M. & Sugiura, M. (1968) *Yakugaku Zasshi* **88**, 1576 & 1591   [3.4.21.14]
1466. Ito, N. & Grisolia, S. (1959) *J. Biol. Chem.* **234**, 242   [5.4.2.1]
1467. Itokawa, Y. & Cooper, J.R. (1968) *Biochim. Biophys. Acta* **158** 180   [2.7.4.15]
1468. Iwasaki, H. & Matsubara, T. (1972) *J. Biochem. (Tokyo)* **71**, 645   [1.7.99.3]
1469. Iwashima, A., Nishino, H. & Nose, Y. (1972) *Biochim. Biophys. Acta* **258**, 333   [2.7.1.89]
1470. Izaki, K. & Strominger, J.L. (1968) *J. Biol. Chem.* **243**, 3193   [3.4.17.8]
1471. Izumori, K., Mitchell, M. & Elbein, A.D. (1976) *J. Biol. Chem.* **126**, 533   [5.3.1.20]
1472. Izumori, K., Rees, A.W. & Elbein, A.D (1975) *J. Biol. Chem.* **250**, 8085   [5.3.1.20]
1473. Jackson C.M. & Hanahan D.J. (1968) *Biochemistry* **7**, 4506   [3.4.21.6]
1474. Jackson, C.M., Johnson, F.F. & Hanahan, D.J. (1968) *Biochemistry* **7**, 4492   [3.4.21.6]
1475. Jackson R.L. & Matsueda G.R. (1970) Methods Enzymol. **19**, 591   [3.4.99.29]
1476. Jackson R.L. & Wolfe R.S. (1968) *J. Biol. Chem.* **243**, 879   [3.4.99.29]
1477. Jacobsen, J.G., Thomas, L.L. & Smith, L.H. Jr. (1964) *Biochim. Biophys. Acta* **85**, 103   [4.1.1.29]
1478. Jacobsen, J.V., Yamaguchi, M., Howard, F.D. & Bernhard, R.A. (1968) *Arch. Biochem. Biophys.* **127**, 252   [4.4.1.4]
1479. Jacobson, B. & Davidson, E.A. (1962) *J. Biol. Chem.* **237**, 638   [5.1.3.12]
1480. Jacobson, K.B. & Kaplan, N.O. (1957) *J. Biol. Chem.* **226**, 427   [3.6.1.9]
1481. Jacoby, G.A. & La Du, B.N. (1964) *J. Biol. Chem.* **239**, 419   [2.6.1.5]
1482. Jaenicke, L. & Brode, E. (1961) *Biochem. Z.* **334**, 108   [6.3.4.3]
1483. Jaffe, W.G. (1943) *J. Biol. Chem.* **149**, 1   [3.4.99.9]
1484. Jaffe, W.G. (1943) *Rev. Brasil Biol.* **3**, 149   [3.4.99.23]
1485. Jagendorf, A.T. (1963) *Methods Enzymol.* **6**, 430   [1.6.99.1]
1486. Jakoby, W.B. (1954) *J. Biol. Chem.* **207**, 657   [3.5.1.9]
1487. Jakoby, W.B. (1963) in The Enzymes, 2nd edn (Boyer, P.D., Lardy, H. & Myrbäck, K., ed.) Vol. 7, p. 203, Academic Press, New York   [1.2.1.1, 1.2.1.11, 1.2.1.16, 1.2.1.18, 1.2.1.19, 1.2.1.3, 1.2.1.4, 1.2.1.5]
1488. Jakoby, W.B. & Bonner, D.M. (1953) *J. Biol. Chem.* **205**, 699 and 709   [3.7.1.3]
1489. Jakoby, W.B. & Bonner, D.M. (1956) *J. Biol. Chem.* **221**, 689   [2.6.1.7]
1490. Jakoby, W.B., Brummond, D.O. & Ochoa, S. (1956) *J. Biol. Chem.* **218**, 811   [2.7.1.19]
1491. Jakoby, W.B. & Fjellstedt, T.A. (1972) in The Enzymes, 3rd edn (Boyer, P.D., ed.) Vol. 7, Chapter 5, p. 199, Academic Press, New York   [1.14.99.8, 3.3.2.3]
1492. Jakoby, W.B. & Fredericks, J. (1961) *Biochim. Biophys. Acta* **48**, 26   [1.1.1.9]
1493. Jakoby, W.B. & Fredericks, J. (1959) *J. Biol. Chem.* **234**, 2145   [1.2.1.19]
1494. Jakoby, W.B., Ohmura, E. & Hayaishi, O. (1956) *J. Biol. Chem.* **222**, 435   [4.1.1.2]
1495. Jakoby, W.B. & Scott, E.M. (1959) *J. Biol. Chem.* **234**, 937   [1.11.1.6, 1.2.1.16]
1496. Jankowski, W., Mańkowski, T. & Chojnacki, T. (1974) *Biochim. Biophys. Acta* **337**, 153   [2.4.1.78]
1497. Janoff A. (1973) *Lab. Investig.* **29**, 458
1498. Jansen, E.F. & Balls, A.K. (1941) *J. Biol. Chem.* **137**, 459   [3.4.21.11, 3.4.22.6]
1499. Jansen, E.F., Nutting, M.-D.F. & Balls, A.K. (1948) *J. Biol. Chem.* **175**, 975   [3.1.1.6]
1500. Jansen, P.L.M. (1974) *Biochim. Biophys. Acta* **338**, 170   [2.4.1.76, 2.4.1.77]
1501. Jansen, P.L.M., Chowdhury, J.R., Fischberg, E.B. & Arias, I.M. (1977) *J. Biol. Chem.* **252**, 2710   [2.4.1.95]
1502. Janssen, F.W. & Ruelius, H.W. (1968) *Biochim. Biophys. Acta* **151**, 330   [1.1.3.13]

474

1503.   Janssen, F.W. & Ruelius, H.W. (1968) *Biochim. Biophys. Acta* **167**, 501   [1.1.3.10]
1504.   Jarkovsky, Z., Marcus, D.M. & Grollman, A.P. (1970) *Biochemistry* **9**, 1123   [2.4.1.65]
1505.   Jayaram, H.N., Ramakrishnan, T. & Vaidyanathan, C.S. (1969) *Indian J. Biochem.* **6**, 106   [2.3.2.7]
1506.   Jean, M. & DeMoss, R.D. (1968) *Can. J. Microbiol.* **14**, 429   [1.1.1.110]
1507.   Jedziniak, J.A. & Lionetti, F.J. (1970) *Biochim. Biophys. Acta* **212**, 478   [4.1.2.4]
1508.   Jeffcoat, R., Hassall, H. & Dagley, S. (1969) *Biochem. J.* **115**, 977   [4.2.1.41]
1509.   Jeffrey, A.M., Knight, M. & Evans, W.C. (1972) *Biochem. J.* **130**, 373   [1.14.13.19]
1510.   Jeffrey J.J. & Gross J. (1970) *Biochemistry* **9**, 268   [3.4.24.7]
1511.   Jeffrey, P.L., Brown, D.H. & Brown, B.I. (1970) *Biochemistry* **9**, 1403   [3.2.1.3]
1512.   Jenkins, W.T., Yphantis, D.A. & Sizer, I.W. (1959) *J. Biol. Chem.* **234**, 51   [2.6.1.1]
1513.   Jequier, E., Robinson, B.S., Lovenberg, W. & Sjoerdsma, A. (1969) *Biochem. Pharmacol.* **18**, 1071   [1.14.16.3, 1.14.16.4]
1514.   Jergil, B. & Dixon, G.H. (1970) *J. Biol. Chem.* **245**, 425   [2.7.1.70]
1515.   Jerina, D.M., Daly, J.W., Witkop, B., Salzman-Nirenberg, P. & Udenfriend, S. (1970) *Biochemistry* **9**, 147   [1.14.99.8]
1516.   Jesty J. & Nemerson Y. (1974) *J. Biol. Chem.* **249**, 509   [3.4.21.21]
1517.   Johansen J.T. & Ottesen M. (1968) *C. R. Trav. Lab. Carlsberg* **36**, 265   [3.4.22.2, 3.4.22.6]
1518.   Johnson, A.H. & Baker, J.R. (1973) *Biochim. Biophys. Acta* **320**, 341   [2.8.2.8]
1519.   Johnson, H.S. (1971) *Biochem. Biophys. Res. Commun.* **43**, 703   [1.1.1.82]
1520.   Johnson, H.S. & Hatch, M.D. (1970) *Biochem. J.* **119**, 273   [1.1.1.82]
1521.   Johnson, L.F. & Tate, M.E. (1969) *Ann. N.Y. Acad. Sci.* **165**, 526   [3.1.3.26, 3.1.3.8]
1522.   Johnson, M.K. (1966) *Biochem. J.* **98**, 44
1523.   Johnson, M., Kaye, M.A.G., Hems, R. & Krebs, H.A. (1953) *Biochem. J.* **54**, 625   [3.6.1.8]
1524.   Johnson, P. & Rees, H.H. (1977) *Biochem J.* **168**, 513   [1.14.99.22]
1525.   Johnson, R.C. & Gilbertson, J.R. (1972) *J. Biol. Chem.* **247**, 6991   [1.2.1.42]
1526.   Jollés, P. (1960) in *The Enzymes*, 2nd edn (Boyer, P.D., Lardy, H. & Myrbäck, K., ed.) Vol. 4, p. 431, Academic Press, New York   [3.2.1.17]
1527.   Jones, A., Faulkner, A. & Turner, J.M. (1973) *Biochem. J.* **134** 959   [4.2.99.7]
1528.   Jones, E.E. & Broquist, H.P. (1966) *J. Biol. Chem.* **241**, 3430   [1.5.1.10]
1529.   Jones, G.H. & Ballou, C.E. (1968) *J. Biol. Chem.* **243**, 2442
1530.   Jones, G.H. & Ballou, C.E. (1969) *J. Biol. Chem.* **244**, 1043 and 1052   [3.2.1.77]
1531.   Jones, M.E. & Spector, L. (1960) *J. Biol. Chem.* **235**, 2897   [6.3.4.16]
1532.   Jones, M.E., Spector, L. & Lipmann, F. (1955) *J. Amer. Chem. Soc.* **77**, 819   [2.7.2.2]
1533.   Jones, N.R. (1955) *Biochem. J.* **60**, 81   [3.4.13.5]
1534.   Jones S.R. & Hofmann T. (1972) *Can. J. Biochem.* **50**, 1297   [3.4.16.1]
1535.   Jonsson, A.G. (1969) *Arch. Biochem. Biophys.* **129**, 62   [3.4.21.16]
1536.   Joshi, J.G. & Handler, P. (1964) *J. Biol. Chem.* **239**, 2741   [2.7.5.1]
1537.   Joshi, J.G. & Handler, P. (1960) *J. Biol. Chem.* **235**, 2981   [2.1.1.7]
1538.   Jost, J.-P. & Bock, R.M. (1969) *J. Biol. Chem.* **244**, 5866   [3.1.1.29]
1539.   Jourdian, G.W., Swanson, A., Watson, D. & Roseman, S. (1966) *Methods Enzymol.* **8**, 205   [3.1.3.29]
1540.   Joy, K.W. & Hageman, R.H. (1966) *Biochem. J.* **100**, 263   [1.7.7.1]
1541.   Joyce, B.K. & Grisolia, S. (1958) *J. Biol. Chem.* **233**, 350   [3.1.3.13]
1542.   Joyce, B.K. & Grisolia, S. (1959) *J. Biol. Chem.* **234**, 1330   [2.7.5.4]
1543.   Joyce, B.K. & Grisolia, S. (1960) *J. Biol. Chem.* **235**, 2278   [3.1.3.2]
1544.   Jukova, N.I., Klunova, S.M. & Philippovich, Yu, B. (1971) in '*Biochemistry of Insects*' *Issue 17, 56. V.I. Lenin State Pedagogical Institute, Moscow*   [1.1.1.167]
1545.   Julian, G.R., Wolfe, R.G. & Reithel, F.J. (1961) *J. Biol. Chem.* **236**, 754   [1.1.1.49]
1546.   Jungerman, K., Thauer, R.F., Leimenstoll, G. & Decker, K. (1973) *Biochim. Biophys. Acta* **305**, 268   [1.18.1.3]
1547.   Jungermann, K., Thauer, R.F., Leimenstoll, G. & Decker, K. (1973) *Biochim. Biophys. Acta* **305**, 268
1548.   Juni, E. (1952) *J. Biol. Chem.* **195**, 715
1549.   Jurasek, L., Fackre, D. & Smillie, L.B. (1969) *Biochem. Biophys. Res. Commun.* **37**, 99   [3.4.21.4]
1550.   Jurtshuk, P. & McManus, L. (1974) *Biochim. Biophys. Acta* **368**, 158   [1.4.3.11]
1551.   Kafatos, F.C., Law, J.H. & Tartakoff, A.M. (1967) *J. Biol. Chem.* **242**, 1488   [3.4.21.4]

1552. Kafatos, F.C., Tartakoff, A.M. & Law, J.H. (1967) *J. Biol. Chem.* **242**, 1477
1553. Kaferstein, H. & Jaenicke, L. (1972) *Hoppe-Seyler's Z. Physiol. Chem.* **353**, 1153 [3.4.22.12]
1554. Kagan, Z.S., Dronov, A.S. & Kretovich, V.L. (1967) *Dokl. Akad. Nauk. S.S.S.R* **175**, 1171; (1968) *Dokl. Akad. Nauk. S.S.S.R.* **179**, 1236 [2.6.1.32, 3.4.22.7]
1555. Kagan, Z.S., Kretovich, V.L. & Polyakov, V.A. (1966) *Biokhimiya* **31**, 355; (1968) *Biokhimiya* **33**, 89; (1969) *Biokhimiya* **34**, 59; (1966) *Enzymologia* **30**, 343 [1.4.1.8]
1556. Kageura, E. & Toki, S. (1974) *Biochim. Biophys. Acta* **341**, 172
1557. Kagi, A. & Yoshihara, O. (1971) *Biochim. Biophys. Acta* **250**, 367 [3.2.1.55]
1558. Kai, M. & Hawthorne, J.N. (1967) *Biochem. J.* **102**, 19P [2.7.1.68]
1559. Kai, M., Salway, J.G. & Hawthorne, J.N. (1968) *Biochem. J.* **106**, 791 [2.7.1.68]
1560. Kai, M., White, G.L. & Hawthorne, J.N. (1966) *Biochem. J.* [2.7.1.67]
1561. Kainuma, K., Wako, K., Kobayashi, A., Nogami, A. & Suzuki, S. (1975) *Biochim. Biophys. Acta* **410**, 333
1562. Kainuma, K., Wako, K., Kobayashi, A., Nogami, A. & Suzuki, S. (1975) *Biochim. Biophys. Acta* **410**, 333 [3.2.1.98]
1563. Kaji, A. & Tagawa, K. (1970) *Biochim. Biophys. Acta* **207**, 456
1564. Kaji, A., Tagawa, K. & Ichimi, T. (1969) *Biochim. Biophys. Acta* **171**, 186 [3.2.1.55]
1565. Kakimoto, T., Kato, J., Shibatani, T., Nishimura, N. & Chibata, I. (1969) *J. Biol. Chem.* **244**, 353 [4.1.1.12]
1566. Kakimoto, Y., Kanazawa, A., Taniguchi, K. & Sano, I. (1968) *Biochim. Biophys. Acta* **156**, 374 [2.6.1.22]
1567. Kakimoto, Y., Taniguchi, K. & Sano, I. (1969) *J. Biol. Chem.* **244**, 335 [2.6.1.40]
1568. Kalckar, H.M. (1953) *Biochim. Biophys. Acta* **12**, 250 [2.7.7.10, 2.7.7.9]
1569. Kalckar, H.M. (1947) *J. Biol. Chem.* **167**, 461 [3.5.4.3, 3.5.4.6]
1570. Kalckar, H.M. (1947) *J. Biol. Chem.* **167**, 477 [2.4.2.1]
1571. Kalckar, H.M. (1951) *Pubbl. Staz. Zool. (Napoli)* **23**, Suppl. 87 [3.2.2.1]
1572. Kalckar, H.M., Braganca, B. & Munch-Petersen, A. (1953) *Nature (London)* **172**, 1038 [2.7.7.12]
1573. Kalckar, H.M., MacNutt, W.S. & Hoff-Jørgensen, E. (1952) *Biochem. J.* **50**, 397 [2.4.2.6]
1574. Kalf, G.F. & Rieder, S.V. (1958) *J. Biol. Chem.* **230**, 691 [3.2.1.28]
1575. Kallio, R.E. (1951) *J. Biol. Chem.* **192**, 371 [4.4.1.2]
1576. Kallio, R.E. & Larson, A.D. (1955) in *A Symposium on Amino Acid Metabolism* (McElroy, W.D. & Glass, H.B., ed.) p. 616, Johns Hopkins Press, Baltimore [5.1.1.2]
1577. Kalman, S.M., Duffield, P.H. & Brzozowski, T. (1966) *J. Biol. Chem.* **241**, 1871
1578. Kalousek, F. & Morris, N.R. (1969) *J. Biol. Chem.* **244**, 1157 [2.1.1.37]
1579. Kalyankar, G.D. & Meister, A. (1959) *J. Biol. Chem.* **234**, 3210 [6.3.2.11]
1580. Kamei, S., Wakabayashi, K. & Shimazono, M. (1964) *J. Biochem. (Tokyo)* **56**, 72 [1.1.1.66]
1581. Kamel, M.Y. & Anderson, R.L. (1967) *Arch. Biochem. Biophys.* **120**, 322 [2.7.1.61]
1582. Kaminskas, E., Kimhi, Y. & Magasanik, B. (1970) *J. Biol. Chem.* **245**, 3536
1583. Kammen, H.O. & Koo, R. (1969) *J. Biol. Chem.* **244**, 4888 [2.7.5.6]
1584. Kanamori, M. & Wixom, R.L. (1963) *J. Biol. Chem.* **238**, 998 [4.2.1.9]
1585. Kanarek, L. & Hill, R.L. (1964) *J. Biol. Chem.* **239**, 4202 [4.2.1.2]
1586. Kaneda M. & Toninaga N. (1975) *J. Biochem. (Tokyo)* **78**, 1287 [3.4.21.25]
1587. Kaneshiro, T. & Law, J.H. (1964) *J. Biol. Chem.* **239**, 1705 [2.1.1.17]
1588. Kanfer, J. & Kennedy, E.P. (1964) *J. Biol. Chem.* **239**, 1720 [2.7.8.8, 4.1.1.65]
1589. Kanfer, J.N., Young, O.M., Shapiro, D. & Brady, R.O. (1966) *J. Biol. Chem.* **241**, 1081 [3.1.4.12]
1590. Kaniuga, Z. (1963) *Biochim. Biophys. Acta* **73**, 550 [1.6.99.3]
1591. Kaplan, B.H. & Stadtman, E.R. (1968) *J. Biol. Chem.* **243**, 1787 [4.3.1.7]
1592. Kaplan, H., Symonds, V.B., Dugas, H. & Whitaker, D.R. (1970) *Can. J. Biochem.* **48**, 649 [3.4.21.12]
1593. Kaplan, M.M. & Flavin, M. (1966) *J. Biol. Chem.* **241**, 4463 [4.2.99.9]
1594. Kaplan, M.M. & Flavin, M. (1966) *J. Biol. Chem.* **241**, 5781 [4.2.99.9]
1595. Kaplan, N.O., Colowick, S.P. & Ciotti, M.M. (1952) *J. Biol. Chem.* **194**, 579 [3.5.4.4]
1596. Karasek, M.A. & Greenberg, D.M. (1957) *J. Biol. Chem.* **227**, 191 [4.1.2.5]
1597. Karpetsky, T.P. & White, E.H. (1973) *Tetrahedron* **29**, 3761 [1.13.12.6]

1598. Karr, D., Tweto, J. & Albersheim, P. (1967) *Arch. Biochem. Biophys.* **121**, 732 [2.1.1.12]

1599. Kasai, K., Uchida, T., Egami, F., Yoshida, K. & Nomoto, M. (1969) *J. Biochem. (Tokyo)* **66**, 389 [3.1.27.3]

1600. Kasai, T., Suzuki, I. & Asai, T. (1962) *Koso Kagaku Shimpojiumii* **17**, 77 [1.2.3.5]

1601. Kassab, R., Pradel, L.A. & Thoai, N.V. (1965) *Biochim. Biophys. Acta* **99**, 397 [2.7.3.4, 2.7.3.5]

1602. Katagiri, H., Yamada, H. & Imai, K. (1959) *J. Biochem. (Tokyo)* **46**, 1119 [2.7.1.42]

1603. Katagiri, M., Ganguli, B.N. & Gunsalus, I.C. (1969) *J. Biol. Chem.* **243**, 3543

1604. Katagiri, M. & Hayaishi, O. (1957) *J. Biol. Chem.* **226**, 439 [2.8.3.6]

1605. Katanuma, N., Matsuda, Y. & Tomino, I. (1964) *J. Biochem. (Tokyo)* **56**, 499 [2.6.1.13]

1606. Katchman, B., Betheil, J.J., Schepartz, A.I. & Sanadi, D.R. (1951) *Arch. Biochem. Biophys.* **34**, 437 [2.7.1.23]

1607. Kato, M. & Ikeda, Y. (1968) *J. Biochem. (Tokyo)* **64**, 321 [3.1.22.2]

1608. Kato, S., Ueda, H., Nonomura, S. & Tatsumi, C. (1968) *Nippon Nogei Kagaku Kaishi* **42**, 596 [3.1.1.38]

1609. Katoh, S. (1971) *Arch. Biochem. Biophys.* **146**, 202 [1.1.1.153]

1610. Katze, J.R. & Konigsberg, W. (1970) *J. Biol. Chem.* **245**, 923 [6.1.1.11]

1611. Katzen, H.M. & Buchanan, J.M. (1965) *J. Biol. Chem.* **240**, 825 [1.1.99.15]

1612. Katzen, H.M., Tietze, F. & Stetten, D.J. (1963) *J. Biol. Chem.* **238**, 1006 [1.8.4.2]

1613. Kaufman, B.T. & Gardiner, R.C. (1966) *J. Biol. Chem.* **241**, 1319 [1.5.1.3]

1614. Kaufman, S. (1955) *J. Biol. Chem.* **216**, 153 [6.2.1.5]

1615. Kaufman, S. (1959) *J. Biol. Chem.* **234**, 2677 [1.14.16.1]

1616. Kaufman, S. (1962) *Methods Enzymol.* **5**, 809 [1.6.99.7]

1617. Kaufman, S. & Alivasatos, S.G.A. (1955) *J. Biol. Chem.* **216**, 141 [6.2.1.5]

1618. Kaufman, S., Gilvarg, C., Cori, O. & Ochoa, S. (1953) *J. Biol. Chem.* **203**, 869 [6.2.1.4]

1619. Kaufman, S., Korkes, S. & del Campillo, A. (1951) *J. Biol. Chem.* **192**, 301 [1.1.1.38]

1620. Kauss, H. & Quader, H. (1976) *Plant Physiol.* **58**, 295 [2.4.1.96]

1621. Kauss, H. & Schubert, B. (1971) *FEBS Lett.* **19**, 131 [2.4.1.96]

1622. Kautsky, M.P. & Hagerman, D.D. (1970) *J. Biol. Chem.* **245**, 1978 [1.1.1.62]

1623. Kawachi, T. & Rudney, H. (1970) *Biochemistry* **9**, 1700 [1.1.1.34]

1624. Kawada, M., Kagawa, Y., Takiguchi, H. & Shimazono, N. (1962) *Biochim. Biophys. Acta* **57**, 404 [3.1.1.31]

1625. Kawahara, F.S. & Talalay, P. (1960) *J. Biol. Chem.* **235**, PC1 [5.3.3.1]

1626. Kazenko, A. & Laskowski, M. (1948) *J. Biol. Chem.* **173**, 217

1627. Kaziro, Y., Ochoa, S., Warner, R.C. & Chen, J. (1961) *J. Biol. Chem.* **236**, 1917 [6.4.1.3]

1628. Kean, E.L. & Bighouse, K.J. (1974) *J. Biol. Chem.* **249**, 7813 [3.1.4.40]

1629. Kean, E.L. & Roseman, S. (1966) *J. Biol. Chem.* **241**, 5643 [2.7.7.43]

1630. Kearney, E.B. (1952) *J. Biol. Chem.* **194**, 747 [2.7.1.26]

1631. Keay L., Moseley M.H., Anderson R.G., O'Connor R.J. & Wildi B.S. (1972) *Biotechnol. Bioeng. Symp.* **3**, 63 [3.4.24.4]

1632. Keay, L. & Moser, P.W. (1969) *Biochem. Biophys. Res. Commun.* **34**, 600 [3.4.21.14]

1633. Keele, B.B., McCord, J.M. & Fridovich, I. (1971) *J. Biol. Chem.* **246**, 2875

1634. Keil B., Gilles A.M., Lecroisey A., Hurion N. & Tong N.-T. (1975) *FEBS-Letters* **56**, 292 [3.4.24.3, 3.4.24.8]

1635. Keil-Dlouha V. (1976) *Biochim. Biophys. Acta* **429**, 239 [3.4.24.8]

1636. Keilin, D. & Hartree, E.F. (1948) *Biochem. J.* **42**, 221 [1.1.3.4]

1637. Keilin, D. & Hartree, E.F. (1952) *Biochem. J.* **50**, 331 [1.1.3.4]

1638. Keilin, D. & Hartree, E.F. (1952) *Biochem. J.* **50**, 341 [5.1.3.3]

1639. Keilin, D. & Hartree, E.F. (1936) *Proc. Roy. Soc. Ser. B* **119**, 141 [1.11.1.6]

1640. Keilin, D. & Hartree, E.F. (1938) *Proc. Roy. Soc. Ser. B* **125**, 171 [1.9.3.1]

1641. Keilin, D. & Hartree, E.F. (1939) *Proc. Roy. Soc. Ser. B* **127**, 167 [1.9.3.1]

1642. Keilin, D. & Mann, T. (1938) *Proc. Roy. Soc. B* **125**, 187

1643. Keilin, D. & Mann, T. (1939) *Nature (London)* **144**, 442 [4.2.1.1]

1644. Keilin, D. & Mann, T. (1939) *Nature (London)* **143**, 23 [1.10.3.2]

1645. Keilley, R.K. (1970) *J. Biol. Chem.* **245**, 4204 [2.7.4.9]

1646. Keilová, H. (1970) *FEBS Lett,* **6**, 312 [3.4.23.5]

1647. Keilová, H., Bláha, K. & Keil, B. (1968) *Eur. J. Biochem.* **4**, 442
1648. Keilová, H. & Keil, B. (1969) *FEBS Lett.* **4**, 295   [3.4.22.1]
1649. Keilová, H. & Lapresle, C. (1970) *FEBS Lett.* **9**, 348   [3.4.23.5]
1650. Keir, H.M., Mathog, R.H. & Carter, C.E. (1964) *Biochemistry* **3**, 1188   [3.1.13.1]
1651. Kellerman, G.M. (1959) *Biochim. Biophys. Acta* **33**, 101   [3.6.1.20]
1652. Kellermeyer, R.W., Allen, S.H.G., Stzernholm, R. & Wood, H.G. (1964) *J. Biol. Chem.* **239**, 2562   [5.4.99.2]
1653. Kelley, R.B., Atkinson, M.R., Huberman, J.A. & Kornberg, A. (1969) *Nature* **224**, 495.   [3.1.11.1]
1654. Kelly, J.J. & Alpers, D.H. (1973) *Biochim. Biophys. Acta* **315**, 113   [3.2.1.3]
1655. Kelly, T.J., Jr. & Smith, H.O. *J. Mol. Biol.* **51**, 393-409 (1970).   [3.1.23.20]
1656. Kennedy, E.P. (1962) *Methods Enzymol.* **5**, 484   [2.7.8.2, 2.7.8.3]
1657. Kennedy, E.P. & Weiss, S.B. (1956) *J. Biol. Chem.* **222**, 193   [2.7.7.14, 2.7.7.15, 2.7.8.1, 2.7.8.2]
1658. Kenney, F.T. (1959) *J. Biol. Chem.* **234**, 2707   [2.6.1.5]
1659. Kenten, R.H. (1957) *Biochem. J.* **67**, 25   [2.5.1.2]
1660. Kenten, R.H. & Mann, P.J.G. (1954) *Biochem. J.* **57**, 347   [1.11.1.7]
1661. Kepler, C.R. & Tove, S.B. (1967) *J. Biol. Chem.* **242**, 5686   [5.2.1.5]
1662. Kern, M. & Racker, E. (1954) *Arch. Biochem. Biophys.* **48**, 235   [1.6.5.4, 4.2.99.10]
1663. Kerr, D. (1971) *Methods Enzymol.* **17B**, 446
1664. Kerr M.A. & Kenny A.J. (1974) *Biochem. J.* **137**, 477   [3.4.24.11]
1665. Kersten, H., Kersten, W. & Staudinger, H. (1957) *Biochim. Biophys. Acta* **24**, 222
1666. Kersten, H., Kersten, W. & Staudinger, H. (1958) *Biochim. Biophys. Acta* **27**, 598   [1.6.5.4]
1667. Kessel, D. (1958) *J. Biol. Chem.* **243**, 4739   [2.7.1.74]
1668. Kessel, D. & Roberts, D.W. (1965) *Biochemistry* **4**, 2631   [1.5.1.3]
1669. Kessler, D.L., Johnson, J.L., Cohen, H.J. & Rajagopalan, K.V. (1974) *Biochim. Biophys. Acta* **334**, 86   [1.8.3.1]
1670. Kessler E. & Yaron A. (1973) *Biochem. Biophys. Res. Commun.* **50**, 405   [3.4.11.13]
1671. Keutel, H.J., Jacobs, H.K., Okabe, K., Yue, R.H. & Kuby, S.A. (1968) *Biochemistry* **7**, 4283   [2.7.3.2]
1672. Khairallah, E.A. & Wolf, G. (1967) *J. Biol. Chem.* **242**, 32   [4.1.1.42]
1673. Khalkhali, Z. & Marshall, R.D. (1975) *Biochem. J.* **146**, 299   [2.4.1.94]
1674. Khalkhali, Z. & Marshall, R.D. (1976) *Carbohydrate Res.* **49**, 455   [2.4.1.94]
1675. Khalkhali, Z., Marshall, R.D., Reuvers, F., Habets-Willems, C. & Boer, P. (1976) *Biochem. J.* **160**, 37   [2.4.1.94]
1676. Khorana, G.H. (1961) in *The Enzymes*, 2nd edn (Boyer, P.D., Lardy, H. & Myrbäck, K., ed.) Vol. 5, p. 79, Academic Press, New York   [3.1.4.1]
1677. Kielley, W.W. (1961) in *The Enzymes*, 2nd edn (Boyer, P.D., Lardy, H. & Myrbäck, K., ed.) Vol. 5, p. 159, Academic Press, New York   [3.6.1.3]
1678. Kielley, W.W. & Bradley, L.B. (1954) *J. Biol. Chem.* **206**, 327   [3.1.2.7]
1679. Kijimoto, S., Ishibashi, T. & Makita, A. (1974) *Biochem. Biophys. Res. Commun.* **56**, 177   [2.4.1.88]
1680. Kikuchi, G., Kimar, A., Talmage, P. & Shemin, D. (1958) *J. Biol. Chem.* **233**, 1214   [2.3.1.37]
1681. Kikuchi, K. & Tsuiki, S. (1973) *Biochim. Biophys. Acta* **327**, 193   [5.1.3.14]
1682. Kikuchi, M., Hayashida, H., Nakano, E. & Sakaguchi, K. (1971) *Biochemistry* **10**, 1222   [3.5.1.43, 3.5.1.44]
1683. Kilgore, W.W. & Starr, M.P. (1959) *J. Biol. Chem.* **234**, 2227   [1.1.1.57, 5.3.1.12]
1684. Kilgore, W.W. & Starr, M.P. (1959) *Nature (London)* **183**, 1412   [1.2.1.35]
1685. Kim, K. (1964) *J. Biol. Chem.* **239**, 783   [2.6.1.29]
1686. Kim, S., Benoiton, L. & Paik, W.K. (1964) *J. Biol. Chem.* **239**, 3790   [1.5.3.4]
1687. Kim, S. & Paik, W.K. (1970) *J. Biol. Chem.* **245**, 1806   [2.1.1.24]
1688. Kim, Y.S., Perdomo, J., Bella, A. & Nordberg, J. (1971) Biochim. Biophys. Acta **244**, 505   [2.4.99.1]
1689. Kimura, T. & Tobari, J. (1963) *Biochim. Biophys. Acta* **73**, 399   [1.1.99.16]
1690. Kindler, S.H. & Gilvarg, C. (1960) *J. Biol. Chem.* **235**, 3532   [3.5.1.18]
1691. King, H.L. Jr. & Wilken, D.R. (1972) *J. Biol. Chem.* **247**, 4096   [1.1.1.168, 1.1.1.169]
1692. King, J. & Waygood, E.R. (1968) *Can. J. Biochem.* **46**, 771   [2.6.1.45]
1693. King, T.E. & Cheldelin, V.H. (1956) *J. Biol. Chem.* **220**, 177   [1.2.1.5]

478

1694. Kingdon, H.S., Shapiro, B.M. & Stadtman, E.R. (1967) *Proc. Nat. Acad. Sci. U.S.* **58**, 1703 [2.7.7.42]

1695. Kinsky, S.C. (1960) *J. Biol. Chem.* **235**, 94   [2.3.1.2]

1696. Kiritani, K., Narise, S. & Wagner, R.P. (1966) *J. Biol. Chem.* **241**, 2047   [1.1.1.86]

1697. Kirkland, R.J.A. & Turner, J.F. (1959) *Biochem. J.* **72**, 716   [2.7.4.6]

1698. Kirschke H., Langner J., Wiederanders B., Ansorge S. & Bohley P. (1977) Eur. J. Biochem **74**, 293   [3.4.22.15]

1699. Kishi, Y., Goto, T., Hirata, Y., Shimomura, O. & Johnson, F.H. (1966) Tetrahedron Lett. **29**, 3427   [1.13.12.6]

1700. Kishida, T. & Yoshimura, S. (1966) *Agr. Biol. Chem. (Tokyo)* **30**, 1183   [3.4.99.8]

1701. Kisiel W. & Davie E.W. (1975) *Biochemistry* **14**, 4928   [3.4.21.21]

1702. Kita, H., Kamimoto, M., Senoh, S., Adachi, T. & Takeda, Y. (1965) *Biochem. Biophys. Res. Commun.* **18**, 66   [1.13.11.7]

1703. Kitahara, K., Ôbayashi, A. & Fukui, S. (1953) *Enzymologia* **15**, 259   [5.1.2.1]

1704. Kitcher, J.P., Trudgill, P.W. & Rees, J.S. (1972) *Biochem. J.* **130**, 121   [1.3.99.8]

1705. Kito, M. & Pizer, L.I. (1969) *J. Biol. Chem.* **244**, 3316   [1.1.1.94]

1706. Kivirikko, K.I., Kishida, Y., Sakakibara, S. & Prockop, J. (1972) *Biochim. Biophys. Acta* **271**, 347   [1.14.11.2]

1707. Kivirikko, K.I. & Prockop, D.J. (1968) *Proc. Nat. Acad. Sci., USA* **60**, 1473   [1.14.11.4]

1708. Kivirikko, K.I. & Prockop, D.J. (1967) *Arch. Biochem. Biophys.* **118**, 611   [1.14.11.2]

1709. Kizer, D.E. & Holman, L. (1974) *Biochim. Biophys. Acta* **350**, 193   [2.7.1.21]

1710. Klee, W.A., Richards, H.H. & Cantoni, G.L. (1961) *Biochim. Biophys. Acta* **54**, 157   [2.1.1.3, 2.1.1.5]

1711. Kleid, D., Humayun, Z., Jeffrey, A. & Ptashne, A. (1976) *Proc. Nat. Acad. Sci. USA* **73**, 293. [3.1.23.25]

1712. Klein, A.O. & Vishniac, W. (1961) *J. Biol. Chem.* **236**, 2544   [3.1.1.14]

1713. Kline, L.K., Fittler, F. & Hall, R.H. (1969) *Biochemistry* **8**, 4361   [2.5.1.8]

1714. Klingman, J.D. & Handler, P. (1958) *J. Biol. Chem.* **232**, 369

1715. Klinman, J.P. & Rose, I.A. (1971) *Biochemistry* **10**, 2253 and 2259   [5.3.3.7]

1716. Klungsoeyr, L. (1966) *Biochim. Biophys. Acta* **122**, 361   [2.7.1.57]

1717. Knappe, J., Blaschkowski, H.P., Gröbner, P. & Schmitt, T. (1974) *Eur. J. Biochem.* **50**, 253

1718. Knappe, J., Brümmer, W. & Biederbick, K. (1963) *Biochem. Z.* **338**, 599   [3.5.1.12]

1719. Knappe, J., Schlegel, H.-G. & Lynen, F. (1961) *Biochem. Z.* **335**, 101   [6.4.1.4]

1720. Kneen, E. & Beckord, L.D. (1946) *Arch. Biochem.* **10**, 41

1721. Knizley, H. Jr. (1967) *J. Biol. Chem.* **242**, 4619   [3.4.14.1]

1722. Knizley, H. Jr. (1967) *J. Biol. Chem.* **242**, 4619   [2.3.1.17]

1723. Knop, J.K. & Hansen, R.G. (1970) *J. Biol. Chem.* **245**, 2499   [2.7.7.9]

1724. Knox, W.E. (1953) *Biochem. J.* **53**, 379   [3.7.1.3]

1725. Knox, W.E. (1946) *J. Biol. Chem.* **163**, 699   [1.2.3.1]

1726. Knox, W.E. (1955) *Methods Enzymol.* **2**, 289   [5.3.2.1]

1727. Knox, W.E. & Edwards, S.W. (1955) *J. Biol. Chem.* **216**, 479   [1.13.11.5]

1728. Knox, W.E. & Pitt, B.M. (1957) *J. Biol. Chem.* **225**, 675   [5.3.2.1]

1729. Koch, A.L. (1956) *J. Biol. Chem.* **223**, 535   [2.4.2.5, 3.2.2.2]

1730. Koch, G.L.E., Shaw, D.C. & Gibson, F. (1970) *Biochim. Biophys. Acta* **212**, 375   [1.3.1.12]

1731. Kochakian, C.D., Carroll, B.R. & Uhri, B. (1957) *J. Biol. Chem.* **224**, 811   [1.1.1.50]

1732. Koelle, G.B. (1953) *Biochem. J.* **53**, 217   [3.1.1.8]

1733. Koen, A.L. & Shaw, C.R. (1966) *Biochim. Biophys. Acta* **128**, 48   [1.1.1.105]

1734. Kohlhaw, G., Leary, T.R. & Umbarger, H.E. (1969) *J. Biol. Chem.* **244**, 2218   [4.1.3.12]

1735. Kohn, L.D. & Jakoby, W.B. (1968) *J. Biol. Chem.* **243**, 2472   [1.1.1.37]

1736. Kohn, L.D. & Jakoby, W.B. (1968) *J. Biol. Chem.* **243**, 2486   [1.1.1.92]

1737. Kohn, L.D. & Jakoby, W.B. (1968) *J. Biol. Chem.* **243**, 2494   [1.1.1.81]

1738. Kohn, L.D. & Jakoby, W.B. (1966) *Methods Enzymol.* **9**, 236   [1.3.1.7]

1739. Kohn, L.D., Packman, P.M., Allen, R.H. & Jakoby, W.B. (1968) *J. Biol. Chem.* **243**, 2479   [1.1.1.93]

1740. Kohnert, K.-D., Hahn, H.-J., Zühlke, H., Schmidt, S. & Fiedler, H. (1974) *Biochim. Biophys. Acta* **338**, 68

1741. Kohno, M. & Yamashina, I. (1972) *Biochim. Biophys. Acta* **258**, 600 [3.5.1.26]
1742. Koida M. & Walter R. (1976) *J. Biol. Chem.* **251**, 7593 [3.4.21.26]
1743. Koida T., Kato H. & Davie E.W. (1976) *Methods Enzymol.* **45**, 65
1744. Koide, A., Kataoka, T. & Matsuoka, Y. (1969) *J. Biochem. (Tokyo)* **65**, 475 [3.4.21.1]
1745. Koide, H., Shishido, T., Nagayama, H. & Shimura, K. (1956) *J. Agr. Chem. Soc. (Japan)* **30**, 283 [2.6.1.44]
1746. Koide, N. & Muramatsu, T. (1974) *J. Biol. Chem.* **249**, 4897 [3.2.1.96]
1747. Koike, M., Shah, P.C. & Reed, L.J. (1960) *J. Biol. Chem.* **235**, 1939
1748. Kojima, Y., Itada, N. & Hayaishi, O. (1961) *J. Biol. Chem.* **236**, 2223 [1.13.11.2]
1749. Kok, P. & Astrup, T. (1969) *Biochemistry* **8**, 79 [3.4.21.31]
1750. Kokubu, T., Akutsu, H., Fujimoto, S., Ueda, E., Hiwada, K. & Yamamura, Y. (1969) *Biochim. Biophys. Acta* **191**, 668 [3.4.99.3]
1751. Kolattukudy, P.E. (1970) *Biochemistry* **9**, 1095 [1.1.1.164]
1752. Koli, A.K., Yearby, C., Scott, W. & Donaldson, K.O. (1969) *J. Biol. Chem.* **244**, 621 [1.6.99.5, 1.6.99.6]
1753. Kolinska, J. & Semenza, G. (1967) *Biochim. Biophys. Acta* **146**, 181 [3.2.1.48]
1754. Komai, H. & Neilands, J.B. (1969) *Biochim. Biophys. Acta* **171**, 311 [4.2.1.24]
1755. Kondo, H. & Ishimoto, M. (1975) *J. Biochem. (Tokyo)* **78**, 317 [4.4.1.12]
1756. Kondo, H., Kagotani, K., Oshima, M. & Ishimoto, M. (1973) *J. Biochem. (Tokyo)* **73**, 1269 [1.4.99.2]
1757. Kono T., (1968) *Biochemistry* **7**, 1106 [3.4.24.3]
1758. Koolman, J. & Karlson, P. (1975) *Hoppe-Seyler's Z. Physiol. Chem.* **356**, 1131 [1.1.3.16]
1759. Koren E., Lukac J. & Milkovic S. (1974) *J. Reprod. Fertil.* **36**, 161 [3.4.99.31]
1760. Koren E. & Milkovic S. (1973) *J. Reprod. Fertil.* **32**, 349 [3.4.99.31]
1761. Koritz, S.B. (1964) *Biochemistry* **3**, 1098 [1.1.1.145]
1762. Kormann, A.W., Hurst, R.O. & Flynn, T.G. (1972) *Biochim. Biophys. Acta* **258**, 40 [1.1.1.72]
1763. Korn, E.D. (1957) *J. Biol. Chem.* **226**, 841
1764. Korn, E.D. & Quigley, T.W. (1957) *J. Biol. Chem.* **226**, 833 [3.1.1.3]
1765. Kornalik F. & Blomback B. (1975) *Thrombosis Research* **6**, 53 [3.4.99.27]
1766. Kornberg, A. (1950) *J. Biol. Chem.* **182**, 805 [2.7.1.23]
1767. Kornberg, A., Kornberg, S.R. & Simms, E.S. (1956) *Biochim. Biophys. Acta* **20**, 215 [2.7.4.1]
1768. Kornberg, A., Lieberman, I. & Simms, E.S. (1954) *J. Amer. Chem. Soc.* **76**, 2844 [2.4.2.10, 2.7.6.1]
1769. Kornberg, A., Lieberman, I. & Simms, E.S. (1955) *J. Biol. Chem.* **215**, 417 [2.4.2.7, 2.4.2.8]
1770. Kornberg, A. & Pricer, W.E. (1950) *J. Biol. Chem.* **182**, 763 [3.6.1.9]
1771. Kornberg, A. & Pricer, W.E. (1951) *J. Biol. Chem.* **189**, 123 [1.1.1.41]
1772. Kornberg, A. & Pricer, W.E. (1951) *J. Biol. Chem.* **191**, 535 [2.7.7.1]
1773. Kornberg, A. & Pricer, W.E. (1951) *J. Biol. Chem.* **193**, 481 [2.7.1.20, 2.7.1.40]
1774. Kornberg, A. & Pricer, W.E. (1953) *J. Biol. Chem.* **204**, 329 [6.2.1.3]
1775. Kornberg, A. & Pricer, W.E. (1953) *J. Biol. Chem.* **204**, 345 [2.3.1.15]
1776. Kornberg, S.R. (1956) *J. Biol. Chem.* **218**, 213 [3.6.1.2]
1777. Kornberg, S.R., Lehman, I.R., Bessman, M.J., Simms, E.S. & Kornberg, A. (1958) *J. Biol. Chem.* **233**, 159 [3.1.5.1]
1778. Kornberg, S.R., Zimmerman, S.B. & Kornberg, A. (1961) *J. Biol. Chem.* **236**, 1487 [2.4.1.26, 2.4.1.27, 2.4.1.28]
1779. Kornfeld, S. & Glaser, L. (1961) *J. Biol. Chem.* **236**, 1791 [2.7.7.24]
1780. Kortt A.A., Hinds J.A. & Zerner B. (1973) *Biochemistry* **13**, 2029 [3.4.22.3]
1781. Korzenovsky, M., Diller, E.R., Marshall, A.C. & Auda, B.M. (1960) *Biochem. J.* **76**, 238 [3.1.1.13]
1782. Kosaka, A., Spivey, H.O. & Gholson, R.K. (1971) *J. Biol. Chem.* **246**, 3277 [2.4.2.11]
1783. Kosicki, G.W. (1968) *Biochemistry* **7**, 4299 [4.1.1.3]
1784. Kosuge, T. & Conn, E.E. (1962) *J. Biol. Chem.* **237**, 1653 [3.1.1.35]
1785. Kosuge, T., Heskett, M.G. & Wilson, E.E. (1966) *J. Biol. Chem.* **241**, 3738 [1.13.12.3]
1786. Koukol, J. & Conn, E.E. (1961) *J. Biol. Chem.* **236**, 2692 [4.3.1.5]
1787. Kovaleva G.G., Shimanskaya M.P. & Stepanov U.M. (1972) *Biochem. Biophys. Res. Commun.* **49**, 1075 [3.4.23.6]

480

1788. Krakow, G. & Barkulis, S.S. (1956) *Biochim. Biophys. Acta* **21**, 593   [4.1.1.47]
1789. Krakow, J.S., Coutsogeorgopoulos, C. & Canellakis, E.S. (1962) *Biochim. Biophys. Acta* **55**, 639   [2.7.7.19, 2.7.7.31]
1790. Kramer D.E. & Whitaker J.R. (1964) *J. Biol. Chem.* **239**, 2178   [3.4.22.3]
1791. Kraus E., Kiltz H.H. & Fembert U.F. (1976) *Hoppe-Seyler's Z. Physiol. Chem.* **357**, 233   [3.4.21.14]
1792. Kream, J. & Chargaff, E. (1952) *J. Amer. Chem. Soc.* **74**, 5157   [3.5.4.1]
1793. Krebs, E.G. (1966) *Methods Enzymol.* **8**, 543   [2.7.1.18]
1794. Krebs, E.G. & Fischer, E.H. (1956) *Biochim. Biophys. Acta* **20**, 150   [2.7.1.38]
1795. Krebs, E.G., Kent, A.B. & Fischer, E.H. (1958) *J. Biol. Chem.* **231**, 73   [2.7.1.38]
1796. Krebs, H.A. & Hems, R. (1953) *Biochim. Biophys. Acta* **12**, 172   [2.7.4.6]
1797. Kredich, N.M. & Tomkins, G.M. (1966) *J. Biol. Chem.* **241**, 4955   [2.3.1.30]
1798. Krehbiel, R. & Darrach, M. (1968) *Can. J. Biochem.* **46**, 1075   [1.1.1.149]
1799. Kreis, W. & Hession, C. (1973) *Cancer Res.* **33**, 1862   [4.4.1.11]
1800. Krell, K. & Eisenberg, M.A. (1970) *J. Biol. Chem.* **245**, 6558   [6.3.3.3]
1801. Krenitsky, T.A., Neil, S.M. & Miller, R.L. (1970) *J. Biol. Chem.* **245**, 2605   [2.4.2.22]
1802. Kretovich, V.L. & Stepanovich, K.M. (1961) *Dokl. Akad. Nauk. S.S.S.R.* **139**, 136   [2.6.1.51]
1803. Kretovich, V.L. & Stepanovich, K.M. (1964) *Dokl. Akad. Nauk. S.S.S.R.* **159**, 449; (1966) *Izv. Akad. Nauk. S.S.S.R. Ser. Biol.* No. 2, 295   [1.4.1.7]
1804. Krieble, V.K. & Wieland, W.A. (1921) *J. Amer. Chem. Soc.* **43**, 164   [4.1.2.10]
1805. Krishman, P.S. (1952) *Arch. Biochem. Biophys.* **37**, 224   [3.6.1.11, 3.6.1.5]
1806. Krishna, R.V., Krishnaswarmy, P.R. & Rajagopal Rao, D. (1971) *Biochem. J.* **124**, 905   [2.3.1.53]
1807. Krygier, V. & Momparler, R.L. (1968) *Biochim. Biophys. Acta* **161**, 578   [2.7.1.76]
1808. Kubowitz, F. & Ott, P. (1943) *Biochem. Z.* **314**, 94   [1.1.1.27]
1809. Kubowitz, F. & Ott, P. (1944) *Biochem. Z.* **317**, 193   [2.7.1.40]
1810. Kuby, S.A. & Lardy, H.A. (1953) *J. Amer. Chem. Soc.* **75**, 890   [3.2.1.23]
1811. Kuby, S.A., Noda, L. & Lardy, H.A. (1954) *J. Biol. Chem.* **209**, 191   [2.7.3.2]
1812. Kuby, S.A. & Noltmann, E.A. (1962) in *The Enzymes*, 2nd edn (Boyer, P.D., Lardy, H. & Myrbäck, K., ed.) Vol. 6, p. 515, Academic Press, New York   [2.7.3.2]
1813. Kuhn R.W., Walsh K.A. & Neurath H. (1974) *Biochemistry* **13**, 3871   [3.4.16.1]
1814. Kulaev, I.S. & Bobyk, M.A. (1971) Biokhimiya **36**, 426   [2.7.4.17]
1815. Kulaev, I.S., Bobyk, M.A., Nikolaev, N.N., Sergeev, N.S. & Uryson, S.O. (1971) *Biokhimiya* **36**, 943   [2.7.4.17]
1816. Kulaev, I.S., Konoshenko, G.I. & Umnov, A.M. (1972) *Biokhimiya* **37**, 227   [3.6.1.25]
1817. Kumagai, H., Matsui, H., Ogata, K. & Yamada, H. (1969) *Biochim. Biophys. Acta* **171**, 1   [1.4.3.9]
1818. Kumagai, H., Nagate, T., Yoshida, H. & Yamada, H. (1972) *Biochim. Biophys. Acta* **258**, 779   [2.1.2.1, 4.1.2.5]
1819. Kumagai, H., Yamada, H., Matsui, H., Ohkishi, H. & Ogata, K. (1970) *J. Biol. Chem.* **245**, 1767   [3.1.1.24, 4.1.99.2]
1820. Kumagai, H., Yamada, H., Matsui, H., Ohkishi, H. & Ogata, K. (1970) *J. Biol. Chem.* **245**, 1773   [4.1.99.2]
1821. Kumar, S.A. & Mahadevan, S. (1963) *Arch. Biochem. Biophys.* **103**, 516   [4.2.1.29]
1822. Kumar, S.A., Rao, N.A. & Vaidyanathan, C.S. (1965) *Arch. Biochem. Biophys.* **111**, 646   [3.6.1.9]
1823. Kumon, A., Matsuoka, Y., Kakimoto, Y., Nakajima, T. & Sano, I. (1970) *Biochim. Biophys. Acta* **200**, 466   [3.4.13.4]
1824. Kun, E. (1952) *J. Biol. Chem.* **194**, 603   [1.1.3.1]
1825. Kundig, W., Ghosh, S. & Roseman, S. (1966) *J. Biol. Chem.* **241**, 5619   [2.7.1.60]
1826. Kundig, W., Ghosh, S. & Roseman, S. (1964) *Proc. Nat. Acad. Sci. USA* **52**, 1067   [2.7.1.69, 2.7.3.9]
1827. Kung, H. F., Cederbaum, S., Tsai, L. & Stadtman, T.C. (1970) *Proc. Nat. Acad. Sci. USA* **65**, 978   [5.4.99.4]
1828. Kung, H.-F. & Stadtman, T.C. (1971) *J. Biol. Chem.* **246**, 3378   [5.3.3.6, 5.4.99.4]
1829. Kung, H.-F. & Wagner, C. (1969) *J. Biol. Chem.* **244**, 4136   [3.5.1.2, 6.3.4.12]
1830. Kunimitsu, D.K. & Yasunobu, K.T. (1967) *Biochim. Biophys. Acta* **139**, 405   [3.4.22.6]
1831. Kuninaka, A. (1957) *Koso Kagaku Shinpojiumu* **12**, 65   [3.2.2.12]
1832. Kunitz, M. (1938) *J. Gen. Physiol.* **22**, 207   [3.4.21.1]

1833. Kunitz, M. (1939) *J. Gen. Physiol.* **22**, 447 [3.4.21.9]
1834. Kunitz, M. (1952) *J. Gen. Physiol.* **35**, 423 [3.6.1.1]
1835. Kunitz, M. (1948) *Science,* **108**, 19 [3.1.21.1]
1836. Kunitz, M. & McDonald, M.R. (1946) *J. Gen. Physiol.* **29**, 393 [2.7.1.1]
1837. Kuno, S., Tashiro, M., Taniuchi, H., Horibata, K., Tsukada, K., Hayaishi, O., Sakan, T., Seno, S. & Tokuyama, T. (1961) *Fed. Proc.* **20**, 3 [1.13.11.10]
1838. Kuo, M.-H. & Blumenthal, H.J. (1961) *Biochim. Biophys. Acta* **52**, 13 [3.1.3.2]
1839. Kuo, T.T. & Kosuge, T. (1969) *J. Gen. Appl. Microbiol.* **15**, 51 [1.13.12.3]
1840. Kuo, T.-T. & Tu, J. (1976) *Nature (London)* **263**, 615 [2.1.1.54]
1841. Kupiecki, F.P. & Coon, M.J. (1957) *J. Biol. Chem.* **229**, 743 [2.6.1.22]
1842. Kurahashi, K., Pennington, R.J. & Utter, M.J. (1957) *J. Biol. Chem.* **226**, 1059 [4.1.1.32]
1843. Kurahashi, K. & Sugimura, A. (1960) *J. Biol. Chem.* **235**, 940 [2.7.7.12]
1844. Kuratomi, K. & Fukunaga, K. (1963) *Biochim. Biophys. Acta* **78**, 617 [4.1.3.16]
1845. Kuratomi, K., Fukunaga, K. & Kobayashi, Y. (1963) *Biochim. Biophys. Acta* **78**, 629 [2.6.1.23]
1846. Kurylo-Borowska, Z. & Abramsky, T. (1972) *Biochim. Biophys. Acta* **264**, 1 [5.4.3.6]
1847. Kushner, I., Rapp, W. & Burtin, P. (1964) *J. Clin. Invest.* **43**, 1983 [3.4.23.1, 3.4.23.2, 3.4.23.3]
1848. Kutzbach, C. & Stokstad, E.L.R. (1971) *Biochim. Biophys. Acta* **250**, 459 [1.1.1.171]
1849. Kutzbach, C. & Stokstad, E.L.R. (1968) *Biochem. Biophys. Res. Commun.* **30**, 111 [1.5.1.6]
1850. Kuwabara, S. (1970) *Biochem. J.* **118**, 457 [3.5.2.6]
1851. La Du, B.N. & Zannoni, V.G. (1956) *J. Biol. Chem.* **219**, 273 [1.13.11.27, 1.14.12.2]
1852. La Nauze, J.M. & Rosenberg, H. (1968) *Biochim. Biophys. Acta* **165**, 438 [2.6.1.37, 3.11.1.1]
1853. LaBelle, E.F. Jr. & Hajra, A.K. (1972) *J. Biol. Chem.* **247**, 5825 [1.1.1.101]
1854. Lache, M., Hearn, W.R., Zyskind, J.W., Tipper, D.J. & Strominger, J.L. (1969) *J. Bacteriol.* **100**, 254 [3.4.99.17]
1855. Lack, C.H. (1957) *J. Clin. Pathol.* **10**, 208 [3.4.24.4]
1856. Lack, L. (1961) *J. Biol. Chem.* **236**, 2835 [5.2.1.2, 5.2.1.4]
1857. Lacko, A.G. & Neurath, H. (1967) *Biochem. Biophys. Res. Commun.* **26**, 272 [3.4.17.1, 3.4.21.4]
1858. Lacks, S., & Greenberg, B. (1975) *J. Biol. Chem.* **250**, 4060-4072 [3.1.23.12]
1859. Lagerkvist, U. (1958) *J. Biol. Chem.* **233**, 143 [6.3.5.2]
1860. Lagerqvist, U., Rymo, L., Lindqvist, O. & Andersson, E. (1972) *J. Biol. Chem.* **247**, 3897 [6.1.1.6]
1861. Lahav, M., Chiu, T.H. & Lennarz, W.J. (1969) *J. Biol. Chem.* **244**, 5890 [2.4.1.54]
1862. Lahle, L. & Tanner, W. (1973) Eur. J. Biochem. **38**, 103 [2.4.1.82]
1863. Lajtha, A., Mela, P. & Waelsch, H. (1953) *J. Biol. Chem.* **205**, 553 [6.3.1.2]
1864. Laki, K. & Gladner, J.A. (1964) *Physiol. Rev.* **44**, 127 [3.4.21.5]
1865. Lamont, H.C., Staudenbauer, W.L. & Strominger, J.L. (1972) *J. Biol. Chem.* **247**, 5103 [5.1.1.13]
1866. LaNauze, J.M., Rosenberg, A. & Shaw, D.C. (1970) *Biochim. Biophys. Acta* **212**, 3321 [3.11.1.1]
1867. Lanchantin G.F., Friedman J.A. & Hart D.W. (1973) *J. Biol. Chem.* **248**, 5956 [3.4.99.28]
1868. Landman, A.D. & Dakshinamurti, K. (1975) *Biochem. J.* **145**, 545 [6.3.4.15]
1869. Landman, O.E. (1957) *Biochim. Biophys. Acta* **23**, 558 [3.2.1.23]
1870. Landmann H. & Markwardt F. (1970) *Experientia* **26**, 145 [3.4.21.31]
1871. Lane, M.D., Halenz, D.R., Kosow, D.P. & Hegre, C.S. (1960) *J. Biol. Chem.* **235**, 3082 [6.4.1.3]
1872. Lane, M.D., Young, D.L. & Lynen, F. (1964) *J. Biol. Chem.* **239**, 2858 [6.3.4.9]
1873. Langan, T.A. (1969) *J. Biol. Chem.* **244**, 5763 [2.7.1.70]
1874. Langer, L.J., Alexander, J.A. & Engel, L.L. (1959) *J. Biol. Chem.* **234**, 2609 [1.1.1.62]
1875. Lapresle, C. & Webb, T. (1962) *Biochem. J.* **84**, 455 [3.4.23.5]
1876. Large, P.J. (1971) *FEBS Lett.* **18**, 297 [4.1.2.32]
1877. Larner, J. (1960) in *The Enzymes,* 2nd edn (Boyer, P.D., Lardy, H. & Myrbäck, K., ed.) Vol. 4, p. 369, Academic Press, New York [3.2.1.10, 3.2.1.20, 3.2.1.21, 3.2.1.4]
1878. Larner, J. & Gillespie, R.E. (1956) *J. Biol. Chem.* **223**, 709 [3.2.1.10]

482

1879. Larner, J., Jackson, W.T., Graves, D.J. & Stamer, J.R. (1956) *Arch. Biochem. Biophys.* **60**, 352 [1.1.1.18]

1880. Larson, M.K. (1969) M.S. Thesis. Univ. Calif. (Davis) [3.4.23.6]

1881. Larsson, A. (1973) *Biochim. Biophys. Acta* **324**, 447 [1.17.4.1]

1882. Larsson, A. & Reichard, P. (1966) *J. Biol. Chem.* **241**, 2533 [1.17.4.1]

1883. Larsson, A. & Reichard, P. (1966) *J. Biol. Chem.* **241**, 2540 [1.17.4.1]

1884. Laskowski, M., Jr. & Sealock, R.W. (1971) In *The Enzymes, 3 ed. (Boyer P.D., ed. Vol.3, p. 375 Academic Press, New York)* [3.4.21.4]

1885. Laskowski, M., Mims, V. & Day, P.L. (1945) *J. Biol. Chem.* **157**, 731

1886. Laskowski, M. Sr. (1966) in *Procedures in Nucleic Acid Research*(Cantoni, G.L. & Davies, D.R., ed.) p. 85, Harper & Row, New York [3.1.15.1]

1887. Laskowski, M., Sr. (1971) in *The Enzymes*. 3rd edu (Boyer, P.D., ed.) Vol. 4, p313, Academic Press, New York. [3.1.21.1]

1888. Laster, L. & Blair, A. (1963) *J. Biol. Chem.* **238**, 3348 [2.4.2.16]

1889. Latt, S.A., Holmquist, B. & Vallee, B.L. (1969) *Biochem. Biophys. Res. Commun.* **37**, 333

1890. Lautenberger, J.A. & Linn, S, (1972) *J. Biol. Chem.* **247**, 6176 [2.1.1.37]

1891. Lazarus, G.S., Daniels, J.R., Brown, R.S., Bladen, H.A. & Fullmer, H.M. (1968) *J. Clin. Invest.* **47**, 2622

1892. Lazzarini, R.A. & Atkinson, D.E. (1961) *J. Biol. Chem.* **236**, 3330 [1.6.6.4]

1893. Lecroisey A., Keil-Dlouha V., Woods D.R., Perrin D. & Keil B. (1975) *FEBS-Letters* **59**, 167 [3.4.24.8]

1894. Leder, I.G. (1957) *J. Biol. Chem.* **225**, 125 [2.7.1.12]

1895. Leder, I.G. (1961) *J. Biol. Chem.* **236**, 3066 [2.5.1.3]

1896. Lee, D. & Ryle, A.P. (1967) *Biochem. J.* **104**, 735 & 742 [3.4.23.1]

1897. Lee, E.Y.C., Carter, J.H., Nielsen, L.D. & Fischer, E.H. (1970) *Biochemistry* **9**, 2347 [3.2.1.33]

1898. Lee, E.Y.C., Smith, E.E. & Whelan, W.J. (1970) in *Miami Winter Symposia* (Whelan, W.J. & Schultz, J., ed.) Vol. 1, p. 139, North Holland, Utrecht [2.4.1.25]

1899. Lee, E.Y.C. & Whelan, W.J. (1972) in *The Enzymes*, 3rd edn, (Boyer, P.D., ed.) Vol. 5, pp. 191 & 471, Academic Press, New York [3.2.1.41]

1900. Lee, H.A. & Abeles, R.H. (1963) *J. Biol. Chem.* **238**, 2367 [3.4.11.6, 4.2.1.28]

1901. Lee, N. & Bendet, I. (1967) *J. Biol. Chem.* **242**, 2043 [2.7.1.16]

1902. Lee, N., Patrick, J.W. & Masson, M. (1968) *J. Biol. Chem.* **243**, 4700 [5.1.3.4]

1903. Lee, Y.-P. (1957) *J. Biol. Chem.* **227**, 987, 993 & 999 [3.5.4.6]

1904. Lee, Y.-P., Sowokinos, J.R. & Erwin, M.J. (1967) *J. Biol. Chem.* **242**, 2264 [3.1.3.23]

1905. Leese, H.J. & Semenza, G. (1973) *J. Biol. Chem.* **248**, 8170 [3.2.1.62]

1906. leGoffic, F. & Martel, A. (1974) *Biochemie* **56**, 893 [2.3.1.55]

1907. Legrand Y., Caen J. Booyse F.M., Rafelson M.E., Robert B. & Robert L. (1973) *Biochim. Biophys. Acta* **309**, 406

1908. Lehman, I.R., Bessman, M.J., Simms, E.S. & Kornberg, A. (1958) *J. Biol. Chem.* **233**, 163 [2.7.7.7, 3.4.21.11]

1909. Lehman, I.R. & Nussbaum, A.L. (1964) *J. Biol. Chem.* **239**, 2628 [3.1.11.1]

1910. Lehman, I.R. & Richardson, C.C. (1964) *J. Biol. Chem.* **239**, 233

1911. Lehninger, A.L. & Greville, G.D. (1953) *Biochim. Biophys. Acta* **12**, 188 [1.1.1.35]

1912. Lehninger, A.L., Sudduth, H.C. & Wise, J.B. (1960) *J. Biol. Chem.* **235**, 2450 [1.1.1.30]

1913. Leibach, F.H. & Binkley, F. (1968) *Arch. Biochem. Biophys.* **127**, 292 [2.3.2.2]

1914. Leibowitz, M.J. & Soffer, R.L. (1969) *Biochem. Biophys. Res. Commun.* **36**, 47 [2.3.2.6]

1915. Leibowitz, M.J. & Soffer, R.L. (1970) *J. Biol. Chem.* **245**, 2066 [2.3.2.6]

1916. Leinweber, F.-J., Greenough, R.C., Schwender, C.F., Kaplan, H.R. & DiCarlo, F.J. (1972) *Xenobiotica* **2**, 191 [1.1.1.160]

1917. Leloir, L.F. (1953) *Advan. Enzymol.* **14**,193 [5.1.3.2]

1918. Leloir, L.F. & Cardini, C.E. (1953) *Biochim. Biophys. Acta* **12**, 15 [5.3.1.19]

1919. Leloir, L.F. & Cardini, C.E. (1956) *Biochim. Biophys. Acta* **20**, 33 [2.7.5.2]

1920. Leloir, L.F. & Cardini, C.E. (1962) in *The Enzymes*, 2nd edn (Boyer, P.D., Lardy, H. & Myrbäck, K., ed.) Vol. 6, p. 317, Academic Press, New York [2.4.1.11]

1921. Leloir, L.F., de Fekete, M.A. & Cardini, C.E. (1961) *J. Biol. Chem.* **236**, 636 [2.4.1.21]

1922. Leloir, L.F. & Goldemberg, S.H. (1960) *J. Biol. Chem.* **235**, 919 [2.4.1.11]

1923. Leloir, L.F., Trucco, R.E., Cardini, C.E., Paladini, A.C. & Caputto, R. (1949) *Arch. Biochem.* **24**, 65 [2.7.1.41]
1924. Lennarz, W.J., Bonsen, P.P.M. & van Deenen, L.L.M. (1967) *Biochemistry* **6**, 2307 [2.3.2.3]
1925. Lenney, J.F. & Dalbec, J.M. (1967) *Arch. Biochem. Biophys.* **120**, 42 [3.4.22.9]
1926. Leuthardt, F. & Nielsen, H. (1952) *Helv. Chim. Acta* **35**, 1196 [2.7.6.2]
1927. Leuthardt, F. & Testa, E. (1951) *Helv. Chim. Acta* **34**, 931 [2.7.1.3]
1928. Leuzinger, W., Baker, A.L. & Cauvin, E. (1968) *Proc. Nat. Acad. Sci. USA* **59**, 620 [3.1.1.7]
1929. Levchuk, T.P. & Orekhovich, V.N. (1963) *Biokhimiya* **28**, 1004 [3.4.23.1]
1930. Levenberg, B. & Buchanan, J.M. (1957) *J. Biol. Chem.* **224**, 1005 & 1018 [6.3.3.1]
1931. Levenberg, B. & Hayaishi, O. (1959) *J. Biol. Chem.* **234**, 955 [3.5.4.11]
1932. Leveson J.E. & Esnouf M.P. (1969) *Brit. J. Haematol.* **17**, 173
1933. Levin, D.H. & Racker, E. (1959) *J. Biol. Chem.* **234**, 2532 [4.1.2.16]
1934. Levin, E.Y. & Coleman, D.L. (1967) *J. Biol. Chem.* **242**, 4248 [4.3.1.8]
1935. Levin, E.Y., Levenberg, B. & Kaufman, S. (1960) *J. Biol. Chem.* **235**, 2080 [1.14.17.1]
1936. Levis, G.M. (1970) *Biochem. Biophys. Res. Commun.* **38**, 470 [1.1.1.98]
1937. Levvy, G.A. & Marsh, C.A. (1960) in *The Enzymes*, 2nd edn (Boyer, P.D., Lardy, H. & Myrbäck, K., ed.) Vol. 4, p. 397, Academic Press, New York [3.2.1.31]
1938. Levvy, G.A. & McAllan, A. (1961) *Biochem. J.* **80**, 435 [3.2.1.51]
1939. Levy, C.C. (1967) *J. Biol. Chem.* **242**, 747 [1.14.13.4]
1940. Levy, C.C. & Frost, P. (1966) *J. Biol. Chem.* **241**, 997 [1.14.13.4]
1941. Levy, C.C. & Goldman, P. (1970) *J. Biol. Chem.* **245**, 3257. [3.1.27.6]
1942. Levy, C.C. & Goldman, P. (1967) *J. Biol. Chem.* **242**, 2933
1943. Levy, C.C. & Goldman, P. (1968) *J. Biol. Chem.* **243**, 3507 [3.4.17.5]
1944. Levy, C.C. & Goldman, P. (1969) *J. Biol. Chem.* **244**, 4467 [3.4.17.6]
1945. Levy, C.C. & Weinstein, G.D. (1964) *Biochemistry* **3**, 1944 [1.3.1.11]
1946. Levy, G.B. & Cook, E.S. (1954) *Biochem. J.* **57**, 50 [5.1.3.3]
1947. Levy, H.R. & Popják, G. (1960) *Biochem. J.* **75**, 417 [2.7.1.36, 2.7.4.2, 2.7.6.1]
1948. Levy, H.R. & Talalay, P. (1957) *J. Amer. Chem. Soc.* **79**, 2658 [1.3.1.3]
1949. Levy, H.R. & Talalay, P. (1959) *J. Biol. Chem.* **234**, 2014 [1.3.99.4, 1.3.99.5]
1950. Levy M.R. & Chou S.C. (1974) *Biochim. Biophys. Acta* **334**, 423 [3.4.23.6]
1951. Lewin, L.M. & Brown, G.M. (1961) *J. Biol. Chem.* **236**, 2768 [2.7.1.49, 2.7.4.7]
1952. Lewis, M. & Weissman, S. (1965) *Arch. Biochem. Biophys.* **109**, 490 [3.6.1.15]
1953. Li, S.-C. & Li, Y.-T. (1970) *J. Biol. Chem.* **245**, 5153 [3.2.1.52]
1954. Li, S.-C. & Li, Y.-T. (1970) *J. Biol. Chem.* **245**, 5153 [3.2.1.30]
1955. Li, Y.-T. (1966) *J. Biol. Chem.* **241**, 1010 [3.2.1.24]
1956. Liao, T.-H. & Barber, G.A. (1972) *Biochim. Biophys. Acta* **276** 85 [4.2.1.47]
1957. Liébecq, C., Lallemand, A. & Degueldre-Guillaume, M.-J. (1963) *Bull. Soc. Chim. Biol. (Paris)* **45**, 573 [3.6.1.5]
1958. Lieberman, I. (1956) *J. Biol. Chem.* **222**, 765 [6.3.4.2]
1959. Lieberman, I. (1956) *J. Biol. Chem.* **223**, 327 [6.3.4.4]
1960. Lieberman, I. & Eto, W.H. (1957) *J. Biol. Chem.* **225**, 899 [3.2.1.20]
1961. Lieberman, I. & Kornberg, A. (1953) *Biochim. Biophys. Acta* **12**, 223 [1.3.1.14]
1962. Lieberman, I. & Kornberg, A. (1954) *J. Biol. Chem.* **207**, 911 [3.5.2.3, 3.5.2.4]
1963. Lieberman, I. & Kornberg, A. (1955) *J. Biol. Chem.* **212**, 909 [3.5.1.7]
1964. Lieberman, I., Kornberg, A. & Simms, E.S. (1955) *J. Biol. Chem.* **215**, 403 [2.4.2.10, 4.1.1.23]
1965. Lieberman, I., Kornberg, A. & Simms, E.S. (1955) *J. Biol. Chem.* **215**, 429 [2.7.4.4]
1966. Light, R.J. (1969) *Biochim. Biophys. Acta* **191**, 431 [4.1.1.52]
1967. Lin C.L., Ohtsuki K. & Hatano H. (1973) *J. Biochem. (Tokyo)* **73**,671 [3.4.23.6]
1968. Lin, E.C.C. (1961) *J. Biol. Chem.* **236**, 31 [1.1.1.11]
1969. Lin, E.C.C. & Magasanik, B. (1960) *J. Biol. Chem.* **235**, 1820 [1.1.1.6]
1970. Lin, T.-Y. & Hassid, W.Z. (1966) *J. Biol. Chem.* **241**, 5284 [2.4.1.33]
1971. Lind, K.E. (1972) *Eur. J. Biochem.* **25**, 560 [1.6.99.7]
1972. Lindahl, T., Gally, J.A. & Edelman, G.M. (1969) *J. Biol. Chem.* **244**, 5014 [3.1.11.2]
1973. Lindahl, T., Gally, J.A. & Edelman, G.M. (1969) *Proc. Nat. Acad. Sci. USA* **62**, 597 [3.1.11.3]
1974. Lindberg, B., Klenow, H. & Hansen, K. (1967) *J. Biol. Chem.* **242**, 350 [2.7.1.20]

484

1975. Lindblad, B. Lindstedt, G. & Lindstedt, S. (1970) *J. Amer. Chem. Soc.* **92**, 7446 [1.13.11.27]

1976. Lindl, T., Kreuzaler, F. & Hahlbrock, K. (1973) *Biochim. Biophys. Acta* **302**, 457 [6.2.1.12]

1977. Lindstedt, G. & Lindstedt, S. (1970) *J. Biol. Chem.* **245**, 4178 [1.14.11.1]

1978. Liner I.E. & Friedenson B. (1970) in Methods in Enzymology **19**, 261 [3.4.22.3]

1979. Lineweaver, H. & Jansen, E.F. (1951) *Advan. Enzymol.* **11**, 267 [3.1.1.11, 3.2.1.15]

1980. Linker, A., Hoffman, P., Meyer, K., Sampson, P. & Korn, E.D. (1960) *J. Biol. Chem.* **235**, 3061 [4.2.2.1]

1981. Linker, A., Meyer, K. & Hoffman, P. (1960) *J. Biol. Chem.* **235**, 924 [3.2.1.36]

1982. Linn, S. & Lehman, I.R. (1965) *J. Biol. Chem.* **240**, 1294

1983. Linn, T.C., Pelley, J.W., Pettit, F.H., Hucho, F., Randall, D.D. & Reed, L.J. (1972) *Arch. Biochem. Biophys.* **148**, 327 [2.7.1.99, 3.1.3.43]

1984. Lipke, H. & Kearns, C.W. (1959) *J. Biol. Chem.* **234**, 2123 and 2129 [4.5.1.1]

1985. Lipmann, F. (1946) *Advan. Enzymol.* **6**, 231 [3.6.1.7]

1986. Liss, M., Horwitz, S.B. & Kaplan, N.O. (1962) *J. Biol. Chem.* **237**, 1342 [1.1.1.140]

1987. Littauer, U.Z. & Kornberg, A. (1957) *J. Biol. Chem.* **226**, 1077 [2.7.7.8]

1988. Little, C. (1972) *Biochim. Biophys. Acta* **284**, 375

1989. Little, C. & Ötnass, A.-B. (1975) *Biochim. Biophys. Acta* **391**, 326 [3.1.4.3]

1990. Little, H.N. (1951) *J. Biol. Chem.* **193**, 347 [1.7.3.1]

1991. Little, J.W. (1967) *J. Biol. Chem.* **242**, 679 [3.1.11.3]

1992. Liu, C.-K., Hsu, C.-A. & Abbott, M.T. (1973) *Arch. Biochem. Biophys.* **159**, 180 [1.14.11.6]

1993. Liu, T.Y. & Elliott, S.D. (1965) *J. Biol. Chem.* **240**, 1138 [3.4.22.10]

1994. Livingston, D.M. & Leder, P. (1969) *Biochemistry* **8**, 435 [3.5.1.27]

1995. Ljungdahl, L.G. & Andreesen, J.R. (1975) *FEBS Lett.* **54**, 279 [1.2.1.43]

1996. Ljungquist, S. & Lindahl, T (1977) *J. Biol. Chem.* **252**, 2808 [3.1.25.2]

1997. Lochmüller, H., Wood, H.G. & Davis, J.J. (1966) *J. Biol. Chem.* **241**, 5678 [4.1.1.38]

1998. Lombardini, J.B., Singer, T.P. & Boyer, P.D. (1969) *J. Biol. Chem.* **244**, 1172 [1.13.11.20]

1999. London, M. & Hudson, P.B. (1956) *Biochim. Biophys. Acta* **21**, 290 [1.7.3.3]

2000. Long, C. & Maguire, M.F. (1954) *Biochem. J.* **57**, 223 [3.1.4.3]

2001. Long, C.W., Levitzki, A., Houston, L.L. & Koshland, D.E. Jr. (1968) *Fed. Proc.* **28**, 342 [6.3.4.2, 6.3.4.3]

2002. Loper, J.C. (1968) *J. Biol. Chem.* **243**, 3264 [1.1.1.23]

2003. Lopez G.W. & Barret T.D. (1974) Biol. Bull. **147**, 489 [3.4.24.12]

2004. Lord, J.M. (1972) *Biochim. Biophys. Acta* **267**, 227 [1.1.99.14]

2005. Lorence, J.H. & Nester, E.W. (1967) *Biochemistry* **6**, 1541 [5.4.99.5]

2006. Lornitzo, F.A. & Goldman, D.S. (1964) *J. Biol. Chem.* **239**2730 [2.4.1.15]

2007. Loughlin, R.E., Elford, H.L. & Buchanan, J.M. (1964) *J. Biol. Chem.* **239**, 2888 [2.1.1.13]

2008. Lovenberg, W., Weissbach, H. & Udenfriend, S. (1962) *J. Biol. Chem.* **237**, 89 [4.1.1.28, 4.1.1.53]

2009. Lowenstein, J.M. & Cohen, P.P. (1956) *J. Biol. Chem.* **220**, 57 [2.1.3.2]

2010. Lu, A.Y.H., Junk, K.W. & Coon, M.J. (1969) *J. Biol. Chem.* **244**, 3714 [1.6.2.4]

2011. Lu, A.Y.H., Kuntzman, S.W., Jacobson, M. & Conney, A.H. (1972) *J. Biol. Chem.* **247**, 1727 [1.14.14.1]

2012. Lu, Y.H., Ryan, D., Jerina, D.M., Daly, J.W. & Levin, W. (1975) J. Biol. Chem. **250**, 8285 [3.3.2.3]

2013. Lueck, J.D. & Nordlie, R.C. (1970) *Biochem. Biophys. Res. Commun.* **39**, 190 [3.1.3.9]

2014. Lukens, L.N. & Buchanan, J.M. (1959) *J. Biol. Chem.* **234**, 1799 [4.1.1.21, 6.3.2.6]

2015. Lukens, L.N. & Herrington, K.A. (1957) *Biochim. Biophys. Acta* **24**, 432 [2.4.2.7, 2.4.2.8]

2016. Lukomskaya, I.S. (1959) *Dokl. Akad. Nauk. S.S.S.R.* **129**, 1172 [2.4.1.25]

2017. Lund, P. & Magasanik, B. (1965) *J. Biol. Chem.* **240**, 4316 [3.5.3.8]

2018. Lynen, F., Agranoff, B.W., Eggerer, H., Henning, V. & Möslein, E.M. (1959) *Angew. Chem.* **71**, 657 [2.5.1.1]

2019. Lynen, F., Knappe, J., Lorch, E., Jutting, G. & Ringelmann, E. (1959) *Angew. Chem.* **71**, 481 [5.3.3.3]

2020. Lynen, F., Knappe, J., Lorch, E., Jütting, G., Ringelmann, E. & Lachance, J.-P. (1961) *Biochem. Z.* **335**, 123   [6.4.1.4]
2021. Lynen, F. & Ochoa, S. (1953) *Biochim. Biophys. Acta* **12**, 299   [2.3.1.9, 2.8.3.5]
2022. Lynn, W.S. & Brown, R.H. (1958) *J. Biol. Chem.* **232**, 1015   [1.1.1.51, 1.1.1.53, 1.14.99.9]
2023. Lynn, W.S. & Perryman, N.C. (1960) *J. Biol. Chem.* **235**, 1912   [3.1.1.3]
2024. Maas, W.K. (1956) *Fed. Proc.* **15**, 305   [6.3.2.1]
2025. Maas, W.K. (1952) *J. Biol. Chem.* **198**, 23   [3.4.24.4, 6.3.2.1]
2026. Maas, W.K., Novelli, G.D. & Lipmann, F. (1953) *Proc. Nat. Acad. Sci. USA* **39**, 1004   [2.3.1.1]
2027. Macdonald, I.A., Mahony, D.E., Jellett, J.F. & Meier, C.E. (1977) *Biochim. Biophys. Acta.* **489**, 466   [1.1.1.176]
2028. Macdonald, I.A., Williams, C.N. & Mahony, D.E. (1973) *Biochim. Biophys. Acta* **309**, 243   [1.1.1.159]
2029. Macdonald, I.A., Williams, C.N., Mahony, D.E. & Christie, W.M. (1975) *Biochim. Biophys. Acta* **384**, 12   [1.1.1.159]
2030. Machinist, J.M., Orme-Johnson, W.H. & Ziegler, D.M. (1966) *Biochemistry* **5**, 2939   [4.1.2.24]
2031. MacKenzie, J.J. & Sorensen, L.B. (1973) *Biochim. Biophys. Acta* **327**, 282   [1.6.6.8]
2032. Macleod, N.J. & Pridham, J.B. (1963) *Biochem. J.* **88**, 45P   [4.3.1.11]
2033. MacLeod, R.M., Farkas, W., Fridovitch. I. & Handler, P. (1961) *J. Biol. Chem.* **236**, 1841   [1.8.3.1]
2034. Macmillan, J.D. & Vaughn, R.H. (1964) *Biochemistry* **3**, 564   [4.2.2.9]
2035. MacNutt, W.S. (1952) *Biochem. J.* **50**, 384   [2.4.2.6]
2036. MacNutt, W.S. & Damle, S.P. (1964) *J. Biol. Chem.* **239**, 4272   [5.5.1.3]
2037. Madan, V.K., Hillmer, P. & Gottschalk, G. (1973) *Eur. J. Biochem.* **32**, 51   [1.1.1.157]
2038. Madyastha, K.M., Guarnaccia, R., Baxter, C. & Coscia, C.J. (1973) *J. Biol. Chem.* **248**, 2497   [2.1.1.50]
2039. Magasanik, B., Moyed, H.S. & Gehring, L.B. (1957) *J. Biol. Chem.* **226**, 339   [1.2.1.14]
2040. Magee, P.T. & Snell, E.E. (1966) *Biochemistry* **5**, 409   [1.1.1.84, 1.2.1.33]
2041. Mager, J. & Magasanik, B. (1960) *J. Biol. Chem.* **235**, 1474   [1.6.6.8]
2042. Magnusson, S. (1965) *Ark. Kem.* **24**, 349   [3.4.21.5, 3.4.21.6]
2043. Magnusson, S. (1965) *Ark. Kem.* **24**, 375
2044. Mahadevan, S. (1963) *Arch. Biochem. Biophys.* **100**, 557   [4.2.1.29]
2045. Mahadevan, S., Ayyoub, N.I. & Roels, O.A. (1966) *J. Biol. Chem.* **241**, 57   [3.1.1.21]
2046. Mahadevan, S. & Sauer, F. (1969) *J. Biol. Chem.* **244**, 4448   [3.1.1.28]
2047. Mahadevan, S. & Tappel, A.L. (1967) *J. Biol. Chem.* **242**, 4568   [3.5.1.26]
2048. Mahlen, A. & Gatenbeck, S. (1968) *Acta Chem. Scand.* **22**, 2617   [4.1.3.23]
2049. Mahler, H.R. (1954) *J. Biol. Chem.* **206**, 13   [1.3.99.2]
2050. Mahler, H.R., Hübscher, G. & Baum, H. (1955) *J. Biol. Chem.* **216**, 625   [1.7.3.3]
2051. Mahler, H.R., Mackler, B., Green, D.E. & Bock, R.M. (1954) *J. Biol. Chem.* **210**, 465   [1.2.3.1]
2052. Mahler, H.R., Raw, I., Molinari, R. & DoAmaral, D.F. (1958) *J. Biol. Chem.* **233**, 230   [1.6.2.2]
2053. Mahler, H.R., Wakil, S.J. & Bock, R.M. (1953) *J. Biol. Chem.* **204**, 453   [6.2.1.2]
2054. Mahony, D.E., Meier, C.E., Macdonald, I.A. & Holdeman, L.V. (1977) *App. Environ. Microbiol.* **34**, 419   [1.1.1.176, 2.4.1.31]
2055. Maier, V.P., Hasegawa, S. & Hera, E. (1969) *Phytochemistry* **8**, 405   [3.1.1.36]
2056. Main, A.R. (1960) *Biochem. J.* **75**, 188   [3.1.1.2]
2057. Maitra, U. & Dekker, E.E. (1964) *J. Biol. Chem.* **239**, 1485   [4.1.3.16]
2058. Majerus, P.W., Alberts, A.W. & Vagelos, P.R. (1965) *J. Biol. Chem.* **240**, 618   [4.2.1.58]
2059. Maki, Y., Yamamoto, S., Nozaki, M. & Hayaishi, O. (1969) *J. Biol. Chem.* **244**, 2942   [1.14.13.5]
2060. Makman, M.H. & Cantoni, G.L. (1965) *Biochemistry* **4**, 1434   [6.1.1.11]
2061. Malamy, M.H. & Horecker, B.L. (1964) *Biochemistry* **3**, 1893   [3.1.3.1]
2062. Malathi, K., Padmanaban, G. & Sarma, P.S. (1970) *Phytochemistry* **9**, 1603   [2.3.1.58]
2063. Malathi, P. & Crane, R.K. (1969) *Biochim. Biophys. Acta* **173**, 245   [3.2.1.62]
2064. Mahlén, A. (1973) *Eur. J. Biochem.* **38**, 32   [3.5.1.2, 4.1.3.29]

2065. Maley, F. & Lardy, H.A. (1956) *J. Amer. Chem. Soc.* **78**, 1393
2066. Malmgren, H. (1952) *Acta Chem. Scand.* **6**, 16 [3.6.1.10, 3.6.1.11]
2067. Malmström, B.G. (1961) in *The Enzymes*, 2nd edn (Boyer, P.D., Lardy, H. & Myrbäck, K., ed.) Vol. 5, p. 471, Academic Press, New York [4.2.1.11]
2068. Malmstrom, B.G., Andreasson, L.-E. & Reinhammar, B. (1975) in *The Enzymes*, 3rd edn (Boyer, P.D.,ed.) Vol. 12, p. 507, Academic Press, New York [1.10.3.2, 1.14.18.1]
2069. Mandeles, S. & Bloch, K. (1955) *J. Biol. Chem.* **214**, 639 [6.3.2.2]
2070. Mandl J., Keller S. & Manahan J. (1964) *Biochemistry* **3**, 1737 [3.4.24.3]
2071. Mann, J.D. & Mudd, S.H. (1963) *J. Biol. Chem.* **238**, 381 [2.1.1.27]
2072. Manners, D.J. (1962) *Advan. Carbohyd. Chem.* **17**, 371 [3.2.1.1, 3.2.1.2]
2073. Manners, D.J. & Rowe, K.L. (1971) *J. Inst. Brew.* **77**, 358
2074. Manners, D.J. & Taylor, D.C. (1967) *Arch. Biochem. Biophys.* **121**, 443 [2.4.1.30, 3.6.1.30]
2075. Mans, R.J. & Walter, T.J. (1971) *Biochim. Biophys. Acta* **247**, 113 [2.7.7.19, 2.7.7.6]
2076. Mansour, T.E. (1966) *Methods Enzymol.* **9**, 430 [2.7.1.11, 3.4.23.2]
2077. Mapson, L.W. & Breslow, E. (1957) *Biochem. J.* **65**29P [1.3.2.3, 3.4.23.2]
2078. Mapson, L.W., Isherwood, F.A. & Chen, Y.T. (1954) *Biochem. J.* **56**, 21 [1.3.2.3]
2079. Mapson, L.W., March, J.F. & Wardale, D.A. (1969) *Biochem. J.* **115**, 653 [2.6.1.41]
2080. Marcus, P.I. & Talalay, P. (1956) *J. Biol. Chem.* **218**, 661 [1.1.1.50, 1.1.1.51]
2081. Maréchal, L.R. (1967) *Biochim. Biophys. Acta* **146**, 417 & 431 [2.4.1.30]
2082. Maréchal, L.R. & Goldemberg, S.H. (1964) *J. Biol. Chem.* **239**, 3163 [2.4.1.34]
2083. Margolies, M.N. & Goldberger, R.F. (1966) *J. Biol. Chem.* **241**, 3262 [5.3.1.16]
2084. Marki, F. & Martius, C. (1960) *Biochem. Z.* **333**, 111 [1.6.99.2]
2085. Markland F.S., Brown D.M. & Smith E.L. (1972) *J. Biol. Chem.* **247**, 5596 [3.4.21.14]
2086. Markland F.S. & Damus P.S. (1971) J. Biol. Chem. **264**, 6460 [3.4.21.30]
2087. Markley, K. & Smallman, E. (1961) *Biochim. Biophys. Acta* **47**, 327 [2.7.1.36]
2088. Markovitz, A. (1964) *J. Biol. Chem.* **239**, 2091 [1.1.1.135]
2089. Marks, F. & Raab, I. (1974) *Biochim. Biophys. Acta* **334**, 368 [3.1.4.35]
2090. Marmur, J. & Hotchkiss, R.D. (1955) *J. Biol. Chem.* **214**, 383 [1.1.1.17]
2091. Marotta, C.A., Levy, C.C., Weissman, S.M. & Varricchio, F. (1973) *Biochemistry* **12**, 2901 [3.1.27.6]
2092. Maroux S., Baratti J. & Desnuelle P. (1971) *J. Biol. Chem.* **246**, 5031 [3.4.21.9]
2093. Marr, A.G. & Wilson, P.W. (1954) *Arch. Biochem. Biophys.* **49**, 424 [5.1.1.1]
2094. Marsh, C.A. (1963) *Biochem. J.* **87**, 82
2095. Marshall, M., Metzenberg, R.L. & Cohen, P.P. (1958) *J. Biol. Chem.* **233**, 102 [6.3.4.16]
2096. Marshall, M., Metzenberg, R.L. & Cohen, P.P. (1961) *J. Biol. Chem.* **236**, 2229 [6.3.4.16]
2097. Marshall, W.E., Manion, R. & Porath, J. (1968) *Biochim. Biophys. Acta* **151**, 414 [3.4.99.16]
2098. Martin, R.G. (1963) *J. Biol. Chem.* **238**, 257 [2.4.2.17]
2099. Martin, R.G. & Goldberger, R.F. (1967) *J. Biol. Chem.* **242**, 1168 [2.6.1.9]
2100. Martin, R.O. & Stumpf, P.K. (1959) *J. Biol. Chem.* **234**, 2548 [1.11.1.3]
2101. Martinez-Carrion, M. & Jenkins, W.T. (1965) *J. Biol. Chem.* **240**, 3538 [2.6.1.21]
2102. Martinez, G., Barker, H.A. & Horecker, B.L. (1963) *J. Biol. Chem.* **238**, 1598 [1.1.1.67]
2103. Martius, C. (1963) in *The Enzymes*, 2nd edn (Boyer, P.D., Lardy, H. & Myrbäck, K., ed.) Vol. 7, p. 517, Academic Press, New York [1.10.3.2, 1.6.99.2]
2104. Maruyama, K., Ariga, N., Tsuda, M. & Deguchi, K. (1978) *J. Biochem. (Tokyo)* **83**, 1125 [1.2.1.45]
2105. Mason, H.S. (1956) *Nature (London)* **177**, 79 [1.10.3.1]
2106. Mason, M. (1957) *J. Biol. Chem.* **227**, 61 [2.6.1.7]
2107. Massaro, E.J. & Lennarz, W.J. (1965) *Biochemistry* **4**, 85 [6.2.1.2]
2108. Massey, V. (1960) *Biochim. Biophys. Acta* **38**, 447 [1.2.4.2]
2109. Massey, V. (1963) in *The Enzymes*, 2nd edn (Boyer, P.D., Lardy, H. & Myrbäck, K., ed.) Vol. 7, p. 275, Academic Press, New York [1.6.4.3]
2110. Massey, V., Gibson, Q.H. & Veeger, C. (1960) *Biochem. J.* **77**, 341 [1.6.4.3]
2111. Massey, V., Palmer, G. & Bennett, R. (1961) *Biochim. Biophys. Acta* **48**, 1 [1.4.3.3]
2112. Masters, B.S.S., Kamin, H., Gibson, Q.H. & Williams, C.H. Jr. (1965) *J. Biol. Chem.* **240**, 921 [1.6.2.4]

2113. Masui, T., Herman, R. & Staple, E. (1966) *Biochim. Biophys. Acta* **117**, 266 [1.1.1.161, 1.2.1.40]

2114. Matalon, R., Cifonelli, J.A. & Dorfman, A. (1971) *Biochem. Biophys. Res. Commun.* **42**, 340 [3.2.1.76]

2115. Mathews, C.K., Brown, F. & Cohen, S.S. (1964) *J. Biol. Chem.* **239**, 2957 [2.1.2.8]

2116. Mathis, J.B. & Brown, G.M. (1970) *J. Biol. Chem.* **245**, 3015 [4.1.2.25]

2117. Matock, P. & Jones, J.G. (1970) *Biochem. J.* **116**, 797 [2.8.2.9]

2118. Matsubara, H., Hagihara, B., Nakai, M., Kemiaki, Y., Yonetani, T. & Okunuki, K. (1958) *J. Biochem. (Tokyo)* **45**, 251 [3.4.21.14]

2119. Matsubara, M., Katoh, S., Akino, M. & Kaufman, S. (1966) *Biochim. Biophys. Acta* **122**, 202 [1.1.1.153]

2120. Matsuda, M. & Ogur, M. (1969) *J. Biol. Chem.* **244**, 3352 [2.6.1.39]

2121. Matsuhashi, M., Matsuhashi, S. & Lynen, F. (1964) *Biochem. Z.* **340**, 263 [6.4.1.2]

2122. Matsuhashi, M., Matsuhashi, S., Numa, S. & Lynen, F. (1964) *Biochem. Z.* **340**, 243 [6.4.1.2]

2123. Matsuhashi, M. & Strominger, J.L. (1966) *J. Biol. Chem.* **241**, 4738 [2.6.1.33]

2124. Matsuhashi, S. (1966) *J. Biol. Chem.* **241**, 4275 [5.1.3.10]

2125. Matsuhashi, S., Matsuhashi, M., Brown, J.G. & Strominger, J.L. (1966) *J. Biol. Chem.* **241**, 4283 [4.2.1.45]

2126. Matsuo, Y. & Greenberg, D.M. (1959) *J. Biol. Chem.* **234**, 507 and 516 [4.4.1.1]

2127. Matsuoka, Y. & Koide, A. (1966) *Arch. Biochem. Biophys.* **114**, 422 [3.4.21.1]

2128. Matsushita, S. & Raacke, I.D. (1968) *Biochim. Biophys. Acta* **166**, 707 [3.6.1.15]

2129. Matsushita, T. & Davis, F.F. (1971) *Biochim. Biophys. Acta* **238**, 165 [4.2.1.70]

2130. Mattenheimer, H. (1956) *Hoppe-Seyler's Z. Physiol. Chem.* **303**, 107 [3.6.1.10]

2131. Mauzerall, D. & Granick, S. (1958) *J. Biol. Chem.* **232**, 1141 [4.1.1.37]

2132. Mavrides, C. & Orr, W. (1975) *J. Biol. Chem.* **250**, 4128 [2.6.1.1, 2.6.1.57]

2133. Maw, G.A. (1956) *Biochem. J.* **63**, 116 [2.1.1.3]

2134. Maw, G.A. (1958) *Biochem. J.* **70**, 168 [2.1.1.3]

2135. Maxwell, E.S. & de Robichon-Szulmajster, H. (1960) *J. Biol. Chem.* **235**, 308 [5.1.3.2]

2136. Maxwell, E.S., Kalckar, H.M. & Strominger, J.L. (1956) *Arch. Biochem. Biophys.* **65**, 2 [1.1.1.22]

2137. Mayer, R.M. & Ginsburg, V. (1965) *J. Biol. Chem.* **240**, 1900 [2.7.7.33]

2138. Mayes, J.S. & Hansen, R.G. (1966) *Methods Enzymol.* **9**, 708 [2.7.7.12]

2139. Mazelis, M. & Creveling, R.K. (1963) *J. Biol. Chem.* **238**, 3358 [3.2.2.7]

2140. Mazelis, M., Levin, B. & Mallinson, N. (1965) *Biochim. Biophys. Acta* **105**, 106 [3.3.1.2]

2141. Mazelis, M. & Vennesland, B. (1957) *Plant Physiol.* **32**, 591 [4.1.1.31]

2142. Mazumder, R., Sanadi, D.R. & Rodwell, W.V. (1960) *J. Biol. Chem.* **235**, 2546 [6.2.1.4]

2143. Mazumder, R., Sasakawa, T., Kaziro, Y. & Ochoa, S. (1962) *J. Biol. Chem.* **237**, 3065 [5.1.99.1]

2144. McCaman, R.E. & Finnerty, W.R. (1968) *J. Biol. Chem.* **243**, 5074 [2.7.7.41]

2145. McClure, W.R., Lardy, H.A. & Kneifel, H.P. (1971) *J. Biol. Chem.* **246**, 3569 [6.4.1.1]

2146. McConn, J.C., Tsuru, D. & Yasunobu, K.T. (1964) *J. Biol. Chem.* **239**, 3706

2147. McCord, J.M. & Fridovich, I. (1969) *J. Biol. Chem.* **244**, 6049 [1.15.1.1, 2.3.2.5]

2148. McCorkindale, J. & Edson, N.L. (1954) *Biochem. J.* **57**, 518 [1.1.1.14]

2149. McCormick, D.B. & Butler, R.C. (1962) *Biochim. Biophys. Acta* **65**, 326 [2.7.1.26]

2150. McCormick, D.B., Gregory, M.E. & Snell, E.E. (1961) *J. Biol. Chem.* **236**, 2076 [2.7.1.35]

2151. McCorquodale, D.J. (1964) *Biochim. Biophys. Acta* **91**, 541 [6.1.1.16]

2152. McCoy, E.E. & Najjar, V.A. (1959) *J. Biol. Chem.* **234**, 3017 [2.7.5.1]

2153. McCready, R.M. & Seegmiller, C.G. (1954) *Arch. Biochem. Biophys.* **50**, 440 [3.2.1.15]

2154. McCroskery P.A., Wood S. Jun. & Harris E.D. Jr. (1973) *Science* **182**, 70 [3.4.24.7]

2155. McCullough, W.G., Piligian, J.T. & Daniel, I.J. (1957) *J. Amer. Chem. Soc.* **79**, 628 [4.1.1.24]

2156. McDonald J.K., Callahan P.X. & Ellis S. (1971) in *Tissue Proteinases (Barrett A.J. & Dingle J.T. eds.) North Holland Publ., Amsterdam,* p. 69 [3.4.14.2]

2157. McDonald J.K. & Ellis S. (1975) *Life Science* **17**, (8) [3.4.16.1]

2158. McDonald J.K., Reilly T.J., Zeitman B.B. & Ellis S. (1968) *J. Biol. Chem.* **243**, 2028 [3.4.14.2]

2159. McDonald J.K., Zeitman B.B. & Ellis S. (1972) *Biochem. Biophys. Res. Commun.* **46**, 62 [3.4.14.1]

2160. McDowall M.A. (1970) *Eur. J. Biochem.* **14**, 214 [3.4.22.14]

2161. McEwen, C.M. Jr. (1965) *J. Biol. Chem.* **240**, 2003 [1.4.3.6]

2162. McFadden, B.A. & Howes, W.V. (1963) *J. Biol. Chem.* **238**, 1737

2163. McGilvery, R.W. & Cohen, P.P. (1948) *J. Biol. Chem.* **174**, 813 [4.1.1.25, 4.1.1.28]

2164. McGilvray, D. & Morris, J.G. (1969) *Biochem. J.* **112**, 657 [2.3.1.29]

2165. McGuire, E.J. & Roseman, S. (1967) *J. Biol. Chem.* **242**, 3745 [2.4.1.41]

2166. McGuire, J.S., Hollis, V.W. & Tomkins, G.M. (1960) *J. Biol. Chem.* **235**, 3112 [1.3.1.4]

2167. McIntosh, E.N., Purko, M. & Wood, W.A. (1957) *J. Biol. Chem.* **228**, 499 [4.1.2.12]

2168. McKenna, E.J. & Coon, M.J. (1970) *J. Biol. Chem.* **245**, 3882 [1.14.15.3]

2169. McKenzie, A. (1936) *Ergeb. Enzymforsch.* **5**, 49 [4.1.2.10]

2170. McKhann, G.M., Levy, R. & Ho, W. (1965) *Biochem. Biophys. Res. Commun.* **20**, 109 [2.8.2.11]

2171. McManus, I.R. (1962) *J. Biol. Chem.* **237**, 1207 [2.1.1.22]

2172. McMullen, A.I. & McSweeney, G.P. (1966) *Biochem. J.* **101**, 42 [2.5.1.20]

2173. Mecke, D., Wulff, K. & Holzer, H. (1966) *Biochem. Biophys. Res. Commun.* **24**, 452 [2.7.7.42]

2174. Mecke, D., Wulff, K. & Holzer, H. (1966) *Biochim. Biophys. Acta* **128**, 559 [2.7.7.42]

2175. Medina, A. & Nicholas, D.J.D. (1957) *Nature (London)* **179**, 533 [1.6.6.6]

2176. Medina, A. & Sols, A. (1956) *Biochim. Biophys. Acta* **19**, 378 [2.7.1.4]

2177. Mehl, E. & Jatzkewitz, H. (1964) *Hoppe-Seyler's Z. Physiol. Chem.* **339**, 260 [3.1.6.8]

2178. Mehler, A.H. & Knox, W.E. (1950) *J. Biol. Chem.* **187**, 431 [3.5.1.9]

2179. Mehler, A.H. & Mitra, S.K. (1967) *J. Biol. Chem.* **242**, 5495 [6.1.1.19]

2180. Mehler, A.H. & Tabor, H. (1953) *J. Biol. Chem.* **201**, 775 [4.3.1.3]

2181. Meigs, R.A. & Ryan, K.J. (1966) *J. Biol. Chem.* **241**, 4011 [1.1.1.147]

2182. Meister, A. (1954) *J. Biol. Chem.* **206**, 587 [2.6.1.13]

2183. Meister, A. (1954) *J. Biol. Chem.* **210**, 17 [2.6.1.15]

2184. Meister, A. (1962) in *The Enzymes*, 2nd edn (Boyer, P.D., Lardy, H. & Myrbäck, K., ed.) Vol. 6, p. 443, Academic Press, New York [6.3.1.2]

2185. Meister, A., Bukenberger, M.W. & Strassburger, M. (1963) *Biochem. Z.* **338**, 217 [4.2.1.48]

2186. Meister, A. & Fraser, P.E. (1954) *J. Biol. Chem.* **210**, 37 [2.6.1.14]

2187. Meister, A. & Greenstein, J.P. (1948) *J. Biol. Chem.* **175**, 573 [3.7.1.2]

2188. Meister, A., Levintow, L., Greenfield, R.E. & Abendschein, P.A. (1955) *J. Biol. Chem.* **215**, 441 [3.5.1.3]

2189. Meister, A., Radhakrishnan, A.N. & Buckley, S.D. (1957) *J. Biol. Chem.* **229**, 789 [1.5.1.1, 1.5.1.2]

2190. Meister, A. & Wellner, D. (1963) in *The Enzymes*, 2nd edn (Boyer, P.D., Lardy, H. & Myrbäck, K., ed.) Vol 7, p. 609, Academic Press, New York [1.4.3.2, 1.4.3.3]

2191. Meister, P.D., Reinecke, L.M., Meeks, R.C., Murray, H.C., Eppstein, S.H., Osborn, H.M.L., Weintraub, A. & Peterson, D.H. (1954) *J. Amer. Chem. Soc.* **76**, 4050 [1.14.99.9]

2192. Mella, K., Volz, M. & Pfleiderer, G. (1967) *Anal. Biochem.* **21**, 219 [3.4.24.1]

2193. Melnick, I. & Buchanan, J.M. (1957) *J. Biol. Chem.* **225**, 157 [6.3.5.3]

2194. Melo, A., Elliott, H. & Glaser, L. (1968) *J. Biol. Chem.* **243**, 1467 [4.2.1.45, 4.2.1.46, 4.2.1.47]

2195. Melo, A. & Glaser, L. (1968) *J. Biol. Chem.* **243**, 1475 [1.1.1.133, 5.1.3.13]

2196. Meloche, H.P. & Wood, W.A. (1964) *J. Biol. Chem.* **239**, 3505 [4.2.1.12]

2197. Meloche, H.P. & Wood, W.A. (1964) *J. Biol. Chem.* **239**, 3515 [4.1.2.14]

2198. Mendicino, J. (1960) *J. Biol. Chem.* **235**, 3347 [2.4.1.14]

2199. Mendicino, J. & Muntz, J.A. (1958) *J. Biol. Chem.* **233**, 178 [3.5.4.6]

2200. Menon, G.K.K., Friedman, D.L. & Stern, J.R. (1960) *Biochim. Biophys. Acta* **44**, 375 [6.2.1.6]

2201. Menon, G.K.K. & Stern, J.R. (1960) *J. Biol. Chem.* **235**, 3393 [2.8.3.5]

2202. Menzies, C.A. & McQuillan, M.T. (1967) *Biochim. Biophys. Acta* **132**, 444 [3.4.16.3]

488

2203. Merell, H., Clark, M.J., Knowles, P.F. & Sprinson, D.B. (1967) J. Biol. Chem. **242**, 82 [2.5.1.19]
2204. Merrett, T.R., Bar-Eli, E. & van Vunakis, H. (1969) *Biochemistry* **8**, 3696 [3.4.23.1]
2205. Merrick, J.M. & Roseman, S. (1966) *Methods Enzymol.* **9**, 657 [4.3.1.9]
2206. Merritt, A.D. & Tomkins, G.M. (1959) *J. Biol. Chem.* **234**, 2778 [1.1.1.1]
2207. Meselson, M. & Yuan, R. (1968) *Nature (London)* **217**, 1110 [3.1.24.2, 3.1.24.3]
2208. Messer, M. & Ottesen, M. (1965) *C.R. Trav. Lab. Carlsberg* **35**, 1
2209. Metrione, R.M., Neves, A. & Fruton, J.S. (1966) *Biochemistry* **5**, 1597 [3.4.14.1]
2210. Metzler, D.E. & Snell, E.E. (1952) *J. Biol. Chem.* **198**, 363 [4.2.1.14]
2211. Meyer-Arendt, E., Beisenherz, G. & Bücher, T. (1953) *Naturwiss.* **40**, 59 [5.3.1.1]
2212. Meyer, K., Hoffman, P. & Linker, A. (1960) in *The Enzymes*, 2nd edn (Boyer, P.D., Lardy, H. & Myrbäck, K., ed.) Vol. 4, p. 447, Academic Press, New York [3.2.1.35, 3.2.1.36]
2213. Meyer, K. & Rapport, M.M. (1952) *Advan. Enzymol.* **13**, 199 [3.2.1.35, 4.2.2.1]
2214. Meyerhof, O. & Beck, L.V. (1944) *J. Biol. Chem.* **156**, 109 [5.3.1.1]
2215. Meyerhof, O., Shatas, R. & Kaplan, A. (1953) *Biochim. Biophys. Acta* **12**, 121 [3.6.1.2]
2216. Michaels, G.B., Davidson, J.T. & Peck, H.D. Jr. (1970) *Biochem. Biophys. Res. Commun.* **39**, 321 [1.8.99.2]
2217. Middleton, J.H., Edgell, M.H. & Hutchison, C.A. III J. Virol. **10**, 42-50 (1972). [3.1.23.17]
2218. Middleton, J.H., Stankus, P.V., Edgell, M.H. & Hutchison, C.A. III, unpublished results [3.1.23.22, 3.1.23.25]
2219. Miflin, B.F. & Lea, P.J. (1974) Nature **251**, 614 [1.4.7.1]
2220. Mikulski, A.J. & Laskowski, M. Sr. (1970) *J. Biol. Chem.* **245**, 5026 [3.1.30.2]
2221. Mill, P.J. & Tuttobello, R. (1961) *Biochem. J.* **79**, 57 [3.2.1.15]
2222. Miller, A. & Waelsch, H. (1957) *J. Biol. Chem.* **228**, 397 [2.1.2.5]
2223. Miller, J.E. & Litwack, G. (1971) J. Biol. Chem. **246**, 3234 [2.6.1.5]
2224. Miller, K.D. & Copeland, W.H. (1956) *Biochim. Biophys. Acta* **22**, 193 [3.2.1.3]
2225. Miller, K.D. & van Vunakis, H. (1956) *J. Biol. Chem.* **223**, 227 [3.4.21.5]
2226. Miller, O.N., Huggins, C.G. & Arai, K. (1953) *J. Biol. Chem.* **202**, 263 [1.1.1.8]
2227. Miller, R.E. & Stadtman, E.R. (1972) J. Biol. Chem. **247**, 7407 [1.4.1.13]
2228. Miller, W.L., Kalafer, M.E., Gaylor, J.L. & Delwicke, C.V. (1967) Biochemistry **6**, 2673 [1.14.99.16]
2229. Millerd, A. & Bonner, J. (1954) *Arch. Biochem. Biophys.* **49**, 343 [6.2.1.1]
2230. Millet J. (1970) J. Appl. Bacteriol **33**, 207 [3.4.24.4]
2231. Millet J. & Acher R. (1968) *Biochim. Biophys. Acta* **151**, 302 [3.4.24.4]
2232. Millet J. & Acher R. (1969) *Eur. J. Biochem.* **9**, 456 [3.4.24.4]
2233. Mills, G.C. (1960) *Arch. Biochem. Biophys.* **86**, 1 [1.11.1.9]
2234. Mills, J.B. (1949) *Biochem. J.* **44**, 302 [3.1.1.11]
2235. Milstien, S. & Goldman, P. (1972) *J. Biol. Chem.* **247**, 6280 [3.5.3.14]
2236. Milstone, J.H. (1964) *Fed. Proc.* **23**, 742 [3.4.21.6]
2237. Minamikawa, T., Jayasankar, N.P., Bohm, B.A., Taylor, I.E.P. & Towers, G.H.N. (1970) *Biochem. J.* **116**, 889 [3.7.1.4]
2238. Minatogawa, Y., Noguchi, T. & Kido, R. (1977) *Hoppe-Seyler's Z. Physiol. Chem.* **358** [2.6.1.58]
2239. Minson, A.C. & Creaser, E.H. (1969) *Biochem. J.* **114**, 49 [3.5.4.19]
2240. Misaka, E. & Nakanishi, K. (1963) *J. Biochem. (Tokyo)* **53**, 465 [1.6.99.2]
2241. Misra, H.P. & Fridovich, I. (1972) *J. Biol. Chem.* **247**, 3410 [1.15.1.1]
2242. Mitchell W.M. (1968) *Biochim. Biophys. Acta* **159**, 554 [3.4.24.3]
2243. Mitchell, W.M. & Harrington, W.F. (1968) *J. Biol. Chem.* **243**, 4683 [3.4.22.8]
2244. Mitoma, C. (1956) *Arch. Biochem. Biophys.* **60**, 477 [1.14.16.1]
2245. Mitoma, C., Posner, H.S., Reitz, H.C. & Udenfriend, S. (1956) *Arch. Biochem. Biophys.* **61**, 431 [1.14.14.1]
2246. Mitoma, C. & Udenfriend, S. (1962) *Methods Enzymol.* **5**, 816 [1.14.14.1]
2247. Mitra, S.K. & Mehler, A.H. (1967) *J. Biol. Chem.* **242**, 5491 [6.1.1.19]
2248. Mitsuhashi, S. & Davis, B.D. (1954) *Biochim. Biophys. Acta* **15**, 54 [4.2.1.10]
2249. Mitsuhashi, S. & Davis, B.D. (1954) *Biochim. Biophys. Acta* **15**, 268 [1.1.1.24, 1.1.1.25, 4.2.1.10]

2250. Mitton, J.R., Scholan, N.A. & Boyd, G.S. (1971) *Eur. J. Biochem.* **20**, 569
       [1.14.13.17]
2251. Mitz, M.A. & Heinrikson, R.L. (1961) *Biochim. Biophys. Acta* **46**, 45   [1.1.1.66]
2252. Miyata, M. & Mori, T. (1969) *J. Biochem. (Tokyo)* **66**, 463   [1.7.2.1]
2253. Miyoshi, T., Sato, H. & Harada, T. (1974) *Biochim. Biophys. Acta* **358**, 231   [1.1.1.165]
2254. Mizugaki, M., Swindell, A.C. & Wakil, S.J. (1968) *Biochem. Biophys. Res. Commun.* **33**,
       520   [4.2.1.59, 4.2.1.61]
2255. Mizugaki, M., Weeks, G., Toomey, R.E. & Wakil, S.J. (1968) *J. Biol. Chem.* **243**, 3661
       [4.2.1.58]
2256. Mizuno, Y. & Ito, E. (1968) *J. Biol. Chem.* **243**, 2665   [6.3.2.13]
2257. Mizusaki, S., Tanabe, Y., Noguchi, M. & Tamaki, E. (1971) *Plant Cell Physiol.* **12**, 633
       [2.1.1.53]
2258. Mizusawa K. & Yoshida F. (1973) *J. Biol. Chem.* **248**, 4417   [3.4.21.14]
2259. Mizushima, S. & Kitahara, K. (1962) *J. Gen. Appl. Microbiol.* **8**, 56   [1.11.1.1]
2260. Moffa, D.J., Lotspeich, F.J. & Krause, R.F. (1970) *J. Biol. Chem.* **245**, 439   [1.2.1.36]
2261. Mokrash, L.C. & McGilvery, R.N. (1956) *J. Biol. Chem.* **221**, 909   [3.1.3.11]
2262. Moldave, K. & Meister, A. (1957) *J. Biol. Chem.* **229**, 463   [2.3.1.14]
2263. Molinari, E. & Hoffmann-Ostenhof, O. (1968) *Hoppe-Seyler's Z. Physiol. Chem.* **349**, 1797
       [2.7.1.64]
2264. Momparler, R.L. & Fischer, G.A. (1968) *J. Biol. Chem.* **243**, 4298   [2.7.1.74]
2265. Monder, C. (1967) *J. Biol. Chem.* **242**, 4603   [1.2.1.23]
2266. Monder, C. & White, A. (1965) *J. Biol. Chem.* **240**, 71   [1.1.1.150, 1.1.1.151]
2267. Mondovi, B., Costa, M.T., Agro, A.F. & Rotilio, G. (1967) *Arch. Biochem. Biophys.* **119**,
       373   [1.4.3.6]
2268. Monod, J. & Cohn, M. (1952) *Advan. Enzymol.* **31**, 67   [3.2.1.23]
2269. Moog, P. & Krisch, K. (1974) *Hoppe-Seyler's Z. Physiol. Chem.* **355**, 529   [3.1.1.10]
2270. Moore, E.C. & Hurlbert, R.B. (1966) *J. Biol. Chem.* **241**, 4802   [1.17.4.1]
2271. Moore, E.C., Reichard, P. & Thelander, L. (1964) *J. Biol. Chem.* **239**, 3445   [1.6.4.5]
2272. Moore, J.H. & Williams, D.L. (1964) *Biochim. Biophys. Acta* **84**, 41   [3.1.1.4]
2273. Moore, J.T. Jr. & Gaylor, J.L. (1969) *J. Biol. Chem.* **244**, 6334   [2.1.1.41]
2274. Moore, M.R., O'Brien, W.E. & Ljungdahl, L.G. (1974) *J. Biol. Chem.* **249**, 5250
       [1.5.1.15]
2275. Moorefield, H.H. (1956) *Contr. Boyce Thompson Inst.* **18**, 303   [4.5.1.1]
2276. Mora, J., Tarrab, R., Martuscelli, J. & Soberón, G. (1965) *Biochem. J.* **96**, 588
       [3.5.3.7]
2277. Moran, F., Nasuno, S. & Starr, M.P. (1968) *Arch. Biochem. Biophys.* **123**, 298
       [4.2.2.1]
2278. Moran, F., Nasuno, S. & Starr, M.P. (1968) *Arch. Biochem. Biophys.* **125**, 734
       [4.2.2.6]
2279. Morell, H., Clark, M.J., Knowles, P.F. & Sprinson, D.B. (1967) *J. Biol. Chem.* **242**, 82
       [4.6.1.4]
2280. Morell, H. & Sprinson, D.B. (1968) *J. Biol. Chem.* **243**, 676   [2.7.1.71]
2281. Morell, P. & Radin, N.S. (1969) *Biochemistry* **8**, 506   [2.4.1.45]
2282. Morgan, L.R. Jr., Weimorts, D.M. & Aubert, C.C. (1965) *Biochim. Biophys. Acta* **100**, 393
       [1.10.3.5]
2283. Morihara, K. (1963) *Biochim. Biophys. Acta* **73**, 113   [3.4.24.4]
2284. Morihara K. & Touzuki H. (1975) *Agr. Biol. Chem.* **39**, 1489   [3.4.21.14]
2285. Morihara, K. & Tsuzuki, H. (1968) *Arch. Biochem. Biophys.* **126**, 971   [3.4.21.4]
2286. Morihara, K. & Tsuzuki, H. (1964) *Biochim. Biophys. Acta* **92**, 351   [3.4.24.4]
2287. Morihara, K., Tsuzuki, H. & Oka, T. (1968) *Arch. Biochem. Biophys.* **123**, 572
       [3.4.24.4]
2288. Morihara K., Tsuzuki H.& Oka T. (1973) *Biochim. Biophys. Acta* **309**, 414   [3.4.24.4]
2289. Morihara K., Tsuzuki H., Oka T., Inoue H. & Ebata M. (1965) *J. Biochem. (Tokyo)* **240**,
       3295   [3.4.24.4]
2290. Moritani, M. (1952) *Hukuoka Acta Med.* **43**, 651 & 731   [1.5.3.2]
2291. Moritani, M., Tung, T.-C., Fujii, S., Mito, H., Izumika, N., Kenmochi, K. & Hirohata,
       R. (1954) *J. Biol. Chem.* **209**, 485   [1.5.3.2]
2292. Morley, C.G.D. & Stadtman, T.C. (1970) *Biochemistry* **9**, 4890   [5.4.3.4]
2293. Morley, N. & Kuksis, A. (1972) *J. Biol. Chem.* **247**, 6389   [3.1.1.34]
2294. Morris, D.R. & Hager, L.P. (1966) *J. Biol. Chem.* **241**, 1763   [1.11.1.10]

2295. Morrison, J.F. (1954) *Biochem. J.* **56**, 99 [4.2.1.3]
2296. Morrison, J.F., Griffiths, D.E. & Ennor, A.H. (1957) *Biochem. J.* **65**, 143 [2.7.3.3]
2297. Morrison, M., Hamilton, H.B. & Stotz, E. (1957) *J. Biol. Chem.* **228**, 767 [1.11.1.7]
2298. Morse, D.E. & Horecker, B.L. (1968) *Advan. Enzymol.* **31**, 125 [4.1.2.13]
2299. Morton, R.K. (1953) *Biochem. J.* **55**, 795 [3.1.3.1]
2300. Moskowitz, G.J. & Merrick, J.M. (1969) *Biochemistry* **8**, 2748 [4.2.1.17, 4.2.1.55]
2301. Motokawa, Y. & Kikuchi, G. (1969) *Arch. Biochem. Biophys.* **135**, 402 [2.1.2.10]
2302. Mounter, L.A. (1960) in *The Enzymes*, 2nd edn. (Boyer, P.D., Lardy, H. & Myrbäck, K., ed.) Vol. 4, p. 541, Academic Press, New York [3.8.2.1]
2303. Moustafa, E. & Wong, E. (1967) *Phytochemistry* **6**, 625 [5.5.1.6]
2304. Moyed, H.S. & Magasanik, B. (1957) *J. Biol. Chem.* **226**, 351 [6.3.4.1]
2305. Moyed, H.S. & Williamson, V. (1967) *J. Biol. Chem.* **242**, 1075 [1.3.1.17]
2306. Moyle, J. & Dixon, M. (1956) *Biochem. J.* **63**, 548 [1.1.1.42]
2307. Mudd, S.H. (1959) *J. Biol. Chem.* **234**, 87 [2.5.1.4]
2308. Mudd, S.H. (1959) *J. Biol. Chem.* **234**, 1784 [2.5.1.4]
2309. Mudd, S.H. & Cantoni, G.L. (1958) *J. Biol. Chem.* **231**, 481 [2.5.1.6]
2310. Mue, S., Tuboi, S. & Kikuchi, G. (1964) *J. Biochem. (Tokyo)* **56**, 545 [6.2.1.9]
2311. Mueller, G.C. & Miller, J.A. (1949) *J. Biol. Chem.* **180** 1125 [1.6.6.7]
2312. Mueller, G.C. & Rumney, G. (1957) *J. Amer. Chem. Soc.* **79** 1004 [1.14.99.11]
2313. Muhammed, A. (1961) *Biochim. Biophys. Acta* **54**, 121 [2.7.4.1]
2314. Munch-Petersen, A. (1955) *Arch. Biochem. Biophys.* **55**, 592 [2.7.7.13]
2315. Murao, S. & Nishino, T. (1974) *Agr. Biol. Chem. (Tokyo)* **38**, 2483 [2.7.6.4]
2316. Murphy, T.A. & Wyatt, G.R. (1965) *J. Biol. Chem.* **240**, 1500 [2.4.1.15]
2317. Murray, K., Hughes, S.G., Brown, J.S. & Bruce, S. (1976) *Biochem. J.* **159**, 317 [3.1.23.3, 3.1.23.4]
2318. Murray, K., Morrison, A., Cooke, H.W. & Roberts, R.J., unpublished observations. [3.1.23.15]
2319. Myers, F.L. & Northcote, D.H. (1959) *Biochem. J.* **71**, 749 [3.2.1.4]
2320. Myers, J.W. (1961) *J. Biol. Chem.* **236**, 1414 [4.2.1.9]
2321. Myers, P.A. & Zatman, L.J. (1971) *Biochem. J.* **121**, 10P [4.1.2.32]
2322. Myrbäck, K. (1960) in *The Enzymes*, 2nd edn (Boyer, P.D., Lardy, H. & Myrbäck, K., ed.) Vol. 4, p. 379, Academic Press, New York [3.2.1.26]
2323. Myrbäck, K. & Örtenblad, B. (1937) *Biochem. Z.* **291**, 61 [3.2.1.28]
2324. Nachmansohn, D. & Wilson, I.B. (1951) *Advan. Enzymol.* **12**, 259 [3.1.1.7, 3.1.1.8]
2325. Nadai, Y. (1958) *J. Biochem. (Tokyo)* **45**, 1011 [3.5.3.10]
2326. Nadanishi T. & Yamamoto T. (1974) *Agr. Biol. Chem. (Tokyo)* **38**, 2391 [3.4.99.11]
2327. Nagai, J. & Bloch, K. (1968) *J. Biol. Chem.* **243**, 4626 [1.14.99.6]
2328. Nagai, S. & Black, S. (1968) *J. Biol. Chem.* **243**, 1942 [1.8.4.4]
2329. Nagai, S. & Flavin, M. (1967) *J. Biol. Chem.* **242**, 3884 [2.3.1.31]
2330. Nagai, Y., Lapiere, C.M. & Gross, J. (1966) *Biochemistry* **5**, 3123 [3.4.24.7]
2331. Naganawa, H., Yagisawa, M., Kondo, S., Takeuchi, T. & Umezawa, H. (1971) *J. Antibiot.* **24**, 913 [2.7.7.46]
2332. Nagatsu, T., Levitt, M. & Udenfriend, S. (1964) *J. Biol. Chem.* **239**, 2910 [1.14.16.2]
2333. Nagayama, H., Muramatsu, M. & Shimura, K. (1958) *Nature (London)* **181**, 417 [2.6.1.47]
2334. Nagel, C.W. & Vaughn, R.H. (1961) *Arch. Biochem. Biophys.* **94**, 328 [4.2.2.2]
2335. Nair, K.G. (1966) *Biochemistry* **5**, 150 [3.1.4.17]
2336. Nair, P.M. (1969) *Arch. Biochem. Biophys.* **133**, 208 [6.3.1.4]
2337. Nair, P.M. & Vaidyanathan, C.S. (1964) *Biochim. Biophys. Acta* **81**, 496 [1.13.11.17]
2338. Nair, P.M. & Vaidyanathan, C.S. (1964) *Biochim. Biophys. Acta* **81**, 507 [1.10.3.4]
2339. Nair, P.M. & Vaidyanathan, C.S. (1965) *Biochim. Biophys. Acta* **110**, 521
2340. Nair, P.M. & Vining, L.C. (1964) *Arch. Biochem. Biophys.* **106**, 422 [1.1.3.14]
2341. Nair, P.M. & Vining, L.C. (1965) *Biochim. Biophys. Acta* **96**, 318 [1.10.3.4]
2342. Nair, P.P., Gordon, M. & Reback, J. (1967) *J. Biol. Chem.* **242**, 7 [3.5.1.24]
2343. Najjar, V.A. (1948) *J. Biol. Chem.* **175**, 281 [2.7.5.1]
2344. Najjar, V.A. (1962) in *The Enzymes*, 2nd edn (Boyer, P.D., Lardy, H. & Myrbäck, K., ed.) Vol. 6, p. 161, Academic Press, New York [2.7.5.1]
2345. Nakada, H.I. (1964) J. Biol. Chem. **239**, 468 [2.6.1.4]
2346. Nakada, H.I. & Sweeny, P.C. (1967) *J. Biol. Chem.* **242**, 845 [4.2.2.3]
2347. Nakada, H.I. & Wolfe, J.B. (1961) *Arch. Biochem. Biophys.* **94**, 244 [4.2.2.5]

2348. Nakagawa, H. & Kimura, H. (1968) *Biochem. Biophys. Res. Commun.* **32**, 209 [4.2.1.22]

2349. Nakagawa, H., Kimura, H. & Suda, M. (1968) in *Symposium on Pyridoxal Enzymes*, 3rd edn. Wiley, New York p. 101 [4.2.1.13]

2350. Nakagawa, H. & Takeda, Y. (1962) *Biochim. Biophys. Acta* **62**, 423 [1.14.13.7]

2351. Nakagawa, Y. & Noltmann, E.A. (1965) *J. Biol. Chem.* **240**, 1877 [5.3.1.9]

2352. Nakajima M. (1974) *Eur. J. Biochem.* **44**, 87 [3.4.24.4]

2353. Nakajima, Y., Fukunaga, N., Sasaki, S. & Usami, S. (1973) *Biochim. Biophys. Acta* **293**, 242 [3.6.1.22]

2354. Nakamatu, T. & Yamanaka, K. (1969) *Biochim. Biophys. Acta* **178**, 156 [5.3.1.4]

2355. Nakamura, H. & Sugino, Y. (1966) *J. Biol. Chem.* **241**, 4917 [2.7.4.6]

2356. Nakamura, K. & Bernheim, F. (1961) *Biochim. Biophys. Acta* **50**, 147 [1.2.1.15]

2357. Nakamura, T. (1958) *Biochim. Biophys. Acta* **30**, 44 & 538 [1.10.3.2]

2358. Nakamura, W., Hosoda, S. & Hayashi, K. (1974) *Biochim. Biophys. Acta* **358**, 251

2359. Nakanishi, K. (1959) *J. Biochem. (Tokyo)* **46**, 1263 [3.4.23.6]

2360. Nakanishi K. (1960) J. Biochem. (Tokyo) **47**, 16 [3.4.23.6]

2361. Nakanishi, N., Hasegawa, H. & Watabe, S. (1977) *J. Biochem. (Tokyo)* **81**, 681 [1.6.99.10]

2362. Nakano, M. (1967) *J. Biol. Chem.* **242**, 73 [2.6.1.24]

2363. Nakano, M. & Danowski, T.S. (1964) *Biochim. Biophys. Acta* **85**, 18 [2.6.1.25]

2364. Nakano, M. & Danowski, T.S. (1966) *J. Biol. Chem.* **241**, 2075 [1.4.3.2]

2365. Nakano, M., Ushijima, Y., Saga, M., Tsutsumi, Y. & Asami, H. (1968) *Biochim. Biophys. Acta* **167**, 9 [1.1.3.15]

2366. Nakayama, S. & Amagase, S. (1968) *Proc. Japan Acad.* **44** 358

2367. Nakayama, T. (1960) *J. Biochem. (Tokyo)* **48**, 812 [1.2.1.4]

2368. Nakazawa, K., Ueda, K., Honjo, T., Yoshihara, K., Nishizuka, Y. & Hayaishi, O. (1968) *Biochem. Biophys. Res. Commun.* **32**, 143 [3.2.2.5]

2369. Nakazawa, T., Hori, K. & Hayaishi, O. (1972) *J. Biol. Chem.* **247**, 3439 [1.13.12.2]

2370. Namba, Y., Yoshizawa, K., Ejima, A., Hayashi, T. & Kaneda, T. (1969) *J. Biol. Chem.* **244**, 4437 [1.2.1.25]

2371. Narahashi, Y., Shibuya, K. & Yanagita, M. (1968) *J. Biochem. (Tokyo)* **64**, 427

2372. Nardacci N.J., Mukhopadhyay S. & Campbell B.J. (1975) *Biochim. Biophys. Acta* **377**, 146 [3.4.15.2]

2373. Narrod, S.A. & Jakoby, W.B. (1964) *J. Biol. Chem.* **239**, 2189 [1.4.3.8]

2374. Narrod, S.A. & Wood, W.A. (1956) *J. Biol. Chem.* **220**, 45 [2.7.1.12]

2375. Nason, A. (1963) in *The Enzymes*, 2nd edn (Boyer, P.D., Lardy, H. & Myrbäck, K., ed.) Vol. 7, p. 587, Academic Press, New York [1.6.6.1, 1.6.6.2, 1.6.6.3, 1.7.99.4]

2376. Nason, A. & Evans, H.J. (1953) *J. Biol. Chem.* **202**, 655 [1.6.6.3]

2377. Nason, A., Wosilait, W.D. & Terrell, A.J. (1954) *Arch. Biochem. Biophys.* **48**, 233 [1.6.5.4]

2378. Nasuno, S. & Starr, M.P. (1967) *Biochem. J.* **104**, 178 [4.2.2.2]

2379. Nathenson, S.G., Ishimoto, N. & Strominger, J.L. (1966) *Methods Enzymol.* **8**, 426 [2.4.1.70]

2380. Nathenson, S.G., Strominger, J.L. & Ito, E. (1964) *J. Biol. Chem.* **239**, 1773 [6.3.2.8]

2381. Neal, D.L. & Kindel, P.K. (1970) *J. Bacteriol.* **101**, 910 [1.1.1.114]

2382. Nebert, D.W. & Gelboin, H.V. (1968) *J. Biol. Chem.* **243**, 6242 [1.14.14.1]

2383. Needham, D.M. & Williams, J.M. (1959) *Biochem. J.* **73**, 171 [3.6.1.3]

2384. Negelein, E. & Wulff, H.-J. (1937) *Biochem. Z.* **293**, 351 [1.1.1.1]

2385. Neilson, N.E. (1955) *Biochim. Biophys. Acta* **17**, 139 [4.2.1.4]

2386. Nelson, D.J. & Carter, C.E. (1969) *J. Biol. Chem.* **244**, 5254 [2.7.4.9]

2387. Nelson, T.E., Kolb, E. & Larner, J. (1969) *Biochemistry* **8**, 1419 [3.2.1.33]

2388. Nemerson Y. & Esnouf M.P. (1972) *Proc. Nat. Acad. Sci. USA* **70**, 310 [3.4.21.21]

2389. Nesterova, M.V., Sashchenko, L.P., Vasiliev, V.Y. & Severin, E.S. (1975) *Biochim. Biophys. Acta* **377**, 271 [2.7.1.37]

2390. Neuberger, A. & Turner, J.M. (1963) *Biochim. Biophys. Acta* **67**, 342 [2.6.1.43]

2391. Neufeld, E.F., Feingold, D.S. & Hassid, W.Z. (1959) *Arch. Biochem. Biophys.* **83**, 96 [2.7.1.43]

2392. Neufeld, E.F., Feingold, D.S. & Hassid, W.Z. (1960) *J. Biol. Chem.* **235**, 906 [2.7.1.46, 2.7.1.6]

2393. Neufeld, E.F., Feingold, D.S., Ilves, S.M., Kessler, G. & Hassid, W.Z. (1961) *J. Biol. Chem.* **236**, 3102 [2.7.1.44]
2394. Neufeld, H.A., Green, L.F., Latterell, F.M. & Weintraub, R.L. (1958) *J. Biol. Chem.* **232**, 1093 [1.8.3.2]
2395. Neuhaus, F.C. (1962) *Fed. Proc.* **21**, 229 [6.3.2.3]
2396. Neuhaus, F.C. & Byrne, W.L. (1959) *J. Biol. Chem.* **234**, 113 [3.1.3.3]
2397. Neujahr, H.Y. & Gaal, A. (1973) *Eur. J. Biochem.* **35**, 386 [1.14.13.7]
2398. Neujahr, H.Y. & Gaal, A. (1975) *Eur. J. Biochem.* **58**, 351 [1.14.13.7]
2399. Neumann, N.P. & Lampen, J.O. (1967) *Biochemistry* **6**, 468 [3.2.1.26]
2400. Neurath, H. (1960) in *The Enzymes*, 2nd edn (Boyer, P.D., Lardy, H. & Myrbäck, K., ed.) Vol. 4, p. 11, Academic Press, New York [3.4.17.1, 3.4.17.2]
2401. Neville, A.M., Orr, J.C. & Engel, L.L. (1968) *Biochem. J.* **107**, 20P [1.1.1.145]
2402. Newton, W.A., Morino, Y. & Snell, E.E. (1965) *J. Biol. Chem.* **240**, 1211 [4.1.99.1]
2403. Nicholas, D.J.D., Medina, A. & Jones, L.T.G. (1960) *Biochim. Biophys. Acta* **37**, 468 [1.6.6.4]
2404. Nicholas, D.J.D. & Nason, A. (1955) *J. Bacteriol.* **69**, 580 [1.6.6.1]
2405. Nicholas, D.J.D. & Nason, A. (1954) *J. Biol. Chem.* **207**, 353 [1.6.6.3]
2406. Nicholls, P. & Schonbaum, G.R. (1963) in *The Enzymes*, 2nd edn (Boyer, P.D., Lardy, H. & Myrbäck, K., ed.) Vol. 8, p. 147, Academic Press, New York [1.11.1.6]
2407. Nickerson, W.J. & Durand, S.C. (1963) *Biochim. Biophys. Acta* **77**, 87 [3.4.99.11]
2408. Niehaus, W.G. Jr., Kisic, A., Torkelson, A., Bednarczyk, D.J. & Schroepfer, G.J. Jr. (1970) *J. Biol. Chem.* **245**, 3790 [4.2.1.53]
2409. Nigg, H.N., Svoboda, J.A., Thompson, M.J., Dutky, S.R., Kaplanis, J.N. & Robbins, W.E. (1976) *Experientia* **32**, 438 [1.14.99.22]
2410. Niimura, T., Tokieda, T. & Yamaha, T. (1974) *J. Biochem. (Tokyo)* **75**, 407 [3.10.1.2]
2411. Nilsson-Ehle, P., Belfrage, P. & Borgström, B. (1971) *Biochim. Biophys. Acta* **248**, 114 [3.1.1.34]
2412. Nimmo-Smith, R.H. (1960) *Biochem. J.* **75**, 284 [3.5.1.13]
2413. Ninjoor V., Taylor S.L. & Tappel A.L. (1974) *Biochim. Biophys. Acta* **370**, 308 [3.4.16.1]
2414. Nirenberg, M.W. & Jakoby, W.B. (1960) *J. Biol. Chem.* **235**, 954 [1.1.1.61, 1.2.1.16]
2415. Nishihara, H. (1971) *Biochemistry* **10**, 1353 [4.1.2.31]
2416. Nishihara, H. & Dekker, E.E. (1972) *J. Biol. Chem.* **247**, 5079 [4.1.2.31, 4.1.3.16]
2417. Nishimura, H. & Maruo, B. (1960) *Biochim. Biophys. Acta* **40**, 355 [3.1.27.2]
2418. Nishimura, J.S. & Greenberg, D.M. (1961) *J. Biol. Chem.* **236**, 2684 [4.2.1.16]
2419. Nishino, H. (1972) *J. Biochem. (Tokyo)* **72**, 1093 [2.7.4.16]
2420. Nishino, T. & Murao, S. (1974) *Agr. Biol. Chem. (Tokyo)* **38**, 2491 [2.7.6.4]
2421. Nishino, T. & Murao, S. (1975) *Agr. Biol. Chem. (Tokyo)* **39**, 1007 [2.7.6.4]
2422. Nisizawa, K. & Hashimoto, Y. (1959) *Arch. Biochem. Biophys.* **81**, 211 [3.2.1.4]
2423. Nisman, B. (1954) *Bacteriol. Rev.* **18**, 16 [1.4.1.2]
2424. Nisman, B. & Mager, J. (1952) *Nature (London)* **169**, 243 [1.4.1.5]
2425. Niyomporn, B., Dahl, J.L. & Strominger, J.L. (1968) *J. Biol. Chem.* **243**, 773 [6.1.1.14]
2426. Noda, L. (1958) *J. Biol. Chem.* **232**, 237 [2.7.4.3]
2427. Noda, L. (1962) in *The Enzymes*, 2nd edn (Boyer, P.D., Lardy, H. & Myrbäck, K., ed.) Vol. 6, p. 139, Academic Press, New York [2.7.4.14, 2.7.4.3, 2.7.4.4]
2428. Noda, L. & Kuby, S.A. (1957) *J. Biol. Chem.* **226**, 541 & 551 [2.7.4.3]
2429. Noltmann, E.A. (1964) *J. Biol. Chem.* **239**, 1545 [5.3.1.9]
2430. Noltmann, E.A., Gubler, C.J. & Kuby, S.A. (1961) *J. Biol. Chem.* **236**, 1225 [1.1.1.49]
2431. Noltmann, E. & Bruns, F.H. (1959) *Biochem. Z.* **331**, 436 [5.3.1.9]
2432. Nomura, J., Nishizuka, Y. & Hayaishi, O. (1963) *J. Biol. Chem.* **238**, 1441 [4.4.1.6]
2433. Nordlie, R.C. (1971) in *The Enzymes*, 3rd edn (Boyer, P.D., ed.) Vol. 4, p. 543, Academic Press, New York
2434. Nordlie, R.C. (1974) *Current Topics Cell. Regulat.* **8**, 33 [3.1.3.9]
2435. Nordlie, R.C. (1971) in *The Enzymes*, 3rd edn (Boyer, P.D., ed.) Vol. 4, p. 543 [3.1.3.9]
2436. Nordlie, R.C. & Arion, W.A. (1964) *J. Biol. Chem.* **239**, 1680 [3.1.3.9]
2437. Nordlie, R.C. & Fromm, H.J. (1959) *J. Biol. Chem.* **234**, 2523 [1.1.1.56]
2438. Nordwig, A. & Strauch, L. (1963) *Hoppe-Seyler's Z. Physiol. Chem.* **330**, 145
2439. Norris, A.T., Matsumura, S. & Bloch, K. (1964) *J. Biol. Chem.* **239**, 3653 [4.2.1.60]

494

2440. Norris, E.R. & Elam, D.W. (1940) *J. Biol. Chem.* **134**, 443 [3.4.23.1]
2441. Norris, E.R. & Mathies, J.C. (1953) *J. Biol. Chem.* **204**, 673 [3.4.23.1]
2442. Northrop, J.H. (1930) *J. Gen. Physiol.* **13**, 739 & 767 [3.4.23.1]
2443. Northrop, J.H. (1933) *J. Gen. Physiol.* **15**, 29 [3.4.23.2]
2444. Northrop, J.H. (1933) *J. Gen. Physiol.* **16**, 615 [3.4.23.1]
2445. Northrop, J.H., Kunitz, M. & Herriott, R.M. (1948) in *Crystalline Enzymes*, 2nd edn, Columbia University Press [2.7.1.1, 3.4.21.4, 3.4.21.9]
2446. Norton, J. & Roth, J.S. (1967) *J. Biol. Chem.* **242**, 2029. [3.1.26.2]
2447. Norton, S.J. (1964) *Arch. Biochem. Biophys.* **106**, 147 [6.1.1.15]
2448. Norton, S.J., Ravel, J.M. & Shive, W. (1963) *J. Biol. Chem.* **238**, 269 [6.1.1.12]
2449. Norum, K.R. (1964) *Biochim. Biophys. Acta* **89**, 95 [2.3.1.21]
2450. Nose, Y. & Lipmann, F. (1958) *J. Biol. Chem.* **233**, 1348 [2.8.2.2, 2.8.2.4]
2451. Nossal, N.G. & Singer, M. (1968) *J. Biol. Chem.* **243**, 913 [3.1.13.1]
2452. Novelli, G.D. (1953) *Fed. Proc.* **12**, 675 [2.7.1.34, 2.7.7.3]
2453. Novogrodsky, A. & Hurwitz, J. (1966) *J. Biol. Chem.* **241**, 2923 [2.7.1.78]
2454. Novogrodsky, A. & Meister, A. (1964) *J. Biol. Chem.* **239**, 879 [4.1.1.12]
2455. Novogrodsky, A., Tal, M., Traub, A. & Hurwitz, J. (1966) *J. Biol. Chem.* **241**, 2933 [2.7.1.78]
2456. Nowotny, E., Sananez, R.D., Nattoro, G., Yantorno, C. & Faillaci, M.G. (1974) *Hoppe-Seyler's Z. Physiol. Chem.* **355**, 716 [4.1.2.30]
2457. Nozaki, M., Kagamiyama, H. & Hayaishi, O. (1963) *Biochem. Z.* **338**, 582 [1.13.11.2]
2458. Nugteren, D.H. & Hazelhof, E. (1973) *Biochim. Biophys. Acta* **326**, 448 [5.3.99.2, 5.3.99.3]
2459. Nugteren, D.H. & van Dorp, D.A. (1965) *Biochim. Biophys. Acta* **98**, 654 [1.14.99.1]
2460. Nurmikko, V., Salo, E., Hakola, H., Mäkinen, K. & Snell, E.E. (1966) *Biochemistry* **5**, 399 [3.5.1.22]
2461. Nuss, D.L. & Furuichi, Y. (1977) *J. Biol. Chem.* **252**, 2815
2462. Nuss, D.L., Furuichi, Y. & Koch, G. (1975) *Cell* **6**, 21 [3.6.1.30]
2463. Nussbaum, J.-L. & Mandel, P. (1972) *J. Neurochem.* **19**, 1789 [2.8.2.13]
2464. Nygaard, A.P. (1961) *J. Biol. Chem.* **236**, 920 [1.1.2.4, 1.1.99.6]
2465. Nygaard, A.P. (1963) in *The Enzymes*, 2nd edn (Boyer, P.D., Lardy, H., & Myrbäck, K., ed.) Vol. 7, p. 557, Academic Press, New York [1.1.2.3, 1.1.2.4]
2466. Nyns, E.J., Zach, D. & Snell, E.E. (1969) *J. Biol. Chem.* **244**, 2601 [3.5.1.29]
2467. Ochoa, S. (1954) *Advan. Enzymol.* **15**, 183 [1.2.4.1, 1.2.4.2]
2468. Ochoa, S., Mehler, A.H. & Kornberg, A. (1948) *J. Biol. Chem.* **174**, 979 [1.1.1.40]
2469. Ochoa, S. & Mii, S. (1961) *J. Biol. Chem.* **236**, 3303 [2.7.7.8]
2470. Ochoa, S., Stern, J.R. & Schneider, M.C. (1951) *J. Biol. Chem.* **193**, 691 [4.1.3.7]
2471. O'Connor, R.J. & Halvorson, H. (1961) *Biochim. Biophys. Acta* **48**, 47 [1.4.1.1]
2472. Odeide, R., Guilloton, M., Dupuis, B., Ravon, D. & Rosenberg, A.J. (1968) *Bull. Soc. Chim. Biol. (Paris)* **50**, 2023 [2.7.1.11]
2473. Oesch. F. (1974) *Biochem. J.* **139**, 77 [3.3.2.3]
2474. Oesch. F. & Daly, J. (1971) *Biochim. Biophys. Acta* **227**, 692 [3.3.2.3]
2475. Oesch. F., Jerina, D.M. & Daly, J. (1971) *Biochim. Biophys. Acta* **227**, 685
2476. Oesch. F., Jerina, D.M. & Daly, J.W. (1971) *Arch. Biochem. Biophys.* **144**, 253
2477. Oesch. F., Kaubisch, N., Jerina, D.M. & Daly, J.W. (1971) *Biochemistry* **10**, 4858
2478. Oeschger, M.P. & Bessman, M.J. (1966) *J. Biol. Chem.* **241** 5452 [2.7.4.8]
2479. Oginsky, E.L. & Gehrig, R.F. (1952) *J. Biol. Chem.* **198**, 799 [3.5.3.6]
2480. Ohashi, H., Matsuhashi, M. & Matsuhashi, S. (1971) *J. Biol. Chem.* **246**, 2325 [2.6.1.59]
2481. Ohesson K. & Olsson I. (1974) *Eur. J. Biochem.* **42**, 519
2482. Ohlsson K. & Ohlsson J. (1973) *Eur. J. Biochem.* **36**, 473 [3.4.21.11]
2483. Ohmura, E. & Hayaishi, O. (1957) *J. Biol. Chem.* **227**, 181 [3.4.24.7, 3.5.1.8]
2484. Ohshima G., Iwanaga S. & Suzuki T. (1971) *Biochim. Biophys. Acta* **250**, 416 [3.4.24.6]
2485. Ohtaka, Y., Uchida, T. & Sakai, T. (1963) *J. Biochem. (Tokyo)* **54**, 322. [3.1.14.1]
2486. Ohya, T., Sawai, T., Uemura, S. & Abe, K. (1978) *Agr. Biol. Chem. (Tokyo)* **42**, 571 [3.2.1.70]
2487. Oishi, M. (1969) *Proc. Nat. Acad. Sci. USA* **64**, 1292. [3.1.11.5]
2488. Oka, T. & Simpson, F.J. (1971) *Biochem. Biophys. Res. Commun.* **43**, 1 [1.13.11.24]
2489. Okamoto, H. & Hayaishi, O. (1967) *Biochem. Biophys. Res. Commun.* **29**, 394 [1.14.13.9]

2490. Okamoto, K. (1963) *J. Biochem. (Tokyo)* **53**, 448   [1.1.1.69]
2491. Okazaki, R. & Kornberg, A. (1964) *J. Biol. Chem.* **239**, 269   [2.7.1.21]
2492. Okuda, K. & Hoshita, N. (1968) *Biochim. Biophys. Acta* **164**, 381   [1.14.13.15]
2493. Olafson, R.W., Drummond, G.I. & Lee, J.F. (1969) *Can. J. Biochem.* **47**, 961
       [3.1.4.16]
2494. Old, R., Murray, K., & Roizes, G. J. Mol. Biol. **92**, 331-339 (1975).   [3.1.23.21]
2495. Oliver, I.T. & Peel, J.L. (1956) *Biochim. Biophys. Acta* **20**, 390   [2.7.4.3]
2496. Olomucki, A., Pho, D.B., Lebar, R., Delcambe, L. & Thoai, N.V. (1968) *Biochim.*
       Biophys. Acta **151**, 353   [1.13.12.1]
2497. Olson, J.A. (1961) in *The Enzymes*, 2nd ed. (Boyer, P.D., Lardy, H. & Myrbäck, K., ed.)
       Vol. 5, p. 387, Academic Press, New York   [4.1.3.1]
2498. Olson, J.A. & Anfinsen, C.B. (1952) *J. Biol. Chem.* **197**, 67   [1.4.1.3]
2499. Olson, M.O.J., Nagabushan, N., Dzwiniel, M., Smillie, L.B. & Whitaker, D.R. (1970)
       Nature (London) **228**, 438   [3.4.21.12]
2500. Omura, T., Sanders, E., Estabrook, R.W., Cooper, D.Y. & Rosenthal, O. (1966) *Arch.*
       Biochem. Biophys. **117**, 660   [1.18.1.2]
2501. Ondarza, R.N., Abney, R. & López-Colomé, A.M. (1969) *Biochim. Biophys. Acta* **191**, 239
       [1.6.4.6]
2502. O'Neil, S.R. & DeMoss, R.D. (1968) *Arch. Biochem. Biophys.* **127**, 369   [2.6.1.27]
2503. Ong P.S. & Gaucher G.M. (1976) Can. J. Microbiol. **22**, 165   [3.4.21.14]
2504. Ono, M., Inoue, H., Suzuki, F. & Takeda, Y. (1972) *Biochim. Biophys. Acta* **284**, 285
       [4.1.1.17]
2505. Orekhovich, V.N. (1968) *Ital. J. Biochem.* **17**, 241   [3.4.15.1]
2506. Orengo, A. (1969) *J. Biol. Chem.* **244**, 2204   [2.7.1.48]
2507. Orlowski, M., Richman, P.G. & Meister, A. (1969) *Biochemistry* **8**, 1049   [2.3.2.4]
2508. Ornston, L.N. (1966) *J. Biol. Chem.* **241**, 3787   [5.5.1.2]
2509. Ornston, L.N. (1966) *J. Biol. Chem.* **241**, 3795   [4.1.1.44, 5.3.3.4, 5.5.1.1]
2510. Orsi, B.A. & Spencer, B. (1964) *J. Biochem. (Tokyo)* **56**, 81   [2.8.2.6]
2511. Ortmann, R., Sutter, A. & Grisebach, H. (1972) *Biochim. Biophys. Acta* **289**, 293
       [2.4.2.25]
2512. Osaki, S. (1966) *J. Biol. Chem.* **241**, 5053   [1.16.3.1]
2513. Osaki, S. & Walaas, O. (1967) *J. Biol. Chem.* **242**, 2653   [1.16.3.1]
2514. Osborn, M.J. & D'Ari, L. (1964) *Biochem. Biophys. Res. Commun.* **16**, 568   [2.4.1.56]
2515. Osborn, M.J. & Huennekens, F.M. (1957) *Biochim. Biophys. Acta* **26**, 646   [1.5.1.5]
2516. Osborn, M.J. & Weiner, I.M. (1968) *J. Biol. Chem.* **243**, 2631   [2.4.1.60]
2517. Osborn, M.J. & Yuan Tze-Yuen, R. (1968) *J. Biol. Chem.* **243**, 5145   [2.7.8.6]
2518. Oshima G., Sato-Ohmori T. & Suzuki T. (1969) *Toxicon* 7, 229   [3.4.24.6]
2519. Oshino, N., Imai, Y. & Sato, R. (1966) Biochim. Biophys. Acta **128**, 13   [1.14.99.5]
2520. Oshino, N., Imai, Y. & Sato, R. (1971) *J. Biochem. (Tokyo)* **69**, 155   [1.14.99.5]
2521. Ota, S., Moore, S. & Stein, W.H. (1964) *Biochemistry* **3**, 180   [3.4.22.4]
2522. Otey, M.L., Birnbaum, S.M. & Greenstein, J.P. (1954) *Arch. Biochem. Biophys.* **49**, 245
2523. Otha, Y., Ogura, Y. & Wada, A. (1966) *J. Biol. Chem.* **241**, 5919   [3.4.24.4]
2524. Otha, Y. & Ribbons, D.W. (1970) *FEBS Lett.* **11**, 189   [1.14.13.6]
2525. Otsuka, K. (1958) *J. Gen. Appl. Microbiol.* **4**, 211   [1.1.1.54]
2526. Otten, L.A.B.M., Vreugdenhil, D. & Schilperoort, R.A. (1977) *Biochim. Biophys. Acta* **485**,
       268   [1.5.1.16]
2527. Ottesen, M. & Rickert, W. (1970) *C.R. Trav. Lab. Carlsberg* 37, 301   [3.4.23.6]
2528. Ottesen M. & Svendsen I. (1970) Methods Enzymol. **19**   [3.4.21.14]
2529. Otto, K. & Bhakdi, S. (1969) *Hoppe-Seyler's Z. Physiol. Chem.* **350**, 1577   [3.4.22.1]
2530. Otto K. & Riesenkönig H. (1975) *Biochim. Biophys. Acta* **379**, 462   [3.4.16.1]
2531. Overath, P., Kellerman, G.M. & Lynen, F. (1962) *Biochem. Z.* **335**, 500
       [5.1.99.1, 5.4.99.2]
2532. Owen W.G. & Jackson C.M. (1973) *Thrombosis Research* 3, 705   [3.4.99.28]
2533. Ozawa, T., Fukuda, M. & Sasaoka, K. (1973) *Biochim. Biophys. Res. Commun.* **52**, 998
       [2.6.1.21]
2534. Pabst, M.J., Habig, W.H. & Jakoby, W.B. (1974) *J. Biol. Chem.* **249**, 7140   [2.5.1.18]
2535. Pacaud M. & Richaud C. (1975) *J. Biol. Chem.* **250**, 7771
2536. Pacaud M., Sibilli L. & Le Bras G. (1976) *Eur. J. Biochem.* **69**, 141   [3.4.21.14]
2537. Packman, P.M. & Jakoby, W.B. (1965) *J. Biol. Chem.* **240**, 4107   [2.4.2.19]
2538. Paege, L.M. & Schlenk, F. (1952) *Arch. Biochem. Biophys.* **40**, 42   [2.4.2.3]

496

2539. Paigen, K. (1958) *J. Biol. Chem.* **233**, 388   [3.1.3.16]
2540. Paik, W.K., Bloch-Frankenthal, L., Birnbaum, S.M., Winitz, M. & Greenstein, J.P. (1957)
       Arch. Biochem. Biophys. **69**, 56   [3.5.1.17]
2541. Paik, W.K. & Kim, S. (1964) *Arch. Biochem. Biophys.* **108**, 221   [2.3.1.32]
2542. Paik, W.K. & Kim, S. (1968) *J. Biol. Chem.* **243**, 2108   [2.1.1.23]
2543. Paik, W.K. & Kim, S. (1970) *J. Biol. Chem.* **245**, 88   [2.1.1.43]
2544. Paladini, A.C., Caputto, R., Leloir, L.F., Trucco, R.E. & Cardini, C.E. (1949) *Arch.*
       Biochem. **23**, 55   [2.7.1.10]
2545. Palekar, A.G., Tate, S.S. & Meister, A. (1970) *Biochemistry* **9**, 2310   [4.1.1.12]
2546. Palleroni, N.J. & Doudoroff, M. (1957) *J. Bacteriol.* **74**, 180   [1.1.1.116, 3.1.1.30]
2547. Palleroni, N.J. & Doudoroff, M. (1956) *J. Biol. Chem.* **218**, 535   [5.3.1.7]
2548. Palmatier, R.D., McCroskey, R.P. & Abbott, M.T. (1970) *J. Biol. Chem.* **245**, 6706
       [4.1.1.66]
2549. Palmer, R.E. & Anderson, R.L. (1972) *J. Biol. Chem.* **247**, 3415   [2.7.1.85]
2550. Palmer, R.E. & Anderson, R.L. (1972) *J. Biol. Chem.* **247**, 3420   [3.2.1.86]
2551. Paltauf, F. & Holasek, A. (1973) J. Biol. Chem. **248**, 1609   [1.14.99.19]
2552. Paneque, A., del Campo, F.F., Ramirez, J.M. & Losada, M. (1965) *Biochim. Biophys.*
       Acta **109**, 79   [1.6.6.2]
2553. Pangburn M.K., Burstein Y., Morgan P.H., Walsh K.A. & Neurath H. (1973) *Biochem.*
       Biophys. Res. Commun. **54**, 371   [3.4.24.4]
2554. Pangburn M.K. & Walsh K.A. (1975) *Biochemistry* **14**, 4050   [3.4.24.4]
2555. Pant, R. (1959) *Biochem. J.* **73**, 30   [2.7.3.5]
2556. Pape, H. & Strominger, J.L. (1969) *J. Biol. Chem.* **244**, 3598   [1.17.1.1]
2557. Park, R.W. & Fox, S.W. (1960) *J. Biol. Chem.* **235**, 3193   [3.5.1.14]
2558. Parks, R.E., Ben-Gershom, E. & Lardy, H.A. (1957) *J. Biol. Chem.* **227**, 231   [2.7.1.3]
2559. Parmeggiani, A., Luft, J.H., Love, D.S. & Krebs, E.G. (1966) *J. Biol. Chem.* **241**, 4625
       [2.7.1.11]
2560. Parsons, S.J. & Burns, R.O. (1969) *J. Biol. Chem.* **244**, 996   [1.1.1.85]
2561. Parvin, R. & Smith, R.A. (1969) *Biochemistry* **8**, 1748   [3.9.1.1]
2562. Parzen, S.D. & Fox, A.S. (1964) *Biochim. Biophys. Acta* **92**, 465   [1.2.1.37]
2563. Paskhina, T.S. & Trapeznikova, S.S. (1967) *Biokhimiya* **32**, 527   [3.4.17.3]
2564. Patel, T.R. & Gibson, D.T. (1974) *J. Bacteriol.* **119**, 879   [1.3.1.29]
2565. Patil, S.S. & Zucker, M. (1965) J. Biol. Chem. **240**, 3938
2566. Patil, S.S. & Zucker, M. (1965) *J. Biol. Chem.* **240**, 3938
2567. Pattabiraman, T.N. & Bachhawat, B.K. (1962) *Biochim. Biophys. Acta* **59**, 681   [2.3.1.4]
2568. Pattabiraman, T.N. & Bachhawat, B.K. (1961) *Biochim. Biophys. Acta* **50**, 129
       [2.7.7.23]
2569. Pattabiraman, T.N. & Bachhawat, B.K. (1961) *Biochim. Biophys. Acta* **54**, 273
       [5.3.1.10]
2570. Patterson E.K. (1976) Methods Enzymol. **45**, 377   [3.4.13.11]
2571. Patterson E.K., Gatmaitan J.S. & Hayman S. (1973) *Biochemistry* **12**, 3701   [3.4.13.11]
2572. Patterson, M.K. Jr. & Orr, G.R. (1968) *J. Biol. Chem.* **243**, 376   [6.3.5.4]
2573. Paul, K.G. (1963) in *The Enzymes*, 2nd edn (Boyer, P.D., Lardy, H. & Myrbäck, K., ed.)
       Vol. 8, p. 227, Academic Press, New York   [1.11.1.7]
2574. Paul, R.C. & Ratledge, C. (1973) *Biochim. Biophys. Acta* **320**, 9   [2.3.1.5]
2575. Paulus, H. & Gray, E. (1967) *J. Biol. Chem.* **242**, 4980   [2.7.2.4]
2576. Paulus, H. & Kennedy, E.P. (1960) *J. Biol. Chem.* **235**, 1303   [2.7.8.11]
2577. Pauly, H.E. & Pfleiderer, G. (1976) *Hoppe-Seyler's Z. Physiol. Chem.* **356**, 1613
       [1.1.1.47]
2578. Pazur, J.H. & Anderson, J.S. (1963) *J. Biol. Chem.* **238**, 3155   [2.7.7.32]
2579. Pazur, J.H. & Okada, S. (1968) *J. Biol. Chem.* **243**, 4732   [2.4.1.25]
2580. Pazur, J.H. & Shuey, E.W. (1961) *J. Biol. Chem.* **236**, 1780   [2.7.7.24]
2581. Peanasky, R.J., Gratecos, D., Baratti, J. & Rovery, M. (1969) *Biochim. Biophys. Acta* **181**,
       82   [3.4.21.2]
2582. Peers, F.G. (1953) *Biochem. J.* **53**, 102
2583. Peisach. J. & Levine, W.G. (1965) *J. Biol. Chem.* **240**, 2284
2584. Peisach, J. & Levine, W.G. (1965) *J. Biol. Chem.* **240**, 2284   [1.14.18.1]
2585. Peraino, C., Bunville, L.G. & Tahmisian, T.N. (1969) *J. Biol. Chem.* **244**, 2241
       [2.6.1.13]
2586. Perry, S.V. (1960) *Biochem. J.* **74**, 94   [3.6.1.3]

2587. Peterofsky, B. & Gilvarg, C. (1961) *J. Biol. Chem.* **236**, 1432 [2.6.1.17]
2588. Peterson, D. & Llaneza, J. (1974) *Arch. Biochem. Biophys.* **162**, 135 [4.2.99.12]
2589. Peterson, J.A., Kusunose, M., Kusunose, E. & Coon, M.J. (1967) *J. Biol. Chem.* **242**, 4334 [1.14.15.3, 1.18.1.1]
2590. Peterson, P.J. & Fowden, L. (1965) *Biochem. J.* **97**, 112 [6.1.1.15]
2591. Petrack, B., Greengard, P., Craston, A. & Sheppy, F. (1965) *J. Biol. Chem.* **240**, 1725 [3.5.1.19]
2592. Petrack, B., Sullivan, L. & Ratner, S. (1957) *Arch. Biochem. Biophys.* **69**, 186
2593. Pettersson, G. (1965) *Acta Chem. Scand.* **19**, 2013 [4.1.1.58]
2594. Petzold, G.L. & Agranoff, B.W. (1967) *J. Biol. Chem.* **242**, 1187 [2.7.7.41]
2595. Pfleger, E.C., Piantadosi, C. & Snyder, F. (1967) Biochim. Biophys. Acta **144**, 633 [1.14.16.5]
2596. Pfleiderer, G. & Celliers, P.G. (1963) *Biochem. Z.* **339**, 186 [3.4.11.2]
2597. Pfleiderer, G., Celliers, P.G., Stanulovic, M., Wachsmuth, E.D., Determann, H. & Braunitzer, G. (1964) *Biochem. Z.* **340**, 552 [3.4.11.2]
2598. Pfleiderer, G. & Kraft, A. (1965) *Biochem. Z.* **342**, 85 [3.4.24.1]
2599. Pfleiderer, G. & Sumyk, G. (1961) *Biochim. Biophys. Acta* **51**, 482 [3.4.21.4, 3.4.24.1, 3.4.99.6]
2600. Pfleiderer, G., Zwilling, R. & Sonneborn, H. (1967) *Hoppe Seyler's Z. Physiol. Chem.* **348**, 1319
2601. Phaff, H.J. & Demain, A.L. (1956) *J. Biol. Chem.* **218**, 875 [1.6.99.3, 3.2.1.15]
2602. Philippovich, Yu. B., Klumova, S.M. & Jukova, N.I. (1974) *Dokl. Akad. Nauk. S.S.S.R.* **217**, 241 [1.1.1.167]
2603. Phillips, A.T. & Wood, W.A. (1965) *J. Biol. Chem.* **240**, 4703 [4.2.1.16]
2604. Piérard, A. & Wiame, J.M. (1960) *Biochim. Biophys. Acta* **37**, 490 [1.4.1.1]
2605. Pieringer, R.A. & Hokin, L.E. (1962) *J. Biol. Chem.* **237**, 653 [2.7.1.94]
2606. Pieringer, R.A. & Kunnes, R.S. (1965) *J. Biol. Chem.* **240**, 2833 [2.7.1.94]
2607. Pierpoint, W.S., Hughes, D.E., Baddiley, J. & Mathias, A.P. (1955) *Biochem. J.* **61**, 368 [2.7.1.33]
2608. Pierre, K.J. & LePage, G.A. (1968) *Proc. Soc. Exp. Biol. Med.* **127**, 432 [2.7.1.73]
2609. Pigman, W.W. (1941) *J. Res. Nat. Bur. Stand.* **26**, 197 [3.2.3.1]
2610. Pincus, P. (1950) *Nature (London)* **66**, 187 [3.1.6.4]
2611. Piquilloud, Y., Reinharz, A. & Roth, M. (1970) *Biochim. Biophys. Acta* **206**, 136 [3.4.15.1]
2612. Pirrotta, V. (1976) *Nucleic Acids Res.* **3**, 1747-1760. [3.1.23.10]
2613. Pizer, L.I. (1962) in *The Enzymes*, 2nd edn (Boyer, P.D., Lardy, H. & Myrbäck, K., ed.) Vol. 6, p. 179, Academic Press, New York [2.7.5.3]
2614. Plager, J.E. & Samuels, L.T. (1953) *Arch. Biochem. Biophys.* **42**, 477 [1.14.99.10]
2615. Plancot M.T. & Han K.K. (1972) *Eur. J. Biochem.* **28**, 327 [3.4.13.11]
2616. Planta, R.J., Gorter, J. & Gruber, M. (1964) *Biochim. Biophys. Acta* **89**, 511 [3.4.14.1]
2617. Plapp, R. & Strominger, J.L. (1970) *J. Biol. Chem.* **245**, 3675 [2.3.2.10]
2618. Plaut, G.W.E. (1955) J. Biol. Chem. **217**, 235 [3.6.1.5]
2619. Plaut, G.W.E. (1963) *J. Biol. Chem.* **238**, 2225 [2.5.1.9]
2620. Plaut, G.W.E. (1963) in *The Enzymes*, 2nd edn (Boyer, P.D., Lardy, H. & Myrbäck, K., ed.) Vol. 7, p. 105, Academic Press, New York [1.1.1.41, 1.1.1.42]
2621. Plaut, G.W.E. & Harvey, R.A. (1971) *Methods Enzymol.* **18B**, 527 [2.5.1.9]
2622. Plaut, G.W.E. & Sung, S.-C. (1954) *J. Biol. Chem.* **207**, 305 [1.1.1.41]
2623. Plaut, R.J., Gorter, J. & Gruber, M. (1955) *J. Biol. Chem.* **217**, 235 [3.6.1.6]
2624. Pliska V., Thorn N.A. & Vilhardt H. (1971) *Acta Endocrinol.* **67**, 12 [3.4.15.2]
2625. Plong, J. & Kjeldgaard, N.O. (1957) *Biochim. Biophys. Acta* **24**, 278
2626. Polakoski K.L. & McRorie R.A. (1973) *J. Biol. Chem.* **248**, 8183 [3.4.21.10]
2627. Polakovski, K.L., Zaneveld, L.J.D. & Williams, W.L. (1972) *Biol. Reprod.* **6**, 23 [3.4.21.10, 3.4.21.31]
2628. Polet, H. & Levine, L. (1975) *J. Biol. Chem.* **250**, 351 [5.3.3.9]
2629. Pollock, M.R. (1960) in *The Enzymes*, 2nd edn (Boyer, P.D., Lardy, H. & Myrbäck, K., ed.) Vol. 4, p. 269, Academic Press, New York [3.5.2.6]
2630. Pollock, M.R., Torriani, A.-M. & Tridgell, E.G. (1956) *Biochem. J.* **62**, 387 [3.5.2.6]
2631. Pomerantz, S. (1967) *J. Biol. Chem.* **242**, 5308 [1.10.3.1]
2632. Pomerantz, S.H. (1963) *J. Biol. Chem.* **238**, 2351 [1.14.18.1]

498

2633.   Pontis, H., Degerstedt, G. & Reichard, P. (1961) *Biochim. Biophys. Acta* **51**, 138
         [2.4.2.3]
2634.   Pontremoli, S., de Flora, A., Grazi, E., Mangiarotti, G., Bonsignore, A. & Horecker, B.L.
         (1961) *J. Biol. Chem.* **236**, 2975   [1.1.1.44]
2635.   Pontremoli, S., Traniello, S., Luppis, B. & Wood, W.A. (1965) *J. Biol. Chem.* **240**, 3459
         [3.1.3.11]
2636.   Pope, J.L., McPherson, J.C. & Tidwell, H.C. (1966) *J. Biol. Chem.* **241**, 2306
         [3.1.1.23]
2637.   Porra, R.J. & Jones, O.T.G. (1963) *Biochem. J.* **87**, 181 & 186   [4.99.1.1]
2638.   Porter, J.W. & Long, R.W. (1958) *J. Biol. Chem.* **233**, 20   [3.1.2.2]
2639.   Porter, W.H. & Mitchell, W.M. (1973) *Mol. Cell. Biochem.* **1**, 95   [3.4.22.8]
2640.   Potts, J.R.M., Weklych, R. & Conn, E.E. (1974) *J. Biol. Chem.* **249**, 5019   [1.14.13.11]
2641.   Poulson, R. (1976) *J. Biol. Chem.* **251**, 3730   [1.3.3.4]
2642.   Poulson, R. & Polglase, W.J. (1975) *J. Biol. Chem.* **250**, 1269   [1.3.3.4]
2643.   Powell, J.F. & Hunter, J.R. (1956) *Biochem. J.* **62**, 381   [3.5.4.4]
2644.   Pradel, L.-A., Kassab, R., Conlay, C. & Thoai, N.V. (1968) *Biochim. Biophys. Acta* **154**,
         305   [2.7.3.1, 3.4.13.11]
2645.   Pradel, L.-A., Kassab, R. & Thoai, N.V. (1964) *Biochim. Biophys. Acta* **81**, 86
         [2.7.3.1]
2646.   Prahl, J.W. & Neurath, H. (1966) *Biochemistry* **5**, 4137   [3.4.17.2]
2647.   Prairie, R.L. & Talalay, P. (1963) *Biochemistry* **2**, 203   [1.14.99.12]
2648.   Prasada Reddy, T.L., Suryanarayana Murthy, P. & Venkitasubramanian, T.A. (1975)
         Biochim. Biophys. Acta **376**, 210
2649.   Pratt, A.G., Crawford, E.J. & Friedkin, M. (1968) *J. Biol. Chem.* **243**, 6367   [3.4.13.7]
2650.   Preddie, E.C. (1969) *J. Biol. Chem.* **244**, 3958   [6.1.1.2]
2651.   Preiss, J. (1964) *J. Biol. Chem.* **239**, 3127   [1.1.1.132]
2652.   Preiss, J. (1966) *Methods Enzymol.* **9**, 602   [5.3.1.17]
2653.   Preiss, J. & Ashwell, G. (1962) *J. Biol. Chem.* **237**, 309   [4.2.2.3]
2654.   Preiss, J. & Ashwell, G. (1962) *J. Biol. Chem.* **237**, 317   [1.1.1.126]
2655.   Preiss, J. & Ashwell, G. (1963) *J. Biol. Chem.* **238**, 1577   [1.1.1.127]
2656.   Preiss, J., Govins, S., Eidels, L., Lammel, C., Greenberg, E., Edelmann, P. & Sabraw, A.
         (1970) in *Miami Winter Symposia* (Whelan, W.J. & Schultz, J., ed.) Vol. 1, p. 122, North
         Holland, Utrecht   [2.4.1.21]
2657.   Preiss, J. & Handler, P. (1957) *J. Biol. Chem.* **225**, 759   [2.4.2.12]
2658.   Preiss, J. & Wood, E. (1964) *J. Biol. Chem.* **239**, 3119   [2.7.7.13]
2659.   Premkumar, R., Subba Rao, P.V., Sreeleela, N.S. & Vaidyanathan, C.S. (1969) *Can. J.
         Biochem.* **47**, 825   [1.14.99.13]
2660.   Prescott, D.J. & Vagelos, P.R. (1972) *Advan. Enzymol.* **36**, 298   [1.1.1.100]
2661.   Prescott, D.J. & Vagelos, P.R. (1972) *Advan. Enzymol.* **36**, 295
         [2.3.1.38, 2.3.1.39, 2.3.1.41]
2662.   Prescott, J.M. & Wilkes, S.H. (1966) *Arch. Biochem. Biophys.* **117**, 328   [3.4.11.10]
2663.   Prescott, P.J. & Vagelos, P.R. (1972) *Advan. Enzymol.* **36**, 284   [2.7.8.7]
2664.   Press, E.M., Porter, R.R. & Cebra, J. (1960) *Biochem. J.* **74**, 501   [3.4.23.5]
2665.   Prottey, C. & Hawthorne, J.N. (1967) *Biochem. J.* **105**, 379   [2.7.8.11]
2666.   Pugh, D., Leaback, D.H. & Walker, P.G. (1957) *Biochem. J.* **65**, 464   [3.2.1.30]
2667.   Putman, E.W., Litt, C.F. & Hassid, W.Z. (1955) *J. Amer. Chem. Soc.* **77**, 4351
         [2.4.1.8]
2668.   Quastel, J.H. & Witty, R. (1951) *Nature (London)* **167**, 556   [2.6.1.13]
2669.   Quayle, J.R. (1963) *Biochem. J.* **89**, 492   [4.1.1.8]
2670.   Quayle, J.R. (1966) *Methods Enzymol.* **9**, 360   [1.2.1.2]
2671.   Quayle, J.R., Keech, D.B. & Taylor, G.A. (1961) *Biochem. J.* **78**, 225   [2.8.3.2]
2672.   Quayle, J.R. & Taylor, G.A. (1961) *Biochem. J.* **78**, 611   [1.2.1.17]
2673.   Qureshi, A.A., Beytia, E. & Porter, J.W. (1973) *J. Biol. Chem.* **248**, 1848 & 1856
         [2.5.1.21]
2674.   Rabier D. & Desmazeaud M.J. (1973) *Biochimie* **55**, 389   [3.4.11.10, 3.4.13.11]
2675.   Rabin, R., Reeves, H.C. & Ajl, S.J. (1963) *J. Bacteriol.* **86**, 937   [4.1.3.10]
2676.   Rabinowitz, J.C. & Barker, H.A. (1956) *J. Biol. Chem.* **218**, 161   [3.5.4.3]
2677.   Rabinowitz, J.C. & Pricer, W.E. (1957) *Fed. Proc.* **16**, 236   [2.1.2.4]
2678.   Rabinowitz, J.C. & Pricer, W.E. (1956) *J. Amer. Chem. Soc.* **78**, 4176   [3.5.4.9]
2679.   Rabinowitz, J.C. & Pricer, W.E. (1956) *J. Biol. Chem.* **222**, 537   [3.5.4.8]

2680. Rabinowitz, J.C. & Pricer, W.E. (1962) *J. Biol. Chem.* **237**, 2898 [6.3.4.3]
2681. Rabinowitz, J.C. & Pricer, W.E. (1956) *J. Amer. Chem.Soc.* **78**, 5702 [2.1.2.4, 4.3.1.4]
2682. Rabinowitz, M. & Lipmann, F. (1960) *J. Biol. Chem.* **235**, 1043 [2.7.1.37]
2683. Race, C., Ziderman, D. & Watkins, W.M. (1968) *Biochem. J.* **107**, 733 [2.4.1.37]
2684. Racker, E. (1952) *Biochim. Biophys. Acta* **9**, 577 [3.1.2.6]
2685. Racker, E. (1947) *J. Biol. Chem.* **167**, 843 [2.7.1.11]
2686. Racker, E. (1949) *J. Biol. Chem.* **177**, 883 [1.2.1.3]
2687. Racker, E. (1951) *J. Biol. Chem.* **190**, 685 [4.4.1.5]
2688. Racker, E. (1952) *J. Biol. Chem.* **196**, 347 [4.1.2.4]
2689. Racker, E. (1955) *J. Biol. Chem.* **217**, 855 [1.6.4.2]
2690. Racker, E. (1955) *J. Biol. Chem.* **217**, 867 [1.8.4.1]
2691. Racker, E. (1961) in *The Enzymes*, 2nd edn (Boyer, P.D., Lardy, H. & Myrbäck, K., ed.) Vol. 5, p. 397, Academic Press, New York [2.2.1.1]
2692. Racker, E. (1961) in *The Enzymes*, 2nd edn (Boyer, P.D., Lardy, H. & Myrbäck, K., ed.) Vol. 5, p. 407, Academic Press, New York [2.2.1.2]
2693. Racker, E. (1962) *Methods Enzymol.* **5**, 270 [3.1.3.37]
2694. Radcliffe, B.C. & Nicholas, D.J.D. (1970) *Biochim. Biophys. Acta* **205**, 273 [1.7.99.4]
2695. Radcliffe R. & Nemerson Y. (1975) *J. Biol. Chem.* **250**, 388 [3.4.21.21]
2696. Raetz, C.R.H., Hirschberg, B., Dowhan, W., Wickner, W.T. & Kennedy, E.P. (1972) *J. Biol. Chem.* **247**, 2245 [3.6.1.26]
2697. Rafter, G.W. (1960) *J. Biol. Chem.* **235**, 2475 [3.6.1.1]
2698. Ragot, J. (1967) *C.R. Acad. Sci. Ser. D(Paris)* **265**, 965
2699. Rahim, M.A. & Sih, C.J. (1965) *J. Biol. Chem.* **241**, 3615 [1.14.99.4]
2700. Rahimtula, A.D. & Gaylor, J.L. (1972) *J. Biol. Chem.* **247**, 9 [1.1.1.170]
2701. Raijman, L., Grisolia, S. & Edelhoch, H. (1960) *J. Biol. Chem.* **235**, 2340 [3.6.1.7.]
2702. Raison, J.K., Henson, G. & Rienits, K.G. (1966) *Biochim. Biophys. Acta* **118**, 285 [1.2.1.29]
2703. Rajagopalan, K.V. & Handler, P. (1967) *J. Biol. Chem.* **242**, 4097 [1.2.1.37]
2704. Rajagopalan, T.G., Stein, W.H. & Moore, S. (1966) *J. Biol. Chem.* **241**, 4940 [3.4.23.1]
2705. Rall, T.W., Wosilait, W.D. & Sutherland, E.W. (1956) *Biochim. Biophys. Acta* **20**, 69 [2.7.1.38, 3.1.3.17]
2706. Ramakrishnan, C.V. & Martin, S.M. (1955) *Arch. Biochem. Biophys.* **55**, 403 [1.1.1.41]
2707. Ramakrishnan, T. & Campbell, J.J.R. (1955) *Biochim. Biophys. Acta* **17**, 122 [1.1.99.3]
2708. Raman, P.B., Sharma, D.C. & Dorfman, R.I. (1966) *Biochemistry* **5**, 1795 [1.14.15.5]
2709. Ramasarma, T. & Giri, K.V. (1956) *Arch. Biochem. Biophys.* **62**, 91 [5.3.1.9]
2710. Ramasastri, B.V. & Blakley, R.T. (1962) *J. Biol. Chem.* **237**, 1982 [1.5.1.5]
2711. Ramaswamy, N.K. & Nair, P.M. (1973) *Biochim. Biophys. Acta* **293**, 269 [2.3.1.37]
2712. Ramírez, J.M., Del Campo, F.F., Paneque, A. & Losada, M. (1966) *Biochim. Biophys. Acta* **118**, 58 [1.7.7.1]
2713. Randall, D.D. & Tolbert, N.E. (1971) *J. Biol. Chem.* **246**, 5510 [3.1.23.19, 3.1.3.38]
2714. Ranjan, S., Patnaik, K.K. & Laloraya, M.M. (1961) *Naturwiss.* **48**, 406 [5.1.2.5]
2715. Rao, D.R. & Greenberg, D.M. (1961) *J. Biol. Chem.* **236**, 1758 [3.5.2.7]
2716. Rao, D.R., Hariharan, K. & Vijayalakshmi, K.R. (1969) *Biochem. J.* **114**, 107 [2.6.1.46]
2717. Rao, D.R. & Oesper. P. (1961) *Biochem. J.* **81**, 405 [2.7.2.3]
2718. Rao, G.S. & Breuer, H. (1969) *J. Biol. Chem.* **244**, 5521 [2.4.1.59]
2719. Rao, G.S., Rao, M.L. & Breuer, H. (1970) *Biochem. J.* **118**, 625 [2.4.1.61]
2720. Rao, M.M., Rebello, P.F. & Pogell, B.M. (1969) *J. Biol. Chem.* **244**, 112 [2.1.1.38]
2721. Rapoport, S. & Luebering, J. (1951) *J. Biol. Chem.* **189**, 683 [3.1.3.13]
2722. Rapoport, S. & Luebering, J. (1952) *J. Biol. Chem.* **196**, 583 [2.7.5.4]
2723. Rapport, M.M., Myer, K. & Linker, A. (1951) *J. Amer. Chem. Soc.* **73**, 2416 [3.2.1.35]
2724. Ratliff, R.L., Weaver, R.H., Lardy, H.A. & Kuby, S.A. (1964) *J. Biol. Chem.* **239**, 301 [2.7.4.6]
2725. Ratner, S. (1954) *Advan. Enzymol.* **15**, 319 [3.5.3.6, 6.3.4.5]
2726. Ratner, S. (1962) in *The Enzymes*, 2nd edn (Boyer, P.D., Lardy, H. & Myrbäck, K., ed.) Vol. 6, p. 267, Academic Press, New York [2.1.4.1]
2727. Ratner, S. & Rochovansky, O. (1956) *Arch. Biochem. Biophys.* **63**, 277 & 296 [2.1.4.1]
2728. Ravdin, R.G. & Crandall, D.I. (1951) *J. Biol. Chem.* **189**, 137 [1.13.11.5]

500

2729.   Ravel, J.M., Norton, S.J., Humphreys, J.S. & Shive, W. (1962) *J. Biol. Chem.* **237**, 2845
        [6.3.1.1]
2730.   Ravel, J.M., Wang, S., Heinemeyer, C. & Shive, W. (1965) *J. Biol. Chem.* **240**, 432
        [6.1.1.17, 6.1.1.18]
2731.   Ravindranath, S.D. & Appaji Rao, N. (1969) *Arch. Biochem. Biophys.* **133**, 54
        [3.6.1.18]
2732.   Ray, W.J. & Roscelli, G.A. (1964) *J. Biol. Chem.* **239**, 1228   [2.7.5.1]
2733.   Reay, P.F. & Conn, E.E. (1974) *J. Biol. Chem.* **249**, 5826   [2.4.1.85]
2734.   Recknagel, R.O. (1957) *J. Biol. Chem.* **227**, 273   [1.1.1.53]
2735.   Reddi, K.K. (1967) *Methods Enzymol.* **12A** 257
2736.   Reddi, K.K. & Mauser, L.J. (1965) *Proc. Nat. Acad. Sci. USA* **53**, 607   [3.1.27.1]
2737.   Reddy, C.C. & Vaidyanathan, C.S. (1975) *Biochim. Biophys. Acta* **384**, 46   [1.14.13.12]
2738.   Reeck, G.R., Winter, W.P. & Neurath, H. (1970) *Biochemistry* **9**, 1398   [3.4.21.4]
2739.   Reed, L.J. & Cox, D.J. (1970) in *The Enzymes, 3rd edn. (Boyer, P.D., ed.) Vol.* **1**, p. 213,
        Academic Press, New York   [2.3.1.61]
2740.   Reem, G.H. (1968) *J. Biol. Chem.* **243**, 5695   [6.3.4.7]
2741.   Rees, D.A. (1961) *Biochem. J.* **80**, 449   [2.5.1.5]
2742.   Rees, D.A. (1961) *Biochem. J.* **81**, 347   [2.5.1.5]
2743.   Rees, H.H., Goad, L.J. & Goodwin, T.W. (1969) *Biochim. Biophys. Acta* **176**, 892
        [5.4.99.8]
2744.   Rees, M. (1968) *Biochemistry* **7**, 353   [1.7.3.4]
2745.   Reese, E.T. (1965) *Can. J. Microbiol.* **11**, 167   [3.2.1.78]
2746.   Reese, E.T. & Avigad, G. (1966) *Biochim. Biophys. Acta* **113**, 79   [2.4.1.10]
2747.   Reese, E.T. & Mandels, M. (1959) *Can. J. Microbiol.* **5**, 173   [3.2.1.39, 3.2.1.6]
2748.   Reese, E.T., Parrish, F.W. & Mandels, M. (1961) *Can. J. Microbiol.* **7**, 309   [3.2.1.71]
2749.   Reese, E.T., Parrish, F.W. & Mandels, M. (1962) *Can. J. Microbiol.* **8**, 327   [3.2.1.75]
2750.   Reeves, H.C. & Ajl, S.J. (1962) *J. Bacteriol.* **84**, 186   [4.1.3.9]
2751.   Reeves, R.E. (1968) *J. Biol. Chem.* **243**, 3202   [2.7.9.1]
2752.   Reeves, R.E. (1971) *Biochem. J.* **125**, 531   [2.7.9.1]
2753.   Reeves, R.E. & Guthrie, J.D. (1975) *Biochem. Biophys. Res. Commun.* **66**, 1389
        [2.7.2.12]
2754.   Reeves, R.E., Menzies, R.A. & Hsu, D.S. (1968) *J. Biol. Chem.* **243**, 5486   [2.7.9.1]
2755.   Reeves, R.E., Montalvo, F.E. & Lushbaugh, T.S. (1971) *Int. J. Biochem.* **2**, 55
        [1.1.1.2]
2756.   Reeves, R.E., Montalvo, F.E. & Lushbaugh, T.S. *Private Communication*   [1.1.1.80]
2757.   Reeves, R.E. & South, D.J. (1974) *Biochem. Biophys. Res. Commun.* **58**, 1053
        [2.7.2.10]
2758.   Reeves, R.E., South, D.J., Blytt, H.J. & Warren, L.G. (1974) *J. Biol. Chem.* **249**, 7737
        [2.7.1.90]
2759.   Reeves, R.E., Warren, L.G. & Hsu, D.S. (1966) *J. Biol. Chem.* **241**, 1257   [2.7.1.56]
2760.   Reeves, R.E., Warren, L.G., Susskind, B. & Lo, H.-S. (1977) *J. Biol. Chem.* **252**, 726
        [1.2.3.6, 6.2.1.13]
2761.   Reglero, A. & Cabezas, J.A. (1976) *Eur. J. Biochem.* **66**, 379
2762.   Reichard, P. & Hanshoff, G. (1956) *Acta Chem. Scand.* **10**, 548   [2.1.3.2]
2763.   Reichelt D., Jacobsohn E. & Haschen R.J. (1974) *Biochim. Biophys. Acta* **341**, 15
        [3.4.23.5]
2764.   Reiner, A.M. (1972) *J. Biol. Chem.* **247**, 4960   [1.3.1.25]
2765.   Reiser, J. & Yuan, R. (1977) *J. Biol. Chem.* **252**, 451-456   [3.1.24.4]
2766.   Reissig, J.L. (1956) *J. Biol. Chem.* **219**, 753   [2.7.5.2]
2767.   Reissig, J.L. & Leloir, L.F. (1966) *Methods Enzymol.* **8**, 175   [2.7.5.2]
2768.   Reitz, M.S. & Rodwell, V.W. (1970) *J. Biol. Chem.* **245**, 3091   [3.5.1.30]
2769.   Rembold, H. & Simmersbach, F. (1969) *Biochim. Biophys. Acta* **184**, 589   [3.5.4.11]
2770.   Remold H., Fasold H. & Staib F. (1968) *Biochim. Biophys. Acta* **167**, 399   [3.4.23.6]
2771.   Remy, C.N., Remy, W.T. & Buchanan, J.M. (1955) *J. Biol. Chem.* **217**, 885
        [2.4.2.8, 2.7.6.1]
2772.   Renart, M.F. & Sillero, A. (1974) *Biochim. Biophys. Acta* **341**, 178   [1.6.6.8]
2773.   Rendina, G. & Coon, M.J. (1957) *J. Biol. Chem.* **225**, 523   [3.1.2.4]
2774.   Rendina, G. & Singer, T.P. (1959) *J. Biol. Chem.* **234**, 1605   [1.1.99.1]
2775.   Rensen, J.F., Matsushita, T., Chirikjian, J.G. & Davis, F.F. (1972) *Biochim. Biophys. Acta*
        281, 481   [4.2.1.70]

2776. Renwick, A.G.C. & Engel, L.L. (1967) *Biochim. Biophys. Acta* **146**, 336 [1.1.1.148]
2777. Reusch, V.M. & Neuhaus, F.C. (1971) *J. Biol. Chem.* **246**, 6136 [6.3.2.16]
2778. Riazuddin, S. & Grossman, L. (1977) *J. Biol. Chem.* **252**, 6287. [3.1.25.1]
2779. Ribbons, D.W. (1966) *Biochem. J.* **99**, 30P [1.13.11.14]
2780. Ribbons, D.W. & Evans, W.C. (1960) *Biochem. J.* **76**, 310 [4.1.1.55]
2781. Richardson, C.C. & Kornberg, A. (1964) *J. Biol. Chem.* **239**, 242 [3.1.11.2]
2782. Richardson, C.C., Lehman, I.R. & Kornberg, A. (1964) *J. Biol. Chem.* **239**, 251 [3.1.11.2]
2783. Richardson, K.E. & Tolbert, N.E. (1961) *J. Biol. Chem.* **236**, 1280 [1.1.3.1]
2784. Richardson, K.E. & Tolbert, N.E. (1961) *J. Biol. Chem.* **236**, 1285 [3.1.3.18]
2785. Richey, D.P. & Brown, G.M. (1971) *Methods Enzymol.* **18B**, 765 [2.7.6.3]
2786. Richey, D.P. & Brown, G.M. (1969) *J. Biol. Chem.* **244**, 1582 [2.5.1.15, 2.7.6.3]
2787. Richmond, V., Tang, J., Wolf, S., Trucco, R. & Caputto, R. (1958) *Biochim. Biophys. Acta* **29**, 453 [3.4.23.3]
2788. Rieske, J.S. (1976) *Biochim. Biophys. Acta* **456**, 195 [1.10.2.2]
2789. Riethmüller, G. & Tuppy, H. (1964) *Biochem. Z.* **340**, 413 [4.99.1.1]
2790. Rigo, L.U., Nakano, M., Veiga, L.A. & Feingold, D.S. (1976) *Biochim. Biophys. Acta* **445**, 286 [1.1.1.173]
2791. Riklis, E. & Rittenberg, D. (1961) *J. Biol. Chem.* **236**, 2526 [1.12.2.1]
2792. Riley, W.O. & Snell, E.E. (1968) *Biochemistry* **7**, 3520 [4.1.1.22]
2793. Rilling, H.C. & Coon, M.J. (1960) *J. Biol. Chem.* **235**, 3087 [5.3.3.3, 6.4.1.4]
2794. Rimington, C. (1958) *Rev. Pure Appl. Chem.* **8**, 129 [4.99.1.1]
2795. Rindler-Ludwig R. & Braunsteiner H. (1975) *Biochim. Biophys. Acta* **379**, 606 [3.4.21.20]
2796. Ringler, R.L. (1961) *J. Biol. Chem.* **236**, 1192 [1.1.99.5]
2797. Rippon, J.W. (1968) *Biochim. Biophys. Acta* **159**, 147 [3.4.24.3]
2798. Rippon, J.W. (1968) *J. Bacteriol.* **95**, 43 [3.4.24.9]
2799. Robbins, K.C., Barnett, E.L. & Grant, N.H. (1955) *J. Biol. Chem.* **216**, 27 [1.7.3.3]
2800. Robbins K.C., Summaria L., Hsieh B. & Shah K.J. (1967) *J. Biol. Chem.* **242**, 2333 [3.4.21.31]
2801. Robbins, P.W. (1962) in *The Enzymes*, 2nd edn (Boyer, P.D., Lardy, H. & Myrbäck, K., ed.) Vol. 6, p. 363, Academic Press, New York [2.8.2.1, 2.8.2.2, 2.8.2.3, 2.8.2.5]
2802. Robbins, P.W. & Lipmann, F. (1957) *J. Biol. Chem.* **229**, 837 [2.7.1.25, 2.7.7.5]
2803. Robbins, P.W. & Lipmann, F. (1958) *J. Biol. Chem.* **233**, 681 & 686 [2.7.7.4, 2.7.7.5, 2.8.2.1]
2804. Robert-Baudouy, J.M. & Stoeber, F.R. (1973) *Biochim. Biophys. Acta* **309**, 473 [4.2.1.8]
2805. Roberts, D.W.A. (1956) *J. Biol. Chem.* **222**, 259 [3.5.4.5]
2806. Roberts, E. (1954) *Arch. Biochem. Biophys.* **48**, 395 [2.6.1.8]
2807. Roberts, E. (1960) in *The Enzymes*, 2nd edn (Boyer, P.D., Lardy, H. & Myrbäck, K., ed.) Vol. 4, p. 285, Academic Press, New York
2808. Roberts, E. & Frankel, S. (1951) *J. Biol. Chem.* **190**, 505 [4.1.1.15]
2809. Roberts, J., Holcenberg, J.S. & Dolowy, W.C. (1972) *J. Biol. Chem.* **247**, 84 [3.5.1.38]
2810. Roberts, R.J., Breitmeyer, J.B., Tabachnik, N.F. & Meyers, P.A. (1975) *J. Mol. Biol.* **91**, 121-123. [3.1.23.16]
2811. Roberts, R.J., Myers, P.A. Morrison, A., & Murray, K. (1976) *J. Mol. Biol.,* **102**, 157-165. [3.1.23.1]
2812. Roberts, R.J., Myers, P.A., Morrison, A., & Murray, K. (1976) *J. Mol. Biol.,* **103**, 199-208.
2813. Roberts, R.J. & Myers, P.A., unpublished observations. [3.1.23.16, 3.1.23.40]
2814. Roberts, R.J., Wilson, G.A., & Young, F.E. (1977) *Nature* **265**, 82-84. [3.1.23.6]
2815. Roberts, R.M. (1971) J. Biol. Chem. **246**, 4995 [2.7.7.44]
2816. Robertson, H.D., Altman, S. & Smith, J.D. (1972) *J. Biol. Chem.* **247**, 5243. [3.1.26.5]
2817. Robertson, H.D., Webster, R.E. & Zinder, N.D. (1968) *J. Biol. Chem.* **243**, 82 [3.1.26.3]
2818. Robinson, J.R., Klein, S.M. & Sagers, R.D. (1973) *J. Biol. Chem.* **248**, 5319 [2.1.2.10]
2819. Robinson, W.G. & Coon, M.J. (1957) *J. Biol. Chem.* **225**, 511 [1.1.1.31]
2820. Robyt, J.F. & Ackerman, R.J. (1971) *Arch. Biochem. Biophys.* **145**, 105 [3.2.1.60]
2821. Rocca, E. & Ghiretti, F. (1958) *Arch. Biochem. Biophys.* **17**, 336 [1.4.3.7]
2822. Rocha e Silva, M. (1963) *Ann. N.Y. Acad. Sci.* **104**, 190 [3.4.21.8]

502

2823. Roche, B. & Azoulay, E. (1969) *Eur. J. Biochem.* **8**, 426   [1.1.1.73]
2824. Roche, J., Lacombe, G. & Girard, H. (1950) *Biochim. Biophys. Acta* **6**, 210
        [3.5.3.2, 3.5.3.3]
2825. Rock, C.O. & Snyder, F. (1974) *J. Biol. Chem.* **249**, 5382   [2.7.1.93]
2826. Rodriguez, P., Bass, S.T. & Hansen, R.G. (1968) *Biochim. Biophys. Acta* **167**, 201
        [3.6.1.21]
2827. Rodwell, V.W., Towne, J.C. & Grisolia, S. (1957) *J. Biol. Chem.* **228**, 875   [2.7.5.3]
2828. Roe, C.R. & Kaplan, N.O. (1969) *Biochemistry* **8**, 5093   [1.1.1.152]
2829. Rohm K.H., (1974) *Hoppe-Seyler's Z. Physiol. Chem.* **255**, 675   [3.4.13.11]
2830. Rohr, M. (1968) *Mh Chem.* **99**, 2255   [3.5.1.32]
2831. Rohr, M. (1968) *Mh Chem.* **99**, 2278   [3.5.1.32]
2832. Romano, A.H. & Nickerson, W.J. (1954) *J. Biol. Chem.* **208**, 409   [1.6.4.1]
2833. Roncari G., Stoll E. & Zuber H. (1976) *Methods Enzymol.* **45**, 522   [3.4.11.12]
2834. Roncari G. & Zuber H. (1969) *Intern. J. Protein. Res.* **1**, 45   [3.4.11.12]
2835. Roon, R.J. & Barker, H.A. (1972) *J. Bacteriol.* **109**, 44   [2.1.3.6]
2836. Roon, R.J. & Levenberg, B. (1972) *J. Biol. Chem.* **247**, 4107   [3.5.1.45]
2837. Roon, R.J. & Levenberg, B. (1968) *J. Biol. Chem.* **243**, 5213   [6.3.4.6]
2838. Rose, I.A. (1962) in *The Enzymes*, 2nd edn (Boyer, P.D., Lardy, H. & Myrbäck, K., ed.)
        Vol. 6, p. 115, Academic Press, New York   [2.7.2.1]
2839. Rose, I.A., Grunberg-Manago, M., Korey, S. & Ochoa, S. (1955) *J. Biol. Chem.* **211**, 737
        [2.7.2.1]
2840. Rose, S.P.R. & Heald, P.J. (1961) *Biochem. J.* **81**, 339   [3.1.3.16]
2841. Rose, Z.B. & Racker, E. (1966) *Methods Enzymol.* **9**, 357   [1.2.1.1]
2842. Roseman, S. (1957) *J. Biol. Chem.* **226**, 115   [3.5.1.33]
2843. Roseman, S., Jourdian, G.W., Watson, D. & Rood, R. (1961) *Proc. Nat. Acad. Sci. USA*
        47, 958   [4.1.3.20]
2844. Rosen, O.M., Hoffee, P. & Horecker, B.L. (1965) *J. Biol. Chem.* **240**, 1517   [4.1.2.4]
2845. Rosenbaum, N. & Gefter, M.L. (1972) *J. Biol. Chem.* **247**, 5675   [2.5.1.8]
2846. Rosenberg, A. (1960) *Biochim. Biophys. Acta* **45**, 297   [3.4.13.3]
2847. Rosenberg, L.L. & Arnon, D.I. (1955) *J. Biol. Chem.* **217**, 361   [1.2.1.13, 1.2.1.9]
2848. Rosenfeld, E. & Wiederschein, G. (1965) *Bull. Soc. Chim. Biol. (Paris)* **47**, 1433
        [3.2.1.40]
2849. Rosenthaler, J., Guirard, B.M., Chang, G.W. & Snell, E.E. (1965) *Proc. Nat. Acad. Sci.
        U.S.* **54**, 152   [4.1.1.22]
2850. Ross, G.W. & Boulton, M.G. (1973) *Biochim. Biophys. Acta* **309**, 430   [3.5.2.6]
2851. Rossi, C.R. & Gibson, D.M. (1964) *J. Biol. Chem.* **239**, 1694   [6.2.1.10]
2852. Rosso, J.-P., Forget, P. & Pichinoty, F. (1973) *Biochim. Biophys. Acta* **321**, 443
        [1.7.99.4]
2853. Rotenberg, S.L. & Sprinson, D.B. (1970) *Proc. Nat. Acad. Sci. USA* **67**, 1669   [4.6.1.3]
2854. Rothberg, S. & Hayaishi, O. (1957) *J. Biol. Chem.* **229**, 897
2855. Rothe, W., Pfleiderer, G. & Zwilling, R. (1970) *Hoppe Seyler's Z. Physiol. Chem.* **351**, 629
        [3.4.24.2]
2856. Rothfield, L., Osborn, M.J. & Horecker, B.L. (1964) *J. Biol. Chem.* **239**, 2788
        [2.4.1.58]
2857. Rothschild, H.A. & Barron, E.S.G. (1954) *J. Biol. Chem.* **209**, 511   [1.2.1.8]
2858. Rotini, O.F. (1956) *Ric. Sci.* **26**, 2786   [3.5.5.3]
2859. Roush, A.H. & Betz, R.F. (1958) *J. Biol. Chem.* **233**, 261   [2.4.2.6]
2860. Rowbury, R.J. & Woods, D.D. (1964) *J. Gen. Microbiol.* **36**, 341   [2.3.1.46]
2861. Rowen, J.W. & Kornberg, A. (1951) *J. Biol. Chem.* **193**, 497   [2.7.1.22]
2862. Rowley, B. & Tucci, A.F. (1970) *Arch. Biochem. Biophys.* **141**, 499   [1.1.1.155]
2863. Rowsell, E.V. (1956) *Biochem. J.* **64**, 235   [2.6.1.5, 2.6.1.6]
2864. Rowsell, E.V. (1956) *Biochem. J.* **64**, 246   [2.6.1.12]
2865. Roy, A.B. (1960) *Advan. Enzymol.* **22**, 205   [3.1.6.1, 3.1.6.2, 3.1.6.3]
2866. Roy, A.B. (1960) *Biochem. J.* **74**, 49   [2.8.2.3]
2867. Roy, A.B. (1954) *Biochim. Biophys. Acta* **15**, 300   [3.1.6.2]
2868. Rozenfel'd, E.L. & Lukomskaya, I.S. (1956) *Biokhimiya* **21**, 412   [3.2.1.11]
2869. Rudney, H. (1957) *J. Biol. Chem.* **227**, 363   [4.1.3.5]
2870. Ruelius, H.W., Kerwin, R.M. & Janssen, F.W. (1968) *Biochim. Biophys. Acta* **167**, 493
        [1.1.3.10]
2871. Ruffner, B.W. Jr. & Anderson, E.P. (1969) *J. Biol. Chem.* **244**, 5994   [2.7.4.14]

2872. Ruis, H. & Hoffmann,-Ostenhof, O. (1969) *Eur. J. Biochem.* **7**, 442
[1.1.1.142, 1.1.1.143]
2873. Ruiz-Herrera, J., Amezcua-Ortega, R. & Trujillo, A. (1968) *J. Biol. Chem.* **243**, 4083
[1.6.4.2]
2874. Runnegar, M.T.C., Scott, K., Webb, E.C. & Zerner, B. (1969) *Biochemistry* **8**, 2013
[3.1.1.1]
2875. Russell, D.W. & Conn, E.E. (1967) *Arch. Biochem. Biophys.* **122**, 256   [1.14.13.11]
2876. Rutter, W.J. & Lardy, H.A. (1958) *J. Biol. Chem.* **233**, 374   [1.1.1.40]
2877. Ryan, C.A. (1965) *Arch. Biochem. Biophys.* **110**, 169   [3.4.21.1, 3.4.21.4]
2878. Ryan, C.A., Clary, J.J. & Tominatsu, Y. (1965) *Arch. Biochem. Biophys.* **110**, 175
2879. Ryan, K.J. & Engel, L.L. (1957) *J. Biol. Chem.* **225**, 103   [1.14.99.10]
2880. Rychlik, I. (1966) *Biochim. Biophys. Acta* **114**, 425   [2.3.2.12]
2881. Rychlík, I., Černá, J., Chládek, S., Žemlička, J. & Haladová, Z. (1969) J. Mol. Biol. **43**,
13   [2.3.2.12]
2882. Ryle, A.P. (1960) *Biochem. J.* **75**, 145   [3.4.23.2, 3.4.23.3]
2883. Ryle, A.P. (1965) *Biochem. J.* **96**, 6
2884. Ryle, A.P. & Hamilton, M.P. (1966) *Biochem. J.* **101**, 176   [3.4.23.3]
2885. Ryle, A.P. & Porter, R.R. (1959) *Biochem. J.* **73**, 75
2886. Sable, H.Z. & Guarino, A.J. (1952) *J. Biol. Chem.* **196**, 395   [2.7.1.12]
2887. Sabo, D.J. & Orlando, J.A. (1968) *J. Biol. Chem.* **243** 3742   [1.6.2.5]
2888. Sacks, W. & Jensen, C.O. (1951) *J. Biol. Chem.* **192**, 231   [4.2.1.31]
2889. Sadana, J.C. & McElroy, W.D. (1957) *Arch. Biochem. Biophys.* **67**, 16   [1.9.6.1]
2890. Sadana, J.C. & Morey, A.V. (1961) *Biochim. Biophys. Acta* **50**, 153   [1.12.2.1]
2891. Sadowski, P.D. & Hurwitz, J. (1969) *J. Biol. Chem.* **244**, 6182   [3.1.21.2]
2892. Sagers, R.D., Beck, J.V., Gruber, W. & Gunsalus, I.C. (1956) *J. Amer. Chem. Soc.* **78**,
694   [2.1.2.4]
2893. Saier, M.H. Jr., Freucht, B.U. & Roseman, S. (1971) *J. Biol. Chem.* **246**, 7819
[2.7.1.98]
2894. Saini P.K. & Rosenberg I.H. (1974) *J. Biol. Chem.* **249**, 5131   [3.4.22.12]
2895. Saito, K. & Hanahan, D.J. (1962) *Biochemistry* **1**, 521   [3.1.1.4]
2896. Saito, K. & Komamine, A. (1976) *Eur. J. Biochem.* **68**, 237   [1.13.11.29, 1.13.11.30]
2897. Saito, K. & Komamine, A. (1978) *Eur. J. Biochem.* **82**, 385   [1.13.11.29, 1.13.11.30]
2898. Saito, S., Ozutsumi, M. & Kurahashi, K. (1967) *J. Biol. Chem.* **242**, 2362   [2.7.7.12]
2899. Saito, Y., Hayaishi, O. & Rothberg, S. (1957) *J. Biol. Chem.* **229**, 921   [1.14.13.9]
2900. Sakaguchi, K. & Murao, S. (1950) *J. Agr. Chem. Soc. (Japan)* **23**, 411   [3.5.1.11]
2901. Sakano, Y., Masuda, N. & Kobayashi, T. (1971) *Agr. Biol. Chem. (Tokyo)* **35**, 971
[3.2.1.57]
2902. Sallach, H.J. (1956) *J. Biol. Chem.* **223**, 1101   [2.6.1.12, 2.6.1.51]
2903. Salway, J.G., Harewood, J.L., Kai, M., White, G.L. & Hawthorne, J.N. (1968) *J.
Neurochem.* **15**, 221   [2.7.8.11]
2904. Samuelsson, B. (1965) *J. Amer. Chem. Soc.* **87**, 3011   [1.14.99.1]
2905. Sanadi, D.R., Gibson, D.M. & Ayengar, P. (1954) *Biochim. Biophys. Acta* **14**, 434
[6.2.1.4]
2906. Sanadi, D.R., Littlefield, J.W. & Bock, R.M. (1952) *J. Biol. Chem.* **197**, 851   [1.2.4.2]
2907. Sanders, H.K., Becker, G.E. & Nason, A. (1972) *J. Biol. Chem.* **247**, 2015   [1.4.2.1]
2908. Sano, K., Amemura, A. & Harada, T. (1975) Biochim. Biophys. Acta **377**, 410
[3.2.1.21]
2909. Sanwal, B.D. & Zink, M.W. (1961) *Arch. Biochem. Biophys.* **94**, 430   [1.4.1.9]
2910. Sapico, V. & Anderson, R.L. (1969) *J. Biol. Chem.* **244**, 6280   [2.7.1.56]
2911. Sarda, L. & Desnuelle, P. (1958) *Biochim. Biophys. Acta* **30**, 513   [3.1.1.3]
2912. Sarid, S., Berger, A. & Katchalski, E. (1962) *J. Biol. Chem.* **237**, 2207
[3.4.11.5, 3.4.13.8]
2913. Sarma, D.S.R., Rajalakshmi, S. & Sarma, P.S. (1964) *Biochim. Biophys. Acta* **81**, 311
[3.5.1.19]
2914. Saroja, K., Venkataraman, R. & Giri, K.V. (1955) *Biochem. J.* **60**, 399   [2.4.1.24]
2915. Sasaoka, K. & Kito, M. (1964) *Agr. Biol. Chem. (Tokyo)* **28**, 313   [6.3.1.6]
2916. Sasaoka, K., Kito, M. & Inagaki, H. (1963) *Agr. Biol. Chem. (Tokyo)* **27**, 467
[6.3.1.6]
2917. Sasaoka, K., Kito, M. & Onishi, Y. (1965) Agr. Biol. Chem. (Tokyo) **29**, 984   [6.3.1.6]
2918. Sastry, L.V.S. & Ramakrishnan, T. (1961) *J. Sci. Indian Res.* **20C**, 277   [2.6.1.4]

2919. Sato, K., Yamada, Y., Aida, K. & Uemura, T. (1969) *J. Biochem. (Tokyo)* **66**, 521 [1.1.99.12]
2920. Sato, S., Hutchison, C.A. & Harris, J.I. (1977) *Proc. Nat. Acad. Sci. USA* **74**, 542-546. [3.1.23.39]
2921. Satyanarayana, T. & Radhakrishnan, A.N. (1965) *Biochim. Biophys. Acta* **110**, 380 [1.1.1.86]
2922. Saunders, P.P. & Broquist, H.P. (1966) *J. Biol. Chem.* **241**, 3435 [1.5.1.7]
2923. Saunders, P.P., Wilson, B.A. & Saunders, G.F. (1969) *J. Biol. Chem.* **244**, 3691 [2.4.2.2]
2924. Savage, N. (1957) *Biochem. J.* **67**, 146 [1.6.4.3]
2925. Sawada, H. & Takagi, Y. (1964) *Biochim. Biophys. Acta* **92**, 26 [4.1.2.19]
2926. Sawada, J. (1963) *Agr. Biol. Chem. (Tokyo)* **27**, 677 [3.4.23.6]
2927. Sawada J. (1966) *Agr. Biol. Chem. (Tokyo)* **30**, 393 [3.4.23.6]
2928. Sawada, Y., Ohyama, T. & Yamazaki, I. (1972) *Biochim. Biophys. Acta* **268**, 305 [1.15.1.1]
2929. Sawai, T., Toriyama, K. & Yano, K. (1974) *J. Biochem. (Tokyo)* **75**, 105 [3.2.1.94]
2930. Sawai, T., Yamaki, T. & Ohya, T. (1976) *Agr. Biol. Chem. (Tokyo)* **40**, 1293 [3.2.1.70]
2931. Sawyer, C.H. (1945) *Science* **101**, 385 [3.1.1.8]
2932. Saz, H.J. & Hubbard, J.A. (1957) *J. Biol. Chem.* **225**, 921 [1.1.1.39]
2933. Scarano, E. (1960) *J. Biol. Chem.* **235**, 706 [3.5.4.12]
2934. Scarano, E., Bonaduce, L. & de Petrocellis, B. (1960) *J. Biol. Chem.* **235**, 3556 [3.5.4.12]
2935. Scarano, E. Quoted by H.M. Kalckar (1953) *Biochim. Biophys. Acta* **12**, 250 [2.7.1.18]
2936. Schabort, J.C. & Potgieter, D.J.J. (1971) *Biochim. Biophys. Acta* **250**, 329 [1.3.99.9]
2937. Schabort, J.C. & Potgieter, D.J.J. (1968) *Biochim. Biophys. Acta* **151**, 47 [1.3.1.5]
2938. Schabort, J.C., Potgieter, D.J.J. & de Villiers, V. (1968) *Biochim. Biophys. Acta* **151**, 33 [1.3.1.5]
2939. Schabort, J.C., Wilkins, D.C., Holzapfel, C.W., Potgieter, D.J.J. & Neitz, A.W. (1971) *Biochim. Biophys. Acta* **250**, 311 [1.3.99.9]
2940. Schachman, H.K., Adler, J., Radding, C.M., Lehman, I.R. & Kornberg, A. (1960) *J. Biol. Chem.* **235**, 3242 [2.7.7.7]
2941. Schachter, D. & Taggart, J.V. (1954) *J. Biol. Chem.* **208**, 263 [2.3.1.13]
2942. Schachter, H., Sarney, J., McGuire, E.J. & Roseman, S. (1969) *J. Biol. Chem.* **244**, 4785 [1.1.1.122]
2943. Schack P. (1967) *C.R. Trav. Lab. Carlsberg* **36**, 67 [3.4.22.6]
2944. Schacter, M. (1969) *Physiol. Rev.* **49**, 509 [3.4.21.8]
2945. Schaub, M.C. & Strauch, L. (1968) *Hoppe-Seyler's Z. Physiol. Chem.* **349**, 809
2946. Schauer, R. (1970) *Hoppe-Seyler's Z. Physiol. Chem.* **351**, 595 & 749 [2.3.1.44, 2.3.1.45]
2947. Schauer, R. (1970) *Hoppe-Seyler's Z. Physiol. Chem.* **351**, 783 [1.14.99.18]
2948. Schauer, R. & Wember, M. (1971) *Hoppe-Seyler's Z. Physiol. Chem.* **352**, 1282 [1.14.99.18]
2949. Schechter, I. (1974) *Biochim. Biophys. Acta* **362**, 233 [2.7.4.18]
2950. Scher, B.M. & Horecker, B.L. (1966) *Arch. Biochem. Biophys.* **116**, 117 [1.1.1.115, 1.1.1.21]
2951. Schiffman S., Theodore I. & Rappaport S.I. (1969) *Biochemistry* **8**, 1397 [3.4.21.23]
2952. Schirch, L.V. & Gross, T. (1968) *J. Biol. Chem.* **243**, 5651 [2.1.2.1]
2953. Schiwara, V.H.W. & Domagk, G.F. (1968) *Hoppe-Seyler's Z. Physiol. Chem.* **349**, 1321 [1.1.1.120]
2954. Schiwara, V.H.W., Domschke, W. & Domagk, G.F. (1968) *Hoppe Seyler's Z. Physiol. Chem.* **349**, 1575 [1.1.1.115, 1.1.1.116]
2955. Schlender, K.K., Wei, S.H. & Villar-Palasi, C. (1969) *Biochim. Biophys. Acta* **191**, 272 [2.7.1.37]
2956. Schlesinger, M.J. & Coon, M.J. (1961) *J. Biol. Chem.* **236**, 2421 [1.1.1.32]
2957. Schleuning W.D., Kolb H.J., Hell R. & Fritz H. (1975) *Hoppe-Seyler's Z. Physiol. Chem.* **356**, 1923 [3.4.21.10]
2958. Schlossberg, M.A., Bloom, R.J., Richert, D.A. & Westerfeld, W.W. (1970) *Biochemistry* **9**, 1148 [4.1.3.15]
2959. Schlossberg, M.A., Richert, D.A., Bloom, R.J. & Westerfeld, W.W. (1968) *Biochemistry* **7**, 333 [4.1.3.15]

2960. Schlossmann, K., Brüggemann, J. & Lynen, F. (1962) *Biochem. Z.* **336**, 258 [4.2.1.22]
2961. Schmidt, A. & Trebst, A. (1969) *Biochim. Biophys. Acta* **180**, 529 [1.8.7.1]
2962. Schmidt, G. (1961) in *The Enzymes*, 2nd edn (Boyer, P.D., Lardy, H. & Myrbäck, K., ed.) Vol. 5, p. 37, Academic Press, New York [3.1.3.19]
2963. Schmidt W. & Havermann K. (1974) Hoppe-Seyler's *Z. Physiol. Chem.* **355**, 1077 [3.4.21.20]
2964. Schmit, J.C. & Zalkin, H. (1969) *Biochemistry* **8**, 175 [4.2.1.51]
2965. Schmitt, A., Bottke, I. & Siebert, G. (1966) *Hoppe-Seyler's Z. Physiol. Chem.* **347**, 18 [4.1.1.3]
2966. Schneider, K. & Schlegel, H.G. (1976) *Biochem. Biophys. Acta* **452**, 66 [1.12.1.2]
2967. Schneider, Z., Larsen, E.G., Jacobsen, G., Johnson, B.C. & Pawelkiewicz, J. (1970) *J. Biol. Chem.* **245**, 3388 [4.2.1.30]
2968. Schneider, Z. & Pawelkiewicz, J. (1966) *Act. Biochim. Pol.* **13**, 311 [4.2.1.30]
2969. Scholnick, P.L., Hammaker, L.E. & Marver, H.S. (1972) *J. Biol. Chem.* **247**, 4126 & 4132 [2.3.1.37]
2970. Schöpp, W., Sorger, H., Kleber, H.-P. & Aurich, H. (1969) *Eur. J. Biochem.* **10**, 56 [1.1.1.108]
2971. Schramm, M., Klybas, V. & Racker, E. (1958) *J. Biol. Chem.* **233**, 1283 [4.1.2.22, 4.1.2.9]
2972. Schreiber, G., Eckstein, M., Oeser, A. & Holzer, H. (1964) *Biochem. Z.* **340**, 13 [2.6.1.1]
2973. Schreker, A.W. & Kornberg, A. (1950) *J. Biol. Chem.* **182**, 795 [2.7.7.2]
2974. Schroeder, D.D. & Shaw, E. (1968) *J. Biol. Chem.* **243**, 2943 [3.4.21.4]
2975. Schubert-Wright C., Alden R.A. & Kraut J. (1969) *Nature* **221**, 235 [3.4.21.14]
2976. Schuberth, J. (1966) *Biochim. Biophys. Acta* **122**, 470 [2.3.1.6]
2977. Schuegraf, A., Ratner, S. & Warner, R.C. (1960) *J. Biol. Chem.* **235**, 3597 [6.3.4.5]
2978. Schultz, J. & Mosbach, K. (1971) *Eur. J. Biochem.* **22**, 153 [3.1.1.40]
2979. Schulz, A.R. & Oliner, L. (1967) *Life Sci.* **6**, 873 [4.1.1.53]
2980. Schulze, H., Gallenkamp, H. & Staudinger, H. (1970) *Hoppe Seyler's Z. Physiol. Chem.* **351**, 809 [1.6.5.4]
2981. Schweet, R.S. & Allen, E.H. (1958) *J. Biol. Chem.* **223**, 1104 [6.1.1.1]
2982. Schwimmer, S. (1953) *Arch. Biochem. Biophys.* **43**, 108 [2.4.1.19]
2983. Schwimmer, S. & Balls, A.K. (1949) *J. Biol. Chem.* **179**, 1063 [3.2.1.1]
2984. Sciaky, D. & Roberts, R.J., unpublished results [3.1.23.8]
2985. Scocca J. & Lee Y.N. (1969) *J. Biol. Chem.* **244**, 4852 [3.4.22.4]
2986. Scott, D.B.M. & Cohen, S.S. (1953) *Biochem. J.* **55**, 23 [1.1.1.44]
2987. Scott, D.B.M. & Cohen, S.S. (1957) *Biochem. J.* **641**, 686 [1.1.1.44]
2988. Scott, E.M. & Jakoby, W.B. (1959) *J. Biol. Chem.* **234**, 932 [2.6.1.19]
2989. Scriba, P. & Holzer, H. (1961) *Biochem. Z.* **334**, 473 [1.2.4.1]
2990. Scrutton, M.C., Young, M.R. & Utter, M.F. (1970) *J. Biol. Chem.* **245**, 6220 [6.4.1.1]
2991. Seegmiller, J.E. (1953) *J. Biol. Chem.* **201**, 629 [1.2.1.4]
2992. Seely, M.K., Criddle, R.S. & Conn, E.E. (1966) *J. Biol. Chem.* **241**, 4457 [4.1.2.11]
2993. Segal, H.L. & Brenner, B.M. (1960) *J. Biol. Chem.* **235**, 471 [3.1.3.5]
2994. Seidman, M.M., Toms, A. & Wood, J.M. (1969) *J. Bacteriol.* **97**, 1192 [1.13.11.22, 3.4.23.1]
2995. Seifter S. & Harper E. (1971) in *The Enzymes (Boyer P. ed.) Acad. Press New York* [3.4.24.7]
2996. SeifterR S., Gallop M., Klein L. & Meilman E. (1957) *J. Biol. Chem.* **234**, 285 [3.4.24.3]
2997. Seijffers, M.J., Miller, L.L. & Segal, H.L. (1964) *Biochemistry* **3**, 1203
2998. Seiler, N. & Al-Therib, M.J. (1974) *Biochim. Biophys. Acta* **354**, 206 [2.3.1.57]
2999. Sekeris, C.E. (1963) *Hoppe-Seyler's Z. Physiol. Chem.* **332**, 70 [4.1.1.28]
3000. Sekine H. (1976) *Agr. Biol. Chem. (Tokyo)* **40**, 703 [3.4.24.4]
3001. Sekizawa, Y., Maragoudakis, M.E., King, T.E. & Cheldelin, V.H. (1966) *Biochemistry* **5**, 2392 [4.1.3.13]
3002. Sellinger, O.Z. & Miller, O.N. (1957) *Fed. Proc.* **16**, 245 [2.7.1.29]
3003. Sellinger, O.Z. & Miller, O.N. (1959) *J. Biol. Chem.* **234**, 1641 [1.1.1.7]
3004. Seltzer, S. (1973) *J. Biol. Chem.* **248**, 215 [5.2.1.2]
3005. Semenza, G. (1957) *Biochim. Biophys. Acta* **24**, 401 [3.4.13.6]
3006. SentheShanmuganathan, S. (1960) *Biochem. J.* **77**, 619 [2.6.1.5]

506

3007. Sergott, R.C., Debeer, L.J. & Bessman, M.J. (1971) *J. Biol. Chem.* **246**, 7755 [3.5.4.12]
3008. Serif, G.S. & Kirkwood, S. (1958) *J. Biol. Chem.* **233**, 109 [1.11.1.8]
3009. Setlow, J.K. & Bollum, F.J. (1968) *Biochim. Biophys. Acta* **157**, 233 [4.1.99.3]
3010. Setten, M.R. (1964) *J. Biol. Chem.* **239**, 3576 [3.1.3.9]
3011. Seubert, W. & Fass, E. (1964) *Biochem. Z.* **341**, 23 [4.1.3.26, 4.2.1.57]
3012. Seubert, W., Fass, E. & Remberger, U. (1963) *Biochem. Z.* **338**, 265 [6.4.1.5]
3013. Seubert, W., Lamberts, I., Kramer, R. & Ohly, B. (1968) *Biochim. Biophys. Acta* **164**, 498 [1.3.1.8]
3014. Seubert, W. & Remberger, U. (1961) *Biochem. Z.* **334**, 401 [6.4.1.1]
3015. Sgoutas, D.S. (1970) *Biochemistry* **9**, 1826 [2.3.1.26]
3016. s'Gravenmade, E.J., van der Drift, C. & Vogels, G.D. (1971) Biochim. Biophys. Acta **198**, 339 [3.5.2.6]
3017. Shaffer, P.M., McCroskey, R.P., Palmatier, R.D., Midgett, R.J. & Abbott, M.T. (1968) *Biochem. Biophys. Res. Commun.* **33**, 806 [1.14.11.3]
3018. Shannon, L.M. & Marcus, A. (1962) *J. Biol. Chem.* **237**, 3342 [4.1.3.17]
3019. Shapiro, B. (1953) *Biochem. J.* **53**, 663
3020. Shapiro, B.M. (1969) *Biochemistry* **8**, 659 [3.1.4.15]
3021. Shapiro, B.M. & Stadtman, E.R. (1968) *J. Biol. Chem.* **243**, 3769 [2.7.7.42, 3.1.4.15]
3022. Shapiro, S.K. (1958) *Biochim. Biophys. Acta* **29**, 405 [2.1.1.10]
3023. Shapiro, S.K. & Yphantis, D.A. (1959) *Biochim. Biophys. Acta* **36**, 241 [2.1.1.10]
3024. Sharma, H.K. & Vaidyanathan, C.S. (1975) *Eur. J. Biochem.* **56**, 163 [1.13.11.28]
3025. Sharp, P.A., Sugden, B. & Sambrook, (1973) *J. Biochemistry* **12**, 3055. [3.1.23.23, 3.1.23.24]
3026. Shaw, D.C. & Wells, J.R.E. (1967) *Biochem. J.* **104**, 5C [3.4.16.1, 3.4.22.9]
3027. Shaw, D.C.& Wells, J.R.E. (1972) *Biochem. J.* **128**, 229 [3.4.15.2]
3028. Shaw, D.R.D. (1956) *Biochem. J.* **64**, 394 [1.1.1.15, 1.1.1.16]
3029. Shaw, D.R.D. (1962) *Biochem. J.* **82**, 297 [2.7.7.39, 2.7.7.40]
3030. Shaw, E. & Glover, G., (1970) *Arch. Biochem. Biophys.* **139**, 298 [3.4.21.4]
3031. Shaw, E., Mares-Guia, M. & Cohen, W., (1965) *Biochemistry* **4**, 2219 [3.4.21.4]
3032. Shaw, P.D. (1967) *Biochemistry* **6**, 2253 [1.3.1.16]
3033. Shaw, W.V. (1967) *J. Biol. Chem.* **242**, 687 [2.3.1.28]
3034. Shaw, W.V. & Brodsky, R.F. (1968) *J. Bacteriol.* **95**, 28 [2.3.1.28]
3035. Shaw, W.V., Tsai, L. & Stadtman, E.R. (1966) *J. Biol. Chem.* **241**, 935 [2.1.1.21]
3036. Shedlarski, J.G. & Gilvarg, C. (1970) *J. Biol. Chem.* **245**, 1362 [4.2.1.52]
3037. Shefer, S., Hauser, S. & Mosbach, E.H. (1966) *J. Biol. Chem.* **241**, 946 [1.3.1.22]
3038. Sheldon, R., Jurale, C. & Kates, J. (1972) *Proc. Nat. Acad. Sci. U.S.* **69**, 417 [2.7.7.19, 2.7.7.6]
3039. Shen, L. & Preiss, J. (1965) *J. Biol. Chem.* **240**, 2334 [2.7.7.27]
3040. Shepherdson, M. & Pardee, A.B. (1960) *J. Biol. Chem.* **235**, 3233 [2.1.3.2]
3041. Sherman, W.R., Stewart, M.A. & Zinbo, M. (1969) *J. Biol. Chem.* **244**, 5703 [5.5.1.4]
3042. Sherry, S. & Alkjaersig, N. (1957) *Ann. N.Y. Acad. Sci.* **68**, 52 [3.4.21.31]
3043. Sheth, K. & Alexander, J.K. (1969) *J. Biol. Chem.* **244**, 457 [2.4.1.49]
3044. Shibahara, M., Moody, J.A. & Smith, L.L. (1970) *Biochim. Biophys. Acta* **202**, 172 [1.14.99.14]
3045. Shiio, I., Shiio, T. & McFadden, B.A. (1965) *Biochim. Biophys. Acta* **96**, 114 [4.1.3.1]
3046. Shikita, M., Inano, H. & Tamaoki, B. (1967) *Biochemistry* **6**, 1760 [1.1.1.149]
3047. Shimazono, N., Mano, Y., Tanaka, R. & Kaziro, Y. (1959) *J. Biochem. (Tokyo)* **46**, 959 [2.7.6.2]
3048. Shimomura, O. & Johnson, F.H. (1968) *Biochemistry* **7**, 1734 [1.14.99.21]
3049. Shimomura, O. & Johnson, F.H. (1975) *Proc. Nat. Acad. Sci. USA* **72**, 1546 [1.13.12.5]
3050. Shimomura, O., Johnson, F.H. & Kohama, Y. (1972) *Proc. Nat. Acad. Sci. USA* **69**, 2086 [1.14.99.21]
3051. Shimono, H. & Sugino, Y. (1971) *Eur. J. Biochem.* **19**, 256 [2.7.4.8]
3052. Shin, M., Tagawa, K. & Arnon, D.I. (1963) *Biochem. Z.* **338**, 84 [1.18.1.2]
3053. Shinano S. & Fukushima K. (1971) *Agr. Biol. Chem. (Tokyo)* **35**, 1488 [3.4.23.13]
3054. Shiokawa, H. & Noda, L. (1970) *J. Biol. Chem.* **245**, 669 [3.6.1.7]
3055. Shiota, T., Baugh, C.M., Jackson, R. & Dillard, R. (1969) *Biochemistry* **8**, 5022 [2.5.1.15, 2.7.6.3]

3056. Shipolini R.A., Callewaert G.L., Cottrell R.C. & Vernon C.A. (1974) *Eur. J. Biochem.* **48**, 465 [3.4.99.32]

3057. Shishido, K., Kato, M. & Ikeda, Y (1968) *J. Biochem. (Tokyo)* **65**, 479. [3.1.22.2]

3058. Shizuta, Y., Nakazawa, A., Tokushige, M. & Hayaishi, O. (1969) *J. Biol. Chem.* **244**, 1883 [4.2.1.16]

3059. Short, E.C. & Koerner, J.F. (1969) *J. Biol. Chem.* **244**, 1487

3060. Shrawder, E. & Martinez-Carrion, M. (1972) *J. Biol. Chem.* **247**, 2486 [2.6.1.1]

3061. Shug, A.L., Wilson, P.W., Green, D.E. & Mahler, H.R. (1954) *J. Amer. Chem. Soc.* **76**, 3355 [1.18.3.1]

3062. Shuster, C.W. (1966) *Methods Enzymol.* **9**, 524 [4.1.2.21]

3063. Shuster, L. & Kaplan, N.O. (1953) *J. Biol. Chem.* **201**, 535 [3.1.3.6]

3064. Sidbury, J.B., Rosenberg, L.L. & Najjar, V.A. (1956) *J. Biol. Chem.* **222**, 89 [2.7.1.41]

3065. Siebert, G., Dubuc, J., Warner, R.C. & Plaut, G.W.E. (1957) *J. Biol. Chem.* **226**, 965 [1.1.1.42]

3066. Siegel, L., Foote, J.L. & Coon, M.J. (1965) *J. Biol. Chem.* **240**, 1025 [6.3.4.10]

3067. Siegel, L.M., Murphy, M.J. & Kamin, H. (1973) *J. Biol. Chem.* **248**, 251 [1.8.1.2]

3068. Siffert, O., Emöd, I. & Keil, B. (1976) *FEBS Letters* **66**, 114 [3.4.21.4, 3.4.22.8]

3069. Silber, R., Malathi, V.G. & Hurwitz, J. (1972) *Proc. Nat. Acad. Sci. USA* **69**, 3009

3070. Silink M., Reddel R., Berthel M. & Rowe P.B. (1975) *J. Biol. Chem.* **250**, 5982 [3.4.22.12]

3071. Sillero, M.A.G., Sillero, A. & Sols, A. (1969) *Eur. J. Biochem.* **10**, 345 [2.7.1.28]

3072. Sillero, M.A.G., Villalba, R., Moreno, A., Quintanilla, M., Lobatón, C.D. & Sillero, A. *Eur. J. Biochem.* (in press) [3.6.1.29]

3073. Silverman, M., Keresztesy, J.C., Koval, G.J. & Gardiner, R.C. (1957) *J. Biol. Chem.* **226**, 83 [2.1.2.6]

3074. Silverstein R.M. (1974) *Anal. Biochem.* **62**, 478 [3.4.22.4]

3075. Silverstein R.M. & Kezdy F.J. (1975) *Arch. Biochem. Biophys.* **167**, 678 [3.4.22.4]

3076. Silverstein, R., Voet, J., Reed, D. & Abeles, R.H. (1967) *J. Biol. Chem.* **242**, 1338 [2.4.1.7]

3077. Simon, D., Hoshino, J. & Kröger, H. (1973) *Biochim. Biophys. Acta* **321**, 361 [4.2.1.13]

3078. Simpson, F.J. (1966) *Methods Enzymol.* **9**, 454 [2.7.1.17]

3079. Simpson, F.J., Wolin, M.J. & Wood, W.A. (1958) *J. Biol. Chem.* **230**, 457 [2.7.1.16, 3.1.23.16]

3080. Singer, M.F. & Fruton, J.S. (1957) *J. Biol. Chem.* **229**, 111 [3.9.1.1]

3081. Singer, T.P. & Hofstee, B.H.J. (1948) *Arch. Biochem.* **18**, 229 & 245 [3.1.1.3]

3082. Singer, T.P. & Kearney, E.B. (1963) in *The Enzymes*, 2nd edn (Boyer, P.D., Lardy, H. & Myrbäck, K., ed.) Vol. 7, p. 383, Academic Press, New York [1.3.99.1]

3083. Singer, T.P., Kearney, E.B. & Bernath, P. (1956) *J. Biol. Chem.* **223**, 599 [1.3.99.1]

3084. Singer, T.P. & Pensky, J. (1952) *J. Biol. Chem.* **196**, 375 [4.1.1.1]

3085. Singh, J. (1974) *Biochim. Biophys. Acta* **333**, 28 [1.9.3.2]

3086. Singh, M., Böttger, B., Stewart, C., Brooks, G.C. & Srere, P.A. (1973) *Biochem. Biophys. Res. Commun.* **53**, 1 [2.3.1.49]

3087. Singh, M., Böttger, B., Stewart, C., Brooks, G.C. & Srere, P.A. (1973) *Biochem. Biophys. Res. Commun.* **53**, 1

3088. Singh, R.M.M. & Adams, E. (1965) *J. Biol. Chem.* **240**, 4344 & 4352 [3.5.4.22]

3089. Singleton, J.W. & Laster, L. (1965) *J. Biol. Chem.* **240**, 4780 [1.3.1.24]

3090. Sinha, A.K. & Chatterjee, G.C. (1968) *Biochem. J.* **107**, 165 [3.5.4.20]

3091. Sipe, J.E., Anderson, W.M. Jr., Remy, C.N. & Love, S.H. (1972) *J. Bacteriol.* **110**, 81 [2.1.1.48]

3092. Sistrom, W.R. & Stanier, R.Y. (1954) *J. Biol. Chem.* **210**, 821 [1.13.11.1, 5.5.1.1]

3093. Sivak, A. & Hoffmann-Ostenhof, O. (1961) *Biochim. Biophys. Acta* **53**, 426 [1.1.1.19]

3094. Sjöholm, I. (1967) *Acta Pharm. Suecica* **4**, 81 [3.4.11.3]

3095. Sjöholm, I. & Yman, L. (1967) *Acta Pharm. Suecica* **4**, 65

3096. Sköld, O. (1960) *J. Biol. Chem.* **235**, 3273 [2.7.1.48]

3097. Slein, M.W. (1955) *J. Amer. Chem. Soc.* **77**, 1663 [2.7.1.17, 5.3.1.5]

3098. Slein, M.W. (1950) *J. Biol. Chem.* **186**, 753 [5.3.1.8]

3099. Sluyterman, L.A.E. (1964) *Biochim. Biophys. Acta* **85**, 305 & 316 [3.4.22.2]

3100. Sly, W.S. & Stadtman, E.R. (1963) *J. Biol. Chem.* **238**, 2632 [3.1.2.10]

3101. Sly, W.S. & Stadtman, E.R. (1963) *J. Biol. Chem.* **238**, 2639 [2.7.2.6]

508

3102. Small, G.D. & Cooper, C. (1966) *Biochemistry* **5**, 14 [3.6.1.14]
3103. Smeby, R.R. & Bumpus, F.M. (1970) *Methods Enzymol.* **19** 699 [3.4.99.19]
3104. Smiley, J.D. & Ashwell, G. (1960) *J. Biol. Chem.* **235**, 1571 [4.2.1.7]
3105. Smiley, J.D. & Ashwell, G. (1961) *J. Biol. Chem.* **236**, 357 [1.1.1.45, 4.1.1.34]
3106. Smiley, K.L. & Sobolov, M. (1962) *Arch. Biochem. Biophys.* **97**, 538 [4.2.1.30]
3107. Smillie, L.B., Enenkel, A.G. & Kay, C.M. (1966) *J. Biol. Chem.* **241**, 2097 [3.4.21.1]
3108. Smith, D.L., Blattner, F.R., & Davies, (1976) *J. Nuc. Acid Res.* **3**, 343 [3.1.23.26, 3.1.23.31]
3109. Smith, E.E.B. & Mills, G.T. (1955) *Biochim. Biophys. Acta* **18**, 152 [2.7.7.12, 2.7.7.9]
3110. Smith, E.L. (1951) *Advan. Enzymol.* **12**, 191 [3.4.13.3]
3111. Smith, E.L. (1948) *J. Biol. Chem.* **176**, 9 [3.4.13.11]
3112. Smith, E.L. (1955) *Methods Enzymol.* **2**, 83 [3.4.11.4, 3.4.13.3]
3113. Smith, E.L. & Hill, R.L. (1960) in *The Enzymes*, 2nd edn (Boyer, P.D., Lardy, H. & Myrbäck, K., ed.) Vol. 4, p. 37, Academic Press, New York
3114. Smith, E.L. & Kimmel, J.R. (1960) in *The Enzymes*, 2nd edn (Boyer, P.D., Lardy, H. & Myrbäck, K., ed.) Vol. 4, p. 133, Academic Press, New York [3.4.22.2]
3115. Smith G.D., Murray M.A., Nichol L.W. & Trikojus V.M. (1969) *Biochim. Biophys. Acta* **171**, 288 [3.4.23.11]
3116. Smith, H.O. & Wilcox, K.W. (1970) *J. Mol. Biol.* **51**, 379 [3.1.23.20]
3117. Smith, I.K. (1973) *Biochim. Biophys. Acta* **321**, 156 [2.6.1.45]
3118. Smith, I.K. & Thompson, J.F. (1971) *Biochim. Biophys. Acta* **227**, 288 [2.3.1.30]
3119. Smith, I.K. & Thompson, J.F. (1969) *Biochim. Biophys. Acta* **184**, 130 [4.2.99.10]
3120. Smith, L. (1954) *Bacteriol. Rev.* **18**, 106 [1.9.3.2]
3121. Smith, M.E. & Greenberg, D.M. (1956) *Nature (London)* **177**, 1130 [1.5.1.2]
3122. Smith, R.A. & Gunsalus, I.C. (1957) *J. Biol. Chem.* **229**, 305 [4.1.3.1]
3123. Smith, R.A. & Thiesen, M.C. (1966) *Methods Enzymol.* **9**, 403 [2.7.1.62]
3124. Smith, R.L. & Shaw, E. (1969) *J. Biol. Chem.* **244**, 4704 [3.4.21.4]
3125. Smith, S.T., Rajagopalan, K.V. & Handler, P. (1967) *J. Biol. Chem.* **242**, 4108
3126. Smith, S.W., Weiss, S.B. & Kennedy, E.P. (1957) *J. Biol. Chem.* **228**, 915 [3.1.3.4]
3127. Smith, T.A. (1969) *Phytochemistry* **8**, 2111 [3.5.3.12]
3128. Smith, T.E. & Mitoma, C. (1962) *J. Biol. Chem.* **237**, 1177 [1.1.1.104]
3129. Snell, E.E., Smucker, A.A., Ringelmann, E. & Lynen, F. (1964) *Biochem. Z.* **341**, 109 [4.1.1.51]
3130. Snoke, J.E. (1956) *J. Biol. Chem.* **223**, 271 [1.4.1.3]
3131. Snoke, J.E., Yanari, S. & Bloch, K. (1953) *J. Biol. Chem.* **201**, 573 [6.3.2.2]
3132. Snyder, F., Malone, B. & Piantadosi, C. (1973) *Biochim. Biophys. Acta* **316**, 259 [1.14.16.5]
3133. Snyder, S.H., Silva, O.L. & Kies, M.W. (1961) *J. Biol. Chem.* **236**, 2996 [3.5.2.7]
3134. Soda, K. & Misono, H. (1968) *Biochemistry* **7**, 4110 [2.6.1.36]
3135. Soda, K., Misono, H. & Yamamoto, T. (1968) *Biochemistry* **7**, 4102 [2.6.1.36]
3136. Soda, K. & Moriguchi, M. (1969) *Biochem. Biophys. Res. Commun.* **34**, 34 [4.1.1.18]
3137. Soda, K. & Osumi, T. (1969) *Biochem. Biophys. Res. Commun.* **35**, 363 [5.1.1.10]
3138. Sodek, J. & Hofmann, T. (1970) *Can. J. Biochem.* **48**, 1014 [3.4.23.6]
3139. Sodek, J. & Hofmann, T. (1968) *J. Biol. Chem.* **243**, 450 [3.4.23.6]
3140. Soffer, R.L. (1970) *J. Biol. Chem.* **245**, 731 [2.3.2.8]
3141. Soffer, R.L. (1973) *J. Biol. Chem.* **248**, 2918 [2.3.2.8]
3142. Soffer, R.L. (1973) *J. Biol. Chem.* **248**, 8424 [2.3.2.6]
3143. Soffer, R.L., Hechtman, P. & Savage, M. (1973) *J. Biol. Chem.* **248**, 1224
3144. Soffer, R.L. & Horinishi, H. (1969) *J. Mol. Biol.* **43**, 163 [2.3.2.8]
3145. Sokatch. J.R., Sanders, L.E. & Marshall, V.P. (1968) *J. Biol. Chem.* **243**, 2500 [1.2.1.27]
3146. Solomon, L.R. & Breitman, T.R. (1971) *Biochem. Biophys. Res. Commun.* **44**, 299 [2.7.1.83]
3147. Sols, A. & Salas, M.L. (1966) *Methods Enzymol.* **9**, 425 [2.7.1.11]
3148. Somack, R. & Costilow, R.N. (1973) *Biochemistry* **12**, 2597 [5.1.1.12, 5.4.3.5]
3149. Somack, R. & Costilow, R.N. (1973) *J. Biol. Chem.* **248**, 385 [1.4.1.12]
3150. Sonneborn, H.-H., Pfleiderer, G. & Ishay, J. (1969) *Hoppe Seyler's Z. Physiol. Chem.* **350**, 389 [3.4.21.1]
3151. Sonneborn, H.-H., Zwilling, R. & Pfleiderer, G. (1969) *Hoppe Seyler's Z. Physiol. Chem.* **350**, 1097 [3.4.99.6]

3152. Sonnino, S., Carminatti, H. & Cabib, E. (1966) *Arch. Biochem. Biophys.* **116**, 26 [3.2.1.42]

3153. Soodsma, J.F., Piantadosi, C. & Snyder, F. (1972) *J. Biol. Chem.* **247**, 3923 [1.14.16.5]

3154. Sörbo, B.H. (1957) *Biochim. Biophys. Acta* **24**, 324 [2.8.1.2]

3155. Sörbo, B.H. (1953) *Acta Chem. Scand.* **7**, 1129 & 1137 [2.8.1.1]

3156. Soucek, A., Michalec, C. & Couckova, A. (1971) *Biochim. Biophys. Acta.* **227**, 116 [3.1.4.41]

3157. Sova, V.V., Elyakova, L.A. & Vaskovsky, V.E. (1970) *Biochim. Biophys. Acta* **212**, 111 [3.2.1.6]

3158. Spahr, P.F. (1964) *J. Biol. Chem.* **239**, 3716 [3.1.13.1]

3159. Spahr, P.F. & Gesteland, R.F. (1970) *Eur. J. Biochem.* **12**, 270 [3.6.1.24]

3160. Sparrow, L.G., Ho, P.P.K., Sundaram, T.K., Zach, D., Nyns, E.J. & Snell, E.E. (1969) *J. Biol. Chem.* **244**, 2590 [1.14.12.4, 1.14.12.5]

3161. Spector, T. & Massey, V. (1972) *J. Biol. Chem.* **247**, 4679, 5632 and 7123 [1.14.13.2]

3162. Spencer, D. (1959) *Aust. J. Biol. Sci.* **12**, 181 [1.6.6.1]

3163. Spencer, R.L. & Preiss, J. (1967) *J. Biol. Chem.* **242**, 385 [6.3.1.5]

3164. Speranza, M.L., Ronchi, S. & Minchiotti, L. (1973) *Biochim. Biophys. Acta* **327**, 274 [1.6.4.5]

3165. Spiekerman A.M., Fredericks K.K., Wagner F.W. & Prescott J.M. (1973) *Biochim. Biophys. Acta* **293**, 464

3166. Spiro, M.H. & Spiro, R.G. (1968) *J. Biol. Chem.* **243**, 6520 [2.4.99.1]

3167. Spiro, M.H. & Spiro, R.G. (1968) *J. Biol. Chem.* **243**, 6529 [2.4.1.38]

3168. Sporn, M.B., Lazarus, H.M., Smith, J.M. & Henderson, W.R. (1969) *Biochemistry* **8**, 1698 [3.1.13.1]

3169. Sprossler B., Heilman H.D., Grumpp E.& Ulig H. (1971) *Hoppe-Seyler's Z. Physiol. Chem.* **352**, 1524 [3.4.16.1]

3170. Sprössler, B. & Lingens, F. (1970) *Hoppe-Seyler's Z. Physiol. Chem.* **351**, 448 [5.4.99.5]

3171. Sreeleela, N.S., SubbaRao, P.V., Premkumar, R. & Vaidyanathan, C.S. (1969) *J. Biol. Chem.* **244**, 2993 [1.14.12.2]

3172. Srere, P.A. & Lipmann, F. (1953) *J. Amer. Chem. Soc.* **75**, 4874 [4.1.3.8]

3173. Srere, P.A., Seubert, W. & Lynen, F. (1959) *Biochim. Biophys. Acta* **33**, 313 [3.1.2.2]

3174. Sribney, M. (1966) *Biochim. Biophys. Acta* **125**, 542 [2.3.1.24]

3175. Sribney, M. & Kennedy, E.P. (1958) *J. Biol. Chem.* **233**, 1315 [2.7.8.3]

3176. Sridhara, S. & Wu, T.T. (1969) *J. Biol. Chem.* **244**, 5233 [1.2.1.22]

3177. Srinivasan, P.R., Rothschild, J. & Sprinson, D.B. (1963) *J. Biol. Chem.* **238**, 3176 [4.6.1.3]

3178. Srinivasan, P.R. & Sprinson, D.B. (1959) *J. Biol. Chem.* **234**, 716 [4.1.2.15]

3179. Sriprakash, K.S., Luudh, N., Noo-On Huh, M. & Radding, C.M. (1975) *J. Biol. Chem.* **250**, 5438.

3180. Stachow, C.S., Stevenson, I.L. & Day, D. (1967) *J. Biol. Chem.* **242**, 5294 [1.2.1.7]

3181. Stadtman, E.R. (1952) *Fed. Proc.* **11**, 291 [2.8.3.1]

3182. Stadtman, E.R. (1955) *J. Amer. Chem. Soc.* **77**, 5765 [4.3.1.6]

3183. Stadtman, E.R. (1952) *J. Biol. Chem.* **196**, 527 [2.3.1.8]

3184. Stadtman, E.R. (1955) *Methods. Enzymol.* **1**, 596 [2.3.1.8]

3185. Stadtman, E.R., Novelli, G.D. & Lipmann, F. (1951) *J. Biol. Chem.* **191**, 365 [2.3.1.54]

3186. Stadtman, T.C. (1956) *Biochem. J.* **62**, 614 [1.4.1.6]

3187. Stadtman, T.C. (1961) in *The Enzymes*, 2nd edn (Boyer, P.D., Lardy, H. & Myrbäck, K., ed.) Vol. 5, p. 55, Academic Press, New York [3.1.3.1]

3188. Stadtman, T.C. (1973) *Advanc. Enzymol.* **38**, 441 [1.4.1.12]

3189. Stadtman, T.C., Cherkes, A. & Anfinsen, C.B. (1954) *J. Biol. Chem.* **206**, 511 [1.1.3.6]

3190. Stadtman, T.C. & Elliott, P. (1957) *J. Biol. Chem.* **228**, 983 [1.4.1.6, 5.1.1.4]

3191. Stadtman, T.C. & Renz, P. (1968) *Arch. Biochem. Biophys.* **125**, 226 [5.4.3.3]

3192. Stadtman, T.C. & Tsai, L. (1967) *Biochem. Biophys. Res. Commun.* **28**, 920 [5.4.3.4]

3193. Stafford, H.A., Magaldi, A. & Vennesland, B. (1954) *J. Biol. Chem.* **207**, 621 [1.1.1.29]

3194. Stambaugh, R. & Buckley, J. (1971) *Fed. Proc.* **30**, 1184 [3.4.21.10]

3195. Stambaugh, R. & Buckley, J. (1968) *Science* **161**, 585

3196. Stambaugh R. & Smith M. (1974) *Science* **186**, 745 [3.4.21.10]

510

3197. Stanier, R.Y. & Ingraham, J.L. (1954) *J. Biol. Chem.* **210**, 799   [1.13.11.3]
3198. Stark, G.R. & Dawson, C.R. (1963) in *The Enzymes*, 2nd edn (Boyer, P.D., Lardy, H. & Myrbäck, K., ed.) Vol. 8, p. 297, Academic Press, New York   [1.10.3.3]
3199. Starkey P.M. & Barrett A.J. (1976) *Biochem. J.* **155**, 255   [3.4.21.20]
3200. Starnes W.L. & Behal F.J. (1974) *Biochemistry* **13**, 3221   [3.4.11.14, 3.4.21.11]
3201. Starr, J.L. & Goldthwait, D.A. (1963) *J. Biol. Chem.* **238**, 682   [2.7.7.25]
3202. Staub, M. & Dénes, G. (1966) *Biochim. Biophys. Acta* **128**, 82   [2.3.1.35]
3203. Staudinger, H., Krisch, K. & Leonhäuser, S. (1961) *Ann. N.Y. Acad. Sci.* **92**, 195   [1.6.5.4]
3204. Stavrianopoulos, J.G. & Chargaff, E. (1973) *Proc. Nat. Acad. Sci. (USA)* **70**, 1959.   [3.1.26.4]
3205. Steinman, C.R. & Jakoby, W.B. (1967) *J. Biol. Chem.* **242**, 5019   [1.2.1.5]
3206. Stenesh, J.J. & Winnick, T. (1960) *Biochem. J.* **77**, 575   [3.4.24.4, 6.3.2.11]
3207. Stern, A.I. & Avron, M. (1966) *Biochim. Biophys. Acta* **118**, 577   [2.7.7.35]
3208. Stern, J.R. (1957) *Biochim. Biophys. Acta* **26**, 448   [1.1.1.35]
3209. Stern, J.R. (1957) *Biochim. Biophys. Acta* **26**, 641   [4.2.1.17]
3210. Stern, J.R. (1961) in *The Enzymes*, 2nd edn (Boyer, P.D., Lardy, H. & Myrbäck, K., ed.) Vol. 5, p. 367, Academic Press, New York   [4.1.3.7]
3211. Stern, J.R., Coon, M.J. & del Campillo, A. (1953) *Nature (London)* **171**, 28   [2.3.1.16]
3212. Stern, J.R., Coon, M.J., del Campillo, A. & Schneider, M.C. (1956) *J. Biol. Chem.* **221**, 15   [2.8.3.5]
3213. Stern, J.R. & del Campillo, A. (1955) *J. Amer. Chem. Soc.* **77**, 1073   [5.1.2.3]
3214. Stern, J.R. & Drummond, G.I. (1956) *Fed. Proc.* **15**, 363
3215. Stern, J.R., Drummond, G.I., Coon, M.J. & del Campillo, A. (1960) *J. Biol. Chem.* **235**, 313   [2.3.1.9]
3216. Stern, J.R. & O'Brien, R.W. (1969) *J. Bacteriol.* **98**, 147   [1.1.1.83]
3217. Stern, J.R. & Ochoa, S. (1951) *J. Biol. Chem.* **191**, 161   [2.7.2.1]
3218. Stern, N. & Tietz, A. (1973) *Biochim. Biophys. Acta* **296**, *136*   [*2.4.1.84*]
3219. Stern, R. & Mehler, A.H. (1965) *Biochem. Z.* **342**, 400   [6.1.1.6]
3220. Stetten, M.R. (1970) *Biochim. Biophys. Acta* **208**, 394   [2.7.1.79]
3221. Stevens, A. & Hilmoe, R.J. (1960) *J. Biol. Chem.* **235**, 3016 & 3023   [3.1.30.2]
3222. Stevens R.L., Micalizzi E.R., Fessler D.C. & Pals D.T. (1972) *Biochemistry* **11**, 2999   [3.4.15.1]
3223. Stevenson K.J. & Gaucher G.M. (1975) *Biochem. J.* **151**, 527   [3.4.21.14]
3224. Stewart, P.R. & Quayle, J.R. (1967) *Biochem. J.* **102**, 885   [4.1.3.15]
3225. Steyn-Parvé, E.P. (1952) *Biochim. Biophys. Acta* **8**, 310   [2.7.6.2]
3226. Still, J.L., Buell, M.V., Knox, W.E. & Green, D.E. (1949) *J. Biol. Chem.* **179**, 831   [1.4.3.1]
3227. Still, J.L. & Sperling, E. (1950) *J. Biol. Chem.* **182**, 585   [1.4.3.1]
3228. Stinson, R.A. & Spencer, M.S. (1969) *Biochem. Biophys. Res. Commun.* **34**, 120   [2.6.1.18]
3229. Stitch, S.R., Halkerston, I.D.K. & Hillman, J. (1956) *Biochem. J.* **63**, 705   [3.1.6.2]
3230. Stocker K. & Barlow G.H. (1976) *Methods Enzymol.* **45**, 214   [3.4.21.29]
3231. Stoffel, W., Bauer, E. & Stahl, J. (1974) *Hoppe-Seyler's Z. Physiol. Chem.* **355**, 61   [2.7.1.91]
3232. Stoffel, W. & Därr, W. (1974) Hoppe-Seyler's Z. *Physiol. Chem.* **355**, 54   [1.3.1.27]
3233. Stoffel, W., Ditzer, R. & Caesar, H. (1964) *Hoppe-Seyler's Z. Physiol. Chem.* **339**, 167   [5.3.3.8]
3234. Stoffel, W., Heimann, G. & Hellenbroich, B. (1973) *Hoppe-Seyler's Z. Physiol. Chem.* **354**, 562   [2.7.1.91]
3235. Stoffel, W., Le Kim, D. & Sticht, G. (1969) *Hoppe-Seyler's Z. Physiol. Chem.* **350**, 1233   [4.1.2.27]
3236. Stoffel, W., Le Kim, D. & Sticht, G. (1968) *Hoppe-Seyler's Z. Physiol. Chem.* **349**, 664   [1.1.1.102, 2.3.1.50]
3237. Stoffel, W., Le Kim, D. & Sticht, G. (1968) *Hoppe-Seyler's Z. Physiol. Chem.* **349**, 1637   [1.1.1.102]
3238. Stoffel, W., LeKim, D. & Heyn, G. (1970) *Hoppe-Seyler's Z. Physiol. Chem.* **351**, 875   [1.1.1.164]
3239. Stoolmiller, A.C. & Abeles, R.H. (1966) *J. Biol. Chem.* **241**, 5764   [4.2.1.43]

3240. Stoolmiller, A.C., Horwitz, A.L. & Dorfman, A. (1972) *J. Biol. Chem.* **247**, 3525 [2.4.2.26]
3241. Størmer, F.C. (1967) *J. Biol. Chem.* **242**, 1756 [4.1.1.5]
3242. Størmer, F.C., Solberg, Y. & Hovig, T. (1969) *Eur. J. Biochem.* **10**, 251 [4.1.3.18]
3243. Stouthamer, A.H. (1961) *Biochim. Biophys. Acta* **48**, 484 [2.7.1.58]
3244. Stransky, H. & Amberger, A. (1973) *Z. Pflanzenphysiol.* **70**, 74 [4.2.1.69]
3245. Strassman, M. & Ceci, L.N. (1967) *Arch. Biochem. Biophys.* **119**, 420 [4.1.3.10]
3246. Strassman, M. & Ceci, L.N. (1964) *Biochem. Biophys. Res. Commun.* **14**, 262 [4.1.3.21]
3247. Strassman, M. & Ceci, L.N. (1965) *J. Biol. Chem.* **240**, 4357 [1.1.1.87]
3248. Strassman, M. & Ceci, L.N. (1966) *J. Biol. Chem.* **241**, 5401 [4.2.1.36]
3249. Straub, F.B. (1939) *Biochem. J.* **33**, 787 [1.6.4.3]
3250. Straub, F.B. (1940) *Biochem. J.* **34**, 483 [1.1.1.27]
3251. Straub, F.B. (1942) *Hoppe-Seyler's Z. Physiol. Chem.* **275**, 63 [1.1.1.37]
3252. Strauch, L. & Grossmann, W. (1966) *Hoppe-Seyler's Z. Physiol. Chem.* **344**, 140 [3.4.24.3]
3253. Strauch, L. & Vencelj, H. (1967) *Hoppe-Seyler's Z. Physiol. Chem.* **348**, 465
3254. Strecker, H.J. (1953) *Arch. Biochem. Biophys.* **46**, 128 [1.4.1.3]
3255. Strecker, H.J. (1960) *J. Biol. Chem.* **235**, 3218 [1.5.1.12]
3256. Strecker, H.J. (1965) *J. Biol. Chem.* **240**, 1225 [2.6.1.13]
3257. Strecker, H.J. & Harary, I. (1954) *J. Biol. Chem.* **211**, 263 [1.1.1.4, 1.1.1.5]
3258. Strecker, H.J. & Korkes, S. (1952) *J. Biol. Chem.* **196**, 769 [1.1.1.47]
3259. Strelitz, F. (1944) *Biochem. J.* **38**, 86 [3.1.1.8]
3260. Strickland, R.G. (1959) *Biochem. J.* **73**, 646 & 654 [1.1.1.40]
3261. Strickland, S. & Massey, V. (1973) *J. Biol. Chem.* **248**, 2944 and 2953 [1.14.13.4]
3262. Strittmatter, P. (1963) in *The Enzymes*, 2nd edn (Boyer, P.D., Lardy, H. & Myrbäck, K., ed.) Vol. 8, p. 113, Academic Press, New York [1.6.2.2]
3263. Strittmatter, P. & Ball, E.G. (1955) *J. Biol. Chem.* **213**, 445 [1.2.1.1]
3264. Strittmatter, P. & Velick, S.F. (1957) *J. Biol. Chem.* **228**, 785 [1.6.2.2]
3265. Strobel, G.A. & Kosuge, T. (1965) *Arch. Biochem. Biophys.* **109**, 622 [1.1.1.138]
3266. Strominger, J.L. (1955) *Biochim. Biophys. Acta* **16**, 616 [2.7.1.40]
3267. Strominger, J.L. & Mapson, L.W. (1957) *Biochem. J.* **66**, 567 [1.1.1.22]
3268. Strominger, J.L., Maxwell, E.S., Axelrod, J. & Kalckar, H.M. (1957) *J. Biol. Chem.* **224**, 79 [1.1.1.22]
3269. Strominger, J.L. & Smith, M.S. (1959) *J. Biol. Chem.* **234**, 1822 [2.7.7.23]
3270. Struve, W.G., Sinha, R.K. & Neuhaus, F.C. (1966) *Biochemistry* **5**, 82 [2.7.8.13]
3271. Stulberg, M.P. (1967) *J. Biol. Chem.* **242**, 1060 [6.1.1.20]
3272. Stumpf, P.K. & Horecker, B.L. (1956) *J. Biol. Chem.* **218**, 753 [2.7.1.17, 5.1.3.1]
3273. Su, J.-C., Li, C.-C. & Ting, C.C. (1966) *Biochemistry* **5**, 536 [3.5.4.17]
3274. Subba Rao, P.V., Moore, K. & Towers, G.H.N. (1967) *Arch. Biochem. Biophys.* **122**, 466 [4.1.1.46]
3275. Subba Rao, P.V. & Vaidyanathan, C.S. (1967) *Arch. Biochem. Biophys.* **118**, 388 [1.10.3.4]
3276. Subramanian, A.R. & Kalnitsky, G. (1964) *Biochemistry* **3** 1861 & 1868 [3.4.21.14]
3277. Subramanian, S.S. & Raghavendra Rao, M.R. (1968) *J. Biol. Chem.* **243**, 2367 [4.2.1.35]
3278. Suda, T., Robinson, J.C. & Fjellstedt, T.A. (1976) *Arch. Biochem. Biophys.* **176**, 610 [1.1.1.172]
3279. Sudo, T., Kikuno, M. & Kurihara, T. (1972) Biochim. Biophys. Acta **255**, 640 [3.1.4.37]
3280. Sugimoto, E. & Pizer, L.I. (1968) *J. Biol. Chem.* **243**, 2081 [1.1.1.95]
3281. Sugiura, M., Ito, A. & Yamaguchi, T. (1974) *Biochim. Biophys. Acta* **350**, 61 [3.2.1.95]
3282. Sugiyama, S., Yano, K. & Arima, K. (1960) *Bull. Agr. Chem. Soc. (Japan)* **24**, 243 & 249 [1.13.11.4]
3283. Suhara, K., Takemori, S. & Katagiri, M. (1969) *Arch. Biochem. Biophys.* **130**, 422 [1.1.1.90]
3284. Sukanya, N.K. & Vaidyanathan, C.S. (1964) *Biochem. J.* **92**, 594 [2.6.1.28]
3285. Suld, H.M. & Herbut, P.A. (1970) *J. Biol. Chem.* **245**, 2797 [3.5.1.1]
3286. Sulkowski, E. & Laskowski, M., Sr. (1966) *J. Biol. Chem.* **241**, 4386
3287. Sullivan, J.D. & Ikawa, M. (1973) *Biochim. Biophys. Acta* **309**, 11 [1.1.3.5]
3288. Sullivan J.J. & Jago G.R. (1970) Aus. *J. Dairy Technol.* **25**, 141 [3.4.11.8]

512

3289. Sumizu, K. (1962) *Biochim. Biophys. Acta* **63**, 210   [1.8.1.3]
3290. Sumner, J.B. (1951) in *The Enzymes* (Sumner, J.B. & Myrbäck, K., ed.) Vol. 1, p. 873, Academic Press, New York   [3.5.1.5]
3291. Sumner, J.B. & Dounce, A.L. (1937) *J. Biol. Chem.* **121**, 417   [1.11.1.6]
3292. Sund, H. & Theorell, H. (1963) in *The Enzymes*, 2nd edn (Boyer, P.D., Lardy, H. & Myrbäck, K., ed.) Vol. 7, p. 25, Academic Press, New York   [1.1.1.1]
3293. Sundarajan, T.A. & Sarma, P.S. (1959) *Biochem. J.* **71**, 537   [3.1.3.16, 3.9.1.1]
3294. Sundaram, T.K. & Snell, E.E. (1969) *J. Biol. Chem.* **244**, 2577   [1.1.3.12, 1.1.99.9]
3295. Sung, C.-P. & Johnstone, R.M. (1967) *Biochem. J.* **105**, 497   [2.7.1.82]
3296. Sussman, M. & Osborn, M.J. (1964) *Proc. Nat. Acad. Sci. USA* **52**, 81   [2.4.1.74]
3297. Sutherland, E.W., Cohn, M., Posternak, T. & Cori, C.F. (1949) *J. Biol. Chem.* **180**, 1285   [2.7.5.1]
3298. Sutherland, E.W., Posternak, T. & Cori, C.F. (1949) *J. Biol. Chem.* **181**, 153   [2.7.5.3]
3299. Sutter, A. & Grisebach, H. (1973) *Biochim. Biophys. Acta* **309**, 289   [2.4.1.91]
3300. Sutter, A., Ortmann, R. & Grisebach. H. (1972) *Biochim. Biophys. Acta* **258**, 71   [2.4.1.81]
3301. Sutton, C.R. & King, H.K. (1962) *Arch. Biochem. Biophys.* **96**, 360   [4.1.1.14]
3302. Sutton, W.B. (1957) *J. Biol. Chem.* **226**, 395   [1.13.12.4]
3303. Sutton, W.D. (1971) *Biochim Biophys Acta* **240**, 522   [3.1.30.1]
3304. Suzuki, H., Li, S.-C. & Li, Y.-T. (1970) *J. Biol. Chem.* **245**, 781   [3.2.1.22]
3305. Suzuki, I. & Silver, M. (1966) *Biochim. Biophys. Acta* **122**, 22   [1.13.11.18]
3306. Suzuki, I. & Werkman, C.H. (1960) *Biochem. J.* **74**, 359   [1.6.4.2]
3307. Suzuki, K., Mano, Y. & Shimazono, N. (1960) *J. Biochem. (Tokyo)* **48**, 313   [1.1.1.20]
3308. Suzuki, K., Takemori, S. & Katagiri, M. (1969) *Biochim. Biophys. Acta* **191**, 77   [1.14.13.1]
3309. Suzuki, S. & Strominger, J.L. (1960) *J. Biol. Chem.* **235**, 257, 267 & 274   [2.8.2.5]
3310. Suzuki, S., Trenn, R.H. & Strominger, J.L. (1961) *Biochim. Biophys. Acta* **50**, 169   [2.8.2.12]
3311. Suzuki, T. & Hochater, R.M. (1966) *Can. J. Biochem.* **44**, 259   [4.2.1.70]
3312. Swai, T. & Niwa, Y. (1975) *Agr. Biol. Chem. (Tokyo)* **39**, 1007   [3.2.1.94]
3313. Swaine, D. (1969) *Biochim. Biophys. Acta* **178**, 609   [4.2.1.49]
3314. Swaninathan, N. & Radhakrishnan, A.N. (1969) *Biochim. Biophys. Acta* **191**, 322   [3.2.1.23]
3315. Swartz, M.N., Kaplan, N.O. & Lamborg, M.F. (1958) *J. Biol. Chem.* **232**, 1051   [3.6.1.9]
3316. Sweat, M.L., Samuels, L.T. & Lumry, R. (1950) *J.Biol. Chem.* **185**, 75   [1.1.1.63, 1.1.1.64]
3317. Sweet, B., McNicol, G.P. & Douglas, A.S. (1965) *Clin. Sci.* **29**, 375
3318. Swell, L. & Treadwell, C.R. (1950) *J. Biol. Chem.* **185**, 349   [3.1.1.13]
3319. Swick, R.W. & Wood, H.G. (1960) *Proc. Nat. Acad. Sci. U.S.* **46**, 28   [2.1.3.1]
3320. Switzer, R.L. (1969) *J. Biol. Chem.* **244**, 2854   [2.7.6.1]
3321. Szörényi, E.T., Dvornikova, P.D. & Degtyar, P.G. (1949) *Dokl. Akad. Nauk. S.S.S.R.* **67**, 341   [2.7.3.3]
3322. Szulmajster, J. (1958) *Biochim. Biophys. Acta* **30**, 154   [3.5.4.21]
3323. Szymona, M. (1962) *Acta Biochim. Pol.* **9**, 165   [2.7.1.63]
3324. Szymona, M. & Ostrowski, W. (1964) *Biochim. Biophys. Acta* **85**, 283   [2.7.1.63]
3325. Tabor, C.W. (1962) *Methods Enzymol.* **5**, 756   [4.1.1.50]
3326. Tabor, C.W. (1962) *Methods Enzymol.* **5**, 761   [2.5.1.16]
3327. Tabor, C.W. & Kellogg, P.D. (1970) *J. Biol. Chem.* **245**, 5425   [1.5.99.6]
3328. Tabor, H., Mehler, A.H. & Stadtman, E.R. (1953) *J. Biol. Chem.* **204**, 127   [2.3.1.5]
3329. Tabor, H. & Tabor, C.W. (1972) *Advances Enzymol.* **36**, 225   [1.5.99.6]
3330. Tabor, H. & Tabor, C.W. (1972) *Advances Enzymol.* **36**, 203   [2.5.1.16]
3331. Tabor, H. & Wyngarden, L. (1959) *J. Biol. Chem.* **234**, 1830   [2.1.2.5, 3.5.4.9]
3332. Tabuchi, T. & Šatoh, T. (1976) *Agr. Biol. Chem. (Tokyo)* **40**, 1863   [4.1.3.30]
3333. Tabuchi, T. & Satoh T. (1977) *Agr. Biol. Chem. (Tokyo)* **41**, 169   [4.1.3.30]
3334. Tack, B.F., Chapman, P.J. & Dagley, S. (1972) *J. Biol. Chem.* **247**, 6444   [4.1.3.17]
3335. Tagawa, K. & Arnon, D.I. (1962) *Nature (London)* **195**, 537   [1.18.3.1]
3336. Tagawa, K. & Kagi, A. (1969) Carbohydr. Res. **11**, 293   [3.2.1.55]
3337. Tagawa, K., Shin, M. & Okunuki, K. (1959) *Nature (London)* **183**, 111   [1.11.1.7]
3338. Tager, J.M. & Rautanen, N. (1955) *Biochim. Biophys. Acta* **18**, 111   [1.8.3.1]

3339. Tai, T., Yamashita, K., Ogata-Arakawa, M., Koide, N., Muramatsu, T., Iwashita, S., Inoue, Y. & Kobata, A. (1975) *J. Biol. Chem.* **250**, 8569 [3.2.1.96]
3340. Tait, G.H. (1973) Biochem. J. **131**, 389 [2.3.1.37]
3341. Takagi, Y. (1962) *Agr. Biol. Chem. (Tokyo)* **26**, 719 [1.1.1.128]
3342. Takagi, Y. & Horecker, B.L. (1956) *J. Biol. Chem.* **225**, 77 [3.2.2.1]
3343. Takahashi, K. (1961) *J. Biochem. (Tokyo)* **49**, 1 [3.1.27.3]
3344. Takahashi, N. & Egami, F. (1961) *Biochem. J.* **80**, 384 [3.1.6.7]
3345. Takahashi T. & Ohsaka A. (1970) *Biochim. Biophys. Acta* **198**, 293 [3.4.24.6]
3346. Takanami, M. (1973) *FEBS Letters* **34**, 318-322 [3.1.23.18]
3347. Takasaki, Y. & Tanabe, O. (1962) *Hakko Kyokaishi* **20**, 449
3348. Takebe, I. (1961) *J. Biochem. (Tokyo)* **50**, 245 [3.1.6.6.]
3349. Takeda, H. & Hayaishi, O. (1966) *J. Biol. Chem.* **241**, 2733 [1.13.12.2]
3350. Takeda, H., Yamamoto, S., Kojima, Y. & Hayaishi, O. (1969) *J. Biol. Chem.* **244**, 2935 [1.13.12.2, 3.5.1.30]
3351. Takemori, S., Yasuda, H., Mihara, K., Suzuki, K. & Katagiri, M. (1969) *Biochim. Biophys. Acta* **191**, 69 [1.14.13.1]
3352. Takemori, S., Yasuda, H., Mihara, K., Suzuki, K. & Katagiri, M. (1969) *Biochim. Biophys. Acta* **191**, 58 [1.14.13.1]
3353. Takeuchi, T., Weinbach, E.C. & Diamond, L.S. (1975) *Biochem. Biophys. Res. Commun.* **65**, 591 [1.2.3.6]
3354. Taku, A. & Anwar, R.A. (1973) J. Biol. Chem. **248**, 4971 [1.1.1.158]
3355. Taku, A., Gunetileke, K.G. & Anwar, R.A. (1970) *J. Biol. Chem.* **245**, 5012 [1.1.1.158]
3356. Talalay, P. & Dobson, M.M. (1953) *J. Biol. Chem.* **205**, 823 [1.1.1.51]
3357. Talalay, P. & Wang, V.S. (1955) *Biochim. Biophys. Acta* **18**, 300 [5.3.3.1]
3358. Tamir, H. (1971) *Methods Enzymol.* **17B**, 134 [1.3.1.26]
3359. Tanaka, K., Budd, M.A., Efron, M.L. & Isselbacher, K.J. (1966) *Proc. Nat. Acad. Sci. USA* **56**, 236 [1.3.99.10]
3360. Tanaka, K., Nakano, T., Noguchi, S. & Pigman, W. (1968) *Arch. Biochem. Biophys.* **126**, 624 [3.2.1.51]
3361. Tanaka, T. & Knox, W.E. (1959) *J. Biol. Chem.* **234**, 1162 [1.13.11.11]
3362. Tanenbaum, S.W. (1956) *Biochim. Biophys. Acta* **21**, 335 [1.1.1.2, 1.2.1.5]
3363. Tanenbaum, S.W. & Shemin, D. (1950) *Fed. Proc.* **9**, 236 [2.6.1.6]
3364. Tang, J. & Tang, K. (1963) *J. Biol. Chem.* **238**, 606 [3.4.23.3]
3365. Tangen, O., Fonnum, F. & Haavaldsen, R. (1965) *Biochim. Biophys. Acta* **96**, 82 [2.6.1.27]
3366. Tani, Y., Ukita, M. & Ogata, K. (1972) *Agr. Biol. Chem. (Tokyo)* **36**, 181 [2.6.1.54]
3367. Taniguchi, S., Asano, A., Iida, K., Kono, M., Omachi, K. & Egami, F. (1958) *Proc. Int. Symp. Enzyme Chem. (Tokyo & Kyoto)* **2**, 238 [1.7.99.1]
3368. Taniguchi, S., Mitsui, H., Nakamura, K. & Egami, F. (1955) *Ann. Acad. Sci. Fenn. Ser. A II* **60**, 200 [1.6.6.3, 1.6.6.4, 1.7.99.1]
3369. Taniuchi, H., Hatanaka, M., Kuno, S., Hayaishi, O., Nakajima M. & Kurihara, N. (1964) *J. Biol. Chem.* **239**, 2204 [1.14.12.1]
3370. Taniuchi, H. & Hayaishi, O. (1963) *J. Biol. Chem.* **238**, 283 [1.14.99.2, 1.3.1.18]
3371. Tanner, W. (1967) *Ber. Deut. Bot. Ges.* **80**, 111 [2.4.1.67]
3372. Tarentino, A.L. & Maley, F. (1969) *Arch. Biochem. Biophys.* **130**, 295 [3.5.1.26]
3373. Tarr, H.L.A. (1955) *Biochem. J.* **59**, 386 [3.2.2.1, 3.2.2.2]
3374. Tarr, H.L.A. (1973) *J. Fish Res. Bd (Can.)* **30**, 1861 [3.2.2.13]
3375. Taussig, A. (1960) *Biochim. Biophys. Acta* **44**, 510 [3.5.5.3]
3376. Taussig, A. (1965) *Can. J. Biochem.* **43**, 1063 [3.5.5.3]
3377. Taylor, E.S. & Gale, E.F. (1945) *Biochem. J.* **39**, 52 [4.1.1.17, 4.1.1.19]
3378. Taylor, M.B. & Juni, E. (1960) *Biochim. Biophys. Acta* **39**, 448 [1.1.1.4, 1.1.1.76, 5.1.2.4]
3379. Taylor, R.T. (1971) *Biochim. Biophys. Acta* **242**, 355 [2.1.1.13, 3.2.1.51]
3380. Taylor, R.T. & Jenkins, W.T. (1966) *J. Biol. Chem.* **241**, 4396 [2.6.1.42]
3381. Taylor S.L. & Tappel A.L. (1974) *Biochim. Biophys. Acta* **341**, 99 [3.4.16.2]
3382. Taylor S.L. & Tappel A.L. (1974) *Biochim. Biophys. Acta* **341**, 112 [3.4.16.1]
3383. Taylor, S.S. & Heath, E.C. (1969) *J. Biol. Chem.* **244**, 6605 [2.3.1.40]
3384. Taylor, W.H. (1962) *Physiol. Rev.* **42**, 519 [3.4.23.1, 3.4.23.3]

514

3385. Taylor, W.H., Taylor, M.L. & Eames, D.F. (1966) *J. Bacteriol.* **91**, 2251
       [1.3.1.15, 1.3.3.1]
3386. Tchen, T.T. (1958) *J. Biol. Chem.* **233**, 1100   [2.7.1.36]
3387. Tchen, T.T. & Bloch, K. (1957) *J. Biol. Chem.* **226**, 921   [1.14.99.7]
3388. Telegdi, M. (1968) *Biochim. Biophys. Acta* **159**, 227   [1.1.1.8]
3389. Tempest, D.W., Meers, J.L. & Brown, C.M. (1970) *Biochem. J.* **117**, 405   [1.4.1.13]
3390. Tenhunen, R., Marver, H.S. & Schmid, R. (1969) *J. Biol. Chem.* **244**, 6388   [1.14.99.3]
3391. Terada, M., Tatibana, M. & Hayaishi, O. (1967) *J. Biol. Chem.* **242**, 5578   [3.2.2.8]
3392. Tezuka, T. & Tonomura, K. (1976) *J. Biochem. (Tokyo)* **80**, 79   [4.99.1.2]
3393. Thanassi, N.M. & Nakada, H.I. (1967) *Arch. Biochem. Biophys.* **118**, 172   [3.2.1.44]
3394. Thangamani A. & Hofmann T. (1966) *Can. J. Biochem.* **44**, 579   [3.4.23.6]
3395. Thelander, L. (1973) *J. Biol. Chem.* **248**, 4591   [1.17.4.1]
3396. Theorell, H. (1958) *Advan. Enzymol.* **20**, 31   [1.1.1.1]
3397. Theorell, H. (1943) *Ark. Kem. Min. Geol.* **16A**, No. 2   [1.11.1.7]
3398. Theorell, H. (1935) *Biochem. Z.* **278**, 263   [1.6.99.1]
3399. Theorell, H. & Akesson, Ã (1956) *Arch. Biochem. Biophys.* **65**, 439   [1.6.99.1]
3400. Theorell, H., Holman, R.T. & Akesson, Ã (1947) *Acta Chem. Scand.* **1**, 571
       [1.13.11.12]
3401. Thimann, K.V. & Mahadevan, S. (1964) *Arch. Biochem. Biophys.* **105**, 133   [3.5.5.1]
3402. Thoai, N.V. (1957) *Bull. Soc. Chim. Biol. (Paris)* **39**, 197   [2.7.3.1, 2.7.3.4]
3403. Thoai, N.V., di Jeso, F., Robin, Y. & der Terrossian, E. (1966) *Biochim. Biophys. Acta*
       113, 542   [2.7.3.7, 3.1.23.6]
3404. Thoai, N.V., Huc, C., Pho, D.B. & Olomucki, A. (1969) *Biochim. Biophys. Acta* **191**, 46
3405. Thoai, N.V. & Olomucki, A. (1962) *Biochim. Biophys. Acta* **59**, 533   [1.13.12.1]
3406. Thoai, N.V. & Olomucki, A. (1962) *Biochim. Biophys. Acta* **59**, 545   [1.13.12.1]
3407. Thoai, N.V. & Robin, Y. (1961) *Biochim. Biophys. Acta* **52**, 221   [1.5.1.11]
3408. Thoai, N.V., Robin, Y. & Pradel, L. (1963) *Biochim. Biophys. Acta* **73**, 437
       [2.7.3.4, 2.7.3.6]
3409. Thoai, N.V., Thome-Beau, F. & Olomucki, A. (1966) *Biochim. Biophys. Acta* **115**, 73
       [3.5.3.7]
3410. Thoma, R.W. & Peterson, W.H. (1954) *J. Biol. Chem.* **210**, 569   [3.5.1.12]
3411. Thomas, J.H. & Tudball, N. (1967) *Biochem. J.* **105**, 467   [4.3.1.10]
3412. Thomas, P.J. (1968) *Biochem. J.* **109**, 695   [2.3.1.27]
3413. Thompson, J.F. & Moore, D.P. (1968) *Biochem. Biophys. Res. Commun.* **31**, 281
       [4.2.99.8]
3414. Thompson, J.S. & Richardson, K.E. (1967) *J. Biol. Chem.* **242**, 3614   [2.6.1.12]
3415. Thompson, R.E. & Carper, W.R. (1970) *Biochim. Biophys. Acta* **198**, 397   [1.1.1.47]
3416. Thompson, W. & Dawson, R.M.C. (1964) *Biochem. J.* **91**, 237   [3.1.4.11]
3417. Thorne, C.B. (1955) in *A Symposium on Amino Acid Metabolism*(McElroy, W.D. & Glass,
       B., ed.) p. 41, Johns Hopkins Press, Baltimore   [2.6.1.21]
3418. Thorne, C.B., Gómez, C.G. & Housewright, R.D. (1955) *J. Bacteriol.* **69**, 357
       [2.6.1.21]
3419. Thorne, C.B. & Molnar, D.M. (1955) *J. Bacteriol.* **70**, 420
3420. Tietz, A., Lindberg, M. & Kennedy, E.P. (1964) *J. Biol. Chem.* **239**, 4081   [1.14.16.5]
3421. Tietz, A. & Ochoa, S. (1958) *Arch. Biochem. Biophys.* **78**, 477   [2.7.1.40]
3422. Tigerstrom, M.V. & Tener, G.M. (1967) *Can. J. Biochem.* **45**, 1067   [6.1.1.21]
3423. Ting, S.-M., Miller, O.N. & Sellinger, O.Z. (1965) *Biochim. Biophys. Acta* **97**, 407
       [1.1.1.78]
3424. Ting, S.-M., Sellinger, O.Z. & Miller, O.N. (1964) *Biochim. Biophys. Acta* **89**, 217
       [1.1.1.77]
3425. Tipton, K.F. (1962) *Biochim. Biophys. Acta* **92**, 341   [3.4.99.2]
3426. Tipton, K.F. (1968) *Biochim. Biophys. Acta* **159**, 451   [1.4.3.4]
3427. Tissiéres, A. (1951) *Biochem. J.* **50**, 279   [1.9.3.2]
3428. Titani K., Ericsson L.H., Neurath H. & Walsh K.A. (1975) *Biochemistry* **14**, 1358
       [3.4.21.4]
3429. Titani K., Fujikawa K., Enfield D.L., Ericsson L.H., Walsh K.A. & Neurath H. (1975)
       Proc. Nat. Acad. Sci. USA **72**, 3082
3430. Titani K., Hermodson M.A., Ericsson L.H., Walsh K.A. & Neurath H.(1972) Nature
       New Biology **238**, 35   [3.4.21.8, 3.4.24.4]

3431. Titani K., Hermodson M.A., Fujikawa K., Ericsson L.H., Walsh K.A., Neurath H. & Davie E.W. (1972) *Biochemistry,* **11**, 4899

3432. Tobe S., Takami T., Ikeda S. & Mitsugi K. (1976) *Agr. Biol. Chem. (Tokyo)* **40**, 1087 [3.4.21.14]

3433. Toews, C.J. (1967) *Biochem. J.* **105**, 1067 [1.1.1.72]

3434. Tokieda, T., Niimura, T., Takamura, F. & Yamaha, T. (1977) *J. Biochem. (Tokyo)* **81**, 851 [1.4.3.12]

3435. Tolosa, E.A., Chepurnova, N.K., Khomutov, R.M. & Severin, E.S. (1969) *Biochim. Biophys. Acta* **171**, 369 [4.4.1.10]

3436. Tomášek, V., Šorm, F., Zwilling, R. & Pfleiderer, G. (1970) *FEBS Lett.* **6**, 229 [3.4.21.4]

3437. Tomio, J.M., Garcia, R.C., San Martin de Viale, L.C. & Grinstein, M. (1970) *Biochim. Biophys. Acta* **198**, 353 [4.1.1.37]

3438. Tomita, F. & Takahashi, I. (1969) *Biochim. Biophys. Acta* **179**, 18 [3.5.4.13]

3439. Tomita, K., Mocha, C.-J. & Lardy, H.A. (1964) *J. Biol. Chem.* **239**, 1202 [2.1.1.26]

3440. Tomizawa, H.H. & Halsey, Y.D. (1959) *J. Biol. Chem.* **234**, 307

3441. Tomkins, G.M. (1956) *J. Biol. Chem.* **218**, 437 [1.1.1.50]

3442. Tomkins, G.M. (1957) *J. Biol. Chem.* **225**, 13 [1.3.1.3]

3443. Tomkins, G.M., Michael, P.J. & Curran, J.F. (1957) *Biochim. Biophys. Acta* **23**, 655 [1.14.15.4]

3444. Tomoda, K. & Shimazono, H. (1964) *Agr. Biol. Chem.* (Tokyo) **28**, 770

3445. Tomoeda, M. & Kitamura, R. (1977) *Biochim. Biophys. Acta* **480**, 315 [5.2.1.6]

3446. Tookey, H.L. & Balls, A.K. (1956) *J. Biol. Chem.* **218**, 213 [3.1.4.4]

3447. Toomey, R.E. & Wakil, S.J. (1966) *Biochim. Biophys. Acta* **116**, 189 [1.1.1.100]

3448. Toomey, R.E. & Wakil, S.J. (1966) *J. Biol. Chem.* **241**, 1159 [2.3.1.41]

3449. Topham, R.W. & Gaylor, J.L. (1970) *J. Biol. Chem.* **245**, 2319 [4.2.1.62]

3450. Toriyama, N. (1965) *Nichidai Igaku Zasshi* **24**, 423 [1.6.6.10]

3451. Toro-Goyco, E. & Matos, M. (1966) *Nature (London)* **210**, 527 [3.4.99.18]

3452. Touster, O., Reynolds, V.H. & Hutcheson, R.M. (1956) *J. Biol. Chem.* **221**, 697 [1.1.1.10]

3453. Toyama, S., Misono, H. & Soda, K. (1972) *Biochem. Biophys. Res. Commun.* **46**, 1374 [2.6.1.55]

3454. Tracey, M.V. (1955) *Biochem. J.* **61**, 579 [3.2.1.14]

3455. Traniello, S., Calcagno, M. & Pontremoli, S. (1971) *Arch. Biochem. Biophys.* **146**, 603 [3.1.3.37]

3456. Traniello, S. & Vescia, A. (1964) *Arch. Biochem. Biophys.* **105**, 465 [3.4.13.11]

3457. Traut, R.R. & Monro, R.E. (1964) *J. Mol. Biol.* **10**, 63 [2.3.2.12]

3458. Travis, J. (1968) *Biochem. Biophys. Res. Commun.* **30**, 730 [3.4.21.4]

3459. Travis, J. & Roberts, R.C. (1969) *Biochemistry* **8**, 2884 [3.4.21.4]

3460. Trijbels, F. & Vogels, G.D. (1966) *Biochim. Biophys. Acta* **118**, 387 [3.5.3.4]

3461. Trijbels, F. & Vogels, G.D. (1967) *Biochim. Biophys. Acta* **132**, 115 [4.3.2.3]

3462. Trilling, D.M. & Aposhian, H.V. (1968) *Proc. Nat. Acad. Sci. U.S.* **60**, 214 [3.1.11.4]

3463. Trippett, S., Dagley, S. & Stopher, D.A. (1960) *Biochem. J.* **76**, 9P [1.13.11.8]

3464. Troll, W., Sherry, S. & Wachman, J. (1954) *J. Biol. Chem.* **208**, 85 [3.4.21.7]

3465. Trop, M. & Birk, Y. (1968) *Biochem. J.* **109**, 475 [3.4.21.4]

3466. Trufanov, A.F. & Kirsanova, J.A. (1946) *Byull. Eksp. Biol. Med.* **22**, No. 6 [2.7.1.35]

3467. Tsai, S.-C. & Gaylor, J.L. (1966) *J. Biol. Chem.* **241**, 4043 [3.1.7.1]

3468. Tschesche H. & Kupfer S. (1972) *Eur. J. Bioch.* **26**, 33 [3.4.16.1]

3469. Tsuboi, K.K., Estrada, J. & Hudson, P.B. (1958) *J. Biol. Chem.* **231**, 19 [5.3.1.9]

3470. Tsuboi, K.K. & Hudson, P.B. (1956) *Arch. Biochem. Biophys.* **61**, 197 [3.1.3.2, 3.1.3.21]

3471. Tsuboi, K.K. & Hudson, P.B. (1957) *J. Biol. Chem.* **224**, 879 [2.4.2.1]

3472. Tsuboi, K.K., Wiener, G. & Hudson, P.B. (1957) *J. Biol. Chem.* **224**, 621 [3.1.3.19, 3.1.3.2]

3473. Tsuda, Y. & Friedman, H.C. (1970) *Fed. Proc.* **29**, 597

3474. Tsuda, Y. & Friedmann, H.C. (1970) *J. Biol. Chem.* **245**, 5914 [1.4.1.12]

3475. Tsuji, F.I., Lynch, R.V. & Stevens, C.L. (1974) *Biochemistry* **13**, 5204 [1.13.12.6]

3476. Tsujisaka, Y., Fukimoto, J. & Yamamoto, T. (1958) *Nature (London)* **181**, 770 [3.2.1.3]

3477. Tsukada, K. (1966) *J. Biol. Chem.* **241**, 4522 [1.4.99.1]

3478. Tsunazawa, S., Narita, K. & Ogata K. (1975) *J. Biochem. (Tokyo)* **77**, 89 [3.4.14.3]
3479. Tsuru, D., Hattori, A., Yamamoto, T. & Fukumoto, J. (1969) *Agr. Biol. Chem. (Tokyo)* **33**, 1419 [3.4.23.6]
3480. Tsuru, D., Oka, I. & Yoshimoto, T. (1976) *Agr. Biol. Chem. (Tokyo)* **40(5)**, 1011 [3.5.2.10]
3481. Tsuru K., Yamamoto T. & Fukumoto J. (1966) *J. Agr. Biol. Chem. (Tokyo)* **30**, 651 [3.4.24.4]
3482. Tsusué, M. (1971) *J. Biochem. (Tokyo)* **69**, 781 [3.4.23.6, 3.5.4.24]
3483. Tu, C-P.D., Roychoudhury, R., & Wu, R. (1976) *Biochem. Biophys. Res. Commun.* **72**, 355-362
3484. Tubbs, P.K. & Greville, G.D. (1959) *Biochim. Biophys. Acta* **34**, 290 [1.1.99.6]
3485. Tuboi, S. & Kikuchi, G. (1965) *Biochim. Biophys. Acta* **96**, 148 [4.1.3.24]
3486. Tung, K. & Nordin, J.H. (1968) *Biochim. Biophys. Acta* **158**, 154 [3.2.1.61]
3487. Tuppy, H. & Staudenbauer, W.L. (1966) *Biochemistry* **5**, 1742 [3.2.1.49]
3488. Turková, J., Mikeš, O., Hayashi, K., Danno, G. & Polgár, L. (1972) *Biochim. Biophys. Acta* **257**, 257 [3.4.21.14]
3489. Turner, D.H. & Turner, J.F. (1960) *Biochem. J.* **74**, 486 [3.1.3.10]
3490. Turner, D.H. & Turner, J.F. (1961) *Biochem. J.* **79**, 143 [3.5.4.6]
3491. Turner, J.F. & King, J.E. (1961) *Biochem. J.* **79**, 147 [1.2.1.14]
3492. Turner, J.M. (1966) *Biochem. J.* **99**, 427 [1.1.1.75]
3493. Turner, J.M. (1967) *Biochem. J.* **104**, 112 [1.1.1.75]
3494. Twarog, R. & Wolfe, R.S. (1962) *J. Biol. Chem.* **237**, 2474 [2.7.2.7]
3495. Tyler, T.R. & Leatherwood, J.M. (1967) *Arch. Biochem. Biophys.* **119**, 363 [5.1.3.11]
3496. Uchida, T. & Egami, F. (1967) *J. Biochem. (Tokyo)* **61**, 44 [3.1.27.1]
3497. Uchida, T. & Eqami, F. (1971) in *The Enzymes.* 3rd edn (Boyer, P.D., ed.) Vol.4, p205, Academic Press, New York. [3.1.27.4]
3498. Uchino, F., Kurano, Y. & Doi, S. (1967) *Agr. Biol. Chem. (Tokyo)* **31**, 428 [3.4.23.6]
3499. Uchiyama, H. & Tabuchi, T. (1976) *Agr. Biol. Chem. (Tokyo)* **40**, 1411 [4.1.3.31]
3500. Uchiyama, T., Niwa, S. & Tanaka, K. (1973) *Biochim. Biophys. Acta* **315**, 412 [2.4.1.93]
3501. Udenfriend, S. & Cooper, J.R. (1952) *J. Biol. Chem.* **194**, 503 [1.14.16.1]
3502. Ueda, T., Lode, E.T. & Coon, M.J. (1972) *J. Biol. Chem.* **247**, 2109 & 5010
3503. Uehara, K. (1952) *J. Chem. Soc. (Japan), Pure Chem. Sect.* **73**, 311
3504. Uehara, K. & Takeda, M. (1962) *J. Biochem. (Tokyo)* **52**, 461 [1.1.1.113]
3505. Uehara, K., Tanimoto, T. & Sato, H. (1974) *J. Biochem. (Tokyo)* **75**, 333 [1.1.1.162]
3506. Ukada, S. & Vennesland, B. (1962) *J. Biol. Chem.* **237**, 2018 [1.3.1.15]
3507. Umbreit, W.W. & Heneage, P. (1953) *J. Biol. Chem.* **201**, 15 [4.1.1.16]
3508. Umezawa, H., Takahashi, Y., Fujii, A., Saino, T., Shirai, T. & Takita, T. (1974) *J. Antibiot.* **26**, 117 [3.5.1.40]
3509. Umnov, A.M., Egorov, S.N., Mansurova, S.E. & Kulaev, I.S. (1974) *Biokhimiya* **39**, 373 [3.6.1.25]
3510. Unemoto, T. & Hayashi, M. (1969) *Biochim. Biophys. Acta* **171**, 89 [3.1.4.16]
3511. Unemoto, T., Hayashi, M., Miyaki, K. & Hayashi, M. (1965) *Biochim. Biophys. Acta* **110**, 319 [1.6.6.9]
3512. Uotila, L. (1973) *Biochemistry* **12**, 3944 [3.1.2.6]
3513. Uotila, L. (1973) *Biochemistry* **12**, 3938 [3.1.2.12, 3.1.2.13]
3514. Uotila, L. (1976) Personal communication [3.1.2.13]
3515. Uotila, L. & Koivusalo, M. (1974) *J. Biol. Chem.* **249**, 7683 [1.2.1.1]
3516. Uotila, L. & Koivusalo, M. (1974) *J. Biol. Chem.* **249**, 7664 [3.1.2.12, 3.1.2.6]
3517. Urich, K. (1968) *Z. Naturforsch. B.* **23**, 1508 [1.4.3.7]
3518. Utter, M.F. & Keech, D.B. (1963) *J. Biol. Chem.* **238**, 2603 [6.4.1.1]
3519. Uyeda, K. & Kurooka, S. (1970) *J. Biol. Chem.* **245**, 3315 [2.7.1.11]
3520. Uyeda, K. & Rabinowitz, J.C. (1971) *J. Biol. Chem.* **246**, 3111 and 3120 [1.2.7.1]
3521. Uziel, M. & Hanahan, D.J. (1957) *J. Biol. Chem.* **226**, 789 [5.4.1.1]
3522. Vachek, H. & Wood, J.L. (1972) *Biochim. Biophys. Acta* **258**, 133 [2.8.1.2]
3523. Vagelos, P.R. & Larrabee, A.R. (1967) *J. Biol. Chem.* **242**, 1776 [3.1.4.14]
3524. Valentine, R.C. & Wolfe, R.S. (1960) *J. Biol. Chem.* **235**, 1948 [2.3.1.19]
3525. Vallee, B.L. & Neurath, H. (1954) *J. Amer. Chem. Soc.* **76**, 5006 [3.4.17.1]
3526. Vallejo, C.M., Lobatón, C.D., Quintanilla, M., Sillero, A. & Sillero, M.A.G. (1976) *Biochim. Biophys. Acta* **438**, 304

3527. van den Bosch, H., Aarsman, A.J., De Jong, J.G.N. & van Deenen, L.L.M. (1973) *Biochim. Biophys. Acta* **296**, 94   [3.1.1.5]

3528. van den Bosch, H., van Golde, L.M.G., Eibl, H. & van Deenen, L.L.M. (1967) *Biochim. Biophys. Acta* **144**, 613   [2.3.1.23]

3529. van den Hamer, C.J.A., Morell, A.G. & Scheinberg, H.I. (1967) *J. Biol. Chem.* **242**, 2514   [2.8.1.2]

3530. van der Drift, C., van Helvoort, P.E.M. & Vogels, G.D. (1971) *Arch. Biochem. Biophys.* **145**, 465   [1.1.1.154]

3531. van der Drift, L., Vogels, G.D. & van der Drift, C. (1975) *Biochim. Biophys. Acta* **391**, 240   [5.1.99.3]

3532. van der Werf, P., Orlowski, M. & Meister, A. (1971) *Proc. Nat. Acad. Sci. USA* **68**, 2982   [3.5.2.9]

3533. van Eys, J. & Kaplan, N.O. (1957) *J. Amer. Chem. Soc.* **79**, 2782   [1.1.1.1]

3534. van Heyningen, R. & Pirie, A. (1953) *Biochem. J.* **53**, 436   [1.6.4.2]

3535. van Melle, P.J., Lewis, S.H., Samsa, E.G. & Westfall, R.J. (1963) *Enzymologia* **26**, 133

3536. van Tamelen, E.E., Willett, J.D., Clayton, R.B. & Lord, K.E. (1966) *J. Amer. Chem. Soc.* **88**, 4752   [1.14.99.7, 3.4.21.4]

3537. van Thoai, N., Robin, Y. & Guillou, Y. (1972) *Biochemistry* **11**, 3890   [2.7.3.5]

3538. Vance, D.E., Mituhashi, O. & Bloch, K. (1973) J. Biol. Chem. **248**, 2303   [2.3.1.38]

3539. Vance, P.G., Keele, B.B. & Rajagopalan, K.V. (1972) J. Biol. Chem. **247**, 4782   [1.15.1.1]

3540. Vanderwinkel, E., Furmanski, P., Reeves, H.C. & Ajl, S.J. (1968) *Biochem. Biophys. Res. Commun.* **33**, 902   [2.8.3.8]

3541. Varner, J.E. (1960) in *The Enzymes*, 2nd edn (Boyer, P.D., Lardy, H. & Myrbäck, K., ed.) Vol. 4, p. 247, Academic Press, New York   [3.5.1.5]

3542. Vaughan, P.F.T. & Butt, V.S. (1969) *Biochem. J.* **113**, 109   [1.14.17.2]

3543. Veerkamp, J.H. (1974) *Biochim. Biophys. Acta* **348**, 23   [2.4.1.46]

3544. Veibel, S. (1951) in *The Enzymes* (Sumner, J.B. & Myrbäck, K., ed.) Vol. 1, p. 621, Academic Press, New York   [3.2.1.22]

3545. Velick, S.F. & Furfine, C. (1963) in *The Enzymes*, 2nd edn (Boyer, P.D., Lardy, H. & Myrbäck, K., ed.) Vol. 7, p. 243, Academic Press, New York   [1.2.1.12]

3546. Venkataraman, R. & Racker, E. (1961) *J. Biol. Chem.* **236**, 1876 and 1883   [2.2.1.2]

3547. Verachtert, H., Rodriguez, P., Bass, S.T. & Hansen, R.G. (1966) *J. Biol. Chem.* **241**, 2007   [2.7.7.28]

3548. Verma, I.M. (1975) *J. Virology* **15**, 843.   [3.1.13.2]

3549. Vessey, D.A. & Zakim, D. (1973) *Biochim. Biophys. Acta* **315**, 43   [2.4.1.75]

3550. Vesterberg K. & Vesterberg O. (1972) *J. Med. Microbiol.* **5**, 441   [3.4.24.4]

3551. Vidal-Lieria, M. & van Uden, N. (1973) *Biochim. Biophys. Acta* **293**, 295   [1.1.1.18]

3552. Villee, C.A. & Spencer, J.M. (1960) *J. Biol. Chem.* **235**, 3615   [1.1.1.63, 1.1.1.64]

3553. Villemez, C.L., Swanson, A.L. & Hassid, W.Z. (1966) *Arch. Biochem. Biophys.* **116**, 446   [2.4.1.43]

3554. Virden, R., Watts, D.C. & Baldwin, E. (1965) *Biochem. J.* **94**, 536   [2.7.3.3]

3555. Virtanen, A.I. (1962) *Arch. Biochem. Biophys. Suppl.* **1**, 200   [5.99.1.1]

3556. Visuri K., Kikola J. & Enori T.-M. (1969) *Eur. J. Biochem.* **7**, 193   [3.4.16.1]

3557. Vitols, E., Walker, G.A. & Huennekens, F.M. (1966) *J. Biol. Chem.* **241**, 1455   [2.5.1.17]

3558. Vogel, G. & Lynen, F. (1970) *Naturwiss.* **57**, 664   [4.1.1.52]

3559. Vogel, H.J. (1953) *Proc. Nat. Acad. Sci. U.S.* **39**, 578   [2.6.1.11, 3.5.1.16]

3560. Vogel, H.J. & Bonner, D.M. (1956) *J. Biol. Chem.* **218**, 97   [3.5.1.16]

3561. Vogel, H.J. & McLellan, W.L. (1970) *Methods Enzymol.* **17A**, 251   [2.7.2.8]

3562. Vogels, G.D. (1966) *Biochim. Biophys. Acta* **113**, 277   [3.5.3.9]

3563. Vogt, V.M. (1973) *Eur. J. Biochem.* **33**, 192.   [3.1.30.1]

3564. Volk, W.A. (1960) *J. Biol. Chem.* **235**, 1550   [5.3.1.13]

3565. Volk, W.A. (1962) *J. Biol. Chem.* **237**, 19   [2.7.1.54]

3566. Volk, W.A. & Larsen, J.L. (1962) *J. Biol. Chem.* **237**, 2454   [1.1.1.130]

3567. Voll, M.J., Appella, E. & Martin, R.G. (1967) *J. Biol. Chem.* **242**, 1760   [2.4.2.17]

3568. Voynick, I.M. & Fruton, J.S. (1968) *Biochemistry* **7**, 40   [3.4.14.1]

3569. Wachsman, J.T. (1956) *J. Biol. Chem.* **223**, 19   [1.4.1.4]

3570. Wachsmuth D., Fritze I. & Pfleiderer G. (1966) *Biochemistry* **5**, 169   [3.4.11.2]

3571. Wacker H. (1974) *Biochim. Biophys. Acta* **334**, 417   [3.4.11.2]

518

3572. Wacker, H., Harvey, R.A., Winestock, C.H. & Plaut, G.W.E. (1964) *J. Biol. Chem.* **239**, 3493 [2.5.1.9]
3573. Wada, H. & Snell, E.E. (1961) *J. Biol. Chem.* **236**, 2089 [1.4.3.5]
3574. Wada, H. & Snell, E.E. (1962) *J. Biol. Chem.* **237**, 127 [2.6.1.31]
3575. Wada, H. & Snell, E.E. (1962) *J. Biol. Chem.* **237**, 133 [2.6.1.30]
3576. Wagner, C., Lusty, S.M. Jr., Kung, H.-F. & Rogers, N.L. (1967) *J. Biol. Chem.* **242**, 1287 [2.1.1.19]
3577. Wagner, I., Hofmann, H. & Hoffmann-Ostenhof, O. (1969) *Hoppe Seyler's Z. Physiol. Chem.* **350**, 1460 [2.1.1.40]
3578. Wahba, A.J. & Friedkin, M. (1962) *J. Biol. Chem.* **237**, 3794 [2.1.1.45]
3579. Wählby, S. (1968) *Biochim. Biophys. Acta* **151**, 409 [3.4.21.14]
3580. Wählby, S. & Engström, L. (1968) *Biochim. Biophys. Acta* **151**, 402 [3.4.21.4]
3581. Wainio, W.W., Eichel, B. & Gould, A. (1960) *J. Biol. Chem.* **235**, 1521 [1.9.3.1]
3582. Wakabayashi, M. & Fishman, W.H. (1961) *J. Biol. Chem.* **236**, 996 [3.2.1.31]
3583. Wakil, S.J. (1955) *Biochim. Biophys. Acta* **18**, 314 [5.1.2.3]
3584. Wakil, S.J. (1958) *J. Amer. Chem. Soc.* **80**, 6465 [6.4.1.2]
3585. Wakil, S.J. & Bressler, R. (1962) *J. Biol. Chem.* **237**, 687 [1.1.1.36]
3586. Wakil, S.J., Green, D.E., Mii, S. & Mahler, H.R. (1954) *J. Biol. Chem.* **207**, 631 [1.1.1.35]
3587. Waku, K. & Lands, W.E.M. (1968) *J. Biol. Chem.* **243**, 2654 [2.3.1.25]
3588. Walker, D.A. (1960) *Biochem. J.* **74**, 216 [1.1.1.40]
3589. Walker, G.A. & Kilgour, G.L. (1965) *Arch. Biochem. Biophys.* **111**, 534 [1.11.1.1]
3590. Walker, G.A., Murphy, S. & Huennekens, F.M. (1969) *Arch. Biochem. Biophys.* **134**, 95 [1.6.99.8, 1.6.99.9]
3591. Walker, G.C. & Nicholas, D.J.D. (1961) *Biochim. Biophys. Acta* **49**, 350 [1.7.99.3]
3592. Walker, G.C. & Nicholas, D.J.D. (1961) *Biochim. Biophys. Acta* **49**, 361 [1.7.99.1]
3593. Walker, G.J. & Pulkownik, A. (1973) *Carbohyd. Res.* **29**, 1 [3.2.1.70]
3594. Walker, G.J. & Whelan, W.J. (1959) *Nature (London)* **183**, 46 [2.4.1.25]
3595. Walker, J.B. (1956) *J. Biol. Chem.* **218**, 549 [2.1.4.1]
3596. Walker, J.B. (1957) *J. Biol. Chem.* **224**, 57 [2.1.4.1]
3597. Walker, J.B. (1971) *Lloydia* **34**, 363 [2.6.1.56, 2.7.1.65]
3598. Walker, J.B. & Skorvaga, M. (1973) *J. Biol. Chem.* **248**, 2435 [2.7.1.72, 2.7.1.87, 2.7.1.88]
3599. Walker, J.B. & Skorvaga, M. (1973) *J. Biol. Chem.* **248**, 2441 [3.1.3.39]
3600. Walker, J.B. & Walker, M.S. (1967) *Biochemistry* **6**, 3821 [2.7.1.65]
3601. Walker, J.B. & Walker, M.S. (1969) *Biochemistry* **8**, 763 [2.6.1.50, 2.6.1.56]
3602. Walker, J.B. & Walker, M.S. (1967) *Biochim. Biophys. Acta* **148**, 335 [2.7.1.72]
3603. Walker, M.S. & Walker, J.B. (1966) *J. Biol. Chem.* **241**, 1262 [2.1.4.2]
3604. Walker, M.S. & Walker, J.B. (1971) *J. Biol. Chem.* **246**, 7034 [3.1.3.39, 3.1.3.40]
3605. Wallach, D.P. & Grisolia, S. (1957) *J. Biol. Chem.* **226**, 277 [3.5.2.2]
3606. Wallenfels, K. & Malhotra, O.P. (1960) in *The Enzymes*, 2nd edn (Boyer, P.D., Lardy, H. & Myrbäck, K., ed.) Vol. 4, p. 409, Academic Press, New York [3.2.1.23]
3607. Walsh, D.A., Perkins, J.P. & Krebs, E.G. (1968) *J. Biol. Chem.* **243**, 3763 [2.7.1.37]
3608. Walter R.(1976) *Biochim. Biophys. Acta* **422**, 138 [3.4.21.26]
3609. Wang, C.C. & Barker, H.A. (1969) *J. Biol. Chem.* **244**, 2516 [4.2.1.34]
3610. Wang, S.-F. & Gabriel, O. (1969) *J. Biol. Chem.* **244**, 3430 [4.2.1.46]
3611. Wang, T.P. & Kaplan, N.O. (1954) *J. Biol. Chem.* **206**, 311 [2.7.1.23, 2.7.1.24]
3612. Wang, T.P., Sable, H.Z. & Lampen, J.O. (1950) *J. Biol. Chem.* **184**, 17 [3.5.4.5]
3613. Warburg, O. & Christian, W. (1939) *Biochem. Z.* **303**, 40 [1.2.1.12]
3614. Warner, A.H., Beers, P.C. & Huang, F.L. (1974) *Can. J. Biochem.* **52**, 231 [2.7.7.45]
3615. Warner, A.H. & Finamore, F.J. (1965) *Biochemistry* **4**, 1568 [3.6.1.17]
3616. Warner, H.R. & Lands, W.E.M. (1961) *J. Biol. Chem.* **236**, 2404 [3.3.2.2]
3617. Warnick, G.R. & Burnham, B.F. (1971) J. Biol. Chem. **246**, 6880 [2.3.1.37]
3618. Warren, L. & Buchanan, J.M. (1957) *J. Biol. Chem.* **229**, 613 [2.1.2.2]
3619. Warringa, M.G.P.J. & Giuditta, A. (1958) *J. Biol. Chem.* **230**, 111 [1.3.99.1]
3620. Watanabe, Y., Konishi, S. & Shimura, K. (1957) *J. Biochem. (Tokyo)* **44**, 299 [2.7.1.39]
3621. Watkins, W.M. & Hassid, W.Z. (1962) *J. Biol. Chem.* **237**, 1432 [2.4.1.22, 3.4.11.2]
3622. Watson, D.R., Jourdian, G.W. & Roseman, S. (1966) *J. Biol. Chem.* **241**, 5627 [4.1.3.20]

3623. Watson, G.H., Houghton, C. & Cain, R.B. (1974) *Biochem. J.* **140**, 277 [3.7.1.5]
3624. Weaver, R.F., Blatti, S.P. & Rutter, W.J. (1971) *Proc. Nat. Acad. Sci. U.S.* **68**, 2994 [2.7.7.6]
3625. Weaver, R.H. & Herbst, E.J. (1958) *J. Biol. Chem.* **231**, 647 [1.4.3.4]
3626. Webb, E.C. & Morrow, P.F.W. (1959) *Biochem. J.* **73**, 7 [3.1.6.1]
3627. Webster, G.C. (1961) *Biochim. Biophys. Acta* **49**, 141 [6.1.1.7]
3628. Webster, G.C. & Varner, J.E. (1955) *J. Biol. Chem.* **215**, 91 [6.3.1.1]
3629. Webster, G.R., Marples, E.A. & Thompson, R.H.S. (1957) *Biochem. J.* **65**, 374 [3.1.4.2]
3630. Webster, L.T. & Davie, E.W. (1961) *J. Biol. Chem.* **236**, 479 [6.1.1.11]
3631. Webster, L.T. Jr., Gerowin, L.D. & Rakita, L. (1965) *J. Biol. Chem.* **240**, 29 [6.2.1.2]
3632. Webster, M.E. (1970) in *Handbook of Experimental Pharmacology* (Erdö, E.G., ed.) Vol. 25, p. 659, Springer-Verlag, Berlin
3633. Webster, M.E. & Pierce, J.V. (1963) *Ann. N.Y. Acad. Sci. US* **104**, 91
3634. Webster, R.E. & Gross, S.R. (1965) *Biochemistry* **4**, 2309 [3.4.21.8, 4.1.3.12]
3635. Wechter, W.J., Mikulski, A.J. & Laskowski, M. Sr. (1968) *Biochem. Biophys. Res. Commun.* **30**, 318 [3.1.30.2]
3636. Weeks, G. & Wakil, S.J. (1968) *J. Biol. Chem.* **243**, 1180 [1.3.1.10, 1.3.1.9]
3637. Wegman, J. & DeMoss, J.A. (1965) *J. Biol. Chem.* **240**, 3781 [2.4.2.18]
3638. Weigl, J. & Yashe, W. (1966) *Can. J. Microbiol.* **12**, 939 [3.2.1.83]
3639. Weil-Malherbe, H. (1937) *Biochem. J.* **31**, 2080 [1.1.99.2]
3640. Weil-Malherbe, W. & Green, R.H. (1955) *Biochem. J.* **61**, 218 [3.5.4.6]
3641. Weimberg, R. & Doudoroff, M. (1955) *J. Biol. Chem.* **217**, 607 [1.1.1.46, 3.1.1.15, 4.2.1.25, 4.2.1.5]
3642. Weinhold, P.A. & Rethy, V.B. (1972) *Biochim. Biophys. Acta* **276**, 143 [2.7.1.82]
3643. Weiss, B. & Richardson, C.C. (1967) *Proc. Nat. Acad. Sci. U.S.* **57**, 1021 [6.5.1.1]
3644. Weiss, S.B., Kennedy, E.P. & Kiyasu, J.Y. (1960) *J. Biol. Chem.* **235**, 40 [2.3.1.20]
3645. Weiss, S.B., Smith, S.W. & Kennedy, E.P. (1958) *J. Biol. Chem.* **231**, 53 [2.7.8.2]
3646. Weissbach, H., Bogdanski, D.F., Redfield, B.G. & Udenfriend, S. (1957) *J. Biol. Chem.* **227**, 617 [4.1.1.28]
3647. Weissbach, H., Redfield, B.G. & Axelrod, J. (1961) *Biochim. Biophys. Acta* **54**, 190 [2.3.1.5]
3648. Weissbach, H., Toohey, J. & Barker, H.A. (1959) *Proc. Nat. Acad. Sci. U.S.* **45**, 521 [5.4.99.1]
3649. Weissmann, B. (1955) *J. Biol. Chem.* **216**, 783 [3.2.1.35]
3650. Weissmann, B. & Hinrichsen, D.F. (1969) *Biochemistry* **8**, 2034 [3.2.1.49]
3651. Weissmann, B., Rowen, G., Marshall, J. & Friederici, D. (1967) *Biochemistry* **6**, 207 [3.2.1.50]
3652. Welch, G.R., Cole, K.W. & Gaertner, F.H. (1974) *Arch. Biochem. Biophys.* **165**, 505 [4.6.1.4]
3653. Welker N.E. & Campbell L.L. (1967) *J. Bacteriol.* **94**, 1124 [3.4.21.14]
3654. Weller, M., Virmaux, N. & Mandel, P. (1975) *Proc. Nat. Acad. Sci., USA* **72**, 381 [2.7.1.97]
3655. Wellner, D. & Meister, A. (1960) *J. Biol. Chem.* **235**, 2013 [1.4.3.2]
3656. Wells, J.R.E. (1965) *Biochem. J.* **97**, 228 [3.4.16.1]
3657. Welton R.L. & Woods D.R. (1975) *Biochim. Biophys. Acta* **384**, 228 [3.4.24.8]
3658. Wengenmayer, H., Ebel, J. & Grisebach, H. (1974) *Eur. J. Biochem.* **50**, 135 [2.1.1.46]
3659. Wengenmeyer, H., Ebel, J. & Grisebach, H. (1976) *Eur. J. Biochem.* **65**, 529 [1.2.1.44]
3660. Wenger, D.A., Petipas, J.W. & Pieringer, R.A. (1968) *Biochemistry* **7**, 3700 [2.4.1.46]
3661. Werries, E., Wollek, E., Gottschalk, A. & Buddecke, E. (1969) *Eur. J. Biochem.* **10**, 445 [3.2.1.49, 3.2.1.50]
3662. Westhead, E.W. & McLain, G. (1964) *J. Biol. Chem.* **239**, 2464 [4.2.1.11]
3663. Westlake, D.W.S. (1963) *Can. J. Microbiol.* **9**, 211 [3.2.1.66]
3664. Westley, J. & Green, J.R. (1959) *J. Biol. Chem.* **234**, 2325 [2.8.1.1]
3665. Whistler, R.L. & Masek, E. (1955) *J. Amer. Chem. Soc.* **77**, 1241 [3.2.1.8]
3666. Whitaker, D.R., Hanson, K.R. & Datta, P.K. (1963) *Can. J. Biochem. Physiol.* **41**, 671 [3.2.1.4]
3667. Whitaker, D.R. & Roy, C. (1967) *Can. J. Biochem.* **45**, 911 [3.4.24.4]

3668. Whitaker, D.R., Roy, C., Tsai, C.S. & Jurasek, L. (1965) *Can. J. Biochem.* **43**, 1961 [3.4.24.4]

3669. White, E.H., McCapra, F., Field, G.F. & McElroy, W.D. (1963) *J. Amer. Chem. Soc.* **83**, 2402 [1.13.12.7]

3670. White, E.H., Rapaport, E., Hopkins, T.A. & Seliger, H.H. (1969) *J. Amer. Chem. Soc.* **91**, 2178 [1.13.12.7]

3671. White, G.A. & Krupka, R.M. (1965) *Arch. Biochem. Biophys.* **110**, 448 [1.13.11.13]

3672. White, R.J. & Pasternak, C.A. (1967) *Biochem. J.* **105**, 121 [3.5.1.25]

3673. White, W.F., Barlow, G.H. & Mozen, M.M. (1966) *Biochemistry* **5**, 2160 [3.4.21.31]

3674. Whiteley, H.R., Osborn, M.J. & Huennekens, F.M. (1959) *J. Biol. Chem.* **234**, 1538 [6.3.4.3]

3675. Whitfield, C.D., Steers, E.J. Jr. & Weissbach. H. (1970) *J. Biol. Chem.* **245**, 390 [2.1.1.14]

3676. Whiting, G.C. & Coggins, R.A. (1974) *Biochem. J.* **141**, 35 [1.1.1.166]

3677. Wickner, R.B. & Tabor, H. (1971) *Method. Enzymol.* **17B**, 80 [3.5.3.13]

3678. Wickremasinghe, R., Hedegaard, J. & Roche, J. (1967) *C.R. Soc. Biol.* **161**, 1891 [2.6.1.38]

3679. Wiebers, J.L. & Garner, H.R. (1967) *J. Biol. Chem.* **242**, 12 [4.2.99.9]

3680. Wiebers, J.L. & Garner, H.R. (1967) *J. Biol. Chem.* **242**, 5644 [4.2.99.9]

3681. Wiederschain, G. (1973) *Dokl. Akad. Nauk. S.S.S.R.* **211**, 974 [3.2.1.38]

3682. Wiederschain, G. (1974) *Dokl. Akad. Nauk. S.S.S.R.* **214**, 462

3683. Wiederschain, G. & Beyer, E. (1976) Dokl. Akad. Nauk. S.S.S.R. **231**, 486 [3.2.1.22]

3684. Wiederschain, G. & Prokopenkov, A. (1973) *Arch. Biochem. Biophys.* **158**, 539 [3.2.1.38]

3685. Wieland, O. & Suyter, M. (1957) *Biochem. Z.* **329**, 320 [2.7.1.30]

3686. Wilcox, H.G. & Fried, M. (1963) *Biochem. J.* **87**, 192

3687. Wilkinson, J.F. (1949) *Biochem. J.* **44**, 460 [2.7.1.6]

3688. Williams-Ashman, H.G. & Banks, J. (1954) *Arch. Biochem. Biophys.* **50**, 513 [1.1.1.14]

3689. Williams-Ashman, H.G. & Banks, J. (1956) *J. Biol. Chem.* **223**, 509 [2.7.7.15]

3690. Williams, C.H. Jr. & Kamin, H. (1962) *J. Biol. Chem.* **237**, 587 [1.6.2.4]

3691. Williams, F.R. & Hager, L.P. (1966) *Arch. Biochem. Biophys.* **116**, 168 [1.2.3.3, 4.1.1.15]

3692. Williams, F.R. & Hager, L.P. (1961) *J. Biol. Chem.* **236**, PC36 [1.2.2.2]

3693. Williams, W.J., Litwin, J. & Thorne, C.B. (1955) *J. Biol. Chem.* **212**, 427 [2.3.2.1]

3694. Williamson, I.P. & Wakil, S.J. (1966) *J. Biol. Chem.* **241**, 2326 [2.3.1.38, 2.3.1.39]

3695. Willingham, A.K. & Matschiner, J.T. (1974) *Biochem. J.* **140**, 435 [1.14.99.20]

3696. Willis, J.E. & Sallach, H.J. (1964) *Biochim. Biophys. Acta* **81**, 39 [1.1.1.95]

3697. Wilson, D.B. & Hogness, D.S. (1964) *J. Biol. Chem.* **239**, 2469 [5.1.3.2]

3698. Wilson, D.G., King, K.W. & Burris, R.H. (1954) *J. Biol. Chem.* **208**, 863 [2.6.1.12, 2.6.1.2, 2.6.1.4]

3699. Wilson, D.M. & Ajl, S. (1957) *J. Bacteriol.* **73**, 415 [2.7.1.5]

3700. Wilson, E.M. & Kornberg, H.L. (1963) *Biochem. J.* **88**, 578 [4.1.1.12]

3701. Wilson, E.M. & Snell, E.E. (1962) *J. Biol. Chem.* **237**, 3171 [2.1.2.7]

3702. Wilson, G.A. & Young, F.E. (1975) J. Mol. Biol. **97**, 123-126

3703. Wilson, G.A. & Young, F.E. in *Microbiology 1976. D. Schlessinger, ed., American Society of Microbiology, Washington, D.C., pp.350-357* [3.1.23.10, 3.1.23.9]

3704. Wilton, D.C., Rahimtula, A.D. & Akhtar, M. (1969) *Biochem. J.* **114**, 71 [5.3.3.5]

3705. Wingard M., Matsueda G. & Wolfe R.S. (1972) J. Bacteriol. **112**, 940 [3.4.99.30]

3706. Winkelman, J. & Lehninger, A.L. (1958) *J. Biol. Chem.* **233**, 794 [3.1.1.19]

3707. Winkler, N.W. & Markovitz, A. (1971) *J. Biol. Chem.* **246**, 5868 [1.1.1.135]

3708. Wintersberger, E., Cox, D. & Neurath, H. (1962) *Biochemistry* **1**, 1069 [3.4.17.2]

3709. Wishnick, M., Lane, M.D., Scrutton, M.C. & Mildvan, A.S. (1969) *J. Biol. Chem.* **244**, 5761 [4.1.1.39]

3710. Wiss, O. & Weber, F. (1956) *Hoppe-Seyler's Z. Physiol. Chem.* **304**, 232 [3.7.1.3]

3711. Witheiler J. & Wilson D.B. (1972) *J. Biol. Chem.* **247**, 2217 [3.4.21.24]

3712. Wittenberg, J. & Kornberg, A. (1953) *J. Biol. Chem.* **202**, 431 [2.7.1.32]

3713. Wittliff, J.L. & Airth, R.L. (1968) *Biochemistry* **7**, 736 [2.5.1.2]

3714. Wolf, D., Ebner, E. & Hinze, H. (1972) *Eur. J. Biochem.* **25**, 239 [2.7.7.42]

3715. Wolf, W.A. & Brown, G.M. (1969) *Biochim. Biophys. Acta* **192**, 468 [3.5.4.16]

3716. Wolfe, R.G. & Nielands, J.B. (1956) *J. Biol. Chem.* **221**, 61 [1.1.1.37]

3717. Wolff, J.B., Britton, B.B. & Nakada, H.I. (1957) *Arch. Biochem. Biophys.* **66**, 333 [5.3.1.10]
3718. Wolff, J.B. & Kaplan, N.O. (1956) *J. Biol. Chem.* **218**, 849 [1.1.1.17]
3719. Wolff, J.B. & Kaplan, N.O. (1955) *Methods Enzymol.* **1**, 346 [1.1.1.17]
3720. Wolin, M.J., Simpson, F.J. & Wood, W.A. (1958) *J. Biol. Chem.* **232**, 559 [5.1.3.4]
3721. Wong, K.K., Meister, A. & Moldove, K. (1959) *Biochim. Biophys. Acta* **36**, 531 [6.1.1.2]
3722. Wood, B.J.B. & Rainbow, C. (1961) *Biochem. J.* **78**, 204 [2.4.1.8]
3723. Wood, J.L. & Cavallini, D. (1966) *Arch. Biochem. Biophys.* **119**, 368 [1.13.11.19]
3724. Wood, W.A. (1955) *Methods Enzymol.* **2**, 212 [5.1.1.1]
3725. Wood, W.A. (1972) in *The Enzymes, 3rd ed.* (Boyer, P.D., ed.) Vol. 7, p. 281, Academic Press, New York [4.1.3.16, 4.1.3.17]
3726. Wood, W.A. & Gunsalus, I.C. (1951) *J. Biol. Chem.* **190**, 403 [5.1.1.1]
3727. Wood, W.A., McDonough, M.J. & Jacobs, L.B. (1961) *J. Biol. Chem.* **236**, 2190 [1.1.1.11, 1.1.1.56]
3728. Woodin, T.S. & Nishioka, L. (1973) *Biochim. Biophys. Acta* **309**, 211 [5.4.99.5]
3729. Woolfolk, C.A., Shapiro, B. & Stadtman, E.R. (1966) *Arch. Biochem. Biophys.* **116**, 177 [6.3.1.2]
3730. Woollen, J.W., Heyworth, R. & Walker, P.G. (1961) *Biochem. J.* **78**, 111 [3.2.1.30]
3731. Woolley D.E., Tueker J.S., Green G. & Evanson J.M. (1976) *Biochem. J.* **153**, 119 [3.4.24.7]
3732. Wosilait, W.D. (1960) *J. Biol. Chem.* **235**, 1196 [1.6.99.2]
3733. Wren, A. & Massey, V. (1966) *Biochim. Biophys. Acta* **122**, 436
3734. Wright, A., Dankert, M., Fennessen, P. & Robbins, P.W. (1967) *Proc. Nat. Acad. Sci. U.S.* **57**, 1798 [2.7.8.6]
3735. Wright, M., Buttin, G. & Hurwitz, J. (1971) *J. Biol. Chem.* **246**, 6543. [3.1.11.5]
3736. Wu, H.L.C. & Mason, M. (1964) *J. Biol. Chem.* **239**, 1492 [2.6.1.31]
3737. Wurster, B. & Hess, B. (1972) *FEBS Lett.* **23**, 341 [5.1.3.15]
3738. Wykle, R.L., Blank, M.L., Malone, B. & Snyder, F. (1972) *J. Biol. Chem.* **247**, 5442 [1.14.99.19]
3739. Wykle, R.L. & Schremmer, J.M. (1974) *J. Biol. Chem.* **249**, 1742 [3.1.4.39]
3740. Yagi, T., Honya, M. & Tamiya, N. (1968) *Biochim. Biophys. Acta* **153**, 699 [1.12.2.1, 4.2.1.27]
3741. Yamada, E.W. (1964) *Can. J. Biochem.* **42**, 317 [2.4.2.23]
3742. Yamada, E.W. (1961) *J. Biol. Chem.* **236**, 3043 [2.4.2.15]
3743. Yamada, E.W. & Jakoby, W.B. (1958) *J. Biol. Chem.* **233**, 707 [4.2.1.72]
3744. Yamada, E.W. & Jakoby, W.B. (1959) *J. Biol. Chem.* **234**, 941 [4.2.1.26, 4.2.1.71]
3745. Yamada, E.W. & Jakoby, W.B. (1960) *J. Biol. Chem.* **235**, 589 [1.2.1.18]
3746. Yamada, H. (1971) Method. Enzymol. **17B**, 726 [1.4.3.10]
3747. Yamada, H., Adachi, O. & Ogata, K. (1965) *Agr. Biol. Chem. (Tokyo)* **29**, 649, 863 & 912 [1.4.3.6]
3748. Yamada, H., Okamoto, K., Kodoma, K., Noguchi, F. & Tanaka, S. (1961) *J. Biochem. (Tokyo)* **49**, 404 [3.1.3.22]
3749. Yamada, H. & Yasunobu, K.T. (1962) *J. Biol. Chem.* **237**, 1511 & 3077 [1.4.3.4]
3750. Yamada, M. & Kurahashi, K. (1969) *J. Biochem. (Tokyo)* **66**, 529 [5.1.1.11]
3751. Yamada, Y., Aida, K. & Uemura, T. (1967) *J. Biochem. (Tokyo)* **61**, 636 [1.1.99.11]
3752. Yamada, Y., Iizuka, K., Aida, K. & Uemura, T. (1967) *J. Biochem. (Tokyo)* **62**, 223 [1.1.3.11]
3753. Yamafuji, K. & Eto, M. (1954) *Enzymologia* **16**, 247 [2.6.3.1]
3754. Yamafuji, K., Omura, H. & Miura, K. (1953) *Enzymologia* **16**, 75 [2.6.3.1]
3755. Yamafuji, K., Shimamura, M. & Omura, H. (1956) *Enzymologia* **17**, 359 [2.6.3.1]
3756. Yamagata, S. & Takeshima, K. (1976) *J. Biochem. (Tokyo)* **80**, 777 & 787 [4.2.99.10]
3757. Yamagata, S., Takeshima, K. & Naiki, N. (1974) *J. Biochem (Tokyo)* **75**, 1221 [4.2.99.10]
3758. Yamagata, T., Saito, H., Habuchi, O. & Suzuki, S. (1968) *J. Biol. Chem.* **243**, 1523 & 1536 [3.1.6.10, 3.1.6.9, 4.2.2.4]
3759. Yamaguchi, I., Shibata, H., Seto, H. & Misato, T. (1975) *J. Antibiot.* **28**, 7 [3.5.4.23]
3760. Yamamoto, S. & Bloch, K. (1970) *J. Biol. Chem.* **245**, 1670 [1.14.99.7]
3761. Yamamoto, S., Katagiri, M., Maeno, H. & Hayaishi, O. (1965) *J. Biol. Chem.* **240**, 3408 [1.14.13.1]

3762. Yamanaka, K. (1968) *Biochim. Biophys. Acta* **151**, 670   [5.3.1.5]
3763. Yamanaka, K., Gino, M. & Kaneda, R. (1977) *Agr. Biol. Chem. (Tokyo)* **41**, 1493 [1.1.1.175]
3764. Yamanaka, T., Kijimoto, S., Okunuki, K. & Kusai, K. (1962) *Nature (London)* **194**, 759 [1.9.3.2]
3765. Yamanaka, T. & Okunuki, K. (1970) *Biochim. Biophys. Acta* **220**, 354   [1.11.1.5]
3766. Yamasaki, H. & Moriyama, T. (1971) *Biochim. Biophys. Acta* **227**, 698   [4.2.1.24]
3767. Yamasaki, M. & Arima, K (1967) *Biochim. Biohys. Acta* **139**, 202.   [3.1.27.2]
3768. Yamasaki, M. & Arima, K. (1969) *Biochem. Biophys. Res. Commun.* **37**, 430   [3.1.27.2]
3769. Yamashina, I. (1956) *Ark. Kem.* **9**, 225   [3.4.21.9]
3770. Yamashita, S., Hosaka, K. & Numa, S. (1973) *Eur. J. Biochem.* **38**, 25 [2.3.1.51, 2.3.1.52]
3771. Yanagawa, H. & Egami, F. (1975) *Biochim. Biophys. Acta* **384**, 342   [1.6.4.7]
3772. Yanagita, T. & Foster, J.W. (1956) *J. Biol. Chem.* **221**, 593   [3.5.99.1]
3773. Yang, H., Tani, Y. & Ogata, K. (1970) *Agr. Biol. Chem. (Tokyo)* **34**, 1748   [6.3.3.3]
3774. Yang, H.Y.T. & Erdös, E.G. (1967) *Nature (London)* **215**, 1402   [3.4.15.1]
3775. Yang, H.Y.T., Erdös, E.G. & Chiang, T.S. (1968) *Nature (London)* **218**, 1224 [3.4.16.2]
3776. Yang, H.Y.T., Erdös, E.G., Chiang, T.S., Jenssen, T.A. & Rodgers, J.G. (1970) *Biochem. Pharmacol.* **19**, 1201   [3.4.16.2]
3777. Yang, H.Y.T., Erdös, E.G., Jenssen, T.A. & Levin, Y. (1970) *Fed. Proc.* **29**, 281 [3.4.15.1]
3778. Yang, H.Y.T., Jenssen, T.A. & Erdös, E.G. (1970) *Clin. Res.* **18**, 88
3779. Yaniv, H. & Gilvarg, C. (1955) *J. Biol. Chem.* **213**, 787   [1.1.1.25]
3780. Yaron A., (1976) *Methods Enzymol.* **45**, 599   [3.4.15.1]
3781. Yaron A. & Berger A. (1970) in *Methods in Enzymology* **19**, 522   [3.4.11.9]
3782. Yaron, A. & Mlynar, D. (1968) *Biochem. Biophys. Res. Commun.* **32**, 658   [3.4.11.9]
3783. Yates, M.G. (1969) *Biochim. Biophys. Acta* **171**, 299   [2.4.99.2, 3.5.4.17]
3784. Yates, M.G. & Nason, A. (1966) *J. Biol. Chem.* **241**, 4872   [1.9.99.1]
3785. Yefimochkina, E.F. & Braunstein, A.E. (1959) *Arch. Biochem. Biophys.* **83**, 350 [6.3.4.4]
3786. Yeh, Y.-C. & Greenberg, D.M. (1965) *Biochim. Biophys. Acta* **105**, 279   [1.5.1.5]
3787. Yip, G.B. & Dain, J.A. (1970) *Biochim. Biophys. Acta* **206**, 252   [2.4.1.62]
3788. Yip, M.C.M. (1973) *Biochim. Biophys. Acta* **306**, 298
3789. Yip, M.C.M. & Dain, J.A. (1970) *Biochem. J.* **118**, 247   [2.4.1.62]
3790. Yip, M.C.M. & Knox, W.E. (1970) *J. Biol. Chem.* **245**, 2199   [6.3.5.5]
3791. Yman, L. (1970) *Acta Pharm. Suecica* **7**, 75   [3.4.11.3]
3792. Yokobayashi, K., Misaki, A. & Harada, T. (1970) *Biochim. Biophys. Acta* **212**, 458 [3.2.1.68]
3793. Yokoyama S., Oobayashi A., Tanabe O. & Ichishima E. (1975) *Agr. Biol. Chem. (Tokyo)* **39**, 1211   [3.4.16.1]
3794. Yokoyama S., Oobayashi A., Tanabe O., Sugawara S., Araki E. & Ichishima E. (1974) *Appl. Microbiol. USA* **27**, 953   [3.4.16.1]
3795. Yonetani, T. (1970) *Advan. Enzymol.* **33**, 309   [1.11.1.5]
3796. Yonetani, T. (1960) *J. Biol. Chem.* **235**, 3138   [1.9.3.1]
3797. Yonetani, T. (1961) *J. Biol. Chem.* **236**, 1680   [1.9.3.1]
3798. Yorifuji, T., Ogata, K. & Soda, K. (1969) *Biochem. Biophys. Res. Commun.* **34**, 760 [5.1.1.9]
3799. York, J.L., Grollman, A.P. & Bublitz, C. (1961) *Biochim. Biophys. Acta* **47**, 298 [1.1.1.19]
3800. Yoshida, A. & Freese, E. (1965) *Biochim. Biophys. Acta* **96**, 248   [1.4.1.1]
3801. Yoshida, A. & Liu, H. (1972) *J. Biol. Chem.* **247**, 952   [3.4.14.3]
3802. Yoshida E. & Noda H. (1965) *Biochim. Biophys. Acta* **105**, 562   [3.4.24.3]
3803. Yoshida, F. & Ichishima, E. (1961) *Agr. Biol. Chem. (Tokyo)* **25**, 102   [3.4.23.6]
3804. Yoshida, F. & Nagasawa, M. (1956) *Bull. Agr. Chem. Soc. (Japan)* **20**, 262   [3.4.23.6]
3805. Yoshida, T., Takahashi, S. & Kikuchi, J. (1974) *J. Biochem. (Tokyo)* **75**, 1187 [1.14.99.3]
3806. Yoshikawa, K. & Adachi, K. (1973) *Biochim. Biophys. Acta* **315**, 333   [2.7.1.37]
3807. Yoshimori, R.N. Ph.D. Thesis (1971) University of California, San Francisco [3.1.23.13, 3.1.23.14]

3808. Yoshimoto, A., Nakamura, T. & Sato, R. (1967) *J. Biochem. (Tokyo)* **62**, 756 [1.8.99.1]
3809. Yoshimoto, A. & Sato, R. (1968) *Biochim. Biophys. Acta* **153**, 555 [1.8.1.2]
3810. Yoshimoto, T., Oka, I. & Tsuru, D. (1976) *J. Biochem. (Tokyo)* **79**, 1381 [3.5.3.3]
3811. Yoshimura, S. & Danno, G. (1964) *Agr. Chem. Soc. (Japan)* **38**, ·178 [3.4.99.20]
3812. Young, F.E. (1966) *J. Biol. Chem.* **241**, 3462 [3.4.17.7]
3813. Young, I.G. & Gibson, F. (1969) *Biochim. Biophys. Acta* **177**, 401 [1.3.1.28, 3.3.2.1, 5.4.99.6]
3814. Young, M.R. & Neish, A.C. (1966) *Phytochemistry* **5**, 1121 [4.3.1.5]
3815. Yourno, J. & Ino, I. (1968) *J. Biol. Chem.* **243**, 3273 [1.1.1.23]
3816. Yu, C.A. & Gunsalus, I.C. (1969) *J. Biol. Chem.* **244**, 6149 [1.14.15.2]
3817. Yu R.J., Grappel S.F. & Blank F. (1972) *Experientia* **28**, 886 & 1512 [3.4.24.10]
3818. Yu, R.J., Harmon, S.R. & Blank, F. (1968) *J. Bacteriol.* **96**, 1435 [3.4.24.10]
3819. Yu, R.J., Harmon, S.R. & Blank, F. (1969) *J. Invest. Dermatol.* **53**, 166 [3.4.24.10]
3820. Yuen, R. & Schachter, H. (1972) *Can. J. Biochem.* **50**, 798 [4.2.1.68]
3821. Yugari, Y. & Gilvarg, C. (1965) *J. Biol. Chem.* **240**, 4710 [4.2.1.52]
3822. Yura, T. & Vogel, H.J. (1959) *J. Biol. Chem.* **234**, 335 [1.5.1.2]
3823. Yura, T. & Vogel, H.J. (1959) *J. Biol. Chem.* **234**, 339 [1.1.1.3]
3824. Yurewicz, E.C., Ghalambor, M.A., Duckworth, D.H. & Heath, E.C. (1971) *J. Biol. Chem.* **246**, 5607 [3.2.1.87]
3825. Yurewicz, E.C., Ghalambor, M.A. & Heath, E.C. (1971) *J. Biol. Chem.* **246**, 5596 [3.2.1.87]
3826. Zabeau, M., Greene, R., Myers, P.A. & Roberts, R.J., unpublished observations [3.1.23.29]
3827. Zabin, I. (1963) *J. Biol. Chem.* **238**, 3300 [2.3.1.18]
3828. Zabin, I., Kepes, A. & Monod, J. (1962) *J. Biol. Chem.* **237**, 253 [2.3.1.18]
3829. Zain, B.S. & Roberts, R.J., (1977) *J. Mol. Biol.* **45** , 249 [3.1.23.41]
3830. Zakrzewski, S.F. & Nichol, C.A. (1960) *J. Biol. Chem.* **235**, 2984 [1.5.1.3]
3831. Zalkin, H. & Kling, D. (1968) *Biochemistry* **7**, 3566 [4.1.3.27]
3832. Zancan, G.T., Recondo, E.F. & Leloir, L.F. (1964) *Biochim. Biophys. Acta* **92**, 125 [3.1.3.28]
3833. Zannoni, V.G. & Weber, W.W. (1966) *J. Biol. Chem.* **241**, 1340 [1.1.1.96]
3834. Zappia, V. & Barker, H.A. (1970) *Biochim. Biophys. Acta* **207**, 505 [5.4.3.2]
3835. Zatman, L.J., Kaplan, N.O. & Colowick, S.P. (1953) *J. Biol. Chem.* **200**, 197 [3.2.2.6]
3836. Zatman, L.J., Kaplan, N.O., Colowick, S.P. & Ciotti, M.M. (1953) *J. Amer. Chem. Soc.* **75**, 3293 [3.2.2.6]
3837. Zechmeister, L. & Toth, G. (1939) *Enzymologia* **7**, 165 [3.2.1.14, 3.2.1.30]
3838. Zelitch, I. (1953) *J. Biol. Chem.* **201**, 719 [1.1.1.26]
3839. Zelitch, I. (1955) *J. Biol. Chem.* **216**, 553 [1.1.1.26]
3840. Zelitch, I. & Ochoa, S. (1953) *J. Biol. Chem.* **201**, 707 [1.1.3.1]
3841. Zeller, E.A. (1963) in *The Enzymes*, 2nd edn (Boyer, P.D., Lardy H. & Myrbäck, K., ed.) Vol. 8, p. 313, Academic Press, New York [1.4.3.4, 1.4.3.6]
3842. Zenk, M.H. & Schmitt, J. (1964) *Naturwiss.* **51**, 510 [2.3.1.34]
3843. Zenk, M.H. & Schmitt, J.H. (1965) *Biochem. Z.* **342**, 54 [2.3.1.36]
3844. Zerner, B., Coutts, S.M., Lederer, F., Waters, H.H. & Westheimer, F.H. (1966) *Biochemistry* **5**, 813 [4.1.1.4]
3845. Ziegler, D.M. & Pettit, F.H. (1966) *Biochemistry* **5**, 2932 [1.14.13.8]
3846. Zimmerman, D.C. (1970) *Lipids* **5**, 392 [1.13.11.12]
3847. Zimmerman, D.C. & Vick, B.A. (1970) *Plant Physiol.* **46**, 445 [5.3.99.1]
3848. Zimmerman, M. & Seidenberg, J. (1964) *J. Biol. Chem.* **239**, 2618 & 2622 ·[2.4.2.4]
3849. Zimmerman, S.B. & Kornberg, A. (1961) *J. Biol. Chem.* **236**, 1480 [3.6.1.12]
3850. Zimmerman, S.B., Little, J.W., Oshinsky, C.K. & Gellert, M. (1967) *Proc. Nat. Acad. Sci. U.S.* **57**, 1841 [6.5.1.2]
3851. Zink, M.W. & Sanwal, B.D. (1962) *Arch. Biochem. Biophys.* **99**, 72 [1.4.1.9]
3852. Zittle, C.A., Dellamonica, E.S., Custer, J.H. & Krikorian, R. (1955) *Arch. Biochem. Biophys.* **56**, 469 [3.1.1.7]
3853. Zonneveld, B.J.M. (1972) *Biochim. Biophys. Acta* **258**, 541 [3.2.1.84]
3854. Zuber, H. (1964) *Nature (London)* **201**, 613 [3.4.16.1]
3855. Zuber, H. & Matilt P. (1968) *Z. Naturforsch.* **B 23**, 663 [3.4.16.1]
3856. Zuidweg, M.H.J. (1968) *Biochim. Biophys. Acta* **152**, 144 [1.14.15.4]

524

3857.   Zumft, W.G. & Mortenson, L.E. (1975) *Biochim. Biophys. Acta* **416**, 1
        [1.18.2.1, 1.18.3.1, 1.19.2.1]
3858.   Zumft, W.G., Paneque, A., Aparicio, P.J. & Losada, M. (1969) *Biochem. Biophys. Res. Commun.* **36**, 980   [1.7.7.1]
3859.   Zwilling, R. (1968) *Hoppe-Seyler's Z. Physiol. Chem.* **349**, 326   [3.4.21.18, 3.4.21.4]
3859.   Zwilling, R. (1968) *Hoppe-Seyler's Z. Physiol. Chem.* **349**, 326

INDEX TO THE ENZYME LIST

Recommended names are printed in bold roman letters; other names are printed in ordinary roman. The *systematic names*, used as a basis of classification, are in italics. Names of deleted entries are shown in small roman letters.

525

Acetyl-CoA:acetyl-CoA *C-acetyltransferase*	2.3.1.9
Acetyl-CoA acetyltransferase	2.3.1.9
Acetyl-CoA acylase	3.1.2.1
Acetyl-CoA:[acyl-carrier-protein] *S-acetyltransferase*	2.3.1.38
Acetyl-CoA acyltransferase	2.3.1.16
Acetyl-CoA:D-amino-acid *N-acetyltransferase*	2.3.1.36
Acetyl-CoA:arylamine N-acetyltransferase	2.3.1.5
Acetyl-CoA:L-aspartate *N-acetyltransferase*	2.3.1.17
Acetyl-CoA:carbon-dioxide ligase *(ADP-forming)*	6.4.1.2
Acetyl-CoA carboxylase	6.4.1.2
Acetyl-CoA:carnitine O-acetyltransferase	2.3.1.7
Acetyl-CoA:chloramphenicol *3-O-acetyltransferase*	2.3.1.28
Acetyl-CoA:choline O-acetyltransferase	2.3.1.6
Acetyl-CoA:cortisol O-acetyltransferase	2.3.1.27
Acetyl-CoA deacylase	3.1.2.1
Acetyl-CoA:dihydrolipoamide *S-acetyltransferase*	2.3.1.12
Acetyl-CoA:formate C-acetyltransferase	2.3.1.54
Acetyl-CoA:galactoside *6-O-acetyltransferase*	2.3.1.18
Acetyl-CoA:gentamicin-C *N³-acetyltransferase*	2.3.1.60
Acetyl-CoA:gentamicin-C₁ₐ *N²'-acetyltransferase*	2.3.1.59
Acetyl-CoA:L-glutamate *N-acetyltransferase*	2.3.1.1
Acetyl-CoA:glycine C-acetyltransferase	2.3.1.29
Acetyl-CoA:L-histidine N-acetyltransferase	2.3.1.33
Acetyl-CoA:histone acetyltransferase	2.3.1.48
Acetyl-CoA:L-homoserine *O-acetyltransferase*	2.3.1.31
Acetyl-CoA:hydrogen-sulphide *S-acetyltransferase*	2.3.1.10
Acetyl-CoA hydrolase	3.1.2.1
Acetyl-CoA hydrolase	3.1.2.1
Acetyl-CoA:imidazole N-acetyltransferase	2.3.1.2
Acetyl-CoA:kanamycin *N⁶'-acetyltransferase*	2.3.1.55
Acetyl-CoA:malonate CoA-transferase	2.8.3.3
Acetyl-CoA:N-acetylneuraminate *4-O-acetyltransferase*	2.3.1.44
Acetyl-CoA:N-acetylneuraminate *7 (or 8)-O-acetyltransferase*	2.3.1.45
Acetyl-CoA:orthophosphate *acetyltransferase*	2.3.1.8
Acetyl-CoA:L-phenylalanine *N-acetyltransferase*	2.3.1.53

Acetyl-CoA:propionate CoA-transferase	2.8.3.1
Acetyl-CoA:putrescine N-acetyltransferase	2.3.1.57
Acetyl-CoA:L-serine O-acetyltransferase	2.3.1.30
Acetyl-CoA synthetase	6.2.1.1
Acetyl-CoA synthetase (ADP-forming)	6.2.1.13
Acetyl-CoA:thioethanolamine *S-acetyltransferase*	2.3.1.11
Acetyl-CoA:D-tryptophan *N-acetyltransferase*	2.3.1.34
Acetyl-CoA:2-amino-2-deoxy-D-glucose *N-acetyltransferase*	2.3.1.3
Acetyl-CoA:2-amino-2-deoxy-D- *glucose-6-phosphate N-acetyltransferase*	2.3.1.4
Acetylenedicarboxylate hydrase	4.2.1.72
Acetylenemonocarboxylate hydrase	4.2.1.71
Acetylesterase	3.1.1.6
β-N-Acetyl-D-galactosaminidase	3.2.1.53
α-N-Acetyl-D-galactosaminidase	3.2.1.49
N-Acetylglucosamine deacetylase	3.5.1.33
N-Acetyl-D-glucosamine kinase	2.7.1.59
Acetylglucosaminephosphate isomerase	5.3.1.11
Acetylglucosamine phosphomutase	2.7.5.2
N-**Acetylglucosamine-6-phosphate** **deacetylase**	3.5.1.25
α-N-Acetyl-D-glucosaminidase	3.2.1.50
β-N-Acetyl-D-glucosaminidase	3.2.1.30
trans-N-Acetylglucosaminosylase	2.4.1.16
β-N-Acetylglucosaminylsaccharide **fucosyltransferase**	2.4.1.65
6-O-Acetyl-D-glucose acetylhydrolase	3.1.1.33
6-O-Acetylglucose deacetylase	3.1.1.33
Acetylglutamate kinase	2.7.2.8
N-*Acetyl-L-glutamate-5-semialdehyde:* *NADP⁺ oxidoreductase (phosphorylating)*	1.2.1.38
N-**Acetyl-γ-glutamyl-phosphate reductase**	1.2.1.38
β-N-Acetyl-D-hexosaminidase	3.2.1.52
N-*Acetyl-L-histidine amidohydrolase*	3.5.1.34
Acetylhistidine deacetylase	3.5.1.34
O-Acetyl-L-homoserine acetate-lyase *(adding methanethiol)*	4.2.99.10
O-**Acetylhomoserine (thiol)-lyase**	4.2.99.10
Acetylindoxyl oxidase	1.7.3.2
N-Acetylindoxyl:oxygen oxidoreductase	1.7.3.2
N-**Acetyllactosamine synthase**	2.4.1.90
Acetylmethylcarbinol racemase	5.1.2.4
N-**Acetylmuramoyl-L-alanine amidase**	3.5.1.28
N-*Acetylneuraminate,hydrogen-* *donor:oxygen oxidoreductase (N-acetyl-* *hydroxylating)*	1.14.99.18
N-**Acetylneuraminate lyase**	4.1.3.3
N-**Acetylneuraminate monooxygenase**	1.14.99.18

N-*Acetylneuraminate pyruvate-lyase*	4.1.3.3
N-*Acetylneuraminate pyruvate-lyase (pyruvate-phosphorylating)*	4.1.3.19
N-**Acetylneuraminate synthase**	4.1.3.19
N-**Acetylneuraminate 4-*O*-acetyltransferase**	2.3.1.44
N-**Acetylneuraminate 7 (or 8)-*O*-acetyltransferase**	2.3.1.45
N-Acetylneuraminic acid aldolase	4.1.3.3
Acetylornithine aminotransferase	2.6.1.11
Acetylornithine deacetylase	3.5.1.16
*Acetyl-phosphate:*L-*lysine N⁶-acetyltransferase*	2.3.1.32
S-Acetylphosphopantetheine: deacetyl-[citrate-oxaloacetate-lyase (pro-3S-CH₂COO→acetate)] S-acetyltransferase	2.3.1.49
*O-Acetyl-*L-*serine acetate-lyase (adding hydrogen-sulphide)*	4.2.99.8
D-**Acetylserine (thiol)· lyase**	4.2.99.8
Acetylserotonin methyltransferase	2.1.1.4
Achromobacter iophagus **collagenase**	3.4.24.8
Acid:CoA ligase (AMP-forming)	6.2.1.3
Acid:CoA ligase (GDP-forming)	6.2.1.10
Acid maltase	3.2.1.3
Acid phosphatase	3.1.3.2
Acid phosphomonoesterase	3.1.3.2
Aconitase	4.2.1.3
cis-Aconitate carboxy-lyase	4.1.1.6
Aconitate decarboxylase	4.1.1.6
Aconitate hydratase	4.2.1.3
Aconitate Δ²—Δ³-isomerase	5.3.3.7
Aconitate Δ-isomerase	5.3.3.7
Acrosin	3.4.21.10
Actinidia anionic protease	3.4.22.14
Actinidin	3.4.22.14
Actinomycin lactonase	3.1.1.39
Actinomycin lactonohydrolase	3.1.1.39
Activated Christmas Factor	3.4.21.22
Acyl-activating enzyme	6.2.1.3
Acyl-activating enzyme	6.2.1.1
Acyl-activating enzyme	6.2.1.2
Acyl-[(acyl-carrier-protein)] desaturase	1.14.99.6
Acyl-[(acyl-carrier-protein)],hydrogen-donor:oxygen oxidoreductase	1.14.99.6
Acyl-[acyl-carrier-protein]:malonyl-[acyl-carrier-protein] C-acyltransferase (decarboxylating)	2.3.1.41
Acyl-[acyl-carrier-protein]:NAD⁺ oxidoreductase	1.3.1.9
Acyl-[acyl-carrier-protein]:NADP⁺ oxidoreductase	1.3.1.10
*Acyl-[acyl-carrier-protein]:O-(2-acyl-*sn-*glycero-3-phospho)-ethanolamine O-acyltransferase*	2.3.1.40
Acyl-[acyl-carrier-protein]-phospholipid acyltransferase	2.3.1.40
Acylagmatine amidase	3.5.1.40
Acylamidase	3.5.1.4
Acylamide amidohydrolase	3.5.1.4
*2-Acylamido-2-deoxy-*D-*glucose 2-epimerase*	5.1.3.8
*2-Acylamido-2-deoxy-*D-*glucose-6-phosphate-2-epimerase*	5.1.3.9
N-*Acylamino-acid amidohydrolase*	3.5.1.14
Acylamino-acid-releasing enzyme	3.4.14.3
N-Acylaminoacyl-peptide hydrolase	3.4.14.3
Acylase	3.5.1.4
*N-Acyl-*L-*aspartate amidohydrolase*	3.5.1.15
O-Acylcarnitine acylhydrolase	3.1.1.28
Acylcarnitine hydrolase	3.1.1.28
[Acyl-carrier-protein] acetyltransferase	2.3.1.38
[Acyl-carrier-protein] malonyltransferase	2.3.1.39
[Acyl-carrier-protein] phosphodiesterase	3.1.4.14
[Acyl-carrier-protein] 4′-pantetheinephosphohydrolase	3.1.4.14
Acylcholine acylhydrolase	3.1.1.8
Acyl-CoA:(acceptor) oxidoreductase	1.3.99.3
Acyl-CoA:acetate CoA-transferase	2.8.3.8
Acyl-CoA:acetyl-CoA C-acyltransferase	2.3.1.16
Acyl-CoA:cholesterol O-acyltransferase	2.3.1.26
Acyl-CoA dehydrogenase	1.3.99.3
Acyl-CoA dehydrogenase (NADP⁺)	1.3.1.8
Acyl-CoA desaturase	1.14.99.5
Acyl-CoA:dihydroxyacetone-phosphate O-acyltransferase	2.3.1.42
Acyl-CoA:glycine N-acyltransferase	2.3.1.13
Acyl-CoA,hydrogen-donor:oxygen oxidoreductase	1.14.99.5
Acyl-CoA:NADP⁺ oxidoreductase	1.3.1.8
Acyl-CoA:O-1-alk-1-enyl-glycero-3-phosphocholine O-acyltransferase	2.3.1.25
*Acyl-CoA:*sn-*glycerol-3-phosphate O-acyltransferase*	2.3.1.15
Acyl-CoA:sphingosine N-acyltransferase	2.3.1.24
Acyl-CoA synthetase	6.2.1.3
Acyl-CoA synthetase (GDP-forming).	6.2.1.10
Acyl-CoA:1-acylglycero-3-phosphocholine O-acyltransferase	2.3.1.23
*Acyl-CoA:1-acyl-*sn-*glycerol-3-phosphate O-acyltransferase*	2.3.1.51

Acyl-CoA:1,2-diacylglycerol *O-acyltransferase*	2.3.1.20
Acyl-CoA:2-acyl-sn-glycerol-3-phosphate *O-acyltransferase*	2.3.1.52
Acyl-CoA:2-acyl-sn-glycero-3- *phosphocholine O-acyltransferase*	2.3.1.62
Acyl dehydrogenase	1.3.99.3
Acylglucosamine 2-epimerase	5.1.3.8
Acylglucosamine-6-phosphate 2-epimerase	5.1.3.9
S-Acylglutathione hydrolase	3.1.2.7
Acylglycerol palmitoyltransferase	2.3.1.22
2-Acylglycerophosphate acyltransferase	2.3.1.52
1-Acylglycerophosphate acyltransferase	2.3.1.51
2-Acylglycerophosphocholine **acyltransferase**	2.3.1.62
Acyl-lysine deacylase	3.5.1.17
*N-**Acyl-D-mannosamine kinase***	2.7.1.60
Acylmuramoylalaninase	3.4.17.7
Acylmuramoyl-alanine carboxypeptidase	3.4.17.7
N-*Acylmuramoyl-L-alanine hydrolase*	3.4.17.7
Acylneuraminate cytidylyltransferase	2.7.7.43
*N-**Acylneuraminate-9-phosphatase***	3.1.3.29
N-*Acylneuraminate-9-phosphate* *phosphohydrolase*	3.1.3.29
N-*Acylneuraminate-9-phosphate* *pyruvate-lyase (pyruvate-phosphorylating)*	4.1.3.20
*N-**Acylneuraminate-9-phosphate synthase***	4.1.3.20
Acylneuraminyl hydrolase	3.2.1.18
Acylphosphatase	3.6.1.7
Acyl-phosphate—hexose **phosphotransferase**	2.7.1.61
Acyl-phosphate:D-hexose *phosphotransferase*	2.7.1.61
Acylphosphate phosphohydrolase	3.6.1.7
5′-Acylphosphoadenosine acylhydrolase	3.6.1.20
5′-Acylphosphoadenosine hydrolase	3.6.1.20
Acylpyruvate acylhydrolase	3.7.1.5
Acylpyruvate hydrolase	3.7.1.5
N-*Acylsphingosine amidohydrolase*	3.5.1.23
Acylsphingosine deacylase	3.5.1.23
Adenase	3.5.4.2
Adenine aminase	3.5.4.2
Adenine aminohydrolase	3.5.4.2
Adenine deaminase	3.5.4.2
Adenine phosphoribosyltransferase	2.4.2.7
Adenosinase	3.2.2.7
Adenosine aminohydrolase	3.5.4.4
Adenosine deaminase	3.5.4.4
Adenosine diphosphatase	3.6.1.5
Adenosine kinase	2.7.1.20
Adenosine nucleosidase	3.2.2.7

Adenosine (phosphate) aminohydrolase	3.5.4.17
Adenosine (phosphate) deaminase	3.5.4.17
Adenosine ribohydrolase	3.2.2.7
Adenosinetetraphosphatase	3.6.1.14
Adenosinetetraphosphate phosphohydrolase	3.6.1.14
Adenosinetriphosphatase	3.6.1.3
Adenosinetriphosphatase (Mg-activated)	3.6.1.4
Adenosinetriphosphatase (Na+,K+-activated)	3.6.1.3
Adenosine-3′,5′-bisphosphate *3′-phosphohydrolase*	3.1.3.7
Adenosylhomocysteinase	3.3.1.1
S-Adenosyl-L-homocysteine *homocysteinylribohydrolase*	3.2.2.9
S-Adenosyl-L-homocysteine hydrolase	3.3.1.1
Adenosylhomocysteine nucleosidase	3.2.2.9
S-Adenosyl-L-methionine alkyltransferase *(cyclizing)*	2.5.1.4
S-Adensyl-L-methionine:carnosine *N-methyltransferase*	2.1.1.22
S-Adenosyl-L-methionine carboxy-lyase	4.1.1.50
S-Adenosyl-L-methionine:catechol *O-methyltransferase*	2.1.1.6
S-Adenosylmethionine cleaving enzyme	3.3.1.2
Adenosylmethionine cyclotransferase	2.5.1.4
Adenosylmethionine decarboxylase	4.1.1.50
S-Adenosyl-L-methionine:DNA *(cytosine-5-)-methyltransferase*	2.1.1.37
S-Adenosyl-L-methionine:fatty acid *O-methyltransferase*	2.1.1.15
S-Adenosyl-L-methionine:glycine *methyltransferase*	2.1.1.20
S-Adenosyl-L-methionine:guanidinoacetate *N-methyltransferase*	2.1.1.2
S-Adenosyl-L-methionine:histamine *N-methyltransferase*	2.1.1.8
S-Adenosyl-L-methionine:L-homocysteine *S-methyltransferase*	2.1.1.10
S-Adenosyl-L-methionine hydrolase	3.3.1.2
Adenosylmethionine hydrolase	3.3.1.2
S-Adenosyl-L-methionine:indolepyruvate *C-methyltransferase*	2.1.1.47
S-Adenosyl-L-methionine:isoflavone *4′-O-methyltransferase*	2.1.1.46
S-Adenosyl-L-methionine:loganate *11-O-methyltransferase*	2.1.1.50
S-Adenosyl-L-methionine:magnesium- *protoporphyrin O-methyltransferase*	2.1.1.11
S-Adenosyl-L-methionine:L-methionine *S-methyltransferase*	2.1.1.12
S-Adenosyl-L-methionine:myo-inositol *1-methyltransferase*	2.1.1.39

S-*Adenosyl*-L-*methionine:myo-inositol 3-methyltransferase*	2.1.1.40
S-*Adenosyl*-L-*methionine:N-acetylserotonin O-methyltransferase*	2.1.1.4
S-*Adenosyl*-L-*methionine:nicotinamide N-methyltransferase*	2.1.1.1
S-*Adenosyl*-L-*methionine:nicotinate N-methyltransferase*	2.1.1.7
S-*Adenosyl*-L-*methionine:N$^\alpha$,N$^\alpha$-dimethyl-*L-*histidine N$^\alpha$-methyltransferase*	2.1.1.44
S-*Adenosyl*-L-*methionine: O-demethylpuromycin O-methyltransferase*	2.1.1.38
S-*Adenosyl*-L-*methionine:phenol O-methyltransferase*	2.1.1.25
S-*Adenosyl*-L-*methionine:phenylethanolamine N-methyltransferase*	2.1.1.28
S-*Adenosyl*-L-*methionine:phosphatidyl-ethanolamine N-methyltransferase*	2.1.1.17
S-*Adenosyl*-L-*methionine:protein (arginine) N-methyltransferase*	2.1.1.23
S-*Adenosyl*-L-*methionine:protein (lysine) N-methyltransferase*	2.1.1.43
S-*Adenosyl*-L-*methionine:protein O-methyltransferase*	2.1.1.24
S-*Adenosyl*-L-*methionine:putrescine N-methyltransferase*	2.1.1.53
S-*Adenosyl*-L-*methionine:rRNA (adenine-6-)-methyltransferase*	2.1.1.48
S-*Adenosyl*-L-*methionine:rRNA (guanine-1-)-methyltransferase*	2.1.1.51
S-*Adenosyl*-L-*methionine:rRNA (guanine-2-)-methyltransferase*	2.1.1.52
S-*Adenosyl*-L-*methionine:thiol S-methyltransferase*	2.1.1.9
S-*Adenosyl*-L-*methionine:tRNA (adenine-1-)-methyltransferase*	2.1.1.36
S-*Adenosyl*-L-*methionine:tRNA (cytosine-5-)-methyltransferase*	2.1.1.29
S-*Adenosyl*-L-*methionine:tRNA (guanine-1-)-methyltransferase*	2.1.1.31
S-*Adenosyl*-L-*methionine:tRNA (guanine-2-)-methyltransferase*	2.1.1.32
S-*Adenosyl*-L-*methionine:tRNA (guanine-7-)-methyltransferase*	2.1.1.33
S-*Adenosyl*-L-*methionine:tRNA (guanosine-2'-)-methyltransferase*	2.1.1.34
S-*Adenosyl*-L-*methionine:tRNA (purine-2- or -6-)-methyltransferase*	2.1.1.30
S-*Adenosyl*-L-*methionine:tRNA (uracil-5-)-methyltransferase*	2.1.1.35
S-*Adenosyl*-L-*methionine:tryptamine N-methyltransferase*	2.1.1.49
S-*Adenosyl*-L-*methionine:tyramine N-methyltransferase*	2.1.1.27
S-*Adenosyl*-L-*methionine:unsaturated-phospholipid methyltransferase*	2.1.1.16
S-*Adenosyl*-L-*methionine:zymosterol methyltransferase*	2.1.1.41
S-*Adenosyl*-L-*methionine:1,4-α-*D-*glucan 6-O-methyltransferase*	2.1.1.18
S-*Adenosyl*-L-*methionine:2-iodophenol methyltransferase*	2.1.1.26
S-*Adenosyl*-L-*methionine: 5,7,3',4'-tetrahydroxyflavone 3'-O-methyltransferase*	2.1.1.42
Adenylate cyclase	4.6.1.1
Adenylate kinase	2.7.4.3
Adenyl cyclase	4.6.1.1
Adenylic acid deaminase	3.5.4.6
Adenylosuccinase	4.3.2.2
Adenylosuccinate AMP-lyase	4.3.2.2
Adenylosuccinate lyase	4.3.2.2
Adenylosuccinate synthase	6.3.4.4
Adenylosuccinate synthetase	6.3.4.4
Adenylpyrophosphatase	3.6.1.3
Adenylyl cyclase	4.6.1.1
*Adenylyl-[*L-*glutamate:ammonia ligase (ADP-forming)] adenylylhydrolase*	3.1.4.15
Adenylyl-[glutamine-synthetase] hydrolase	3.1.4.15
Adenylylsulphatase	3.6.2.1
Adenylylsulphate kinase	2.7.1.25
Adenylylsulphate reductase	1.8.99.2
Adenylylsulphate sulphohydrolase	3.6.2.1
ADP aminohydrolase	3.5.4.7
ADPase	3.6.1.5
ADP deaminase	3.5.4.7
ADPglucose pyrophosphorylase	2.7.7.27
ADPglucose—starch glucosyltransferase	2.4.1.21
*ADPglucose:1,4-α-*D-*glucan 4-α-*D-*glucosyltransferase*	2.4.1.21
ADPphosphoglycerate phosphatase	3.1.3.28
ADPphosphoglycerate phosphohydrolase	3.1.3.28
ADPribose phosphorylase	2.7.7.35
ADPribose pyrophosphatase	3.6.1.13
ADPribose ribophosphohydrolase	3.6.1.13
*ADP:*D-*ribose-5-phosphate adenylyltransferase*	2.7.7.35
ADPsugar phosphorylase	2.7.7.36
*Acyl-CoA:1-alkyl-*sn-*glycero-3-phosphate O-acyltransferase*	2.3.1.63

ADPsugar pyrophosphatase	3.6.1.21
ADPsugar sugarphosphohydrolase	3.6.1.21
ADP:sugar-1-phosphate adenylyltransferase	2.7.7.36
ADP:sulphate adenylyltransferase	2.7.7.5
ADP-sulphurylase	2.7.7.5
Adrenalin oxidase	1.4.3.4
Adrenodoxin reductase	1.18.1.2
Aerobacter-capsular-polysaccharide galactohydrolase	3.2.1.87
Aeromonas proteolytica **aminopeptidase**	3.4.11.10
Aeromonas proteolytica **neutral proteinase**	3.4.24.4
A-esterase	3.1.1.2
Agarase	3.2.1.81
Agaritine γ-**glutamyltransferase**	2.3.2.9
Agarose 3-glycanohydrolase	3.2.1.81
Agavain	3.4.99.2
Agkistrodon **serine proteinase**	3.4.21.28
Agmatinase	3.5.3.11
Agmatine amidinohydrolase	3.5.3.11
Agmatine deiminase	3.5.3.12
Agmatine iminohydrolase	3.5.3.12
D-Alanine:D-alanine ligase (ADP-forming)	6.3.2.4
D-Alanine:alanyl-poly(glycerophosphate) ligase (ADP-forming)	6.3.2.16
Alanine aminotransferase	2.6.1.2
D-Alanine aminotransferase	2.6.1.21
D-Alanine aminotransferase	2.6.1.10
Alanine carboxypeptidase	3.4.17.6
Alanine dehydrogenase	1.4.1.1
L-Alanine:glyoxylate aminotransferase	2.6.1.44
Alanine—glyoxylate aminotransferase	2.6.1.44
L-Alanine:malonate-semialdehyde aminotransferase	2.6.1.18
D-Alanine:membrane-acceptor ligase	6.3.2.16
L-Alanine:NAD+ oxidoreductase (deaminating)	1.4.1.1
Alanine—oxo-acid aminotransferase	2.6.1.12
β-Alanine—oxoglutarate aminotransferase	2.6.1.19
L-Alanine:oxomalonate aminotransferase	2.6.1.47
Alanine—oxomalonate aminotransferase	2.6.1.47
D-Alanine:poly(phosphoribitol) ligase (AMP-forming)	6.1.1.13
β-**Alanine—pyruvate aminotransferase**	2.6.1.18
Alanine racemase	5.1.1.1
Alanine racemase	5.1.1.1
D-Alanine-sRNA synthetase	6.1.1.8
L-Alanine:tRNA^Ala ligase (AMP-forming)	6.1.1.7

L-*Alanine:2-oxo-acid aminotransferase*	2.6.1.12
D-*Alanine:2-oxoglutarate aminotransferase*	2.6.1.21
L-*Alanine:2-oxoglutarate aminotransferase*	2.6.1.2
L-*Alanine:4,5-dioxovalerate aminotransferase*	2.6.1.43
D-Alanylalanine synthetase	6.3.2.4
D-Alanyl-alanyl-poly (glycerophosphate) synthetase	6.3.2.16
β-*Alanyl-CoA ammonia-lyase*	4.3.1.6
β-**Alanyl-CoA ammonia-lyase**	4.3.1.6
O-**Alanylphosphatidylglycerol synthase**	2.3.2.11
D-Alanyl-poly (phosphoribitol) synthetase	6.1.1.13
Alanyl-tRNA:phosphatidylglycerol alanyltransferase	2.3.2.11
Alanyl-tRNA synthetase	6.1.1.7
L-*Alanyl-tRNA:UDP-N-acetylmuramoyl-L-alanyl-D-glutamyl-L-lysyl-D-alanyl-D-alanine N^6-alanyltransferase*	2.3.2.10
Alcohol:(acceptor) oxidoreductase	1.1.99.8
Alcohol dehydrogenase	1.1.1.1
Alcohol dehydrogenase (acceptor)	1.1.99.8
Alcohol dehydrogenase (NAD(P)+)	1.1.1.71
Alcohol dehydrogenase (NADP+)	1.1.1.2
Alcohol:NAD+ oxidoreductase	1.1.1.1
Alcohol:NADP+ oxidoreductase	1.1.1.2
Alcohol:NAD(P)+ oxidoreductase	1.1.1.71
Alcohol oxidase	1.1.3.13
Alcohol:oxygen oxidoreductase	1.1.3.13
Aldehyde dehydrogenase	1.2.1.3
Aldehyde dehydrogenase (NADP+)	1.2.1.4
Aldehyde dehydrogenase (NAD(P)+)	1.2.1.5
Aldehyde:NAD+ oxidoreductase	1.2.1.3
Aldehyde:NAD(P)+ oxidoreductase	1.2.1.5
Aldehyde:NADP+ oxidoreductase	1.2.1.4
Aldehyde oxidase	1.2.3.1
Aldehyde:oxygen oxidoreductase	1.2.3.1
Aldehyde reductase	1.1.1.1
Alditol:NADP+ 1-oxidoreductase	1.1.1.21
D-*Aldohexoside:(acceptor) 5-oxidoreductase*	1.1.99.13
Aldoketomutase	4.4.1.5
Aldolase	4.1.2.13
L-3-Aldonate dehydrogenase	1.1.1.45
Aldonolactonase	3.1.1.18
D-Aldopantoate dehydrogenase	1.2.1.33
Aldose dehydrogenase	1.1.1.121
Aldose mutarotase	5.1.3.3
D-*Aldose.NAD+ 1-oxidoreductase*	1.1.1.121
Aldose reductase	1.1.1.21
Aldose 1-epimerase	5.1.3.3

Aldose 1-epimerase	5.1.3.3
Alginase	3.2.1.16
Alginate lyase	4.2.2.3
Alginate synthase	2.4.1.33
Ali-esterase	3.1.1.1
Alkaline phosphatase	3.1.3.1
Alkaline phosphomonoesterase	3.1.3.1
Alkane,reduced-rubredoxin:oxygen 1-oxidoreductase	1.14.15.3
Alkane 1-hydroxylase	1.14.15.3
Alkane 1-monooxygenase	1.14.15.3
2-Alkenal reductase	1.3.1.27
Alkenyl-glycerophosphocholine hydrolase	3.3.2.2
1-(1-Alkenyl)-glycero-3-phosphocholine aldehydohydrolase	3.3.2.2
Alkylacylglycerophosphoethanolamine desaturase	1.14.99.19
Alkylamidase	3.5.1.39
3-Alkylcatechol 2,3-dioxygenase	1.13.11.25
S-Alkyl-L-cysteine alkylthiol-lyase (deaminating)	4.4.1.6
S-Alkylcysteine lyase	4.4.1.6
Alkyldihydroxyacetone kinase	2.7.1.84
Alkylglycerol kinase	2.7.1.93
1-Alkylglycerophosphate acyltransferase	2.3.1.63
Alkylglycerophosphoethanolamine phosphodiesterase	3.1.4.39
Alkylhalidase	3.8.1.1
Alkyl-halide halidohydrolase	3.8.1.1
Alkylmercury lyase	4.99.1.2
Alkylmercury mercuric-lyase	4.99.1.2
1-Alkyl-sn-glycerol,tetrahydropteridine: oxygen oxidoreductase	1.14.16.5
1-Alkyl-sn-glycero-3-phosphoethanolamine ethanolaminehydrolase	3.1.4.39
O-1-Alkyl-2-acyl-sn-glycero-3-phosphoethanolamine,hydrogen-donor: oxygen oxidoreductase	1.14.99.19
2-Alkyn-1-ol dehydrogenase	1.1.1.165
Allantoate amidinohydrolase	3.5.3.4
Allantoate amidinohydrolase (decarboxylating)	3.5.3.9
Allantoate deiminase	3.5.3.9
Allantoicase	3.5.3.4
Allantoin amidohydrolase	3.5.2.5
Allantoinase	3.5.2.5
Allantoin racemase	5.1.99.3
Allantoin racemase	5.1.99.3
Alliin alkyl-sulphenate-lyase	4.4.1.4
Alliinase	4.4.1.4

Alliin lyase	4.4.1.4
Allokinase	2.7.1.55
Allose kinase	2.7.1.55
Allothreonine aldolase	4.1.2.6
Allyl-alcohol dehydrogenase	1.1.1.54
Allyl-alcohol:NADP+ oxidoreductase	1.1.1.54
Allylic-terpene-diphosphate: isopentenyldiphosphate terpenoid-allyltransferase	2.5.1.11
Alternaria **serine proteinase**	3.4.21.16
Altronate dehydratase	4.2.1.7
D-*Altronate hydro-lyase*	4.2.1.7
D-*Altronate:NAD+ 3-oxidoreductase*	1.1.1.58
Amidase	3.5.1.4
ω-**Amidase**	3.5.1.3
Amidinoaspartase	3.5.3.14
N-*Amidino-L-aspartate amidinohydrolase*	3.5.3.14
ω-*Amidodicarboxylate amidohydrolase*	3.5.1.3
Amidophosphoribosyltransferase	2.4.2.14
Amine dehydrogenase	1.4.99.3
Amine oxidase	1.4.3.4
Amine oxidase (copper-containing)	1.4.3.6
Amine oxidase (flavin-containing)	1.4.3.4
Amine oxidase (Pyridoxal containing)	1.4.3.6
Amine:oxygen oxidoreductase (deaminating) (copper-containing)	1.4.3.6
Amine:oxygen oxidoreductase (deaminating) (flavin-containing)	1.4.3.4
D-*Amino-acid:(acceptor) oxidoreductase (deaminating)*	1.4.99.1
D-Amino-acid acetyltransferase	2.3.1.36
Amino-acid acetyltransferase	2.3.1.1
L-Amino-acid dehydrogenase	1.4.1.5
D-Amino-acid dehydrogenase	1.4.99.1
L-*Amino-acid:NAD+ oxidoreductase (deaminating)*	1.4.1.5
L-Amino-acid oxidase	1.4.3.2
D-Amino-acid oxidase	1.4.3.3
L-*Amino-acid:oxygen oxidoreductase (deaminating)*	1.4.3.2
D-*Amino-acid:oxygen oxidoreductase (deaminating)*	1.4.3.3
Amino-acid racemase	5.1.1.10
Amino-acid racemase	5.1.1.10
Aminoacylase	3.5.1.14
Aminoacylase II	3.5.1.15
α-*Aminoacyl-dipeptide hydrolase*	3.4.11.4
Aminoacyl-histidine dipeptidase	3.4.13.3
Aminoacyl-L-histidine hydrolase	3.4.13.3
Aminoacyl-L-lysine (-L-arginine) hydrolase	3.4.13.4
Aminoacyl-lysine dipeptidase	3.4.13.4

Aminoacyl-methylhistidine dipeptidase	3.4.13.5
α-*Aminoacyl-peptide hydrolase*	3.4.11.11
α-*Aminoacyl-peptide hydrolase*	3.4.11.14
α-*Aminoacyl-peptide hydrolase*	3.4.11.3
α-*Aminoacyl-peptide hydrolase*	3.4.11.12
α-*Aminoacyl-peptide hydrolase*	3.4.11.13
α-*Aminoacyl-peptide hydrolase* (Aeromonas proteolytica)	3.4.11.10
α-*Aminoacyl-peptide hydrolase (cytosol)*	3.4.11.1
α-*Aminoacyl-peptide hydrolase (microsomal)*	3.4.11.2
Aminoacylproline aminopeptidase	3.4.11.9
*Aminoacyl-*L*-proline hydrolase*	3.4.13.9
Aminoacylprolyl-peptide hydrolase	3.4.11.9
*Aminoacyl-prosmethyl-*L*-histidine hydrolase*	3.4.13.5
Aminoacyl-tRNA aminoacylhydrolase	3.1.1.29
Aminoacyl-tRNA hydrolase	3.1.1.29
2-Aminoadipate aminotransferase	2.6.1.39
Aminoadipate-semialdehyde dehydrogenase	1.2.1.31
L*-2-Aminoadipate:2-oxoglutarate aminotransferase*	2.6.1.39
L*-2-Aminoadipate-6-semialdehyde: NAD(P)*$^+$ *oxidoreductase*	1.2.1.31
Aminobenzoate carboxy-lyase	4.1.1.24
Aminobenzoate decarboxylase	4.1.1.24
Aminobutyraldehyde dehydrogenase	1.2.1.19
4-Aminobutyraldehyde:NAD$^+$ *oxidoreductase*	1.2.1.19
Aminobutyrate aminotransferase	2.6.1.19
4-Aminobutyrate:2-oxoglutarate aminotransferase	2.6.1.19
N^2*-(4-Aminobutyryl)-*L*-lysine hydrolase*	3.4.13.4
Aminocarboxymuconate-semialdehyde decarboxylase	4.1.1.45
Aminodeoxygluconate dehydratase	4.2.1.26
(2-Aminoethyl) phosphonate aminotransferase	2.6.1.37
(2-Aminoethyl) phosphonate:pyruvate aminotransferase	2.6.1.37
Aminoimidazolase	3.5.4.8
4-Aminoimidazole aminohydrolase	3.5.4.8
L*-3-Aminoisobutyrate aminotransferase*	2.6.1.22
D-3-Aminoisobutyrate—pyruvate aminotransferase	2.6.1.40
L*-3-Aminoisobutyrate:2-oxoglutarate aminotransferase*	2.6.1.22
Aminolaevulinate aminotransferase	2.6.1.43
Aminolaevulinate dehydratase	4.2.1.24
5-Aminolaevulinate hydro-lyase (adding 5-aminolaevulinate and cyclizing)	4.2.1.24
δ-Aminolaevulinate synthase	2.3.1.37

Aminomalonate decarboxylase	4.1.1.10
Aminomuconate-semialdehyde dehydrogenase	1.2.1.32
2-Aminomuconate-6-semialdehyde:NAD$^+$ *oxidoreductase*	1.2.1.32
5-Amino-n-valeramide amidohydrolase	3.5.1.30
Aminopeptidase	3.4.11.11
Aminopeptidase (cytosol)	3.4.11.1
Aminopeptidase (human liver)	3.4.11.14
Aminopeptidase (microsomal)	3.4.11.2
Aminopeptidase P	3.4.11.9
o-**Aminophenol oxidase**	1.10.3.4
o-Aminophenol:oxygen oxidoreductase	1.10.3.4
L-**Aminopropanol dehydrogenase**	1.1.1.75
D-Aminopropanol dehydrogenase	1.1.1.74
L*-1-Aminopropan-2-ol:NAD*$^+$ *oxidoreductase*	1.1.1.75
Aminopropyltransferase	2.5.1.16
5-Aminovaleramidase	3.5.1.30
5-Aminovalerate aminotransferase	2.6.1.48
5-Aminovalerate:lipoate oxidoreductase (cyclizing)	1.4.4.1
5-Aminovalerate:NAD$^+$ *oxidoreductase (cyclizing)*	1.4.1.6
5-Aminovalerate:2-oxoglutarate aminotransferase	2.6.1.48
*2-Amino-2-deoxy-*D*-gluconate ammonia-lyase (isomerizing)*	4.3.1.9
*2-Amino-2-deoxy-*D*-gluconate hydro-lyase (deaminating)*	4.2.1.26
*2-Amino-2-deoxy-*D*-glucose-6-phosphate ketol-isomerase (amino-transferring)*	5.3.1.19
*2-Amino-2-deoxy-*D*-glucose-6-phosphate ketol-isomerase (deaminating)*	5.3.1.10
1-(4-Amino-2-methylpyrimid-5-ylmethyl)-3-(β-hydroxyethyl)-2-methylpyridinium-bromide aminohydrolase	3.5.4.20
2-Amino-4-hydroxypteridine aminohydrolase	3.5.4.11
*2-Amino-4-hydroxy-6-(*D*-erythro-1',2',3'-trihydroxypropyl)-7,8-dihydropteridine glycolaldehyde-lyase*	4.1.2.25
2-Amino-4-hydroxy-6-hydroxymethyldihydropteridine pyrophosphokinase	2.7.6.3
2-Amino-4-hydroxy-6-hydroxymethyl-7,8-dihydropteridine-diphosphate: 4-aminobenzoate 2-amino-4-hydroxydihydropteridine-6-methenyltransferase	2.5.1.15
(R)-3-Amino-2-methylpropionate·pyruvate aminotransferase	2.6.1.40
(*R*)-3-Amino-2-methylpropionate—pyruvate aminotransferase	2.6.1.40

Ammonia:(acceptor) oxidoreductase	1.7.99.1
Ammonia:ferredoxin oxidoreductase	1.7.7.1
Ammonia kinase	2.7.3.8
AMP aminase	3.5.4.6
AMP aminohydrolase	3.5.4.6
AMP deaminase	3.5.4.6
AMP nucleosidase	3.2.2.4
AMP phosphoribohydrolase	3.2.2.4
AMP:pyrophosphate phosphoribosyltransferase	2.4.2.7
AMP pyrophosphorylase	2.4.2.7
AMP,sulphite:(acceptor) oxidoreductase	1.8.99.2
Amygdalase	3.2.1.21
γ-Amylase	3.2.1.3
α-Amylase	3.2.1.1
β-Amylase	3.2.1.2
Amyloglucosidase	3.2.1.3
Amylopectin-1,6-glucosidase	3.2.1.9
Amylopectin 6-glucanohydrolase	3.2.1.41
Amylopectin 6-glucanohydrolase	3.2.1.69
Amylophosphorylase	2.4.1.1
Amylosucrase	2.4.1.4
Amylo-1,6-glucosidase	3.2.1.33
Ancrod	3.4.21.28
Androstene-3,17-dione hydroxylase	1.14.99.12
4-Androstene-3,17-dione monooxygenase	1.14.99.12
Androst-4-ene-3,17-dione, hydrogen-donor:oxygen (13-oxidoreductase, lactonizing)	1.14.99.12
Angiotensinase	3.4.99.3
Angiotensinase C	3.4.16.2
Angiotensin converting enzyme	3.4.15.1
Anserinase	3.4.13.5
Anthranilate hydroxylase	1.14.12.2
Anthranilate hydroxylase	1.14.12.1
Anthranilate,NAD(P)H:oxygen oxidoreductase (1,2-hydroxylating, deaminating, decarboxylating)	1.14.12.1
Anthranilate,NADPH:oxygen oxidoreductase (2,3-hydroxylating, deaminating)	1.14.12.2
Anthranilate phosphoribosyltransferase	2.4.2.18
Anthranilate synthase	4.1.3.27
Anthranilate,tetrahydropteridine:oxygen oxidoreductase (3-hydroxylating)	1.14.16.3
Anthranilate 1,2-dioxygenase (deaminating, decarboxylating)	1.14.12.1
Anthranilate 2,3-dioxygenase (deaminating)	1.14.12.2
Anthranilate 3-hydroxylase	1.14.16.3
Anthranilate 3-monooxygenase	1.14.16.3
AP I, II, III	3.4.11.12

D-Apiitol:NAD+ 1-oxidoreductase	1.1.1.114
D-Apiose reductase	1.1.1.114
APS-kinase	2.7.1.25
Apyrase	3.6.1.5
Aquacobalamin reductase	1.6.99.8
Aquacob(I)alamin adenosyltransferase	2.5.1.17
L-Arabinitol dehydrogenase	1.1.1.12
D-Arabinitol dehydrogenase	1.1.1.11
L-Arabinitol dehydrogenase (ribulose-forming)	1.1.1.13
L-Arabinitol:NAD+ 2-oxidoreductase (L-ribulose-forming)	1.1.1.13
D-Arabinitol:NAD+ 4-oxidoreductase	1.1.1.11
L-Arabinitol:NAD+ 4-oxidoreductase (L-xylulose-forming)	1.1.1.12
α-L-Arabinofuranosidase	3.2.1.55
α-L-Arabinofuranoside arabinofuranohydrolase	3.2.1.55
α-L-Arabinofuranoside hydrolase	3.2.1.79
Arabinogalactanase	3.2.1.90
Arabinogalactanase	3.2.1.89
Arabinogalactan 3-β-D-galactanohydrolase	3.2.1.90
Arabinogalactan 4-β-D-galactanohydrolase	3.2.1.89
L-Arabinokinase	2.7.1.46
D-Arabinokinase	2.7.1.54
L-Arabinonate dehydratase	4.2.1.25
Arabinonate dehydratase	4.2.1.5
L-Arabinonate hydro-lyase	4.2.1.25
D-Arabinonate hydro-lyase	4.2.1.5
Arabinonolactonase	3.1.1.15
D-Arabinonolactonase	3.1.1.30
D-Arabinono-γ-lactone lactonohydrolase	3.1.1.30
L-Arabinono-γ-lactone lactonohydrolase	3.1.1.15
D-Arabinose dehydrogenase	1.1.1.116
L-Arabinose dehydrogenase	1.1.1.46
D-Arabinose dehydrogenase (NAD(P)+)	1.1.1.117
L-Arabinose isomerase	5.3.1.4
Arabinose isomerase	5.3.1.3
L-Arabinose ketol-isomerase	5.3.1.4
D-Arabinose ketol-isomerase	5.3.1.3
D-Arabinose:NAD(P)+ 1-oxidoreductase	1.1.1.117
L-Arabinose:NAD+ 1-oxidoreductase	1.1.1.46
D-Arabinose:NAD+ 1-oxidoreductase	1.1.1.116
Arabinosephosphate isomerase	5.3.1.13
D-Arabinose-5-phosphate ketol-isomerase	5.3.1.13
β-L-Arabinosidase	3.2.1.88
Arabinosidase	3.2.1.55
β-L-Arabinoside arabinohydrolase	3.2.1.88
Arene monooxygenase (epoxidizing)	1.14.99.8
Arene-oxide hydratase	3.3.2.3

Arginase	3.5.3.1
D-Arginase	3.5.3.10
Arginine amidinase	3.5.3.1
L-*Arginine amidinohydrolase*	3.5.3.1
D-*Arginine amidinohydrolase*	3.5.3.10
Arginine aminopeptidase	3.4.11.6
L-*Arginine carboxy-lyase*	4.1.1.19
Arginine carboxypeptidase	3.4.17.3
Arginine decarboxylase	4.1.1.19
Arginine deiminase	3.5.3.6
Arginine dihydrolase	3.5.3.6
L-*Arginine:glycine amidinotransferase*	2.1.4.1
L-*Arginine iminohydrolase*	3.5.3.6
Arginine kinase	2.7.3.3
L-*Arginine:oxygen 2-oxidoreductase (decarboxylating)*	1.13.12.1
Arginine racemase	5.1.1.9
Arginine racemase	5.1.1.9
L-*Arginine:tRNA^{Arg} ligase (AMP-forming)*	6.1.1.19
L-*Arginine:1-amino-1-deoxy-*scyllo-*inositol-4-phosphate amidinotransferase*	2.1.4.2
Arginine 2-monooxygenase	1.13.12.1
Argininosuccinase	4.3.2.1
L-*Argininosuccinate arginine-lyase*	4.3.2.1
Argininosuccinate lyase	4.3.2.1
Argininosuccinate synthetase	6.3.4.5
L-*Arginyl (*L-*lysyl)-peptide hydrolase*	3.4.11.6
Arginyltransferase	2.3.2.8
L-*Arginyl-tRNA:protein arginyltransferase*	2.3.2.8
Arginyl-tRNA synthetase	6.1.1.19
Armillaria mellea neutral proteinase	3.4.99.32
Aromatic-amino-acid aminotransferase	2.6.1.57
Aromatic-L-*amino-acid carboxy-lyase*	4.1.1.28
Aromatic-L-amino-acid decarboxylase	4.1.1.28
Aromatic-amino-acid-glyoxylate aminotransferase	2.6.1.60
Aromatic-amino-acid:glyoxylate aminotransferase	2.6.1.60
Aromatic-amino-acid:2-oxoglutarate aminotransferase	2.6.1.57
Aromatic-hydroxylamine acetyltransferase	2.3.1.56
Arthrobacter serine proteinase	3.4.21.14
Aryl acylamidase	3.5.1.13
Aryl-acylamide amidohydrolase	3.5.1.13
Aryl-alcohol dehydrogenase	1.1.1.90
Aryl-alcohol dehydrogenase (NADP$^+$)	1.1.1.91
Aryl-alcohol:NAD$^+$ oxidoreductase	1.1.1.90
Aryl-alcohol:NADP$^+$ oxidoreductase	1.1.1.91
Aryl-alcohol oxidase	1.1.3.7

Aryl-alcohol:oxygen oxidoreductase	1.1.3.7
Aryl-aldehyde dehydrogenase	1.2.1.29
Aryl-aldehyde dehydrogenase (NADP$^+$)	1.2.1.30
Aryl-aldehyde:NAD$^+$ oxidoreductase	1.2.1.29
Aryl-aldehyde:NADP$^+$ oxidoreductase (ATP-forming)	1.2.1.30
Arylamine acetylase	2.3.1.5
Arylamine acetyltransferase	2.3.1.5
Arylamine sulphotransferase	2.8.2.3
Arylesterase	3.1.1.2
Aryl-ester hydrolase	3.1.1.2
Aryl-formylamine amidohydrolase	3.5.1.9
Arylsulphatase	3.1.6.1
Aryl-sulphate sulphohydrolase	3.1.6.1
Aryl sulphotransferase	2.8.2.1
Aryl 4-hydroxylase	1.14.14.1
Aryl 4-monooxygenase	1.14.14.1
Asclepain	3.4.22.7
Ascorbase	1.10.3.3
L-Ascorbate—cytochrome b_5 reductase	1.10.2.1
L-*Ascorbate:ferricytochrome b_5 oxidoreductase*	1.10.2.1
Ascorbate oxidase	1.10.3.3
L-*Ascorbate:oxygen oxidoreductase*	1.10.3.3
Ascorbate:oxygen 2,3-oxidoreductase (bond-cleaving)	1.13.11.13
Ascorbate 2,3-dioxygenase	1.13.11.13
Asparaginase	3.5.1.1
Asparaginase II	3.5.1.1
L-*Asparagine amidohydrolase*	3.5.1.1
L-*Asparagine hydro-lyase*	4.2.1.65
L-*Asparagine:hydroxylamine γ-aspartyltransferase*	2.3.2.7
Asparagine→hydroxylamine transaspartase	2.3.2.7
Asparagine—oxo-acid aminotransferase	2.6.1.14
Asparagine synthetase	6.3.1.1
Asparagine synthetase (ADP-forming)	6.3.1.4
Asparagine synthetase (glutamine-hydrolysing)	6.3.5.4
L-*Asparagine:tRNA ligase (AMP-forming)*	6.1.1.22
L-*Asparagine:2-oxo-acid aminotransferase*	2.6.1.14
Asparaginyl-tRNA synthetase	6.1.1.22
Asparagusate dehydrogenase	1.6.4.7
Asparagusate reductase (NADH)	1.6.4.7
Aspartase	4.3.1.1
Aspartate acetyltransferase	2.3.1.17
Aspartate aminopeptidase	3.4.11.7
D-Aspartate aminotransferase	2.6.1.21
Aspartate aminotransferase	2.6.1.1

L-*Aspartate:ammonia ligase (ADP-forming)*	6.3.1.4
L-*Aspartate:ammonia ligase (AMP-forming)*	6.3.1.1
L-*Aspartate ammonia-lyase*	4.3.1.1
Aspartate ammonia-lyase	4.3.1.1
Aspartate carbamoyltransferase	2.1.3.2
Aspartate carboxypeptidase	3.4.17.5
L-*Aspartate:L-glutamine amido-ligase (AMP-forming)*	6.3.5.4
Aspartate kinase	2.7.2.4
D-**Aspartate oxidase**	1.4.3.1
D-*Aspartate:oxygen oxidoreductase (deaminating)*	1.4.3.1
Aspartate racemase	5.1.1.13
Aspartate racemase	5.1.1.13
Aspartate-semialdehyde dehydrogenase	1.2.1.11
L-*Aspartate-β-semialdehyde hydro-lyase (adding pyruvate and cyclizing)*	4.2.1.52
L-*Aspartate-β-semialdehyde:NADP+ oxidoreductase (phosphorylating)*	1.2.1.11
Aspartate transcarbamylase	2.1.3.2
L-*Aspartate:tRNA^Asp ligase (AMP-forming)*	6.1.1.12
Aspartate 1-decarboxylase	4.1.1.11
L-*Aspartate:2-oxoglutarate aminotransferase*	2.6.1.1
L-*Aspartate 4-carboxy-lyase*	4.1.1.12
Aspartate 4-decarboxylase	4.1.1.12
Aspartic oxidase	1.4.3.1
Aspartoacylase	3.5.1.15
Aspartokinase	2.7.2.4
β-**Aspartylacetylglucosaminidase**	3.2.2.11
β-L-*Aspartyl-amino acid hydrolase*	3.4.13.10
β-**Aspartyldipeptidase**	3.4.13.10
Aspartylglucosyl-aminase	3.5.1.26
Aspartylglucosylamine deaspartylase	3.5.1.26
L-α-*Aspartyl (L-α-glutamyl)-peptide hydrolase*	3.4.11.7
4-L-*Aspartylglycosylamine amidohydrolase*	3.5.1.37
β-Aspartyl peptidase	3.4.13.10
Aspartyltransferase	2.3.2.7
Aspartyl-tRNA synthetase	6.1.1.12
1-β-*Aspartyl-2-acetamido-1,2-dideoxy-D-glucosylamine-L-asparagino-hydrolase*	3.2.2.11
Aspergillopeptidase A	3.4.23.6
Aspergillus **alkaline proteinase**	3.4.21.14
Aspergillus **Deoxyribonuclease K₁**	3.1.22.2
Aspergillus DNase K₂	3.1.21.2
Aspergillus DNase K₁	3.1.22.2

Aspergillus niger var. macrosporus **carboxyl proteinase**	3.4.23.6
Aspergillus nuclease S₁	3.1.30.1
Aspergillus oryzae **carboxyl proteinase**	3.4.23.6
Aspergillus oryzae **neutral proteinase**	3.4.24.4
Aspergillus oryzae ribonuclease	3.1.27.3
Aspergillus proteinase B	3.4.21.14
Aspergillus saitoi **carboxyl proteinase**	3.4.23.6
Aspergillus sojae DNase	3.1.21.2
Assimilatory nitrate reductase	1.6.6.2
Assimilatory nitrate reductase	1.6.6.1
ATP:acetate phosphotransferase	2.7.2.1
ATP:adenosine 5'-phosphotransferase	2.7.1.20
ATP:adenylylsulphate 3'-phosphotransferase	2.7.1.25
ATP:D-allose 6-phosphotransferase	2.7.1.55
ATP aminohydrolase	3.5.4.18
ATP:ammonia phosphotransferase	2.7.3.8
ATP:AMP phosphotransferase	2.7.4.3
ATP:L-arabinose 1 phosphotransferase	2.7.1.46
ATP:D-arabinose 5-phosphotransferase	2.7.1.54
ATP:L-arginine Nω-phosphotransferase	2.7.3.3
ATPase	3.6.1.8
ATP:L-aspartate 4-phosphotransferase	2.7.2.4
ATP:butyrate phosphotransferase	2.7.2.7
ATP:carbamate phosphotransferase	2.7.2.2
ATP:cellobiose 6-phosphotransferase	2.7.1.85
ATP:choline phosphotransferase	2.7.1.32
ATP:citrate oxaloacetate-lyase (pro-3S-CH₂COO^Δ→acetyl-CoA; ATP-dephosphorylating)	4.1.3.8
ATP citrate (pro-3S)-lyase	4.1.3.8
ATP:CMP phosphotransferase	2.7.4.14
ATP:cob(I)alamin Co-β-adenosyltransferase	2.5.1.17
ATP:creatine N-phosphotransferase	2.7.3.2
ATP:C-55-isoprenoid-alcohol phosphotransferase	2.7.1.66
ATP:(d)AMP phosphotransferase	2.7.4.11
ATP deaminase	3.5.4.18
ATP:deoxyadenosine 5'-phosphotransferase	2.7.1.76
ATP:deoxynucleosidemonophosphate phosphotransferase	2.7.4.13
ATP:dephospho-CoA 3'-phosphotransferase	2.7.1.24
ATP:(d)GMP phosphotransferase	2.7.4.8
ATP:dihydrostreptomycin-6-phosphate 3'α-phosphotransferase	2.7.1.88
ATP-diphosphatase	3.6.1.5
ATP diphosphohydrolase	3.6.1.5

ATP:(d)NMP phosphotransferase	2.7.4.12
ATP:dTMP phosphotransferase	2.7.4.9
ATP:erythritol 4-phosphotransferase	2.7.1.27
ATP:D-erythro-dihydrosphinganine 1-phosphotransferase	2.7.1.91
ATP:ethanolamine O-phosphotransferase	2.7.1.82
ATP:farnesyl-diphosphate phosphotransferase	2.7.4.18
ATP:FMN adenylyltransferase	2.7.7.2
ATP:formate phosphotransferase	2.7.2.6
ATP:D-fructose-1-phosphate 6-phosphotransferase	2.7.1.56
ATP:D-fructose 1-phosphotransferase	2.7.1.3
ATP:D-fructose-6-phosphate 1-phosphotransferase	2.7.1.11
ATP:D-fructose 6-phosphotransferase	2.7.1.4
ATP:L-fuculose 1-phosphotransferase	2.7.1.51
ATP:D-galactose 1-phosphotransferase	2.7.1.6
ATP:D-galacturonate 1-phosphotransferase	2.7.1.44
ATP:D-gluconate 6-phosphotransferase	2.7.1.12
ATP:α-D-glucose-1-phosphate adenylyltransferase	2.7.7.27
ATP:D-glucose-1-phosphate 6-phosphotransferase	2.7.1.10
ATP:D-glucose 6-phosphotransferase	2.7.1.2
ATP:D-glucuronate 1-phosphotransferase	2.7.1.43
ATP:[(L-glutamate:ammonia ligase (ADP-forming)] adenylyltransferase	2.7.7.42
ATP:L-glutamate γ-phosphotransferase	2.7.2.11
ATP:D-glyceraldehyde 3-phosphotransferase	2.7.1.28
ATP:D-glycerate 3-phosphotransferase	2.7.1.31
ATP:glycerol 3-phosphotransferase	2.7.1.30
ATP:guanidinoacetate N-phosphotransferase	2.7.3.1
ATP:guanidinoethyl-methyl-phosphate phosphotransferase	2.7.3.7
ATP:D-hexose 6-phosphotransferase	2.7.1.1
ATP:L-homoserine O-phosphotransferase	2.7.1.39
ATP:hydroxyacetone phosphotransferase	2.7.1.29
ATP:hypotaurocyamine Nω-phosphotransferase	2.7.3.6
ATP:inosine 5'-phosphotransferase	2.7.1.73
ATP:kanamycin 3'-O-phosphotransferase	2.7.1.95
ATP:lombricine Nω-phosphotransferase	2.7.3.5
ATP:mannitol 1-phosphotransferase	2.7.1.57
ATP:D-mannose 6-phosphotransferase	2.7.1.7
ATP:L-methionine S-adenosyltransferase	2.5.1.6
ATP:mevalonate 5-phosphotransferase	2.7.1.36
ATP:monoacylglycerol 3-phosphotransferase	2.7.1.94
ATP monophosphatase	3.6.1.3

ATP:myo-inositol 1-phosphotransferase	2.7.1.64
ATP:N-acetyl-L-glutamate 5-phosphotransferase	2.7.2.8
ATP:NADH 2'-phosphotransferase	2.7.1.86
ATP:NAD+ 2'-phosphotransferase	2.7.1.23
ATP:nicotinatemononucleotide adenylyltransferase	2.7.7.18
ATP:NMN adenylyltransferase	2.7.7.1
ATP:N-ribosylnicotinamide 5'-phosphotransferase	2.7.1.22
ATP:nucleosidediphosphate phosphotransferase	2.7.4.6
ATP:nucleosidemonophosphate phosphotransferase	2.7.4.4
ATP:nucleoside-5'-monophosphate pyrophosphotransferase	2.7.6.4
ATP:O-alkyldihydroxyacetone phosphotransferase	2.7.1.84
ATP:L(or D)-ribulose 5-phosphotransferase	2.7.1.16
ATP:oxaloacetate carboxy-lyase (transphosphorylating)	4.1.1.49
ATP:pantetheine-4'-phosphate adenylyltransferase	2.7.7.3
ATP:pantetheine 4'-phosphotransferase	2.7.1.34
ATP:pantothenate 4'-phosphotransferase	2.7.1.33
ATP:phosphatidylinositol 4-phosphate 5-phosphotransferase	2.7.1.68
ATP:phosphatidylinositol 4-phosphotransferase	2.7.1.67
ATP phosphohydrolase	3.6.1.3
ATP phosphoribosyltransferase	2.4.2.17
ATP:phosphorylase-b phosphotransferase	2.7.1.38
ATP:photo-bleached-rhodopsin phosphotransferase	2.7.1.97
ATP:polynucleotide adenylyltransferase	2.7.7.19
ATP:polyphosphate phosphotransferase	2.7.4.1
ATP:protamine O-phosphotransferase	2.7.1.70
ATP:protein phosphotransferase	2.7.1.37
ATP:pseudouridine 5'-phosphotransferase	2.7.1.83
ATP:pyridoxal 5-phosphotransferase	2.7.1.35
ATP pyrophosphatase	3.6.1.8
ATP pyrophosphate-lyase (cyclizing)	4.6.1.1
ATP pyrophosphohydrolase	3.6.1.8
ATP:[pyruvate dehydrogenase (lipoamide)] phosphokinase	2.7.1.99
ATP:pyruvate,orthophosphate phosphotransferase	2.7.9.1
ATP:pyruvate,water phosphotransferase	2.7.9.2
ATP:pyruvate 2-O-phosphotransferase	2.7.1.40
ATP:L-rhamnulose 1-phosphotransferase	2.7.1.5
ATP:riboflavin 5'-phosphotransferase	2.7.1.26

*ATP:*D*-ribose-5-phosphate pyrophosphotransferase*	2.7.6.1
*ATP:*D*-ribose-5-phosphate 1-phosphotransferase*	2.7.1.18
*ATP:*D*-ribose 5-phosphotransferase*	2.7.1.15
*ATP:*D*-ribulose-5-phosphate 1-phosphotransferase*	2.7.1.19
*ATP:*D*-ribulose 5-phosphotransferase*	2.7.1.47
ATP:sedoheptulose7-phosphotransferase	2.7.1.14
ATP:shikimate 5-phosphotransferase	2.7.1.71
ATP:streptomycin 3″-adenylyltransferase	2.7.7.47
ATP:streptomycin 3″-phosphotransferase	2.7.1.87
ATP:streptomycin 6-phosphotransferase	2.7.1.72
ATP:sulphate adenylyltransferase	2.7.7.4
ATP-sulphurylase	2.7.7.4
ATP:taurocyamine N$^\omega$-phosphotransferase	2.7.3.4
ATP:thiamin-diphosphate phosphotransferase	2.7.4.15
ATP:thiamin-monophosphate phosphotransferase	2.7.4.16
ATP:thiamin phosphotransferase	2.7.1.89
ATP:thiamin pyrophosphotransferase	2.7.6.2
ATP:thymidine5′-phosphotransferase	2.7.1.21
ATP:tRNA adenylyltransferase	2.7.7.25
ATP:uridine 5′-phosphotransferase	2.7.1.48
*ATP:*L*-xylulose 5-phosphotransferase*	2.7.1.53
*ATP:*D*-xylulose 5-phosphotransferase*	2.7.1.17
ATP:1-amino-1-deoxy-scyllo-inositol 4-phosphotransferase	2.7.1.65
ATP:1-O-alkyl-sn-glycerol 3-phosphotransferase	2.7.1.93
*ATP:2-acetamido-2-deoxy-*D*-glucose 6-phosphotransferase*	2.7.1.59
*ATP:2-acylamino-2-deoxy-*D*-mannose 6-phosphotransferase*	2.7.1.60
*ATP:2-amino-2-deoxy-*D*-glucose phosphotransferase*	2.7.1.8
ATP:2-amino-4-hydroxy-6-hydroxymethyl-7,8-dihydropteridine 6′-pyrophosphotransferase	2.7.6.3
*ATP:2-keto-*D*-gluconate 6-phosphotransferase*	2.7.1.13
*ATP:2-keto-3-deoxy-*D*-galactonate phosphotransferase*	2.7.1.58
*ATP:2-keto-3-deoxy-*D*-gluconate 6-phosphotransferase*	2.7.1.45
ATP:2-methyl-4-amino-5-hydroxymethyl-pyrimidine 5-phosphotransferase	2.7.1.49
ATP:2-methyl-4-amino-5-phosphomethyl-pyrimidine phosphotransferase	2.7.4.7
*ATP:3-phospho-*D*-glycerate 1-phosphotransferase*	2.7.2.3

ATP:4-methyl-5-(2′-hydroxyethyl)-thiazole 2′-phosphotransferase	2.7.1.50
ATP:5′-dephosphopolynucleotide 5′-phosphotransferase	2.7.1.78
ATP:5-diphosphomevalonate carboxy-lyase (dehydrating)	4.1.1.33
*ATP:5-keto-2-deoxy-*D*-gluconate 6-phosphotransferase*	2.7.1.92
ATP:5-phosphomevalonate phosphotransferase	2.7.4.2
*ATP:6-deoxy-*L*-galactose 1-phosphotransferase*	2.7.1.52
Atropine acylhydrolase	3.1.1.10
Azobenzene reductase	1.6.6.7
Azotobacter agilis RNase	3.1.27.2
Azotobacter nuclease	3.1.30.2
Bacillus amyloliquefaciens F (BamFI), (BamKI)	3.1.23.6
Bacillus amyloliquefaciens N (BamNI)	3.1.23.6
Bacillus macerans amylase	2.4.1.19
Bacillus stearothermophilus (BstI)	3.1.23.6
Bacillus subtilis: see also B. subtilis	
Bacillus subtilis (Bsul247)	3.1.23.31
Bacillus subtilis **neutral proteinase**	3.4.24.4
Bacillus subtilis X5 (BsuI)	3.1.23.17
Bacillus thermoproteolyticus **neutral proteinase**	3.4.24.4
Baker's yeast proteinase	3.4.22.9
Baker's yeast proteinase A	3.4.23.6
Barbiturase	3.5.2.1
Barbiturate amidohydrolase	3.5.2.1
B. atrox coagulant enzyme	3.4.21.29
Batroxobin	3.4.21.29
Benzaldehyde dehydrogenase	1.2.1.6
Benzaldehyde dehydrogenase (NAD⁺)	1.2.1.28
Benzaldehyde dehydrogenase (NADP⁺)	1.2.1.7
Benzaldehyde:NAD⁺ oxidoreductase	1.2.1.28
Benzaldehyde:NADP⁺ oxidoreductase	1.2.1.7
Benzamidase	3.5.1.14
1,2-Benzenediol:oxygen oxidoreductase	1.10.3.1
Benzenediol:oxygen oxidoreductase	1.10.3.2
Benzene hydroxylase	1.14.12.3
Benzene,NADH:oxygen 1,2-oxidoreductase	1.14.12.3
Benzene 1,2-dioxygenase	1.14.12.3
Benzoate hydroxylase	1.13.99.2
Benzoate,NADPH:oxygen oxidoreductase (4-hydroxylating)	1.14.13.12
Benzoate:oxygen oxidoreductase	1.13.99.2
Benzoate 1,2-dioxygenase	1.13.99.2
Benzoate 4-monooxygenase	1.14.13.12

Benzopyrene 3-monooxygenase	1.14.14.2
Benzopyrene 3-monooxygenase	1.14.14.1
Benzoylagmatine amidohydrolase	3.5.1.40
N-*Benzoylamino-acid amidohydrolase*	3.5.1.32
Benzoylcholinesterase	3.1.1.9
Benzoylcholinesterase	3.1.1.8
Benzoylformate carboxy-lyase	4.1.1.7
Benzoylformate decarboxylase	4.1.1.7
Benzyl-thiocyanate isomerase	5.99.1.1
B-esterase	3.1.1.1
Betaine-aldehyde dehydrogenase	1.2.1.8
Betaine-aldehyde:NAD+ oxidoreductase	1.2.1.8
Betaine—homocysteine methyltransferase	2.1.1.5
Betaine:L-homocysteine S-methyltransferase	2.1.1.5
Bile-salt sulphotransferase	2.8.2.14
Bilirubin-glucuronoside:bilirubin-glucuronoside glucuronosyltransferase	2.4.1.95
Bilirubin-glucuronoside glucuronosyltransferase	2.4.1.95
Bilirubin monoglucuronide transglucuronidase	2.4.1.95
Bilirubin:NAD(P)+ oxidoreductase	1.3.1.24
Biliverdin reductase	1.3.1.24
Biotin-[acetyl-CoA carboxylase] synthetase	6.3.4.15
Biotin-amide amidohydrolase	3.5.1.12
Biotin:apo-[acetyl-CoA:carbon-dioxide ligase (ADP-forming)] ligase (AMP-forming)	6.3.4.15
Biotin:apo-[methylmalonyl-CoA:pyruvate carboxyltransferase] ligase (AMP-forming)	6.3.4.9
Biotin:apo-[propionyl-CoA:carbon-dioxide ligase (ADP-forming)] ligase (AMP-forming)	6.3.4.10
Biotin:apo-[3-methylcrotonoyl-CoA: carbon-dioxide ligase (ADP-forming)] ligase (AMP-forming)	6.3.4.11
Biotin carboxylase	6.3.4.14
Biotin-carboxyl-carrier-protein: carbon-dioxide ligase (ADP-forming)	6.3.4.14
Biotin:CoA ligase (AMP-forming)	6.2.1.11
Biotinidase	3.5.1.12
Biotin—[methylcrotonoyl-CoA-carboxylase] synthetase	6.3.4.11
Biotin—[methylmalonyl-CoA-carboxyltransferase] synthetase	6.3.4.9
Biotin—[propionyl-CoA-carboxylase (ATP-hydrolysing)] synthetase	6.3.4.10
Biotinyl-CoA synthetase	6.2.1.11
Bis(5′-Adenosyl) triphosphatase	3.6.1.29
P¹,P³-Bis(5′-Adenosyl)-triphosphate adenylohydrolase	3.6.1.29
Bisphosphoglycerate phosphatase	3.1.3.13

Bis(5′-Guanosyl) tetraphosphatase	3.6.1.17
P¹,P⁴-Bis(5′-Guanosyl)-tetraphosphate guanylohydrolase	3.6.1.17
2,3-Bisphospho-D-glycerate:2-phospho-D-glycerate phosphotransferase	2.7.5.3
2,3-Bisphospho-D-glycerate 2-phosphohydrolase	3.1.3.13
Bisphosphoglyceromutase	2.7.5.4
Bisphosphoglyceromutase	5.4.2.1
Blasticidin-S aminohydrolase	3.5.4.23
Blasticidin-S deaminase	3.5.4.23
Blood-group-substance α-D-galactosyltransferase	2.4.1.37
Bordetella bronchiseptica (BbrI)	3.1.23.21
Bothrops atrox **serine proteinase**	3.4.21.29
Bovine adrenal cortex RNase	3.1.26.1
Branched-chain-amino-acid aminotransferase	2.6.1.42
Branched-chain-amino-acid:2-oxoglutarate aminotransferase	2.6.1.42
Branched-chain α-keto acid dehydrogenase	1.2.4.4
Branching enzyme	2.4.1.18
Brevibacterium luteum (BluI)	3.1.23.42
Brevibacterium luteum (BluII)	3.1.23.17
Brewer's yeast proteinase	3.4.22.9
Bromelain	3.4.22.4
B. subtilis endonuclease	3.1.21.2
B. subtilis nuclease	3.1.16.1
B. subtilis RNase	3.1.27.3
Bunolol reductase	1.1.1.160
L(+)-Butanediol dehydrogenase	1.1.1.76
D(−)-Butanediol dehydrogenase	1.1.1.4
D(−)-2,3-*Butanediol:NAD+ oxidoreductase*	1.1.1.4
L(+)-2,3-*Butanediol:NAD+ oxidoreductase*	1.1.1.76
DL-5-[(tert-Butylamino)-2′-hydroxypropoxy]-1,2,3,4-tetrahydro-1-naphthol: NADP+ oxidoreductase	1.1.1.160
Butyleneglycol dehydrogenase	1.1.1.4
2-Butyne-1,4-diol:NAD+ 1-oxidoreductase	1.1.1.165
Butyrate:CoA ligase (AMP-forming)	6.2.1.2
Butyrate CoA-transferase	2.8.3.4
Butyrate kinase	2.7.2.7
γ-Butyrobetaine,2-oxoglutarate dioxygenase	1.14.11.1
Butyrylcholine esterase	3.1.1.8
Butyryl-CoA:(acceptor) oxidoreductase	1.3.99.2
Butyryl-CoA dehydrogenase	1.3.99.2
Butyryl-CoA:orthophosphate butyryltransferase	2.3.1.19
Butyryl-CoA synthetase	6.2.1.2
Butyryl dehydrogenase	1.3.99.2

Caffeate 3,4-dioxygenase	1.13.11.22
Calf thymus RNase	3.1.26.3
Callose synthetase	2.4.1.34
Camphor,reduced-putida-ferredoxin:oxygen oxidoreductase (5-hydroxylating)	1.14.15.1
Camphor,reduced-rubredoxin:oxygen oxidoreductase (1,2-lactonizing)	1.14.15.2
Camphor 1,2-monooxygenase	1.14.15.2
Camphor 5-monooxygenase	1.14.15.1
Canavanase	3.5.3.1
Candida albicans **carboxyl proteinase**	3.4.23.6
Candida lipolytica **serine proteinase**	3.4.21.14
CAP	3.4.11.13
Capsular-polysaccharide galactohydrolase	3.2.1.87
Carbamate kinase	2.7.2.2
N-*Carbamoyl β alanine amidohydrolase*	3.5.1.6
N-*Carbamoyl-L-aspartate amidohydrolase*	3.5.1.7
Carbamoylaspartate decarboxylase	4.1.1.13
Carbamoylaspartic dehydrase	3.5.2.3
Carbamoylphosphate:L aspartate carbamoyltransferase	2.1.3.2
Carbamoylphosphate:L-ornithine carbamoyltransferase	2.1.3.3
Carbamoylphosphate:oxamate carbamoyltransferase	2.1.3.5
Carbamoylphosphate:putrescine carbamoyltransferase	2.1.3.6
Carbamoyl-phosphate synthase (ammonia) ·	6.3.4.16
Carbamoyl-phosphate synthase (glutamine)	6.3.5.5
Carbamoyl-phosphate synthetase (ammonia)	6.3.4.16
Carbamoyl-phosphate synthetase (glutamine-hydrolysing)	6.3.5.5
Carbamoylserine ammonia-lyase	4.3.1.13
O-Carbamoyl-L-serine ammonia-lyase (pyruvate-forming)	4.3.1.13
O-Carbamoyl-L-serine deaminase	4.3.1.13
Carbamylaspartotranskinase	2.1.3.2
Carbonate dehydratase	4.2.1.1
Carbonate hydro-lyase	4.2.1.1
Carbon-dioxide:ammonia ligase (ADP-forming, carbamate-phosphorylating)	6.3.4.16
Carbon-dioxide:L-glutamine amido-ligase (ADP-forming, carbamate-phosphorylating)	6.3.5.5
Carbonic anhydrase	4.2.1.1ˉ
Carboxyamidase	3.4.15.2
Carboxycathepsin	3.4.15.i
Carboxy-*cis-cis*-muconate cyclase	5.5.1.5
3-Carboxy-*cis-cis*-muconate cycloisomerase	5.5.1.2

Carboxydismutase	4.1.1.39
3-Carboxyethylcatechol 2,3-dioxygenase	1.13.11.16
N²-(1-Carboxyethyl)-L-arginine:NAD⁺ oxidoreductase (L-arginine-forming)	1.5.1.11
N²-(1-Carboxyethyl)-L-lysine:NADP⁺ oxidoreductase (L-lysine forming)	1.5.1.16
α-Carboxylase	4.1.1.1
Carboxylesterase	3.1.1.1
Carboxylic-ester hydrolase	3.1.1.1
4-Carboxymethylbut-3-enolide(1,4)enol-lactonohydrolase	3.1.1.24
L-*5-Carboxymethylhydantoin amidohydrolase*	3.5.2.4
Carboxymethylhydantoinase	3.5.2.4
Carboxymethyloxysuccinate glycollate-lyase	4.2.99.12
Carboxymethyloxysuccinate lyase	4.2.99.12
4-Carboxymethyl-4-hydroxyisocrotonolactone lyase (decyclizing)	5.5.1.1
4-Carboxymuconolactone carboxy-lyase	4.1.1.44
4-Carboxymuconolactone decarboxylase	4.1.1.44
4-Carboxymuconolactone lyase (decyclizing)	5.5.1.2
3-Carboxymuconolactone lyase (decyclizing)	5.5.1.5
Carboxypeptidase A	3.4.17.1
Carboxypeptidase B	3.4.17.2
'Carboxypeptidase G'	3.4.22.12
Carboxypeptidase N	3.4.17.3
1-(2'-Carboxyphenylamino)-1-deoxyribulose-5-phosphate carboxy-lyase (cyclizing)	4.1.1.48
Carboxypolypeptidase	3.4.17.1
Carnitine acetyltransferase	2.3.1.7
Carnitine carboxy-lyase	4.1.1.42
Carnitine decarboxylase	4.1.1.42
Carnitine dehydrogenase	1.1.1.108
Carnitine:NAD⁺ oxidoreductase	1.1.1.108
Carnitine palmitoyltransferase	2.3.1.21
Carnosinase	3.4.13.3
Carnosine *N*-methyltransferase	2.1.1.22
Carnosine synthetase	6.3.2.11
Carotene oxidase	1.13.11.12
β-Carotene:oxygen 15,15'-oxidoreductase (bond-cleaving)	1.13.11.21
β-Carotene 15,15'-dioxygenase	1.13.11.21
κ-Carrageenanase	3.2.1.83
κ-Carrageenan 4-β-D-glycanohydrolase	3.2.1.83
Catalase	1.11.1.6
Catechol methyltransferase	2.1.1.6
Catechol oxidase	1.10.3.1
Catechol oxidase (dimerizing)	1.1.3.14

Catechol:oxygen oxidoreductase (dimerizing)	1.1.3.14
Catechol:oxygen 1,2-oxidoreductase (decyclizing)	1.13.11.1
Catechol:oxygen 2,3-oxidoreductase (decyclizing)	1.13.11.2
Catechol 1,2-dioxygenase	1.13.11.1
Catechol 2,3-dioxygenase	1.13.11.2
Cathepsin B	3.4.22.1
Cathepsin B₁	3.4.22.1
Cathepsin C	3.4.14.1
Cathepsin D	3.4.23.5
Cathepsin G	3.4.21.20
Cathepsin L	3.4.22.15
CDPabequose epimerase	5.1.3.10
CDPabequose:D-mannosyl-rhamnosyl-galactose-1-diphospholipid:abequosyltransferase	2.4.1.60
CDPabequose 2-epimerase	5.1.3.10
CDPcholine:N-acylsphingosine cholinephosphotransferase	2.7.8.3
CDPcholine:sphingosine cholinephosphotransferase	2.7.8.10
CDPcholine:1,2-diacylglycerol cholinephosphotransferase	2.7.8.2
CDPdiacylglycerol—inositol 3-phosphatidyltransferase	2.7.8.11
CDPdiacylglycerol:myo-inositol 3-phosphatidyltransferase	2.7.8.11
CDPdiacylglycerol phosphatidylhydrolase	3.6.1.26
CDPdiacylglycerol pyrophosphatase	3.6.1.26
CDPdiacylglycerol:L-serine O-phosphatidyltransferase	2.7.8.8
CDPdiacylglycerol:sn-glycerol-3-phosphate phosphatidyltransferase	2.7.8.5
CDP-diglyceride—inositol phosphatidyltransferase	2.7.8.11
CDPdiglyceride pyrophosphorylase	2.7.7.41
CDPdiglyceride—serine O-phosphatidyl-transferase	2.7.8.8
CDPethanolamine:L-serine ethanolaminephosphotransferase	2.7.8.4
CDPethanolamine:1,2-diacylglycerol ethanolaminephosphotransferase	2.7.8.1
CDPglucose pyrophosphorylase	2.7.7.33
CDPglucose 4,6-dehydratase	4.2.1.45
CDPglucose 4,6-hydro-lyase	4.2.1.45
CDPglycerol glycerophosphotransferase	2.7.8.12
CDPglycerol phosphoglycerylhydrolase	3.6.1.16
CDPglycerol:poly (glycerophosphate) glycerophosphotransferase	2.7.8.12
CDPglycerol pyrophosphatase	3.6.1.16

CDPglycerol pyrophosphorylase	2.7.7.39
CDPparatose epimerase	5.1.3.10
CDPribitol pyrophosphorylase	2.7.7.40
CDPribitol:teichoic-acid phosphoribitoltransferase	2.4.1.55
CDP-4-keto-3,6-dideoxy-D-glucose:NAD(P)+ 3-oxidoreductase	1.17.1.1
CDP-4-keto-6-deoxy-D-glucose reductase	1.17.1.1
Cellobiase	3.2.1.21
Cellobiose epimerase	5.1.3.11
Cellobiose:orthophosphate α-D-glucosyltransferase	2.4.1.20
Cellobiose phosphorylase	2.4.1.20
Cellobiose 2-epimerase	5.1.3.11
Cellodextrin phosphorylase	2.4.1.49
Cellulase	3.2.1.4
Cellulose polysulphatase	3.1.6.7
Cellulose-sulphate sulphohydrolase	3.1.6.7
Cellulose synthase (GDP-forming)	2.4.1.29
Cellulose synthase (UDP-forming)	2.4.1.12
C₁ enzyme	3.2.1.91
Cephalosporinase	3.5.2.6
Cephalosporin-C acetylhydrolase	3.1.1.41
Cephalosporin-C deacetylase	3.1.1.41
Ceramide cholinephosphotransferase	2.7.8.3
Cerebroside-sulphatase	3.1.6.8
Cerebroside-3-sulphate 3-sulphohydrolase	3.1.6.8
C-esterase	3.1.1.6
Chalaropsis RNase	3.1.27.3
Chalcone isomerase	5.5.1.6
Chitin amidohydrolase	3.5.1.41
Chitinase	3.2.1.14
Chitin deacetylase	3.5.1.41
Chitin synthase	2.4.1.16
Chitin—UDP acetylglucosaminyltransferase	2.4.1.16
Chitobiase	3.2.1.29
Chitodextrinase	3.2.1.14
Chlamydomonas nuclease	3.1.31.1
Chloramphenicol acetyltransferase	2.3.1.28
Chlorate reductase	1.97.1.1
Chlorella DNase	3.1.21.2
Chloride:hydrogen-peroxide oxidoreductase	1.11.1.10
Chloride peroxidase	1.11.1.10
Chlorite:acceptor oxidoreductase	1.97.1.1
Chlorophyllase	3.1.1.14
Chlorophyll chlorophyllidohydrolase	3.1.1.14
Cholate:CoA ligase (AMP-forming)	6.2.1.7
Cholate thiokinase	6.2.1.7

Cholestanetetraol 26-dehydrogenase	1.1.1.161
Cholestanetriol-26-al 26-dehydrogenase	1.2.1.40
Cholestanetriol 26-monooxygenase	1.14.13.15
5β-Cholestane-3α,7α,12α-triol, NADPH:oxygen oxidoreductase (26-hydroxylating)	1.14.13.15
5β-Cholestane-3α,7α,12α-triol-26-al:NAD+ oxidoreductase	1.2.1.40
5β-Cholestane-3α,7α,12α,26-tetraol:NAD+ 26-oxidoreductase	1.1.1.161
Cholestenol Δ-isomerase	5.3.3.5
Δ7-Cholestenol Δ7—Δ8-isomerase	5.3.3.5
Cholestenone 5α-reductase	1.3.1.22
Cholestenone 5β-reductase	1.3.1.23
Cholesterol acyltransferase	2.3.1.26
Cholesterol esterase	3.1.1.13
Cholesterol,NADPH:oxygen oxidoreductase (7α-hydroxylating)	1.14.13.17
Cholesterol:NADP+ Δ7-oxidoreductase	1.3.1.21
Cholesterol oxidase	1.1.3.6
Cholesterol:oxygen oxidoreductase	1.1.3.6
Cholesterol 20-hydroxylase	1.14.1.9
Cholesterol 7α-monooxygenase	1.14.13.17
5α-Cholest-7-en-3β-ol:oxygen Δ5-oxidoreductase	1.3.3.2
Choline:(acceptor) oxidoreductase	1.1.99.1
Choline acetylase	2.3.1.6
Choline acetyltransferase	2.3.1.6
Choline dehydrogenase	1.1.99.1
Choline esterase I	3.1.1.7
Choline esterase II (unspecific)	3.1.1.8
Choline kinase	2.7.1.32
Choline oxidase	1.1.3.17
Choline:oxygen 1-oxidoreductase	1.1.3.17
Choline phosphatase	3.1.4.4
Cholinephosphate cytidylyltransferase	2.7.7.15
Cholinephosphotransferase	2.7.8.2
Cholinesterase	3.1.1.7
Cholinesterase	3.1.1.8
Cholinesulphatase	3.1.6.6.
Choline-sulphate sulphohydrolase	3.1.6.6.
Choline sulphotransferase	2.8.2.6
Choloyl-CoA synthetase	6.2.1.7
Choloylglycine hydrolase	3.5.1.24
Chondroitin ABC eliminase	4.2.2.4
Chondroitin ABC lyase	4.2.2.4
Chondroitin ABC lyase	4.2.2.4
Chondroitin AC eliminase	4.2.2.5
Chondroitin AC lyase	4.2.2.5
Chondroitin AC lyase	4.2.2.5

Chondroitinase	3.2.1.34
Chondroitinase	4.2.2.4
Chondroitinase	4.2.2.5
Chondroitinase	3.1.6.4
Chondroitinsulphatase	3.1.6.4
Chondroitin sulphate lyase	4.2.99.6
Chondroitin sulphate lyase	4.2.2.5
Chondroitin-sulphate sulphohydrolase	3.1.6.4
Chondroitin sulphotransferase	2.8.2.5
Chondro-4-sulphatase	3.1.6.9
Chondro-6-sulphatase	3.1.6.10
Chorismate mutase	5.4.99.5
Chorismate pyruvate-lyase (amino-accepting)	4.1.3.27
Chorismate pyruvatemutase	5.4.99.5
Chorismate synthase	4.6.1.4
Chymase	3.4.23.4
Chymopapain	3.4.22.6
Chymosin	3.4.23.4
Chymotrypsin	3.4.21.1
Chymotrypsin A and B	3.4.21.1
Chymotrypsin C	3.4.21.2
Cinnamaldehyde:NADP+ oxidoreductase (CoA-cinnamoylating)	1.2.1.44
trans-Cinnamate,NADPH:oxygen oxidoreductase (2-hydroxylating)	1.14.13.14
trans-Cinnamate,NADPH:oxygen oxidoreductase (4-hydroxylating)	1.14.13.11
trans-Cinnamate 2-monooxygenase	1.14.13.14
trans-Cinnamate 4-monooxygenase	1.14.13.11
Cinnamic acid 2-hydroxylase	1.14.13.14
Cinnamoyl-CoA reductase	1.2.1.44
Citraconate hydratase	4.2.1.35
Citramalate CoA-transferase	2.8.3.7
(+)-Citramalate hydro-lyase	4.2.1.34
(−)-Citramalate hydro-lyase	4.2.1.35
Citramalate lyase	4.1.3.22
Citramalate pyruvate-lyase	4.1.3.22
Citramalyl-CoA hydro-lyase	4.2.1.56
Citramalyl-CoA lyase	4.1.3.25
Citramalyl-CoA pyruvate-lyase	4.1.3.25
Citrase	4.1.3.6
Citratase	4.1.3.6
Citrate aldolase	4.1.3.6
Citrate cleavage enzyme	4.1.3.8
Citrate condensing enzyme	4.1.3.7
Citrate dehydratase	4.2.1.4
Citrate hydro-lyase	4.2.1.4
Citrate (isocitrate) hydro-lyase	4.2.1.3

Citrate oxaloacetate-lyase (pro-3R-CH$_2$COO$^-$→acetyl-CoA)	4.1.3.28
Citrate oxaloacetate-lyase (pro-3S-CH$_2$COO$^-$→acetate)	4.1.3.6
Citrate oxaloacetate-lyase (pro-3S-CH$_2$COO$^-$→acetyl-CoA)	4.1.3.7
Citrate (*pro-3S*)-lyase	4.1.3.6
Citrate (*re*)-synthase	4.1.3.28
Citrate (*si*)-synthase	4.1.3.7
Citridesmolase	4.1.3.6
Citritase	4.1.3.6
Citrogenase	4.1.3.7
Citrullinase	3.5.1.20
L-Citrulline:L-aspartate ligase (AMP-forming)	6.3.4.5
L-Citrulline N^5-carbamoyldihydrolase	3.5.1.20
Citrulline phosphorylase	2.1.3.3
Clearing factor lipase	3.1.1.34
Clostridiopeptidase A	3.4.24.3
Clostridiopeptidase B	3.4.22.8
Clostridium histolyticum **aminopeptidase**	3.4.11.13
Clostridium histolyticum **collagenase**	3.4.24.3
Clostridium histolyticum proteinase B′	3.4.22.7
Clostridium histolyticum proteinase B	3.4.22.8
Clostridium oedematiens β- and *γ*-toxins	3.1.4.3
Clostridium welchii α-toxin	3.1.4.3
Clostripain	3.4.22.8
dCMP aminohydrolase	3.5.4.12
dCMP deaminase	3.5.4.12
CMP-N-acetylneuraminate:D-galactosyl-glycoprotein N-acetylneuraminyltransferase	2.4.99.1
CMP-*N*-acetylneuraminate—galactosyl-glycoprotein sialyltransferase	2.4.99.1
CMP-N-acetylneuraminate:D-galactosyl-2-acetamido-2-deoxy-D-galactosyl-(N-acetylneuraminyl)-D-galactosyl-D-glucosylceramide N-acetylneuraminyltransferase	
CMP-*N*-acetylneuraminate—monosialoganglioside sialyltransferase	2.4.99.2
CMP-N-acylneuraminate N-acylneuraminohydrolase	3.1.4.40
CMP-*N*-acylneuraminate phosphodiesterase	3.1.4.40
CMP-sialate hydrolase	3.1.4.40
CMPsialate pyrophosphorylase	2.7.7.43
CMPsialate synthase	2.7.7.43
CMP-3-deoxy-D-*manno*-octulosonate pyrophosphorylase	2.7.7.38
CoA:apo-[acyl-carrier-protein] pantetheinephosphotransferase	2.7.8.7
Coagulant protein of Russell's viper venom	3.4.21.23
Coagulation Factor IXa	3.4.21.22
Coagulation factor VIIa (bovine)	3.4.21.21
Coagulation Factor Xa	3.4.21.6
Coagulation Factor XIa	3.4.21.27
CoAS—Sglutathione reductase (NADPH)	1.6.4.6
Cob (I) alamin adenosyltransferase	2.5.1.17
Cob (II) alamin reductase	1.6.99.9
Cocain esterase	3.1.1.1
Coccus P (proteinase)	3.4.24.4
Coenzyme A:oxidized-glutathione oxidoreductase	1.8.4.3
Coeruloplasmin	1.16.3.1
Colicin E2 and E3	3.1.21.1
Collagenase A	3.4.24.3
Collagenase I	3.4.24.3
Condensing enzyme	4.1.3.7
Conjugase	3.4.22.12
Coproporphyrinogenase	1.3.3.3
Coproporphyrinogen oxidase	1.3.3.3
Coproporphyrinogen:oxygen oxidoreductase (decarboxylating)	1.3.3.3
Correndonuclease II	3.1.25.1
Corticosterone,reduced-adrenal-ferredoxin:oxygen oxidoreductase (18-hydroxylating)	1.14.15.5
Corticosterone 18-hydroxylase	1.14.15.5
Corticosterone 18-monooxygenase	1.14.15.5
Cortisol acetyltransferase	2.3.1.27
Cortisone α-reductase	1.3.1.4
Cortisone reductase	1.1.1.53
Cortisone β-reductase	1.3.1.3
Corynebacterium humiferum (ChuI)	3.1.23.21
4-Coumarate:CoA ligase (AMP-forming)	6.2.1.12
p-Coumarate hydroxylase	1.14.17.2
p-Coumarate 3-monooxygenase	1.14.17.2
4-Coumaroyl-CoA synthetase	6.2.1.12
Crayfish low-molecular-weight proteinase	3.4.99.6
Creatinase	3.5.3.3
Creatine amidinohydrolase	3.5.3.3
Creatine kinase	2.7.3.2
Creatininase	3.5.2.10
Creatinine amidohydrolase	3.5.2.10
Creatinine deiminase	3.5.4.21
Creatinine iminohydrolase	3.5.4.21
Cresolase	1.14.18.1
Crotalase	3.4.21.30
Crotalus adamanteus **serine proteinase**	3.4.21.30

Crotalus atrox **metalloproteinase**	3.4.24.1
Crotonase	4.2.1.17
Crotonoyl-[acyl-carrier-protein] hydratase	4.2.1.58
dCTP aminohydrolase	3.5.4.13
dCTPase	3.6.1.12
CTP:cholinephosphate cytidylyltransferase	2.7.7.15
dCTP deaminase	3.5.4.13
CTP:ethanolaminephosphate cytidylyltransferase	2.7.7.14
CTP:D-glucose-1-phosphate cytidylyltransferase	2.7.7.33
CTP:N-acylneuraminate cytidylyltransferase	2.7.7.43
dCTP nucleotidohydrolase	3.6.1.12
CTP:phosphatidate cytidylyltransferase	2.7.7.41
CTP:D-ribitol-5-phosphate cytidylyltransferase	2.7.7.40
CTP:sn-glycerol-3-phosphate cytidylyltransferase	2.7.7.39
CTP synthetase	6.3.4.2
CTP:tRNA cytidylyltransferase	2.7.7.21
CTP:3-deoxy-D-manno-octulosonate cytidylyltransferase	2.7.7.38
Cucumisin	3.4.21.25
Cucurbitacin Δ²³-reductase	1.3.1.5
Cyanamide hydratase	4.2.1.69
Cyanase	3.5.5.3
Cyanate aminohydrolase	3.5.5.3
Cyanate hydrolase	3.5.5.3
Cyanide hydratase	4.2.1.66
3-Cyanoalanine hydratase	4.2.1.65
β-Cyanoalanine synthase	4.4.1.9
Cyanohydrin glucosyltransferase	2.4.1.85
Cyclamate sulphamatase	3.10.1.2
Cyclamate sulphamidase	3.10.1.2
3′:5′-Cyclic-GMP phosphodiesterase	3.1.4.35
3′:5′-Cyclic-GMP 5′-nucleotidohydrolase	3.1.4.35
1:2-Cyclic-inositol-monophosphate phosphodiesterase	3.1.4.36
3′:5′-Cyclic-nucleotide phosphodiesterase	3.1.4.17
2′:3′-Cyclic-nucleotide 2′-phosphodiesterase	3.1.4.16
2′:3′-Cyclic-nucleotide 3′-phosphodiesterase	3.1.4.37
3′:5′-Cyclic-nucleotide 5′-nucleotidohydrolase	3.1.4.17
Cyclodextrin glucanotransferase	2.4.1.19
Cycloheptaglucanase	3.2.1.12
3,5-Cyclohexadiene-1,2-diol-1-carboxylate dehydrogenase	1.3.1.25
3,5-Cyclohexadiene-1,2-diol-1-carboxylate: NAD⁺ oxidoreductase (decarboxylating)	1.3.1.25
Cyclohexaglucanase	3.2.1.13
Cyclohexan-1,2-diol dehydrogenase	1.1.1.174
trans-Cyclohexan-1,2-diol:NAD⁺ oxidoreductase	1.1.1.174
Cyclohexylamine oxidase	1.4.3.12
Cyclohexylamine:oxygen oxidoreductase (deaminating)	1.4.3.12
Cyclohexylsulphamate sulphamidase	3.10.1.2
Cyclomaltodextrinase	3.2.1.54
Cyclomaltodextrin dextrin-hydrolase (decyclizing)	3.2.1.54
Cyclomaltodextrin glucanotranferase	2.4.1.19
Cyclopentanol dehydrogenase	1.1.1.163
Cyclopentanol:NAD⁺ oxidoreductase	1.1.1.163
Cyclopentanone monooxygenase	1.14.13.16
Cyclopentanone,NADPH:oxygen oxidoreductase (5-hydroxylating, lactonizing)	1.14.13.16
β-Cyclopiazonate:(acceptor) oxidoreductase (cyclizing)	1.3.99.9
β-Cyclopiazonate dehydrogenase	1.3.99.9
β-Cyclopiazonate oxidocyclase	1.3.99.9
Cypridina luciferin:oxygen 2-oxidoreductase (decarboxylating)	1.13.12.6
Cypridina **luciferin 2-monooxygenase**	1.13.12.6
β-Cystathionase	4.4.1.8
Cystathionase	4.4.1.1
L-Cystathionine cysteine-lyase (deaminating)	4.4.1.1
Cystathionine L-homocysteine-lyase (deaminating)	4.4.1.8
Cystathionine β-lyase	4.4.1.8
Cystathionine γ-lyase	4.4.1.1
Cystathionine β-synthase	4.2.1.21
Cystathionine γ-synthase	4.2.99.9
Cystathionine β-synthase	4.2.1.22
Cysteamine dehydrogenase	1.8.1.1
Cysteamine dioxygenase	1.13.11.19
Cysteamine:oxygen oxidoreductase	1.13.11.19
Cysteine aminotransferase	2.6.1.3
Cysteine desulphhydrase, γ-Cystathionase	4.4.1.1
Cysteine dioxygenase	1.13.11.20
L-Cysteine hydrogen-sulphide-lyase (adding HCN)	4.4.1.9
L-Cysteine hydrogen-sulphide-lyase (adding sulphite)	4.4.1.10
Cysteine lyase	4.4.1.10
L-Cysteine:oxygen oxidoreductase	1.13.11.20
L-Cysteine-sulphinate carboxy-lyase	4.1.1.29
Cysteine sulphinate decarboxylase	4.1.1.29
Cysteine synthase	4.2.99.8

L-*Cysteine-tRNA^{Cys} ligase (AMP-forming)*	6.1.1.16
L-*Cysteine:2-oxoglutarate aminotransferase*	2.6.1.3
Cysteinyl-glycine dipeptidase	3.4.13.6
L-*Cysteinyl-glycine hydrolase*	3.4.13.6
Cysteinyl-tRNA synthetase	6.1.1.16
Cystine desulphhydrase	4.4.1.1
Cystine reductase (NADH)	1.6.4.1
Cystyl aminopeptidase	3.4.11.3
Cytidine aminohydrolase	3.5.4.5
Cytidine deaminase	3.5.4.5
Cytidylate kinase	2.7.4.14
Cytochrome a_3	1.9.3.1
Cytochrome b_5 reductase	1.6.2.2
Cytochrome *cd*	1.9.3.2
Cytochrome c_3 hydrogenase	1.12.2.1
Cytochrome *c* oxidase	1.9.3.1
Cytochrome *c* reductase	1.6.99.3
Cytochrome oxidase	1.9.3.1
Cytochrome peroxidase	1.11.1.5
Cytochrome reductase (NADPH)	1.6.2.3
Cytosine aminohydrolase	3.5.4.1
Cytosine deaminase	3.5.4.1
DDT-dehydrochlorinase	4.5.1.1
Deacetyl-[citrate-(*pro-3S*)-lyase] acetyltransferase	2.3.1.49
Deamido-NAD^{+}:ammonia ligase (AMP-forming)	6.3.1.5
Deamido-NAD^{+}:L-glutamine amido-ligase (AMP-forming)	6.3.5.1
Deamido-NAD^{+} pyrophosphorylase	2.7.7.18
Debranching enzyme	3.2.1.68
Debranching enzyme	3.2.1.41
Decapase	3.6.1.30
Decylcitrate synthase	4.1.3.23
Decylhomocitrate synthase	4.1.3.29
7-Dehydrocholesterol reductase	1.3.1.21
Dehydropeptidase II	3.5.1.14
3-Dehydroquinate dehydratase	4.2.1.10
3-Dehydroquinate hydro-lyase	4.2.1.10
3-Dehydroquinate synthase	4.6.1.3
3-Dehydrosphinganine reductase	1.1.1.102
O-**Demethylpuromycin methyltransferase**	2.1.1.38
D-enzyme	2.4.1.25
Deoxyadenosine kinase	2.7.1.76
(Deoxy)adenylate kinase	2.7.4.11
Deoxy-CTPase	3.6.1.12
Deoxycytidine aminohydrolase	3.5.4.14
Deoxycytidine deaminase	3.5.4.14
Deoxycytidine kinase	2.7.1.74

Deoxycytidinetriphosphatase	3.6.1.12
Deoxycytidylate hydroxymethyltransferase	2.1.2.8
Deoxycytidylate kinase	2.7.4.5
Deoxycytidylate kinase	2.7.4.14
Deoxycytidylate methyltransferase	2.1.1.54
*6-Deoxy-*L-*galactose:NAD^{+} 1-oxidoreductase*	1.1.1.122
2-Deoxy-D-**gluconate dehydrogenase**	1.1.1.125
*2-Deoxy-*D-*gluconate:NAD^{+} 3-oxidoreductase*	1.1.1.125
Deoxy-GTPase	3.1.5.1
Deoxyguanylate kinase	2.7.4.8
3-Deoxy-D-*manno*-**octulosonate aldolase**	4.1.2.23
*3-Deoxy-*D-*manno-octulosonate* D-*arabinose-lyase*	4.1.2.23
3-Deoxy-*manno*-octulosonate cytidylyltransferase	2.7.7.38
(Deoxy) nucleosidemonophosphate kinase	2.7.4.13
Deoxynucleosidetriphosphate:DNA deoxynucleotidyltransferase	2.7.7.7
3′-Deoxynucleotidase	3.1.3.34
Deoxynucleotide 3′-phosphatase	3.1.3.34
Deoxyriboaldolase	4.1.2.4
Deoxyribocyclobutadipyrimidine pyrimidine-lyase	4.1.99.3
Deoxyribodipyrimidine photolyase	4.1.99.3
2′-Deoxyriboncleoside-triphosphate: oxidized thioredoxin 2′-oxidoreductase	1.17.4.2
Deoxyribonuclease I	3.1.21.1
Deoxyribonuclease II	3.1.22.1
Deoxyribonuclease S_1 *N. crassa* nuclease 1423	3.1.30.1
2′-Deoxyribonucleoside-diphosphate:oxidized-thioredoxin 2′-oxidoreductase	1.17.4.1
Deoxyribonucleotide 3′-phosphohydrolase	3.1.3.34
Deoxyribose-phosphate aldolase	4.1.2.4
*2-Deoxy-*D-*ribose-5-phosphate acetaldehyde-lyase*	4.1.2.4
4-Deoxy-L-*threo*-**5-hexosulose-uronate ketol-isomerase**	5.3.1.17
*4-Deoxy-*L-*threo-5-hexosulose-uronate ketol-isomerase*	5.3.1.17
Deoxyuridine:orthophosphate deoxyribosyltransferase	2.4.2.23
Deoxyuridine phosphorylase	2.4.2.23
Deoxyuridinetriphosphatase	3.6.1.23
(5′-Deoxy-5′-adenosyl)(3-aminopropyl) methylsulphonium-salt:putrescine 3-aminopropyltransferase	2.5.1.16

Dephospho-CoA kinase	2.7.1.24
Dephospho-CoA pyrophosphorylase	2.7.7.3
Dephosphophosphorylase kinase	2.7.1.38
Desulphinase	4.1.1.12
Desulphoheparin sulphotransferase	2.8.2.8
Dethiobiotin synthetase	6.3.3.3
Dextranase	3.2.1.11
Dextransucrase	2.4.1.5
Dextrin dextranase	2.4.1.2
Dextrin glycosyltransferase	2.4.1.25
Dextrin 6-α-D-glucanohydrolase	3.2.1.10
Dextrin 6-α-D-glucosidase	3.2.1.33
Dextrin 6-glucosyltransferase	2.4.1.2
DFPase	3.8.2.1
DHAP synthase	4.1.2.15
Diacetyl reductase	1.1.1.5
Diacylglycerol acyltransferase	2.3.1.20
Diacylglycerol lipase	3.1.1.34
2,2-Dialkyl-L-amino-acid carboxy-lyase (amino-transferring)	4.1.1.64
Dialkylamino-acid decarboxylase (pyruvate)	4.1.1.64
Diamine aminotransferase	2.6.1.29
Diamine oxidase	1.4.3.6
Diamine:2-oxoglutarate aminotransferase	2.6.1.29
Diamino-acid aminotransferase	2.6.1.8
Diaminobutyrate—pyruvate aminotransferase	2.6.1.46
L-2,4-Diaminobutyrate:pyruvate aminotransferase	2.6.1.46
L-3,6-Diaminohexanoate aminomutase	5.4.3.3
D-2,6-Diaminohexanoate aminomutase	5.4.3.4
L-erythro-3,5-Diaminohexanoate dehydrogenase	1.4.1.11
L-erythro-3,5-Diaminohexanoate:NAD+ oxidoreductase (deaminating)	1.4.1.11
7,8-Diaminononanoate:-carbon-dioxide cyclo-ligase (ADP-forming)	6.3.3.3
Diamino oxhydrase	1.4.3.6
2,4-Diaminopentanoate dehydrogenase	1.4.1.12
2,4-Diaminopentanoate:NAD(P)+ oxidoreductase (deaminating)	1.4.1.12
meso-2,6-Diaminopimelate carboxy-lyase	4.1.1.20
Diaminopimelate decarboxylase	4.1.1.20
Diaminopimelate epimerase	5.1.1.7
2,6-LL-Diaminopimelate 2-epimerase	5.1.1.7
2,3-Diaminopropionate oxalyltransferase	2.3.1.58
2,5-Diaminovalerate:2-oxoglutarate aminotransferase	2.6.1.8
Diaphorase	1.6.4.3
Diastase	3.2.1.2
Diastase	3.2.1.1
3,4-Dicarboxy-3-hydroxytetradecanoate oxaloacetate-lyase (CoA-acylating)	4.1.3.23

4,5-Dicarboxy-4-hydroxypentadecanoate 2-oxoglutarate-lyase (CoA-acylating)	4.1.3.29
N6-(1,3-Dicarboxypropyl)-L-lysine:NAD+ oxidoreductase (L-glutamate-forming)	1.5.1.9
N6-(1,3-Dicarboxypropyl)-L-lysine:NAD+ oxidoreductase (L-lysine-forming)	1.5.1.7
N6-(1,3-Dicarboxypropyl)-L-lysine:NADP+ oxidoreductase (L-glutamate-forming)	1.5.1.10
N6-(1,3-Dicarboxypropyl)-L-lysine:NADP+ oxidoreductase (L-lysine-forming)	1.5.1.8
1,2-Didehydropipecolate reductase	1.5.1.14
Diglyceride acyltransferase	2.3.1.20
Diglyceride lipase	3.1.1.34
Diguanosinetetra-phosphatase	3.6.1.17
o-Dihydric phenol methyltransferase	2.1.1.42
trans-**1,2-Dihydrobenzene-1,2-diol dehydrogenase**	1.3.1.20
cis-**1,2-Dihydrobenzene-1,2-diol dehydrogenase**	1.3.1.19
cis-1,2-Dihydrobenzene-1,2-diol:NAD+ oxidoreductase	1.3.1.19
trans-1,2-Dihydrobenzene-1,2-diol:NADP+ oxidoreductase	1.3.1.20
7,8-Dihydrobiopterin:NADP+ oxidoreductase	1.1.1.153
Dihydrobunolol dehydrogenase	1.1.1.160
4,5α-Dihydrocortisone:NADP+ Δ4-oxidoreductase	1.3.1.4
4,5β-Dihydrocortisone:NADP+ Δ4-oxidoreductase	1.3.1.3
Dihydrocoumarin hydrolase	3.1.1.35
Dihydrocoumarin lactonohydrolase	3.1.1.35
23,24-Dihydrocucurbitacin:NAD(P)+ Δ23-oxidoreductase	1.3.1.5
(1S,3R,4S)-3,4-Dihydroxycyclohexane-1-carboxylate:NAD+ 3-oxidoreductase	1.1.1.166
Dihydrodipicolinate reductase	1.3.1.26
Dihydrodipicolinate synthase	4.2.1.52
Dihydrofolate dehydrogenase	1.5.1.4
Dihydrofolate reductase	1.5.1.3
Dihydrofolate synthetase	6.3.2.12
Dihydrolipoamide acetyltransferase	2.3.1.12
Dihydrolipoamide reductase (NAD+)	1.6.4.3
Dihydrolipoamide succinyltransferase	2.3.1.61
Dihydroneopterin aldolase	4.1.2.25
Dihydro-orotase	3.5.2.3
L-5,6-Dihydro-orotate amidohydrolase	3.5.2.3
L-5,6-Dihydroorotate:NAD+ oxidoreductase	1.3.1.14
L-5,6-Dihydroorotate:NADP+ oxidoreductase	1.3.1.15
Dihydroorotate oxidase	1.3.3.1
L-5,6-Dihydroorotate:oxygen oxidoreductase	1.3.3.1

Dihydropteridine reductase	1.6.99.7
Dihydropteridine reductase (NADH)	1.6.99.10
*Dihydropteroate:*L-*glutamate ligase (ADP-forming)*	6.3.2.12
Dihydropteroate pyrophosphorylase	2.5.1.15
Dihydropteroate synthase	2.5.1.15
Dihydropyrimidinase	3.5.2.2
5,6-Dihydropyrimidine amidohydrolase	3.5.2.2
D-*erythro-Dihydrosphingosine:NADP⁺ 3-oxidoreductase*	1.1.1.102
Dihydrosphingosine kinase	2.7.1.91
Dihydrosphingosine-1-phosphate aldolase	4.1.2.27
Dihydrosphingosine-1-phosphate palmitaldehyde-lyase	4.1.2.27
Dihydrostreptomycin-6-phosphate 3'α-kinase	2.7.1.88
Dihydrouracil dehydrogenase	1.3.1.1
Dihydrouracil dehydrogenase (NADP⁺)	1.3.1.2
5,6-Dihydrouracil:NAD⁺ oxidoreductase	1.3.1.1
5,6-Dihydrouracil:NADP⁺ oxidoreductase	1.3.1.2
Dihydroxyacetone-phosphate acyltransferase	2.3.1.42
Dihydroxyacetone-phosphate phospho-lyase	4.2.99.11
Dihydroxyacetone reductase	1.1.1.156
Dihydroxyacetone-transferase	2.2.1.2
Dihydroxyacid dehydratase	4.2.1.9
2,3-Dihydroxyacid hydro-lyase	4.2.1.9
2,3-Dihydroxybenzoate carboxy-lyase	4.1.1.46
2,3-Dihydroxybenzoate:oxygen 2,3-oxidoreductase (decyclizing)	1.13.11.28
2,3-Dihydroxybenzoate:oxygen 3,4-oxidoreductase (decyclizing)	1.13.11.14
*2,3-Dihydroxybenzoate:*L-*serine ligase*	6.3.2.14
2,3-Dihydroxybenzoate 2,3-dioxygenase	1.13.11.28
2,3-Dihydroxybenzoate 3,4-dioxygenase	1.13.11.14
2,3-Dihydroxybenzoylserine synthetase	6.3.2.14
Dihydroxyfumarate carboxy-lyase	4.1.1.54
Dihydroxyfumarate decarboxylase	4.1.1.54
2,3-Dihydroxyindole:oxygen 2,3-oxidoreductase (decyclizing)	1.13.11.23
2,3-Dihydroxyindole 2,3-dioxygenase	1.13.11.23
Dihydroxyisovalerate dehydrogenase (isomerizing)	1.1.1.86
Dihydroxyisovalerate dehydrogenase (isomerizing)	1.1.1.89
2,3-Dihydroxy-isovalerate:NADP⁺ oxidoreductase (isomerizing)	1.1.1.86
7,8-Dihydroxy-kynurenate oxygenase	1.13.11.10
7,8-Dihydroxykynurenate:oxygen 8,8a-oxidoreductase (decyclizing)	1.13.11.10
7,8-Dihydroxykynurenate 8,8a-dioxygenase	1.13.11.10

3,4-Dihydroxyphenylacetate:oxygen 2,3-oxidoreductase (decyclizing)	1.13.11.15
3,4-Dihydroxyphenyl-acetate:oxygen 3,4-oxidoreductase (decyclizing)	1.13.11.7
3,4-Dihydroxyphenylacetate 2,3-dioxygenase	1.13.11.15
3,4-Dihydroxyphenylacetate 3,4-dioxygenase	1.13.11.7
Dihydroxyphenylalanine aminotransferase	2.6.1.49
Dihydroxyphenylalanine ammonia-lyase	4.3.1.11
*3,4-Dihydroxy-*L-*phenylalanine ammonia-lyase*	4.3.1.11
3,4-Dihydroxyphenylalanine:oxygen 3,4-oxidoreductase (recyclizing)	1.13.11.30
3,4-Dihydroxyphenylalanine:oxygen 4,5-oxidoreductase (recyclizing)	1.13.11.29
*3,4-Dihydroxy-*L-*phenylalanine: 2-oxoglutarate aminotransferase*	2.6.1.49
3,4-Dihydroxyphenylethyl-amine,ascorbate: oxygen oxidoreductase (β-hydroxylating)	1.14.17.1
β-(2,3-Dihydroxyphenyl) propionate: oxygen 1,2-oxidoreductase (decyclizing)	1.13.11.16
4,5-Dihydroxyphthalate carboxy-lyase	4.1.1.55
4,5-Dihydroxyphthalate decarboxylase	4.1.1.55
2,6-Dihydroxypyridine,NADH:oxygen oxidoreductase (3-hydroxylating)	1.14.13.10
2,5-Dihydroxypyridine oxygenase	1.13.11.9
2,5-Dihydroxypyridine:oxygen 5,6-oxidoreductase (decyclizing)	1.13.11.9
2,6-Dihydroxypyridine 3-monooxygenase	1.14.13.10
2,5-Dihydroxypyridine 5,6-dioxygenase	1.13.11.9
3,4-Dihydroxy-trans-cinnamate:oxygen 3,4-oxidoreductase (decyclizing)	1.13.11.22
11α,15-Dihydroxy-9-oxoprost-13-enoate: NAD⁺ 15-oxidoreductase	1.1.1.141
3,4-Dihydroxy-9,10-secoandrosta-1,3,5(10)-triene-9,17-dione:oxygen 4,5-oxidoreductase (decyclizing)	1.13.11.25
3,4-Dihydroxy-9,10-secoandrosta-1,3,5(10)-triene-9,17-dione 4,5-dioxygenase	1.13.11.25
cis-**1,2-Dihydro-1,2-dihydroxynaphthalene dehydrogenase**	1.3.1.29
cis-1,2-Dihydro-1,2-dihydroxynaphthalene: NAD⁺ 1,2-oxidoreductase	1.3.1.29
2,3-Dihydro-2,3-dihydroxybenzoate dehydrogenase	1.3.1.28
2,3-Dihydro-2,3-dihydroxybenzoate:NAD⁺ oxidoreductase	1.3.1.28
2,3-Dihydro-2,3-dihydroxy-benzoate synthase	3.3.2.1
7,8-Dihydro-7,8-dihydroxykynurenate: NAD⁺ oxidoreductase	1.3.1.18
Diiodophenylpyruvate reductase	1.1.1.96

Diiodotyrosine aminotransferase	2.6.1.24
3,5-Diiodo-L-tyrosine:2-oxoglutarate	
aminotransferase	2.6.1.24
β-(3,5-Diiodo-4-hydroxyphenyl)-lactate:	
NAD+ oxidoreductase	1.1.1.96
Di-isopropyl-fluorophosphatase	3.8.2.1
Di-isopropyl-fluorophosphate	
fluorohydrolase	3.8.2.1
Diisopropylfluorophosphonate	
halogenase	3.8.2.1
Di-isopropyl phosphorofluoridase	3.8.2.1
2,5-Diketocamphane lactonizing enzyme	1.14.15.2
β-Diketonase	3.7.1.2
Dimethylallyldiphosphate:	
isopentenyldiphosphate	
dimethylallyltransferase	2.5.1.1
Dimethylallyltransferase	2.5.1.1
Dimethylaniline monooxygenase	
(N-oxide-forming)	1.14.13.8
N,N-Dimethylaniline,NADPH:oxygen	
oxidoreductase (N-oxide-forming)	1.14.13.8
Dimethylaniline-N-oxide aldolase	4.1.2.24
N,N-Dimethylaniline-N-oxide	
formaldehyde-lyase	4.1.2.24
Dimethylaniline oxidase	1.14.13.8
Dimethylglycine dehydrogenase	1.5.99.2
Dimethylhistidine methyltransferase	2.1.1.44
Dimethylmalate dehydrogenase	1.1.1.84
3,3,-Dimethyl-D-malate:NAD+	
oxidoreductase (decarboxylating)	1.1.1.84
Dimethylpropiothetin dethiomethylase	4.4.1.3
S-Dimethyl-β-propiothetin	
dimethyl-sulphide-lyase	4.4.1.3
Dimethylthetin—homocysteine	
methyltransferase	2.1.1.3
Dimethylthetin:L-homocysteine	
S-methyltransferase	2.1.1.3
4,4-Dimethyl-5α-cholest-7-en-3β-ol,	
hydrogen-donor:oxygen oxidoreductase	1.14.99.16
6,7-Dimethyl-8-(1'-D-ribityl) lumazine:	
6,7-dimethyl-8-(1'-D-ribityl) lumazine	
2,3-butanediyltransferase	2.5.1.9
3,5-Dinitrotyrosine aminotransferase	2.6.1.26
Dinucleoside-triphosphatase	3.6.1.29
Dinucleotide nucleotidohydrolase	3.6.1.9
2,3-Di-O-acyl-1-O(β-D-galactosyl)-	
D-glycerol acylhydrolase	3.1.1.26
2,4-Dioxotetrahydropyrimidine	
nucleotide:pyrophosphate	
phosphoribosyltransferase	2.4.2.20
Dioxotetrahydropyrimidine	
phosphoribosyltransferase	2.4.2.20
Dioxotetrahydropyrimidineribonucleotide	
pyrophosphorylase	2.4.2.20

2,5-Dioxovalerate dehydrogenase	1.2.1.26
2,5-Dioxovalerate:NADP+ oxidoreductase	1.2.1.26
Dipeptidase	3.4.13.11
Dipeptidase M	3.4.13.12
Dipeptide hydrolase	3.4.13.11
Dipeptidyl-amino-peptidase I	3.4.14.1
Dipeptidyl-aminopeptidase II	3.4.14.2
Dipeptidyl carboxypeptidase	3.4.15.1
Dipeptidyl peptidase I	3.4.14.1
Dipeptidyl peptidase II	3.4.14.2
Dipeptidylpeptide hydrolase	3.4.14.2
Dipeptidylpeptide hydrolase	3.4.14.1
Dipeptidyl transferase	3.4.14.1
Diphenol oxidase,*o*-Diphenolase	1.10.3.1
Diphosphoinositide kinase	2.7.1.68
Diplococcus pneumoniae (Dpn II)	3.1.23.27
Disproportionating enzyme	2.4.1.25
Disulphoglucosamine-6-sulphatase	3.1.6.11
DNA (cytosine-5-)-methyltransferase	2.1.1.37
DNA joinase	6.5.1.2
DNA joinase	6.5.1.1
DNA ligase	6.5.1.1
DNA ligase	6.5.1.2
DNA nucleotidylexotransferase	2.7.7.31
DNA nucleotidyltransferase	2.7.7.7
DNA polymerase	2.7.7.7
DNA repair enzyme	6.5.1.1
DNA repair enzyme	6.5.1.2
DNase	3.1.21.1
DNase II	3.1.22.1
DNA 5'-dinucleotidohydrolase	3.1.11.4
Dodecenoyl-CoA	
Δ³-cis—Δ²-trans-isomerase	5.3.3.8
Dodecenoyl-CoA Δ-isomerase	5.3.3.8
Donor:hydrogen-peroxide oxidoreductase	1.11.1.7
DOPA decarboxylase	4.1.1.28
DOPA decarboxylase	4.1.1.26
Dopamine β-hydroxylase	1.14.17.1
Dopamine β-monooxygenase	1.14.17.1
DPNase	3.2.2.5
DPN hydrolase	3.2.2.5
DPN kinase	2.7.1.23
DT-diaphorase	1.6.99.2
Ecdysone,hydrogen-donor:oxygen	
oxidoreductase (20-hydroxylating)	1.14.99.22
Ecdysone oxidase	1.1.3.16
Ecdysone:oxygen 3-oxidoreductase	1.1.3.16
Ecdysone 20-monooxygenase	1.14.99.22
Echarin	3.4.99.27
Echis carnatus **prothrombin activating**	
proteinase	3.4.99.27

E. coli, see also *Escherichia coli*	
E. coli endonuclease II	3.1.22.3
E.coli endonuclease I "Nicking" nuclease of calf thymus	3.1.21.1
E. coli endonuclease IV DNase V (mammalian)	3.1.21.2
E. coli endonucleases III and V	3.1.25.1
E. coli exonuclease III. *Haemophilus influenzae* exonuclease	3.1.11.2
E. coli exonuclease I. Mammalian DNase III	3.1.11.1
E. coli exonuclease V. *H. influenzae* ATP-dependent *DNase*	3.1.11.5
E. coli exonuclease VII. *Micrococcus luteus* exonuclease	3.1.11.6
E. coli protease I	3.4.21.14
E. coli RNase	3.1.26.1
E. coli RNase I	3.1.27.1
8,11,14-Eicosatrienoate, hydrogen-donor:oxygen oxidoreductase	1.14.99.1
Elastase	3.4.21.11
Endodeoxyribonuclease *Alu*I	3.1.23.1
Endodeoxyribonuclease (apurinic or apyrimidinic)	3.1.25.2
Endodeoxyribonuclease *Asu*I	3.1.23.2
Endodeoxyribonuclease (ATP- and S-adenosyl-methionine-dependent)	3.1.4.32
Endodeoxyribonuclease (ATP-hydrolysing)	3.1.4.33
Endodeoxyribonuclease *Ava*I	3.1.23.3
Endodeoxyribonuclease *Ava*II	3.1.23.4
Endodeoxyribonuclease *Bal*I	3.1.23.5
Endodeoxyribonuclease *Bam*FI	3.1.23.6
Endodeoxyribonuclease *Bam*HI	3.1.23.6
Endodeoxyribonuclease *Bam*KI	3.1.23.6
Endodeoxyribonuclease *Bam*NI	3.1.23.6
Endodeoxyribonuclease *Bbr*I	3.1.23.21
Endodeoxyribonuclease *Bbv*I	3.1.23.7
Endodeoxyribonuclease *Bcl*I	3.1.23.8
Endodeoxyribonuclease *Bgl*I	3.1.23.9
Endodeoxyribonuclease *Bgl*II	3.1.23.10
Endodeoxyribonuclease *Blu*I	3.1.23.42
Endodeoxyribonuclease *Blu*II	3.1.23.17
Endodeoxyribonuclease *Bpu*I	3.1.23.11
Endodeoxyribonuclease *Bsp*RI	3.1.23.17
Endodeoxyribonuclease *Bst*I	3.1.23.6
Endodeoxyribonuclease *Bsu*I	3.1.23.31
Endodeoxyribonuclease *Bsu*I	3.1.23.17
Endodeoxyribonuclease *Bsu*I247	3.1.23.31
Endodeoxyribonuclease *Chu*I	3.1.23.21
Endodeoxyribonuclease *Chu*II	3.1.23.20
Endodeoxyribonuclease *Dpn*I	3.1.23.12

Endodeoxyribonuclease *Dpn*II	3.1.23.27
Endodeoxyribonuclease *Eco*B	3.1.24.1
Endodeoxyribonuclease *Eco*K	3.1.24.2
Endodeoxyribonuclease *Eco*PI	3.1.24.3
Endodeoxyribonuclease *Eco*P15	3.1.24.4
Endodeoxyribonuclease *Eco*RI	3.1.23.13
Endodeoxyribonuclease *Eco*RII	3.1.23.14
Endodeoxyribonuclease *Hae*I	3.1.23.15
Endodeoxyribonuclease *Hae*II	3.1.23.16
Endodeoxyribonuclease *Hae*III	3.1.23.17
Endodeoxyribonuclease *Hap*II	3.1.23.24
Endodeoxyribonuclease *Hga*I	3.1.23.18
Endodeoxyribonuclease *Hha*I	3.1.23.19
Endodeoxyribonuclease *Hha*II	3.1.23.22
Endodeoxyribonuclease *Hhg*I	3.1.23.17
Endodeoxyribonuclease *Hinb*III	3.1.23.21
Endodeoxyribonuclease *Hind*II	3.1.23.20
Endodeoxyribonuclease *Hind*III	3.1.23.21
Endodeoxyribonuclease *Hine*II	3.1.23.20
Endodeoxyribonuclease *Hinf*I	3.1.23.22
Endodeoxyribonuclease *Hinf*II	3.1.23.21
Endodeoxyribonuclease *Hin*HI	3.1.23.16
Endodeoxyribonuclease *Hpa*I	3.1.23.23
Endodeoxyribonuclease *Hpa*II	3.1.23.24
Endodeoxyribonuclease *Hph*I	3.1.23.25
Endodeoxyribonuclease IV (Phage T4-induced)	3.1.21.2
Endodeoxyribonuclease *Kpn*I	3.1.23.26
Endodeoxyribonuclease *Mbo*I	3.1.23.27
Endodeoxyribonuclease *Mbo*II	3.1.23.28
Endodeoxyribonuclease *Mnl*I	3.1.23.29
Endodeoxyribonuclease *Mnn*I	3.1.23.20
Endodeoxyribonuclease *Mno*I	3.1.23.24
Endodeoxyribonuclease *Mos*I	3.1.23.27
Endodeoxyribonuclease *Ngo*I	3.1.23.16
Endodeoxyribonuclease *Pfa*I	3.1.23.30
Endodeoxyribonuclease *Pol*I	3.1.23.17
Endodeoxyribonuclease *Pst*I	3.1.23.31
Endodeoxyribonuclease *Pvu*I	3.1.23.32
Endodeoxyribonuclease *Pvu*II	3.1.23.33
Endodeoxyribonuclease (pyrimidine dimer)	3.1.25.1
Endodeoxyribonuclease *Sac*I	3.1.23.34
Endodeoxyribonuclease *Sac*II	3.1.23.35
Endodeoxyribonuclease *Sac*III	3.1.23.36
Endodeoxyribonuclease *Sal*I	3.1.23.37
Endodeoxyribonuclease *Sal*PI	3.1.23.31
Endodeoxyribonuclease *Sfa*I	3.1.23.17
Endodeoxyribonuclease *Sgr*I	3.1.23.38
Endodeoxyribonuclease *Sma*I	3.1.23.44

Endodeoxyribonuclease *Sau*3A1	3.1.23.27
Endodeoxyribonuclease *Sst*I	3.1.23.34
Endodeoxyribonuclease *Sst*II	3.1.23.35
Endodeoxyribonuclease *Sst*III	3.1.23.36
Endodeoxyribonuclease *Taq*I	3.1.23.39
Endodeoxyribonuclease *Taq*II	3.1.23.40
Endodeoxyribonuclease *Tgl*I	3.1.23.35
Endodeoxyribonuclease V	3.1.22.3
Endodeoxyribonuclease *Xam*I	3.1.23.37
Endodeoxyribonuclease *Xba*I	3.1.23.41
Endodeoxyribonuclease *Xho*I	3.1.23.42
Endodeoxyribonuclease *Xho*II	3.1.23.43
Endodeoxyribonuclease *Xma*I	3.1.23.44
Endodeoxyribonuclease *Xma*II	3.1.23.31
Endodeoxyribonuclease *Xni*I	3.1.23.45
Endodeoxyribonuclease *Xpa*I	3.1.23.42
Endo-α-*N*-acetylgalactosaminidase	3.2.1.97
Endo-β-*N*-acetylglucosaminidase	3.2.1.96
Endo-β-N-acetyl-glucosaminidase	3.4.99.17
Endonuclease R. *Corynebacterium humiferum* (*Chu*II)	3.1.23.20
Endonuclease S₁ (*Aspergillus*)	3.1.30.1
Endonuclease (*Serratia marcescens*)	3.1.30.2
Endonuclease Z. *Bacillus sphaericus* (*Bsp*RI)	3.1.23.17
Endopolyphosphatase	3.6.1.10
Endoribonuclease H (calf thymus)	3.1.26.4
Endoribonuclease III	3.1.4.24
Endothia **carboxyl proteinase**	3.4.23.6
Endo-1,2-β-D-glucanase	3.2.1.71
Endo-1,3-β-D-galactanase	3.2.1.90
Endo-1,3-α-D-glucanase	3.2.1.59
Endo-1,3-β-D-glucanase	3.2.1.6
Endo-1,3-β-D-glucanase	3.2.1.39
Endo-1,3-β-D-xylanase	3.2.1.32
Endo-1,3(4)-β-D-glucanase	3.2.1.6
Endo-1,4-β-D-galactanase	3.2.1.89
Endo-1,4-β-glucanase	3.2.1.4
Endo-1,4-β-D-mannanase	3.2.1.78
Endo-1,4-β-D-xylanase	3.2.1.8
Endo-1,6-β-D-glucanase	3.2.1.75
Enolase	4.2.1.11
3-Enolpyruvoylshikimate-5-phosphate synthase	2.5.1.19
Enoyl-[acyl-carrier-protein] reductase	1.3.1.9
Enoyl-[acyl-carrier-protein] reductase (NADPH)	1.3.1.10
Enoyl-CoA hydratase	4.2.1.17
Δ³-*cis*-Δ²-*trans* Enoyl-CoA isomerase	5.3.3.7
Enoyl-CoA reductase	1.3.1.8
Enoyl hydrase	4.2.1.17
Enoylpyruvate transferase	2.5.1.7
Enterokinase	3.4.21.9
Enteropeptidase	3.4.21.9
Entomophthora **collagenolytic proteinase**	3.4.21.33
Epoxide hydratase	3.3.2.3
Epoxide hydrolase	3.3.2.3
Epoxide hydrolase	3.3.2.3
trans-Epoxysuccinate hydratase	4.2.1.37
5α-Ergosta-7,22-diene-3β,5-diol 5,6-hydro-lyase	4.2.1.62
Erythritol kinase	2.7.1.27
Erythritol:NADP⁺ oxidoreductase	1.1.1.162
Erythrose isomerase	5.3.1.2
D-Erythrulose reductase	1.1.1.162
Erythrulose-1-phosphate formaldehyde-lyase	4.1.2.2
Erythrulose-1-phosphate synthetase	4.1.2.2
Escherichia coli, see also *E. coli*	
Escherichia coli **periplasmic proteinase**	3.4.21.14
Escherichia freundii **proteinase**	3.4.24.4
Estradiol, See Oestradiol	
Ethanolamine ammonia-lyase	4.3.1.7
Ethanolamine ammonia-lyase	4.3.1.7
Ethanolamine kinase	2.7.1.82
Ethanolamine oxidase	1.4.3.8
Ethanolamine:oxygen oxidoreductase (deaminating)	1.4.3.8
Ethanolaminephosphate cytidylyltransferase	2.7.7.14
Ethanolaminephosphate phospho-lyase	4.2.99.7
Ethanolaminephosphate phospho-lyase (deaminating)	4.2.99.7
Ethanolaminephosphotransferase	2.7.8.1
Ethylene reductase	1.3.99.2
2-Ethylmalate glyoxylate-lyase (CoA-butyrylating)	4.1.3.10
2-Ethylmalate synthase	4.1.3.10
Etiocholanolone 3α-dehydrogenase	1.1.1.152
Euphorbain	3.4.99.7
Exo-cellobiohydrolase	3.2.1.91
Exodeoxyribonuclease I	3.1.11.1
Exodeoxyribonuclease II	3.1.4.26
Exodeoxyribonuclease III	3.1.11.2
Exodeoxyribonuclease (Lambda-induced)	3.1.11.3
Exodeoxyribonuclease (Phage SP3-induced)	3.1.11.4
Exodeoxyribonuclease V	3.1.11.5
Exodeoxyribonuclease VII	3.1.11.6
Exo-β-D-fructosidase	3.2.1.80
Exo-isomaltohydrolase	3.2.1.94

Exo-isomaltotrihydrolase	3.2.1.95
Exo-malthohexaohydrolase	3.2.1.98
Exo-maltotetrahydrolase	3.2.1.60
Exo-β-*N*-acetylmuramidase	3.2.1.92
3'-Exonuclease	3.1.16.1
5'-Exonuclease	3.1.4.1
Exonuclease IV	3.1.11.1
Exopolygalacturonase	3.2.1.67
Exopolygalacturonate lyase	4.2.2.9
Exo-poly-α-D-galacturonosidase	3.2.1.82
Exopolyphosphatase	3.6.1.11
Exoribonuclease	3.1.13.1
Exoribonuclease H	3.1.13.2
Exoribonuclease. *Lactobacillus plantarum* RNase	3.1.13.1
Exo-1,2-1,3-α-D-mannosidase	3.2.1.77
Exo-1,3-α-glucanase	3.2.1.84
Exo-1,3-β-D-glucosidase	3.2.1.58
Exo-1,3-β-D-xylosidase	3.2.1.72
Exo-1,4-α-D-glucosidase	3.2.1.3
Exo-1,4-β-D-glucosidase	3.2.1.74
Exo-1,4-β-D-xylosidase	3.2.1.37
Exo-1,6-α-D-glucosidase	3.2.1.70
FAD nucleotidohydrolase	3.6.1.18
FAD pyrophosphatase	3.6.1.18
FAD pyrophosphorylase	2.7.7.2
Farnesyl-diphosphate:farnesyl-diphosphate farnesyltransferase	2.5.1.21
Farnesyl-diphosphate kinase	2.7.4.18
Farnesylpyrophosphate synthetase	2.5.1.1
Farnesyltransferase	2.5.1.21
Fatty acid desaturase	1.14.99.5
Fatty-acid-hydroperoxide isomerase	5.3.99.1
Fatty acid ω-hydroxylase	1.14.15.3
Fatty acid methyltransferase	2.1.1.15
Fatty-acid peroxidase	1.11.1.3
Fatty acid thiokinase (long chain)	6.2.1.3
Fatty acid thiokinase (medium chain)	6.2.1.2
Fatty acyl-CoA reductase	1.2.1.42
Ferredoxin:H+ oxidoreductase	1.18.3.1
Ferredoxin:NAD+ oxidoreductase	1.18.1.3
Ferredoxin:NADP+ oxidoreductase	1.18.1.2
Ferredoxin-NADP+ reductase	1.18.1.2
Ferredoxin-NAD+ reductase	1.18.1.3
Ferredoxin—nitrate reductase	1.7.7.1
Ferrihaemoprotein *P450* reductase	1.6.2.4
Ferrochelatase	4.99.1.1
Ferrocytochrome c:hydrogen-peroxide oxidoreductase	1.11.1.5
Ferrocytochrome c:iron oxidoreductase	1.9.99.1

Ferrocytochrome c:oxygen oxidoreductase	1.9.3.1
Ferrocytochrome c2:oxygen oxidoreductase	1.9.3.2
Ferrocytochrome:nitrate oxidoreductase	1.9.6.1
Ferroxidase	1.16.3.1
Fibrinase	3.4.21.7
Fibrinogenase	3.4.21.5
Fibrinolysin	3.4.21.7
Ficin	3.4.22.3
Flavanone lyase (decyclizing)	5.5.1.6
Flavokinase	2.7.1.26
Flavoprotein-linked monooxygenase	1.14.14.1
FMN adenylyltransferase	2.7.7.2
Formaldehyde dehydrogenase	1.2.1.1
Formaldehyde:NAD+ oxidoreductase (glutathione-formylating)	1.2.1.1
Formamidase	3.5.1.9
Formamide hydro-lyase	4.2.1.66
Formate acetyltransferase	2.3.1.54
Formate dehydrogenase	1.2.1.2
Formate dehydrogenase (cytochrome)	1.2.2.1
Formate dehydrogenase (NADP+)	1.2.1.43
Formate:ferricytochrome b1 oxidoreductase	1.2.2.1
Formate hydrogenlyase	1.2.1.2
Formate kinase	2.7.2.6
Formate:NAD+ oxidoreductase	1.2.1.2
Formate:NADP+ oxidoreductase	1.2.1.43
Formate:tetrahydrofolate ligase (ADP-forming)	6.3.4.3
Formic dehydrogenase	1.2.1.1
Formiminoaspartate deiminase	3.5.3.5
N-Formimino-L-aspartate iminohydrolase	3.5.3.5
Formiminoglutamase	3.5.3.8
Formiminoglutamate deiminase	3.5.3.13
N-Formimino-L-glutamate formiminohydrolase	3.5.3.8
N-Formimino-L-glutamate iminohydrolase	3.5.3.13
5-Formiminotetrahydrofolate ammonia-lyase (cyclizing)	4.3.1.4
Formiminotetrahydrofolate cyclodeaminase	4.3.1.4
5-Formiminotetrahydrofolate:L-glutamate N-formiminotransferase	2.1.2.5
5-Formiminotetrahydrofolate:glycine N-formimino-transferase	2.1.2.4
Formylase	3.5.1.9
N-Formyl-L-aspartate amidohydrolase	3.5.1.8
Formylaspartate deformylase	3.5.1.8
Formyl-CoA hydrolase	3.1.2.10
Formyl-CoA hydrolase	3.1.2.10

S-Formylglutathione hydrolase	3.1.2.12
S-Formylglutathione hydrolase	3.1.2.12
Formylkynureninase	3.5.1.9
N-*Formyl-*L-*methionine amidohydrolase*	3.5.1.31
Formylmethionine deformylase	3.5.1.31
N-*Formyl-*L-*methionylaminoacyl-tRNA amidohydrolase*	3.5.1.27
N-**Formylmethionylaminoacyl-tRNA deformylase**	3.5.1.27
10-Formyltetrahydrofolate	3.5.1.10
5-Formyltetrahydrofolate cyclo-ligase	6.3.3.2
5-Formyltetrahydrofolate cyclo-ligase (ADP-forming)	6.3.3.2
Formyltetrahydrofolate deformylase	3.5.1.10
Formyltetrahydrofolate dehydrogenase	1.5.1.6
*5-Formyltetrahydrofolate:*L-*glutamate* N-*formyltransferase*	2.1.2.6
10-Formyltetrahydrofolate: L-*methionyl-tRNA* N-*formyltransferase*	2.1.2.9
10-Formyltetrahydrofolate:NADP+ oxidoreductase	1.5.1.6
Formyltetrahydrofolate synthetase	6.3.4.3
10-Formyltetrahydrofolate: 5′-phosphoribosyl-5-amino- 4-imidazolecarboxamide formyltransferase	2.1.2.3
FP synthase	2.4.1.96
*2,6-β-*D-*Fructan fructanohydrolase*	3.2.1.65
*2,1-β-*D-*Fructan fructanohydrolase*	3.2.1.7
*β-*D-*Fructan fructohydrolase*	3.2.1.80
*2,6-β-*D-*Fructan 6-β-*D-*fructofuranosylfructohydrolase*	3.2.1.64
2,6-β-D-**Fructan 6-levanbiohydrolase**	3.2.1.64
β-D-**Fructofuranosidase**	3.2.1.26
*β-*D-*Fructofuranoside fructohydrolase*	3.2.1.26
Fructokinase	2.7.1.4
D-*Fructose:(acceptor) 5-oxidoreductase*	1.1.99.11
Fructose-bisphosphatase	3.1.3.11
Fructose-bisphosphate aldolase	4.1.2.13
D-*Fructose:NADP+ 5-oxidoreductase*	1.1.1.124
Fructose-1-phosphate kinase	2.7.1.56
D-*Fructose-1,6-bisphosphate* D-*glyceraldehyde-3-phosphate-lyase*	4.1.2.13
Fructose-1,6-bisphosphate triosephosphate-lyase	4.1.2.13
D-*Fructose-1,6-bisphosphate 1-phosphohydrolase*	3.1.3.11
D-**Fructose 5-dehydrogenase**	1.1.99.11
Fructose 5-dehydrogenase (NADP+)	1.1.1.124
D-*Fructose-6-phosphate* D-*erythrose- 4-phosphate-lyase (phosphate-acetylating)*	4.1.2.22
Fructose-6-phosphate phosphoketolase	4.1.2.22
*β-h-*Fructosidase	3.2.1.26

Fructuronate reductase	1.1.1.57
Fucoidanase	3.2.1.44
Fucokinase	2.7.1.52
D-**Fuconate dehydratase**	4.2.1.67
L-**Fuconate dehydratase**	4.2.1.68
D-*Fuconate hydro-lyase*	4.2.1.67
L-*Fuconate hydro-lyase*	4.2.1.68
*2-O-α-*L-*Fucopyranosyl-β-*D-*galactoside fucohydrolase*	3.2.1.63
L-**Fucose dehydrogenase**	1.1.1.122
Fucose-1-phosphate guanylyltransferase	2.7.7.30
α-L-**Fucosidase**	3.2.1.51
β-D-**Fucosidase**	3.2.1.38
1,2-α-L-**Fucosidase**	3.2.1.63
*α-*L-*Fucoside fucohydrolase*	3.2.1.51
*β-*D-*Fucoside fucohydrolase*	3.2.1.38
Fucosyl-galactose acetylgalactosaminyltransferase	2.4.1.40
L-**Fuculokinase**	2.7.1.51
L-**Fuculosephosphate aldolase**	4.1.2.17
L-*Fuculose-1-phosphate* L-*lactaldehyde-lyase*	4.1.2.17
Fumarase	4.2.1.2
Fumarate hydratase	4.2.1.2
Fumarate reductase	1.3.99.1
Fumarate reductase (NADH)	1.3.1.6
Fumaric aminase	4.3.1.1
Fumaric hydrogenase	1.3.99.1
Fumarylacetoacetase	3.7.1.2
4-Fumarylacetoacetate fumarylhydrolase	3.7.1.2
2-Furoyl-CoA:(acceptor) oxidoreductase (hydroxylating)	1.3.99.8
2-Furoyl-CoA dehydrogenase	1.3.99.8
Furoyl-CoA hydroxylase	1.3.99.8
Furylfuramide isomerase	5.2.1.6
2-(2-Furyl)-3-(5-nitro-2-furyl) acrylamide cis-trans-*isomerase*	5.2.1.6
Galactanase	3.2.1.90
Galactanase	3.2.1.89
Galactarate dehydratase	4.2.1.42
D-*Galactarate hydro-lyase*	4.2.1.42
Galactinol—raffinose galactosyltransferase	2.4.1.67
Galactinol—sucrose galactosyltransferase	2.4.1.82
Galactitol dehydrogenase	1.1.1.16
Galactitol:NAD+ 3-oxidoreductase	1.1.1.16
Galactokinase	2.7.1.6
Galactolipase	3.1.1.26
Galactonate dehydratase	4.2.1.6
D-*Galactonate hydro-lyase*	4.2.1.6

Galactonolactone dehydrogenase	1.3.2.3
L-*Galactono-γ-lactone:ferricytochrome c oxidoreductase*	1.3.2.3
Galactose dehydrogenase	1.1.1.48
Galactose dehydrogenase (NADP+)	1.1.1.120
D-*Galactose:NADP+ 1-oxidoreductase*	1.1.1.120
D-*Galactose:NAD+ 1-oxidoreductase*	1.1.1.48
Galactose oxidase	1.1.3.9
D-*Galactose:oxygen 6-oxidoreductase*	1.1.3.9
Galactose-1-phosphate thymidylyltransferase	2.7.7.32
Galactose-1-phosphate uridylyltransferase	2.7.7.10
Galactose-1-phosphate uridylyltransferase	2.7.7.10
Galactose-6-sulphatase	2.5.1.5
Galactose-6-sulphate alkyltransferase (cyclizing)	2.5.1.5
Galactose-6-sulphurylase	2.5.1.5
α-D-Galactosidase	3.2.1.22
β-D-Galactosidase	3.2.1.23
Galactoside acetyltransferase	2.3.1.18
α-D-*Galactoside galactohydrolase*	3.2.1.22
β-D-*Galactoside galactohydrolase*	3.2.1.23
Galactosylceramidase	3.2.1.46
Galactosylceramide sulphotransferase	2.8.2.11
Galactosylgalactosylglucosylceramidase	3.2.1.47
D-*Galactosyl-D-galactosyl-D-glucosylceramide galactohydrolase*	3.2.1.47
1-O-α-D-Galactosyl-myo-inositol:raffinose galactosyltransferase	2.4.1.67
1-O-α-D-Galactosyl-myo-inositol:sucrose 6-galactosyltransferase	2.4.1.82
D-*Galactosyl-N-acetamidodeoxy-α-D-galactoside* D-*galactosyl-N-acetamidodeoxy-D-galactohydrolase*	3.2.1.97
D-*Galactosyl-N-acylsphingosine galactohydrolase*	3.2.1.46
Galactowaldenase	5.1.3.2
Galacturonokinase	2.7.1.44
Gallate carboxy-lyase	4.1.1.59
Gallate decarboxylase	4.1.1.59
GDPfucose—galactosyl-glucosaminyl-galactosyl-glucosylceramide α-L-fucosyltransferase	2.4.1.89
*GDPfucose:*D-*galactosyl-(1,4)-2-acetamido-2-deoxy-D-glucosyl-(1,3)-D-galactosyl-(1,4)-D-glucosylceramide α-L-fucosyltransferase*	2.4.1.89
GDPfucose—glycoprotein fucosyltransferase	2.4.1.68
GDPfucose:glycoprotein fucosyltransferase	2.4.1.68
GDPfucose—lactose fucosyltransferase	2.4.1.69
GDPfucose:lactose fucosyltransferase	2.4.1.69

GDPfucose pyrophosphorylase	2.7.7.30
GDPfucose:β-2-acetamido-2-deoxy-D-glucosaccharide 4-α-L-fucosyltransferase	2.4.1.65
GDPglucose glucohydrolase	3.2.1.42
GDPglucose—glucosephosphate glucosyltransferase	2.4.1.36
*GDPglucose:*D-*glucose-6-phosphate α-D-glucosyltransferase*	2.4.1.36
GDPglucose pyrophosphorylase	2.7.7.34
GDPglucose:1,4-β-D-glucan 4-β-glucosyltransferase	2.4.1.29
GDPglucosidase	3.2.1.42
GDPhexose pyrophosphorylase	2.7.7.29
GDPmannose dehydrogenase	1.1.1.132
GDPmannose:dolicholphosphate mannosyltransferase	2.4.1.83
GDPmannose dolicholphosphate mannosyltransferase	2.4.1.83
GDPmannose:glucomannan 1,4-β-D-mannosyltransferase	2.4.1.32
GDPmannose:heteroglycan 2,3-α-D-mannosyltransferase	2.4.1.48
GDPmannose α-D-mannosyltransferase	2.4.1.48
GDPmannose:NAD+ 6-oxidoreductase	1.1.1.132
GDPmannose—phosphatidyl-*myo*-inositol α-D-mannosyltransferase	2.4.1.57
GDPmannose:phosphomannan mannosephosphotransferase	2.7.8.9
GDPmannose phosphorylase	2.7.7.22
GDPmannose—undecaprenyl-phosphate mannosyltransferase	2.4.1.54
GDPmannose:undecaprenyl-phosphate mannosyltransferase	2.4.1.54
*GDP:*D-*mannose-1-phosphate guanylyltransferase*	2.7.7.22
GDPmannose:1-phosphatidyl-myo-inositol α-D-mannosyltransferase	2.4.1.57
GDPmannose 4,6-dehydratase	4.2.1.47
GDPmannose 4,6-hydro-lyase	4.2.1.47
GDPmannuronate:alginate D-mannuronyltransferase	2.4.1.33
GDP-6-deoxy-D-talose dehydrogenase	1.1.1.135
GDP-6-deoxy-D-talose:NAD(P)+ 4-oxidoreductase	1.1.1.135
Gentamicin acetyltransferase I	2.3.1.60
Gentamicin acetyltransferase II	2.3.1.59
Gentamicin 2′-acetyltransferase	2.3.1.59
Gentamicin 2″-nucleotidyltransferase	2.7.7.46
Gentamicin 3-acetyltransferase	2.3.1.60
Gentiobiase	3.2.1.21
Gentisate carboxy-lyase	4.1.1.62
Gentisate decarboxylase	4.1.1.62

Gentisate oxygenase	1.13.11.4
Gentisate:oxygen 1,2-oxidoreductase (decyclizing)	1.13.11.4
Gentisate 1,2-dioxygenase	1.13.11.4
Geranoyl-CoA:carbon-dioxide ligase (ADP-forming)	6.4.1.5
Geranoyl-CoA carboxylase	6.4.1.5
Geranyldiphosphate:isopentenydiphosphate geranyltransferase	2.5.1.10
Geranyltransferase	2.5.1.10
Gliocladium **proteinase**	3 4.99.8
1,4-α-Glucan branching enzyme	2.4.1.18
1,4-β-D-Glucan cellobiohydrolase	3.2.1.91
1,6-β-D-Glucan glucanohydrolase	3.2.1.75
1,3-β-D-Glucan glucanohydrolase	3.2.1.39
1,4-α-D-Glucan glucanohydrolase	3.2.1.1
1,2-β-D-Glucan glucanohydrolase	3.2.1.71
1,4-α-D-Glucan glucohydrolase	3.2.1.3
1,6-α-D-Glucan glucohydrolase	3.2.1.70
1,3-β-D-Glucan glucohydrolase	3.2.1.58
1,4-β-D-Glucan glucohydrolase	3.2.1.74
1,6-α-D-Glucan isomaltohydrolase	3.2.1.94
1,6-α-D-Glucan isomaltotriohydrolase	3.2.1.95
1,4-α-D-Glucan maltohexaohydrolase	3.2.1.98
1,4-α-D-Glucan maltohydrolase	3.2.1.2
1,4-α-D-Glucan maltotetraohydrolase	3.2.1.60
1,3-β-D-Glucan:orthophosphate glucosyltransferase	2.4.1.97
1,4-α-D-Glucan:orthophosphate α-D-glucosyltransferase	2.4.1.1
4-α-D-Glucanotransferase	2.4.1.3
4-α-D-Glucanotransferase	2.4.1.25
1,3-β-D-Glucan phosphorylase	2.4.1.97
1,3-β-D-Glucan synthase	2.4.1.34
1,3-β-D-Glucan—UDPglucosyltransferase	2.4.1.34
1,4-α-D-Glucan:1,4-α-D-glucan (D-glucose) 6-α-D-glucosyltransferase	2.4.1.24
1,4-α-D-Glucan:1,4-α-D-glucan 4-α-D-glycosyltransferase	2.4.1.25
1,4-α-D-Glucan:1,4-α-D-glucan 6-α-D-(1,4-α D-glucano)-transferase	2.4.1.18
1,4-α-D-Glucan:1,6-α-D-glucan 6-α-D-glucosyltransferase	2.4.1.2
1,3-(1,3;1,4)-α-D-Glucan 3-glucanohydrolase	3.2.1.59
1,3-α-D-Glucan 3-glucohydrolase	3.2.1.84
1,3-(1,3;1,4)-β-D-Glucan 3 (4)-glucanohydrolase	3.2.1.6
1,3-1,4-β-D-Glucan 4-glucanohydrolase	3.2.1.73
1,4-(1,3;1,4)-β-D-Glucan 4-glucanohydrolase	3.2.1.4
1,3-1,4-α-D-Glucan 4-glucanohydrolase	3.2.1.61
1,4-α-D-Glucan 4-α-D-(1,4-α-D-glucano)-transferase (cyclizing)	2.4.1.19
1,6-α-D-Glucan 6-glucanohydrolase	3.2.1.11
1,4-α-D-Glucan 6-α-D-glucosyltransferase	2.4.1.24
Glucarate dehydratase	4.2.1.40
D-*Glucarate hydro-lyase*	4.2.1.40
Glucoamylase	3.2.1.3
Glucodextranase	3.2.1.70
Glucoinvertase	3.2.1.20
Glucokinase	2.7.1.2
Glucomannan 4-β-D-mannosyltransferase	2.4.1.32
D-*Gluconate:(acceptor) 2-oxidoreductase*	1.1.99.3
Gluconate dehydratase	4.2.1.39
D-*Gluconate hydro-lyase*	4.2.1.39
D-*Gluconate:NAD(P)⁺ 5-oxidoreductase*	1.1.1.69
Gluconate 2-dehydrogenase	1.1.99.3
Gluconate 5-dehydrogenase	1.1.1.69
Gluconokinase	2.7.1.12
Gluconolactonase	3.1.1.17
D-*Glucono-δ-lactone lactonohydrolase*	3.1.1.17
Glucosaminate ammonia-lyase	4.3.1.9
Glucosamine acetylase	2.3.1.3
Glucosamine acetyltransferase	2.3.1.3
Glucosamine kinase	2.7.1.8
Glucosamine-phosphate acetyltransferase	2.3.1.4
Glucosaminephosphate isomerase	5.3.1.10
Glucosaminephosphate isomerase (glutamine-forming)	5.3.1.19
4-N-(2-β-D-Glucosaminyl)-L-asparaginase	3.5.1.26
D-*Glucose:(acceptor) 1-oxidoreductase*	1.1.99.10
Glucose dehydrogenase	1.1.1.47
Glucose dehydrogenase (acceptor)	1.1.99.10
Glucose dehydrogenase (NAD⁺)	1.1.1.118
Glucose dehydrogenase (NADP⁺)	1.1.1.119
Glucose dehydrogenase (*Aspergillus*)	1.1.99.10
Glucose isomerase	5.3.1.18
β-D-*Glucose:NAD(P)⁺ 1-oxidoreductase*	1.1.1.47
D-*Glucose:NADP⁺ 1-oxidoreductase*	1.1.1.119
D-*Glucose:NAD⁺ 1-oxidoreductase*	1.1.1.118
Glucose oxidase	1.1.3.4
β-D-*Glucose:oxygen 1-oxidoreductase*	1.1.3.4
Glucose oxyhydrase	1.1.3.4
Glucosephosphate isomerase	5.3.1.9
Glucose phosphomutase	2.7.5.1
Glucose phosphomutase	2.7.5.5
Glucose-1-phosphatase	3.1.3.10
Glucose-1-phosphate adenylyltransferase	2.7.7.27
Glucose-1-phosphate cytidylyltransferase	2.7.7.33

553

D-*Glucose-1-phosphate:D-glucose-1-phosphate 6-phosphotransferase*	2.7.1.41
D-*Glucose-1-phosphate:D-glucose 6-phosphotransferase*	2.7.5.5
Glucose-1-phosphate guanylyltransferase	2.7.7.34
Glucose-1-phosphate phosphodismutase	2.7.1.41
D-*Glucose-1-phosphate phosphohydrolase*	3.1.3.10
D-*Glucose-1-phosphate:riboflavin 5'-phosphotransferase*	2.7.1.42
Glucose-1-phosphate thymidylyltransferase	2.7.7.24
Glucose-1-phosphate uridylyltransferase	2.7.7.9
α-D-*Glucose-1,6-bisphosphate:deoxy-D-ribose-1-phosphate phosphotransferase*	2.7.5.6
α-D-*Glucose-1,6-bisphosphate:α-D-glucose-1-phosphate phosphotransferase*	2.7.5.1
Glucose-6-phosphatase	3.1.3.9
Glucose-6-phosphate dehydrogenase	1.1.1.49
Glucose-6-phosphate isomerase	5.3.1.9
D-*Glucose-6-phosphate ketol-isomerase*	5.3.1.9
D-*Glucose-6-phosphate:NADP+ 1-oxidoreductase*	1.1.1.49
D-*Glucose-6-phosphate phosphohydrolase*	3.1.3.9
Glucose-6-phosphate1-epimerase	5.1.3.15
Glucose-6-phosphate 1-epimerase	5.1.3.15
β-D-Glucosidase	3.2.1.21
α-1,3-Glucosidase	3.2.1.27
α-D-Glucosidase	3.2.1.20
trans-N-Glucosidase	2.4.2.6
β-D-*Glucoside glucohydrolase*	3.2.1.21
α-D-*Glucoside glucohydrolase*	3.2.1.20
β-D-Glucoside kinase	2.7.1.85
D-Glucoside 3-dehydrogenase	1.1.99.13
Glucosidosucrase	3.2.1.20
Glucosulphatase	3.1.6.3
Glucosylceramidase	3.2.1.45
3-O-β-D-*Glucosylglucose:orthophosphate glucosyltransferase*	2.4.1.31
D-*Glucosyl-N-acylsphingosine glucohydrolase*	3.2.1.45
Glucuronate isomerase	5.3.1.12
D-*Glucuronate ketol-isomerase*	5.3.1.12
Glucuronate reductase	1.1.1.19
Glucuronate-1-phosphate uridylyltransferase	2.7.7.44
β-D-Glucuronidase	3.2.1.31
β-D-*Glucuronide glucuronosohydrolase*	3.2.1.31
Glucuronokinase	2.7.1.43
D-*Glucurono-δ-lactone lactonohydrolase*	3.1.1.19
Glucuronolactone reductase	1.1.1.20

Glucuronosyl-disulphoglucosamine glucuronidase	3.2.1.56
Δ*4,5-β-D-Glucuronosyl-(1,4)-2-acetamido-2-deoxy-D-galactose-4-sulphate 4-sulphohydrolase*	3.1.6.9
Δ*4,5-β-D-Glucuronosyl-(1,4)-2-acetamido-2-deoxy-D-galactose-6-sulphate 6-sulphohydrolase*	3.1.6.10
1,3-D-Glucuronosyl-2-sulphamido-2-deoxy-6-O-sulpho-β-D-glucose glucuronohydrolase	3.2.1.56
Glutamate acetyltransferase	2.3.1.35
L-*Glutamate:ammonia ligase (ADP-forming)*	6.3.1.2
D-Glutamate cyclase	4.2.1.48
L-*Glutamate:L-cysteine γ-ligase (ADP-forming)*	6.3.2.2
Glutamate decarboxylase	4.1.1.15
Glutamate dehydrogenase	1.4.1.2
Glutamate dehydrogenase (NADP+)	1.4.1.4
Glutamate dehydrogenase (NAD(P)+)	1.4.1.3
L-*Glutamate:ethylamine ligase (ADP-forming)*	6.3.1.6
L-*Glutamate:ferredoxin oxidoreductase (transaminating)*	1.4.7.1
Glutamate formiminotransferase	2.1.2.5
Glutamate formyltransferase	2.1.2.6
D-*Glutamate hydro-lyase (cyclizing)*	4.2.1.48
Glutamate kinase	2.7.2.11
L-*Glutamate:methylamine ligase (ADP-forming)*	6.3.4.12
Glutamate mutase	5.4.99.1
L-*Glutamate:NAD+ oxidoreductase (deaminating)*	1.4.1.2
L-*Glutamate:NAD+ oxidoreductase (transaminating)*	1.4.1.14
L-*Glutamate:NADP+ oxidoreductase (deaminating)*	1.4.1.4
L-*Glutamate:NAD(P)+ oxidoreductase (deaminating)*	1.4.1.3
L-*Glutamate:NADP+ oxidoreductase (transaminating)*	1.4.1.13
L-Glutamate oxidase	1.4.3.11
D-Glutamate oxidase	1.4.3.7
D-*Glutamate:oxygen oxidoreductase (deaminating)*	1.4.3.7
L-*Glutamate:oxygen oxidoreductase (deaminating)*	1.4.3.11
Glutamate racemase	5.1.1.3
Glutamate racemase	5.1.1.3
Glutamate-semialdehyde dehydrogenase	1.2.1.41
L-*Glutamate-γ-semialdehyde:NADP+ oxidoreductase (phosphorylating)*	1.2.1.41

Glutamate synthase (ferredoxin)	1.4.7.1
Glutamate synthase (NADH)	1.4.1.14
Glutamate synthase (NADPH)	1.4.1.13
L-*Glutamate:tRNA^{Glu} ligase (AMP-forming)*	6.1.1.17
L-*Glutamate 1-carboxy-lyase*	4.1.1.15
Glutamic-alanine transaminase	2.6.1.2
Glutamic-aspartic transaminase	2.6.1.1
Glutamic dehydrogenase	1.4.1.3
Glutamic dehydrogenase	1.4.1.2
Glutamic dehydrogenase	1.4.1.4
Glutamic-oxaloacetic transaminase	2.6.1.1
D-Glutamic oxidase	1.4.3.7
Glutamic-pyruvic transaminase	2.6.1.2
Glutaminase	3.5.1.2
D-Glutaminase	3.5.1.35
Glutaminase II	2.6.1.15
Glutamin-(asparagin-)ase	3.5.1.38
D-*Glutamine amidohydrolase*	3.5.1.35
L-*Glutamine amidohydrolase*	3.5.1.2
L-*Glutamine (L-asparagine) amidohydrolase*	3.5.1.38
Glutamine - fructose-6-phosphate aminotransferase	5.3.1.19
Glutamine:D-glutamyl-peptide glutamyltransferase	2.3.2.1
Glutamine—oxo-acid aminotransferase	2.6.1.15
Glutamine phenylacetyltransferase	2.3.1.14
Glutamine phosphoribosylpyrophosphate amidotransferase	2.4.2.14
Glutamine—scyllo-inosose aminotransferase	2.6.1.50
Glutamine synthetase	6.3.1.2
Glutamine-synthetase adenylyltransferase	2.7.7.42
L-*Glutamine:tRNA^{Gln} ligase (AMP-forming)*	6.1.1.18
L-*Glutamine:2-oxo-acid aminotransferase*	2.6.1.15
L-*Glutamine:2,4,6/3,5-pentahydroxycyclohexanone aminotransferase*	2.6.1.50
L-*Glutaminyl-peptide amidohydrolase*	3.5.1.44
Glutaminyl-peptide glutaminase	3.5.1.44
Glutaminyl-peptide γ-glutamyltransferase	2.3.2.13
Glutaminyl-tRNA cyclotransferase	2.3.2.5
L-*Glutaminyl-tRNA γ-glutamyltransferase (cyclizing)*	2.3.2.5
Glutaminyl-tRNA synthetase	6.1.1.18
(5-L-Glutamyl)-L-amino-acid 5-glutamyltransferase (cyclizing)	2.3.2.4
γ-Glutamylcyclotransferase	2.3.2.4
γ-L-*Glutamyl-L-cysteine:glycine ligase (ADP-forming)*	6.3.2.3
γ-Glutamylcysteine synthetase	6.3.2.2
r-Glutamylglutamate carboxypeptidase	3.4.12.13
α-Glutamyl-glutamate dipeptidase	3.4.13.7
2-L-*Glutamyl-L-glutamate hydrolase*	3.4.13.7
γ-Glutamyl hydrolase	3.4.22.12
γ-Glutamylmethylamide synthetase	6.3.4.12
(5-Glutamyl)-peptide:amino-acid 5-glutamyltransferase	2.3.2.2
γ-Glutamylphosphate reductase	1.2.1.41
γ-Glutamyltransferase	2.3.2.2
D-Glutamyltransferase	2.3.2.1
Glutamyl transpeptidase	2.3.2.2
D-Glutamyl transpeptidase	2.3.2.1
Glutamyl-tRNA synthetase	6.1.1.17
Glutarate:CoA ligase (ADP-forming)	6.2.1.6
Glutarate-semialdehyde dehydrogenase	1.2.1.20
Glutarate-semialdehyde:NAD$^+$ oxidoreductase	1.2.1.20
Glutaryl-CoA:(acceptor) oxidoreductase (decarboxylating)	1.3.99.7
Glutaryl-CoA dehydrogenase	1.3.99.7
Glutaryl-CoA synthetase	6.2.1.6
Glutathione—CoAS-SG transhydrogenase	1.8.4.3
Glutathione:cystine oxidoreductase	1.8.4.4
Glutathione—cystine transhydrogenase	1.8.4.4
Glutathione:dehydroascorbate oxidoreductase	1.8.5.1
Glutathione dehydrogenase (ascorbate)	1.8.5.1
Glutathione:homocystine oxidoreductase	1.8.4.1
Glutathione—homocystine transhydrogenase	1.8.4.1
Glutathione:hydrogen-peroxide oxidoreductase	1.11.1.9
Glutathione-insulin transhydrogenase	1.8.4.2
Glutathione peroxidase	1.11.1.9
Glutathione:protein-disulphide oxidoreductase	1.8.4.2
Glutathione reductase (NAD(P)H)	1.6.4.2
Glutathione S-alkyltransferase	2.5.1.12
Glutathione S-alkyltransferase	2.5.1.18
Glutathione S-aralkyltransferase	2.5.1.14
Glutathione S-aralkyltransferase	2.5.1.18
Glutathione S-aryltransferase	2.5.1.13
Glutathione S-aryltransferase	2.5.1.18
Glutathione synthetase	6.3.2.3
Glutathione thiolesterase	3.1.2.7
Glutathione transferase	2.5.1.18
Glyceraldehyde-phosphate dehydrogenase	1.2.1.12

Glyceraldehyde-phosphate dehydrogenase (NADP⁺)	1.2.1.9
Glyceraldehyde-phosphate dehydrogenase (NADP⁺) (phosphorylating)	1.2.1.13
D-*Glyceraldehyde-3-phosphate ketol-isomerase*	5.3.1.1
D-*Glyceraldehyde-3-phosphate:NAD⁺ oxidoreductase (phosphorylating)*	1.2.1.12
D-*Glyceraldehyde-3-phosphate:NADP⁺ oxidoreductase*	1.2.1.9
D-*Glyceraldehyde-3-phosphate:NADP⁺ oxidoreductase (phosphorylating)*	1.2.1.13
Glycerate dehydrogenase	1.1.1.29
Glycerate kinase	2.7.1.31
D-*Glycerate:NAD⁺ oxidoreductase*	1.1.1.29
D-*Glycerate:NAD(P)⁺ oxidoreductase*	1.1.1.60
D-*Glycerate:NAD(P)⁺ oxidoreductase (carboxylating)*	1.1.1.92
D-*Glycerate:NAD(P)⁺ 2-oxidoreductase*	1.1.1.81
Glycerate phosphomutase	2.7.5.4
Glycerate phosphomutase	2.7.5.3
D-*Glycerate-2-phosphate phosphohydrolase*	3.1.3.20
D-*Glycerate-3-phosphate phosphohydrolase*	3.1.3.38
Glycerol dehydratase	4.2.1.30
Glycerol dehydrogenase	1.1.1.6
Glycerol dehydrogenase (NADP⁺)	1.1.1.72
Glycerol hydro-lyase	4.2.1.30
Glycerol kinase	2.7.1.30
Glycerol-monoester acylhydrolase	3.1.1.23
Glycerol:NADP⁺ oxidoreductase	1.1.1.72
Glycerol:NADP⁺ 2-oxidoreductase (dihydroxyacetone-forming)	1.1.1.156
Glycerol:NAD⁺ 2-oxidoreductase	1.1.1.6
Glycerol-1-phosphatase	3.1.3.21
Glycerol-1-phosphate phosphohydrolase	3.1.3.21
Glycerol 2-dehydrogenase (NADP⁺)	1.1.1.156
Glycerol-2-phosphatase	3.1.3.19
Glycerol-2-phosphate phosphohydrolase	3.1.3.19
sn-*Glycerol-3-phosphate:(acceptor) oxidoreductase*	1.1.99.5
Glycerol-3-phosphate cytidylyltransferase	2.7.7.39
Glycerol-3-phosphate dehydrogenase	1.1.99.5
Glycerol-3-phosphate dehydrogenase (NAD⁺)	1.1.1.8
sn-**Glycerol-3-phosphate dehydrogenase (NAD(P)⁺)**	1.1.1.94
sn-*Glycerol-3-phosphate:NAD(P)⁺ 2-oxidoreductase*	1.1.1.94
sn-*Glycerol-3-phosphate:NAD⁺ 2-oxidoreductase*	1.1.1.8
Glycerophosphatase	3.1.3.1
Glycerophosphatase	3.1.3.2

Glycerophosphate acyltransferase	2.3.1.15
Glycerophosphate phosphatidyltransferase	2.7.8.5
Glycerophosphocholine cholinephosphodiesterase	3.1.4.38
L-*3-Glycerophosphocholine cholinephosphohydrolase*	3.1.4.38
L-*3-Glycerophosphocholine glycerophosphohydrolase*	3.1.4.2
Glycerophosphocholine phosphodiesterase	3.1.4.2
Glyceryl ether cleaving enzyme	1.14.16.5
Glyceryl-ether monooxygenase	1.14.16.5
Glycinamide ribonucleotide synthetase	6.3.4.13
Glycine acetyltransferase	2.3.1.29
Glycine acyltransferase	2.3.1.13
Glycine amidinotransferase	2.1.4.1
Glycine aminotransferase	2.6.1.4
Glycine carboxypeptidase	3.4.17.4
Glycine—cytochrome *c* reductase	1.4.2.1
Glycine dehydrogenase	1.4.1.10
Glycine dehydrogenase (cytochrome)	1.4.2.1
Glycine:ferricytochrome c oxidoreductase (deaminating)	1.4.2.1
Glycine formiminotransferase	2.1.2.4
Glycine methyltransferase	2.1.1.20
Glycine:NAD⁺ oxidoreductase (deaminating)	1.4.1.10
Glycine:oxaloacetate aminotransferase	2.6.1.35
Glycine—oxaloacetate aminotransferase	2.6.1.35
Glycine synthase	2.1.2.10
Glycine:tRNA^Gly ligase (AMP-forming)	6.1.1.14
Glycine:2-oxoglutarate aminotransferase	2.6.1.4
Glycocholase	3.5.1.24
Glycocyaminase	3.5.3.2
Glycogenase	3.2.1.1
Glycogenase	3.2.1.2
Glycogen (starch) synthase	2.4.1.11
Glycogen synthase *a* kinase	2.7.1.37
Glycogen-synthase-D phosphatase	3.1.3.42
Glycogen 6-glucanohydrolase	3.2.1.68
Glycolaldehyde dehydrogenase	1.2.1.21
Glycolaldehyde:NAD⁺ oxidoreductase	1.2.1.21
Glycolaldehydetransferase	2.2.1.1
Glycolate:NAD⁺ oxidoreductase	1.1.1.26
Glycollate:(acceptor) oxidoreductase	1.1.99.14
Glycollate dehydrogenase	1.1.99.14
Glycollate:NADP⁺ oxidoreductase	1.1.1.79
Glycollate oxidase	1.1.3.1
Glycollate:oxygen oxidoreductase	1.1.3.1
Glycoprotein β-D-galactosyltransferase	2.4.1.38
Glycosulphatase	3.1.6.3

Glycosylceramidase	3.2.1.62
Glycosyl-N-acylsphingosine glycohydrolase	3.2.1.62
Glycyl-glycine endopeptidase	3.4.99.17
Glycyl-tRNA synthetase	6.1.1.14
Glyoxalase I	4.4.1.5
Glyoxalase II	3.1.2.6
Glyoxylate carbo-ligase	4.1.1.47
Glyoxylate carboxy-lyase (dimerizing)	4.1.1.47
Glyoxylate dehydrogenase (acylating)	1.2.1.17
Glyoxylate:NADP+ oxidoreductase (CoA-oxalylating)	1.2.1.17
Glyoxylate oxidase	1.2.3.5
Glyoxylate:oxygen oxidoreductase	1.2.3.5
Glyoxylate reductase	1.1.1.26
Glyoxylate reductase (NADP+)	1.1.1.79
Glyoxylate transacetylase	4.1.3.2
GMP reductase	1.6.6.8
GMP synthetase	6.3.4.1
GMP synthetase (glutamine-hydrolysing)	6.3.5.2
dGTPase	3.1.5.1
GTP cyclohydrolase	3.5.4.16
GTP:fucose-1-phosphate guanylyltransferase	2.7.7.30
GTP:α-D-glucose-1-phosphate guanylyltransferase	2.7.7.34
GTP:GTP guanylyltransferase	2.7.7.45
GTP:α-D-hexose-1-phosphate guanylyltransferase	2.7.7.29
GTP—mannose-1-phosphate guanylyltransferase	2.7.7.13
GTP:α-D-mannose-1-phosphate guanylyltransferase	2.7.7.13
GTP:oxaloacetate carboxy-lyase (transphosphorylating)	4.1.1.32
GTP pyrophosphate-lyase (cyclizing)	4.6.1.2
dGTP triphosphohydrolase	3.1.5.1
GTP:3-phospho-D-glycerate 1-phosphotransferase	2.7.2.10
GTP:5-hydroxy-L-lysine O-phosphotransferase	2.7.1.81
GTP 7,8-8,9-dihydrolase	3.5.4.16
Guanase	3.5.4.3
Guanidinoacetate amidinohydrolase	3.5.3.2
Guanidinoacetate kinase	2.7.3.1
Guanidinoacetate methyltransferase	2.1.1.2
Guanidinobutyrase	3.5.3.7
4-Guanidinobutyrate amidinohydrolase	3.5.3.7
Guanidinodeoxy-*scyllo*-inositol-4-phosphatase	3.1.3.40
1-Guanidino-1-deoxy-scyllo-inositol-4-phosphate 4-phosphohydrolase	3.1.3.40

1D-1-Guanidino-3-amino-1,3-dideoxy-*scyllo*-inositol aminotransferase	2.6.1.56
1D-1-Guanidino-3-amino-1,3-dideoxy-scyllo-inositol:pyruvate aminotransferase	2.6.1.56
Guanine aminase	3.5.4.3
Guanine aminohydrolase	3.5.4.3
Guanine deaminase	3.5.4.3
Guanosine aminase	3.5.4.15
Guanosine aminohydrolase	3.5.4.15
Guanosine deaminase	3.5.4.15
Guanosine:orthophosphate ribosyltransferase	2.4.2.15
Guanosine phosphorylase	2.4.2.15
Guanosinetriphosphate guanylyltransferase	2.7.7.45
Guanylate cyclase	4.6.1.2
Guanylate kinase	2.7.4.8
Guanyl cyclase	4.6.1.2
Guanyloribonuclease	3.1.27.3
Guanylyl cyclase	4.6.1.2
L-Gulonate dehydrogenase	1.1.1.45
L-Gulonate:NADP+ 1-oxidoreductase	1.1.1.19
L-Gulonate:NAD+ 3-oxidoreductase	1.1.1.45
L-Gulono-γ-lactone lactonohydrolase	3.1.1.18
L-Gulono-γ-lactone:NADP+ 1-oxidoreductase	1.1.1.20
L-Gulonolactone oxidase	1.1.3.8
L-Gulono-γ-lactone:oxygen 2-oxidoreductase	1.1.3.8
m7G(5′)pppN pyrophosphatase	3.6.1.30
Haem,hydrogen-donor:oxygen oxidoreductase (α-methene-oxidizing, hydroxylating)	1.14.99.3
Haemophilus aprophilus (HapII)	3.1.23.24
Haemophilus haemoglobinophilus (HhgI) and Streptococcus faecalis (SfaI)	3.1.23.17
Haemophilus haemolyticus (HhaII)	3.1.23.22
Haemophilus influenzae HI (Hin HI) and Neisseria gonorrhoea (NgoI)	3.1.23.16
Haemophilus influenzae serotype b (HinbIII)	3.1.23.21
Haemophilus influenzae serotype c (HincII)	3.1.23.20
Haemophilus suis (HsuI)	3.1.23.21
Haem oxygenase (decyclizing)	1.14.99.3
Haemphilus influenzae serotype f (HinfII)	3.1.23.21
Haloacetate dehalogenase	3.8.1.3

Haloacetate halidohydrolase	3.8.1.3		L-*Histidine:2-oxoglutarate*	
2-Haloacid dehalogenase	3.8.1.2		*aminotransferase*	2.6.1.38
2-Haloacid halidohydrolase	3.8.1.2		**Histidinol dehydrogenase**	1.1.1.23
Halogenase	3.8.1.1		L-*Histidinol:NAD⁺ oxidoreductase*	1.1.1.23

Haloacetate halidohydrolase 3.8.1.3
2-Haloacid dehalogenase 3.8.1.2
2-Haloacid halidohydrolase 3.8.1.2
Halogenase 3.8.1.1
Hatching enzyme 3.4.24.12
HeLa cell RNase 3.1.26.1
Heparinase 3.2.1.19
Heparinase 4.2.2.7
Heparin eliminase 4.2.2.7
Heparin lyase 4.2.2.7
Heparin lyase 4.2.2.7
Heparin-sulphate eliminase 4.2.2.8
Heparin-sulphate lyase 4.2.2.8
Heparitinsulphate lyase 4.2.2.8
Heparitin sulphotransferase 2.8.2.12
Heptulokinase 2.7.1.14
Heterophosphatase 2.7.1.1
Hexadecanal dehydrogenase (acylating) 1.2.1.42
Hexadecanal:NAD⁺ oxidoreductase (CoA-acylating) 1.2.1.42
Hexadecanal:NADP⁺ oxidoreductase 1.3.1.27
Hexadecanol dehydrogenase 1.1.1.164
Hexadecanol:NAD⁺ oxidoreductase 1.1.1.164
2-Hexadecenal reductase 1.3.1.27
Hexokinase 2.7.1.1
Hexosediphosphatase 3.1.3.11
Hexose oxidase 1.1.3.5
D-*Hexose:oxygen 1-oxidoreductase* 1.1.3.5
Hexosephosphate aminotransferase 5.3.1.19
Hexosephosphate isomerase 5.3.1.9
Hexose-1-phosphate guanylyltransferase 2.7.7.29
Hexose-1-phosphate uridylyltransferase 2.7.7.12
Hippurate hydrolase 3.5.1.32
Hippuricase 3.5.1.14
Histaminase 1.4.3.6
Histamine methyltransferase 2.1.1.8
Histidase 4.3.1.3
Histidinase 4.3.1.3
Histidine acetyltransferase 2.3.1.33
L-*Histidine:β-alanine ligase (AMP-forming)* 6.3.2.11
Histidine aminotransferase 2.6.1.38
Histidine ammonia-lyase 4.3.1.3
L-*Histidine ammonia-lyase* 4.3.1.3
L-*Histidine carboxy-lyase* 4.1.1.22
Histidine α-deaminase 4.3.1.3
Histidine decarboxylase 4.1.1.22
L-*Histidine:tRNA^His ligase (AMP-forming)* 6.1.1.21

L-*Histidine:2-oxoglutarate aminotransferase* 2.6.1.38
Histidinol dehydrogenase 1.1.1.23
L-*Histidinol:NAD⁺ oxidoreductase* 1.1.1.23
Histidinol phosphatase 3.1.3.15
Histidinol-phosphate aminotransferase 2.6.1.9
Histidinol-phosphate phosphohydrolase 3.1.3.15
L-*Histidinol-phosphate:2-oxoglutarate aminotransferase* 2.6.1.9
Histidyl-tRNA synthetase 6.1.1.21
Histone acetyltransferase 2.3.1.48
Histone kinase 2.7.1.70
Histozyme 3.5.1.14
HLA 3.4.11.14
Hog kidney phosphodiesterase 3.1.15.1
Holo-[acyl-carrier-protein] synthase 2.7.8.7
Homoaconitate hydratase 4.2.1.36
Homocitrate synthase 4.1.3.21
Homocysteine desulphhydrase 4.4.1.2
L-*Homocysteine hydrogen-sulphide-lyase (deaminating)* 4.4.1.2
Homocysteine methyltransferase 2.1.1.10
Homogentisate oxygenase 1.13.11.5
Homogentisate:oxygen 1,2-oxidoreductase (decyclizing) 1.13.11.5
Homogentisate 1,2-dioxygenase 1.13.11.5
Homogentisicase 1.13.11.5
Homoisocitrate dehydrogenase 1.1.1.155
Homoprotocatechuate oxygenase 1.13.11.7
Homoserine acetyltransferase 2.3.1.31
Homoserine deaminase 4.4.1.1
Homoserine dehydratase 4.2.1.15
Homoserine dehydratase 4.4.1.1
Homoserine dehydrogenase 1.1.1.3
Homoserine kinase 2.7.1.39
L-*Homoserine:NAD(P)⁺ oxidoreductase* 1.1.1.3
Homoserine O-transsuccinylase 2.3.1.46
Homoserine succinyltransferase 2.3.1.46
Human acid DNase of gastric mucosa and cervix 3.1.22.1
Human placenta endonuclease 3.1.22.3
Hurain 3.4.99.9
Hyaluronate lyase 4.2.2.1
Hyaluronate lyase 4.2.2.1
Hyaluronate 3-glycanohydrolase 3.2.1.36
Hyaluronate 4-glycanohydrolase 3.2.1.35
Hyaluronidase 3.2.1.35
Hyaluronidase 3.2.1.36
Hyaluronidase (but *cf* EC 3.2.1.35 and 3.2.1.36) 4.2.2.1

Hyaluronoglucosaminidase	3.2.1.35
Hyaluronoglucosidase	3.2.1.35
Hyaluronoglucuronidase	3.2.1.36
Hybrid nuclease	3.1.4.34
Hydantoinase	3.5.2.2
Hydrogenase	1.18.3.1
Hydrogenase	1.12.1.2
Hydrogenase	1.12.2.1
Hydrogenase	1.18.3.1
Hydrogen dehydrogenase	1.12.1.2
Hydrogen:ferricytochrome c₃ oxidoreductase	1.12.2.1
Hydrogenlyase	1.18.3.1
Hydrogen:NAD⁺ oxidoreductase	1.12.1.2
Hydrogen-peroxide:hydrogen-peroxide oxidoreductase	1.11.1.6
Hydrogen sulphide:(acceptor) oxidoreductase	1.8.99.1
Hydrogen-sulphide acetyltransferase	2.3.1.10
Hydrogen-sulphide:ferredoxin oxidoreductase	1.8.7.1
Hydrogen-sulphide:NADP⁺ oxidoreductase	1.8.1.2
Hydroperoxide isomerase	5.3.99.1
D-*2-Hydroxyacid:(acceptor) oxidoreductase*	1.1.99.6
D-2-Hydroxyacid dehydrogenase	1.1.99.6
L-2-Hydroxyacid oxidase	1.1.3.15
L-*2-Hydroxyacid:oxygen oxidoreductase*	1.1.3.15
Hydroxyacid racemase, Lacticoracemase	5.1.2.1
3-Hydroxyacrylate hydro-lyase	4.2.1.71
D-*3-Hydroxyacyl-[acyl-carrier-protein]: NADP⁺ oxidoreductase*	1.1.1.100
3-Hydroxyacyl-CoA dehydrogenase	1.1.1.35
L-*3-Hydroxyacyl-CoA hydro-lyase*	4.2.1.17
L-*3-Hydroxyacyl-CoA:NAD⁺ oxidoreductase*	1.1.1.35
D-*3-Hydroxyacyl-CoA:NADP⁺ oxidoreductase*	1.1.1.36
β-Hydroxyacyl dehydrogenase	1.1.1.35
Hydroxyacylglutathione hydrolase	3.1.2.8
Hydroxyacylglutathione hydrolase	3.1.2.6
S-*2-Hydroxyacylglutathione hydrolase*	3.1.2.6
2-Hydroxyadipate:NAD⁺ oxidoreductase	1.1.1.172
S-(Hydroxyalkyl) glutathione lyase	4.4.1.7
S-*(Hydroxyalkyl) glutathione lyase*	2.5.1.18
L-Hydroxyaminoacid dehydratase	4.2.1.13
D-Hydroxyaminoacid dehydratase	4.2.1.14
3-Hydroxyanthranilate oxidase	1.10.3.5
3-Hydroxyanthranilate oxygenase	1.13.11.6

3-Hydroxyanthranilate:oxygen oxidoreductase	1.10.3.5
3-Hydroxyanthranilate:oxygen 3,4-oxidoreductase (decyclizing)	1.13.11.6
3-Hydroxyanthranilate 3,4-dioxygenase	1.13.11.6
3-Hydroxyaspartate aldolase	4.1.3.14
erythro-**3-Hydroxyaspartate dehydratase**	4.2.1.38
erythro-3-Hydroxy-L₅-aspartate glyoxylate-lyase	4.1.3.14
erythro-3-Hydroxy-L₅-aspartate hydro-lyase (deaminating)	4.2.1.38
4-Hydroxybenzoate carboxy-lyase	4.1.1.61
4-Hydroxybenzoate decarboxylase	4.1.1.61
3-Hydroxybenzoate,hydrogen-donor:oxygen oxidoreductase (4-hydroxylating)	1.14.99.13
p-Hydroxybenzoate hydroxylase	1.14.13.2
4-Hydroxybenzoate,NADPH:oxygen oxidoreductase (3-hydroxylating)	1.14.13.2
4-Hydroxybenzoate 3-monooxygenase	1.14.13.2
3-Hydroxybenzoate 4-hydroxylase	1.14.99.13
3-Hydroxybenzoate 4-monooxygenase	1.14.99.13
3-Hydroxybenzyl-alcohol dehydrogenase	1.1.1.97
3-Hydroxybenzyl-alcohol:NADP⁺ oxidoreductase	1.1.1.97
3-Hydroxybutyrate dehydrogenase	1.1.1.30
4-Hydroxybutyrate dehydrogenase	1.1.1.61
Hydroxybutyrate-dimer hydrolase	3.1.1.22
4-Hydroxybutyrate:NAD⁺ oxidoreductase	1.1.1.61
D-*3-Hydroxybutyrate:NAD⁺ oxidoreductase*	1.1.1.30
D-*3-Hydroxybutyryl-[acyl-carrier-protein] hydro-lyase*	4.2.1.58
D-3-Hydroxybutyryl-CoA dehydratase	4.2.1.55
3-Hydroxybutyryl-CoA dehydrogenase	1.1.1.157
3 Hydroxybutyryl-CoA epimerase	5.1.2.3
D-*3-Hydroxybutyryl-CoA hydro-lyase*	4.2.1.55
L-*3-Hydroxybutyryl-CoA:NADP⁺ oxidoreductase*	1.1.1.157
3-Hydroxybutyryl-CoA 3-epimerase	5.1.2.3
3-D-(3-D-Hydroxybutyryloxy)-butyrate hydroxybutyrohydrolase	3.1.1.22
3α-Hydroxycholanate dehydrogenase	1.1.1.52
25-Hydroxycholecalciferol, NADPH:oxygen oxidoreductase (1-hydroxylating)	1.14.13.13
25-Hydroxy-cholecalciferol 1-hydroxylase	1.14.13.13
25-Hydroxycholecalciferol 1-monooxygenase	1.14.13.13
4-Hydroxycinnamate,ascorbate:oxygen oxidoreductase (3-hydroxylating)	1.14.17.2

Hydroxycyclohexanecarboxylate dehydrogenase	1.1.1.166
2-Hydroxycyclohexanone 2-monooxygenase	1.14.12.6
2-Hydroxycyclohexan-1-one	*1.14.12.6*
ω-Hydroxydecanoate dehydrogenase	1.1.1.66
10-Hydroxydecanoate:NAD+ oxidoreductase	*1.1.1.66*
D-3-Hydroxydecanoyl-[acyl-carrier-protein] dehydratase	4.2.1.60
D-3-Hydroxydecanoyl-[acyl-carrier-protein] hydro-lyase	*4.2.1.60*
2-Hydroxyethylene-dicarboxylate hydro-lyase	*4.2.1.72*
Hydroxyethylthiazole kinase	2.7.1.50
L-2-Hydroxy-fatty-acid dehydrogenase	1.1.1.99
D-2-Hydroxy-fatty-acid dehydrogenase	1.1.1.98
Hydroxyglutamate decarboxylase	4.1.1.16
4-Hydroxyglutamate transaminase	2.6.1.23
3-Hydroxy-L-glutamate 1-carboxy-lyase	*4.1.1.16*
4-Hydroxy-L-glutamate:2-oxoglutarate aminotransferase	*2.6.1.23*
L-2-Hydroxyglutarate:(acceptor) oxidoreductase	*1.1.99.2*
2-Hydroxyglutarate dehydrogenase	1.1.99.2
2-Hydroxyglutarate glyoxylate-lyase (CoA-propionylating)	*4.1.3.9*
2-Hydroxyglutarate synthase	4.1.3.9
3-Hydroxyisobutyrate dehydrogenase	1.1.1.31
3-Hydroxyisobutyrate:NAD+ oxidoreductase	*1.1.1.31*
3-Hydroxyisobutyryl-CoA hydrolase	*3.1.2.4*
3-Hydroxyisobutyryl-CoA hydrolase	3.1.2.4
Hydroxylamine oxidase	1.7.3.4
Hydroxylamine:oxygen oxidoreductase	*1.7.3.4*
Hydroxylamine reductase	1.7.99.1
Hydroxylamine reductase (NADH)	1.6.6.11
ω-Hydroxylase	1.14.15.3
Hydroxylysine kinase	2.7.1.81
Hydroxymalonate dehydrogenase	1.1.1.167
Hydroxymalonate:NAD+ oxidoreductase	*1.1.1.167*
4-Hydroxymandelonitrile hydroxybenzaldehyde-lyase	*4.1.2.11*
Hydroxymandelonitrile lyase	4.1.2.11
Hydroxymethylglutaryl-CoA hydrolase	3.1.2.5
Hydroxymethylglutaryl-CoA lyase	4.1.3.4
Hydroxymethylglutaryl-CoA reductase	1.1.1.88
Hydroxymethylglutaryl-CoA reductase (NADPH)	1.1.1.34
Hydroxymethylglutaryl-CoA synthase	4.1.3.5
Hydroxymethylpyrimidine kinase	2.7.1.49

5-Hydroxymethyluracil,2-oxoglutarate dioxygenase	1.14.11.5
6-Hydroxynicotinate reductase	1.3.7.1
6-Hydroxy-L-nicotine oxidase	1.5.3.5
6-Hydroxy-D-nicotine oxidase	1.5.3.6
6-Hydroxy-L-nicotine:oxygen oxidoreductase	*1.5.3.5*
6-Hydroxy-D-nicotine:oxygen oxidoreductase	*1.5.3.6*
5-Hydroxy-*N*-methylpyroglutamate synthase	3.5.1.36
3-Hydroxyoctanoyl-[acyl-carrier-protein] dehydratase	4.2.1.59
3-Hydroxyoctanoyl-[acyl-carrier-protein] hydro-lyase	*4.2.1.59*
3-Hydroxypalmitoyl-[acyl-carrier-protein] dehydratase	4.2.1.61
3-Hydroxypalmitoyl-[acyl-carrier-protein] hydro-lyase	*4.2.1.61*
4-Hydroxyphenylacetate	*1.14.13.18*
4-Hydroxyphenylacetate,NADH:oxygen oxidoreductase (3-hydroxylating)	*1.14.13.3*
4-Hydroxyphenyl acetate 1-hydroxylase	1.14.13.18
4-Hydroxyphenylacetate 1-monooxygenase	1.14.13.18
p-Hydroxyphenyl acetate 3-hydroxylase	1.14.13.3
4-Hydroxyphenylacetate 3-monooxygenase	1.14.13.3
2-Hydroxyphenyl propionate hydroxylase	1.14.13.4
3-(2-Hydroxyphenyl)-propionate,NADH:oxygen oxidoreductase (3-hydroxylating)	*1.14.13.4*
2-Hydroxyphenylpropionate:NAD+ oxidoreductase	*1.3.1.11*
4-Hydroxyphenylpyruvate dioxygenase	1.13.11.27
4-Hydroxyphenylpyruvate:oxygen oxidoreductase (hydroxylating, decarboxylating)	*1.13.11.27*
17α-Hydroxyprogesterone acetaldehyde-lyase	*4.1.2.30*
17α-Hydroxyprogesterone aldolase	4.1.2.30
Hydroxyproline epimerase	5.1.1.8
4-Hydroxy-L-proline:NAD+ oxidoreductase	*1.1.1.104*
Hydroxyproline 2-epimerase	*5.1.1.8*
3-Hydroxypropionate dehydrogenase	1.1.1.59
3-Hydroxypropionate:NAD+ oxidoreductase	*1.1.1.59*
15-Hydroxyprostaglandin dehydrogenase	1.1.1.141
Hydroxypyruvate carboxy-lyase	*4.1.1.40*
Hydroxypyruvate decarboxylase	4.1.1.40
Hydroxypyruvate reductase	1.1.1.81

D-*2-Hydroxystearate:NAD+* *oxidoreductase*	1.1.1.98
L-*2-Hydroxystearate:NAD+* *oxidoreductase*	1.1.1.99
*10-*D-*Hydroxystearate10-hydro-lyase*	4.2.1.53
β-Hydroxysteroid dehydrogenase	1.1.1.51
20β-Hydroxysteroid dehydrogenase	1.1.1.53
20α-Hydroxysteroid dehydrogenase	1.1.1.149
12α-Hydroxysteroid dehydrogenase	1.1.1.176
21-Hydroxysteroid dehydrogenase	1.1.1.150
3α-Hydroxysteroid dehydrogenase	1.1.1.50
11β-Hydroxysteroid dehydrogenase	1.1.1.146
16α-Hydroxysteroid dehydrogenase	1.1.1.147
7α-Hydroxysteroid dehydrogenase	1.1.1.159
3β-Hydroxy-Δ⁵-steroid dehydrogenase	1.1.1.145
21-Hydroxysteroid dehydrogenase (NADP+)	1.1.1.151
16-Hydroxysteroid epimerase	5.1.99.2
3α-Hydroxysteroid:NAD(P)+ *oxidoreductase*	1.1.1.50
3(or 17) β-Hydroxysteroid:NAD(P)+ *oxidoreductase*	1.1.1.51
11β-Hydroxysteroid:NADP+ *11-oxidoreductase*	1.1.1.146
16α-Hydroxysteroid:NAD(P)+ *16-oxidoreductase*	1.1.1.147
17α-Hydroxysteroid:NAD(P)+ *17-oxidoreductase*	1.1.1.148
17β-Hydroxysteroid:NADP+ *17-oxidoreductase*	1.1.1.64
20α-Hydroxysteroid:NAD(P)+ *20-oxidoreductase*	1.1.1.149
21-Hydroxysteroid:NADP+ *21-oxidoreductase*	1.1.1.151
12α-Hydroxysteroid:NAD+ *12α-oxidoreductase*	1.1.1.176
17β-Hydroxysteroid:NAD+ *17-oxidoreductase*	1.1.1.63
21-Hydroxysteroid:NAD+ *21-oxidoreductase*	1.1.1.150
3β-Hydroxy-Δ⁵-steroid:NAD+ *3-oxidoreductase*	1.1.1.145
7α-Hydroxysteroid:NAD+ *7-oxidoreductase*	1.1.1.159
3β-Hydroxysteroid sulphotransferase	2.8.2.2
16-Hydroxysteroid 16-epimerase	5.1.99.2
5α-Hydroxysterol dehydratase	4.2.1.62
Hydroxytryptophan decarboxylase	4.1.1.28
1-Hydroxy-1,2,4-butanetricarboxylate:NAD+ *oxidoreductase (decarboxylating)*	1.1.1.155
3-Hydroxy-2-methylpyridine-4,5-dicarboxylate 4-carboxy-lyase	4.1.1.51

3-Hydroxy-2-methyl-pyridine-4,5-dicarboxylate 4-decarboxylase	4.1.1.51
2-Hydroxy-2-methyl-3-oxobutyrate carboxy-lyase	4.1.1.5
4-Hydroxy-2-oxoglutarate aldolase	4.1.3.16
4-Hydroxy-2-oxoglutarate glyoxylate-lyase	4.1.3.16
2-Hydroxy-3-carboxyadipate dehydrogenase	1.1.1.87
2-Hydroxy-3-carboxyadipate hydro-lyase	4.2.1.36
2-Hydroxy-3-carboxyadipate:NAD+ *oxidoreductase (decarboxylating)*	1.1.1.87
3-Hydroxy 3 carboxyadipate 2-oxoglutarate-lyase (CoA-acetylating)	4.1.3.21
3-Hydroxy-3-isohexenylglutaryl-CoA hydro-lyase	4.2.1.57
3-Hydroxy-3-isohexenylglutaryl-CoA lyase	4.1.3.26
3-Hydroxy-3-isohex-3-enylglutaryl-CoA isopentenylacetoacetyl-CoA-lyase	4.1.3.26
3-Hydroxy-3-methylglutaryl-CoA acetoacetate-lyase	4.1.3.4
3-Hydroxy-3-methylglutaryl-CoA acetoacetyl-CoA-lyase (CoA-acetylating)	4.1.3.5
3-Hydroxy-3-methylglutaryl-CoA hydrolase	3.1.2.5
3-Hydroxy-3-methylglutaryl-CoA hydro-lyase	4.2.1.18
2-Hydroxy-3-oxoadipate carboxylase	4.1.3.15
2-Hydroxy-3-oxoadipate glyoxylate-lyase (carboxylating)	4.1.3.15
D-*2-Hydroxy-3,3-dimethyl-3-formylpropionate: NAD+ 4-oxidoreductase*	1.2.1.33
N-*Hydroxy-4-acetylaminobiphenyl:* N-*hydroxy-4-aminobiphenyl* O-*acetyltransferase*	2.3.1.56
2-Hydroxy-4-carboxy-cis-cis-muconate-6-semialdehyde:NADP+ oxidoreductase	1.2.1.45
(+)-4-Hydroxy-4-carboxymethylisocrotonolactone Δ²—Δ³-isomerase	5.3.3.4
2-Hydroxy-4-carboxymuconate-6-semialdehyde dehydrogenase	1.2.1.45
3β-Hydroxy-4β-methylcholestenoate dehydrogenase (decarboxylating)	1.1.1.170
4-Hydroxy-4-methyl-2-oxoglutarate aldolase	4.1.3.17
4-Hydroxy-4-methyl-2-oxoglutarate pyruvate-lyase	4.1.3.17
2-Hydroxy-4-methyl-3-carboxyvalerate hydro-lyase	4.2.1.33
2-Hydroxy-4-methyl-3-carboxyvalerate:NAD+ oxidoreductase	1.1.1.85

3-Hydroxy-4-methyl-3-carboxyvalerate 2-oxo-3-methylbutyrate-lyase (CoA-acetylating)	4.1.3.12
3β-Hydroxy-4β-methyl-5α-cholest-7-en-4α-oate:NAD+ oxidoreductase (decarboxylating)	1.1.1.170
3α-Hydroxy-5β-cholanate:NAD+ oxidoreductase	1.1.1.52
3α-Hydroxy-5β-steroid:NAD+ 3-oxidoreductase	1.1.1.152
Hyponitrite reductase	1.6.6.6
Hypotaurine dehydrogenase	1.8.1.3
Hypotaurine:NAD+ oxidoreductase	1.8.1.3
Hypotaurocyamine kinase	2.7.3.6
Hypoxanthine oxidase	1.2.3.2
Hypoxanthine phosphoribosyltransferase	2.4.2.8
D-Iditol dehydrogenase	1.1.1.15
L-Iditol dehydrogenase	1.1.1.14
L-Iditol:NAD+ 5-oxidoreductase	1.1.1.14
D-Iditol:NAD+ 5-oxidoreductase	1.1.1.15
L-Idonate dehydrogenase	1.1.1.128
L-Idonate:NADP+ 2-oxidoreductase	1.1.1.128
L-Iduronidase	3.2.1.76
Imidazoleacetate,NADH:oxygen oxidoreductase (hydroxylating)	1.14.13.5
Imidazoleacetate 4-monooxygenase	1.14.13.5
Imidazoleacetate: 5′-phosphoribosyldiphosphate ligase (ADP and pyrophosphate-forming)	6.3.4.8
Imidazole acetylase	2.3.1.2
Imidazole acetyltransferase	2.3.1.2
Imidazoleglycerol-phosphate dehydratase	4.2.1.19
D-erythro-Imidazoleglycerol-phosphate hydro-lyase	4.2.1.19
Imidazolonepropionase	3.5.2.7
4-Imidazolone-5-propionate amidohydrolase	3.5.2.7
4-Imidazolone-5-propionate hydro-lyase	4.2.1.49
Imidazolylacetol-phosphate aminotransferase	2.6.1.9
Imidazol-5-yl-lactate dehydrogenase	1.1.1.111
Imidazol-5-yl-lactate:NAD(P)+ oxidoreductase	1.1.1.111
Iminodipeptidase	3.4.13.8
Iminodipeptidase	3.4.13.9
IMP:L-aspartate ligase (GDP-forming)	6.3.4.4
IMP cyclohydrolase	3.5.4.10
IMP dehydrogenase	1.2.1.14
IMP:NAD+ oxidoreductase	1.2.1.14
IMP:pyrophosphate phosphoribosyltransferase	2.4.2.8
IMP pyrophosphorylase	2.4.2.8

IMP 1,2-hydrolase (decyclizing)	3.5.4.10
Indanol dehydrogenase	1.1.1.112
1-Indanol:NAD(P)+ oxidoreductase	1.1.1.112
Indoleacetaldoxime dehydratase	4.2.1.29
3-Indoleacetaldoxime hydro-lyase	4.2.1.29
Indoleglycerolphosphate aldolase	4.1.2.8
Indolelactate dehydrogenase	1.1.1.110
Indolelactate:NAD+ oxidoreductase	1.1.1.110
Indole:oxygen 2,3-oxidoreductase (decyclizing)	1.13.11.17
Indolepyruvate methyltransferase	2.1.1.47
Indole 2,3-dioxygenase	1.13.11.17
Indole-3-glycerol-phosphate synthase	4.1.1.48
Indophenolase	1.9.3.1
Indophenol oxidase Atmungsferment	1.9.3.1
Inorganic pyrophosphatase	3.6.1.1
*scyllo-***Inosamine kinase**	2.7.1.65
Inosamine-phosphate amidinotransferase	2.1.4.2
Inosinase	3.2.2.2
Inosinate nucleosidase	3.2.2.12
5′-Inosinate phosphoribohydrolase	3.2.2.12
Inosine kinase	2.7.1.73
Inosine nucleosidase	3.2.2.2
Inosine phosphorylase	2.4.2.1
Inosine ribohydrolase	3.2.2.2
myo-Inositol-hexakisphosphate 3-phosphohydrolase	3.1.3.8
myo-Inositol-hexakisphosphate 6-phosphohydrolase	3.1.3.26
myo-Inositol:NAD+ 2-oxidoreductase	1.1.1.18
*myo-***Inositol oxygenase**	1.13.99.1
myo-Inositol:oxygen oxidoreductase	1.13.99.1
*myo-***Inositol 1-kinase**	2.7.1.64
*myo-***Inositol 1-methyltransferase**	2.1.1.39
1L-*myo-*Inositol-1-phosphatase	3.1.3.25
1L-myo-Inositol-1-phosphate lyase (isomerizing)	5.5.1.4
1L-myo-Inositol-1-phosphate phosphohydrolase	3.1.3.25
*myo-***Inositol-1-phosphate synthase**	5.5.1.4
D-myo-Inositol-1:2-cyclic-phosphate 2-inositolphosphohydrolase	3.1.4.36
*myo-***Inositol 2-dehydrogenase**	1.1.1.18
*myo-***Inositol 3-methyltransferase**	2.1.1.40
*myo-***Inosose-2 dehydratase**	4.2.1.44
Insulinase	3.4.22.11
Insulinase	3.4.99.10
Insulin reductase	1.8.4.2
Inulase	3.2.1.7
Inulase II	2.4.1.93

2-Keto-3-deoxy-D-gluconate:NAD(P)+ 5-oxidoreductase	1.1.1.127
2-Keto-3-deoxy-D-gluconate:NADP+ 6-oxidoreductase	1.1.1.126
2-Keto-3-deoxygluconokinase	2.7.1.45
2-Keto-3-deoxy-L-pentonate aldolase	4.1.2.18
2-Keto-3-deoxy-D-pentonate aldolase	4.1.2.28
2-Keto-3-deoxy-D-pentonate glycollaldehyde-lyase	4.1.2.28
2-Keto-3-deoxy-L-pentonate glycollaldehyde-lyase	4.1.2.18
5-Keto-4-deoxy-D-glucarate dehydratase	4.2.1.41
5-Keto-4-deoxy-D-glucarate hydro-lyase (decarboxylating)	4.2.1.41
Kidney brush border neutral proteinase	3.4.24.11
Kinase II	3.4.15.1
Kininogenase	3.4.21.8
Kininogenin	3.4.21.8
Kynurenate,hydrogen-donor:oxygen oxidoreductase (hydroxylating)	1.14.99.2
Kynurenate-7,8-dihydrodiol dehydrogenase	1.3.1.18
Kynurenate 7,8-hydroxylase	1.14.99.2
Kynureninase	3.7.1.3
Kynurenine aminotransferase	2.6.1.7
Kynurenine formamidase	3.5.1.9
L-Kynurenine hydrolase	3.7.1.3
L-Kynurenine,NADPH:oxygen oxidoreductase (3-hydroxylating)	1.14.13.9
L-Kynurenine:2-oxoglutarate aminotransferase (cyclizing)	2.6.1.7
Kynurenine 3-hydroxylase	1.14.13.9
Kynurenine 3-monooxygenase	1.14.13.9
Laccase	1.10.3.2
Lactaldehyde dehydrogenase	1.2.1.22
D-Lactaldehyde dehydrogenase	1.1.1.78
L-Lactaldehyde:NAD+ oxidoreductase	1.2.1.22
D-Lactaldehyde:NAD+ oxidoreductase	1.1.1.78
Lactaldehyde reductase	1.1.1.77
Lactaldehyde reductase (NADPH)	1.1.1.55
β-Lactamase I	3.5.2.6
β-Lactamase II	3.5.2.6
Lactase	3.2.1.23
Lactate dehydrogenase	1.1.1.27
D-Lactate dehydrogenase	1.1.1.28
D-Lactate dehydrogenase (cytochrome)	1.1.2.4
Lactate dehydrogenase (cytochrome)	1.1.2.3
L-Lactate:ferricytochrome c oxidoreductase	1.1.2.3
D-Lactate:ferricytochrome c oxidoreductase	1.1.2.4

Lactate—malate transhydrogenase	1.1.99.7
L-Lactate:NAD+ oxidoreductase	1.1.1.27
D-Lactate:NAD+ oxidoreductase	1.1.1.28
Lactate:oxaloacetate oxidoreductase	1.1.99.7
Lactate oxidative decarboxylase	1.13.12.4
L-Lactate:oxygen 2-oxidoreductase (decarboxylating)	1.13.12.4
Lactate racemase	5.1.2.1
Lactate racemase	5.1.2.1
Lactate 2-monooxygenase	1.13.12.4
Lactic acid dehydrogenase	1.1.1.28
Lactic acid dehydrogenase	1.1.2.3
Lactic acid dehydrogenase	1.1.2.4
Lactic acid dehydrogenase	1.1.1.27
Lactobacillus exonuclease	3.1.15.1
γ-Lactonase	3.1.1.25
Lactonase	3.1.1.17
γ-Lactone hydroxyacylhydrolase	3.1.1.25
Lactose synthase	2.4.1.22
Lactoyl-CoA dehydratase	4.2.1.54
Lactoyl-CoA hydro-lyase	4.2.1.54
Lactoyl-glutathione lyase	4.4.1.5
S-D-Lactoyl-glutathione methylglyoxal-lyase (isomerizing)	4.4.1.5
Lambda exonuclease. T4	3.1.11.3
Laminaribiose phosphorylase	2.4.1.31
Laminarinase	3.2.1.6
Laminarinase	3.2.1.39
Laminarin phosphorylase	2.4.1.97
Lathosterol oxidase	1.3.3.2
Latia luciferin,hydrogen-donor:oxygen oxidoreductase (demethylating)	1.14.99.21
Latia luciferin monooxygenase (demethylating)	1.14.99.21
Lecanorate hydrolase	3.1.1.40
Lecithin acyltransferase	2.3.1.43
Lecithinase A	3.1.1.4
Lecithinase B	3.1.1.5
Lecithinase C	3.1.4.3
Lecithinase D	3.1.4.4
Lecithin:cholesterol acyltransferase	2.3.1.43
Lecithin—cholesterol acyltransferase	2.3.1.43
Lens neutral proteinase	3.4.24.5
Leucine amino-peptidase ('LAP')	3.4.11.1
Leucine aminotransferase	2.6.1.6
Leucine dehydrogenase	1.4.1.9
L-Leucine:NAD+ oxidoreductase (deaminating)	1.4.1.9
L-Leucine:tRNA^Leu ligase (AMP-forming)	6.1.1.4
L-Leucine:2-oxoglutarate aminotransferase	2.6.1.6

Leucostoma **neutral proteinase**	3.4.24.6
Leucyltransferase	2.3.2.6
L-*Leucyl-tRNA:protein leucyltransferase*	2.3.2.6
Leucyl-tRNA synthetase	6.1.1.4
Levanase	3.2.1.65
Levansucrase	2.4.1.10
Lichenase	3.2.1.73
Licheninase	3.2.1.5
Limit dextrinase	3.2.1.41
Limit dextrinase	3.2.1.10
Limonin-D-ring-lactonase	3.1.1.36
Limonoate-D-ring-lactone lactonohydrolase	3.1.1.36
Linamarin synthase	2.4.1.63
Linoleate Δ^{12}-cis-Δ^{11}-trans-isomerase	5.2.1.5
Linoleate isomerase	5.2.1.5
Linoleate:oxygen oxidoreductase	1.13.11.12
Lipase	3.1.1.3
Lipoamide dehydrogenase (NADH), Lipoamide reductase (NADH)	1.6.4.3
Lipoate acetyltransferase	2.3.1.12
Lipophosphodiesterase I	3.1.4.3
Lipophosphodiesterase II	3.1.4.4
Lipoprotein lipase	3.1.1.34
Lipoxidase	1.13.11.12
Lipoxygenase	1.13.11.12
Lipoyl dehydrogenase	1.6.4.3
Liver acid DNase	3.1.22.1
Loganate methyltransferase	2.1.1.50
Lohmann's enzyme	2.7.3.2
Lombricine kinase	2.7.3.5
Lotus **carboxyl proteinase**	3.4.23.13
Luciferin sulphotransferase	2.8.2.10
Luteolin methyltransferase	2.1.1.42
Lysine acetyltransferase	2.3.1.32
L-*Lysine carboxy-lyase*	4.1.1.18
Lysine decarboxylase	4.1.1.18
Lysine dehydrogenase	1.4.1.15
Lysine hydroxylase	1.14.11.4
β-Lysine mutase	5.4.3.3
D-α-Lysine mutase	5.4.3.4
L-*Lysine:NAD$^+$ oxidoreductase (deaminating, cyclizing)*	1.4.1.15
L-*Lysine:oxygen 2-oxidoreductase (decarboxylating)*	1.13.12.2
Lysine racemase	5.1.1.5
Lysine racemase	5.1.1.5
L-*Lysine:tRNA^{Lys} ligase (AMP-forming)*	6.1.1.6
Lysine 2-monooxygenase	1.13.12.2
Lysine,2-oxoglutarate dioxygenase	1.14.11.4

L-*Lysine:2-oxoglutarate 6-aminotransferase*	2.6.1.36
Lysine-2-oxo glutaryl reductase	1.5.1.7
Lysine 2,3-aminomutase	5.4.3.2
L-*Lysine 2,3-aminomutase*	5.4.3.2
D-**Lysine 5,6-aminomutase**	5.4.3.4
β-**Lysine 5,6-aminomutase**	5.4.3.3
L-**Lysine 6-aminotransferase**	2.6.1.36
Lysolecithin acylhydrolase	3.1.1.5
Lysolecithin acylmutase	5.4.1.1
Lysolecithin acyltransferase	2.3.1.23
Lysolecithinase	3.1.1.5
Lysolecithin migratase	5.4.1.1
Lysolecithin 2,3-acylmutase	5.4.1.1
Lysophospholipase	3.1.1.5
Lysophospholipase D	3.1.4.39
D-**Lysopine dehydrogenase**	1.5.1.16
Lysosomal carboxypeptidase C	3.4.16.2
Lysosomal α-glucosidase	3.2.1.3
Lysozyme	3.2.1.17
Lysyltransferase	2.3.2.3
L-*Lysyl-tRNA:phosphatidylglycerol 3'-lysyltransferase*	2.3.2.3
Lysyl-tRNA synthetase	6.1.1.6
D-*Lyxose ketol-isomerase*	5.3.1.15
D-*Lyxose ketol-isomerase*	5.3.1.15
Magnesium-protoporphyrin methyltransferase	2.1.1.11
L-*Malate:(acceptor) oxidoreductase*	1.1.99.16
Malate:CoA ligase (ADP-forming)	6.2.1.9
Malate condensing enzyme	4.1.3.2
Malate dehydrogenase	1.1.1.37
Malate dehydrogenase (acceptor)	1.1.99.16
D-**Malate dehydrogenase (decarboxylating)**	1.1.1.83
Malate dehydrogenase (decarboxylating)	1.1.1.39
Malate dehydrogenase (NADP$^+$)	1.1.1.82
Malate dehydrogenase (oxaloacetate-decarboxylating)	1.1.1.38
Malate dehydrogenase (oxaloacetate-decarboxylating) (NADP$^+$)	1.1.1.40
L-*Malate glyoxylate-lyase (CoA-acetylating)*	4.1.3.2
D-*Malate hydro-lyase*	4.2.1.31
L-*Malate hydro-lyase*	4.2.1.2
L-*Malate:NAD$^+$ oxidoreductase*	1.1.1.37
L-*Malate:NAD$^+$ oxidoreductase (decarboxylating)*	1.1.1.39
D-*Malate:NAD$^+$ oxidoreductase (decarboxylating)*	1.1.1.83

L-*Malate:NAD+ oxidoreductase* (*oxaloacetate-decarboxylating*)	1.1.1.38
L-*Malate:NADP+ oxidoreductase*	1.1.1.82
L-*Malate:NADP+ oxidoreductase* (*oxaloacetate-decarboxylating*)	1.1.1.40
Malate oxidase	1.1.3.3
L-*Malate:oxygen oxidoreductase*	1.1.3.3
Malate synthase	4.1.3.2
Malate synthetase	4.1.3.2
Maleate cis-trans-isomerase	5.2.1.1
Maleate hydratase	4.2.1.31
Maleate isomerase	5.2.1.1
4-Maleylacetoacetate cis-trans-isomerase	5.2.1.2
Maleylacetoacetate isomerase	5.2.1.2
3-Maleylpyruvate cis-trans-isomerase	5.2.1.4
Maleylpyruvate isomerase	5.2.1.4
Malic dehydrogenase	1.1.1.37
'Malic' enzyme	1.1.1.40
'Malic' enzyme	1.1.1.38
'Malic' enzyme	1.1.1.39
Malonate CoA-transferase	2.8.3.3
Malonate-semialdehyde dehydratase	4.2.1.27
Malonate-semialdehyde dehydrogenase	1.2.1.15
Malonate-semialdehyde dehydrogenase (acetylating)	1.2.1.18
Malonate-semialdehyde hydro-lyase	4.2.1.27
Malonate-semialdehyde:NAD(P)+ oxidoreductase	1.2.1.15
Malonate-semialdehyde:NAD(P)+ oxidoreductase (*decarboxylating, CoA-acetylating*)	1.2.1.18
Malonyl-CoA:[acyl-carrier-protein] S-malonyltransferase	2.3.1.39
Malonyl-CoA carboxyltransferase	2.1.3.4
Malonyl-CoA carboxy-lyase	4.1.1.9
Malonyl-CoA decarboxylase	4.1.1.9
Maltase	3.2.1.20
Maltose:orthophosphate 1-β-D-glucosyltransferase	2.4.1.8
Maltose phosphorylase	2.4.1.8
Maltose 3-glycosyltransferase	2.4.1.6
Malyl-CoA glyoxylate-lyase	4.1.3.24
Malyl-CoA lyase	4.1.3.24
Malyl-CoA synthetase	6.2.1.9
Mammalian DNase IV	3.1.11.3
Mandelate racemase	5.1.2.2
Mandelate racemase	5.1.2.2
Mandelonitrile benzaldehyde-lyase	4.1.2.10
Mandelonitrile lyase	4.1.2.10
Mannanase	3.2.1.25
1,4-β-D-Mannan mannanohydrolase	3.2.1.78

1,2-1,3-α-D-Mannan mannohydrolase	3.2.1.77
Mannase	3.2.1.25
Mannitol dehydrogenase	1.1.1.67
Mannitol dehydrogenase (cytochrome)	1.1.2.2.
Mannitol dehydrogenase (NADP+)	1.1.1.138
D-*Mannitol:ferricytochrome 2-oxidoreductase*	1.1.2.2
Mannitol kinase	2.7.1.57
D-*Mannitol:NADP+ 2-oxidoreductase*	1.1.1.138
D-*Mannitol:NAD+ 2-oxidoreductase*	1.1.1.67
Mannitol-1-phosphatase	3.1.3.22
Mannitol-1-phosphate dehydrogenase	1.1.1.17
D-*Mannitol-1-phosphate:NAD+ 2-oxidoreductase*	1.1.1.17
D-*Mannitol-1-phosphate phosphohydrolase*	3.1.3.22
Mannokinase	2.7.1.7
Mannonate dehydratase	4.2.1.8
D-**Mannonate dehydrogenase (NAD(P)+)**	1.2.1.34
D-*Mannonate hydro-lyase*	4.2.1.8
D-*Mannonate:NAD(P)+ 6-oxidoreductase*	1.1.1.131
D-*Mannonate:NAD(P)+ 6-oxidoreductase* (D-*mannuronate-forming*)	1.2.1.34
D-*Mannonate:NAD+ 5-oxidoreductase*	1.1.1.57
Mannose isomerase	5.3.1.7
D-*Mannose ketol-isomerase*	5.3.1.7
Mannosephosphate isomerase	5.3.1.8
Mannose-1-phosphate guanylyltransferase	2.7.7.13
Mannose-1-phosphate guanylyltransferase (GDP)	2.7.7.22
Mannose-6-phosphate isomerase	5.3.1.8
D-*Mannose-6-phosphate ketol-isomerase*	5.3.1.8
α-D-Mannosidase	3.2.1.24
β-D-Mannosidase	3.2.1.25
α-D-*Mannoside mannohydrolase*	3.2.1.24
β-D-*Mannoside mannohydrolase*	3.2.1.25
Mannosyl-glycoprotein 1,4-N-acetamidodeoxy-β-D-glycohydrolase	3.2.1.96
Mannuronate reductase	1.1.1.131
Melibiase	3.2.1.22
Melilotate dehydrogenase	1.3.1.11
Melilotate hydroxylase	1.14.13.4
Melilotate 3-monooxygenase	1.14.13.4
Menadione reductase	1.6.99.2
3-Mercaptopyruvate:cyanide sulphurtransferase	2.8.1.2
3-Mercaptopyruvate sulphurtransferase	2.8.1.2
Mesaconate hydratase	4.2.1.34
Metaphosphatase	3.6.1.11
Metaphosphatase	3.6.1.10
Metapyrocatechase	1.13.11.2

Methylitaconate Δ-isomerase	5.3.3.6
Methylitaconate Δ²—Δ³-isomerase	5.3.3.6
Methylmalonate-semialdehyde dehydrogenase (acylating)	1.2.1.27
Methylmalonate-semialdehyde:NAD⁺ oxidoreductase (CoA-propionylating)	1.2.1.27
Methylmalonyl-CoA carboxyltransferase	2.1.3.1
(R)-Methylmalonyl-CoA carboxy-lyase	4.1.1.41
Methylmalonyl-CoA CoA-carbonylmutase	5.4.99.2
Methylmalonyl-CoA decarboxylase	4.1.1.41
S-Methylmalonyl-CoA mutase	5.4.99.2
Methylmalonyl-CoA:pyruvate carboxyltransferase	2.1.3.1
Methylmalonyl-CoA racemase	5.1.99.1
Methylmalonyl-CoA racemase	5.1.99.1
Methylmethionine-sulphonium-salt hydrolase	3.3.1.2
5-O-Methyl-myo-inositol:NAD⁺ oxidoreductase	1.1.1.143
3-Methyloxindole:NADP⁺ oxidoreductase	1.3.1.17
6-Methylsalicylate carboxy-lyase	4.1.1.52
6-Methylsalicylate decarboxylase	4.1.1.52
2-Methylserine hydroxymethyltransferase	2.1.2.7
Methylsterol monooxygenase	1.14.99.16
5-Methyltetrahydrofolate:(acceptor) oxidoreductase	1.1.99.15
5-Methyltetrahydrofolate:NADP⁺ oxidoreductase	1.1.1.171
5-Methyltetrahydropteroyl-L-glutamate: L-homocysteine S-methyltransferase	2.1.1.13
5-Methyltetrahydropteroyl-tri-L-glutamate:L-homocysteine S-methyltransferase	2.1.1.14
Methylthiophosphoglycerate phosphatase	3.1.3.14
1-Methylthio-3-phospho-D-glycerate phosphohydrolase	3.1.3.14
N-Methyl-2-oxoglutaramate hydrolase	3.5.1.36
N-Methyl-2-oxoglutaramate methylamidohydrolase	3.5.1.36
2-Methyl-3-hydroxypyridine-5-carboxylate, NAD(P)H:oxygen oxidoreductase (decyclizing)	1.14.12.4
2-Methyl-4-amino-5-hydroxymethyl-pyrimidinediphosphate:4-methyl-5-(2'-phosphoethyl)-thiazole 2-methyl-4-aminopyrimidine-5-methenyltransferase	2.5.1.3
Metridium proteinase A	3.4.21.3
Mevaldate reductase	1.1.1.32
Mevaldate reductase (NADPH)	1.1.1.33
Mevalonate kinase	2.7.1.36
Mevalonate:NAD⁺ oxidoreductase	1.1.1.32
Mevalonate:NAD⁺ oxidoreductase (CoA-acylating)	1.1.1.88
Mevalonate:NADP⁺ oxidoreductase	1.1.1.33

Mevalonate:NADP⁺ oxidoreductase (CoA-acylating)	1.1.1.34
Mexicanain	3.4.99.14
Microbial carboxyl proteinases	3.4.23.6
Microbial metalloproteinases	3.4.24.4
Microbial RNase I	3.1.27.3
Microbial RNase II	3.1.27.1
Microbial serine proteinases	3.4.21.14
Micrococcal endonuclease	3.1.31.1
Micrococcus caseolyticus **neutral proteinase**	3.4.24.4
Mixed-function oxidase	1.14.14.1
Monoacylglycerol kinase	2.7.1.94
Monoacylglycerol lipase	3.1.1.23
Monoamine oxidase	1.4.3.4
Monobutyrase	3.1.1.1
Monodehydroascorbate reductase (NADH)	1.6.5.4
Monoglyceride acyltransferase	2.3.1.22
Monophenol,dihydroxyphenyl alanine:oxygen oxidoreductase	1.14.18.1
Monophenol monooxygenase	1.14.18.1
Monophosphatidylinositol inositolphosphohydrolase	3.1.4.10
Monophosphatidylinositol phosphodiesterase	3.1.4.10
Moraxella nonliquefaciens (MnnI)	3.1.23.20
Moraxella nonliquefaciens(Mno I)	3.1.23.24
Moraxella osloensis (MosI)	3.1.23.27
Mouse nuclear RNase	3.1.13.1
Mucinase	3.2.1.35
Mucinase	3.2.1.36
Mucinase	4.2.2.1
Muconate cycloisomerase	5.5.1.1
Muconolactone Δ-isomerase	5.3.3.4
Mucopeptide amidohydrolase	3.5.1.28
Mucopeptide glycohydrolase	3.2.1.17
Mucopeptide N-acetylmuramoylhydrolase	3.2.1.17
Mucopolysaccharide α-L-iduronohydrolase	3.2.1.76
Mucopolysaccharide β-N-acetylmuramoylexohydrolase	3.2.1.92
Mucor pusillus **carboxyl proteinase**	3.4.23.6
Mucor rennin	3.4.23.6
Mung bean nuclease	3.1.30.1
Muramidase	3.2.1.17
Muramoyl-pentapeptide carboxypeptidase	3.4.17.8
Muscle phosphorylase *a* and *b*	2.4.1.1
Mutarotase	5.1.3.3
Mycodextranase	3.2.1.61
Myokinase	2.7.4.3
Myrosinase	3.2.3.1

Myxobacter **AL-1 proteinase I**	3.4.99.29
Myxobacter **AL-1 proteinase II**	3.4.99.30
Myxobacter **α-lytic proteinase**	3.4.21.12
Myxobacter **β-lytic proteinase**	3.4.24.4
N²-Acetyl-L-ornithine amidohydrolase	3.5.1.16
N²-Acetyl-L-ornithine:L-glutamate N-acetyltransferase	2.3.1.35
N²-Acetyl-L-ornithine:2-oxoglutarate aminotransferase	2.6.1.11
N⁶-Acyl-L-lysine umidohydrolase	3.5.1.17
NADase	3.2.2.5
NAD⁺ glycohydrolase	3.2.2.5
NADH:(acceptor) oxidoreductase	1.6.99.3
NADH:aquacob(III)alamin oxidoreductase	1.6.99.8
NADH:asparagusate oxidoreductase	1.6.4.7
NADH:cob(II)alamin oxidoreductase	1.6.99.9
NADH:L-cystine oxidoreductase	1.6.4.1
NADH dehydrogenase	1.6.99.3
NADH dehydrogenase (quinone)	1.6.99.5
NADH:ferricytochrome b₅ oxidoreductase	1.6.2.2
NADH:hydrogen-peroxide oxidoreductase	1.11.1.1
NADH:hydroxylamine oxidoreductase	1.6.6.11
NADH:hyponitrite oxidoreductase	1.6.6.6
NADH kinase	2.7.1.86
NADH:lipoamide oxidoreductase	1.6.4.3
NADH:monodehydroascorbate oxidoreductase	1.6.5.4
NADH:nitrate oxidoreductase	1.6.6.1
NADH:(quinone-acceptor) oxidoreductase	1.6.99.5
NADH:trimethylamine-N-oxide oxidoreductase	1.6.6.9
NADH.6,7-dihydropteridine oxidoreductase	1.6.99.10
NAD⁺ kinase	2.7.1.23
NAD⁺ nucleosidase	3.2.2.5
NAD⁺ peroxidase	1.11.1.1
NAD(P)⁺ glycohydrolase	3.2.2.6
NADPH:(acceptor) oxidoreductase	1.6.99.1
NADPH:CoAS— S glutathione oxidoreductase	1.6.4.6
NADPH—cytochrome c₂ reductase	1.6.2.5
NADPH-cytochrome *c* reductase	1.6.2.4
NADPH-cytochrome reductase	1.6.2.4
NADPH dehydrogenase	1.6.99.1
NADPH dehydrogenase (quinone)	1.6.99.6
NAD(P)H dehydrogenase (quinone)	1.6.99.2
NADPH diaphorase	1.6.99.1
NADPH:dimethylaminoazobenzene oxidoreductase	1.6.6.7
NADPH:ferricytochrome c₂ oxidoreductase	1.6.2.5
NADPH:ferricytochrome oxidoreductase	1.6.2.4
NADPH:GMP oxidoreductase (deaminating)	1.6.6.8
NADPH:hydrogen-peroxide oxidoreductase	1.11.1.2
NADPH:NAD⁺ oxidoreductase	1.6.1.1
NADPH:nitrate oxidoreductase	1.6.6.3
NAD(P)H:nitrate oxidoreductase	1.6.6.2
NAD(P)H:nitrite oxidoreductase	1.6.6.4
NAD⁺ phosphohydrolase	3.6.1.22
NAD(P)H:oxidized-glutathione oxidoreductase	1.6.4.2
NADPH:oxidized-thioredoxin oxidoreductase	1.6.4.5
NAD(P)H:oxygen oxidoreductase (1-hydroxylating)	1.14.13.18
NADPH:oxygen 2-oxidoreductase (1,2-lactonizing)	1.14.12.6
NAD(P)H:protein-disulphide oxidoreductase	1.6.4.4
NAD(P)H:(quinone-acceptor) oxidoreductase	1.6.99.2
NADPH:(quinone-acceptor) oxidoreductase	1.6.99.6
NAD(P)H:4-nitroquinoline-N-oxide oxidoreductase	1.6.6.10
NADPH:6,7-dihydropteridine oxidoreductase	1.6.99.7
NAD(P)⁺ nucleosidase	3.2.2.6
NADP⁺ peroxidase	1.11.1.2
NAD(P)⁺ transhydrogenase	1.6.1.1
NAD⁺ pyrophosphatase	3.6.1.22
NAD⁺ pyrophosphorylase	2.7.7.1
NAD⁺ synthetase	6.3.1.5
NAD⁺ synthetase (glutamine-hydrolysing)	6.3.5.1
Naphthalene,hydrogen-donor:oxygen oxidoreductase (1,2-epoxidizing)	1.14.99.8
NDPhexose pyrophosphorylase	2.7.7.28
NDPsugar phosphorylase	2.7.7.37
NDP:sugar-1-phosphate nucleotidyltransferase	2.7.7.37
Neomycin-kanamycin phosphotransferase	2.7.1.95
Nepenthes **carboxyl proteinase**	3.4.23.12
N⁵-Ethyl-L-glutamine synthetase	6.3.1.6
Neuraminidase	3.2.1.18
N³-(γ-L-Glutamyl)-4-hydroxymethylphenylhydrazine: (acceptor) γ-glutamyltransferase	2.3.2.9
Nicotinamidase	3.5.1.19

Nicotinamide amidohydrolase	3.5.1.19
Nicotinamide methyltransferase	2.1.1.1
Nicotinamidenucleotide amidase	3.5.1.42
Nicotinamidenucleotide amidohydrolase	3.5.1.42
Nicotinamidenucleotide phosphoribohydrolase	3.2.2.14
Nicotinamidenucleotide:pyrophosphate phosphoribosyltransferase	2.4.2.12
Nicotinamide phosphoribosyltransferase	2.4.2.12
Nicotinate dehydrogenase	1.5.1.13
Nicotinate methyltransferase	2.1.1.7
Nicotinatemononucleotide adenylyltransferase	2.7.7.18
Nicotinatemononucleotide pyrophosphorylase (carboxylating)	2.4.2.19
Nicotinate:NADP+ 6-oxidoreductase (hydroxylating)	1.5.1.13
Nicotinatenucleotide:dimethylbenzimidazole phosphoribosyltransferase	2.4.2.21
Nicotinatenucleotide— dimethylbenzimidazole phosphoribosyltransferase	2.4.2.21
Nicotinatenucleotide:pyrophosphate phosphoribosyltransferase	2.4.2.11
Nicotinatenucleotide:pyrophosphate phosphoribosyltransferase (carboxylating)	2.4.2.19
Nicotinate phosphoribosyltransferase	2.4.2.11
Nicotine:(acceptor) 6-oxidoreductase (hydroxylating)	1.5.99.4
Nicotine dehydrogenase	1.5.99.4
Nitrate-ester reductase	1.8.6.1
Nitrate reductase	1.7.99.4
Nitrate reductase (cytochrome)	1.9.6.1
Nitrate reductase (NADH)	1.6.6.1
Nitrate reductase (NAD(P)H)	1.6.6.2
Nitrate reductase (NADPH)	1.6.6.3
Nitric-oxide:(acceptor) oxidoreductase	1.7.99.3
Nitric-oxide:ferricytochrome c oxidoreductase	1.7.2.1
Nitric-oxide reductase	1.7.99.2
Nitrilase	3.5.5.1
Nitrile aminohydrolase	3.5.5.1
Nitrite:(acceptor) oxidoreductase	1.7.99.4
Nitrite reductase	1.7.99.3
Nitrite reductase (cytochrome)	1.7.2.1
Nitrite reductase (NAD(P)H)	1.6.6.4
β-Nitroacrylate reductase	1.3.1.16
Nitroethane oxidase	1.7.3.1
Nitroethane:oxygen oxidoreductase	1.7.3.1
Nitrogen:(acceptor) oxidoreductase	1.7.99.2
Nitrogenase	1.18.2.1
Nitrogenase (flavodoxin)	1.19.2.1

p-Nitrophenol conjugating enzyme	2.4.1.75
4-Nitrophenylphosphatase	3.1.3.41
4-Nitrophenylphosphate phosphohydrolase	3.1.3.41
3-Nitropropionate:NADP+ oxidoreductase	1.3.1.16
Nitroquinoline-*N*-oxide reductase	1.6.6.10
N⁶-Methyl-lysine oxidase	1.5.3.4
N⁶-Methyl-L-lysine:oxygen oxidoreductase (demethylating)	1.5.3.4
NMN adenylyltransferase	2.7.7.1
NMNase	3.2.2.14
NMN nucleosidase	3.2.2.14
NMN pyrophosphorylase	2.4.2.12
N,N-Dimethylglycine:(acceptor) oxidoreductase (demethylating)	1.5.99.2
Noradrenalin *N*-methyltransferase	2.1.1.28
Notatin	1.1.3.4
NTP:deoxycytidine 5′-phosphotransferase	2.7.1.74
NTP:gentamicin 2″-nucleotidyltransferase	2.7.7.46
NTP:hexose-1-phosphate nucleotidyltransferase	2.7.7.28
NTP polymerase	2.7.7.19
Nucleosidase	3.2.2.1
Nucleoside deoxyribosyltransferase	2.4.2.6
Nucleosidediphosphatase	3.6.1.6
Nucleosidediphosphate kinase	2.7.4.6
Nucleosidediphosphate phosphohydrolase	3.6.1.6
Nucleosidemonophosphate kinase	2.7.4.4
Nucleoside phosphoacylhydrolase	3.6.1.24
Nucleoside phosphotransferase	2.7.1.77
Nucleoside:purine (pyrimidine) deoxyribosyltransferase	2.4.2.6
Nucleoside:purine (pyrimidine) ribosyltransferase	2.4.2.5
Nucleoside ribohydrolase	3.2.2.8
Nucleoside ribosyltransferase	2.4.2.5
Nucleoside triphosphatase	3.6.1.15
Nucleosidetriphosphate—adenylate kinase	2.7.4.10
Nucleosidetriphosphate:AMP phosphotransferase	2.7.4.10
Nucleosidetriphosphate:DNA deoxynucleotidylexotransferase	2.7.7.31
Nucleosidetriphosphate-hexose-1-phosphate nucleotidyltransferase	2.7.7.28
Nucleosidetriphosphate pyrophosphatase	3.6.1.19
Nucleosidetriphosphate pyrophosphohydrolase	3.6.1.19
Nucleosidetriphosphate:RNA nucleotidyltransferase	2.7.7.6
Nucleoside-2′:3′-cyclic-phosphate 2′-nucleotidohydrolase	3.1.4.37
Nucleoside-2′:3′-cyclic-phosphate 3′-nucleotidohydrolase	3.1.4.16

Nucleoside-5′-phosphoacylate acylhydrolase	3.6.1.24
3′-Nucleotidase	3.1.3.6
Nucleotidase	3.1.3.31
5′-Nucleotidase	3.1.3.5
Nucleotid phosphohydrolase	3.1.3.31
Nucleotide pyrophosphatase	3.6.1.9
Nucleotide pyrophosphokinase	2.7.6.4
Nucleotide:3′-deoxynucleoside 5′-phosphotransferase	2.7.1.77
Octanol dehydrogenase	1.1.1.73
Octanol:NAD⁺ oxidoreductase	1.1.1.73
Octopine dehydrogenase	1.5.1.11
17β-Oestradiol—UDP glucuronyltransferase	2.4.1.59
Oestradiol 17β-dehydrogenase	1.1.1.62
Oestradiol 17α-dehydrogenase	1.1.1.148
Oestradiol-17β,hydrogen-donor:oxygen oxidoreductase (6β-hydroxylating)	1.14.99.11
Oestradiol-17β:NAD⁺ 17-oxidoreductase	1.1.1.62
Oestradiol 6β-hydroxylase	1.14.99.11
Oestradiol 6β-monooxygenase	1.14.99.11
Oestriol—UDP 16α-glucuronyltransferase	2.4.1.61
Oestriol—UDP 17β-glucuronyltransferase	2.4.1.42
Oestriol 2-hydroxylase	1.14.1.11
Oestrone sulphotransferase	2.8.2.4
'Old Yellow enzyme'	1.6.99.1
Oleate hydratase	4.2.1.53
Oligodeoxyribonucleate exonuclease	3.1.4.29
Oligogalacturonide lyase	4.2.2.6
Oligogalacturonide lyase	4.2.2.6
Oligoglucan-branching glycosyltransferase	2.4.1.24
1,3-β-D-Oligoglucan:orthophosphate glucosyltransferase	2.4.1.30
1,4-β-D-Oligoglucan:orthophosphate α-D-glucosyltransferase	2.4.1.49
β-1,3-Oligoglucan:orthophosphate glucosyltransferase II	2.4.1.30
1,3-β-D-Oligoglucan phosphorylase	2.4.1.30
Oligonucleate 5′-nucleotidohydrolase	3.1.4.1
Oligonucleotidase	3.1.13.3
Oligoribonuclease of *E. coli*	3.1.13.1

Oligo-1,6-glucosidase	3.2.1.10
Opheline kinase	2.7.3.7
Ophio-amino-acid oxidase (for the snake enzyme only)	1.4.3.2
Opsin kinase	2.7.1.97
Orcinol hydroxylase	1.14.13.6
Orcinol,NADH:oxygen oxidoreductase (2-hydroxylating)	1.14.13.6
Orcinol 2-monooxygenase	1.14.13.6
L-*Ornithine ammonia-lyase (cyclizing)*	4.3.1.12
Ornithine carbamoyltransferase	2.1.3.3
L-*Ornithine carboxy-lyase*	4.1.1.17
Ornithine cyclodeaminase	4.3.1.12
Ornithine decarboxylase	4.1.1.17
Ornithine—oxo-acid aminotransferase	2.6.1.13
Ornithine racemase	5.1.1.12
Ornithine racemase	5.1.1.12
Ornithine transcarbamylase	2.1.3.3
L-*Ornithine:2-oxo-acid aminotransferase*	2.6.1.13
D-**Ornithine 4,5-aminomutase**	5.4.3.5
D-*Ornithine 4,5-aminomutase*	5.4.3.5
Ornithine 4,5-aminomutase	5.4.3.1
Orotate phosphoribosyltransferase	2.4.2.10
Orotate reductase	1.3.1.14
Orotate reductase (NADPH)	1.3.1.15
Orotidine-5′-phosphate carboxy-lyase	4.1.1.23
Orotidine-5′-phosphate decarboxylase	4.1.1.23
Orotidine-5′-phosphate:pyrophosphate phosphoribosyltransferase	2.4.2.10
Orotidine-5′-phosphate pyrophosphorylase	2.4.2.10
Orotidylic acid phosphorylase	2.4.2.10
Orsellinate carboxy-lyase	4.1.1.58
Orsellinate decarboxylase	4.1.1.58
Orsellinate-depside hydrolase	3.1.1.40
Orsellinate-depside hydrolase	3.1.1.40
Orthophosphate:oxaloacetate carboxy lyase (phosphorylating)	4.1.1.31
Orthophosphoric-monoester phosphohydrolase (acid optimum)	3.1.3.2
Orthophosphoric-monoester phosphohydrolase (alkaline optimum)	3.1.3.1
Oxalacetate β-decarboxylase	4.1.1.3
Oxalate carboxy-lyase	4.1.1.2
Oxalate:CoA ligase (AMP-forming)	6.2.1.8

Oxalate CoA-transferase	2.8.3.2
Oxalate decarboxylase	4.1.1.2
Oxalate oxidase	1.2.3.4
Oxalate:oxygen oxidoreductase	1.2.3.4
Oxaloacetase	3.7.1.1
Oxaloacetate acetylhydrolase	3.7.1.1
Oxaloacetate carboxy-lyase	4.1.1.3
Oxaloacetate decarboxylase	4.1.1.3
Oxaloacetate keto—enol-isomerase	5.3.2.2
Oxaloacetate tautomerase	5.3.2.2
Oxaloacetate transacetase	4.1.3.7
Oxaloglycollate reductase (decarboxylating)	1.1.1.92
3-Oxalomalate glyoxylate-lyase	4.1.3.13
Oxalomalate lyase	4.1.3.13
Oxalyl-CoA carboxy-lyase	4.1.1.8
Oxalyl-CoA decarboxylase	4.1.1.8
Oxalyl-CoA synthetase	6.2.1.8
Oxalyl-CoA:L-2,3-diaminopropionate N³-oxalyltransferase	2.3.1.58
Oxalyldiaminopropionate synthase	2.3.1.58
Oxamate carbamoyltransferase	2.1.3.5
Oxamic transcarbamylase	2.1.3.5
2,3-Oxidosqualene cycloartenol-cyclase	5.4.99.8
2,3-Oxidosqualene lanosterol-cyclase	5.4.99.7
2,3-Oxidosqualene mutase (cyclizing, cycloartenol-forming)	5.4.99.8
2,3-Oxidosqualene mutase (cyclizing, lanosterol-forming)	5.4.99.7
Oximinotransferase	2.6.3.1
2-Oxo-acid carboxy-lyase	4.1.1.1
3-Oxoacyl-[acyl-carrier-protein] reductase	1.1.1.100
3-Oxoacyl-[acyl-carrier-protein] synthase	2.3.1.41
3-Oxoadipate CoA-transferase	2.8.3.6
3-Oxoadipate enol-lactonase	3.1.1.24
2-Oxoaldehyde dehydrogenase	1.2.1.23
2-Oxoaldehyde:NAD(P)⁺ oxidoreductase	1.2.1.23
2-Oxobutyrate:ferredoxin oxidoreductase (CoA-propionylating)	1.2.7.2
2-Oxobutyrate synthase	1.2.7.2
(2-Oxoethyl) phosphonate phosphonohydrolase	3.11.1.1
Oxoglutarate dehydrogenase	1.2.4.2
2-Oxoglutarate:ferredoxin oxidoreductase (CoA-succinylating)	1.2.7.3
2-Oxoglutarate:lipoamide oxidoreductase (decarboxylating and acceptor-succinylating)	1.2.4.2
2-Oxoglutarate synthase	1.2.7.3
Oxoisomerase	5.3.1.9

2-Oxoisovalerate dehydrogenase (acylating)	1.2.1.25
2-Oxoisovalerate dehydrogenase (lipoamide)	1.2.4.4
2-Oxoisovalerate:lipoamide oxidoreductase (decarboxylating and acceptor-isobutyrylating)	1.2.4.4
2-Oxoisovalerate:NAD⁺ oxidoreductase (CoA-isobutyrylating)	1.2.1.25
3-Oxolaurate carboxy-lyase	4.1.1.56
3-Oxolaurate decarboxylase	4.1.1.56
2-Oxopantoate formaldehyde-lyase	4.1.2.12
2-Oxopantoate reductase	1.1.1.169
2-Oxopantoyl-lactone reductase	1.1.1.168
5-Oxoprolinase (ATP-hydrolysing)	3.5.2.9
5-Oxo-L-proline amidohydrolase (ATP-hydrolysing)	3.5.2.9
4-Oxoproline reductase	1.1.1.104
(3′-Oxo-prop-2′-enyl)-2-amino-but-2-ene-dioate carboxy-lyase	4.1.1.45
3-Oxosteroid:(acceptor) Δ¹-oxidoreductase	1.3.99.4
3-Oxosteroid Δ¹-dehydrogenase	1.3.99.4
3-Oxosteroid Δ⁵—Δ⁴-isomerase	5.3.3.1
6-Oxotetrahydro-nicotinate dehydrogenase	1.3.7.1
2-Oxo-4-hydroxyglutarate aldolase	4.1.2.31
2-Oxo-4-hydroxyglutarate glyoxylate-lyase	4.1.2.31
2-Oxo-4-hydroxyglutarate lyase	4.1.2.1
3-Oxo-5α-steroid:(acceptor) Δ⁴-oxidoreductase	1.3.99.5
3-Oxo-5β-steroid:(acceptor) Δ⁴-oxidoreductase	1.3.99.6
3-Oxo-5β-steroid Δ⁴-dehydrogenase	1.3.99.6
3-Oxo-5α-steroid Δ⁴-dehydrogenase	1.3.99.5
3-Oxo-5β-steroid:NADP⁺ Δ⁴-oxidoreductase	1.3.1.23
3-Oxo-5α-steroid:NADP⁺ Δ⁴-oxidoreductase	1.3.1.22
7-Oxo-8-aminononanoate synthase	2.3.1.47
Oxytocinase	3.4.11.3
Oxyuranus scutellatus prothrombin-activating proteinase	3.4.99.28
Paecilomyces varioti carboxyl proteinase	3.4.23.6
Palmitate:hydrogen-peroxide oxidoreductase	1.11.1.3
Palmitoyl-CoA:acylglycerol O-palmitoyltransferase	2.3.1.22
Palmitoyl-CoA:L-carnitine O-palmitoyltransferase	2.3.1.21
Palmitoyl-CoA hydrolase	3.1.2.2
Palmitoyl-CoA hydrolase	3.1.2.2

*Palmitoyl-CoA:*L-*serine* C-*palmitoyltransferase (decarboxylating)*	2.3.1.50
Palmitoyldihydroxyacetone-phosphate **reductase**	1.1.1.101
1-Palmitoylglycerol-3-phosphate:NADP+ *oxidoreductase*	1.1.1.101
Pancreatic DNase	3.1.21.1
Pancreatic DNase II,Crab testes DNase	3.1.22.1
Pancreatic RNase	3.1.27.5
Pantetheine kinase	2.7.1.34
Pantetheinephosphate adenylyltransferase	2.7.7.3
Pantoate activating enzyme	6.3.2.1
L-*Pantoate:β-alanine ligase* *(AMP-forming)*	6.3.2.1
Pantoate dehydrogenase	1.1.1.106
D-*Pantoate:NADP+ 2-oxidoreductase*	1.1.1.169
D-*Pantoate:NAD+ 4-oxidoreductase*	1.1.1.106
Pantothenase	3.5.1.22
Pantothenate amidohydrolase	3.5.1.22
Pantothenate kinase	2.7.1.33
Pantothenate synthetase	6.3.2.1
N-*(*L-*Pantothenoyl)-*L-*cysteine* *carboxy-lyase*	4.1.1.30
Pantothenoylcysteine decarboxylase	4.1.1.30
Pantoyl lactone:NADP+ oxidoreductase	1.1.1.168
Papain	3.4.22.2
Papainase	3.4.22.2
Paraoxonase	3.1.1.2
Particle-bound aminopeptidase	3.4.11.2
Pectate lyase	4.2.2.2
Pectate transeliminase	4.2.2.2
Pectinase	3.2.1.15
Pectin demethoxylase	3.1.1.11
Pectin depolymerase	3.2.1.15
Pectinesterase	3.1.1.11
Pectin lyase	4.2.2.10
Pectin methoxylase	3.1.1.11
Pectin methylesterase	3.1.1.11
Pectin pectylhydrolase	3.1.1.11
Penicillin amidase	3.5.1.11
Penicillin amidohydrolase	3.5.1.11
Penicillin amido-β-lactamhydrolase	3.5.2.6
Penicillinase	3.5.2.8
Penicillinase	3.5.2.6
Penicillium citrium nuclease P₁	3.1.30.1
Penicillium janthinellum **carboxyl** **proteinase**	3.4.23.6
Penicillium notatum **extracellular** **proteinase**	3.4.99.16
Penicillium roqueforti **neutral proteinase**	3.4.24.4
Penicillopepsin	3.4.23.6
2,4,6/3,5-Pentahydroxycyclohexanone *hydro-lyase*	4.2.1.44
Pentosealdolase	4.1.2.3
P-enzyme	2.4.1.1
PEP carboxyphosphotransferase	4.1.1.38
Pepsin	3.4.23.1
Pepsin A	3.4.23.1
Pepsin B	3.4.23.2
Pepsin C	3.4.23.3
Peptidase a	3.4.11.11
Peptidase A	3.4.23.6
Peptidase P	3.4.15.1
Peptidoglutaminase I	3.5.1.43
Peptidoglutaminase II	3.5.1.44
Peptidoglycan endopeptidase	3.4.99.17
*Peptidyl-*L-*alanine hydrolase*	3.4.17.6
*Peptidyl-*L-*amino-acid hydrolase*	3.4.16.1
*Peptidyl-*L-*amino-acid hydrolase*	3.4.17.1
Peptidylaminoacylamide hydrolase	3.4.15.2
*Peptidyl-*L-*arginine hydrolase*	3.4.17.3
*Peptidyl-*L-*aspartate hydrolase*	3.4.17.5
Peptidyl carboxyamidase	3.4.15.2
Peptidyldipeptide hydrolase	3.4.15.1
Peptidyl-glutaminase	3.5.1.43
*Peptidyl-*L-*glutamine amidohydrolase*	3.5.1.43
Peptidyl-glycine hydrolase	3.4.17.4
*Peptidyl-*L-*lysine (*L-*arginine) hydrolase*	3.4.17.2
Peptidyllysine,2-oxoglutarate:oxygen *5-oxidoreductase*	1.14.11.4
Peptidylprolyl-amino acid hydrolase	3.4.16.2
Peptidyltransferase	2.3.2.12
Peptidyl-tRNA:aminoacyl-tRNA N-*peptidyltransferase*	2.3.2.12
Peptidyltryptophan:oxygen *2,3-oxidoreductase (decyclizing)*	1.13.11.26
Peptidyltryptophan 2,3-dioxygenase	1.13.11.26
*Peptidyl-*L-*tyrosine hydrolase*	3.4.16.3
Perillyl-alcohol dehydrogenase	1.1.1.144
Perillyl-alcohol:NAD+ oxidoreductase	1.1.1.144
Peroxidase	1.11.1.7
Phage SP3 DNase	3.1.11.4
Phenolase	1.10.3.2
Phenolase,Monophenol oxidase	1.14.18.1
Phenol hydroxylase	1.14.13.7
Phenol,NADPH:oxygen oxidoreductase *(2-hydroxylating)*	1.14.13.7
Phenol O-**methyltransferase**	2.1.1.25
Phenol sulphotransferase	2.8.2.1
Phenol 2-monooxygenase	1.14.13.7

Phenylacetaldehyde dehydrogenase	1.2.1.39
Phenylacetaldehyde:NAD⁺ oxidoreductase	1.2.1.39
Phenylacetyl-CoA:L-glutamine *α-N-phenylacetyltransferase*	2.3.1.14
Phenylalaninase	1.14.16.1
Phenylalanine acetyltransferase	2.3.1.53
L-*Phenylalanine ammonia-lyase*	4.3.1.5
Phenylalanine ammonia-lyase	4.3.1.5
L-*Phenylalanine carboxy-lyase*	4.1.1.53
Phenylalanine decarboxylase	4.1.1.53
Phenylalanine (histidine) aminotransferase	2.6.1.58
L-*Phenylalanine (L-histidine):pyruvate aminotransferase*	2.6.1.58
Phenylalanine racemase (ATP-hydrolysing)	5.1.1.11
Phenylalanine racemase (ATP-hydrolysing)	5.1.1.11
L-*Phenylalanine,tetrahydro-pteridine: oxygen oxidoreductase (4-hydroxylating)*	1.14.16.1
L-*Phenylalanine:tRNA^Phe ligase (AMP-forming)*	6.1.1.20
Phenylalanine 4-hydroxylase	1.14.16.1
Phenylalanine 4-monooxygenase	1.14.16.1
Phenylalanyl-tRNA synthetase	6.1.1.20
Phenylpyruvate carboxy-lyase	4.1.1.43
Phenylpyruvate decarboxylase	4.1.1.43
Phenylpyruvate keto—enol-isomerase	5.3.2.1
Phenylpyruvate tautomerase	5.3.2.1
Phenylserine aldolase	4.1.2.26
L-*threo-3-Phenylserine benzaldehyde-lyase*	4.1.2.26
Phloretin-glucosidase	3.2.1.62
Phloretin hydrolase	3.7.1.4
Phlorizin hydrolase	3.2.1.62
Phosphate acetyltransferase	2.3.1.8
Phosphate butyryltransferase	2.3.1.19
Phosphatidase	3.1.1.4
Phosphatidate cytidylyltransferase	2.7.7.41
Phosphatidate phosphatase	3.1.3.4
L-*α-Phosphatidate phosphohydrolase*	3.1.3.4
Phosphatidate 1-acylhydrolase	3.1.1.32
Phosphatide 2-acylhydrolase	3.1.1.4
Phosphatidolipase	3.1.1.4
Phosphatidylcholine cholinephosphohydrolase	3.1.4.3
Phosphatidylcholine phosphatidohydrolase	3.1.4.4
Phosphatidylethanolamine methyltransferase	2.1.1.17
Phosphatidylglycerophosphatase	3.1.3.27
Phosphatidylglycerophosphate phosphohydrolase	3.1.3.27

Phosphatidyl-inositol-bisphosphate phosphatase	3.1.3.36
Phosphatidylinositol kinase	2.7.1.67
Phosphatidyl-myo-inositol-4,5-bisphosphate phosphohydrolase	3.1.3.36
Phosphatidylserine carboxy-lyase	4.1.1.65
Phosphatidylserine decarboxylase	4.1.1.65
Phosphatidylserine synthase	2.7.8.8
Phosphoacetylglucosamine mutase	2.7.5.2
Phosphoacylase	2.3.1.8
Phosphoadenylate 3′-nucleotidase	3.1.3.7
Phosphoadenylylsulphatase	3.6.2.2
3′-Phosphoadenylylsulphate:arylamine sulphotransferase	2.8.2.3
3′-Phosphoadenylylsulphate:choline sulphotransferase	2.8.2.6
3′-Phosphoadenylylsulphate:chondroitin 4′-sulphotransferase	2.8.2.5
3′-Phosphoadenylylsulphate: galactosylceramide 3′-sulphotransferase	2.8.2.11
3′-Phosphoadenylylsulphate: galactosylsphingosine sulphotransferase	2.8.2.13
3′-Phosphoadenylylsulphate:heparitin N-sulphotransferase	2.8.2.12
3′-Phosphoadenylylsulphate:luciferin sulphotransferase	2.8.2.10
3′-Phosphoadenylylsulphate: N-desulphoheparin N-sulphotransferase	2.8.2.8
3′-Phosphoadenylylsulphate:oestrone 3-sulphotransferase	2.8.2.4
3′-Phosphoadenylylsulphate:phenol sulphotransferase	2.8.2.1
3′-Phosphoadenylylsulphate sulphohydrolase	3.6.2.2
3′- Phosphoadenylylsulphate: taurolithocholate sulphotransferase	2.8.2.14
3′-Phosphoadenylylsulphate: L-tyrosine-methyl-ester sulphotransferase	2.8.2.9
3′-Phosphoadenylylsulphate:UDP-2-acetamido-2-deoxy-D-galactose-4-sulphate 6-sulphotransferase	2.8.2.7
3′-Phosphoadenylylsulphate: 3β-hydroxysteroid sulphotransferase	2.8.2.2
3′-Phosphoadenylylsulphate 3′-phosphatase	3.1.3.30
3′-Phosphoadenylylsulphate 3′-phosphohydrolase	3.1.3.30
Phosphoamidase	3.9.1.1
Phosphoamide hydrolase	3.9.1.1
Phosphodeoxyriboaldolase	4.1.2.4
Phosphodeoxyribomutase	2.7.5.6
Phosphodiesterase I	3.1.4.1

Phosphoenolpyruvate carboxykinase	4.1.1.49
Phosphoenolpyruvate carboxykinase (ATP)	4.1.1.49
Phosphoenolpyruvate carboxykinase (GTP)	4.1.1.32
Phosphoenolpyruvate carboxykinase (pyrophosphate)	4.1.1.38
Phosphoenolpyruvate carboxylase	4.1.1.31
Phosphoenolpyruvate carboxylase	4.1.1.32
Phosphoenolpyruvate carboxylase	4.1.1.38
Phosphoenolpyruvate carboxylase	4.1.1.49
Phosphoenolpyruvate- fructose phosphotransferase	2.7.1.98
Phosphoenolpyruvate:D-fructose 1-phosphotransferase	2.7.1.98
Phosphoenolpyruvate kinase	2.7.1.40
Phosphoenolpyruvate—protein phosphotransferase	2.7.3.9
Phosphoenolpyruvate:protein phosphotransferase	2.7.3.9
Phosphoenolpyruvate:shikimate-5-phosphate enolpyruvoyltransferase	2.5.1.19
Phosphoenolpyruvate synthase	2.7.9.2
Phosphoenolpyruvate: UDP-2-acetamido-2-deoxy-D-glucose 2-enoyl-1-carboxyethyltransferase	2.5.1.7
Phospho*enol* transphosphorylase	2.7.1.40
6-Phosphofructokinase	2.7.1.11
1-Phosphofructokinase	2.7.1.56
6-Phosphofructokinase (pyrophosphate)	2.7.1.90
6-Phospho-β-D-galactosidase	3.2.1.85
6-Phospho-β-D-galactoside 6-phosphogalactohydrolase	3.2.1.85
Phosphoglucoisomerase, Phosphohexoisomerase	5.3.1.9
Phosphoglucokinase	2.7.1.10
Phosphoglucomutase	2.7.5.1
Phosphoglucomutase (glucose-cofactor)	2.7.5.5
Phosphogluconate dehydratase	4.2.1.12
Phosphogluconate dehydrogenase	1.1.1.43
Phosphogluconate dehydrogenase (decarboxylating)	1.1.1.44
6-Phospho-D-gluconate hydro-lyase	4.2.1.12
6-Phospho-D-gluconate-δ-lactone lactonohydrolase	3.1.1.31
6-Phospho-D-gluconate:NAD(P)+ 2-oxidoreductase	1.1.1.43
6-Phospho-D-gluconate:NADP+ 2-oxidoreductase (decarboxylating)	1.1.1.44
Phosphogluconic acid dehydrogenase, 6-Phosphogluconic dehydrogenase, 6-Phosphogluconic carboxylase	1.1.1.44
6-Phosphogluconic dehydrogenase	1.1.1.43
6-Phosphogluconolactonase	3.1.1.31
Phosphoglucosamine acetylase	2.3.1.4
Phosphoglucosamine transacetylase	2.3.1.4
6-Phospho-β-D-glucosidase	3.2.1.86
6-Phospho-β-D-glucosyl-(1,4)-D-glucose glucohydrolase	3.2.1.86
3-Phospho-D-glycerate carboxy-lyase (dimerizing)	4.1.1.39
2-Phosphoglycerate dehydratase	4.2.1.11
Phosphoglycerate dehydrogenase	1.1.1.95
2-Phospho-D-glycerate hydro-lyase	4.2.1.11
Phosphoglycerate kinase	2.7.2.3
Phosphoglycerate kinase (GTP)	2.7.2.10
3-Phosphoglycerate:NAD+ 2-oxidoreductase	1.1.1.95
Phosphoglycerate phosphatase	3.1.3.20
3-Phosphoglycerate phosphatase	3.1.3.38
Phosphoglycerate phosphomutase	5.4.2.1
D-*Phosphoglycerate 2,3-phosphomutase*	5.4.2.1
Phosphoglyceromutase	2.7.5.3
3-Phosphoglyceroyl-phosphate— polyphosphate phosphotransferase	2.7.4.17
3-Phospho-D-glyceroyl-phosphate: polyphosphate phosphotransferase	2.7.4.17
3-Phospho-D-glyceroyl-phosphate: 3-phospho-D-glycerate phosphotransferase	2.7.5.4
Phosphoglycollate phosphatase	3.1.3.18
2-Phosphoglycollate phosphohydrolase	3.1.3.18
Phosphohexoisomerase	5.3.1.8
Phosphohexokinase	2.7.1.11
Phosphohexomutase	5.3.1.8
Phosphohexomutase	5.3.1.9
Phosphohexose isomerase	5.3.1.9
Phosphohistidinoprotein:hexose phosphotransferase	2.7.1.69
Phosphohistidinoprotein—hexose phosphotransferase	2.7.1.69
O-*Phosphohomoserine phospho-lyase (adding water)*	4.2.99.2
Phosphoketolase	4.1.2.9
Phosphoketotetrose aldolase	4.1.2.2
Phospholipase A₁	3.1.1.32
Phospholipase A₂	3.1.1.4
Phospholipase B	3.1.1.5
Phospholipase C	3.1.4.3
Phospholipase D	3.1.4.4
Phosphomannan mannosephosphotransferase	2.7.8.9
Phosphomannose isomerase	5.3.1.8
Phosphomethylpyrimidine kinase	2.7.4.7
Phosphomevalonate kinase	2.7.4.2

Phosphomonoesterase	3.1.3.2
Phosphomonoesterase	3.1.3.1
Phospho-*N*-acetylmuramoyl-pentapeptide-transferase	2.7.8.13
Phosphonoacetaldehyde hydrolase	3.11.1.1
4'-Phospho-N-(L-pantothenoyl)-L-cysteine carboxy-lyase	4.1.1.36
4'-Phospho-L-pantothenate:L-cysteine ligase	6.3.2.5
Phosphopantothenoyl-cysteine decarboxylase	4.1.1.36
Phosphopantothenoylcysteine synthetase	6.3.2.5
Phosphopentokinase	2.7.1.19
Phosphopentomutase	2.7.5.6
Phosphopentosisomerase	5.3.1.6
Phosphoprotein phosphatase	3.1.3.16
Phosphoprotein phosphohydrolase	3.1.3.16
Phosphopyruvate carboxylase	4.1.1.32
Phosphopyruvate carboxylase	4.1.1.38
Phosphopyruvate carboxylase (ATP)	4.1.1.49
Phosphopyruvate hydratase	4.2.1.11
Phosphoramidate—hexose phosphotransferase	2.7.1.62
Phosphoramidate:hexose 1-phosphotransferase	2.7.1.62
Phosphoriboisomerase	5.3.1.6
Phosphoribokinase	2.7.1.18
5-Phospho-α-D-ribose-1-diphosphate: xanthine phosphoribosyltransferase	2.4.2.22
5-Phosphoribosylamine:glycine ligase (ADP-forming)	6.3.4.13
5-Phosphoribosylamine:pyrophosphate phosphoribosyltransferase (glutamate-amidating)	2.4.2.14
5'-Phosphoribosylamine synthetase	6.3.4.7
Phosphoribosylaminoimidazolecarboxamide formyltransferase	2.1.2.3
Phosphoribosylaminoimidazole carboxylase	4.1.1.21
Phosphoribosylaminoimidazole-succinocarboxamide synthetase	6.3.2.6
Phosphoribosylaminoimidazole synthetase	6.3.3.1
Phosphoribosyl-AMP cyclohydrolase	3.5.4.19
Phosphoribosyl-AMP pyrophosphorylase	3.5.4.19
1-N-(5'-Phospho-D-ribosyl)-AMP 1,6-hydrolase	3.5.4.19
N-(5'-Phosphoribosyl)-anthranilate:pyrophosphate phosphoribosyltransferase	2.4.2.18
Phosphoribosyl-anthranilate pyrophosphorylase	2.4.2.18
1-(5'-Phosphoribosyl)-ATP:pyrophosphate phosphoribosyltransferase	2.4.2.17

Phosphoribosyl-ATP pyrophosphohydrolase	3.5.4.19
Phosphoribosyl-ATP pyrophosphorylase	2.4.2.17
Phosphoribosyl-diphosphate 5-amidotransferase	2.4.2.14
N-(5'-Phospho-D-ribosylformimino)-5-amino-1-(5''-phosphoribosyl)-4-imidazolecarboxamide ketol-isomerase	5.3.1.16
N-(5'-Phospho-D-ribosylformimino)-5-amino-1-(5''-phosphoribosyl)-4-imidazolecarboxamide isomerase	5.3.1.16
5'-Phosphoribosylformyl-glycinamide: L-glutamine amido-ligase (ADP-forming)	6.3.5.3
5'-Phosphoribosylformyl-glycinamidine cyclo-ligase (ADP-forming)	6.3.3.1
Phosphoribosylformylglycinamidine synthetase	6.3.5.3
Phosphoribosylglycinamide formyltransferase	2.1.2.2
Phosphoribosylglycinamide synthetase	6.3.4.13
5'-Phosphoribosylimidazoleacetate synthetase	6.3.4.8
5'-Phosphoribosyl-4-carboxy-5-aminoimidazole:L-aspartate ligase (ADP-forming)	6.3.2.6
5'-Phosphoribosyl-5-amino-4-imidazolecarboxylate carboxy-lyase	4.1.1.21
Phosphoribulokinase	2.7.1.19
Phosphoribulose epimerase	5.1.3.1
Phosphorylase	2.4.1.1
Phosphorylase a phosphohydrolase	3.1.3.17
Phosphorylase *b* kinase kinase	2.7.1.37
Phosphorylase kinase	2.7.1.38
Phosphorylase phosphatase	3.1.3.17
Phosphorylcholine-glyceride transferase	2.7.8.2
Phosphorylcholine transferase	2.7.7.15
Phosphorylethanolamine transferase	2.7.7.14
Phosphosaccharomutase	5.3.1.9
Phosphoserine aminotransferase	2.6.1.52
Phosphoserine phosphatase	3.1.3.3
O-Phosphoserine phosphohydrolase	3.1.3.3
O-Phospho-L-serine:2-oxoglutarate aminotransferase	2.6.1.52
Phosphotransacetylase	2.3.1.8
α,α-Phosphotrehalase	3.2.1.93
Phosphotriose isomerase	5.3.1.1
7-Phospho-2-keto-3-deoxy-D-arabinoheptonate D-erythrose-4-phosphate-lyase (pyruvate-phosphorylating)	4.1.2.15
6-Phospho-2-keto-3-deoxy-galactonate aldolase	4.1.2.21
Phospho-2-keto-3-deoxy-gluconate aldolase	4.1.2.14

6-Phospho-2-keto-3-deoxy-D-gluconate D-glyceraldehyde-3-phosphate-lyase	4.1.2.14
Phospho-2-keto-3-deoxy-heptonate aldolase	4.1.2.15
Phospho-2-keto-3-deoxy-octonate aldolase	4.1.2.16
8-Phospho-2-keto-3-deoxy-D-octonate D-arabinose-5-phosphate-lyase (pyruvate-phosphorylating)	4.1.2.16
7-Phospho-3-deoxy-D-arabino-heptulosonate phosphate-lyase (cyclizing)	4.6.1.3
3-Phospho-5-enolpyruvoyl-shikimate phosphate-lyase	4.6.1.4
Phospho-5-keto-2-deoxy-gluconate aldolase	4.1.2.29
6-Phospho-5-keto-2-deoxy-D-gluconate malonate-semialdehyde-lyase	4.1.2.29
Photinus luciferin:oxygen 4-oxidoreductase (decarboxylating, ATP-hydrolysing)	1.13.12.7
Photinus luciferin 4-monooxygenase (ATP-hydrolysing)	1.13.12.7
Photoreactivating enzyme	4.1.99.3
Phylloquinone epoxidase	1.14.99.20
Phylloquinone,hydrogen-donor:oxygen oxidoreductase (2,3-epoxidizing)	1.14.99.20
Phylloquinone monooxygenase (2,3-epoxidizing)	1.14.99.20
Phylloquinone reductase	1.6.99.2
Physarum **carboxyl proteinase**	3.4.23.6
Phytase	3.1.3.26
3-Phytase	3.1.3.8
6-Phytase	3.1.3.26
Phytate 6-phosphatase	3.1.3.26
Pig liver nuclease	3.1.26.1
Pimeloyl-CoA:L-alanine C-pimeloyltransferase (decarboxylating)	2.3.1.47
Pinguinain	3.4.99.18
D-Pinitol dehydrogenase	1.1.1.142
L-Pipecolate:(acceptor) oxidoreductase	1.5.99.3
L-Pipecolate dehydrogenase	1.5.99.3
L-Pipecolate:NADP+ 2-oxidoreductase	1.5.1.14
Plasmalogen synthase	2.3.1.25
Plasmin	3.4.21.7
Plasminogen activator	3.4.21.31
Plasmodium **carboxyl proteinase**	3.4.23.6
Pleospora RNase	3.1.27.4
Poly(deoxyribonucleotide):poly-(deoxyribonucleotide) ligase (AMP-forming)	6.5.1.1

Poly(deoxyribonucleotide):poly-(deoxyribonucleotide) ligase (AMP-forming, NMN-forming)	6.5.1.2
Polydeoxyribonucleotide synthetase (ATP)	6.5.1.1
Polydeoxyribonucleotide synthetase (NAD+)	6.5.1.2
Polygalacturonase	3.2.1.15
Poly(galacturonate) hydrolase	3.2.1.67
Poly-β-glucosaminidase	3.2.1.14
Polyglucuronide lyase	4.2.99.5
Poly(isoprenol)-phosphate galactosephosphotransferase	2.7.8.6
Poly(methoxygalacturonide) lyase	4.2.2.10
1,4-β-Poly-N-acetylglucosaminidase	3.2.1.14
5'-Polynucleotidase	3.1.3.33
2' (3')-Polynucleotidase	3.1.3.32
Polynucleotide adenylyltransferase	2.7.7.19
Polynucleotide ligase	6.5.1.1
Polynucleotide ligase (NAD+)	6.5.1.2
Polynucleotide phosphorylase	2.7.7.8
Polynucleotide 3'-phosphatase	3.1.3.32
Polynucleotide 3'-phosphohydrolase	3.1.3.32
Polynucleotide 5'-hydroxyl-kinase	2.7.1.78
Polynucleotide 5'-phosphatase	3.1.3.33
Polynucleotide 5'-phosphohydrolase	3.1.3.33
Polyol dehydrogenase	1.1.1.14
Polyol dehydrogenase (NADP+)	1.1.1.21
Polyphenol oxidase	1.10.3.2
Polyphosphatase	3.6.1.10
Polyphosphate depolymerase	3.6.1.10
Polyphosphate—glucose phosphotransferase	2.7.1.63
Polyphosphate:D-glucose 6-phosphotransferase	2.7.1.63
Polyphosphate kinase	2.7.4.1
Polyphosphate phosphohydrolase	3.6.1.11
Polyphosphate polyphosphohydrolase	3.6.1.10
Polyphosphorylase	2.4.1.1
Polyribonucleotide nucleotidyltransferase	2.7.7.8
Polyribonucleotide:ortho-phosphate nucleotidyltransferase	2.7.7.8
Poly(ribonucleotide):poly (ribonucleotide) ligase (AMP-forming)	6.5.1.3
Polyribonucleotide synthetase (ATP)	6.5.1.3
Polysaccharide depolymerase	3.2.1.87
Polysaccharide methyltransferase	2.1.1.18
Poly(1,2-α-L-fucoside-4-sulphate) glycanohydrolase	3.2.1.44
Poly(1,4-α-D-galactosiduronate) digalacturonohydrolase	3.2.1.82

Poly(1,4-α-D-galacturonide) exo-lyase	4.2.2.9
Poly(1,4-α-D-galacturonide) galacturonohydrolase	3.2.1.67
Poly(1,4-α-D-galacturonide) glycanohydrolase	3.2.1.15
Poly(1,4-α-D-galacturonide) lyase	4.2.2.2
Poly(1,4-β-D-mannuronide) lyase	4.2.2.3
Poly(1,4-β-(2-acetamido-2-deoxy-D-glucoside)) glycanohydrolase	3.2.1.14
Porphobilinogen ammonia-lyase (polymerizing)	4.3.1.8
Porphobilinogen synthase	4.2.1.24
Porphyran sulphatase	2.5.1.5
Post-proline cleaving enzyme	3.4.21.26
Post-proline endopeptidase	3.4.21.26
Potato nuclease	3.1.30.2
5α-Pregnan-3,20-dione:NADP+ oxidoreductase	1.3.1.30
Prenol-diphosphate pyrophosphohydrolase	3.1.7.1
Prenol pyrophosphatase	3.1.7.1
Prenyltransferase	2.5.1.1
PR-enzyme	4.1.99.3
PR-enzyme	3.1.3.17
Prephenate dehydratase	4.2.1.51
Prephenate dehydrogenase	1.3.1.12
Prephenate dehydrogenase (NADP+)	1.3.1.13
Prephenate hydro-lyase (decarboxylating)	4.2.1.51
Prephenate:NAD+ oxidoreductase (decarboxylating)	1.3.1.12
Prephenate:NADP+ oxidoreductase (decarboxylating)	1.3.1.13
Presqualene synthase	2.5.1.21
Primary amine:acceptor oxidoreductase (deaminating)	1.4.99.3
Procaine esterase	3.1.1.1
Progesterone,hydrogen-donor:oxygen oxidoreductase (hydroxylating)	1.14.99.4
Progesterone,hydrogen-donor:oxygen oxidoreductase (11α-hydroxylating)	1.14.99.14
Progesterone hydroxylase	1.14.99.4
Progesterone monooxygenase	1.14.99.4
Progesterone reductase	1.1.1.145
Progesterone 11α-hydroxylase	1.14.99.14
Progesterone 11α-monooxygenase	1.14.99.14
Progesterone 5α-reductase	1.3.1.30
Prolidase	3.4.13.9
Prolinase	3.4.13.8
Proline carboxypeptidase	3.4.16.2
Proline dipeptidase	3.4.13.9
Proline hydroxylase	1.14.11.2
Proline iminopeptidase	3.4.11.5

L-*Proline:NAD(P)+ 2-oxidoreductase*	1.5.1.1
L-*Proline:NAD(P)+ 5-oxidoreductase*	1.5.1.2
Proline racemase	5.1.1.4
Proline racemase	5.1.1.4
D-Proline reductase	1.4.1.6
D-Proline reductase (dithiol)	1.4.4.1
L-*Proline:tRNA^Pro ligase (AMP-forming)*	6.1.1.15
Proline,2-oxoglutarate dioxygenase	1.14.11.2
L-*Prolyl-amino acid hydrolase*	3.4.13.8
Prolyl dipeptidase	3.4.13.8
L-Prolylglycine dipeptidase	3.4.13.8
Prolyl-glycyl-peptide,2-oxoglutarate:oxygen oxidoreductase	1.14.11.2
L-*Prolyl-peptide hydrolase*	3.4.11.5
Prolyl-tRNA synthetase	6.1.1.15
'Pronase' (component)	3.4.24.4
Propanediol dehydratase	4.2.1.28
1,2-Propanediol hydro-lyase	4.2.1.28
D(or L)-*1,2-Propanediol:NAD+ oxidoreductase*	1.1.1.77
1,2-Propanediol:NADP+ oxidoreductase	1.1.1.55
Propanediol-phosphate dehydrogenase	1.1.1.7
1,2-Propanediol-1-phosphate:NAD+ oxidoreductase	1.1.1.7
2-Propanol:NADP+ oxidoreductase	1.1.1.80
Propionate CoA-transferase	2.8.3.1
Propionyl-CoA:carbon-dioxide ligase (ADP-forming)	6.4.1.3
Propionyl-CoA carboxylase	4.1.1.41
Propionyl-CoA carboxylase (ATP-hydrolysing)	6.4.1.3
3-Propylmalate glyoxylate-lyase (CoA-valerylating)	4.1.3.11
3-Propylmalate synthase	4.1.3.11
P. roqueforti protease II	3.4.24.4
Prostaglandin-A₁ Δ¹⁰— Δ¹¹-isomerase	5.3.3.9
Prostaglandin-A₁ Δ-isomerase	5.3.3.9
Prostaglandin R₂ D-isomerase	5.3.99.2
Prostaglandin R₂ D-isomerase	5.3.99.3
Prostaglandin R₂ D-isomerase	5.3.99.2
Prostaglandin R₂ E-isomerase	5.3.99.3
Prostaglandin synthase	1.14.99.1
Protaminase	3.4.17.2
Protamine kinase	2.7.1.70
Protease III	3.4.24.4
Protein (arginine) methyltransferase	2.1.1.23
Proteinase K	3.4.21.14
Protein disulphide-isomerase	5.3.4.1
Protein disulphide-isomerase	5.3.4.1
Protein—disulphide reductase (glutathione)	1.8.4.2

Protein-disulphide reductase (NAD(P)H)	1.6.4.4
Protein kinase	2.7.1.37
Protein (lysine) methyltransferase	2.1.1.43
Protein methylase I	2.1.1.23
Protein methylase II	2.1.1.24
Protein methylase III	2.1.1.43
Protein *O*-methyltransferase	2.1.1.24
Protein phosphatase	3.1.3.16
Protein—UDP acetylgalactosaminyl-transferase	2.4.1.41
Proteus mirabilis RNase	3.1.27.2
Protoaphin-aglucone dehydratase (cyclizing)	4.2.1.73
Protoaphin-aglucone hydro-lyase (cyclizing)	4.2.1.73
Protocatechuate carboxy-lyase	4.1.1.63
Protocatechuate decarboxylase	4.1.1.63
Protocatechuate oxygenase	1.13.11.3
Protocatechuate:oxygen 3,4-oxidoreductase (decyclizing)	1.13.11.3
Protocatechuate:oxygen 4,5-oxidoreductase (decyclizing)	1.13.11.8
Protocatechuate 3,4-dioxygenase	1.13.11.3
Protocatechuate 4,5-dioxygenase	1.13.11.8
Protocatechuate 4,5-oxygenase	1.13.11.8
Protocollagen hydroxylase	1.14.11.2
Protohaem ferro-lyase	4.99.1.1
Protoporphyrinogen-IX:oxygen oxidoreductase	1.3.3.4
Protoporphyrinogen oxidase	1.3.3.4
Providencia alcalifaciens (PalI)	3.1.23.17
Pseudocholinesterase	3.1.1.8
Pseudomonas aeruginosa **alkaline proteinase**	3.4.24.4
Pseudomonas aeruginosa **neutral proteinase**	3.4.24.4
Pseudomonas **cytochrome oxidase**	1.9.3.2
Pseudomonas serine proteinase	3.4.21.14
Pseudouridine kinase	2.7.1.83
Pseudouridylate synthase	4.2.1.70
Psychosine sulphotransferase	2.8.2.13
Psychosine—UDP galactosyltransferase	2.4.1.23
Pterin deaminase	3.5.4.11
Ptyalin	3.2.1.1
Pullulanase	3.2.1.41
Pullulan 4-glucanohydrolase	3.2.1.57
Pullulan 6-glucanohydrolase	3.2.1.41
Purine nucleosidase	3.2.2.1
Purine-nucleoside:orthophosphate ribosyltransferase	2.4.2.1
Purine-nucleoside phosphorylase	2.4.2.1

Putrescine acetyltransferase	2.3.1.57
Putrescine carbamoyltransferase	2.1.3.6
Putrescine methyltransferase	2.1.1.53
Putrescine oxidase	1.4.3.10
Putrescine:oxygen oxidoreductase (deaminating)	1.4.3.10
Pyranose oxidase	1.1.3.10
Pyranose:oxygen 2-oxidoreductase	1.1.3.10
Pyrazolylalanine synthase	4.2.1.50
Pyridine nucleotide transhydrogenase	1.6.1.1
Pyridoxal dehydrogenase	1.1.1.107
Pyridoxal kinase	2.7.1.35
Pyridoxal:NAD+ oxidoreductase	1.1.1.107
Pyridoxamine:oxaloacetate aminotransferase	2.6.1.31
Pyridoxamine—oxaloacetate transaminase	2.6.1.31
Pyridoxamine-phosphate aminotransferase	2.6.1.54
Pyridoxaminephosphate oxidase	1.4.3.5
Pyridoxaminephosphate:oxygen oxidoreductase (deaminating)	1.4.3.5
Pyridoxamine:pyruvate aminotransferase	2.6.1.30
Pyridoxamine—pyruvate transaminase	2.6.1.30
Pyridoxamine-5'-phosphate:2-oxoglutarate aminotransferase (D-glutamate-forming)	2.6.1.54
5-Pyridoxate dioxygenase	1.14.12.5
5-Pyridoxate,NADPH:oxygen oxidoreductase (decyclizing)	1.14.12.5
5-Pyridoxate oxidase	1.14.12.5
Pyridoxine:(acceptor) 5'-oxidoreductase	1.1.99.9
Pyridoxine dehydrogenase	1.1.1.65
Pyridoxine:NADP+ 4'-oxidoreductase	1.1.1.65
Pyridoxine:oxygen 4'-oxidoreductase	1.1.3.12
Pyridoxine 4-dehydrogenase	1.1.1.65
Pyridoxine 4-oxidase	1.1.3.12
Pyridoxine 5-dehydrogenase	1.1.99.9
Pyridoxin 4-oxidase	1.1.3.12
4-Pyridoxolactonase	3.1.1.27
4-Pyridoxolactone lactonohydrolase	3.1.1.27
Pyridoxol-5-dehydrogenase	1.1.99.9
Pyrimidine deoxyribonucleoside 2'-hydroxylase	1.14.11.3
Pyrimidine-nucleoside:orthophosphate ribosyltransferase	2.4.2.2
Pyrimidine-nucleoside phosphorylase	2.4.2.2
Pyrimidine phosphorylase	2.4.2.4
Pyrimidine phosphorylase	2.4.2.3
Pyrimidine transferase	2.5.1.2
Pyrimidine-5'-nucleotide nucleosidase	3.2.2.10

Pyrimidine-5′-nucleotide phosphoribo – *(deoxyribo)hydrolase*	3.2.2.10
Pyrithiamin deaminase	3.5.4.20
Pyrocatechase	1.13.11.1
*o-***Pyrocatechuate decarboxylase**	4.1.1.46
*o-*Pyrocatechuate oxygenase	1.13.11.14
Pyroglutamase (ATP-hydrolysing)	3.5.2.9
Pyroglutamyl aminopeptidase	3.4.11.8
L-*Pyroglutamyl-peptide hydrolase*	3.4.11.8
Pyrophosphate:acetate phosphotransferase	2.7.2.12
*Pyrophosphate:*D-*fructose-6-phosphate* *1-phosphotransferase*	2.7.1.90
Pyrophosphate—fructose-6-phosphate **1-phosphotransferase**	2.7.1.90
Pyrophosphate—glycerol **phosphotransferase**	2.7.1.79
Pyrophosphate:glycerol *1-phosphotransferase*	2.7.1.79
Pyrophosphate:oxaloacetate carboxy-lyase *(transphosphorylating)*	4.1.1.38
Pyrophosphate phosphohydrolase	3.6.1.1
*Pyrophosphate:*L-*serine* *O-phosphotransferase*	2.7.1.80
Pyrophosphate—serine **phosphotransferase**	2.7.1.80
Pyrophosphomevalonate decarboxylase	4.1.1.33
Pyrrolidone-carboxylate peptidase	3.4.11.8
Pyrroline-2-carboxylate reductase	1.5.1.1
1-Pyrroline-4-hydroxy-2-carboxylate *aminohydrolase (decyclizing)*	3.5.4.22
1-Pyrroline-4-hydroxy-2-carboxylate **deaminase**	3.5.4.22
1-Pyrroline-5-carboxylate dehydrogenase	1.5.1.12
1-Pyrroline-5-carboxylate:NAD⁺ *oxidoreductase*	1.5.1.12
Pyrroline-5-carboxylate reductase	1.5.1.2
Pyrrolooxygenase	1.13.11.26
Pyruvate:carbon-dioxide ligase *(ADP-forming)*	6.4.1.1
Pyruvate carboxylase	6.4.1.1
Pyruvate decarboxylase	4.1.1.1
Pyruvate dehydrogenase (cytochrome)	1.2.2.2
Pyruvate dehydrogenase (lipoamide)	1.2.4.1
[Pyruvate dehydrogenase (lipoamide)] **kinase**	2.7.1.99
[Pyruvate dehydrogenase (lipoamide)]- **phosphatase**	3.1.3.43
[Pyruvate dehydrogenase (lipoamide)]- *phosphate phosphohydrolase*	3.1.3.43
Pyruvate dehydrogenase,Pyruvic dehydrogenase	1.2.2.2

Pyruvate dehydrogenase,Pyruvic dehydrogenase	1.2.4.1
Pyruvate-ferredoxin oxidoreductase *(CoA-acetylating)*	1.2.7.1
Pyruvate:ferricytochrome b₁ *oxidoreductase*	1.2.2.2
Pyruvate formate-lyase	2.3.1.54
Pyruvate kinase	2.7.1.40
Pyruvate:lipoamide oxidoreductase *(decarboxylating and acceptor-acetylating)*	1.2.4.1
Pyruvate,orthophosphate dikinase	2.7.9.1
Pyruvate oxidase	1.2.3.3
Pyruvate oxidase (CoA-acetylating)	1.2.3.6
Pyruvate-oxime:acetone *oximinotransferase*	2.6.3.1
Pyruvate:oxygen oxidoreductase *(CoA-acetylating)*	1.2.3.6
Pyruvate:oxygen oxidoreductase *(phosphorylating)*	1.2.3.3
Pyruvate synthase	1.2.7.1
Pyruvate,water dikinase	2.7.9.2
Pyruvic carboxylase	6.4.1.1
Pyruvic decarboxylase,α-Ketoacid carboxylase	4.1.1.1
Pyruvic-malic carboxylase	1.1.1.38
Pyruvic-malic carboxylase	1.1.1.39
Pyruvic-malic carboxylase	1.1.1.40
Pyruvic oxidase	1.2.3.3
Q-enzyme	2.4.1.18
Quercetin:oxygen 2,3-oxidoreductase *(decyclizing)*	1.13.11.24
Quercetin 2,3-dioxygenase	1.13.11.24
Quercitrinase	3.2.1.66
Quercitrin 3-rhamnohydrolase	3.2.1.66
Quinate dehydrogenase	1.1.1.24
Quinate:NAD⁺ 3-oxidoreductase	1.1.1.24
Quinone reductase	1.6.5.1
Quinone reductase	1.6.99.2
Red cell neutral endopeptidase	3.4.21.24
Reduced ferredoxin:dinitrogen *oxidoreductase (ATP-hydrolysing)*	1.18.2.1
Reduced flavodoxin:dinitrogen *oxidoreductase (ATP-hydrolysing)*	1.19.2.1
Renilla luciferin:oxygen 2-oxidoreductase *(decarboxylating)*	1.13.12.5
Renilla **luciferin 2-monooxygenase**	1.13.12.5
Renin	3.4.99.19
Rennin	3.4.23.4
R-enzyme	3.2.1.41
Reptilase	3.4.21.29
Respiratory nitrate reductase	1.7.99.4
Retinal dehydrogenase	1.2.1.36

Retinal isomerase	5.2.1.3
Retinal:NAD+ oxidoreductase	1.2.1.36
Retinal reductase	1.1.1.71
all-trans-Retinal 11-cis-trans-isomerase	5.2.1.3
Retinene isomerase	5.2.1.3
Retinol dehydrogenase	1.1.1.105
Retinol:NAD+ oxidoreductase	1.1.1.105
Retinol-palmitate esterase	3.1.1.21
Retinol-palmitate palmitohydrolase	3.1.1.21
R-Glutaminyl-peptide:amine γ-glutamyl-yltransferase	2.3.2.13
L-Rhamnofuranose:NAD 11-oxidoreductase	1.1.1.173
L-Rhamnose dehydrogenase	1.1.1.173
L-Rhamnose isomerase	5.3.1.14
L-Rhamnose ketol-isomerase	5.3.1.14
α-L-Rhamnosidase	3.2.1.40
β-L-Rhamnosidase	3.2.1.43
β-L-Rhamnoside rhamnohydrolase	3.2.1.43
α-L-Rhamnoside rhamnohydrolase	3.2.1.40
Rhamnulokinase	2.7.1.5
Rhamnulosephosphate aldolase	4.1.2.19
L-Rhamnulose-1-phosphate L-lactaldehyde-lyase	4.1.2.19
RH hydroxylase	1.14.14.1
Rhizopus **carboxyl proteinase**	3.4.23.6
Rhizopus oligosporus RNase oligosporus RNase.	3.1.27.5
Rhodanese	2.8.1.1
Rhodotorula **carboxyl proteinase**	3.4.23.6
RH,reduced-flavoprotein:oxygen oxidoreductase (RH-hydroxylating)	1.14.14.1
Ribitol dehydrogenase	1.1.1.56
Ribitol:NAD+ 2-oxidoreductase	1.1.1.56
D-Ribitol-5-phosphate cytidylyltransferase	2.7.7.40
Ribitol-5-phosphate dehydrogenase	1.1.1.137
D-Ribitol-5-phosphate:NAD(P)+ 2-oxidoreductase	1.1.1.137
Riboflavinase	3.5.99.1
Riboflavin hydrolase	3.5.99.1
Riboflavin kinase	2.7.1.26
Riboflavin phosphotransferase	2.7.1.42
Riboflavin synthase	2.5.1.9
Ribokinase	2.7.1.15
Ribonuclease alpha	3.1.26.2
Ribonuclease (*Bacillus subtilis*)	3.1.27.2
Ribonuclease (*Enterobacter*)	3.1.27.6
Ribonuclease II	3.1.13.1
Ribonuclease III	3.1.26.3
Ribonuclease II. Plant RNase	3.1.27.1

Ribonuclease I. Venom RNase	3.1.27.5
Ribonuclease P	3.1.26.5
Ribonuclease (pancreatic)	3.1.27.5
Ribonuclease (*Physarum polycephalum*)	3.1.26.1
Ribonuclease T₂	3.1.27.1
Ribonuclease T₁	3.1.27.3
Ribonuclease U₂	3.1.27.4
Ribonucleoside-diphosphate reductase	1.17.4.1
Ribonucleoside-triphosphate reductase	1.17.4.2
5'-Ribonucleotide phosphohydrolase	3.1.3.5
3'-Ribonucleotide phosphohydrolase	3.1.3.6
D-Ribose dehydrogenase (NADP+)	1.1.1.115
Ribose isomerase	5.3.1.20
D-Ribose ketol-isomerase	5.3.1.20
D-Ribose:NADP+ 1-oxidoreductase	1.1.1.115
Ribosephosphate isomerase	5.3.1.6
Ribosephosphate pyrophosphokinase	2.7.6.1
Ribose-5-phosphate adenylyltransferase	2.7.7.35
Ribose-5-phosphate:ammonia ligase (ADP forming)	6.3.4.7
D-Ribose-5-phosphate ketol-isomerase	5.3.1.6
Ribosylhomocysteinase	3.3.1.3
S-Ribosyl-L-homocysteine ribohydrolase	3.3.1.3
Ribosylnicotinamide kinase	2.7.1.22
N-Ribosyl-purine ribohydrolase	3.2.2.1
N-**Ribosylpyrimidine nucleosidase**	3.2.2.8
D-Ribulokinase	2.7.1.47
Ribulokinase	2.7.1.16
Ribulosebisphosphate carboxylase	4.1.1.39
Ribulosephosphate 3-epimerase	5.1.3.1
L-Ribulosephosphate 4-epimerase	5.1.3.4
D-Ribulose-5-phosphate 3-epimerase	5.1.3.1
L-Ribulose-5-phosphate 4-epimerase	5.1.3.4
Ricinine aminohydrolase	3.5.5.2
Ricinine nitrilase	3.5.5.2
RNA adenylating enzyme	2.7.7.19
RNA ligase	6.5.1.3
RNA nucleotidyltransferase	2.7.7.6
RNA polymerase	2.7.7.6
rRNA (adenine-6-)-methyltransferase	2.1.1.48
rRNA (guanine-1-)-methyltransferase	2.1.1.51
rRNA (guanine-2-)-methyltransferase	2.1.1.52
tRNA (adenine-1-)-methyltransferase	2.1.1.36
tRNA adenylyltransferase	2.7.7.20
tRNA adenylyltransferase	2.7.7.25
tRNA CCA-pyrophosphorylase	2.7.7.25
tRNA CCA-pyrophosphorylase	2.7.7.21
tRNA cytidylyltransferase	2.7.7.21
tRNA (cytosine-5-)-methyltransferase	2.1.1.29

tRNA (guanine-1-)-methyltransferase	2.1.1.31
tRNA (guanine-2-)-methyltransferase	2.1.1.32
tRNA (guanine-7-)-methyltransferase	2.1.1.33
tRNA (guanosine-2′-)-methyltransferase	2.1.1.34
tRNA isopentenyltransferase	2.5.1.8
tRNA (purine-2- or -6-)-methyltransferase	2.1.1.30
tRNA (uracil-5-)-methyltransferase	2.1.1.35
RNase	3.1.27.5
RNase I	3.1.27.5
RNase N_2	3.1.27.1
RNase N_1 and N_2 . *N. crassa* RNase N_1 and N_2	3.1.27.3
RNase NU from KB cells	3.1.26.5
RNase U_3	3.1.27.4
RNase U_4	3.1.14.1
Robison ester dehydrogenase	1.1.1.49
Rubber allyltransferase	2.5.1.20
Rubber transferase	2.5.1.20
Rubredoxin:NAD+ oxidoreductase	1.18.1.1
Rubredoxin-NAD+ reductase	1.18.1.1
Rubredoxin reductase	1.18.1.1
RX:glutathione R-transferase	2.5.1.18
Saccharase	3.2.1.26
Saccharogen amylase	3.2.1.2
Saccharomyces carboxyl proteinase	3.4.23.6
Saccharomyces cerevisiae (H_2)	3.1.26.4
Saccharopine dehydrogenase (NAD+, L-glutamate-forming)	1.5.1.9
Saccharopine dehydrogenase (NAD+, lysine-forming)	1.5.1.7
Saccharopine dehydrogenase (NADP+, L-glutamate-forming)	1.5.1.10
Saccharopine dehydrogenase (NADP+, lysine-forming)	1.5.1.8
Salicylate hydroxylase	1.14.13.1
Salicylate,NADH:oxygen oxidoreductase (1-hydroxylating, decarboxylating)	1.14.13.1
Salicylate 1-monooxygenase	1.14.13.1
Salmon testis DNase	3.1.22.1
Salmon testis nuclease	3.1.16.1
Sarcina neutral proteinase	3.4.24.4
Sarcosine:(acceptor) oxidoreductase (demethylating)	1.5.99.1
Sarcosine dehydrogenase	1.5.99.1
Sarcosine oxidase	1.5.3.1
Sarcosine:oxygen oxidoreductase (demethylating)	1.5.3.1
Schardinger enzyme	1.2.3.2

Scopulariopsis proteinase	3.4.99.20
Sea anemone protease A	3.4.21.3
Sealase	6.5.1.1
Sea urchin hatching proteinase	3.4.24.12
Sedoheptulokinase	2.7.1.14
Sedoheptulose-bisphosphatase	3.1.3.37
Sedoheptulose-1,7-bisphosphate 1-phosphohydrolase	3.1.3.37
Sedoheptulose-7-phosphate: D-glyceraldehyde-3-phosphate dihydroxyacetonetransferase	2.2.1.2
Sedoheptulose-7-phosphate: D-glyceraldehyde-3-phosphate glycolaldehyetransferase	2.2.1.1
Sepia proteinase	3.4.24.2
Sepiapterin aminohydrolase	3.5.4.24
Sepiapterin deaminase	3.5.4.24
Sepiapterin reductase	1.1.1.153
Sequoyitol dehydrogenase	1.1.1.143
Serine acetyltransferase	2.3.1.30
Serine aldolase	2.1.2.1
Serine carboxypeptidase	3.4.16.1
Serine deaminase	4.2.1.16
Serine deaminase	4.2.1.13
D-Serine dehydratase	4.2.1.14
L-Serine dehydratase	4.2.1.16
L-Serine dehydratase	4.2.1.13
Serine dehydrogenase	1.4.1.7
Serine-ethanolaminephosphate phosphodiesterase	3.1.4.13
Serine—glyoxylate aminotransferase	2.6.1.45
L-Serine:glyoxylate aminotransferase	2.6.1.45
L-Serine hydro-lyase (adding homocysteine)	4.2.1.22
L-Serine hydro-lyase (adding indoleglycerol-phosphate)	4.2.1.20
L-Serine hydro-lyase (adding pyrazole)	4.2.1.50
D-Serine hydro-lyase (deaminating)	4.2.1.14
L-Serine hydro-lyase (deaminating)	4.2.1.13
Serine hydroxymethylase	2.1.2.1
Serine hydroxymethyltransferase	2.1.2.1
L-Serine:NAD+ oxidoreductase (deaminating)	1.4.1.7
L-Serine-O-sulphate ammonia-lyase (pyruvate-forming)	4.3.1.10
Serine palmitoyltransferase	2.3.1.50
Serine-phosphoethanolamine synthase	2.7.8.4
Serine-phospho-ethanolamine ethanolaminephosphohydrolase	3.1.4.13

L-*Serine:pyruvate aminotransferase*	2.6.1.51
Serine—pyruvate aminotransferase	2.6.1.51
Serinesulphate ammonia-lyase	4.3.1.10
Serine sulphhydrase	4.2.1.22
L-*Serine:tRNA^{Ser} ligase (AMP-forming)*	6.1.1.11
Serratia marcescens **extracellular proteinase**	3.4.24.4
*Serratia marcescens (Sma*I)	3.1.23.44
Seryl-tRNA synthetase	6.1.1.11
Shikimate dehydrogenase	1.1.1.25
Shikimate kinase	2.7.1.71
Shiklimate:NADP^+ 3-oxidoreductase	1.1.1.25
Sialidase	3.2.1.18
Sialyltransferase	2.4.99.1
Silkworm nuclease	3.1.30.2
Single-stranded-nucleate endonuclease	3.1.30.1
Sinigrase	3.2.3.1
Sinigrinase	3.2.3.1
Sinigrin sulphohydrolase, myrosulphatase	3.1.6.5
Snail DNase	3.1.22.1
Solanain	3.4.99.21
Sorbitol dehydrogenase	1.1.1.14
D-**Sorbitol-6-phosphate dehydrogenase**	1.1.1.140
D-*Sorbitol-6-phosphate:NAD^+ 2-oxidoreductase*	1.1.1.140
L-*Sorbose:(acceptor) 5-oxidoreductase*	1.1.99.12
Sorbose dehydrogenase	1.1.99.12
Sorbose dehydrogenase (NADP^+)	1.1.1.123
L-*Sorbose:NADP^+ 5-oxidoreductase*	1.1.1.123
L-**Sorbose oxidase**	1.1.3.11
L-*Sorbose:oxygen 5-oxidoreductase*	1.1.3.11
Sorghum carboxyl proteinase	3.4.23.14
Spermidine:(acceptor) oxidoreductase	1.5.99.6
Spermidine dehydrogenase	1.5.99.6
Spermine oxidase	1.5.3.3
Sphingomyelin ceramide-phosphohydrolase	3.1.4.41
Sphingomyelin cholinephosphohydrolase	3.1.4.12
Sphingomyelin phosphodiesterase	3.1.4.12
Sphingomyelin phosphodiesterase D	3.1.4.41
Sphingosine acyltransferase	2.3.1.24
Sphingosine cholinephosphotransferase	2.7.8.10
Spleen endonuclease	3.1.31.1
Spleen exonuclease	3.1.16.1
Spleen phosphodiesterase	3.1.31.1
Spleen phosphodiesterase. *Lactobacillus acidophilus* nuclease	3.1.16.1
Spreading factor	3.2.1.36
Spreading factor	3.2.1.35
Spreading factor	4.2.2.1
Squalene epoxidase	1.14.99.7

Squalene,hydrogen-donor:oxygen oxidoreductase (2,3-epoxidizing)	1.14.99.7
Squalene hydroxylase	1.14.1.3
Squalene monooxygenase (2,3-epoxidizing)	1.14.99.7
'S-S rearrangase'	5.3.4.1
Staphyloccus aureus (Sau3Al)	3.1.23.27
Staphylococcal proteinase II	3.4.22.13
Staphylococcal serine proteinase	3.4.21.19
Staphylococcal thiol proteinase	3.4.22.13
Staphylococcus aureus **neutral proteinase**	3.4.24.4
Staphylokinase	3.4.24.4
Starch (bacterial glycogen) synthase	2.4.1.21
Steapsin	3.1.1.3
Steroid,hydrogen-donor:oxygen oxidoreductase (17α-hydroxylating)	1.14.99.9
Steroid,hydrogen-donor:oxygen oxidoreductase (21-hydroxylating)	1.14.99.10
Steroid Δ-isomerase	5.3.3.1
Steroid-lactonase	3.1.1.37
Steroid,reduced-adrenal-ferredoxin:oxygen oxidoreductase (11β-hydroxylating)	1.14.15.4
Steroid 11β-hydroxylase	1.14.15.4
Steroid 11β-monooxygenase	1.14.15.4
Steroid 17α-hydroxylase	1.14.99.9
Steroid 17α-monooxygenase	1.14.99.9
Steroid 21-hydroxylase	1.14.99.10
Steroid 21-monooxygenase	1.14.99.10
Steroid 4,5-dioxygenase	1.13.11.25
Sterol-ester acylhydrolase	3.1.1.13
Δ^{24}-Sterol methyltransferase	2.1.1.41
Sterol-sulphatase	3.1.6.2
Sterol-sulphate sulphohydrolase	3.1.6.2
Stipitatonate carboxy-lyase (decyclizing)	4.1.1.60
Stipitatonate decarboxylase	4.1.1.60
Stizolobate synthase	1.13.11.29
Stizolobinate synthase	1.13.11.30
Streptidine kinase	2.7.1.72
thymonuclease. Streptococcal DNase (Streptodornase)	3.1.21.1
Streptococcal proteinase	3.4.22.10
Streptococcus thermophilus **intracellular proteinase**	3.4.24.4
*Streptomyces albus (Sal*PI)	3.1.23.31
Streptomyces **alkalophilic keratinase**	3.4.99.11
Streptomyces griseus **neutral proteinase**	3.4.24.4
*Streptomyces stanford (Sst*I)	3.1.23.34
*Streptomyces stanford (Sst*II)	3.1.23.35
*Streptomyces stanford (Sst*III)	3.1.23.36
Streptomycin 3″-adenylyltransferase	2.7.7.47
Streptomycin 3″-kinase	2.7.1.87

Streptomycin 6-kinase	2.7.1.72
Streptomycin-6-phosphatase	3.1.3.39
Streptomycin-6-phosphate phosphohydrolase	3.1.3.39
Stuart factor	3.4.21.6
Subtilisin	3.4.21.14
Succinate:(acceptor) oxidoreductase	1.3.99.1
Succinate:CoA ligase (ADP-forming)	6.2.1.5
Succinate:CoA ligase (GDP-forming)	6.2.1.4
Succinate dehydrogenase	1.3.99.1
Succinate:NAD+ oxidoreductase	1.3.1.6
Succinate-semialdehyde dehydrogenase	1.2.1.24
Succinate-semialdehyde dehydrogenase (NAD(P)+)	1.2.1.16
Succinate-semialdehyde:NAD+ oxidoreductase	1.2.1.24
Succinate-semialdehyde:NAD(P)+ oxidoreductase	1.2.1.16
Succinic dehydrogenase	1.3.99.1
Succinic thiokinase	6.2.1.4
Succinic thiokinase	6.2.1.5
Succinyl-CoA acylase	3.1.2.3
Succinyl-CoA:citramalate CoA-transferase	2.8.3.7
Succinyl-CoA:dihydrolipoamide S-succinyltransferase	2.3.1.61
Succinyl-CoA:glycine C-succinyltransferase (decarboxylating)	2.3.1.37
Succinyl-CoA:L-homoserine O-succinyltransferase	2.3.1.46
Succinyl-CoA hydrolase	3.1.2.3
Succinyl-CoA hydrolase	3.1.2.3
Succinyl-CoA:oxalate CoA-transferase	2.8.3.2
Succinyl-CoA synthetase (ADP-forming)	6.2.1.5
Succinyl-CoA synthetase (GDP-forming)	6.2.1.4
Succinyl-CoA:3-oxo-acid CoA-transferase	2.8.3.5
Succinyl-CoA:3-oxoadipate CoA-transferase	2.8.3.6
Succinyl-diaminopimelate aminotransferase	2.6.1.17
Succinyl-diaminopimelate desuccinylase	3.5.1.18
S-*Succinylglutathione hydrolase*	3.1.2.13
S-**Succinylglutathione hydrolase**	3.1.2.13
O-*Succinyl-L-homoserine succinate-lyase (adding cysteine)*	4.2.99.9
O-**Succinylhomoserine (thiol)-lyase**	4.2.99.9
Succinyl—β-ketoacyl-CoA transferase	2.8.3.2
N-*Succinyl-LL-2,6-diaminopimelate amidohydrolase*	3.5.1.18
N-*Succinyl-L-2,6-diaminopimelate: 2-oxoglutarate aminotransferase*	2.6.1.17
Sucrase	3.2.1.48
Sucrose-glucan glucosyltransferase	2.4.1.4
Sucrose α-D-glucohydrolase	3.2.1.48
Sucrose α-D-glucohydrolase	3.2.1.48
Sucrose glucosyltransferase	2.4.1.7
Sucrose:orthophosphate α-D-glucosyltransferase	2.4.1.7
Sucrose-phosphatase	3.1.3.24
Sucrose-phosphate synthase	2.4.1.14
Sucrosephosphate—UDP glucosyltransferase	2.4.1.14
Sucrose phosphorylase	2.4.1.7
Sucrose synthase	2.4.1.13
Sucrose—UDP glucosyltransferase	2.4.1.13
Sucrose 1-fructosyltransferase	2.4.1.9
Sucrose:1,4-α-D-glucan 4-α-D-glucosyltransferase	2.4.1.4
Sucrose:1,6-α-D-glucan 6-α-D-glucosyltransferase	2.4.1.5
Sucrose:2,1-β-D-fructan-β-D-fructosyltransferase	2.4.1.9
Sucrose:2,6-β-D-fructan 6-β-D-fructosyltransferase	2.4.1.10
Sucrose 6-fructosyltransferase	2.4.1.10
Sucrose 6-glucosyltransferase	2.4.1.5
Sucrose-6F-phosphate phosphohydrolase	3.1.3.24
Sugar-phosphatase	3.1.3.23
Sugar-phosphate phosphohydrolase	3.1.3.23
Sugar-sulphate sulphohydrolase	3.1.6.3
Sugar-1-phosphate adenylyltransferase	2.7.7.36
Sugar-1-phosphate nucleotidyltransferase	2.7.7.37
Sulfokinase	2.8.2.1
2-Sulphamido-2-deoxy-D-glucose sulphamidase	3.10.1.1
2-Sulphamido-2-deoxy-6-O-sulpho-D-glucose 6-sulphohydrolase	3.1.6.11
Sulphatase	3.1.6.1
Sulphate adenylyltransferase	2.7.7.4
Sulphate adenylyltransferase (ADP)	2.7.7.5
Sulphite dehydrogenase	1.8.2.1
Sulphite:ferricytochrome c oxidoreductase	1.8.2.1
Sulphite oxidase	1.8.3.1
Sulphite:oxygen oxidoreductase	1.8.3.1
Sulphite reductase	1.8.99.1
Sulphite reductase (ferredoxin)	1.8.7.1
Sulphite reductase (NADPH)	1.8.1.2
Sulphoacetaldehyde lyase	4.4.1.12
Sulphoacetaldehyde sulpho-lyase	4.4.1.12
Sulphoglucosamine sulphamidase	3.10.1.1
Sulphur dioxygenase	1.13.11.18
Sulphur:oxygen oxidoreductase	1.13.11.18
Sulphurylase	2.7.7.4
Superoxide dismutase	1.15.1.1

Tyrosine 2,3-aminomutase — 5.4.3.6

Tyrosine 3-hydroxylase — 1.14.16.2

Tyrosine 3-monooxygenase — 1.14.16.2

Tyrosyl-tRNA synthetase — 6.1.1.1

T2 and T4 induced
exodeoxyribonucleases — 3.1.11.1

T4 Endonuclease II — 3.1.21.1

T4 Endonuclease III — 3.1.21.2

T4 endonuclease V — 3.1.25.1

T5 and T7 exonucleases — 3.1.11.3

T7 Endonuclease I — 3.1.21.2

T7 Endonuclease II — 3.1.21.1

Ubiquinol-cytochrome c reductase — 1.10.2.2

*Ubiquinol:ferricytochrome c
oxidoreductase* — 1.10.2.2

Ubiquinone reductase — 1.6.5.3

Uca pugilator collagenolytic proteinase — 3.4.21.32

**UDPacetylgalactosamine—galactosyl-
galactosyl-glucosylceramide
β-N-acetyl-D-galactosaminyltransferase** — 2.4.1.79

**UDPacetylgalactosamine—globoside
α-N-acetyl-D-galactosaminyltransferase** — 2.4.1.88

**UDPacetylgalactosamine—
(N-acetylneuraminyl)-D-galactosyl-
D-glucosylceramide
acetylgalactosaminyltransferase** — 2.4.1.92

**UDPacetylgalactosamine—protein
acetylgalactosaminyltransferase** — 2.4.1.41

**UDPacetylglucosamine—
poly (ribitol-phosphate)
acetylglucosaminyltransferase** — 2.4.1.70

**UDPacetylglucosamine-protein
acetylglucosaminyltransferase** — 2.4.1.94

**UDPacetylglucosamine
pyrophosphorylase** — 2.7.7.23

**UDPacetylglucosamine—steroid
acetylglucosaminyltransferase** — 2.4.1.39

UDPacetylglucosamine 2-epimerase — 5.1.3.14

UDPacetylglucosamine 4-epimerase — 5.1.3.7

**UDPacetylmuramoylpentapeptide lysine
N6-alanyltransferase** — 2.3.2.10

UDPapiose—flavone apiosyltransferase — 2.4.2.25

*UDPapiose:7-O-β-D-glucosyl-5,7,4'-
trihydroxyflavone apiofuranosyltransferase* — 2.4.2.25

UDP-L-arabinose 4-epimerase — 5.1.3.5

UDParabinose 4-epimerase — 5.1.3.5

**UDPgalactose—ceramide
galactosyltransferase** — 2.4.1.62

**UDPgalactose—collagen
galactosyltransferase** — 2.4.1.50

*UDPgalactose:C-55-poly-(isoprenol)-
phosphate galactosephosphotransferase* — 2.7.8.6

**UDPgalactose—galactosyl-glucosaminyl-
galactosyl-glucosylceramide
α-D-galactosyltransferase** — 2.4.1.87

*UDPgalactose:D-galactosyl-(1,4)-2-
acetamido-2-deoxy-D-glucosyl-(1,3)-D-
galactosyl-galactosyl-(1,4)-D-
glucosylceramide α-D-galactosyltransferase* — 2.4.1.87

**UDPgalactose—glucosaminyl-galactosyl-
glucosylceramide
β-D-galactosyltransferase** — 2.4.1.86

**UDPgalactose—glucose
galactosyltransferase** — 2.4.1.22

*UDPgalactose—glycoprotein
galactosyltransferase* — 2.4.1.38

**UDPgalactose—lipopolysaccharide
galactosyltransferase** — 2.4.1.44

*UDPgalactose:lipopolysaccharide
galactosyltransferase* — 2.4.1.44

*UDPgalactose:muco-polysaccharide
galactosyltransferase* — 2.4.1.74

**UDPgalactose—mucopolysaccharide
galactosyltransferase** — 2.4.1.74

**UDPgalactose—N-acylsphingosine
galactosyltransferase** — 2.4.1.47

*UDPgalactose:N-acylsphingosine
galactosyltransferase* — 2.4.1.47

*UDPgalactose:O-α-L-fucosyl-(1,2)-
D-galactose α-D-galactosyltransferase* — 2.4.1.37

*UDPgalactose:sn-glycerol-3-phosphate
α-D-galactosyltransferase* — 2.4.1.96

**UDPgalactose-sn-glycerol-3-phosphate
galactosyltransferase** — 2.4.1.96

**UDPgalactose—sphingosine
β-D-galactosyltransferase** — 2.4.1.23

*UDPgalactose:sphingosine
β-D-galactosyltransferase* — 2.4.1.23

**UDPgalactose—1,2-diacylglycerol
galactosyltransferase** — 2.4.1.46

*UDPgalactose:1,2-diacylglycerol
3-O-galactosyltransferase* — 2.4.1.46

*UDPgalactose:2-acetamido-2-deoxy-
D-galactosyl-(N-acetylneuraminyl)-
D-galactosyl-D-glucosyl-N-acylsphingosine
galactosyltransferase* — 2.4.1.62

*UDPgalactose:2-acetamido-2-deoxy-D-
glucose 4-β-D-galactosyltransferase* — 2.4.1.90

*UDPgalactose:2-acetamido-2-deoxy-D-
glucosyl-glycopeptide galactosyltransferase* — 2.4.1.38

*UDPgalactose:2-acetamido-2-deoxy-D-
glucosyl-(1,3)-D-galactosyl-(1,4)-D-
glucosylceramide β-D-galactosyltransferase* — 2.4.1.86

**UDPgalactose—2-hydroxyacylsphingosine
galactosyltransferase** — 2.4.1.45

*UDPgalactose:2-(2-hydroxyacyl-)
sphingosine galactosyltransferase* — 2.4.1.45

Superoxide:superoxide oxidoreductase — 1.15.1.1

Tabernamontanain — 3.4.99.23

Tabunase — 3.8.2.1

Tagaturonate reductase — 1.1.1.58

Takadiastase — 3.4.23.6

Tannase — 3.1.1.20

Tannin acylhydrolase — 3.1.1.20

Tartrate dehydratase — 4.2.1.32

Tartrate dehydrogenase — 1.1.1.93

meso-**Tartrate dehydrogenase** — 1.3.1.7

Tartrate epimerase — 5.1.2.5

Tartrate epimerase — 5.1.2.5

L(+)-*Tartrate hydro-lyase* — 4.2.1.32

meso-Tartrate hydro-lyase — 4.2.1.37

meso-Tartrate:NAD+ oxidoreductase — 1.3.1.7

Tartrate:NAD+ oxidoreductase — 1.1.1.93

Tartronate-semialdehyde carboxylase — 4.1.1.47

Tartronate-semialdehyde reductase — 1.1.1.60

Tartronate-semialdehyde synthase — 4.1.1.47

*Taurine:(acceptor) oxidoreductase
(deaminating)* — 1.4.99.2

Taurine aminotransferase — 2.6.1.55

Taurine dehydrogenase — 1.4.99.2

Taurine:2-oxoglutarate aminotransferase — 2.6.1.55

Taurocyamine kinase — 2.7.3.4

*Taxifolin,NAD(P)H:oxygen
oxidoreductase (8-hydroxylating)* — 1.14.13.19

Taxifolin 8-monooxygenase — 1.14.13.19

dTDPgalactose pyrophosphorylase — 2.7.7.32

dTDPglucose 4,6-dehydratase — 4.2.1.46

dTDPglucose 4,6-hydro-lyase — 4.2.1.46

**dTDP-4-amino-4,6-dideoxy-D-galactose
aminotransferase** — 2.6.1.59

*dTDP-4-amino-4,6-dideoxy-D-galactose:
2-oxoglutarate aminotransferase* — 2.6.1.59

**dTDP-4-amino-4,6-dideoxy-D-glucose
aminotransferase** — 2.6.1.33

*dTDP-4-amino-4,6-dideoxy-D-glucose:
2-oxoglutarate aminotransferase* — 2.6.1.33

dTDP-4-ketorhamnose reductase — 1.1.1.133

dTDP-4-ketorhamnose 3,5-epimerase — 5.1.3.13

*dTDP-4-keto-6-deoxy-D-glucose
3,5-epimerase* — 5.1.3.13

*dTDP-6-deoxy-L-mannose:NADP+
4-oxidoreductase* — 1.1.1.133

dTDP-6-deoxy-L-talose dehydrogenase — 1.1.1.134

*dTDP-6-deoxy-L-talose:NADP+
4-oxidoreductase* — 1.1.1.134

Teichoic-acid synthase — 2.4.1.55

Tenebrio α-proteinase — 3.4.21.18

T-enzyme — 2.4.1.24

Terminal addition enzyme — 2.7.7.31

Terminal deoxyribonucleotidyl
transferase — 2.7.7.31

Terpenoid-allyltransferase — 2.5.1.11

Testololactone lactonohydrolase — 3.1.1.37

Testosterone 17β-dehydrogenase — 1.1.1.63

**Testosterone 17β-dehydrogenase
(NADP+)** — 1.1.1.64

*2,3,4,5-Tetrahydrodipicolinate:NAD(P)+
oxidoreductase* — 1.3.1.26

Tetrahydrofolate dehydrogenase — 1.5.1.3

*5,6,7,8-Tetrahydrofolate:NADP+
oxidoreductase* — 1.5.1.3

**Tetrahydropteroylglutamate
methyltransferase** — 2.1.1.13

**Tetrahydropteroyltriglutamate
methyltransferase** — 2.1.1.14

*2',4,4',6'-Tetrahydroxy-dehydrochalcone
1,3,5-trihydroxybenzene-hydrolase* — 3.7.1.4

Tetrahydroxypteridine cycloisomerase — 5.5.1.3

Tetrahydroxypteridine lyase (isomerizing) — 5.5.1.3

*1,4,5,6-Tetrahydro-6-oxo-nicotinate:
ferredoxin oxidoreductase* — 1.3.7.1

Tetrahymena **carboxyl proteinase**
and *Tetrahymena pyriformis* — 3.4.23.6
— 3.1.26.4

Theanine synthetase — 6.3.1.6

Thermolysin — 3.4.24.4

Thermomycolase — 3.4.21.14

Thermomycolin — 3.4.21.14

Thermophilic aminopeptidase — 3.4.11.12

**Thermophilic *Streptomyces* serine
proteinase** — 3.4.21.14

Thermopolyspora glauca (Tgl) — 3.1.23.35

Thiaminase — 3.5.99.2

Thiaminase I — 2.5.1.2

Thiaminase II — 3.5.99.2

*Thiamin:base 2-methyl-4-
aminopyrimidine-5-methenyltransferase-* — 2.5.1.2

Thiamindiphosphate kinase — 2.7.4.15

Thiamin hydrolase — 3.5.99.2

Thiamin kinase — 2.7.6.2

Thiamin kinase — 2.7.1.89

Thiamin-monophosphate kinase — 2.7.4.16

Thiamin phosphate pyrophosphorylase — 2.5.1.3

Thiamin pyridinylase — 2.5.1.2

Thiamin pyrophosphokinase — 2.7.6.2

Enzyme	EC number
Thiamin-triphosphatase	3.6.1.28
Thiamin-triphosphate phosphohydrolase	3.6.1.28
Thiobacillus thioparus RNase	3.1.27.5
Thiocyanate isomerase	5.99.1.1
Thioethanolamine acetyltransferase	2.3.1.11
Thiogalactoside acetyltransferase	2.3.1.18
Thioglucosidase	3.2.3.1
Thioglucoside glucohydrolase	3.2.3.1
Thiolase	2.3.1.9
Thiol methyltransferase	2.1.1.9
Thiol oxidase	1.8.3.2
Thiol:oxygen oxidoreductase	1.8.3.2
Thioltransacetylase A	2.3.1.12
Thioltransacetylase B	2.3.1.11
β-Thionase	4.2.1.22
Thioredoxin reductase (NADPH)	1.6.4.5
Thiosulphate:cyanide sulphurtransferase	2.8.1.1
Thiosulphate cyanide transsulphurase	2.8.1.1
Thiosulphate sulphurtransferase	2.8.1.1
L-Threonate dehydrogenase	1.1.1.129
L-Threonate:NAD$^+$ oxidoreductase	1.1.1.129
L-Threonine acetaldehyde-lyase	4.1.2.5
Threonine aldolase	2.1.2.1
Threonine aldolase	4.1.2.5
Threonine deaminase	4.2.1.16
Threonine dehydratase	4.2.1.16
L-Threonine hydro-lyase (deaminating)	4.2.1.16
L-Threonine:NAD$^+$ oxidoreductase	1.1.1.103
Threonine racemase	5.1.1.6
Threonine racemase	5.1.1.6
Threonine synthase	4.2.99.2
L-Threonine:tRNA^{Thr} ligase (AMP-forming)	6.1.1.3
L-Threonine 3-dehydrogenase	1.1.1.103
Threonyl-tRNA synthetase	6.1.1.3
Thrombin	3.4.21.5
Thrombokinase	3.4.21.6
Thymidine kinase	2.7.1.75
Thymidine kinase	2.7.1.21
Thymidine:orthophosphate deoxyribosyltransferase	2.4.2.4
Thymidine phosphorylase	2.4.2.4
Thymidine 2'-hydroxylase	1.14.11.3
Thymidine,2-oxoglutarate dioxygenase	1.14.11.3
Thymidine,2-oxoglutarate:oxygen oxidoreductase (2'-hydroxylating)	1.14.11.3
Thymidylate synthase	2.1.1.45
Thymidylate 5'-nucleotidase	3.1.3.35
Thymidylate 5'-phosphatase	3.1.3.35
Thymidylate 5'-phosphohydrolase	3.1.3.35
Thymine,2-oxoglutarate dioxygenase	1.14.11.6
Thymine,2-oxoglutarate:oxygen oxidoreductase (7-hydroxylating)	1.14.11.6
Thymine 7-hydroxylase	1.14.11.6
Thymus endonuclease	3.1.22.3
Thyroid carboxyl proteinase	3.4.23.11
Thyroid galactosyltransferase	2.4.1.38
Thyroid hormone aminotransferase	2.6.1.26
Thyroid peptidase	3.4.16.3
Thyroid peptide carboxypeptidase	3.4.16.3
Thyroxine aminotransferase	2.6.1.25
Thyroxine:2-oxoglutarate aminotransferase	2.6.1.25
T$_2$-induced deoxynucleotide kinase	2.7.4.12
Tissue endopeptidase degrading collagenase synthetic substrate	3.4.99.31
dTMP kinase	2.7.4.9
TPNH-cytochrome *c* reductase	1.6.2.4
Transaldolase	2.2.1.2
Transaminase A	2.6.1.1
Transcarboxylase	2.1.3.1
Transglutaminase	2.3.2.13
Transhydrogenase	1.6.1.1
Transketolase	2.2.1.1
Transoximinase	2.6.3.1
Transphosphoribosidase	2.4.2.7
Transphosphoribosidase	2.4.2.8
α,α-Trehalase	3.2.1.28
α,α-Trehalose glucohydrolase	3.2.1.28
α,α-Trehalose:orthophosphate β-D-glucosyltransferase	2.4.1.64
Trehalose-phosphatase	3.1.3.12
α,α-Trehalose-phosphate synthase (GDP-forming)	2.4.1.36
α,α-Trehalose-phosphate synthase (UDP-forming)	2.4.1.15
Trehalosephosphate—UDP glucosyltransferase	2.4.1.15
α,α-Trehalose phosphorylase	2.4.1.64
α,α-Trehalose-6-phosphate phosphoglucohydrolase	3.2.1.93
Trehalose-6-phosphate phosphohydrolase	3.1.3.12
Triacetate-lactonase	3.1.1.38
Triacetolactone lactonohydrolase	3.1.1.38
Triacylglycerol acylhydrolase	3.1.1.3
Triacylglycerol lipase	3.1.1.3
Triacylglycero-protein acylhydrolase	3.1.1.34
Tributyrase	3.1.1.3
1,1,1-Trichloro-2,2-bis-(4-chlorophenyl)-ethane hydrogen-chloride-lyase	4.5.1.1
Trichoderma koningi RNase III	3.1.27.4
Trichophyton mentagrophytes **keratinase**	3.4.24.10
Trichophyton schoenleinii **collagenase**	3.4.24.9
Triglyceride lipase	3.1.1.3
17,20β,21-Trihydroxysteroid:NAD$^+$ oxidoreductase	1.1.1.53
3α,7α,12α-Trihydroxy-5β-cholan-24-oylglycine amidohydrolase	3.5.1.24
L-3,5,3'-Triiodothyronine:2-oxoglutarate aminotransferase	2.6.1.26
Trimetaphosphatase	3.6.1.2
Trimetaphosphate hydrolase	3.6.1.2
Trimethylamine:(acceptor) oxidoreductase (demethylating)	1.5.99.7
Trimethylamine dehydrogenase	1.5.99.7
Trimethylamine-N-oxide formaldehyde-lyase	4.1.2.32
Trimethylamine-*N*-oxide reductase	1.6.6.9
Trimethylamine-oxide aldolase	4.1.2.32
4-Trimethylaminobutyrate,2-oxoglutarate: oxygen oxidoreductase (3-hydroxylating)	1.14.11.1
Trimethylsulphonium-chloride: tetrahydrofolate N-methyltransferase	2.1.1.19
Trimethylsulphonium—tetrahydrofolate methyltransferase	2.1.1.19
Triokinase	2.7.1.28
Triosephosphate dehydrogenase	1.2.1.9
Triosephosphate dehydrogenase	1.2.1.12
Triosephosphate dehydrogenase (NADP$^+$)	1.2.1.13
Triosephosphate isomerase	5.3.1.1
Triosephosphate mutase	5.3.1.1
Tripeptide aminopeptidase	3.4.11.4
Triphosphatase	3.6.1.25
Triphosphatase, ATPase	3.6.1.3
Triphosphate phosphohydrolase	3.6.1.25
Triphosphoinositide inositol-trisphosphohydrolase	3.1.4.11
Triphosphoinositide phosphatase	3.1.3.36
Triphosphoinositide phosphodiesterase	3.1.4.11
Tritirachium **alkaline proteinase**	3.4.21.14
Tropinesterase	3.1.1.10
True cholinesterase	3.1.1.7
Trypsin	3.4.21.4
α and β-Trypsin	3.4.21.4
Trypsinogen kinase	3.4.23.6
Tryptamine *N*-methyltransferase	2.1.1.49
D-Tryptophan acetyltransferase	2.3.1.34
Tryptophan aminotransferase	2.6.1.27
Tryptophanase	1.13.11.11
Tryptophanase	4.1.99.1
Tryptophan decarboxylase	4.1.1.28
Tryptophan decarboxylase	4.1.1.27
Tryptophan desmolase	4.2.1.20
L-Tryptophan indole-lyase (deaminating)	4.1.99.1
Tryptophan oxygenase	1.13.11.11
L-Tryptophan:oxygen 2-oxidoreductase (decarboxylating)	1.13.12.3
L-Tryptophan:oxygen 2,3-oxidoreductase (decyclizing)	1.13.11.11
L-Tryptophan:phenylpyruvate aminotransferase	2.6.1.28
Tryptophan—phenylpyruvate aminotransferase	2.6.1.28
Tryptophan pyrrolase	1.13.11.11
Tryptophan synthase	4.2.1.20
L-Tryptophan,tetrahydropteridine:oxygen oxidoreductase (5-hydroxylating)	1.14.16.4
L-Tryptophan:tRNA^{Trp} ligase (AMP-forming)	6.1.1.2
Tryptophanyl-tRNA synthetase	6.1.1.2
Tryptophan 2-monooxygenase	1.13.12.3
L-Tryptophan:2-oxoglutarate aminotransferase	2.6.1.27
Tryptophan 2,3-dioxygenase	1.13.11.11
Tryptophan 5-hydroxylase	1.14.16.4
Tryptophan 5-monooxygenase	1.14.16.4
dTTP:α-D-galactose-1-phosphate thymidylyltransferase	2.7.7.32
dTTP:α-D-glucose-1-phosphate thymidylyltransferase	2.7.7.24
Tyraminase	1.4.3.4
Tyramine *N*-methyltransferase	2.1.1.27
Tyramine oxidase	1.4.3.4
Tyramine oxidase	1.4.3.9
Tyramine:oxygen oxidoreductase (deaminating)	1.4.3.9
Tyrosinase	1.10.3.1
β-Tyrosinase	4.1.99.2
Tyrosinase	1.14.18.1
Tyrosine aminotransferase	2.6.1.5
L-Tyrosine carboxy-lyase	4.1.1.25
Tyrosine carboxypeptidase	3.4.16.3
Tyrosine decarboxylase	4.1.1.25
Tyrosine-ester sulphotransferase	2.8.2.9
Tyrosine phenol-lyase	4.1.99.2
L-Tyrosine phenol-lyase (deaminating)	4.1.99.2
Tyrosine-pyruvate aminotransferase	2.6.1.20
L-Tyrosine,tetrahydropteridine:oxygen oxidoreductase (3-hydroxylating)	1.14.16.2
L-Tyrosine:tRNA^{Tyr} ligase (AMP-forming)	6.1.1.1
L-Tyrosine:2-oxoglutarate aminotransferase	2.6.1.5
L-Tyrosine 2,3-aminomutase	5.4.3.6

UDPgalactose:5-hydroxylysine-collagen
galactosyltransferase 2.4.1.50
UDPgalacturonate
β-galacturonosyltransferase
(acceptor unspecific) 2.4.1.75
UDPgalacturonate—polygalacturonate
α-D-galacturonosyltransferase 2.4.1.43
UDPgalacturonate:1,4-α-poly-D-
galacturonate
4-α-D-galacturonosyltransferase 2.4.1.43
UDPgalacturonosyltransferase 2.4.1.75
UDPglucose—apigenin
β-glucosyltransferase 2.4.1.81
UDPglucose arylamine
glucosyltransferase 2.4.1.71
UDPglucose:arylamine
N-glucosyltransferase 2.4.1.71
UDPglucose—cellulose
glucosyltransferase 2.4.1.12
UDPglucose—ceramide
glucosyltransferase 2.4.1.80
UDPglucose—collagen
glucosyltransferase 2.4.1.66
UDPglucose dehydrogenase 1.1.1.22
UDPglucose:DNA α-D-glucosyltransferase 2.4.1.26
UDPglucose—DNA
α-D-glucosyltransferase 2.4.1.26
UDPglucose—DNA
β-D-glucosyltransferase 2.4.1.27
UDPglucose:DNA β-D-glucosyltransferase 2.4.1.27
UDPglucose—flavonol
glucosyltransferase 2.4.1.91
UDPglucose:flavonol
3-O-glucosyltransferase 2.4.1.91
UDPglucose—fructose
glucosyltransferase 2.4.1.13
UDPglucose—fructosephosphate
glucosyltransferase 2.4.1.14
UDPglucose:D-fructose
2-α-D-glucosyltransferase 2.4.1.13
UDPglucose:D-fructose-6-phosphate
2-α-D-glucosyltransferase 2.4.1.14
UDPglucose:α-D-galactose-1-phosphate
uridylyltransferase 2.7.7.12
UDPglucose:galactosyl-lipopolysaccharide
glucosyltransferase 2.4.1.73
UDPglucose—β-glucan
glucosyltransferase 2.4.1.12
UDPglucose—glucosephosphate
glucosyltransferase 2.4.1.15
UDPglucose:D-glucose-6-phosphate
1-α-D-glucosyltransferase 2.4.1.15
UDPglucose:D-glucosyl-DNA
β-D-glucosyltransferase 2.4.1.28

UDPglucose—glucosyl-DNA
β-D-glucosyltransferase 2.4.1.28
UDPglucose—glycogen
glucosyltransferase 2.4.1.11
[UDPglucose—glycogen
glucosyltransferase-D] phosphohydrolase 3.1.3.42
UDPglucose:glycogen
4-α-D-glucosyltransferase 2.4.1.11
UDPglucose—hexose-1-phosphate
uridylyltransferase 2.7.7.12
UDPglucose:lipopolysaccharide
glucosyltransferase 2.4.1.58
UDPglucose—lipopolysaccharide
glucosyltransferase I 2.4.1.58
UDPglucose—lipopolysaccharide
glucosyltransferase II 2.4.1.73
UDPglucose—luteolin
β-D-glucosyltransferase 2.4.1.81
UDPglucose:N-acylsphingosine
glucosyltransferase 2.4.1.80
UDPglucose:NAD⁺ 6-oxidoreductase 1.1.1.22
UDPglucose:phenol
β-D-glucosyltransferase 2.4.1.35
UDPglucose phosphopolyprenol
glucosyltransferase 2.4.1.78
UDPglucose:phosphopolyprenol
glucosyltransferase 2.4.1.78
UDPglucose—poly (glycerol phosphate)
α-D-glucosyltransferase 2.4.1.52
UDPglucose:poly (glycerol phosphate)
α-D-glucosyltransferase 2.4.1.52
UDPglucose—poly (ribitol phosphate)
β-D-glucosyltransferase 2.4.1.53
UDPglucose:poly (ribitol phosphate)
β-D-glucosyltransferase 2.4.1.53
UDPglucose pyrophosphorylase 2.7.7.9
UDPglucose:(S)-4-hydroxymandelonitrile
β-D-glucosyltransferase 2.4.1.85
UDPglucose—1,3-β-D-glucan
glucosyltransferase 2.4.1.34
UDPglucose:1,4-β-D-glucan
4-β-D-glucosyltransferase 2.4.1.12
UDPglucose:2-hydroxyisobutyronitrile
β-D-glucosyltransferase 2.4.1.63
UDPglucose 4-epimerase 5.1.3.2
UDPglucose 4-epimerase 5.1.3.2
UDPglucose:5-hydroxylysine-collagen
glucosyltransferase 2.4.1.66
UDPglucose:5,7,3',4'-tetrahydroxyflavone
β-D-glucosyltransferase 2.4.1.81
UDPglucosyltransferase 2.4.1.35
UDPglucuronate:bilirubin-glucuronoside
glucuronosyltransferase 2.4.1.77

UDPglucuronate bilirubin-glucuronoside glucuronosyltransferase	2.4.1.77
UDPglucuronate—bilirubin glucuronosyltransferase	2.4.1.76
UDPglucuronate:bilirubin-glucuronosyltransferase	2.4.1.76
UDPglucuronate carboxy-lyase	4.1.1.35
UDPglucuronate decarboxylase	4.1.1.35
UDPglucuronate β-D-glucuronosyltransferase (acceptor-unspecific)	2.4.1.17
UDPglucuronate—oestradiol glucuronosyltransferase	2.4.1.59
UDPglucuronate—oestriol 16α-glucuronosyltransferase	2.4.1.61
UDPglucuronate:oestriol 16α-D-glucuronosyltransferase	2.4.1.61
UDPglucuronate—oestriol 17β-D-glucuronosyltransferase	2.4.1.42
UDPglucuronate→phenol transglucuronidase	2.4.1.17
UDPglucuronate—1,2-diacylglycerol glucuronosyltransferase	2.4.1.84
UDPglucuronate:1,2-diacylglycerol 3-glucuronosyltransferase	2.4.1.84
UDPglucuronate:17β-hydroxysteroid 17β-D-glucuronosyltransferase	2.4.1.42
UDPglucuronate:17β-oestradiol 3-glucuronosyltransferase	2.4.1.59
UDPglucuronate 4-epimerase	5.1.3.6
UDPglucuronate 4-epimerase	5.1.3.6
UDPglucuronate 5′-epimerase	5.1.3.12
UDPglucuronate 5′-epimerase	5.1.3.12
UDPglucuronosyltransferase	2.4.1.17
UDP--*N*--acetylglucosamine dehydrogenase	1.1.1.136
UDP-*N*-acetylenolpyruvoylglucosamine reductase	1.1.1.158
UDP-*N*-acetylgalactosamine-4-sulphate sulphotransferase	2.8.2.7
UDP-*N*-acetylglucosamine—glycoprotein *N*-acetylglucosaminyltransferase	2.4.1.51
UDP-*N*-acetylglucosamine—lipopolysaccharide *N*-acetylglucosaminyltransferase	2.4.1.56
*UDP-N-acetylmuramate:*L-*alanine ligase (ADP-forming)*	6.3.2.8
UDP-N-acetylmuramate:NADP+ oxidoreductase	1.1.1.158
UDP-N-acetylmuramoyl-L-alanine: D-*glutamate ligase (ADP-forming)*	6.3.2.9
UDP-*N*-acetylmuramoylalanine synthetase	6.3.2.8

*UDP-N-acetylmuramoyl-*L-*alanyl-*D-*glutamate:*L-*lysine ligase (ADP-forming)*	6.3.2.7
*UDP-N-acetylmuramoyl-*L-*alanyl-*D-*glutamate:meso-2,6-diaminopimelate ligase (ADP-forming)*	6.3.2.13
UDP-*N*-acetylmuramoyl-L-alanyl-D-glutamate synthetase	6.3.2.9
*UDP-N-acetylmuramoyl-*L-*alanyl-*D-*glutamyl-*L-*lysine:*D-*alanyl-*D-*alanine ligase (ADP-forming)*	6.3.2.10
UDP-*N*-acetylmuramoyl-L-alanyl-D-glutamyl-L-lysine synthetase	6.3.2.7
UDP-*N*-acetylmuramoyl-L-alanyl-D-glutamyl-L-lysyl-D-alanyl-D-alanine synthetase	6.3.2.10
*UDP-N-acetylmuramoyl-*L-*alanyl-*D-γ-*glutamyl-*L-*lysyl-*D-*alanyl-*D-*alanine:undecaprenoid-1-ol-phosphate phospho-N-acetylmuramoyl-pentapeptide-transferase*	2.7.
*UDP-N-acetylmuramoyl-*L-*alanyl-*D-*glutamyl-meso-2,6-diaminopimelate:* D-*alanyl-*D-*alanine ligase (ADP-forming)*	6.3.2.15
UDP-*N*-acetylmuramoyl-L-alanyl-D-glutamyl-*meso*-2,6-diaminopimelate synthetase	6.3.2.13
UDP-*N*-acetylmuramoyl-L-alanyl-D-glutamyl-*meso*-2,6-diaminopimeloyl-D-alanyl-D-alanine synthetase	6.3.2.15
*UDP-N-acetylmuramoyl-tetrapeptidyl-*D-*alanine alanine-hydrolase*	3.4.17.8
UDPxylose:protein xylosyltransferase	2.4.2.26
UDPxylose—protein xylosyltransferase	2.4.2.26
*UDPxylose:1,4-β-*D-*xylan 4-β-*D-*xylosyltransferase*	2.4.2.24
*UDP-2-acetamido-2-deoxy-*D-*galactose:* D-*galactosyl-(1,4)-*D-*galactosyl-(1,4)-* D-*glucosylceramide β-N-acetamidodeoxy-* D-*galactosyltransferase*	2.4.1.79
*UDP-2-acetamido-2-deoxy-*D-*galactose: (N-acetylneuraminyl)-*D-*galactosyl-*D-*glucosylceramide acetamidodeoxygalactosyltransferase*	2.4.1.92
*UDP-2-acetamido-2-deoxy-*D-*galactose:O-α-*L-*fucosyl-(1,2)-*D-*galactose acetamidodeoxygalactosyltransferase*	2.4.1.40
*UDP-2-acetamido-2-deoxy-*D-*galactose:protein acetamidodeoxygalactosyl-transferase*	2.4.1.41
*UDP-2-acetamido-2-deoxy-*D-*galactose: 2-acetamido-2-deoxy-*D-*galactosyl-(1,3)-*D-*galactosyl-(1,4)-*D-*galactosyl-(1,4)-*D-*glucosylceramide α-N-acetamidodeoxy-*D-*2.4.1.88 galactosyltransferase*	
*UDP-2-acetamido-2-deoxy-*D-*glucose:chitin 4-β-acetamidodeoxy-*D-*glucosyltransferase*	2.4.1.16

UDP-2-acetamido-2-deoxy-D-glucose:glycoprotein 2-acetamido-2-deoxy-D-glucosyltransferase	2.4.1.51
UDP-2-acetamido-2-deoxy-D-glucose:poly(ribitol-phosphate) 2-acetamido-2-deoxyglucosyltransferase	2.4.1.70
UDP-2-acetamido-2-deoxy-D-glucose:protein β-N-acetamidodeoxy-D-glucosyltransferase	2.4.1.94
UDP-2-acetamido-2-deoxy-D-glucose:17α-hydroxysteroid-3-D-glucuronoside 17α-acetamidodeoxyglucosyltransferase	2.4.1.39
UDP-2-acetamido-2-deoxy-D-glucose 2-epimerase	5.1.3.14
UDP-2-acetamido-2-deoxy-D-glucose 4-epimerase	5.1.3.7
UDP-4-amino-2-acetamido-2,4,6-trideoxyglucose aminotransferase	2.6.1.34
UDP-4-amino-2-acetamido-2,4,6-trideoxyglucose:2-oxoglutarate aminotransferase	2.6.1.34
UMP:pyrophosphate phosphoribosyltransferase	2.4.2.9
UMP pyrophosphorylase	2.4.2.9
Unsaturated acyl-CoA hydratase	4.2.1.17
Unsaturated acyl-CoA reductase	1.3.99.2
Unsaturated-phospholipid methyltransferase	2.1.1.16
Unspecific diphosphate phosphohydrolase	3.6.1.15
Uracil:(acceptor) oxidoreductase	1.2.99.1
Uracil dehydrogenase	1.2.99.1
Uracil hydro-lyase (adding D-ribose 5-phosphate)	4.2.1.70
Uracil phosphoribosyltransferase	2.4.2.9
Uracil-5-carboxylate carboxy-lyase	4.1.1.66
Uracil-5-carboxylate decarboxylase	4.1.1.66
Urate oxidase	1.7.3.3
Urate:oxygen oxidoreductase	1.7.3.3
Urateribonucleotide:orthophosphate ribosyltransferase	2.4.2.16
Urateribonucleotide phosphorylase	2.4.2.16
Urea amidohydrolase	3.5.1.5
Urea amidohydrolase (ATP-hydrolysing)	3.5.1.45
Urea:carbon-dioxide ligase (ADP-forming) (decarboxylating, deaminating)	6.3.4.6
Urea carboxylase (hydrolysing)	6.3.4.6
Urea hydro-lyase	4.2.1.69
Urease	3.5.1.5
Urease (ATP-hydrolysing)	3.5.1.45
Ureidoglycollate dehydrogenase	1.1.1.154
Ureidoglycollate lyase	4.3.2.3
(S)-Ureidoglycollate:NAD(P)+ oxidoreductase	1.1.1.154
(—)-Ureidoglycollate urea-lyase	4.3.2.3
β-Ureidopropionase	3.5.1.6
Ureidosuccinase	3.5.1.7
Uricase	1.7.3.3
Uridine kinase	2.7.1.48
Uridine nucleosidase	3.2.2.3
Uridine:orthophosphate ribosyltransferase	2.4.2.3
Uridine phosphorylase	2.4.2.3
Uridine ribohydrolase	3.2.2.3
Uridyl transferase	2.7.7.12
Urishiol oxidase	1.10.3.2
Urocanase	4.2.1.49
Urocanate hydratase	4.2.1.49
Urokinase	3.4.99.26
Urokinase	3.4.21.31
Uronate dehydrogenase	1.2.1.35
Uronate:NAD+ 1-oxidoreductase	1.2.1.35
Uronic isomerase	5.3.1.12
Uronolactonase	3.1.1.19
Uroporphyrinogen decarboxylase	4.1.1.37
Uroporphyrinogen III carboxy-lyase	4.1.1.37
Uroporphyrinogen I synthase	4.3.1.8
Ustilago maydis	3.1.26.4
Ustilago sphaerogena RNase	3.1.27.3
UTP:ammonia ligase (ADP-forming)	6.3.4.2
dUTPase	3.6.1.23
UTP:α-D-galactose-1-phosphate uridylyltransferase	2.7.7.10
UTP:α-D-glucose-1-phosphate uridylyltransferase	2.7.7.9
dUTP nucleotidohydrolase	3.6.1.23
UTP:α-D-xylose-1-phosphate uridylyltransferase	2.7.7.11
UTP:1-phospho-α-D-glucuronate uridylyltransferase	2.7.7.44
UTP:2-acetamido-2-deoxy-α-D-glucose-1-phosphate uridylyltransferase	2.7.7.23
Vaccinia virus DNase VI	3.1.21.2
L-Valine carboxy-lyase	4.1.1.14
Valine decarboxylase	4.1.1.14
Valine dehydrogenase (NADP+)	1.4.1.8
ValineΔisoleucine aminotransferase	2.6.1.32
L-Valine:NADP+ oxidoreductase (deaminating)	1.4.1.8
L-Valine:tRNA^Val ligase (AMP-forming)	6.1.1.9
Valine—3-methyl-2-oxovalerate aminotransferase	2.6.1.32
L-Valine:3-methyl-2-oxovalerate aminotransferase	2.6.1.32
Valyl-tRNA synthetase	6.1.1.9
Venom exonuclease	3.1.15.1
Vertebrate collagenase	3.4.24.7

Vinylacetyl-CoA Δ³—Δ²-*isomerase*	5.3.3.3
Vinylacetyl-CoA Δ-isomerase	5.3.3.3
Vipera russelli **proteinase**	3.4.21.23
Vitamin A esterase	3.1.1.12
Vitamin B₁₂ᵣ reductase	1.6.99.9
Xanthine dehydrogenase	1.2.1.37
Xanthine:NAD⁺ oxidoreductase	1.2.1.37
Xanthine oxidase	1.2.3.2
Xanthine:oxygen oxidoreductase	1.2.3.2
Xanthine phosphoribosyltransferase	2.4.2.22
Xanthomonas amaranthicola (XamI)	3.1.23.37
Xanthomonas malvacearum (XmaII)	3.1.23.31
Xanthomonas papavericola (XpaI)	3.1.23.42
Xanthosine-5'-phosphate:ammonia ligase (AMP-forming)	6.3.4.1
Xanthosine-5'-phosphate:L-glutamine amido-ligase (AMP-forming)	6.3.5.2
Xenopus laevis RNase	3.1.27.5
Xylanase	3.2.1.32
1,4-β-D-Xylan synthase	2.4.2.24
1,4-β-D-Xylan xylanohydrolase	3.2.1.8
1,3-β-D-Xylan xylanohydrolase	3.2.1.32
1,4-β-D-Xylan xylohydrolase	3.2.1.37
1,3-β-D-Xylan xylohydrolase	3.2.1.72

Xylitol:NADP⁺ 4-oxidoreductase (L-xylulose-forming)	1.1.1.10
Xylitol:NAD⁺ 2-oxidoreductase (D-xylulose-forming)	1.1.1.9
Xylobiase	3.2.1.37
L-Xylose dehydrogenase	1.1.1.113
D-Xylose dehydrogenase	1.1.1.175
Xylose isomerase	5.3.1.5
D-Xylose ketol-isomerase	5.3.1.5
L-Xylose:NADP⁺ 1-oxidoreductase	1.1.1.113
D-Xylose:NAD⁺ 1-oxidoreductase	1.1.1.175
Xylose-1-phosphate uridylyltransferase	2.7.7.11
β-Xylosidase	3.2.1.37
Xylulokinase	2.7.1.17
L-Xylulokinase	2.7.1.53
D-Xylulose reductase	1.1.1.9
L-Xylulose reductase	1.1.1.10
D-Xylulose-5-phosphate D-*glyceraldehyde-3-phosphate-lyase (phosphate-acetylating)*	4.1.2.9
Yeast carboxypeptidase	3.4.17.4
Yeast DNase	3.1.21.2
Yeast proteinase B	3.4.22.9
Yeast ribonuclease	3.1.14.1
Zwischenferment	1.1.1.49
Zymohexase	4.1.2.13

APPENDIX

NOMENCLATURE OF ELECTRON-TRANSFER PROTEINS

1. General introduction

The processes of oxidation in living cells are catalysed by enzyme systems that transfer hydrogen atoms or electrons in successive steps from an initial donor to a final acceptor.

Examples of such systems are:

a. Oxidation of intermediary metabolites by molecular oxygen in the mitochondria of animal, plant and protist cells, and also in the protoplasmic membranes of those protists whose cells do not contain mitochondria. This enzyme system is commonly referred to as the respiratory chain.

b. Light-driven oxidation of water in chloroplasts of green plants and in membranes of blue-green protists and the light-driven oxidation of water or other suitable reductants in membranes of certain protists. These enzyme systems are commonly referred to as the photosynthetic chain.

c. The oxygenation of compounds by the introduction of one or both of the atoms of molecular oxygen.

d. The reduction of cytidine disphosphate to deoxycytidine diphosphate by NADH.

The initial and final steps of electron transfer in these chains are catalysed by discrete enzymes that can be separated from the other components and studied by conventional methods of enzymology. The initial dehydrogenation of a substrate such as succinate, or a reduced coenzyme such as NADH, is catalysed by a dehydrogenase that is relatively specific for the hydrogen donor but not for the acceptor. Thus, artificial acceptors such as ferricyanide, phenazine methosulphate and methylene blue, or even naturally occurring electron acceptors with which the enzyme does not react in vivo, are often used in the study of dehydrogenases. For example, in the mitochondrial respiratory chain, electrons from NADH are transferred to ubiquinone-10 by an enzyme that could be given the systematic name NADH: ubiquinone oxidoreductase. However, this is hardly appropriate, since, although this enzyme complex (Complex 1) can be separated from the rest of the respiratory chain, the isolated complex reacts sluggishly with added Q-10. It is true that it is possible to separate from Complex 1 a fragment that catalyses the reduction by NADH of ubiquinone homologues of low molecular weight, but, since steps involved in the reaction between NADH and ubiquinone in the intact membrane are by-passed in the reaction catalysed by the isolated fragment, naming it NADH: ubiquinone oxidoreductase would be misleading. For these reasons, the enzyme responsible for the primary dehydrogenation of NADH in the mitochondrial respiratory chain is named simply NADH dehydrogenase (systematic name, NADH: (acceptor) oxidoreductase, EC 1.6.99.3).

The subsequent steps of electron transfer in the mitochondrial respiratory chain are catalysed by electron-transfer carriers, both nonprotein (ubiquinone) and protein. Although each protein carrier, being a catalytically active protein, satisfies the most all-embracing definition of an enzyme, many do not readily fit in with the scheme of enzyme nomenclature, since they catalyse hydrogen or electron transfer from another enzyme to yet a third enzyme. Moreover, since much is known about the electron-carrying centre of these enzymes, it is more appropriate to classify them on the basis of chemical structure of the prosthetic groups and the manner of their attachment to the protein. Six types of electron-transfer proteins have been identified: flavoproteins, proteins containing reducible disulphide groups, cytochromes, iron-sulphur proteins, cuproproteins and molybdoproteins.

Many of these electron-transfer proteins are subunits of enzymes that have been classified under the oxidoreductases. For example, QH_2:ferricytochrome c oxidoreductase (EC 1.10.2.2), contains cytochromes b-562, b-566, and c_1, and an iron-sulphur protein, in addition to subunits in which no electron carrying centre has been identified.

2. Flavoproteins

Flavoproteins contain two types of prosthetic group, FMN *(e.g.* NADH dehydrogenase, EC 1.6.99.3) and FAD. The FMN is non-covalently bound in all known cases. FAD may be non-covalently bound *(e.g.* in lipoamide dehydrogenase (NADH), EC 1.6.4.3) or covalently bound by a methylene bridge between the benzene ring of the isoalloxazine and an amino acid residue in the protein *(e.g.* succinate dehydrogenase, EC 1.3.99.1).

Most flavoproteins are known to catalyse well-defined chemical reactions and are classified in the appropriate subgroup of the list of enzymes. There are two exceptions. One is the so-called electron-transfer flavoprotein that catalyses the transfer of electrons from another enzyme, namely butyryl-CoA dehydrogenase (EC 1.3.99.2), acyl-CoA dehydrogenase (EC 1.3.99.3), sarcosine dehydrogenase (EC 1.5.99.1) or dimethylglycine hydrogenase (EC 1.5.99.2), to the respiratory chain. The second is flavodoxin, a group of flavoproteins of low potential that catalyse electron transfer between two other redox proteins as part of photosynthetic, nitrogen- or sulphate-reducing, or hydrogen-evolving systems.

Flavoproteins that form part of electron-transfer systems are non-autoxidizable. In addition, flavoprotein oxidases, catalysing the direct oxidation of substrates by oxygen are known *e.g.* D-amino-acid oxidase, EC 1.4.3.3.

3. Proteins containing reducible disulphide

Lipoylproteins containing lipoic acid covalently bound by an amide link between its carboxyl group and the 6-amino group of a lysine residue in the protein are involved in the oxidation of both pyruvate and 2-oxoglutarate. The disulphide group of the lipoic acid is both reduced and acylated by pyruvate dehydrogenase (lipoate) (EC 1.2.4.1) and oxoglutarate dehydrogenase (EC 1.2.4.2). Thus, lipoylproteins act as both hydrogen and acyl acceptors. They are subunits of what are often termed the pyruvate or 2-oxoglutarate dehydrogenase complexes.

The flavoproteins catalysing the reduction of lipoyl compounds, oxidized glutathione and thioredoxin by reduced nicotinamide-adenine nucleotide (EC 1.6.4.3, 1.6.4.2 and 1.6.4.5, respectively) contain reducible cystine residues that are involved in the electron-transfer reaction. Thioredoxin, which is required for the reduction of cytidine diphosphate to deoxycytidine diphosphate, is itself an electron-transferring protein with a cystine residue as the electron-transferring centre.

4. Cytochromes

Since their discovery by Keilin, it has been customary to assign the cytochromes to the groups a, b and c according to the nature and mode of binding of the prosthetic haem moieties; group d was introduced by the IUB Enzyme Commission in 1961. Recommendations for the further classification within the four groups were issued in 1961 (1), 1964 (2) and 1972 (3), and have helped to develop widely accepted designations for the various cytochromes.

Since issuance of the 1972 report (3), much has been learned about the primary and tertiary structure of the "classical" cytochromes, and knowledge has accumulated on other cytochromes. It is clear, however, that at the present time the basis for a completely self-consistent chemical classification scheme does not yet exist and that, as stated in the original version, "it is premature to do more than attempt to coordinate and, hopefully, to monitor present practices in classification and nomenclature of cytochromes." The present recommendations follow the same lines as the previous ones. Obsolete portions of the 1972 text (3) have been replaced by appropriate new material and the criteria for assignment of groups that had been specified in an earlier report (1) and omitted from the 1972 version, are again included.

4.1 *Definitions*

The term "haem"[1] remains understood as *any tetrapyrrolic chelate of iron*. The terms "ferrohaem" and "ferrihaem"[2] refer to Fe(II) and Fe(III) oxidation states in haem. A haemochrome is defined as a low-spin compound of haem in which the fifth and sixth coordination places are occupied by strong field ligands[3]. Finally, the term "haemoprotein" refers to a protein containing haem as a prosthetic group.

The classical definition of cytochrome is retained: a cytochrome is a haemoprotein whose characteristic mode of action involves transfer of reducing equivalents associated with a reversible change in oxidation state of the prosthetic group. Formally, this redox change involves a single-electron reversible equilibrium between the Fe(II) and Fe(III) states of the central iron atom.

4.2 *Cytochrome Groups*

Four major groups of cytochromes are currently recognized:

1. Cytochromes *a*. Cytochromes in which the haem prosthetic group contains a formyl side-chain, i.e., haem *a*.

2. Cytochromes *b*. Cytochromes with protohaem or a related haem (without formyl group) as prosthetic group, not covalently bound to protein. This group includes cytochrome "P-450" and cytochrome "o" (see below).

3. Cytochromes *c*. Cytochromes with covalent linkages between the haem side-chains and protein. This group includes *all* cytochromes with prosthetic groups linked in this way, not only those with thioether linkages.

4. Cytochromes *d*. Cytochromes with a tetrapyrollic chelate of iron as prosthetic group in which the degree of conjugation of double bonds is less than in porphyrin, e.g., dihydroporphyrin (chlorin).

Use of the small unprimed italicised letter implies that the haem prosthetic group is in a haemochrome linkage. To indicate that in both the oxidized and reduced forms the haem prosthetic group is not in a haemochrome linkage, a primed small italicised letter, e.g., *c'*, is used.

---

[1] The European spelling "haem", as apposed to "heme", is used here although both are acceptable.

[2] Ferrihaem is sometimes referred to as "haematin", a usage still sanctioned by tradition. However, this term should not be used for the compound crystallized as the chloride or other salt; such compounds are customarily termed "haemin"

[3] The traditional definition that the fifth and sixth coordination positions are occupied by nitrogen atoms was obviously too restrictive as borne out by the case of cytochrome c, wherein one of these sites is occupied by the sulphur atom of methionine.

In the case of a cytochrome having two or more different haem groups attached to a specific protein, each different haem should be indicated, e.g. *Pseudomonas aeruginosa* cytochrome *cd*. In the case of a cytochrome having two or more of the same haem groups attached to a specific protein but in different environments, so that one or more is in a haemochrome linkage and one or more in a non-haemochrome linkage, both types of linkage should be indicated by using both the unprimed and primed small italicised letter, appropriate for the haem in question. As an example, cytochrome *c* oxidase (EC 1.9.3.1) is considered to contain both a haem *a*, in haemochrome-type linkage and a haem *a* (called "*a₃*") in a non-haemochrome-type linkage. By the suggested convention, this cytochrome should be called "cytochrome *aa'*," but current usage still clings to "cytochrome *aa₃*." In general, the name of a cytochrome will not indicate the number of identical molecules per mole of cytochrome.

The main practical tests to be adopted as criteria in determining the group to which a cytochrome belongs should be (1) the position of the α band of the pyridine Fe(II) haemochrome and (2) the ether solubility of the haemin after treatment of the cytochrome with acidified acetone, or acidified methyl ethyl ketone, as shown in Table I.

Table I

Practical Criteria for Determining the Cytochrome Group

Cytochrome	α band of pyridine ferro-haemochrome in alkali	Solubility of product from treatment of cytochrome with acetone-HCl in ether
Group a	580-590 nm	soluble
Group b	556-558 nm	soluble
Group c	549-551 nm	insoluble
Group d	600-620 nm	soluble

It is recommended that for groups a, b, and d, the position of the α band of the pyridine Fe(II) haemochrome should be determined after the acidified acetone-cleaved haemin has been extracted into ether and then extracted from the ether by dilute sodium hydroxide.[4]

Once the nature of the prosthetic haem group and its mode of linkage have been determined, so that the group to which a cytochrome belongs can be stated, the procedure for naming it can follow the steps recommended below:

    a. At the first mention in a publication, the name should be expanded to the systematic name that will include the source in parentheses, e.g. cytochrome *b*[1] (Bacterium X). Since the need has arisen to refer to different *b* and *c* type cytochromes present in the same cell or in the same organelle of a given cell type, it will often be necessary to add distinguishing characteristics, such as "microsomal," "tetrahaem," "high potential," etc.

---

[4]In addition, a number of chemical reactions that can be carried out with small amounts can also be used for the establishment of the group. Oxime formation of the isolated haemin with hydroxylamine combined with conversion of the oxime into a nitrile, and a number of spectroscopically observable reactions of the haematin or porphyrin moiety with sodium bisulphite, or with Dimedon, prove the presence of a haem with a formyl side-chain (group *a*). The reduction of unsaturated side-chain with hydrazine-HI, or Pt-H_2, together with the criterion of α-band position shows the presence of protohaem (group *b*). The splitting of the thioether bond with silver, or mercurous sulphate is a test for group *c*. Group *d* is characterized by the presence of an absorption band of the haemin in acetic acid at about 605 nm, and by the formation of a chlorin with a strong absorption band in the red by removal of iron at room temperature.

b. The names of the already well-established cytochromes with consecutive subscript numbering, listed below, are retained. All cytochromes not fitting in this category should be given a name based upon the α-band wavelength (nm) and written thus: cytochrome c-554.

c. The α-band wavelength used should be determined at room temperature, not liquid-air temperature, and should, if possible be obtained from absolute absorption spectra of the purified protein under carefully defined conditions. Since an error of 1 nm in assigning the position of the band alters the name, care should be taken to standardize the spectrophotometer with standard lines, e.g. with those given by Nd (III). Failing this, the absorption maximum can be determined by calibration against the standard mitochondrial cytochrome c.

Frequently, assignments are made solely on the location of the α-band maximum, but this is a practice that should be discouraged. Reviewers and editors of reports dealing with discovery and description of cytochromes should insist that the above procedures, based on chemical characterization wherever possible, be followed. However, these rules need elaboration whenever solubilization of cytochromes is not possible without denaturation, as in some membrane-bound proteins. The rules given are applied readily only to whole cell systems that have a simple composition, e.g. c-type proteins only. Problems arise when bound cytochromes c occur together with frequently encountered b-type complexes whose α-band maxima overlap those of the c-type proteins. In such cases, attempts should be made to separate adequately the membrane from the soluble fractions in a whole-cell system. Electron-spin resonance spectroscopy is rapidly coming into use to aid in resolving complex cytochrome and haem protein systems which include mixtures of high and low spin components. Use of redox buffers combined with spectroscopy can also be used occasionally to characterize cytochrome components but redox potentials measured at ambient temperatures should not be applied to spectral moieties determined at the low temperature associated with EPR spectroscopy.

Purification of soluble proteins to homogeneity after separation from membrane-bound components should be effected by all available chromatographic, gel-filtration and electrophoretic procedures, together with conventional salting-out techniques and, hopefully, crystallization before assay of physico-chemical properties.

In application of the naming procedures based on location of the α-wavelength maximum, any asymmetry or splitting of the α-peak should be noted, *viz.*, cytochrome c-555 (550) indicates a minor peak or shoulder in parenthesis.

4.3 *Variance in Cytochrome Groups*

The variations in spectroscopic and functional character of cytochromes found, especially as a result of work with prokaryotic systems, led to an undesirable proliferation of subscripts to distinguish sub-groups of uncertain character. It appears that this unhappy tendency has been halted, but adequate characterization of sub-groups remains a problem. Presumably, the primary sequence may provide eventually a rational basis for proper assignments to sub-groups, but the suggestion that a phylogenetic basis for such classification based on homologies may exist is still premature. However, homology can be invoked to reclassify a number of prokaryotic cytochromes c, as is indicated in the revised list of cytochromes given below.

Certain variant cytochromes remain in limbo. Thus, cytochrome "P-450," which originally received its name in a casual manner (location of Soret peak of reduced CO compound) not in conformity with any of the recommendations given herewith or previously, has been characterized as a class of proteins

597

with activity as a monoxygenase involved in hydroxylation associated with electron transfer. Its chemical nature as a *b*-type cytochrome with atypical non-nitrogenous ligands is well established in a number of systems, notably that of the pseudomonads which effect hydroxylation of terpenes. However, no consensus as to a rational terminology is yet in evidence, hence, no new suggestions are submitted as alternative to either the continued usage of P-450, or of other terms ("cytochrome m" has been urged by some investigators).

Another class that is also a sub-group of cytochrome *b*, is cytochrome "o", a type of protohaem oxidase found in prokaryotes. It is recommended that this terminology be abandoned and all examples known be reclassified according to procedures given above. Helicorubin also belongs to the *b* group.

4.4 *List of Cytochromes*

4.4.1 *Cytochrome a group*

Cytochrome *aa*₃	Identical with cytochrome *c* oxidase of eukaryotes. The α-band of the reduced cytochromes is at 605 nm, the γ (Soret) at 445 nm. The reduced cytochrome combines with CO with a shift of the α-band to 590 nm and the γ-band to 430 nm; it also combines with cyanide with a shift of the α-band to 590 nm with little effect on the γ-band. The reduced form is autoxidizable. In the presence of cyanide, one-half of the haem molecule is autoxidizable. It also contains two Cu atoms. It catalyzes oxidation of mitochondrial cytochrome *c* by O_2.
Cytochrome *a*₁	Present in certain bacteria (e.g. *Acetobacter pasteurianum*). Its reduced form is autoxidizable and combines with cyanide without appreciable change in the position of the α-band (590 nm). It is thought to function as a terminal oxidase, but no purified preparation has been reported.

4.4.2 *Cytochrome b group*

Cytochrome *b*	Present in mitochondria of eucaryotes and in chloroplasts. Different *b* species are indicated by the position of the α-band (in nm) of the reduced species, e.g. *b*-563, *b*-566, and also, when possible, by redox potential. Chloroplast cytochrome *b*-563, also known as cytochrome *b*₆, appears to function in one of green plant photosynthetic electron transfer systems while another, "*b*-554" is also present.
Cytochrome *b*₁	Present in certain bacteria (e.g. *Escherichia coli*). Its characteristic absorption band is the α-band of the reduced form at 557-560 nm.
Cytochrome *b*₂	Present in yeast. It contains one molecule of FMN as a second prosthetic group and acts as a lactate dehydrogenase (EC 1.1.2.3).
Cytochrome *b*₃	Present in microsomal material from non-photosynthetic plant tissues. Its characteristic absorption band is the α-band of reduced form at 559 nm.
Cytochrome *b*₅	Present mainly in animal microsomes. It is reduced by NADH in the presence of cytochrome *b*₅ reductase (EC 1.6.2.2). Its characteristic band is the α-band at 554 nm with a shoulder at about 560 nm. It may function in electron transfer associated with desaturation of higher fatty acids and hydroxylation (detoxification).

| Cytochrome b_7 | Present in spadices of various *Arum* species. Its characteristic absorption is the α-band of the ferro form at 560 nm. It is autoxidizable. |

| Cytochrome "o" | A prokaryotic terminal oxidase with protohaem as prosthetic group. (It has also been reported in some protozoans.) The reduced CO compound has absorption peaks at 557-567 nm (α) and 532-537 nm (β). Its absorption spectrum indicates it is a high-spin haem protein. The term, cytochrome o, is discouraged. |

| Cytochrome "P-450" | A *b*-type cytochrome, found in both prokaryotic and eukaryotic systems that effects a two equivalent reduction of oxygen accompanied by oxygenation of organic substrates (monooxygenase). In *Pseudomonas putida*, it selectively performs 5-*exo*-hydroxylation of a number of bicyclic monoterpenes. The ferri form exhibits a mixture of high-spin and low-spin states with a Soret peak at 417 nm. On binding to substrate, the peak shifts to 391 nm. Reduction gives an absorption maximum at 408 nm and a form that rapidly binds oxygen (418 nm) reversibly. |

4.4.3 *Cytochrome c group*

| Cytochrome c | Present in eukaryotic mitochondria where it functions as the substrate for the terminal oxidase (EC 1.9.3.1) in oxidative phosphorylation. It is a soluble, low-spin, monohaem protein with 103-112 residues. Its midpoint redox potential over most of the physiological range is 250 mv. In its reduced form, the α-band maximum is at 550 nm, the β at 520 nm and the Soret peak at 415 nm. |

| Cytochrome c_1 | The membrane-bound *c*-type protein of mitochondria with reduced α peak at 553 nm. On solubilization with detergents this α maximum remains unchanged. It functions as electron donor to cytochrome c in the mitochondrial respiratory chain. |

| Cytochrome c_2 | Soluble, low-spin cytochrome with the same tertiary folding and binding of haem as in mitochondrial cytochrome c, but with limited ability to replace it as a substrate for cytochrome c oxidase (EC 1.9.3.1). It is a monohaem protein with midpoint redox potential usually some 100 mV more oxidizing than that of mitochondrial c, and with much different pH redox profile. It is found mainly in non-sulfur purple photosynthetic bacteria where it functions at the high potential end of the photophosphorylation chain as electron donor to oxidized reaction center bacteriochlorophyll. It appears to be present also in some non-photosynthetic bacteria. |

| Cytochrome c_3 | Low potential low-spin cytochrome with some thioether binding to side chains of haem as in mitochondrial c, but with very different tertiary structure and no homology to the eukaryotic protein. It exists in multihaem form as a monomer with some 80-90 residues and is found in the strictly anaerobic sulfate-reducing bacteria where it participates in sulfate respiration coupled to phosphorylation. It may also exist in other bacteria. It exhibits no reactivity with mitochondrial reductase or oxidase. |

| Cytochrome c_5 | As isolated, it is a low-spin dimeric monohaem protein, usually with 85-90 residues per monomer exhibiting same thioether binding and extraplanar ligands as in mitochondrial cytochrome c. It is unreactive with the cytochrome c reductase and oxidase, but seems to function as an |

599

intermediate in nitrate respiration of facultative anaerobic pseudomonads where it occurs. It appears to be present also in the strictly anaerobic nitrogen-fixing *Azotobacter*. Its midpoint redox potential is somewhat higher than that of mitochondrial c and its alpha peak is red shifted to about 554-555 nm.

Cytochrome c_6 More usually called "cytochrome f:1p", it is a monohaem monomer, low-spin cytochrome which in algae has a molecular weight of about 10,000 and the same binding and extraplanar ligands as in mitochondrial c. Its midpoint redox potential is usually about 100 mV higher than that of the mitochondrial protein. Its reduced α peak is asymmetric and red-shifted to about 555 nm. It functions like cytochrome c_2 in that it mediates electron transfer at the high potential terminus of the photophosphorylation chain in chloroplasts and algae. It is unreactive with mitochondrial cytochrome c reductase or oxidase.

"Pseudomonas" Cytochrome c-551   Another prokaryotic, 80 to 90 residue monomer monohaem cytochrome, unreactive with mitochondrial cytochrome c reductase or oxidase, although possessing similar midpoint redox potential, same extraplanar ligands and haem thioether binding. It is distributed like cytochrome c_5 and apparently functions in nitrite and nitrate respiration in pseudomonads, but it is also found in other bacteria.

Green Bacterial Cytochrome c-555   A low-spin monohaem cytochrome of 80 to 90 residues found exclusively in the green photosynthetic bacteria (*Chlorobium*) with intermediate midpoint redox potential (about 150 mV) and variable reactivities with mitochondrial cytochrome c reductase and oxidase. However, like other prokaryotic cytochromes c mentioned above, it possesses the characteristic thioether haem binding and the same extraplanar ligands. Its reduced α-peak is asymmetric and red-shifted (to about 555 nm) as in cytochrome f.

Cytochrome c A high-spin variant cytochrome c (originally known as ("RHP")) with haem binding through side chain thioether linkage as in mitochondrial cytochrome c but, like some groups mentioned above, unreactive with the mitochondrial c reductase or oxidase. It occurs usually as a dimer, with monomer molecular weights of about 14000, and has a midpoint redox potential at pH 7 close to zero. It lacks haemochrome linkages to extraplanar ligands but exhibits a low-spin haemochrome EPR spectrum at pH values higher than 12. It is found in purple photosynthetic bacteria as well as in nitrate-reducing pseudomonads. When soluble, it reacts only with NO and CO in its reduced form and with NO in its oxidized form. It is unreactive with compounds that provide the usual ligands (CN^-, F^-, N_3^-, etc) for high-spin haem proteins. Its reduced α peak is broad and centered between 440 and 560 nm. Various other forms are as yet insufficiently characterized to place into sub-groups. Two examples are the dihaem cytochrome c_4 (found in *Azotobacter* and related possibly to c_2 on the basis of partial homology) and flavo-cytochrome c (found in purple sulfur photosynthetic bacteria). An important bound form, functional in photometabolism, is "purple-sulfur cytochrome c-556, 552" that has one haem group of high potential and one of low potential.

Cytochrome *d*	Earlier known as cytochrome a_2. Present in many bacteria, its characteristic absorption bands are at about 645-650 nm oxidized and 625-630 nm reduced. Not as yet isolated in pure, soluble form. It is lipophilic; an example is found in *Klebsiella aerogenes*. Another example, found in *Acetobacter peroxydans*, has its α-peak in the reduced form at 612 nm (formerly termed "a_4").
Cytochrome *cd*	This dihaem cytochrome functions as a dissimilatory nitrite reductase in pseudomonads, terminal to a respiratory chain including the *Pseudomonas* cytochrome *c*-551. The haem *d* component at pH 5.6 has its α-peak in reduced form at 625 nm. In *Pseudomonas aeruginosa* cytochrome *cd* appears as a very hydrophilic protein insoluble in ether.

5. Iron-sulphur proteins [5]

Recommendations for the nomenclature of iron-sulphur proteins were formulated in 1971 and published in 1973 (4). Since that time, a number of major contributions to the field have reported structure determination by X-ray crystallography, elaboration of a series of model compounds, and the discovery of novel examples of these proteins. On the basis of the new information, it became desirable to revise the nomenclature of the iron-sulphur proteins and thereby take into account the more precise description of known species while retaining flexibility for future discoveries. The present account reflects the advice of a group of experts given over a four year period.

Wherever possible, terms that have gained wide acceptance have been retained and particular attention has been given to the development of a useful shorthand notation. However, in order to comply with the established rules for the nomenclature of inorganic compounds (5), it was necessary to make some changes from what is presently used in the biochemical literature.[6]

5.1 Proteins containing iron are divided into three groups: hemoproteins, iron-sulphur proteins, and other iron-containing proteins (Scheme I). The last group includes ferritin, transferrin and the oxygenases. The term "iron-sulphur proteins" refers only to those proteins in which a non-heme iron is ligated with inorganic sulphur or cysteine sulphur.

5.2. The iron-sulphur proteins (abbreviation: Fe-S proteins) are in two major categories: *simple* iron-sulphur proteins and *complex* iron-sulphur proteins. "*Simple*" need only be used when the difference from complex iron-sulphur proteins is emphasized. Simple iron-sulphur proteins contain only one or more Fe-S clusters[6] whereas the complex proteins bear such additional active groups as flavin or heme.

5.3 Simple iron-sulphur proteins fall into three groups: rubredoxins, ferredoxins, and other (simple iron-sulphur proteins) (See Scheme I).

---

[5]Based on the 1978 Recommendations of the Nomenclature Committee of the International Union of Biochemistry

[6]According to these rules (5), the Fe_2S_2 or Fe_4S_4 unit of a 2-iron or 4-iron ferredoxin, respectively, is to be considered as the iron-sulphur cluster, thereby excluding the cysteine residues linked to the iron. In this sense then, rubredoxin does not contain an iron-sulphur cluster, merely an iron-centre. When the cluster charge is calculated, the exclusion of the cysteine ligands makes it necessary to digress from present widespread use in biochemistry in that the charge of a 2-iron ferredoxin, e.g., spinach ferredoxin, in its normally isolated oxidized state is now $(+2)$ instead of (-2) and that of reduced spinach ferredoxin, $(+1)$ instead of (-3). In inorganic nomenclature a "centre" is understood as a single atom to which ligands are attached, as it exists, for example, in rubredoxin. In the recent biochemical literature, iron-sulphur clusters of complex iron-sulphur proteins have been called "iron-sulphur centres" in the sense of enzymatic or redox- "active centres" of these proteins. Although this use deviates from that established in inorganic chemistry, the use of "centre 1, 2, etc." of complex proteins in the sense of "active centre" as used in enzymology is sufficiently entrenched and reasonable that it may be continued colloquially. However, the correct nomenclature would be "cluster 1, 2, etc.".

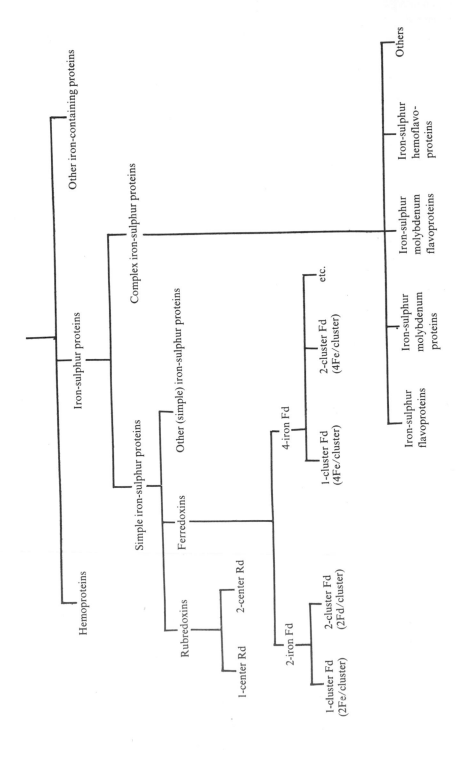

Scheme 1. Iron-containing proteins

5.3.1 Rubredoxins (abbreviation: Rd). This group comprises those iron-sulphur proteins without acid-labile sulphur that are characterized by having iron in a typical mercaptide coordination, *i.e.* an iron center[6] surrounded by four cysteine residues or sulphur-containing ligands. More than one iron center of this type may exist in the molecule. Oxidized rubredoxin has a distinctive electron paramagnetic resonance (EPR) spectrum with a line at $g = 4.3$, whereas the reduced form gives no discernible EPR signal. Only negative redox potentials at pH 7 have been noted for those rubredoxins presently characterized.

The full name should be listed as follows: (source) rubredoxin (function), *e.g. Pseudomonas oleovorans* rubredoxin, alkane ω-hydroxylation.

5.3.2 Ferredoxins (abbreviation: Fd). This group comprises those iron-sulphur proteins that contain an equal number of iron and labile sulphur atoms and that exclusively display electron-carrier activity but not classical enzyme function. Thus, since hydrogenases (iron-sulphur proteins containing one or more Fe-S clusters) are enzymes, they are not classified as ferredoxins.

The criterion that ferredoxins must have negative oxidation-reduction midpoint potentials at pH 7 has been abandoned; there is no need to distinguish between ferredoxins and the previously designated "high potential iron-sulphur proteins"[7]as exemplified by an iron-sulphur protein from *Chromatium vinosum* (9). Ferredoxins may contain one or more clusters of two or four iron and labile-sulphur atoms.

5.3.3 All simple iron-sulphur proteins that are neither rubredoxins nor ferredoxins fall into the category of "other iron-sulphur proteins".

II. Specific Recommendations, Abbreviations and Symbols

5.4 If any iron-sulphur protein has been given another name previously, this should be stated to minimize confusion.

5.5 The term, "high potential iron-sulphur protein" (abbreviated Hipip), may continue to be used for the original iron-sulphur protein of photosynthetic bacteria that had been given this name initially. Otherwise, the use of the terms Hipip or "high potential Fe-S protein" is discouraged; there are ferredoxins occurring naturally at the oxidation level of oxidized and reduced spinach ferredoxin that have oxidation-reduction potentials as high as the original Hipip of bacteria.

5.6 At least once in a report, the source of the protein should precede the term rubredoxin, ferredoxin, or iron-sulphur protein. Similarly, for iron-sulphur clusters, or "centers"[6] of complex iron-sulphur proteins, the proper designation of the parent protein or enzyme should be given, preferably along with the source, *e.g.* beef-heart NADH dehydrogenase Fe-S cluster 1. Thereafter, the designation, cluster 1, may be used. The proliferation of such other abbreviations as center N-1, S-1, or bc-1 is discouraged. With

---

[7]The group of "high potential iron-sulphur proteins" has been deleted on the following basis: a) The three-dimensional structure of the active center of this protein has been found to be essentially the same as that of ferredoxins (6,7). The magnetic properties differ because the Fe-S clusters may assume three different oxidation levels (6, 8). Thus, an oxidized ferredoxin (diamagnetic) corresponds to the reduced *Chromatium vinosum* high-potential iron-sulphur protein (diamagnetic); the extreme oxidation levels, i.e. those of reduced ferredoxin and oxidized *C. vinosum* high potential iron-sulphur protein, have one electron more or one electron less, respectively, than this oxidation level and are paramagnetic. b) After the original study (9) with the *Chromatium* high potential iron-sulphur protein, it became apparent that other iron-sulphur proteins may have the unusual oxidation level of the *Chromatium* protein without a high midpoint oxidation-reduction potential (10). Conversely, ferredoxins with the usual oxidation level and magnetic properties, may have high-midpoint potentials (11). The expression, "high potential Fe-S protein" (or Hipip) has been used increasingly in the literature to indicate the ability to form the higher oxidation level of the *Chromatium* protein. This practice is inappropriate since it uses a term applying to a redox potential for the description of a magnetic property or an oxidation level.

regard to more complex systems, including the complex iron-sulphur proteins, it appears to be neither suitable nor desirable to present designations for Fe-S clusters that are inadequately characterized.

5.7 Iron-sulphur proteins from the same source that have the same type of Fe-S cluster are numbered sequentially with Roman numerals. A newly isolated iron-sulphur protein that is not fully characterized should be called "iron-sulphur protein" and given the lowest unused numeral. By analogy, the various iron-sulphur clusters, or "centres" of an iron-sulphur protein, should be designated Fe-S centre 1 (or Fe-S cluster 1), Fe-S centre 2 (cluster 2), and so on in the order of discovery. Arabic numerals should be used for the different clusters of the same protein or complex, since iron-sulphur proteins and the complexes of the respiratory chain are designated by Roman numerals.

5.8 The designation of a cluster[6] in an iron-sulphur protein containing labile sulphur atoms, should consist of square brackets about the number of iron and labile-sulphur atoms. Thus, [2Fe-2S] represents a two-iron, two-labile-sulphur cluster and [4Fe-4S], a four-iron, four-labile-sulphur cluster. The protein incorporating such a cluster may be called a "two-iron-two-sulphur" or "four-iron-four-sulphur" ferredoxin[8] or iron-sulphur protein. For ferredoxins, it is sufficient simply to refer to a "two-iron" or "four-iron" ferredoxin.

> Comment: The use of hyphens and parentheses is firmly codified in the nomenclature of inorganic chemistry (5). In order to avoid confusion with established practices in coordination chemistry, a short hyphen is used to write Fe-S and the term is placed between square brackets instead of parentheses.

5.9 The presence of several clusters, as in clostridial ferredoxin, is indicated as follows: 2[4Fe-4S]. This may be called a two-cluster four-iron, four-sulphur ferredoxin or iron-sulphur protein or, simply, a two-cluster-four-iron ferredoxin.

5.10 When the formal charge of the cluster is calculated, the sulphur atoms of the bound cysteine residues are *not* included in the calculation, contrary to the widespread practice in the current literature, thus, for the oxidized and reduced forms, respectively, we have:

Spinach ferredoxin: $\qquad$ $[2Fe-2S]^{2+}$; $[2Fe-2S]^{1+}$

Bacillus polymyxa ferredoxin: $\qquad$ $[4Fe-4S]^{2+}$; $[4Fe-4S]^{1+}$

Clostridium pasteurianum ferredoxin: $\qquad$ $2[4Fe-4S]^{2+}$; $2[Fe-4S]^{1+}$

> Comment: The charges referred to are those within the entire cluster. This is in contrast with usage in inorganic chemistry where a charge shown in a formula refers to the whole compound (5). There is also a possibility for misunderstanding the codified use of the term, oxidation state, which always characterizes a single atom but never a group of atoms (5). In order to avoid confusion, the term, "oxidation level" is used to refer to the cluster.

5.11 For ferredoxins at the oxidation level typical of *Chromatium* Hipip, the expression, $[4Fe-4S]^{3+}$, will differentiate between what was formerly called Hipip and a ferredoxin.

5.12 If the oxidation levels in which a ferredoxin can occur are known, this may be indicated as follows for a [4Fe-4S] ferredoxin that is generally obtained on isolation at the (2+) level: $[4Fe-4S]^{2+(3+;2+;1+)}$. This designation implies that the ferredoxin can occur at all three possible oxidation levels. However, these designations should only be used to refer to oxidation levels that can be reached in a

---

[8]In English they would be pronounced as "two-ef-ee-two-es" and "four-ef-ee-four-es."

biological milieu, *i.e.*, in the absence of agents denaturing the protein, even though artificial oxidants may be used to attain such oxidation levels. By this designation, the highest oxidation levels normally found in *Chromatium* Hipip and the protein from *Bacillus polymyxa*, both [4Fe-4S] ferredoxins, may be differentiated:

Chromatium vinosum Hipip: $[4Fe-4S]^{2+(3+:2+)}$

Bacillus polymyxa ferredoxin: $[4Fe-4S]^{2+(2+:1+)}$

This shorthand denotes that *Chromatium* ferredoxin occurs in the reduced (2+) state but also can be found at the (3+) state whereas *B. polymyxa* ferredoxin occurs in the oxidized (2+) state, has not been found in the (3+) but can exist in the (1+) state. The (2+) state is diamagnetic, whereas both the (3+) and (1+) states are paramagnetic and detectable by EPR measurements. It is recommended that the designation for the oxidation level, whether (3+), (2+) or (1+), when used in a publication, should be that occurring in the experiments described. In spoken language, such terms as reduced, oxidized, and super-reduced may be used during a transition period; the formal charge designations should be used when the new recommendations become familiar.

5.13 Rubredoxins are treated in an analogous fashion, except that the basic centre is designated [Rd] since there is no ambiguity concerning numbers of metal atoms involved. Rubredoxins with multiple clusters are denoted as n[Rd]. The formal charges of rubredoxins are $[Rd]^{3+}$ and $[Rd]^{2+}$ for the oxidized and reduced forms, respectively.

5.14 It is useful to present midpoint redox potentials, light absorption and EPR characteristics, particularly when an iron-sulphur protein is first mentioned in a publication.

5.15 The examples given in Table II illustrate and contrast the new designations with those previously applied.

Iron-sulphur proteins with clusters of different types may be designated in a similar manner. For instance, a protein from *Azotobacter vinelandii* (8) has two Fe-S clusters which assume the same range of oxidation levels, although one is found "reduced" and the other "oxidized": $[4Fe-4S]^{2+(3+:\,2+)}$ $[4Fe-4S]^{3+(3+:2+)}$ Fd I.

Table II
Designation of Iron-Sulphur Proteins as Generally Obtained on Isolation

Previous designation	Recommended designation
Spinach chloroplast ferredoxin	Spinach chloroplast $[2Fe-2S]^{2+}$ Fd or, in a specific context: $[2Fe-2S]^{2+(2+:1+)}$ Fd
Azotobacter vinelandii iron–sulphur protein I	*Azotobacter vinelandii* $[2Fe-2S]^{2+}$ Fe-S protein I
Chromatium vinosum high potential iron-sulphur protein	*Chromatium vinosum* $[4Fe-4S]^{2+}$ Fd or, in a specific context: $[4Fe-4S]^{2+(3+:2+)}$ Fd
Clostridium pasteurianum ferredoxin	*Clostridium pasteurianum* $2[4Fe-4S]^{2+}$ Fd or, in a specific context: $2[4Fe-4S]^{2+(2+:1+)}$ Fd

605

6. Other metalloproteins

6.1 Cuproproteins

Oxidases containing copper appear in the List of Enzymes *e.g.* monophenol monooxygenase, EC 1.14.18.1. Cytochrome *c* oxidase, classified both as an enzyme (EC 1.9.3.1) and a cytochrome (cytochrome aa_3), also contains copper. An electron-transfer protein, plastocyanine, is part of the photosynthetic chain in chloroplasts. Azurin and stellacyanin are also electron-transfer proteins.

6.2 Molybdoproteins

Molybdenum is present in xanthine oxidase (EC 1.2.3.2) which also contains FAD and iron-sulphur centres. Aldehyde oxidase (EC 1.2.3.1) also contains molybdenum as well as iron and flavin. Nitric oxide reductase (EC 1.7.99.2) contains molybdenum as well as iron and inorganic sulphur, but not flavin. Nitrate reductase (NADPH) (EC 1.6.6.3) contains molybdenum and iron.

REFERENCES

1. International Union of Biochemistry (1971) *Report of the Commission on Enzymes.* Pergammon Press, New York.
2. International Union of Biochemistry (1965) *Recommendations (1964) on the Nomenclature and Classification of Enzymes.* Elsevier, Amsterdam.
3. International Union of Pure and Applied Chemistry and the International Union of Biochemistry (1973) *Enzyme Nomenclature Recommendations (1972).* Elsevier, Amsterdam.
4. IUPAC-IUB Recommendations: Nomenclature of Iron-Sulphur Proteins (1973), *Eur. J. Biochem.* **35**, 1-2; *J. Biol. Chem.* **248**, 5907-5908
5. International Union of Pure and Applied Chemistry (1970) *Nomenclature of Inorganic Chemistry* 2nd ed., Butterworths, London.
6. Carter, C. W., Jr., Kraut, J., Freer, S. T., Alden, R. A., Sieker, L. C., Adman, E., and Jensen, L. H. (1972), *Proc. Natl. Acad. Sci., USA* **69**, 3526-3529.
7. Carter, C. W., Jr., Kraut, J., Freer, S. T., and Alden, R. A. (1974), *J. Biol. Chem.* **249**, 6339-6346.
8. Herskovitz, T., Averill, B. A., Holm, R. H., Ibers, J. A., Phillips W. D., and Weiher, J. F., (1972), *Proc. Natl. Acad. Sci., USA* **69**, 2347-2441.
9. Bartsch, R. G. (1963) in *Bacterial Photosynthesis* (Gest, H., San Pietro, A., and Vernon, L. P., eds.) p. 315, Antioch Press, Yellow Springs, Ohio.
10. Sweeny, W. V., Rabinowitz, J. C., and Yoch, D. C. (1975), *J. Biol. Chem.* **250**, 7842-7847.
11. Leigh, J. S., Jr., and Erecinska, M. (1975) *Biochim. Biophys. Acta* **387**, 95-106.